Boreal Forest and Climate Change

ADVANCES IN GLOBAL CHANGE RESEARCH

VOLUME 34

For other titles published in this series, go to
www.springer.com/series/5588

Pertti Hari · Liisa Kulmala

Boreal Forest
and Climate Change

Springer

Editors
Dr. Pertti Hari
FI-00014 University of Helsinki
University of Helsinki
Finland

M.Sc. Liisa Kulmala
FI-00014 University of Helsinki
University of Helsinki
Finland

QH
541.5
.T3
B67
2008

ISBN 978-1-4020-8717-2   e-ISBN 978-1-4020-8718-9

Library of Congress Control Number: 2008930996

Printed on acid-free paper

9 8 7 6 5 4 3 2 1

springer.com

B244307

Preface

The Forest Primary Production Research Group was born in the Department of Silviculture, University of Helsinki in the early 1970s. Intensive field measurements of photosynthesis and growth of forest vegetation and use of dynamic models in the interpretation of the results were characteristic of the research in the group. Electric instrumentation was based on analogue techniques and the analysis of the obtained measurements was based on self-written programs.

Joint research projects with the Research Group of Environmental Physics at the Department of Physics, lead by Taisto Raunemaa (1939–2006) started in the late 1970s. The two research groups shared the same quantitative methodology, which made the co-operation fruitful.

Since 1980 until the collapse of the Soviet Union the Academy of Finland and the Soviet Academy of Sciences had a co-operation program which included our team. The research groups in Tartu, Estonia, lead by Juhan Ross (1925–2002) and in Petrozawodsk, lead by Leo Kaipiainen (1932–2004) were involved on the Soviet side. We had annual field measuring campaigns in Finland and in Soviet Union and research seminars. The main emphasis was on developing forest growth models.

The research of Chernobyl fallout started a new era in the co-operation between forest ecologists and physicists in Helsinki. The importance of material fluxes was realized and introduced explicitly in the theoretical thinking and measurements.

The measuring techniques developed slowly during the 1970s and 1980s. The system was computerized, but the utilization of the new possibilities was rather limited. SMEAR I measuring station (station for measuring ecosystem-atmosphere relations) was a clear step forward in utilizing the possibilities of the digital techniques. In addition, the measuring system was planned and constructed to fulfill the research needs of forest ecologists and physicists. Markku Kulmala was leading the physicists in their research of environmental physics, especially the behavior of aerosols.

SMEAR II was planned and constructed to measure all relevant mass and energy fluxes and processes generating the fluxes in a forest ecosystem in southern Finland. The new fully digitalized and automatic measurements opened a new comprehensive source of information for research. A large number of papers dealing with the gas exchange of trees, soil behavior, gas and energy fluxes between ecosystem and atmosphere, and aerosol formation and growth has been published in scientific journals indicating the benefits of the interaction between forest ecologists and physicists.

Our book is a comprehensive summary of the research done within the long lasting co-operation between forest ecologists and physicists at Helsinki University. Several generations of researchers and students have contributed to the work presented here. Despite the fact that the number of contributors is large, a great number of people who contributed to our book remain unnamed. We want to thank them for their contribution.

John Grace from Edinburgh, U.K. read our manuscript and made a large number of constructive and clarifying comments that clearly improved the book. We acknowledge John Grace for his advice and support.

Helsinki   *Pertti Hari and Liisa Kulmala*
June 3 2008

Contents

viii   Contents

Definitions

ABL – Atmospheric boundary layer, also known as the planetary boundary layer. The lower part of the troposphere, where friction on the Earth's surface influences the air flow.

Absolute humidity – Water vapour concentration in the air.

Absorptivity – A property of a material, the ratio of the absorbed radiation to the incident one; in equilibrium absorptivity and emissivity are the same.

Acclimations – Changes in the biochemical regulations system, in functional substances and in fine structure matching vegetation with irregular changes in the environment.

Accuracy – Magnitude of systematic error in measurements.

Adaptation – Results of evolution to a specific environment.

Adhesion – The tendency of certain dissimilar molecules to cling together due to attractive forces.

Aerosol – The combination of solid or liquid aerosol particles and the gaseous medium they are suspended in.

Air seeding – Gas can be sucked from embolised conduits to water conducting ones through pores in the membranes separating the conduits when water tension rises through increasing transpiration demand.

Albedo – The fraction of solar radiation reflected by a surface of an object, often expressed as a percentage.

Allocation to something – The share of available resources used to the growth of the specified component.

Anemometer – Device for measuring wind speed.

Annual cycle of environmental factors – Exact annual cycle in all environmental factors caused by globes orbiting around the sun, seasonality.

Annual cycle of metabolism – Changes in the biochemical regulations system, in functional substances and in fine structure matching vegetation with the annual cycle of environment.

Aquaporins – Integral membrane proteins enabling selective penetration of water molecules across the membrane.

ATP – Adenosine-5′-triphosphate transports chemical energy within cells for metabolism.

Attenuation of light – Reduction of light due to absorption.

Base saturation – The proportion of cation exchange sites occupied by the so-called basic cations (Na^+, K^+, Mg^{2+}, Ca^{2+}) rather than acidic cations (H^+, Al^{3+}).

Biochemical regulation system – Formed by substances that control the functional substances.

Biome – A biotic community of plants and animals.

Black body – A perfectly efficient emitter and absorber of radiation.

Boreal forest – Set of coniferous forest ecosystems that can survive in northern, specifically subarctic, regions.

Boreal zone – Terrestrial division consisting of the northern coniferous parts.

Bulk flow – Ordinary flow of fluid, but "bulk" stresses the contrast to molecular motion.

BVOC – Biogenic volatile organic compounds; a diverse group of volatile compounds evaporating from plant tissues; e.g. isoprenoids, acetone, methanol.

Cambium – Lateral meristem tissue which gives rise to secondary thickness growth; vascular cambium forms phloem and xylem tissues, and cork cambium forms cork tissues.

Carbon balance – Carbon fixed in photosynthesis during time unit is used to the maintenance and growth.

Carbon reactions (dark reactions, Calvin cycle) – Join a carbon atom to five-carbon sugar in chloroplast stroma utilising energy released from ATP and NADPH.

Carbon sequestration – Amount of carbon fixed in vegetation and soil organic matter during a year.

CBL – Convective planetary boundary layer, planetary boundary layer where heating and evapotranspiration at the surface creates thermal instability and thus generates turbulence. The CBL is typical in summer hemisphere during daytime, when turbulence tends to mix air properties and quantities such as conservative concentrations and potential temperature are close to constant.

Cell – Basic structural and functional unit in all living organisms.

Coagulation (aerosol dynamic process) – The process where two particles collide and coalesce, forming a new larger particle.

Cohesion – A physical property of a substance. Cohesion is caused by the intermolecular attraction between similar molecules within a body or substance that acts to unite them. Water, for example, is strongly cohesive.

Cohesion theory – The hypothesis that water is pulled upward along a pressure gradient during transpiration by the electric forces between water molecules. According to this theory, the evaporation of water at the leaf surfaces pulls a continuous water column against a gravitational gradient through a continuous pathway of xylem conduits reaching all the way down to the roots.

Condensation – The phase transition where a vapour is changing to liquid; it is accompanied by the release of (latent) heat.

Conduction – Molecular transport mechanism of energy (heat) arising from random motion of molecules.

Convection – Bulk flow of the fluid. Flowing gas or liquid carries mass, heat and momentum associated to it. In meteorology, convection refers to the vertical bulk flow of air.

Decomposition of soil organic matter – A process where organic macromolecules are disassembled by soil micro-organisms, leads to release of nutrients and CO_2.

Denitrification – The respiratory bacterial reduction of nitrate (NO_3^-) or nitrite (NO_2^-) to nitrogen oxides or molecular nitrogen.

Density of mass – Amount of plant material in a volume element divided with the volume of the element.

Deposition – (aerosol dynamic process) – The process of aerosol particles being transported and deposited onto existing surfaces and thus away from the airborne phase.

Dermal tissues – The outermost layer of cells facing the atmosphere or soil.

Diffusion – Transport of atoms and molecules by random thermal movement.

Dormancy – Period in an organism's annual cycle when growth, development, and possible physical activity is temporarily suspended.

Dynamic model – Mathematical model having time in a central role.

Ecosystem – A spatial unit composed of interacting biotic (animals, plants, micro-organisms) and abiotic (non-living physical and chemical) factors.

Eddy – Flow pattern in the turbulent flow with some coherent structure.

Eddy covariance – Meteorological technique to measure exchange of mass, energy and momentum between the surface and the atmosphere, based on the direct determination of the turbulent transport.

Embolism – Introduction of air bubbles into the water transporting xylem tissue, as individual water-conducting vessels are filled by water vapor or air, and they become embolised. Cause the water column to break.

Emergent property – Feature that cannot be directly derived from lower level phenomena.

Emission scenario – A plausible representation of the future development of emissions of substances that are potentially radiatively active (e.g., greenhouse gases, aerosols), based on a coherent and internally consistent set of assumptions about driving forces (such as demographic and socioeconomic development, technological change) and their key relationships. In general, climate model simulations use some of the so-called SRES scenarios (Nakicenovic and Swart, 2000).

Emissivity – A property of a material, the ratio of the emitted radiation by the material to that of a black body (perfectly efficient) radiator.

Enthalpy – A thermodynamic property of a system related to the internal energy of the system, its pressure and its volume; enthalpy difference between vapour and liquid phases corresponds to the latent heat released in condensation or consumed in evaporation.

Environmental factor – Such feature in the environment that effects on some process.

Enzymes – Large protein molecules acting as biological catalysts in all metabolic processes that occur in living tissue.

Evaporation – The phase transition where liquid is changing to vapour; it is accompanied by the consumption of (latent) heat.

Evapotranspiration – The sum of evaporation and transpiration.

Evolution – New metabolic processes, structures and biochemical regulations and improvements in the existing ones emerge and these novel features are tested against the existing ones in the process of Natural Selection. If the novel feature contributes to the success of the species, then it has the tendency to become more common. This process results in slow evolution of the species as it becomes adapted to the environment of its location.

Extra cellular enzyme – Enzyme emitted by microbes outside the cell to split macromolecules.

FAD$^+$ – Flavin adenine dinucleotide.

Flux density – Amount of flow of material or energy through a plane element during time element devidid with the product of the area and the length of the duration of the time element.

Forced convection – Convection (bulk flow) arising from any other reasons excluding density (temperature) deviations in the fluid, especially flow arising from pressure difference.

Free convection – Natural convection; convection (bulk flow) arising from density deviations in the fluid, especially flow arising from temperature differences.

Freezing tolerance – Ability to survive freezing temperatures.

Functional balance – The uptake of carbon and nitrogen are balanced with each other.

Functional substance – Enable metabolic processes in living organisms. Enzymes, membrane pumps and pigments are examples of functional substances.

GCM – General Circulation Models or General Climatic Models.

Global radiation – The total of direct solar radiation and diffuse sky radiation received by a unit horizontal surface.

GPP – Gross primary production of an ecosystem, i.e., photosynthesis of the ecosystem.

Greenhouse effect – Greenhouse gases absorb thermal infrared radiation, emitted by the Earth's surface, by the atmosphere itself due to the same gases, and by clouds. As a result of this, only a small part of the thermal radiation emitted by the Earth's surface and the lowest atmospheric layers escapes directly to space. Although greenhouses gases by themselves emit thermal radiation to all directions, the upward emitted radiation that finally escapes to space originates from air layers that are much colder than the surface, and the intensity of this radiation is therefore smaller. Therefore, the Earth's surface can maintain an average temperature of about $+14°C$, which is $33°C$ warmer than the temperature $(-19°C)$ that would be observed without the greenhouse effect. As the concentrations of greenhouse gases in the atmosphere are increasing, the greenhouse effect is becoming stronger and this is expected to lead to an increase in the Earth's surface temperature.

Greenhouse gas – Greenhouse gases are gases that absorb and emit radiation at wavelengths within the spectrum of thermal infrared radiation emitted by the Earth's surface, the atmosphere itself, and by clouds. Water vapour (H_2O), carbon dioxide (CO_2), nitrous oxide (N_2O), methane (CH_4) and ozone (O_3) are the primary greenhouse gases in the Earth's atmosphere. In addition, there are a number of entirely human-made greenhouse gases in the atmosphere, such as halocarbons including CFC-11 and CFC-12.

Ground tissues – Present throughout the plant body and formed of parenchyma, collenchyma (supporting) or sclerenchyma (protective and storage) cells.

Growth respiration – Processes supplying energy for formation of new tissue; CO_2 released.

Hierarchy – Ordered structure.

High pressure systems (also Anticyclones) – Characterised by clockwise circulation in the Northern Hemisphere, with high pressure in the middle of the weather system, divergence of flow close to surface and corresponding vertical subsidence.

Humidity – See Relative humidity and Absolute humidity.

Interception of light – Absorption of light in vegetation.

Isoprenoids – A large and diverse class of naturally-occurring organic chemicals similar to terpenes. Isoprenoids are derived from five-carbon isoprene units assembled and modified in thousands of ways. See 'Terpenes'.

LAI – Leaf Area Index, ratio of total upper leaf surface of vegetation divided by the surface area of the land on which the vegetation grows.

Laminar flow – A form of fluid flow in which fluid flows in an ordered manner with regular paths; laminar flow occurs when the flow velocities are low and always close to surfaces, otherwise the flow is turbulent.

Latent heat – Heat absorbed or released when a substance changes its physical state (phase), here especially related to condensation/evaporation and freezing/melting.

Latent heat flux – The flux of heat from the Earth's surface to the atmosphere by evapotranspiration.

Light reactions – Processes in chloroplast thylakoids converting absorbed light quanta into chemical energy (ATP, NADPH) and liberating O_2 from H_2O.

Longwave radiation – Radiation emitted from Earth, typical wavelengths of 4–30 µm.

Low pressure systems (Cyclones) – Counterclockwise circulation in the Northern Hemisphere, low pressure in the middle of the weather system and flow convergence close to surface. The rising air in the middle of the system experiences expansion cooling and therefore supports cloud formation and subsequent precipitation.

LUE – Light use efficiency.

Macrofauna – Soil or benthic organisms which are at least one millimeter in length.

Macromolecules – Large molecules, commonly formed of more than 1,000 atoms, e.g. nucleotides, lipids, proteins and carbohydrates.

Maintenance respiration – Processes supplying energy for cellular household tasks such as protein turnover, active transport or seasonal adjustment of molecular structures; CO_2 released.

Mass density – Mass of vegetation elements in a volume element divided by the volume of the element.

Mass specific process rate – Amount of products produced in a volume element during short time element divided with product of the volume of the space element with the length of the time element.

Mean (arithmetic mean) – The sum of a property over all members in a population divided by the number of items in the population.

Measuring noise – Random error in measurements.

Membrane pumps – Intrinsic proteins which facilitate the movement of polar molecules through membranes either against their electrochemical gradient (active pumps) or in the direction of the electrochemical gradient (channels and carriers). Active pumps need energy provided by ATP for their operation.

Meristematic tissue – Undifferentiated cells in places where new tissue is formed.

Mesophyll – Leaf parenchymatous tissue situated between dermal and vascular tissues, functions in photosynthesis.

Metatheory – Theory of theories.

Methanogenesis – Acetate or CO_2 are converted into CH_4 and CO_2, while hydrogen is consumed.

Microfauna – Microscopic or very small animals (usually including protozoans and very small animals such as rotifers).

Mineralization – A process in which organic molecules in soil are broken down into inorganic forms, such as CO_2 and H_2O, and often accompanied with a release of ions such as NH_4^+, K^+ and Ca^{2+}.

Mitochondria – Cell organelles housing the essential respiratory machinery which generates ATP by the citric acid cycle and electron transfer chain.

Mixed layer (ML) – A layer in which active turbulence has homogenized some range of depths. The depth of the atmospheric mixed layer is known as **the mixing height**. ML is frequently used as a synonym for CBL.

Molecular transport – Transport of mass, heat and momentum arising from random motion of molecules in the presence of gradients. Leads to concepts of diffusion, conduction and viscosity.

Momentum – The product of the mass and velocity of an object.

Momentum absorption – Momentum absorbed from the flowing fluid on the surface.

Momentum flux – Transport rate of momentum in a fluid.

Münch hypothesis – Water is drawn in from the xylem to the phloem at the top of the tree and is pushed back into the xylem at the bottom.

Mycorrhiza – A symbiotic association between fungal hyphae and plant roots, which greatly improves the plant water and nutrient uptake. The ectomycorrhizal hyphae do not penetrate root cells, whereas in endomycorrhizal symbiosis the fungal hyphae pass the root cell wall.

NAD⁺, NADH – Nicotinamide adenine dinucleotide, reduced nicotinamide adenine dinucleotide.

NADP⁺, NADPH – Nicotinamide adenine dinucleotide phosphate, reduced nicotinamide adenine dinucleotide phosphate.

Near infrared radiation (NIR) – Radiation in the wavelength band 750–3,000 nm that contributes to about half of the incoming solar radiation,

NEE – Net Ecosystem Exchange, photosynthesis minus respiration in an ecosystem.

Net radiation – The sum of the downward and upward solar and thermal radiation.

Nitrification – Oxidation of ammonium (NH_4^+) or ammonia (NH_3) to nitrite (NO_2^-) and nitrate (NO_3^-).

Nitrogen balance – All nitrogen taken up by roots and released from the senescencing structures during a year are used for growth during that year.

Nitrogen fixation – a biological process in which nitrogen gas (N_2) from the atmosphere is converted to ammonium form.

NPP – Net primary production, formation of new organic matter in an ecosystem.

NS – Navier-Stokes equation.

Nucleation event – Regional formation and growth of new atmospheric aerosol particles created and grown by nucleation and condensation of atmospheric vapours.

Organelle – A cellular subunit enclosed in a membrane, e.g. plastid, mitochondria, nucleus.

Osmosis – Diffusion through a membrane, which is partly or fully impermeable to dissolved solutes but permeable to water.

Parameter – Constant in a model.

Particle formation (aerosol dynamic process) – The formation of new secondary aerosol particles from vapour phase, which is observed as an increase in the measured nanoparticle concentrations.

Penumbra – Occasion when light coming from the solar disk is partially covered by vegetation.

Proportion of explained variance (PEV) – Commonly used as a measure of the goodness of fit of the model. It gives the proportion of the variance of the measured response that can be explained by the model with the explanatory factors.

Phase transitions – A change in a feature that characterizes a physical system; here especially changes from solid to liquid and liquid to vapour and the reverse changes.

Phenology – Study of timing of phenomena associated with the annual cycle of vegetation.

Phloem – Living vascular tissue where translocation of photoassimilates (sap) takes place.

Photorespiration – Analogous to carbon reactions (Calvin cycle); catalyzed by Rubisco but using O_2 as a substrate instead of CO_2; should not be mixed with "true" respiration although it eventually leads into CO_2 release analogously to respiration.

Photosynthesis – Process where organic carbon molecules (sugars) are formed from CO_2 and H_2O in chloroplasts of green plants using solar radiation as energy source.

Photosynthetic efficiency – Parameter in the optimal stomatal control model of photosynthesis.

Photosynthetic pigments – A group of macromolecules which participate in absorbing selective wavelengths of light quanta and transferring the excitation into processes where it is chemically bound; e.g. chlorophylls, carotenoids, and xanthophylls.

Photosynthetic reaction center – A complex of several proteins, pigments and other macromolecules; the site where molecular excitations originating from sunlight are transformed into a series of electron-transfer reactions.

Photosynthetically active radiation (PAR) – Number of light quanta in the visible range (400–700 nm) of solar radiation.

Phytoelement – Vegetation element.

Pipe model theory – Constant relationship between the cross-sectional sapwood area and foliage area above.

Plastids – Cellular organelles functioning in photosynthesis (chloroplasts), other biosynthetic processes (leucoplasts) and energy storage (amyloplasts).

Podsol – soils typically developing under Boreal coniferous forests. Podzols are characterized by the leaching of acidic organic material and inorganic ions from the A and E horizons to the spodic B horizon characterized by the enrichment of sesquioxides and humus.

Podsolization – Formation of podsol soil.

Potential temperature – The temperature that the parcel or fluid would acquire if adiabatically brought to a standard reference pressure, usually sea level pressure (1,013 millibars).

Precision – Random variation in the model or measurement.

Pressure – force per unit area of solid materia, liquid or gas.

Process – Converts material and/or energy into a new form or moves material through a membrane.

Proteins – Large macromolecules built from amino acid units and folded in a complex three-dimensional structure.

Radiation (electro-magnetic radiation) – The third mode of heat transport beside conduction and bulk flow (convection). Transports heat by the speed of the light in a vacuum.

Radiation energy – Energy carried in the form of electro-magnetic radiation by photons.

Radiative forcing – Radiative forcing is the change in the net, downward minus upward, irradiance (expressed in $W\ m^{-2}$) at the tropopause due to a change in an external driver of climate change, such as, for example, a change in the concentration of carbon dioxide or the output of the Sun. Generally, radiative forcing is defined as the global and annual average change in net radiation resulting from changes in external factors relative to the year 1750.

Regulating substances – Compounds that control the activation, biosynthesis and catabolism of functional substances; e.g. many metabolites, hormones, ions.

Residual – The difference between measured and modeled value.

Respiration – Formation of ATP from ADP through the oxidation of organic substances by metabolic processes of living cells.

Relative humidity (RH) – The ratio, expressed as a percentage, of water vapour content in the air to that if it were saturated with water at the same temperature.

Rhizosphere – The zone that surrounds the roots of plants.

Root exudates – Molecules or ions emitted by roots.

Residual sum of squares (RSS) – The sum of squares of the residuals.

Rubisco – Ribulose-1, 5-bisphosphate carboxylase/oxygenase (RuBisCO) is an enzyme that is used in the Calvin cycle to catalyze the first major step of carbon fixation.

Runoff – The amount of water per unit of space and time leaving an ecosystem as liquid.

Sapwood – Youngest and outermost xylem tissue in stems, consists of living cells involved in conducting water.

SBL – Stably stratified atmospheric boundary layer is the ABL where negative buoyancy flux at the surface damps the turbulence.

Senescence – The process of aging of an organelle, cell, organ or whole plant individual.

Sensible heat – Heat flux associated to the movement of warmer/cooler air.

Shortwave radiation – A term used to describe the radiant energy in the visible, near-ultraviolet, and near-infrared wavelengths. Shortwave radiation may be as broadly defined as between 0.1 and 5.0 µm.

SMEAR – Station for measuring ecosystem-atmosphere relations.

Soil horizons – soil layers differing in characteristics from the adjacent layers as a result of soil forming processes.

Soil textural classes – size groupings of soil particles (such as sandy, loamy and clayey soils) based on the relative proportions of various particle size fractions (clay, silt, sand, gravel and stones).

Soil water storage – Amount of water in top layer up to 1 m of soil per 1 m^2 (units liters m^{-2} or mm).

Solar constant – Amount of the Sun's incoming electromagnetic radiation per unit horizontal area, at the outer surface of Earth's atmosphere. The Solar constant includes all types of Solar radiation, not just the visible light. It is roughly 1,366 W m^{-2} though this fluctuates during a year.

SOM – Soil organic matter denotes all living and dead biologically derived organic material at various stages of decomposition in the soil or on the soil surface, excluding the above-ground portion of living plants.

Space element – So small spatial volume that environmental factors do not vary within it.

Spectrum – Distribution of electro-magnetic radiation as a function of wavelength or frequency.

Stability of atmosphere – At equilibrium state an air parcel has no tendency to either rise or fall and then the atmosphere is at the neutral state; in stable atmosphere upward or downward motions are inhibited leading to weak mixing, whereas in unstable atmosphere the mixing is strong.

Stable – System or property that does not change.

State of functional substances – Regularities in the functional substances generated by the action of regulation system.

Stomatal cavity – Air space in mesophyll tissue below a stoma.

Stomatal conductance – Parameter in models of photosynthesis, transpiration and other gaseous fluxes between leaf and atmosphere introducing diffusion of gas molecules.

Structure of living organisms – Stable carbon compounds in living organisms form its structure.

Subtheory – Theory within the metatheory.

Succession – Orderly changes in the composition or structure of an ecological community. Succession that occurs after the initial succession has been disrupted is called secondary succession.

Synoptic scale – Scale where meteorological processes occur in the order of 1,000 km and 1 day.

Systematic measuring error – Bias in measurements.

Taiga – Earlier referred to coniferous forest by Siberians, nowadays a more general meaning, referring to different forests of the Northern Hemisphere, boreal forest.

Tensiometer – Device used to determine matrix water tension in the soil.

TER – Total ecosystem respiration.

Terpenes – A large group of volatile and non-volatile isoprenoids; the major components of resin (turpentine) in coniferous trees. See 'Isoprenoids'.

Terrestrial radiation – Long-wave radiation emitted by the Earth's surface and its atmosphere.

Thermal radiation – Electromagnetic radiation emitted from the surface of an object which is due to the object's temperature.

Thylakoids – Membrane structures in chloroplasts, housing the pigment system used in capturing solar radiation; grana thylakoids are stacked while stroma thylakoids are unstacked.

Time element – So short time interval that environmental factors do not vary during it.

Tissue – A group of closely connected cells having similar function within an organ.

Trace gas – Gas or gases which make up less than 1% by volume of the earth's atmosphere.

Transpiration – Flow of water vapour through stomata.

Transport – Moves material or energy in space.

Troposphere – Lowest layer of the atmosphere starting at the surface and going up to about 10 km at the poles and 15 km at the equator.

Turbulent flow – A form of fluid flow in which fluid flows in a disordered manner in irregular paths; turbulent flow occurs when the flow velocities are high, otherwise the flow is laminar.

Turgor pressure – Hydrostatic pressure of water within the cells.

Unstable – System or property that does change.

UV – Ultraviolet, electromagnetic radiation with a wavelength between X-rays and visible light (UVA = 315–400 nm, UVB = 280–315 nm).

Variance – Average of the squared distance of random variable from the population mean.

Vascular tissues – The tissues where substances (e.g. water, photoassimilates, mineral nutrients) are transported longitudinally over long-distance; xylem, phloem.

Visible light – Electromagnetic radiation with a wavelength between 380–750 nm.

VOC – Volatile organic chemical compounds.

Water tension – force per unit surface area of water (negative pressure, suction).

Xeromorphic plants – Plants structurally adapted to dry environments, can withstand prolonged drought and have wilting point lower than –1.5 MPa.

Xylem – Vascular tissue where transport of water and solutes occurs.

Contributors

Aakala, Tuomas
University of Helsinki, Department of Forest Ecology, P.O. Box 27, 00014
University of Helsinki, Finland

Aalto, Juho
University of Helsinki, Department of Forest Ecology, P.O. Box 27, 00014
University of Helsinki, Finland

Aalto, Pasi
University of Helsinki, Department of Physics, P.O. Box 64, 00014 University
of Helsinki, Finland

Altimir, Nuria
University of Helsinki, Department of Forest Ecology, P.O. Box 27, 00014
University of Helsinki, Finland

Bäck, Jaana
University of Helsinki, Department of Forest Ecology, P.O. Box 27, 00014
University of Helsinki, Finland

Bonn, Boris
Frankfurt University, Institute of Atmospheric and Environmental Sciences,
Altenhöferallee 1, 60438 Frankfurt / Main, Germany

Boy, Michael
University of Helsinki, Department of Physics, P.O. Box 64, 00014 University
of Helsinki, Finland

Dal Maso, Miikka
(a) ICG-2: Troposphäre, Forschungszentrum Jülich, 52425 Jülich, Germany,
(b) University of Helsinki, Department of Physics, P.O. Box 64, 00014 University
of Helsinki, Finland

Duursma, Remko
(a) University of Western Sydney, Centre for Plant and Food Science, Locked
Bag 1797, Penrith South DC, NSW 1797 Australia, (b) Macquarie University,
Department of Biological Science, N.S.W. 2109, Australia

Garcia, Heli
University of Helsinki, Department of Applied Chemistry and Microbiology,
P.O. Box 27, 00014 University of Helsinki, Finland

Grönholm, Tiia
University of Helsinki, Department of Physics, P.O. Box 64, 00014 University
of Helsinki, Finland

Häkkinen, Risto
Finnish Forest Research Institute, Unioninkatu 40 A, 00170 Helsinki, Finland

Hänninen, Heikki
University of Helsinki, Department of Biological and Environmental Sciences,
P.O. Box 56, 00014 University of Helsinki, Finland

Hari, Pertti
University of Helsinki, Department of Forest Ecology, P.O. Box 27, 00014
University of Helsinki, Finland

Havimo, Mikko
University of Helsinki, Department of Forest Resource Management, P.O. Box 27,
00014 University of Helsinki, Finland

Hellén, Heidi
Finnish Meteorological Institute, P.O. Box 503, 00101 Helsinki, Finland

Hiltunen, Veijo
Hyytiälä Forestry Field Station, University of Helsinki, Hyytiäläntie 124, 35500
Korkeakoski, Finland

Höltta, Teemu
University of Helsinki, Department of Forest Ecology, P.O. Box 27, 00014
University of Helsinki, Finland

Ilvesniemi, Hannu
Finnish Forest Research Institute, P.O. Box 18, 01301 Vantaa, Finland

Jõgiste, Kalev
Estonian University of Life Sciences, Institute of Forestry and Ryral Engineering,
Kreutzwaldi 5, 51014 Tartu, Estonia

Juurola, Eija
University of Helsinki, Department of Forest Ecology, P.O. Box 27, 00014
University of Helsinki, Finland

Kähkönen, Mika
University of Helsinki, Department of Applied Chemistry and Microbiology,
P.O. Box 27, 00014 University of Helsinki, Finland

Kangur, Ahto
Estonian University of Life Sciences, Institute of Forestry and Ryral Engineering,
Kreutzwaldi 5, 51014 Tartu, Estonia

Keronen, Petri
University of Helsinki, Department of Physics, P.O. Box 64, 00014 University
of Helsinki, Finland

Kivekäs, Roope
University of Helsinki, Department of Forest Ecology, P.O. Box 27, 00014
University of Helsinki, Finland

Kolari, Pasi
University of Helsinki, Department of Forest Ecology, P.O. Box 27, 00014
University of Helsinki, Finland

Köster, Kajar
Estonian University of Life Sciences, Institute of Forestry and Ryral Engineering,
Kreutzwaldi 5, 51014 Tartu, Estonia

Kulmala, Liisa
University of Helsinki, Department of Forest Ecology, P.O. Box 27, 00014
University of Helsinki, Finland

Kulmala, Markku
University of Helsinki, Department of Physics, P.O. Box 64, 00014 University
of Helsinki, Finland

Kurola, Jukka
(a) University of Helsinki, Department of Ecological and Environmental Sciences,
Niemenkatu 73, 15140 Lahti, Finland, (b) Department of Applied Chemistry and
Microbiology, University of Helsinki, 00014, Finland

Laakso, Lauri
University of Helsinki, Department of Physics, P.O. Box 64, 00014 University
of Helsinki, Finland

Launiainen, Samuli
University of Helsinki, Department of Physics, P.O. Box 64, 00014 University
of Helsinki, Finland

Lahti, Tapani
Niinikuja 2, 33900 Tampere, Finland

Linkosalo, Tapio
University of Helsinki, Department of Forest Ecology, P.O. Box 27, 00014
University of Helsinki, Finland

Liski, Jari
Finnish Environment Institute, Research Department, Research Programme
for Global Change, P.O. Box 140, 00251 Helsinki, Finland

Mäkelä, Annikki
University of Helsinki, Department of Forest Ecology, P.O. Box 27, 00014
University of Helsinki, Finland

Mönkkönen, Petteri
Värriö Subartic Research Station, University of Helsinki, 98800 Savukoski, Finland

Nikinmaa, Eero
University of Helsinki, Department of Forest Ecology, P.O. Box 27, 00014
University of Helsinki, Finland

Nilsson, Douglas
Stockholm University, Department of Applied Environmental Science, 10691
Stockholm, Sweden

Palva, Lauri
University of Technology, Applied Electronics Laboratory, P.O. Box 3000, 02015
HUT, Finland

Perämäki, Martti
University of Helsinki, Department of Forest Ecology, P.O. Box 27, 00014
University of Helsinki, Finland

Petäjä, Tuukka
(a) University of Helsinki, Department of Physics, P.O. Box 64, 00014 University of
Helsinki, Finland, (b) Earth and Sun Systems Laboratory, Division of Atmospheric
Chemistry, National Center for Atmospheric Research, P.O. Box 3000, Boulder,
CO 80307-5000 USA

Pihlatie, Mari
University of Helsinki, Department of Physics, P.O. Box 64, 00014 University of
Helsinki, Finland

Pohja, Toivo
Hyytiälä Forestry Field Station, University of Helsinki, Hyytiäläntie 124, 35500
Korkeakoski, Finland

Pohjonen, Veli
Värriö Subarctic Research Station, University of Helsinki, 98800 Savukoski,
Finland

Porcar-Castell, Albert
University of Helsinki, Department of Forest Ecology, P.O. Box 27, 00014
University of Helsinki, Finland

Pulkkinen, Minna
University of Helsinki, Department of Forest Ecology, P.O. Box 27, 00014
University of Helsinki, Finland

Pumpanen, Jukka
University of Helsinki, Department of Forest Ecology, P.O. Box 27, 00014
University of Helsinki, Finland

Räisänen, Jouni
University of Helsinki, Department of Physics, P.O. Box 64, 00014 University
of Helsinki, Finland

Raivonen, Maarit
University of Helsinki, Department of Forest Ecology, P.O. Box 27, 00014
University of Helsinki, Finland

Rannik, Üllar
University of Helsinki, Department of Physics, P.O. Box 64, 00014 University
of Helsinki, Finland

Riipinen, Ilona
University of Helsinki, Department of Physics, P.O. Box 64, 00014 University
of Helsinki, Finland

Rita, Hannu
University of Helsinki, Department of Forest Resource Management, P.O. Box 27,
00014 University of Helsinki, Finland

Salkinoja-Salonen, Mirja
University of Helsinki, Department of Applied Chemistry and Microbiology,
P.O. Box 27, 00014 University of Helsinki, Finland

 Schiestl, Pauliina
University of Helsinki, Department of Forest Ecology, P.O. Box 27, 00014
University of Helsinki, Finland

Siivola, Erkki
University of Helsinki, Department of Physics, P.O. Box 64, 00014 University
of Helsinki, Finland

Simojoki, Asko
University of Helsinki, Department of Applied Chemistry and Microbiology,
P.O. Box 27, 00014 University of Helsinki, Finland

Smolander, Sampo
University of Helsinki, Department of Physics, P.O. Box 64, 00014 University
of Helsinki, Finland

Sogacheva, Larisa
University of Helsinki, Department of Physics, P.O. Box 64, 00014 University
of Helsinki, Finland

Tuomenvirta, Heikki
Finnish Meteorological Institute, P.O. Box 503, 00101 Helsinki, Finland

Väänänen, Päivi
University of Helsinki, Department of Forest Ecology, P.O. Box 27, 00014
University of Helsinki, Finland

Vesala, Timo
University of Helsinki, Department of Physics, P.O. Box 64, 00014 University
of Helsinki, Finland

Chapter 1
Introduction

Pertti Hari

1.1 Background

When the earth was born, about 4.54 billion years ago, regularities in the behaviour of material and energy led to development of the properties of the atmosphere, land surface and oceans. Energy input by solar radiation soon became the dominating external influence on the planet. All material emits thermal radiation and its intensity increases with temperature. If the solar radiation remains unchanged for a prolonged period, and if the thermal radiation by the globe is smaller than the solar energy input, then the global temperature increases until it reaches the temperature at which the energy input from the sun and output of thermal radiation by the earth are equal. In principle, the same happens if the output by thermal radiation is larger than solar input, but the global temperature decreases in this case.

Solar energy input generates and drives a complex circulation of energy and material on the globe. This circulation of energy includes several rather different phenomena such as: absorption and emission of radiation, evaporation and condensation of water, heat exchange between solid and liquid material and gas, and convection by mass flow. The atmosphere buffers the diurnal exchange of energy and so reduces the temperature difference between day and night.

The solar energy input also generates a complex water cycle between the surface of the globe and the atmosphere. Some of this energy is expended in causing water to evaporate from oceans, lakes, and other bodies of water or from wet surfaces. Wind transports this water and its latent heat energy over long distances. Water vapour condenses on aerosol particles in the air and subsequently falls as rain to the globe's surface. The water that is not absorbed into the ground or evaporated flows away as rivers into the oceans.

In the beginning, energy and water cycles dominated the globe for long periods until the emergence of photosynthesis. Thereafter the carbon cycle became

P. Hari
University of Helsinki, Department of Forest Ecology, Finland

P. Hari, L. Kulmala (eds.) *Boreal Forest and Climate Change,*
© Springer Science+Business Media B.V., 2009

important to the global system. Photosynthesis extracted CO_2 from a rich supply in the atmosphere and allowed it to be absorbed into vegetation. Plant material became the source of energy and raw material for microbes and animals. Large amounts of organic material were deposited onto the ocean floor and into the ground. In later years, these old repositories became a source of fossil fuels. In the meantime, the process resulted in a reduced atmospheric CO_2 concentration, which started to slow down photosynthesis and formation of organic matter until a relatively stable and low atmospheric CO_2 concentration was reached.

The flows of energy, water and carbon are strongly interconnected. For example, CO_2 concentration has an effect on photosynthesis and on the emission of thermal radiation into space, photosynthesis is connected with transpiration, and volatile organic compounds emitted by vegetation contribute to the growth of aerosol particles after chemical reactions in the air. Thus any major change in one of the circulation systems is reflected in the others. Solar radiation has been rather stable for millions years and circulation systems on the globe have reached relatively stable states in which only small variations occur around stable mean values. The flows of energy, water and carbon, and the operation of the water and carbon cycles, determine to a great extent the vegetation and climate characteristic in each location. Local vegetation and climate respond to any changes in the circulation of energy, water and carbon in the atmosphere.

Photosynthesis enabled the evolution of plant, microbe and animal species. A large number of very different plant, microbe and animal species emerged, each of them specialized to utilise some feature on land or in oceans. Evolution is a very slow process: for example, the evolution of man required millions of years. At least thousands of generations are normally needed to develop a new species, although there are examples of rather fast evolution. Most forest trees live around a few hundreds of years, but some clonal species, such as *Populus tremula*, may live thousands of years. Thus, the evolution of trees is slow, as was the case with humans.

Trees, ground vegetation, microbes and animals on a specific site interact and impose pressures on each other, often developing together in evolution. This is how characteristic ecosystems and vegetation types have emerged to cope with specific environments. Rain forests are characterized by a wet and warm climate, deserts by a dry and warm climate and boreal forests by a cool and wet climate at high latitudes. Coniferous trees dominate the northern forests; the ground vegetation is rich and large quantities of organic matter have accumulated in the soil since the last Ice Age.

We have very little information about conditions on the globe in the time scale of million years. However, some materials can be found which can be dated and which reflect air temperature or gas concentrations characteristic of their formation period. The most valuable sources of information on the time scale of million years are ice-cores taken from the Antarctica or Greenland. Small air bubbles in the ice enable the determination of atmospheric CO_2 concentrations when the ice was formed from accumulating snow. In addition, the ratio of stable oxygen isotopes reflects the temperature conditions of the ocean around the Arctic and Antarctic regions.

Ice-cores that have been extracted from Antarctica and Greenland by research workers indicate that the atmospheric CO_2 concentration has varied in a range from 200 to 300 ppm during the last million years. Changes in atmospheric concentration have been slow; they happen on a time scale of 10,000 years. This is understandable because the amount of carbon dioxide needed for such changes is large and natural processes can generate these amounts only slowly, requiring from a millennium to 10,000 years. In contrast to these large variations, the atmospheric CO_2 concentration has been stable at about 280 ppm during the past 10,000 years (Fig. 1.1.1A).

The stable isotopes of oxygen indicate that temperature has varied simultaneously with the atmospheric CO_2 concentration. Periods of low atmospheric carbon dioxide concentration have been cooler and those of high concentration warmer than the mean temperatures. In addition, during the past 10,000 years the temperature has been relatively stable and warmer than average according to the ice-core data (Fig. 1.1.1B).

The recent stable period of 10,000 years has been long enough for ecosystems to cope with the climate, and a rather clear pattern of vegetation types has emerged. In the northern hemisphere, land areas north from 70°N are tundra; between 70° and 50° N they support coniferous boreal forests. Hardwood forests are typically found further south, from 50°N latitude.

The human population was very small 10,000 years ago and increases in the size of the population were slow until the development of modern medicine and technology. During recent centuries, societies have developed according to their climate

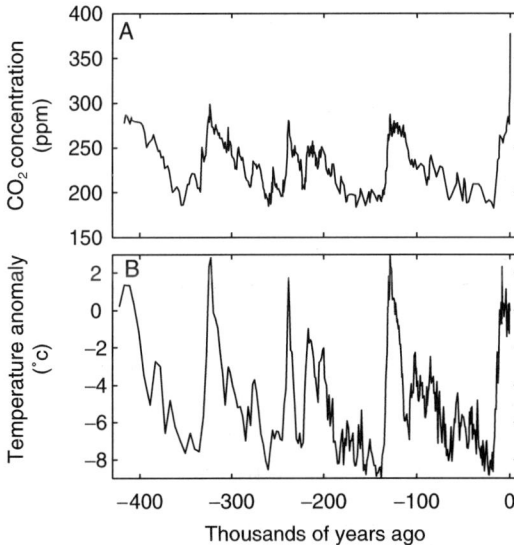

Fig. 1.1.1 (A) Concentration of CO_2 and (B) development of temperature during the last 420,000 years. The concentration data is compiled from three sources (i) Vostok ice cores (Barnola et al., 2003), (ii) Siple station ice core data (Friedli et al., 1986) and (iii) atmospheric measurements (Keeling and Worf, 2005). The temperatures are determined from Vostok ice core isotopes as difference from present values (Jouzel et al., 1987, 1993 and 1996; Petit et al., 1999)

conditions. Each region has developed its own methods of producing food for its population. Crop species and agricultural methods have been chosen to effectively utilize local features of the prevailing climate.

The development of agriculture enabled the growth of human population; fields for crops and animals were able to produce much larger yields per hectare than could be found through gathering and hunting. An increasing population required the clearing of more land for agriculture. For millennia, the solution was to burn forests and to turn them into fields. Thus, much of Europe's temperate forest was already destroyed by medieval times. However, the magnitude of agriculture before the year 1700 was so small that the carbon cycle did not clearly respond to this increasing destruction of forests. After the late 1700s, as a result of the Industrial Revolution, a slowly increasing trend in atmospheric CO_2 concentration can be seen. And since that time, industrialization has been intense and almost always based on energy released from fossil fuels. The atmosphere has responded very clearly; the CO_2 concentration in the atmosphere has increased by 100 ppm since those times.

Thus human activities, use of fossil fuels and destruction of forests, have disrupted the previously stable circulation of carbon from the atmosphere to vegetation, topsoil and the surface layer of oceans. The circulation of energy and water is closely connected with the carbon cycle, thus with changing carbon circulation, the flows of energy and water are responding, and we are facing climate change. There are, however, several phenomena involved in this change that are poorly understood.

The forests are reacting and will continue to react to the changes in atmospheric CO_2 concentration, temperature and availability of nitrogen and water. The reactions of forests generate positive and negative feedbacks into the climate change either accelerating or damping it down. In this book, we analyse the metabolism and growth of forests, and apply our resulting understanding to the interaction between forests and climate change. As an example, we use boreal forests since (i) they are one of the dominating forest vegetation types, (ii) we have a long tradition in their research and (iii) we have versatile measurements from two research stations in Finland, known as SMEAR I and SMEAR II (Section 2.5).

1.2 Theoretical Tools

We urgently need understanding and causal explanations of the effects of climate change on boreal forest ecosystems. Theoretical knowledge has been successfully utilised in physics since the days of Galileo and Newton. Similarly, we should understand the fundamental regularities underlying the growth and development of boreal forests. Theory is the tool for systematising our knowledge and expressing it in a condensed form. The theory should first be formulated and thereafter tested with data from field and laboratory. The observed shortcomings should be improved until the theory passes the tests well enough. Thereafter the theory can be applied to the analysis of the effects of present climate change on boreal forests and to the effects of boreal forests on climate change. Our methodological approach follows the ideas presented by Tuomivaara et al. (1994) and Bunge (1996, pp. 6–7, 30–32, 49, 68–72, 79, 124–126, 179–180).

The starting point of theory formulation is to recognize that there are permanent and general regularities which we do not know. The aim of the research is to discover the permanent and general regularities that govern causal relationships

$$Y = F(X), \qquad (1.2.1)$$

where Y is the feature to be studied, X is an unknown factor and F is an unknown function that links X and Y. Since all factors and the function in the above equation are unknown, they must be specified before any statements can be made.

The theme of the present book comprises a wide range of scales from tissues via individuals and ecosystems to boreal forests and very different phenomena such as photosynthesis, formation of tree structure, carbon sequestration in boreal forests and aerosol formation due to emissions from forests. Proper flow of information across scales between different phenomena is the key to understand the natural development of forests and also of the disturbances being caused by the present climate change. A metatheory is constructed to provide general outlines for the theories of the phenomena involved. The metatheory consists of the basic concepts and ideas. We construct theories for each phenomenon at its scale under the guidance of the metatheory. In this way, the analyses of different phenomena at various levels speak the same language, allowing proper information flow between phenomena across their scales.

The basic concepts of the metatheory define the fundamental properties of the system. Our analysis is focussed on the features and phenomena that the basic concepts cover; those features that they do not cover are omitted as not important. In this way, the basic concepts draw boundaries for the theories involved in the analysis.

Coherent basic concepts in different theories are not sufficient to guarantee the proper flow of information; instead the same types of ideas have to be applied to analyse different phenomena at various levels. The basic ideas of the metatheory are formulated using previously defined concepts and words of common language. These ideas tie together the basic concepts of the metatheory and describe the most important biological, chemical and physical features of phenomena at different levels and utilize knowledge from other disciplines.

Basic Concepts

The metatheory includes ten basic concepts. They define the most essential features of the studied phenomena from tissues to boreal forests. The definition of basic concepts in the metatheory begins with the introduction of process. By definition, a **process** converts material and/or energy into a new form or moves material through a membrane. Photosynthesis forms sugars from carbon dioxide and water using solar energy. Thus it is a process according to the definition in the metatheory.

Similarly, formation of new tissues, i.e., growth, nitrogen uptake by roots and aerosol growth are processes. The processes are quantified with the amounts of material converted or penetrating through a membrane. Processes are also characterised by their rate: the rate is the flux of material or energy generated by the process under consideration.

Environmental factors, by definition, are those features in the environment that have effect on the processes. Light intensity, temperature and carbon dioxide concentration have effects on photosynthesis. Thus at least these three are environmental factors. There are many other factors which are essentially concentrations that affect processes: water vapour in the air, nitrogen ions and water in the soil.

Functional substances enable the metabolic processes in living organisms. Most of the structure of plant cells is formed by cell walls and membranes, which are passive skeletons for the metabolism. All biochemical reactions have specific catalysing enzymes, membrane pumps actively transport material through membranes, and pigments capture light quanta. Thus enzymes, membrane pumps, and pigments form the functional compounds.

Stable carbon compounds form the **structure** of living organisms, and cells are their basic structural units. Similar cells form tissues; they build up organisms and finally organisms form ecosystems. The structure involves several levels that have their characteristic processes and phenomena. Geometrical features, chemical composition and mass characterise the structure.

An ordered structure is called a **hierarchy**. The different levels of structure and time form the hierarchy of ecosystems.

Transport moves material or energy in space. The long distances to be traversed by materials within trees have intrigued scientists for a long time, and transport phenomena are still being actively researched. The transport is characterised by fluxes, i.e., material or energy transported per a unit of area during a unit time.

Living organisms have **biochemical regulation systems** that control the functional substances. New substances are synthesised or activated, existing functional substances are decomposed or inactivated. The biochemical regulation system and its effects on the metabolism has received less attention than it deserves.

New metabolic processes, structures and biochemical regulations and improvements in the existing ones emerge in **evolution**. These novel features are tested against the existing ones in the process of Natural Selection. If the novel feature contributes to the success of the species, then it has the tendency to become more common. This process results in slow '*improvements*' of the species as it becomes adapted to the environment of its location.

Evolution has developed the biochemical regulation system to cope with the very regular and strong annual cycle of light and temperature that is a characteristic feature of the environmental factors in boreal forests. The **annual cycle of metabolism** covers the changes in the biochemical regulation system, in functional substances and in structure matching vegetation with the annual cycle of environment. In boreal ecosystems the annual cycle is quite extreme, and the adaptations shown by plants and animals need to be correspondingly quite pronounced.

Besides the annual cycle, there are slow (i.e. during the life time of individuals) and irregular changes in the environment. **Acclimations** are changes in the biochemical regulation system, in functional substances and in fine structure matching vegetation with slow and irregular changes in environment. In contrast to adaptations, which occur over evolutionary time-scales, acclimations occur over days, weeks and months and do not involve alterations in the genetic make-up.

The basic ideas link concepts with each other. These rather general relationships will be specified when theories of specific phenomena are constructed. In this way, they outline the treatment in each chapter.

Basic idea 1: Environmental factors can be well defined only in sufficiently small space and time elements due to their great variability. The properties of the environment determine the most detailed level in the hierarchy. The great variability is characteristic for light, but also some concentrations and temperatures vary strongly in space and time.

Basic idea 2: Environmental factors and the functional substances determine the conversion of material and energy in processes. The relationships between processes and environmental factors can be determined only at the level of space and time elements due to the great variability of environment.

Basic idea 3: Transport is a physical phenomenon. Physical factors such as pressure difference or thermal movement of molecules generate material and energy fluxes and they should be analysed utilising physical knowledge.

Basic idea 4: Transport and processes effect on the environmental factors. The temporal and spatial properties of environmental factors should be analysed. The development of atmospheric CO_2 concentration, for example, should be known in the past and the predicted increase should be taken into consideration when evaluating development of boreal forests in the future.

Basic idea 5: The spatial levels of hierarchy are space elements of tissue, individual, ecosystem and boreal forests; the temporal levels of hierarchy are time element, year and rotation period. Proper utilization of information gained on a more detailed level and analysis of emergent properties are the keys for transition between the levels.

Basic idea 6: Efficient processes, structures and biochemical regulations have developed during evolution. The emergence of novel features and their test against the existing ones result in effective metabolic, structural and regulation solutions; we can assume that they are close to the best possible ones. Efficient solutions can, at least in some cases, be found as solutions of optimisation problems.

Basic idea 7: The annual cycle of metabolism has been well tested during evolution and it enables individuals to utilise the very regular alternation of warm and freezing periods in the climate of boreal forests.

Basic idea 8: The acclimations, involving processes and biochemical regulation, have been tested during evolution only if the environmental factors generating the acclimation have varied during the lifetime of individuals. Thus most acclimations benefit the acclimating individual, such as changes caused by increasing shading or water deficit in the soil. The atmospheric CO_2 concentration has varied considerably during the last million years. However, the changes have been so slow that during the lifetime of individuals the concentration has been stable. The present increase is outside the range tested during evolution. Thus acclimation to CO_2 can benefit or harm the individual and we cannot predict the acclimations of boreal forests to increasing CO_2.

Basic idea 9: Conservation of mass, energy and momentum is a fundamental feature in nature. Chemical compounds and energy form may change but the number of atoms and the total amount of energy do not change in processes or in transport.

The definition of metatheory:

A theory T of the domain D is a metatheory if the domain D is a set of theories (http://en.org/wiki/Metatheory). Thus metatheory deals with theories, not with phenomena in nature. In our case the domain of our metatheory is theories dealing with boreal forests or climate change in wide sense. Each theory has its own specific concepts but they must be special cases of the concepts in the metatheory. In addition, the theories have their own ideas, but they have to be special cases of the ideas in our metatheory or the theory may have its specific ideas, which do not contradict, with the basic ideas in our metatheory.

The following example may illuminate the difference between metatheory and theory and how the metatheory is utilized to structure the book. Photosynthesis forms sugars using atmospheric CO_2 and solar radiation energy as raw material. Thus photosynthesis is a process and CO_2 concentration and light intensity are environmental factors. There are pigments, membrane pumps and enzymes involved in photosynthesis, which enable the process. These pigments, membrane pumps and enzymes are functional substances. The formation of sugars depend on light intensity, CO_2 concentrations in mesophyll and on functional substances (Basic idea 2). Diffusion transports CO_2 inside the stoma (Basic idea 4). The operation of the stoma is very effective in controlling CO_2 flow into and H_2O flow out of it (Basic idea 6). Thus the fundamental features of the theory dealing with photosynthesis (Section 6.3.2.3) are outlined in the metatheory, but specific features of photosynthesis are introduced with more precise concepts and relationships between them.

Our book includes a very versatile collection of theories dealing with vegetation, forest soil, aerosols and climate change. The flows of material and energy have important role in the theories and the conservation of mass and energy is widely utilised when modelling different phenomena in instrumentation, vegetation and atmosphere. All theories are formulated within the metatheory, which enables communication between the theories.

Theories will be constructed within the above outlined metatheory. Then the phenomenon in consideration is first analysed, the processes and transport phenomena are identified, and the relationships between each other and the environment is analysed. The semantic presentation thus derived is visualised with symbols referring to basic concepts, influences and material/energy amounts and flows (Fig. 1.2.1). These symbols will be used throughout the book.

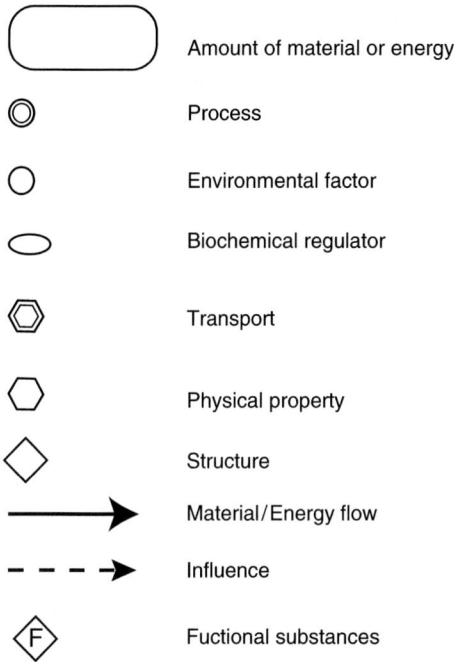

	Amount of material or energy
	Process
	Environmental factor
	Biochemical regulator
	Transport
	Physical property
	Structure
	Material/Energy flow
	Influence
	Fuctional substances

Fig. 1.2.1 Symbols describing the basic concepts to be used in the visualisation of theoretical ideas

The particular regularities in boreal forests and in the interaction between climate change and boreal forests can now be analysed with the basic concepts; process, transport, environmental factors, hierarchy, biochemical regulation, annual cycle and acclimation. The basic concepts allow approximate description of the reality, they do not totally correspond to the phenomena in boreal forests, but they cover the most essential features of the phenomenon under consideration. Thus we can introduce theoretical concepts Y_{th} and X_{th} to treat the phenomenon.

The basic ideas cover the most important connections between the basic concepts. They have an important role together with results from other sciences, especially physics and chemistry, in formulating the theoretical model to describe the relationship F assumed to exist in reality (Eq. 1.2.1). In this way, we can develop the theoretical model f_{th}

$$Y_{th} = f_{th}(X_{th}, a) + \varepsilon_{th}, \tag{1.2.2}$$

where Y_{th} is the response and X_{th} denotes the explaining factors in the theory, ε_{th} is the difference between the real and theoretical response and a denotes the parameters in the model. The quantity ε_{th} is assumed to be a random variable.

Measurements of the response and explaining factors are needed to evaluate the performance of the theoretical model. This introduces an additional complication since measurements always include systematic and random errors. Let $\varepsilon_{m,y}$ and $\varepsilon_{m,x}$ be the random measuring errors of Y_{th} and X_{th} and y_{th} and x_{th} the measured values of Y_{th} and X_{th}. If we omit the systematic error and assume that the errors are additive with the signal to be measured, then

$$Y_{th} = y_{th} + \varepsilon_{m,y} \qquad (1.2.3)$$

and

$$X_{th} = x_{th} + \varepsilon_{m,x} \qquad (1.2.4)$$

When the above equations are introduced into the theoretical model we get

$$y_{th} = f_{th}(x + \varepsilon_{m,x}) + \varepsilon_{th} + \varepsilon_{m,y} \qquad (1.2.5)$$

The measuring noise of the explaining factors generates noise in the modelled response. Let $\varepsilon_{m,f(x)}$ denote the random variation in the modelled response generated by the measuring noise of x_{th}. We get

$$y = f_{th}(x) + \varepsilon_{th} + \varepsilon_{m,y} + \varepsilon_{m,f(x)} \qquad (1.2.6)$$

Thus when we are analysing field data we always have four components in the model. The variation in observed responses, which is explained by the theory and is deterministic, and three random components which are caused by the shortcomings of the theory and measurement noises.

Testing with field measurements of the assumed regularities, especially causal explanations, is a complicated process and it includes random elements. However, testing with measurements enables evaluation of the performance of the theory and evaluation of the reliability of predictions based on the theory.

1.3 The Aim of the Book

Understanding of the effects of present climate change on boreal forests and of the feedbacks from boreal forests to climate change is urgently needed. The changing climate is reflected in the metabolism of boreal forests; for example increasing atmospheric CO_2 concentration accelerates photosynthesis. These changes in metabolism generate responses throughout boreal forest ecosystems. This book is an attempt to bridge this need for knowledge across different scales utilising theoretical thinking within the same metatheory and modern quantitative research technology. The aims of this book are:

(I) To formulate theories within the metatheory at various levels of boreal forests utilising understanding of phenomena at a more detailed level
(II) To quantify and test the theories
(III) To apply the obtained results to analyse and to predict the effects of climate change and of forestry on boreal forests
(IV) To analyse and predict the feedbacks of boreal forest on climate change

The book consists of 11 parts. After this introduction, the second chapter deals with the quantitative methods applied, both modelling and measurements. Thereafter the basic concepts, environmental factors, structure, transport and process are treated separately. Processes, transport and structure are combined in Chapter 7 to deal with trees, ecosystems and atmosphere. The connections between the processes, transport and structure are analysed on tree level in Chapter 8. The obtained results are utilised in the construction of a forest ecosystem model, called MicroForest. In Chapter 9, the resulting model is tested with data obtained from the southern and northern borders of boreal forests. The constructed model is applied to the analysis of the response of boreal forests to climate change and to the effects of boreal forests on climate change in Chapter 10. The book closes with some remarks by summarizing a number of important points and conjecturing some developments for the future.

Chapter 2
Methodologies

2.1 Background

Pertti Hari

University of Helsinki, Department of Forest Ecology, Finland

Any boreal environment is a dynamic system that interacts with its natural and anthropogenic surroundings over different scales of time. A boreal forest system can be divided into several sub-systems, which affect each other via energy and matter flows. When studying a complex multicomponent system like a boreal forest, one can approach the interactions and the system as a whole from several directions. Physical quantities that describe the forest in its current state can be measured, as well as their variability in time. If the processes affecting the system and their interactions are known well enough, one can model the system and predict future behaviour of selected physical quantities. In a post-processing phase the modelled and measured parameters can be intercompared with the aid of statistical methods.

Isaac Newton laid down the foundation of physics for several centuries, and in addition, he established a dynamic modelling tradition that is still very vital. The use of Newtonian methods was very limited in ecology until the beginning of the 20th century. The only application of dynamic models in ecology with a long history is that of the Lotka-Volterra equations which describe the predator-prey interaction. The limited measuring possibilities and great variation evidently hindered any larger scale use of dynamic models.

Statistical methods were developed during the first half of the 20th century. Agricultural field experiments needed methods for proper conclusions under the practical irregularities arising from spatial and temporal variation as well as heterogeneity of material and management. Experimental designs and statistical tests were thus first developed for field crops. Forest inventories utilised very early the statistical methods that had been developed for sampling populations and drawing conclusions about their features. These ideas are still used in the determination of forest properties on various scales.

P. Hari, L. Kulmala (eds.) *Boreal Forest and Climate Change,*
© Springer Science+Business Media B.V., 2009

Before extensive use of computers, i.e., until the 1970s, the required calculations strongly limited the use of statistical methods. The rapid development of computers and creation of statistical software enabled massive calculations without previous practical limitations. This resulted at first in mechanical and non-critical use of statistical methods without real understanding of the basis of the calculations. An opinion shared by many was that ecological research consists of field measurements together with statistical analysis with computers made by professionals. The importance of the link between the research problem and the analysis of measurements was missing to a great extent, and the results obtained from even well-done statistical analysis were of limited scientific value. The situation has improved during recent decades, but there is still a need for improved links between research problem, the underlying theory, and the statistical analysis.

Dynamic modelling in forest ecology benefited from the systems analysis developed for the treatment of dynamics of complicated systems. Although the application of models based on differential or difference equations is increasing, their usage is still rather limited. The main reason is evidently the rather tedious programming effort that is often needed to carry out the calculations. Efficient software, however, was developed during the 1980s and programs like Stella^{TM} and PowerSim^{TM} have made at least simple applications easy to run; in addition, they are useful in teaching.

2.2 Dynamic Modelling

Pertti Hari and Annikki Mäkelä

University of Helsinki, Department of Forest Ecology, Finland

Growth and development are essential concepts in forest ecology; the research focus is thus predominantly on changes over time, not in the static state of the forests. This is especially evident when we consider the response of forests to climate change, for example the effects of increasing CO_2 concentration. The mainstream statistical methods concentrate on analysis of simultaneous properties, such as relationships between various properties of the forest ecosystem, for example, the effect of differing doses of nitrogen on tree growth or the volume of trees in a large area. In certain application areas these are relevant aspects. As the theme of the present book deals with changes in forests, dynamic modelling has a key role in the development of theoretical ideas and, consequently, in the related field work.

The forest ecosystem can be seen as a system of processes that generate flows of material and energy between the pools in the ecosystem. The basic tool in modelling such processes is a container that has inflow and outflow of material or energy (Fig. 2.2.1). Let $X(t)$ denote the amount of material or energy in the container at the moment t and i and o the inflow to and outflow from the container, respectively. The construction of the dynamic model is based on the basic idea of conservation of mass and energy (Section 1.2: Basic idea 9). The amount of material or energy in

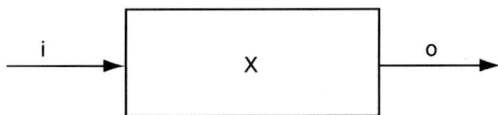

Fig. 2.2.1 Material or energy flows in (*i*) and out (*o*) from a container

the container $X(t + \Delta t)$ after a time interval of length Δt is determined using $X(t)$ and the flows *i* and *o* as follows:

$$X(t + \Delta t) = X(t) + (i - o)\Delta t \qquad (2.2.1)$$

With standard algebraic methods we get

$$\frac{X(t + \Delta t) - X(t)}{\Delta t} = i - o \qquad (2.2.2)$$

When $\Delta t \to 0$, we get

$$\frac{dX}{dt} = i - o \qquad (2.2.3)$$

The differential Eq. 2.2.3 describes the amount of the material or energy in the container. Although the model is very simple, it is a powerful tool in the analysis of phenomena in boreal forests and it makes operational the basic idea 9 (Section 1.2).

The metabolic, physical and chemical processes and transport of material or energy generate flows within living organisms, between organisms and their environment and within the environment. Models for the flows generated by processes or transport will be constructed and they will be combined using Eq. 2.2.3. The models describing the dependence of the flux on environmental factors, enzymes and structure of the system are always approximations and they can be formulated in different ways.

The models include parameters whose values have to be estimated. A standard statistical method, i.e., minimising the residual sum of squares, will be applied. The estimation is often rather problematic since the environmental factors are strongly interconnected and various parameters often have similar effects on the behaviour of the model. These technical problems stress the importance of a critical attitude in the estimation.

A hierarchical structure is characteristic of our approach, from tissue element to boreal forests. The analysis of phenomena at each level should be based on the knowledge from a more detailed level and on emergent properties characteristic for the phenomenon in consideration. The metatheory, which lies behind the models at different levels in the hierarchy, enables combination of the obtained results. The metatheory itself can not be tested due to its generality, and we have to accept or reject it as the basis for the theory. The main criterion in evaluating a metatheory is its demonstrated usefulness.

2.3 Statistical Methods

Pertti Hari[1] and Hannu Rita[2]
[1] University of Helsinki, Department of Forest Ecology, Finland
[2] University of Helsinki, Department of Forest Resource Management, Finland

Deterministic thinking dominates dynamic modelling, as it does in all of the so-called exact sciences, whereas random variation is a characteristic of the realm of statistical thinking. Stochastic behaviour is either a property of the phenomenon itself or is an interpretation of deterministic properties as random variation (noise) to enable their statistical analysis. Noise due to a measuring apparatus is the most evident stochastic element in research. Randomisation procedures in sampling are used to convert deterministic features to random variation. Mathematical statistics has developed efficient tools to evaluate the role of distorting noise in conclusions. In this book, we utilize sampling to gather data from the population and estimation to obtain the values of the parameters in our models. An essential part of modelling is evaluation of the goodness of fit of the model.

Sampling problems deal with the characterization of populations, such as the average length of needles in a forest or the volume of trees in a country. The result can, in principle, be obtained by measuring all individuals in the population. This is, however, impossible in practice due to the labour needed for measurements. Sampling provides tools to characterise a population by measuring only a limited number of individuals. The measurements are then transformed into population level results using statistical methods.

Let H denote a population consisting of N individuals. The property, x, of the individuals in the population follow some distribution having mean μ and variance σ^2. The sampling problem is to estimate the values of the mean and variance using only a small number of individuals measured from the population. In statistical generalisation, the precision of the estimates of population mean μ and variance σ^2 is evaluated using, e.g., confidence intervals.

The key to evaluation of the precision of the values of the relevant parameters is to convert the deterministic property to a random variable. The collection of individuals to be measured is called, by definition, a random sample, S, if each individual in the population H has the same probability of being selected for measurement. The statistical theory has been constructed for random sampling and it is valid only for properly selected subsets of individuals. Systematic sampling is also frequently used, but it has no clear statistical justification. However, systematic sampling when the individuals are selected with some predetermined deterministic rule, seems to give good estimates of the characteristics of the population when wisely used. National forest inventories are good examples of successful systematic sampling.

Let x_i be the observed value in the ith measurement, $\bar{x}$ and s^2 the sample mean and variance, correspondingly. They can be obtained as follows:

$$\bar{x} = \frac{1}{n} \sum_{i=1}^{n} x_i \tag{2.3.1}$$

$$s^2 = \frac{1}{(n-1)} \sum_{i=1}^{n} (x_i - \bar{x})^2 \tag{2.3.2}$$

According to the Central Limit Theorem the distribution of $\bar{x}$ can be approximated with a normal distribution having mean μ and variance σ^2/n. The parameters μ and σ^2 can be estimated by sample mean $\bar{x}$ and variance s^2. The crucial assumption in this strong result is that the measured values are independent of each other. This means that the value of x_i provides no information on x_{i+1}. Randomisation provides justification for the use of statistical analysis and it is the key for statistical generalisation.

The Central Limit Theorem (CLT) enables generalization of the estimated mean and variance to a population by providing confidence intervals for the obtained estimates. The length of this interval is a measure of the precision of the estimates.

Confidence interval for the mean is calculated simply as $\bar{x} \pm \frac{ks^2}{\sqrt{n}}$. The coefficient k is a percentile of the normal distribution; for 95% coverage, it roughly equals 2. The CLT-approximation that is used here works well for samples as small as 30, especially if the measurements come from a symmetrically distributed population. The formula is valid even for smaller samples as soon as they come from a normally distributed population. In this case the percentile k has to be taken from Student's t-distribution, which gives values that exceed 2. Small samples from non-normal populations call for non-parametric procedures.

Confidence intervals for the variance σ^2 can be calculates as well. Because σ^2 can not be negative, the resulting intervals, contrary to the mean, are not symmetric with respect to the estimated value. Both confidence intervals share, however, another important property: the larger the sample size n is, the shorter is the confidence interval. This is consistent with the intuition that larger samples give more precise results. It is important to remember that the bias caused by non-random sampling can not be compensated by increasing sample size.

Usually the models include parameters whose values are unknown. Data are used to estimate the values of the parameters. The sum of squares of the differences between modelled and observed values across all data points is a good measure of the agreement of the model with given parameter values and the data.

Let $y = f_{th}(x, a)$ be the model to be applied, where a denotes the parameter vector. The estimation problem is to determine the parameter values that are relevant with respect to the data available. Let y_i denote the measured response in the ith measurement and x_i the measured value of the explaining factors. The observed responses y_i are assumed to consist of two components, the deterministic one caused by the explaining variables and a stochastic error term, ε, which is often assumed to be normally distributed with zero mean and constant variance

$$y_i = f_{th}(x_i, a) + \varepsilon \tag{2.3.3}$$

The residual sum of squares, RSS, is defined as

$$RSS = \sum_{i=1}^{n} [y_i - f_{th}(x_i, a)]^2 \tag{2.3.4}$$

Among the potential criteria to estimate a, minimising the residual sum of squares is perhaps the most common: the value of a that minimizes RSS

$$\underset{a}{Min}\{RSS\} \tag{2.3.5}$$

gives the solution for the estimation problem. The solution of the minimizing problem is easy in the case of linear regression since it can be found with simple matrix operations. Nonlinear models, common in ecology, call for more advanced numerical methods. The potential local minima may confuse numerical methods when looking for a global minimum: the local minimum residual sum of squares may be rather large when compared with the best possible solution of the minimising problem. Careful inspection of the model behaviour and data should be used to reveal the local minima solutions.

The existence of missing explanatory factors in the model as well as the form of the dependence of included factors can be analysed using residuals. Any systematic behaviour in the residuals ε is a sign of weaknesses in the applied model and enables improvement of the model. Traditionally, statistical analysis of the fit of the model concentrates on the detection of any systematic features in the residuals. This often results in improvements in the model and additional understanding is gained.

The proportion of explained variance, PEV, is commonly used as a measure of the goodness of fit of the model. It gives the proportion of the variance of the measured response that can be explained by the variation of the explanatory factors using the model. The rest, i.e., 100–PEV (or 1–PEV if percentages are not used) gives the proportion of variation that can not be explained by the model. The variance of the residuals determines this variation component. The PEV is calculated for a data set as follows:

$$PEV = 100 \left(1 - \frac{\sum\limits_{i=1}^{n} (f_{th}(x_i, a) - y_i)^2}{\sum\limits_{i=1}^{n} (y_i - \bar{y})^2} \right) \tag{2.3.6}$$

where $\bar{y}$ is the mean of responses in the data.

The proportion of explained variance, PEV, is typically calculated using the same data set from which the values of the parameters have been estimated. In this case, the estimated values will potentially be adjusted for such specific features in the data that are not directly connected with the phenomenon itself. The resulting value of PEV is thus biased towards larger values (i.e., 100%) and is not appropriate for evaluation of the model performance. This bias is great if the number of estimated parameters is large with respect to the number of observations. To get a realistic estimate of PEV, only such data sets that have not been used in the estimation of the parameter values should be used in its calculation.

The residuals have two main sources: they either reflect weaknesses in the model structure or are due to measuring errors. The theoretical model f_{th} is theory-derived assumption dealing with the link between explaining variables and the response. This relationship differs, however, from the real relationship existing in the nature. The difference between the real response and the theoretically modelled response gives rise to error ε_{th}, where the subscript 'th' stands for theory. The measuring noises $\varepsilon_{m,y}$ and $\varepsilon_{m,x}$ of response and explaining factors, respectively, generate further error components that contribute to the residuals. As the measuring noise $\varepsilon_{m,x}$ serves as an input to the model, it has to be converted into an error variation component $\varepsilon_{m,f(x)}$ in the response, generated by measuring errors in the explaining factors. This may take place using, e.g., simulations. Summing up these components we obtain the observed residual error term

$$\varepsilon = \varepsilon_{th} + \varepsilon_{m.y} + \varepsilon_{m,f(x)} \tag{2.3.7}$$

Assuming that the three residual components are independent we obtain

$$Var(\varepsilon) = Var(\varepsilon_{th}) + Var(\varepsilon_{m.y}) + Var(\varepsilon_{m,f(x)}) \tag{2.3.8}$$

The residual variance has thus been split into three components. In Section 2.5.2, it will be shown that reasonable estimates of measuring noise components can be obtained with special measurements. Thereafter, the information embedded in the residuals can be utilised in the conclusions and the weakest component in the empirical research can be identified.

The separation of the residuals into three stochastic components (Eq. 2.3.8) is seldom used. This separation helps to evaluate the quality of measurements. If measuring noises dominate the residuals, the primary task is to improve the measurements, if possible, in order to proceed with the empirical part of the research. The analysis of the role of measuring noise is continued in Sections 2.5.2 and 6.3.2.3.

Statistical analysis is not a lonely island; it is even less only a technical exercise. Various (although not too many) authors have immersed data analysis into a more general framework of empirical inference, one of these approaches being *Scientific Method for Ecological Research* by David Ford (2000).

Design of experiments is perhaps that part of statistics where the connection between theory and data can be seen as its clearest. This ideal has been represented in a clear way in *Experiments in ecology: their logical design and interpretation using analysis of variance* by A.J. Underwood (1997).

Statistical thinking, i.e. heuristic understanding of random phenomena is an efficient tool data analysis. An example of such approach is *Understanding Statistics* by Ott and Mendenhall (1995).

2.4 On Field Measurements

Tuukka Petäjä[1,2], Erkki Siivola[1], Toivo Pohja[3], Lauri Palva[4], and Pertti Hari[5]

[1]University of Helsinki, Department of Physics, Finland
[2]Earth and Sun Systems Laboratory, Division of AtmosphericChemistry, National Center for Atmospheric Research, Boulder, CO, USA
[3]Hyytiälä Forestry Field Station, Finland
[4]University of Technology, Applied Electronics Laboratory, Finland
[5]University of Helsinki, Department of Forest Ecology, Finland

2.4.1 Background

The purpose of measurements is to provide information about physical quantities. A generalized measurement system consists of sensing elements, signal conditioning and processing and data presentation elements. All these steps affect the measurements and have an effect on the final result. Let us take a simple quantity like air temperature as an example. Air temperature can be measured, e.g., with a resistance temperature detector, where in an ideal case the resistance depends linearly as a function of temperature. The change in the resistance is converted to a voltage signal in a bridge circuit. This analog signal is amplified and converted into digital form which is then stored in a computer and can be processed further to take into account, e.g., external calibration factors. The true value of air temperature is thus only indirectly sensed at a single point in space and the measurement signal is processed in a sequence of steps, which all affect the signal in some way.

The development of thermometer and temperature measurements by Celsius, Fahrenheit and Reamur established measurements that can be utilised in ecophysiological studies. Systematic temperature monitoring started in Uppsala, St Petersburg and Edinburgh over 250 years ago. Systematically monitored temperatures covering the whole world have been available for only slightly more than 100 years (Section 10.1.2).

Mechanical methods and sensors to describe the chemical and physical properties of the environment were slowly developed during the 19th and early 20th century. These methods and instruments were often tedious to use and almost impossible for fieldwork. They, however, developed several principles and technological solutions that are still in use.

The development of electric sensors was rapid during the 20th century when the existing principles of mechanical sensors were converted to electric signals and displayed with analogue output or on a chart recorder. This technical development enabled monitoring in field conditions and the measuring accuracy and precision improved. The obtained data was converted manually to digital form for analysis of the results with pen and paper and from 1970 with primitive computers. Large data sets were impossible to analyze properly and they were avoided.

Digitalization of measurements began slowly in the early 1970s. First the output was converted to digital form; later processors became also vital components of the measuring systems. Several previous technical limitations were overcome and novel measuring principles were devised. However, we know from our own experience that the old measuring and analysis traditions limit the utilization of such novel possibilities, and we believe that new instruments can be designed and implemented for ecological measurements.

Ecophysiological measurements have three different goals, which are interconnected but should be clearly recognised when planning and implementing measuring systems. The goal of the first ecological temperature measurements was to describe nature, and this remains a very important goal. We need exact information on environmental factors and processes in the boreal forests, especially on their development during annual cycles.

Testing of models is important for the development of theories within the metatheory. Thus production of data sets for planned tests of models is an important second goal for an ecophysiological measuring system. In a test of a model, we have response variables and explaining variables. For a proper test, we need measurements of the corresponding quantities. The instrumentation should produce these measurements with sufficient accuracy and precision.

Hierarchy is characteristic for boreal forests according to the metatheory. Understanding transitions between levels requires specific information. The third goal of the measurements is to monitor all information needed for proper transition to more aggregated levels in the hierarchy. The three goals of eco-physiological measurements overlap with each other to some extent, but each of them generates some special requirements for the measurements and this is why the three different goals must be clearly recognised.

The planning of instrumentation to obtain data for some well-defined research problem involves several choices which determine the information content of the measurements. The needs of research have to be adjusted to the possibilities available and the constraints imposed by the measuring environment and physical properties of the sensors. Spatial and temporal averaging occurs often in sensors and it is often also needed to limit the amount of data.

The rates of the processes generating the fluxes vary considerably in different time scales. The annual cycle can be seen in all processes and the daily pattern is also very evident. Movement of clouds in the sky generates strong variation in irradiance and is less pronounced for temperature, which is clearly reflected in the processes. The time scale of the movement of individual clouds is a minute or less, and that of a turbulence is 0.1 s.

Variation in solar radiation, temperature and changes in concentrations of gases generate changes in the rates of the processes. If the duration of the measurement of a flux is so long that the factors affecting the rate of the process generating the flux change during the measurement, then the relationship between the flux and explaining factors is smeared. This interferes with the analysis of the data. Thus the duration of the measurement should be so short that the explaining factors remain stable.

Irradiance, especially inside a canopy, has a clear spatial character. If the response area of a monitoring device is so large that it includes considerable spatial variation in irradiance, then the relationship between the rate of the process to be measured and irradiance is smeared by the spatial integration in the measuring system. Thus the flux sensor should be so small that there is only minimal spatial variation in light within the sensor. However, to cover a full range of conditions and to describe the variability, many such sensors will be required.

Technical planning is also important for proper measurements. Rather small details may be crucial for obtaining informative data. For example, proper grounding of instruments reduces electric noise considerably in automated systems.

2.4.2 Measuring Strategy

The old forest ecological measuring traditions do not cover testing of models, although it is commonly the aim of the measurements. The quality of data determines to a great extent the severity of the test and this is why the data should be as informative as possible. There is always variation in the structure and in the processes of living organisms. This variation is problematic for the testing of models using a statistical sampling approach.

Often the theories are tested with the measurements taken from several objects in a population. Thus we want to test the model $y = f_{th}(x, a)$ with the data measured in the field. An additional complication is encountered if the parameters have object-specific values. If we take a random sample from a population, then the variation in the values of the parameters a is converted into random variation. Thus the parameter value for the ith object a_i is

$$a_i = a + \varepsilon \tag{2.4.1}$$

Let x_{ij} denote the explaining variables, y_{ij} the response of the ith object ($i = 1, 2, \ldots, n$) and jth measurement ($j = 1, 2, \ldots, m$). Then we get for the residual sum of squares, RSS,

$$RSS = \sum_{j=1}^{m} \sum_{i=1}^{n} (y_{ij} - f(x_{ij}, a + \varepsilon))^2 \tag{2.4.2}$$

The variation between objects is often large due to different history of the objects and due to genetic variation. This generates additional residual variation and reduces the severity of tests. Evidently, data formed by several measurements of the same object should be preferred, whenever possible, in the testing of a theory.

The forest ecosystems are diverse, both in space and time. Often we need information on both the temporal behaviour and the spatial variation. Measuring temporal development requires frequent observations. To be able to capture the changes in the quantity of interest, measurements should be so frequent that there are only small changes between successive measurements. This requirement may result in frequent measurements, even over ten times in a second, as in an eddy-covariance method.

The number of measuring points of temporal development is often too limited due to high costs, difficult maintenance and huge data sets.

The spatial coverage can be obtained with several measurements. Since the boreal forests are spatially rather inhomogeneous, a large number of monitoring points are needed. If the temporal development should also be observed, then often we meet practical limitations and we have to compromise. A combined strategy is often a good solution; then temporal development is monitored with a few sensors at sufficiently high frequency and spatial variation is measured with campaigns using portable instrumentation.

2.4.3 Accuracy and Precision

Electrical instruments emit a signal, usually voltage or current, that is converted from a measured quantity. This procedure includes several steps, which generate disturbances in the measurement. The obtained result is always contaminated by errors that can be systematic or random.

The standards used in calibration are not exactly correct but fall within some specified limits. The tolerances of several standards in ecophysiological research are rather wide. For example, the cheapest and most commonly used CO_2 standard is only within $\pm 1\%$ of the concentration. Small differences are determined when the CO_2 exchange is measured. Then the poor accuracy of the control gas gives rise to rather large systematic errors.

The properties of the electrical components depend on temperature, thus all measuring devices include a temperature dependent systematic error. This disturbance is often corrected, but still the problem remains although its magnitude is reduced. Light measurements are vulnerable to reflections and to dirt deposition onto the measuring devices. Condensation and evaporation of water and other gasses generate systematic errors in chamber measurements.

Measuring accuracy describes the magnitude of the systematic error. The error is rather easy to determine for a single measuring device, but for large measuring systems it is often very problematic. Especially the surface phenomena in the chamber measurements are difficult to include in the accuracy evaluation.

When a same object is measured close to the detection limit of the instrument in question, the result varies, usually randomly around a proper result. This phenomenon is evident when electrical devices are used, because electrical disturbances generate oscillation in the system. The random variation smears the result and it hinders detection of effects, especially small ones. A measuring precision is the measure of the noise. The standard deviation of the noise component is a practical characteristic of the measuring precision. It should be determined for each set of instrumentation.

Measuring accuracy and precision obtains very little attention in present day ecological fieldwork. This attitude reduces the quality of data, which disturbs proper conclusions based on the field data. Despite natural variability of the measured quantities, the results should be reproducible. In an ideal case this means that the detected phenomenon is measured simultaneously with several different instruments relying on different methods of detection. These kinds of measurements provide information on the instrument accuracy, precision, signal to noise ratio, offset and hysteresis behaviours. A longer time series provides statistics for the observable and also enables correlation analysis with other measured quantities. During long-term measurements, a new additional important issue emerges: the stability and maintenance of the measurement devices.

2.4.4 Measurement Setup

In this section some basic measurement concepts and selected special applications are overviewed, focusing on continuously running field measuring systems. In nature nothing remains stable but is in a constant transient state. The information desired is not in a single value of measurement but in a set of varying samples with properly determined temporal and spatial resolution. Since the object is large, it is bound to have gradients and profiles in the measured quantities. This has to be recognized and taken into account both in the measurements and in the data analysis phase.

All of the above means that the dynamic properties of measurement systems are of increasing importance. Higher sampling rates are desired even at the expense of accuracy. Averaging should be avoided throughout the data collecting chain, in order to maintain the original dynamics and statistics of the target. This is essential when studying nonlinear systems – the usual case.

2.4.4.1 Temperature

Temperature effects most atmospheric and soil processes directly or indirectly and it is by definition also a measure of thermal energy. A wide collection of temperature measurement technology is used in ecological studies, also embedded in instruments for totally different purposes in a more or less visible manner.

In a practical and technical sense, measuring temperature with an uncertainty less than $0.1°C$ is considered a trivial task in a historical perspective on evolution of technology, and even more precise results could be expected. But the reality is worse; the phrase above only reflects the general quality, i.e., technical performance level, of commercial sensing elements, which is only the first link in a practical measuring system. In a free air temperature measurement, radiation introduces larger systematic error than the sensor uncertainty, if it is operated without a proper shielding. Figure 2.4.1 presents an aspirated temperature sensor assembly used in SMEAR II station. Also the spatial and temporal variability is larger than the measurement accuracy. In other words, in environmental studies the character or definition of the object of the measurement is not trivial, at least far from a solid rock on a laboratory

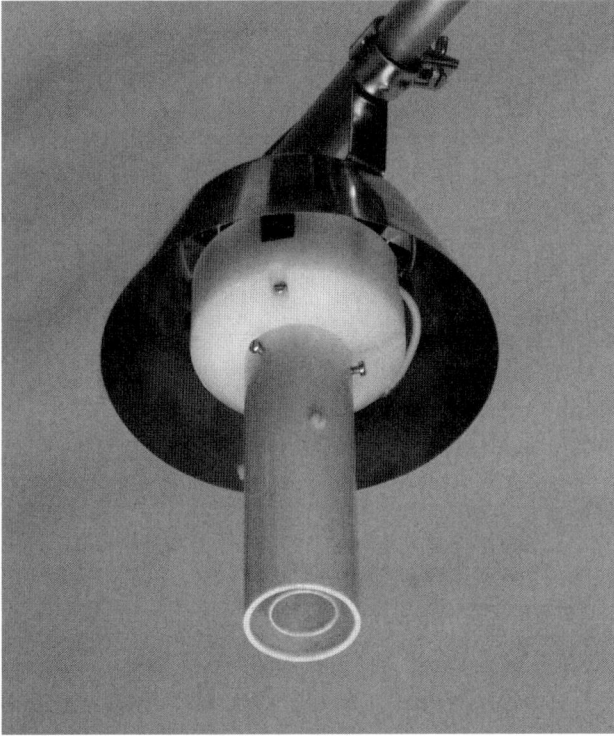

Fig. 2.4.1 Aspirated thermometer used in SMEAR II-station in a vertical profiling assembly. An electrical fan feeds the air through a painted fibreglass double tube structure. The speed of the air passing a centered and shaded PT100-sensor is 4 m/s. Compared to a classic wooden thermometer screen the radiation error is reduced from approximately 1°C to 0.1°C level and the thermal response time is shortened from minutes to some 10 s, which in practice corresponds to the response time of the sensor used. Outer tube diameter is 50 mm (Manufactured by Toivo Pohja)

table. Soil temperature is another good example. Measuring the temperature of a single point in soil presents no technical challenge: variation is slow and limited in range. But the closer to the soil surface the measuring point is, the more sensitive the result is to the displacement, which is hard to define quantitatively in inhomogeneous soil in a repetitive manner. The problem of geometrical definition of the setup is emphasized in a multisensor system where the secondary aim is to calculate thermal fluxes using differences in measured spatial profiles.

In some areas practical improvements on sensor technology are still to be expected. At a shoot or leaf level, small target size poses the principal problem: the instruments tend to disturb the object under study directly or indirectly by shading target from light. In the applications, where a fast time response is desired, there is more room for the technical development. A eddy covariance method is used in micrometeorological gas and thermal flux analysis. It still lacks a fast sensor with less than 0.1 s response time free air that can withstand momentary frosting or icing conditions with a combined acoustic anemometer.

Resistance Temperature Detectors

Several electrical conductor materials are useful as resistance type temperature detectors, which rely on a principle that a resistance depends on surrounding temperature. The first criterion in most applications is a long term stability of a known thermoelectrical resistance characteristic. This is best met by pure metallic resistor materials, like platinum, nickel or copper. Low sensitivity and nonlinearity are secondary issues, and are easily compensated by additional post-processing.

Platinum is known as a chemically inactive metal. So even open thin wire sensors are realizable in applications where fast response in a millisecond range is needed. A usual construction is a coiled wire on a ceramic base, covered with a steel tube resulting in a thermal time constant of several seconds in air.

Mechanical and thermal impulses slowly cause defects in a metallic crystal structure, thus introducing errors, but these effects are typically not important for environmental monitoring purposes. For example, in a standard thermometer screen a platinum sensor can be used for some years with expected drift of less than 0.2°C. When properly packaged and stored, a selected sensor serves as a high quality calibration reference.

For platinum a resistance temperature coefficient is $0.385\,\Omega°C^{-1}$ for a standard PT100 type with $100\,\Omega$ at 0°C. This imposes only moderate requirements on the signal processing system and field installations. A nearly standard level resistance measurement circuitry is adequate, but pulsed excitation supply and 4-wire sensor cabling is preferred. Platinum thermistors as standardized PT100 type variants are widely available, which helps the maintenance of various continuously running systems.

Newly developed platinum thermistors are based on a sputtered or printed film conductor pattern on a ceramic substrate providing a smaller size with faster and cheaper components. For an expanding automotive market, new semiconductor based, but still passive, PTC-resistor sensors have been developed. In order to serve their original target area, they are economical, rugged and slightly more sensitive than metallic thermistors. They are applicable in slow response multipoint systems like soil temperature profiling.

NTC-thermistors are built on bulk metal oxide resistors with a small concentration of impurities to adjust their electrical conductivity. The resulting sensitivity is 5 to 10 times higher than for the platinum sensors. The resistance-temperature function is nearly exponential. Hundreds of types for different purposes with variable precision are commercially available but their interchangeability between products of different manufacturers is poor.

Thermocouples

An electrical circuit connected with junctions between thermoelectrically different conductors forms a simple and robust temperature difference sensor. A voltage is generated between two wires, when the setup is exposed to a temperature difference. This is called a thermoelectric or Seebeck-effect. The output voltage from a thermocouple is only some ten microvolts per degree Celsius in case of copper and constantan. Proper amplification can overcome this low sensitivity, but the accuracy is still limited to 1°C level if standard thermocouple wiring is used. If this is sufficient, thermocouples are economical and easily maintained in multi-channel field installations.

As the thermocouple is an inherently differential and offset-error free sensor, it is ideal for special applications where small local differences are measured excluding absolute target temperature variations. Exemplary setups, where thermocouples are used successfully include soil temperature flux sensors, thermopile pyranometers and peltier psychrometers.

Sensors Based on Integrated Circuits

Modern semiconductor technology offers a variety of sensors where sensing and signal conditioning functions are integrated onto a same component. In the integrated circuit (IC) sensors, the signal is first amplified and then digitized and pre-processed. This simplifies the sensor interfacing with data collecting systems.

The sensing is usually based on the natural temperature-dependent behaviour of a semiconductor junction voltage. As this is not sufficiently controlled by standard semiconductor processes, individual calibration functions are programmed onto the digital part of the circuit during post calibration.

Semiconductors are sensitive to environmental stresses like moisture penetrating plastic packages. Supply voltage needed for active circuits exceeds some volts. For a long term use the semiconductor devices are vulnerable to electrolytic deterioration. This in combination with spurious over-voltages tends to decrease their long term reliability. These setups need additional shielding and transient voltage protection. This easily multiplies the total task of practical system installation. Also communication interfaces and protocols of modern intelligent components tend to be manufacturer specific, which delays application of newly developed systems into large measurement networks.

2.4.4.2 Solar and Terrestrial Radiation

Solar radiation energy is almost a sole driving force for different processes on earth. Radiation here is considered as a multidimensional field variable with a spatial and spectral character connected to a temporal behaviour. This involves a mixed collection of relatively specific instruments adapted to different wavelengths and tasks.

In order to measure energy balance of a boreal forest, one has to monitor both solar (short wavelength) and terrestrial (long-wave, IR) radiation. Incoming solar radiation can reach a measurement point directly, or diffuse to a detector. Some of the radiation is absorbed and part is reflected back into space. All these effects have a wavelength dependency. An example of a construction for diffuse radiation measurements is shown in Fig. 2.4.2.

Fig. 2.4.2 Installation for the diffuse solar radiation at SMEAR II measuring station

In forest ecological studies, radiation methodology can be seen from two basic viewpoints, that of a energetic and that of a process level. From the energetic viewpoint the goal is to monitor the radiant energy flux through a known surface. From the process viewpoint certain discrete wavelength ranges and radiation dynamics are the quantities of interest. For example, infrared-range radiation is closely connected to temperature and thus energy transport, whereas visible radiation utilized in photosynthesis is of a high importance (Section 6.3.2.3), when studying land-atmosphere interactions in a boreal forest environment (Section 7.2). Ultraviolet radiation is a driving factor in many atmospheric photochemical processes (Section 6.3.2.7).

Calorimetric Methods

Wide spectrum solar radiation is traditionally measured by calorimetric methods, i.e., measuring temperature differences caused by the absorbed radiation energy on a specific target. This principle is easily adapted to electronic instrumentation by a thermoelectric converter, a thermopile. The thermopile is essentially a thermocouple circuit whose sensitivity is increased by connecting multiple pairs electrically in series between the two targets of different temperature. This results in a nearly ideal multiplication of sensitivity virtually without an offset error. A simplified structure of the thermopile pyranometer is presented in Fig. 2.4.3, where the upward facing sensing elements are covered with a glass dome. A net radiometer can be constructed in a similar manner except that the dark reference base is replaced with a downfacing absorber. In this particular case the IR-blocking glass dome is then discarded and typically replaced with a polyethylene overall shield.

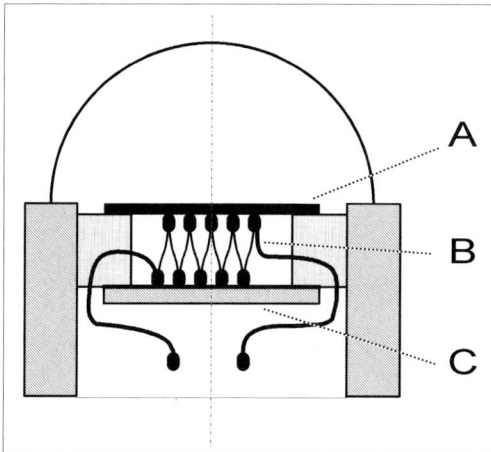

Fig. 2.4.3 A schematic structure of a thermopile pyranometer for global solar radiation measurements. A chain of thermocouples (B) is connected between a black absorbing surface (A) and a reference base (C). A flat sensing plane naturally yields a desired cosine shaped angular sensitivity characteristics but boundary effects and the glass dome tend to distort it at lower angles of incidence. Most commercial moderate quality pyranometers (e.g. CM3, Kipp&Zonen B.V, The Netherlands SK08, Middleton Solar, Australia, TP-3, Reeman, Estonia) follow the cosine function down to 80° in solar zenith angles well within an errorband of ±2%, which is a WMO (World Meteorological Organization) specification for a good quality pyranometer and ISO9060 first class specification

By utilizing the flexibility of the thermopile structure, practical sensors of different angular sensitivity properties can be constructed. Primarily spectral properties depend on the coating material of the absorbing surface. Absorbance variation of less than 2% within 0.1−40 μm range is realized in standard instruments. Wavelength dependency of several commercially available sensors is presented in Fig. 2.4.4.

Although a hardwired thermopile is relatively stable and weather resistant, a shielding cover is used in most field instruments. Practical covering materials such as quartz and germanium or silicon determine their spectral properties and split the instruments into two useful categories: direct solar and far infrared models usually called pyranometer and pyrgeometer respectively. Crossover wavelength is around 2 μm. Thermopile sensors are constantly used as well defined and reliable field instruments and high quality calibration references in solar measurements where sensitivity, speed and size are less important issues.

Another example of radiation measurement devices is presented in Fig. 2.4.5, which represents an absolute tube radiometer (pyrheliometer, aktinometer). High accuracy of the sensitivity factor is based on a balanced differential (potentiometric) principle using electrically controlled resistive power as an absolute reference. A key factor to the high accuracy is the equivalence of the receivers e.g. they should respond thermally equally to both radiant and resistive incident powers. Other sources of errors are mainly eliminated by the principle or relatively easy to

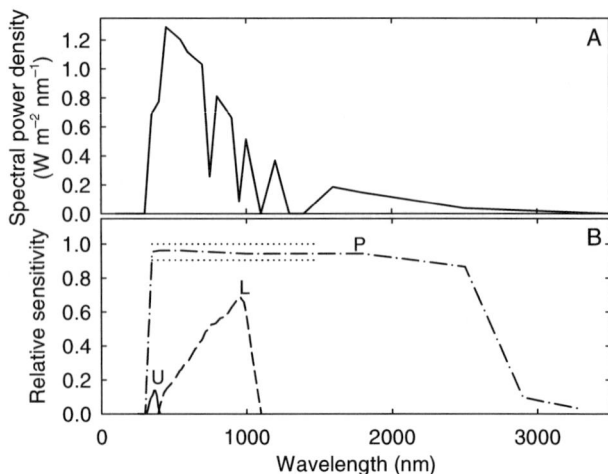

Fig. 2.4.4 Wavelength dependency of solar radiation at sea level (A) and spectral sensitivity of several instruments (B). The solar radiation spectrum is smoothly filtered (Adapted from Monteith and Unsworth, 1990). In the panel B, P is Glass-domed pyranometer (CM3, Kipp&Zonen B.V, The Netherlands) combined with ISO9060 first class ±5%-tolerance band (dashed line). L presents LI-200SA silicon diode pyranometer (LI-COR Ltd, NE, USA) and U shows Ultraviolet-A-sensor (UV-Biometer, model 510, Solar Light Co, Ltd, PA, USA). Note that the spectral power density scale in upper panel represents that of the solar irradiance. The sensor curves (B) are arbitrarily scaled for clarity on the relative sensitivity axis

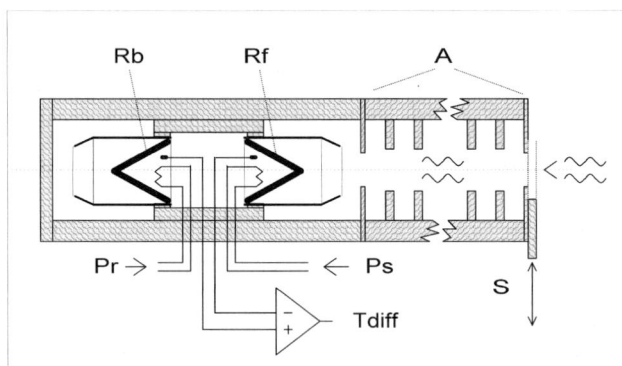

Fig. 2.4.5 A schematic view of an absolute tube radiometer (pyrrheliometer, aktinometer), which is a basic instrument in radiometry. Black absorbing cone receivers, Rf and Rb, are identical and equipped with heating resistors and temperature sensors. Back receiver acts as a dark reference with a constant resistance heating power, Pr. If a temperature difference, Tdiff, is driven to zero with a feedback controlled power supply, Ps, the radiant power entering the front receiver is directly the difference of Pr and Ps. Reference accuracy is still improved by sequentially closing the shutter, S. The dark phase power, Ps is then used as Pr in the continuous mode. The front tube with the apertures (A) defines the effective measurement area and the field of view, usually 1° for direct solar radiation measurements

characterize. A series of instruments, designated as PMO6, based on this principle is build at Physikalisch-Meteorologisches Obsevarorium Davos (Brusa and Fröhlich, 1986) for various ground and space measuring projects. Their uncertainty is determined by different independent methods to be less than 0.2%.

Recent innovations in the IR-sensing field, such as matrix sensors for IR-video imaging or thin film thermopiles used in single spot IR-thermometers, offer new possibilities for measuring traditionally difficult targets like leaf or even needle temperature in field conditions. In addition to being a small and sensitive target, it involves a complex heat and material exchange process driven by radiation by two means – heating and PAR – which lies within a metrological cross-country of temperature and radiation measurements (Kolari et al., 2007).

Photoelectric Methods

A direct photoelectric conversion (Einstein, 1905) is realised in a semiconductive junction, the basic building block of modern integrated circuits. By proper materials and processing the conversion is nearly ideal in many respects necessary in radiation sensors: high sensitivity, wide dynamic and linear range, even 9 decades, small size and high speed.

Modern semiconductor industry offers a wide collection of optoelectrical components as very low cost, mass volume products, or customer specific tailored devices. Commercial photodiodes provide promising economical means for multichannel (light measuring) systems of various geometrical structures needed in forest light measurements. Much attention should be paid to a construction of the optical and mechanical interface around the original sensing component. Angular and spectral characteristics are reshaped and environmental durability needs further improvements. This all should be realised with minimal increase of mechanical size to avoid shading, especially in multipoint field layouts (Aaslyng et al., 1999).

Semiconductors always suffer from environmental stress like moisture, high temperature and hazardous voltages. These present additional challenges in the systems where the optical interface is the main design target. An essential factor of a radiation sensor is its spectral sensitivity characteristic. In case of the semiconductors it is rather complex or discrete if compared with the thermoelectrical sensors. This is due to the fact that only a limited set of base and doping materials are useful in practical semiconductor manufacturing processes.

Consequently, photodiodes are more feasible in certain narrowband applications than for direct wideband radiometer replacements. For example a silicon photodiode naturally covers a spectral range of 0.2–1 μm but it still needs additional filtering in order to roughly approximate pyranometer characteristics.

A photosynthetically active radiation (PAR) sensor is the most common photodiode application used in photosynthesis studies. This arises from the fact that the sensor is sensitive to the radiation at the wavelength range ($\lambda = 400$–700 nm), which corresponds to the wavelengths most important to the photosynthesis. Its small size and speed are essential in shoot level or even cell level model testing in the field. The

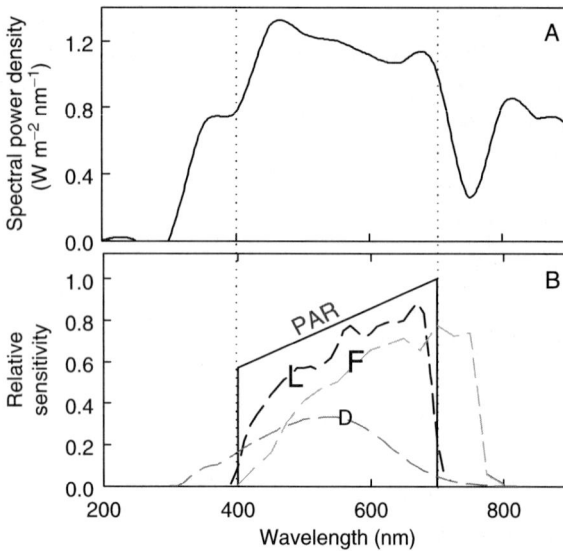

Fig. 2.4.6 A spectral overview of selected practical applications for photosynthetically active radiation (PAR) measurements. Solar radiation spectrum at sea level (A) is adapted from Monteith and Unsworth (2007), power spectral density axis. In the panel B, PAR represents the spectral sensitivity of an ideal PAR quantum sensor and is limited between wavelengths from 400 to 700 nm. On the sensitivity scale the ideal quantum sensor characteristic appears as a linearly rising slope. L is a commonly used LI-190-diode sensor (LI-COR Ltd, NE, USA). F is a multipoint fiberoptic matrix sensor described in Palva et al. (1998a). D is a commercial silicon diode component with an integrated IR-blocking filter (VTB9412B, EG&G Vactec Optoelectronics, USA) (Palva et al., 2001). L, F and D are derived from measurements presented in Palva et al. (1998a and 2001). The sensor curves are arbitrarily scaled on the relative sensitivity axis

desired sampling rate, around 10 Hz, is easily reached (Palva et al., 1998a, 2001). The wavelength dependency of solar radiance and several PAR sensors are presented in Fig. 2.4.6. A series of small sensors are measuring PAR below the canopy in Fig. 2.4.7.

2.4.4.3 Wind Velocity and Direction

Wind is a vector quantity, which has both a scalar component (speed) and direction. Both of them need to be measured. Wind direction is presented in degrees. In meteorology the wind direction is the direction from which the wind is blowing.

Mean wind speed can be measured with several types of anemometers. A cup anemometer (Fig. 2.4.8) typically consists of three or four hemispherical cups mounted on horizontal rods. Rotation of the cup anemometer is proportional to horizontal wind speed. Wind direction on the other hand can be measured with a wind vane, which is a plate, which turns along the wind direction. Both the cup anemometer and the wind vane do not always perform well when the wind speed

Fig. 2.4.7 A series of small PAR sensors measuring the interception of light at SMEAR II measuring station

Fig. 2.4.8 A cup anemometer above SMEAR II site

is low. In particular, wintertime measurements are challenging as snow accumulation to the wind vanes and anemometer cups increase the inertia of the instruments. Their bearings also need periodic maintenance. In a hot-wire anemometer a piece

Fig. 2.4.9 A frozen sonic anemometer above SMEAR II site

of wire is heated up above ambient temperature with a constant current or kept at a constant temperature. Either way, there is a heat flux away from the wire, that is, during high winds the wire cools down faster than during calm periods. One can then relate the change in the resistance or in the amount of current needed to maintain the temperature and relate these to wind velocity around the hot-wire anemometer.

Wind velocity measurements with ultrasonic anemometers rely on the fact that speed of sound travels slower against the wind than it does downwind. This setup consists of a set of sound transmitters and receivers, where the sound travels a known distance (Fig. 2.4.9). The time it takes for the sound to travel from the transmitter to the receiver tells the wind velocity. The effect of temperature to the speed of sound is eliminated by measuring the traveling time both ways between a given pair of transmitters and receivers. This, on the other hand, enables measurement of air temperature. The transmitters and receivers are aligned so that three dimensional wind field is obtained. The ultrasonic anemometers provide a fast response measurement of the wind velocity, which is needed in micrometeorological flux measurements. A typical time resolution is 20 Hz.

Within a boundary layer, the wind velocity changes as a function of height. This is due to drag inflected to the wind by the obstacles on the ground. Also the wind direction changes as function of height. In order to capture these features one needs to make wind velocity measurements at different heights, e.g., from a tall mast. Typically the wind speed is logarithmically proportional to the height. This logarithmic wind law is local and varies from place to place as is proportional to roughness elements around the measurement mast as well as to atmospheric stability.

2.4.4.4 Rain Measurements

Precipitation has been one of the most important quantities to measure during the course of history, since it crucially affects agriculture and crop yields. Both accumulated rain amount and precipitation rate are important parameters to measure. Typically the rain amount is measured with various weighing methods, where rain water is collected in a vessel and the accumulated water is continuously weighed or measured using a tipping bucket setup. Also optical detection of water droplets to a collection funnel with a laser diode and a photodetector are available. With the modern weighing rain gauges, precipitation rates down to 0.01 mm h^{-1} with accuracy of 0.04 mm are possible. Nowadays also continuous measurements of droplet size (distrometer) are possible.

Accurate measurement of rain has proven to be a rather difficult task. Wind affects the collection of rain and snow in a vessel. Also some of the droplets are collected on the sides of the vessel instead of the detector. To increase collection efficiency in the collection vessel, a wind shield has to be used. A commonly used Tretjakov wind shield performs better with respect to snow than did a Wild wind shield used before the 1980s. Evaporation affects slow collection methods. In a complex terrain, e.g., inside a boreal forest, one has to bear in mind the effects caused by the surrounding trees. Some of the rain droplets will be collected by the canopy. Some of this water falls down due to winds shaking the branches of the trees and some will dribble down the tree trunk. Also the rain can have a sharp spatial and temporal variability which affects the measurements and has to be taken into account in the data interpretation. One kind of a rain collecting system is shown in the Fig. 2.4.10.

Fig. 2.4.10 Rain collector below canopy at SMEAR II site

2.4.4.5 Soil Water

Water plays an important role in the biological and chemical processes occurring in the soil. Its one of the major environmental factor together with soil temperature which affects carbon and nitrogen cycles in the forest ecosystem.

Water is bound in the soil by several physical and chemical mechanisms. Most of these like surface tension and capillary force have an electrical origin based on the electrical dipole character of water molecules. These molecular interactions together with the soil pore structure result in the macroscopic quantity, soil water tension or suction, usually called matrix tension (Section 5.4.3). Water tension is expressed in pressure units.

In continuous field soil water measurements two methods are dominating: time domain reflectometry (TDR) measuring water content and tensiometer measuring water tension.

Tensiometer

Soil water tension or suction is traditionally measured with a tensiometer where the suction pressure in liquid phase is directly compared against ambient pressure through a porous cup in a target. This pressure measurement within a 100 kPa range is easily electronically automated and extended to a multipoint field system. The performance of the system is limited by the measurement object and the porous interface properties like nonhomogeneity and thermal transient artefacts, water leakages and bubbling more than by the pressure sensing device.

Hygrometric or psychrometric techniques can be used if the sensor contacts and equilibrates with the soil in water vapor phase. Hygrometric methods are seldom used in the field because of their practical problems in saturated conditions and thermal transient sensitivity although their measuring range could reach extreme dryness.

Time Domain Reflectometry (TDR)

TDR is a direct electrical method: electrical permittivity or apparent dielectric constant of a sample volume is measured with an electrode arrangement with an aid of an excitation signal. The electrical permittivity is directly related to the water content of the soil although the relationship varies slightly with soil type and texture.

This is a general impedance measurement task which could be solved with properly selected constant frequency excitation signal. The TDR is considered as a wide spectrum method where the components of the complex electrical impedance, permittivity and conductivity are detected practically independently.

Originally in soil science, around 1980s, practical applications were build around cable tester instruments where the term TDR refers to (Topp et al., 1980). In the TDR method a step signal is transmitted towards a transmission cable end forming an open transmission line buried in the soil and the reflecting waveform is recorded. The electrical permittivity is evaluated from the record by comparing electrical length, i.e. signal travel time, with the physical length of the sensor.

As an additional by-product the method results in the electrical conductivity of the target zone which is related to salinity of the soil water. The conductivity is calculated from the amplitude of the reflecting step response which represents resistive losses in the target zone (Dalton and Van Genuchten, 1986).

In an ideal case the reflecting waveform is square formed but in practice it is heavily smoothed due to non-homogeneities of the target soil and the sensor geometry. The effective sample volume is not strictly uniformly restricted around the sensor electrodes. The final result is solely sensitive to pulse analysis methods, especially in manual operation. Repeatability, essential in a time series analysis, is better in automated systems using numerical signal analysis algorithms.

A TDR measurement is fairly easily applied to a multipoint system but it requires high frequency multiplexers and low loss cabling. Sensors from distances of several tenths of meters can be connected to one central unit. The system should handle signal frequencies up to some GHz. In contrast the buried part of the instrument is passive simple and rugged.

Constant frequency instruments are used and developed more and more. With these instruments the measuring frequency is around 100 MHz which leads to a relatively simple construction of electronics although impedance measurement of structure sizes near the measuring wavelength is not a trivial task. Usually the active electronics is integrated with the buried electrodes. High frequency cabling is thus avoided. These devices usually provide only one output signal representing the water content which is more easily influenced by the present electrical conductivity than in the TDR systems.

Water tension can be measured using the TDR principle when the sensor electrodes are embedded in a porous ceramic block of known hydrodynamic properties, which determine the dependency between the sensed water content of the block and tension of the surrounding soil (Equitensiometer by Delta-T Devices). This combinational method overcomes the theoretical 100 kPa limit of a normal tensiometer but is more prone to errors caused by impurities and poor calibration.

Calibration of a soil water field measuring setup is a laborious and slow sequence of sample drying treatments in laboratory first to determine the relationship between soil volumetric water content and soil water tension of several samples, possibly including chemical analysis.

2.4.4.6 Trace Gas Measurements

The atmosphere is composed mainly of nitrogen, oxygen and several noble gases. Also water vapour in variable concentration is an important component. There are, however, a lot of other gases present in minuscule quantities. These compounds are called trace gases. Despite their minute concentrations they are important in both local and global scales. Their concentrations are typically given in mixing ratio, i.e., as the ratio of trace gas molar concentration and total molar concentration of air. With aid of the ideal gas law, the mixing ratio can be converted to a ratio of trace gas vapour pressure to total pressure. Depending on the compound, the mixing ratios are parts per million by volume (ppmv, 10^{-6}), parts per billion (ppb, 10^{-9}) or parts per trillion (ppt, 10^{-12}). At a given pressure and temperature the mixing ratio of a given compound can also be presented in units of $\mu g\,m^{-3}$.

Several methods are available for trace gas measurements. For a comprehensive overview of trace gas measurements, the reader is referred to Heard (2006). Some of the trace gasses absorb strongly infra-red radiation (e.g., CO_2 and H_2O). This property can be utilized in concentration measurements, as the absorption of a known infra-red source light is a function of gas concentration in the ambient air. As CO_2 absorbs infra-red radiation, it is a key element in the global warming perspective. Its concentration also provides information on photosynthesis. Water vapour concentration can be measured with the same principle as CO_2, only the wave length has to be changed and the measuring cuvette in the analyser has to be smaller due to higher concentration. A dew point mirror is an alternative measuring principle for water vapour, where a mirror is cooled until condensation of water commences. Formation of liquid layer onto the mirror takes place at dew point temperature. Water vapour concentration can be then calculated from dew point temperature and ambient dry temperature.

Wavelengths varying from UV to visible light can be used in differential optical absorption spectroscopy (DOAS). This enables remote sensing of compounds with a specific absorbance in a short wavelength band. A lot of different trace gasses can be detected with this technique, but due to its high cost this is not a typical air quality monitoring method.

As a gas molecule absorbs UV-radiation, it becomes excited. As it subsequently relaxes to a lower energy state, it spontaneously emits light of a distinct wavelength corresponding to the energy gap. This phenomenon is called fluorescence. It can be utilized in measurements of OH, several halocarbons, SO_2, CO and NO and NO_2. The UV-source can be either a lamp or a laser.

Some compounds are detected after they have undergone chemical conversion. The most important of these chemical methods is chemiluminence, where the reactions produce light which is measured. For example O_3 and NO can be detected with the aid of chemiluminence. Detection of ozone with a variety of dyes is nowadays fast, so it can be even applied to eddy covariance flux measurements (e.g., Keronen et al., 2003). Chemiluminescent reaction of NO and O_3 can be utilized in the measurements of NO. Nitric dioxide (NO_2) on the other hand can be measured after catalytic conversion or using a photolytic technique to transform NO_2 to NO. Conversion is typically not NO_2 specific, therefore other reactive nitrogen compounds (e.g., NO_3, HONO, HNO_3, PAN) are also transformed to NO as well.

Also mass spectrometric methods can be used to measure trace gas concentrations. This method relies on the detection of charged ions in vacuum based on their mass to charge ratios. First the sample gas molecules have to be ionized, e.g., with electron ionization or chemical ionization. Ions are then separated based on their differences in mass to charge (m z^{-1}) ratios with a mass filter, which can be a quadrupole, a time of flight system, a magnetic sector or an ion trap. Finally the detection of m/z separated ions with a faraday cup, secondary electron multiplier or a channeltron. Mass spectrometric methods are used to measure, e.g., concentration of sulphuric acid (Eisele and Tanner, 1993) and OH (Mauldin et al., 1998). Volatile organic compounds can be detected with a Proton Transfer Reaction Mass Spectrometer (PTR-MS, Lindinger et al., 1998) given that their proton affinity is higher than that of water vapour.

2.4.4.7 Aerosol Particles

In addition to various gases, the atmosphere contains ubiquitous numbers of particles suspended in the air. By definition, aerosol is a mixture of solid or liquid aerosol particles suspended in a gaseous medium (Hinds, 1999). Atmospheric aerosol particles affect the global climate directly by scattering incoming solar radiation and indirectly by acting as cloud condensation nuclei (CCN). Regionally, suspended particulate matter degrades visibility (Cabada et al., 2004) and has negative effects on human health (Brunekreef and Holgate, 2002; Von Klot et al., 2005). The net effects depend on the number of particles, their chemical composition, and their physical size.

The size of the particles ranges from nanometers to several tens of micrometers in diameter. Figure 2.4.11 depicts a typical aerosol number, surface area and volume size distributions (modified from Mäkelä et al., 2000a; Saarikoski et al., 2005) in Boreal environment, where three modes below 1 μm can be identified, namely nucleation, Aitken and accumulation modes, which dominate the number concentration. A coarse mode in super-micron size, on the other hand, contributes more to the volume (mass) distribution.

Condensation Particle Counter (CPC)

The total concentration of ambient aerosol particles can be monitored with Condensation Particle Counters (CPCs, e.g., McMurry, 2000). In a CPC the initial size of the sampled aerosol particles are increased by condensing a supersaturated vapour onto the surface of the particles. The particles become large enough to be detected with optical methods. This is a very powerful method to measure concentration of particles that contain only a small amount of mass and are therefore difficult to detect by any other means (McMurry, 2000).

The properties of the condensing vapour (and particles) and the degree of supersaturation determine the smallest size of a particle that can be activated inside the CPCs (Mertes et al., 1995; Sem, 2002; Petäjä et al., 2006). The most frequently used substances as the condensing vapour are water and alcohols. Several methods for producing supersaturation have been suggested including conductive cooling (Bricard et al., 1976), turbulent mixing of cool and warm saturated air (Kousaka et al., 1982), and adiabatic expansion (Aitken, 1897; Kürten et al., 2005). The most

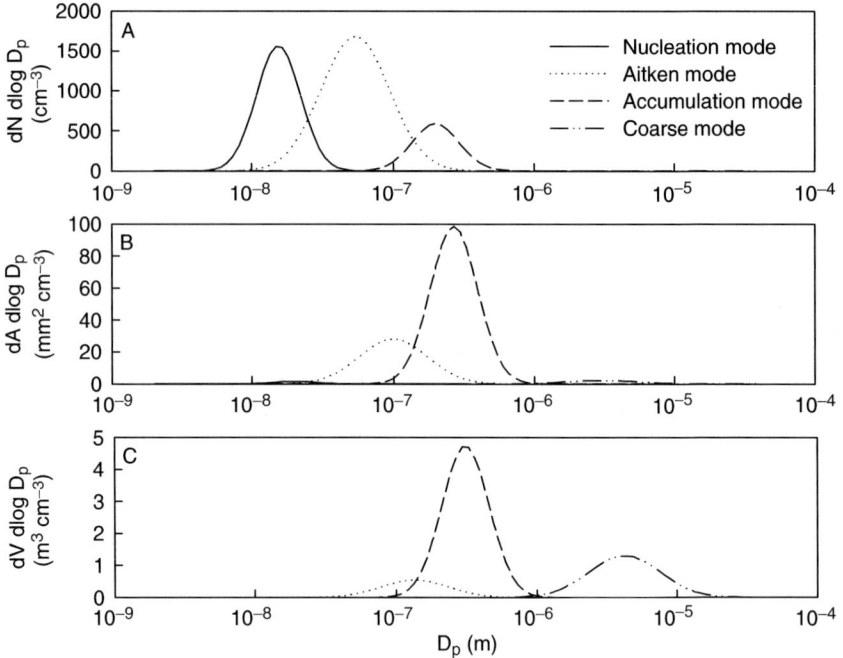

Fig. 2.4.11 A typical aerosol particle number (A), surface (B) and volume (C) size distribution in a Boreal Forest (Adapted from Mäkelä et al., 2000a; Saarikoski et al., 2005)

widely used are conductive cooling type CPCs since they can be operated together with analyzers operating in a continuous flow. In a conductive cooling CPC, the sampled particles are first saturated with a vapour at a higher temperature and then brought into a cooled condenser, where the supersaturated vapour condenses onto the particles. Subsequently the droplets are counted with optical methods.

The most important property for the condensation particle counters its detection efficiency as a function of the particle size. A more practical property is a cut-off size, D_{50}. It is a diameter at which 50% of the sampled particles are successfully counted with the instrument. The cut-off size depends on aerosol losses in the inlet of the CPC, efficiency of the optical system, and particle activation efficiency. However, Stolzenburg and McMurry (1991) showed that the cut-size is mostly a function of activation efficiency alone. In other words, detection efficiency is determined by the amount of supersaturation generated inside the instrument. Mertes et al. (1995) showed that in a conductive cooling CPC operating on butanol condensation, the detection efficiency can be fine-tuned by applying a different temperature difference between the saturator and the condenser of the instrument. Petäjä et al. (2006) showed that also with a recently developed TSI 3785 continuous flow water-based CPC (WCPC, Hering et al., 2005) the detection efficiency depends only on the temperature difference in the system.

Differential Mobility Analyzer (DMPS)

Ambient sub-micron aerosol particle size distribution can be measured with a Differential Mobility Particle Sizer (DMPS, e.g., Aalto, 2004). A twin-DMPS consists of two parallel systems dedicated to different particle size ranges. The combined size distribution represents ambient aerosol size distribution from 3 up to 1,000 nm in electrical mobility equivalent diameters. In short, a DMPS system consists of a Differential Mobility Analyzer (DMA, e.g., Winklmayr et al., 1991), which classifies particles according to their electrical mobility. Size segregated particles are then subsequently counted with a CPC. Since the DMA can only classify charged particles, the particle population is brought to a known charge distribution before sampling with, e.g., a radioactive neutralizer (e.g., Ni-63, Kr-85, Am-241). Particles absorb water vapour in sub-saturated conditions and their size increases (and consequently electrical mobility decreases). To ensure reliable size classification, the sample air is typically dried. Time resolution of the DMPS is typically approximately 10 min. The total sub-micron aerosol number concentration can be obtained from the integrated size distributions. Number size distribution can be obtained also with a Scanning Mobility Particle Sizer (SMPS), where the classifying voltage of the DMA is ramped in a continuous manner while the concentrations are also monitored continuously (Wang and Flagan, 1990). This leads to faster time response. A historical perspective on the electrical aerosol measurements can be found from Flagan (1998).

Chemical Analysis

Atmospheric aerosol particles consist of a mixture of various components from multitude of sources. Major components in the primary particles (directly emitted as particles) include soil-related material (e.g. Fe, Si, Ca, Mg) and organic matter (e.g. pollen, spores). Secondary particles contain nitrates, sulphates, organic carbon and elemental carbon (soot). Mass concentration of particulate matter (PM) is typically collected onto filters. Several mass fractions can be analyzed as the different sized particles are extracted from the ambient population: particle mass below $10 \mu m$ aerodynamic diameter (PM10), below 2.5 μm or below 1 μm (PM1). Historically the chemical analysis has been done off-line, analyzing the composition of the collected PM in the laboratory with various chemical techniques. A thorough overview of available methods is presented in Baron and Willeke (2001).

In the recent years several methods has been developed, which enable chemical characterization in almost real time semi-continuously. These methods include mass spectrometric methods, where particle material is first ionized and then analyzed with mass spectrometric methods (e.g. Jayne et al., 2000). An other method collects particles directly into liquid solution (Particle Into Liquid Solution, PILS, Weber et al., 2001) followed by subsequent analysis with an ion chromatograph. This method provides composition information of water soluble components in a semi-continuous manner. In a thermal desorption chemical ionization mass spectrometer

(TD-CIMS, Voisin et al., 2003; Smith et al., 2005) particles are collected onto a wire. As the wire is heated, evaporated particles are ionized and analyzed with a quadrupole mass spectrometer.

Chemical analysis of nucleation mode particles has proven to be extremely difficult owing to their minuscule mass. Recently the sensitivity of the analysis has improved, and even speciation of single nucleation mode particles in the laboratory has been determined (Wang et al., 2006). In atmospheric conditions, analysis is possible during a growth period of freshly formed particles in the sub-20 nm size range (Smith et al., 2005).

Species in the nucleation mode particles can be inferred indirectly from their hygroscopicity, i.e., how much they absorb water vapour in sub-saturated conditions. This can be done with a Hygroscopic Tandem Differential Mobility Analyzer (HTDMA, e.g., McMurry and Stolzenburg, 1989; Hämeri et al., 2000). In short, this method first selects a narrow size ample from a polydispersed population and then exposes these particles to a controlled amount of water vapour. Depending on their chemical composition they absorb water and their size increases. This is subsequently detected with a DMA. Inorganic salts readily absorb water vapour whereas, e.g., many organic constituents are less hygroscopic or hydrophobic (e.g., soot). Another option to study chemical composition indirectly is to expose particles to higher temperatures. As an example, at 300°C sulphate, nitrate, ammonium, and at least part of organic carbon evaporate from the aerosol phase. The remaining compounds are non-volatile soot, trace metals, fly ash, crustal material, sea salt and carbonaceous compounds (Wehner et al., 2002). By varying the temperature the particles can be classified according to their volatility.

2.4.4.8 Chamber Measurements

Gas fluxes between leaves and atmosphere can be measured with chambers (Fig. 2.4.12). Concentrations inside the chamber change due to the flux between leaves and air. The flux is determined from concentration changes during the closure and air fluxes between the chamber and its surroundings (Fig. 2.4.13).

Let g denote the flux to be measured, C_c the gas concentration inside the chamber, C_r the gas concentration in the air replacing air drawn into the analyser (often ambient air) and, q the flow rate of air into the analyser and V_c the volume of the chamber. The mass balance equation (cf. Eq. 2.4.3) for the gas in the chamber is

$$\frac{d(V_c C_c)}{dt} = g - q C_c + q C_r \qquad (2.4.3)$$

The above mass balance equation can be used in two ways, either assuming steady state conditions and omitting the derivative term or studying concentration changes after closure of the chamber when the derivative term is important. Our measuring system is based on the concentration changes after closure of the chamber. Portable

Fig. 2.4.12 An example of an automatic chamber measuring a Scots pine shoot at SMEAR II station

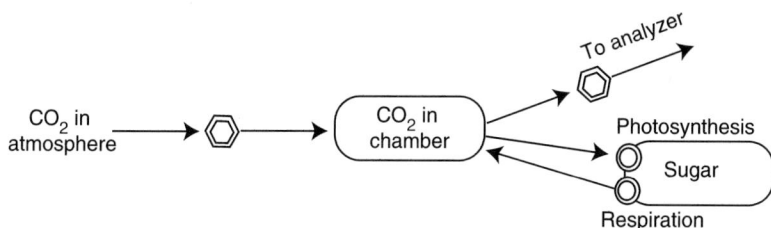

Fig. 2.4.13 CO_2 fluxes in a measuring chamber. The CO_2 exchange, i.e., photosynthetic consumption minus respiratory production can be solved from the mass balance Eq. 2.4.3. The symbols are introduced in the Fig. 1.2.1

gas exchange measuring systems utilise the steady state approach. Thus the derivative in the Eq. 2.4.3 is assumed to be zero and the flux is solved from the simplified equation.

2.4.4.9 Eddy Covariance Measurements

The eddy-covariance method is a rather new method that is based on the novel possibilities opened up by digital techniques. It is designed for the measurement of gas and energy fluxes between the forest ecosystem and the atmosphere. This method utilises the transport of gasses and energy by eddies above the canopy. The three dimensional wind speed, gas concentrations and temperature are measured at high frequency, at least 10 Hz. The micrometeorological fluxes of momentum, heat, CO_2 and H_2O can be calculated as 30 min average co-variances between the scalars (or horizontal wind speed) and vertical wind speed according to commonly accepted procedures (Aubinet et al., 2000, Section 4.2). In principle, the instanta-neous amounts of the scalars transported by turbulent eddies are determined and the mean flux during the period is the average over all transport events. The fluxes should be corrected for system frequency response limitations and low-frequency underestimation based on empirical transfer functions and co-spectral transfer char-acteristics (Rannik et al., 2004) and any other instrument-specific systematic biases such as correction due to density fluctuations (Webb et al., 1980).

Comprehensive handbooks by Ian Strangeways (2003) or by Robert Pearcy et al. (1989) on environmental measuring techniques and field instrumentation are sug-gested for background and further reading.

2.4.5 Planning of Measurements

Field measurements have two main goals; either to describe some system or to test a theoretical model or hypothesis. In addition, sometimes special measurements are needed for proper transition between different levels in the hierarchy. These goals have to be considered in planning of the measurements. There are, however, other aspects such as measuring accuracy and precision, practical arrangements, available money and labour which put strong constraints for the measurements. The actual measuring setup is always a compromise between the conflicting requirements. All the relevant alternatives should be carefully analysed before construction of the mea-suring system. The performance of the measuring system should be analysed after the first measurements and proper improvements should be made.

The two main goals of the measurements result in rather different requirements for the measuring system and there are differences in crucial aspects of the planning. This is why we treat them separately.

Description. Let Z be a property in an ecosystem that we want to describe with our measuring system; it may be temperature, mass, flux, etc. The property varies in space and time, thus

$$Z = Z(t, x) \tag{2.4.4}$$

where t denotes time and x three-dimensional space coordinates. The description problem is to characterize $Z(t, x)$. This can be done in several ways, for example, with some functions. In practice, however, the characterization is done with some statistical parameters, usually with mean and variance. The description is done by taking a sample from all possible values, called population, and with a generaliza-tion from the sample to the population (Section 2.3).

The first step is to choose a description of the property $Z(t, x)$. The separation of time and space is often useful in ecological measurements, since the annual and daily cycles in the environment dominate the temporal behaviour

$$Z(t, x) = Z_t(t) Z_x(x) \tag{2.4.5}$$

where Z_t describes the temporal development and Z_x represents the spatial properties of the quantity Z.

Weather stations have successfully monitored temperature for over 200 years, although global temperature records go back only about 100 years. The spatial variation is a problem during low wind velocities and when the mixing in the lower atmosphere is weak. However, the temporal development of the quantity Z_t can often be measured with the properly planned measurements.

The spatial properties in ecosystems vary strongly and irregularly. Statistical methods are useful to describe the spatial properties. Then we assume that Z_x follows some known distribution in the population, usually a normal distribution. Now the problem is to determine the mean, μ, and standard deviation, σ, in the population.

Statistical methods assume that we have a random or systematic sample from the population. Thus several measuring points should be chosen in the ecosystem. They should be measured in such a way that the temporal changes in the quantity Z do not disturb the results; evidently measurements should be simultaneous. Thereafter statistical methods provide the mean and standard deviation and their confidence intervals (Section 2.3).

The spatial properties of the quantity Z determine the needed measuring precision and accuracy. The measuring precision should be so good that the spatial variation can be measured by the instrument. Let s_I be the standard deviation of the noise of the instrument. If s_I is considerably smaller than the standard deviation in the measured population, σ, then the instrument is able to pick up all the relevant spatial variations. In practise, if $s_I < \sigma/10$, then the instrument is sufficient for the planned application.

The length of the confidence intervals depends on the variation in the measurements and on the number of measurements; a great variation can be compensated with an increasing number of measurements. The amount of labour is proportional to the number of measurements. Thus the proper extent of field work should be determined from the desired length of the confidence intervals.

Testing models. The basic ideas in the planning of the measurements to test a theory, or a theoretical model, is that a measuring noise of a response and explaining factors do not smear out the response to be tested and that the measuring arrangements enable us to see properly the response in the field conditions.

An analysis of the disturbances caused by the measuring noise resulted in Eqs. 2.3.3 and 2.3.7

$$y = f_{th}(x) + \varepsilon_{th} + \varepsilon_{m,y} + \varepsilon_{m,f(x)} \tag{2.4.6}$$

where y is a response, x are explaining variables, $\varepsilon_{m,y}$ are a random measuring error of the response, $\varepsilon_{m,f(x)}$ is a random variation in a modelled response generated by the measuring noise of explaining factors x_{th} and ε_{th} is a random variation caused by shortcomings of the theoretical model. The measuring noises should be so small that they generate only a minimal variation when compared with the shortcomings of the theoretical model. Thus the variance of $\varepsilon_{m,y} + \varepsilon_{m,f(x)}$ should be clearly smaller than the variance of ε_{th}.

A needed precision is an important criterion in a selection of instruments, since the noise level depends strongly on the measuring principle and on the quality of electronics. Shortcomings in the grounding of the measuring system easily generate additional variation in the results. Thus a proper knowledge is needed to eliminate this additional variation component.

An environment is often so variable that it can be well determined only in sufficiently small space and time elements. Thus a relationship between the environment and a process can be determined only at the level of the space and time elements. This spatial and temporal resolution is very difficult to achieve with the present instrumentation and special arrangements are needed to solve the discrepancy between the scales of the phenomena in nature and the instrumentation. One solution to the problem of great variability of the environmental factors is to measure in places where the variation is small, for example, in the top shoots of the canopy (Section 6.3.2.3). If the variability can not be eliminated with practical arrangements, then the environment has to be measured in the same volume and time period as the process is measured (Palva et al., 1998a, b).

Systematic measuring errors are a serious problem when testing models. We have no means to separate the effects of the shortcomings of the theoretical model and the systematic measuring errors from each other. This fact is a major methodological problem when testing the models with field data.

Instrumentation, especially chambers, disturb the object to be measured. These artefacts should be carefully analysed and eliminated if possible. For example, the flows of air are crucial for the accuracy and precision of soil efflux measurements (Pumpanen et al., 2001).

2.5 SMEAR Network

Pertti Hari[1], Eero Nikinmaa[1], Timo Vesala[2], Toivo Pohja[3], Erkki Siivola[2], Tapani Lahti[4], Pasi Aalto[2], Veijo Hiltunen[3], Hannu Ilvesniemi[5], Petri Keronen[2], Pasi Kolari[1], Tiia Grönholm[2], Lauri Palva[6], Jukka Pumpanen[1], Tuukka Petäjä[2,7], Üllar Rannik[2], and Markku Kulmala[2]

[1] University of Helsinki, Department of Forest Ecology, Finland
[2] University of Helsinki, Department of Physics, Finland
[3] Hyytiälä Forestry Field Station, Hyytiäläntie, Korkeakoski, Finland
[4] Niinikuja, Tampere, Finland
[5] Finnish Forest Research Institute, Vantaa, Finland
[6] University of Technology, Applied Electronics Laboratory, Finland
[7] Earth and Sun Systems Laboratory, Division of AtmosphericChemistry, National Center for Atmospheric Research, Boulder, CO, USA

2.5.1 Description of Stations

A forest ecosystem and atmosphere form a complex system which has to be structured before a proper monitoring system can be planned and implemented. Trees are the functional units in a forest stand. They photosynthesize with their needles, take up water and nutrients from the soil, transport water in the woody components, form litter and emit volatile organic vapors. Microbes break down the litter in the soil, releasing CO_2 and other organic vapors into the air. Organic vapors in the atmosphere participate in chemical reactions as well as in aerosol formation and growth. Solar radiation is the primary source of energy for several processes, such as photosynthesis, turbulent mixing, snow melting and many chemical reactions in the atmosphere.

The processes and properties in a forest and atmosphere are strongly connected with each other. For example, the atmospheric CO_2 concentration influences photosynthesis and VOC emissions by trees affect aerosol formation. Simultaneous measurements of several phenomena in a forest ecosystem enables the combination of different processes and analysis of connections between the components of the system, including: (i) carbon fluxes which connect the atmosphere, trees and forest soil with each other, (ii) volatile carbon compounds emitted by trees or forest soil, being important to the formation of aerosol particles, and (iii) rainfall, transpiration, evaporation and run off which together connect the atmosphere, trees and soil.

Recently, a rapid development of measuring techniques has enabled versatile field measurements. New trace compounds can be measured under the field conditions, the accuracy and precision of the measurements are increasing, and the required response time is decreasing. For example, more than ten gasses can be monitored with the present instrumentation using a chamber technique, which requires a response time of a few minutes. In addition, the large measuring systems can be automated with digital techniques providing easy and useful data management.

SMEAR I (Station for Measuring Ecosystem – Atmosphere Relations) was constructed in 1991 in eastern Lapland (67°46′N, 29°35′E) 200 km beyond the Polar Circle 400 m a.s.l. at the timber line near the Värriö research station (Fig. 2.5.1). A mission of the station is to analyse the effects of emissions from the Kola Peninsula on nature in an extremely remote place. The nearest road ends 8 km from the station and the Russian border is 5 km east from the station. The nearest permanently inhabited place is the border guard base at 8 km and nearest civilians live 15 km away from the station. Normally air is very clean but when the wind is blowing from the smelters in Kola, the air quality is poor.

SMEAR II (Station for Measuring Forest Ecosystem – Atmosphere Relations) is located in a rather homogenous Scots pine (*Pinus sylvestris L.*) stand (Figs. 2.5.2 and 2.5.3) on a flat terrain at Hyytiälä Forestry Field Station of the University of Helsinki (61°51′N, 24°17′E, 181 m above sea level) 220 km NW from Helsinki (Fig. 2.5.1). The managed stand was established in 1962 by sowing after the area had been treated with prescribed burning and light soil preparation. The biggest city near the SMEAR II station is Tampere, which is about 60 km from the measurement site with about 200,000 inhabitants.

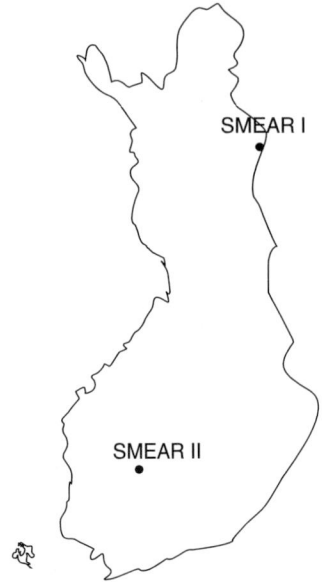

Fig. 2.5.1 SMEAR I and
SMEAR II stations in Finland

Fig. 2.5.2 The study site at SMEAR II is comprised of a 45-year-old Scots pine stand. A view
from the top of the measuring mast

The atmosphere below free troposphere can be described by two compartments;
outer layer and air inside the canopy. These two interact by material, energy and
momentum flows. Trees take carbon from the atmosphere and circulate it back or
feed the carbon pool in the soil. Aerosols, reactive gasses and inert gasses can be
found in the outer layer, inside canopy and soil. Biological, physical and chemical

Fig. 2.5.3 Wintery aspect at SMEAR II

processes generate concentration and temperature differences and fluxes between the components and within them. The main pools and flows are outlined in the Fig. 2.5.4. The SMEAR II measuring station was planned and implemented to monitor the fluxes presented in the Fig. 2.5.4 and to determine relationships between the processes generating fluxes and the environment.

SMEAR II includes four main components: (i) an instrumented 73 m tall mast, (ii) systems to monitor aerosols, (iii) instrumentation to monitor tree functions, and (iv) two instrumented mini catchments. The mast monitors CO_2, H_2O, CO, O_3, SO_2, NO, NO_2, temperature and wind speed profiles, properties of solar and thermal radiation of the stand and fluxes of CO_2, H_2O, O_3 aerosols and several volatile organic compounds between the canopy and the atmosphere. The mast measurements are usually reported as 30 min means. For more details see also (Vesala et al., 1998; Kulmala et al., 2001a; Hari and Kulmala, 2005).

The aerosol and ion size distributions are measured in order to be able to detect ion, cluster and aerosol dynamics. In aerosol dynamics the formation and subsequent growth of fresh atmospheric aerosols is in our focus. Also dry deposition as well as wet scavenging can be investigated. With size distribution measurements, using the growth rate of nucleation mode aerosols and condensation and coagulations sink, the concentration and source rate of condensable vapours can also be estimated (Kulmala et al., 2001b). The aerosol mass composition is also determined. The relations between aerosol dynamics and photochemistry can also be investigated, e.g., analyzing radiation fields, and OH concentrations. Recently Boy et al. (2005) were able to find a closure between calculated and measured sulphuric acid concentrations. The calculated concentrations were based on gas measurements performed at the SMEAR II station.

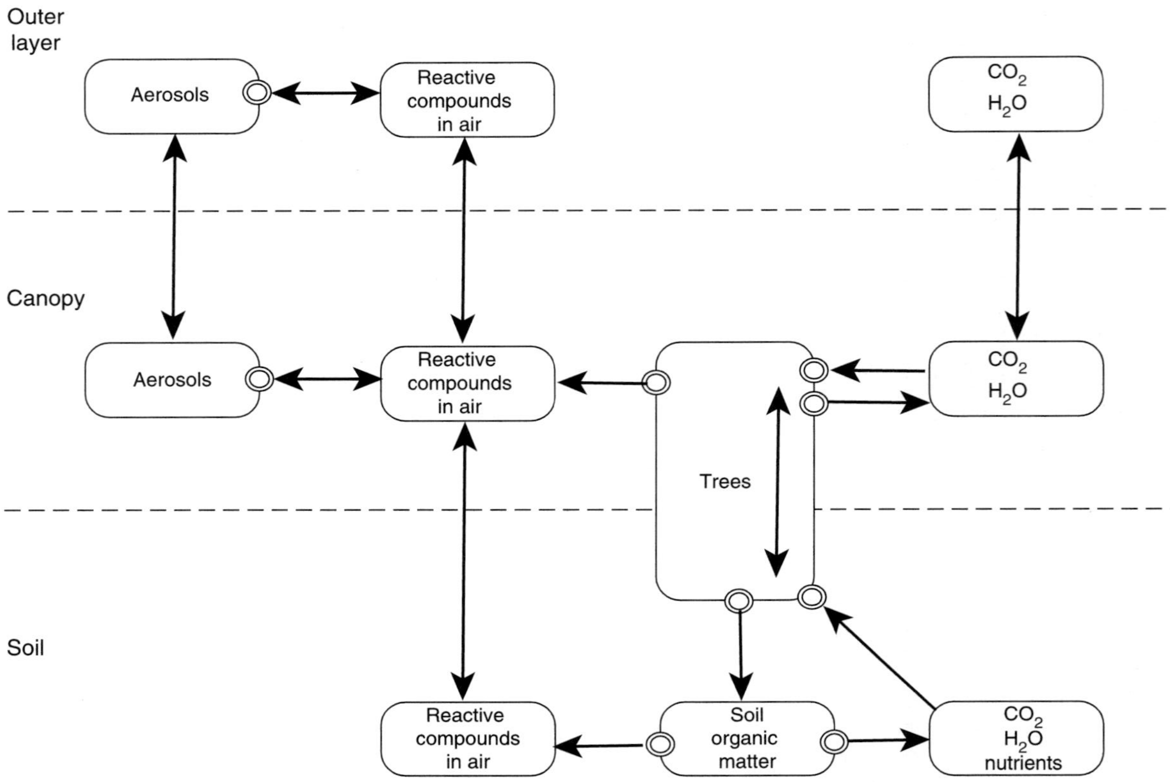

Fig. 2.5.4 Pools and main fluxes in a soil-forest-atmosphere continuum. Symbols have been explained in details in Fig. 1.2.1, arrows indicate flows and boxes amounts

The chamber technique is used to monitor tree processes generating fluxes between trees or soil and atmosphere. The most relevant processes are: photosynthesis, respiration, transpiration, NO_x emission, NO_x and O_3 deposition and emission of volatile organic compounds. Several different versions of the chamber are used depending on the focus of the measurements; needle gas exchange (Hari et al., 1999; Altimir et al., 2002) or stem CO_2 and soil CO_2 efflux (Pumpanen et al., 2001), each using their own modifications of a basic chamber structure. Placing most of the chambers on the top branches in the canopy reduces the disturbances due to temporal and spatial variation in photosynthetically active radiation.

Fluxes between the canopy air space and atmosphere are monitored with micrometeorological methods. In the eddy-covariance method the three-dimensional wind speed and concentrations of CO_2, H_2O, O_3 and aerosols are measured at the frequency of 10 Hz or higher (see, e.g., Suni et al., 2003). In the relaxed-eddy-covariance (see, e.g., Gaman et al., 2004) method the three-dimensional wind measurements control the sampling of air into two containers in such a way that during upward movement the air is collected in the upward bottle and during downward movement into the downward vessel. The concentrations in both of the volumes are measured in regular intervals. A total flux between the canopy space and the atmosphere in the both methods is obtained as a difference between the fluxes down and upwards.

The CO_2 and H_2O fluxes are measured at the SMEAR II station in Hyytiälä continuously by the eddy covariance (EC) technique, located at about 23 m height, roughly 10 m above the forest canopy. The systems include a Solent ultrasonic anemometer (Solent Research 1012R2 and HS1199, Gill Instruments Ltd, Lymington, Hampshire, England) to measure three wind speed components and sonic temperature, and a closed-path infrared gas analyser (LI-6262, Li-Cor Inc., Lincoln, Nebraska, USA) that measures CO_2 and H_2O concentrations. These trace gas flux instruments are typical of those used in the EC flux systems of CO_2 and H_2O (e.g., Aubinet et al., 2000). A distance of the anemometer from the mast is 3.5 m and a 7 m long heated Teflon tube (inner diameter 4 mm) is used to sample air near the anemometer (through a tube inlet fixed below the sensing head of the anemometer, about 15 cm from the center) to the gas analyser situated in an enclosure on the outer side of the mast below the measurement level. The flow rate of 6.3 l/min is used to produce a turbulent flow in the tube. The response time of the LICOR gas analyser to a step-change in concentration is 0.1 s. However, the first order response time of the complete eddy- covariance system for CO_2 was determined to be 0.3 s (Rannik et al., 2004).

A thickness of the soil on the bedrock is very low, only 5–150 cm, due to the ice age. The bedrock is very solid and there is a layer of silt with very high clay content in it, thus we can say that the bedrock is water proof sealed with clay. There are two mini catchments (900 and 300 m^2) near by the measuring cottage (Fig. 2.5.5). These catchments are closed with a dam and the run off from the area is measured. A leakage of substances with the run off is monitored by taking samples for chemical analysis. The water content and tension, CO_2 and temperature profiles are measured.

Fig. 2.5.5 Measured values are stored and monitored in a cottage at SMEAR II station

Solar radiation is the source of energy for several processes in trees and atmosphere. This is why irradiance, diffuse irradiance, photosynthetically active radiation and radiation balance are measured above the canopy. The light distributions within the canopy are monitored with 200 sensors (Palva et al., 2001, Fig. 2.4.7) and within the gas exchange chamber with 400 sensors Palva et al. (1998a, b). The rainfall is measured above and below the canopy. The diameters of the trees are measured annually. The shoot elongation is, however, measured daily in the early summer. On the other hand, stem diameter changes are monitored both above and under bark continuously with a precision of less than 1 μm. The water flow in the wood and in the phloem is calculated from the diameter changes (Perämäki et al., 2001). The litter fall on soil is monitored in two-week intervals.

The system operates year around; only components that may be damaged by freezing water are turned off during the winter. In addition, the growth of the trees around the measuring station has been measured retrospectively to the age of 3 years.

2.5.2 Measuring Accuracy and Precision at SMEAR I and II

An accuracy of sensors and measurements varies and this should be taken into consideration when interpreting the results. A precision of the measurements varies according to the quantity to be measured, but it is often sufficient for interpretation of the results.

Light measurements are vulnerable to systematic errors due to the hard conditions and sensitivity of the sensors. Water and any dirt on the sensor gives rise to too low values and thus biased results. In addition, at low solar angles a horizontal sensor reflects light. The properties of electrical components depend on temperature and also the PAR-sensor and pyranometer have clear temperature dependence. It is strange that the sensor is often mounted onto black aluminium which increases the temperature of the sensor in sunshine.

PAR, temperature, CO_2 and water vapour sensors are read at $10\,s$ intervals at SMEARI during each closure of a chamber. The chamber is closed automatically 180 times during a day. These measurements enable the study of the precision of the measurements in field conditions. If we assume that PAR does not change during a period of $10\,s$, then we have 180 pairs of measurements of same PAR intensity for each day. The regression between the first and second measurement during a closure is rather close. A standard deviation of the difference of two normally distributed random variables having a standard deviation σ is $\sqrt{2}\,\sigma$. This mathematical fact enables calculation of the standard deviation of the measuring noise from the differences of the first and third measurement during the closure. Thus, the standard deviation is $1.2\,\mu\text{mol m}^{-2}\,\text{s}^{-1}$ for PAR, and $0.05\,\text{g m}^{-3}$ for water vapour.

Solar radiation causes the biggest biases in temperature measurements. Complicated radiation shields are used in standard weather stations to get a well-defined environment for the measurements. The temperature increase due to the solar radiation reduces the measuring accuracy considerably and this problem is still more pronounced in ecological measurements than in meteorology and we cannot avoid systematic errors of several degrees Celsius. In addition, a signal of the thermocouples is so small that it may include a systematic bias.

The precision of the temperature measurements can be analysed similarly as light measurements above. The regression between the first and second temperature measurements during a chamber closure is also rather close indicating good precision. The standard error of the noise in temperature measurements based on the differences of the two temperature readings gives us the precision, $0.1°C$.

The chamber measurements of CO_2 exchange is rather complicated and it involves several sources of systematic measuring error. The system can, however, be calibrated by generating exactly known CO_2 flux into the chamber and measuring the flux with the system. In principle, this procedure is similar to those used to calibrate light sensors or gas analysers. The large tolerance of the calibration gasses of the CO_2 analyser (1%) proved to be the weakest link in the measurement of CO_2 exchange with our chambers. The obtained accuracy was 2% (Hari et al., 1999)

The precision of the CO_2 exchange measurements can be determined in an analogous way as previously for light and temperature. The CO_2 exchange of the small shoot in the chamber can be determined for each 10-s period during the closure. The difference of the second and forth closure results in $5\,\mu\text{g}\,(CO_2)\,\text{m}^{-2}\,\text{s}^{-1}$ standard deviation of the $10\,s$ measurement of CO_2 exchange. We use $30\,s$ for determining CO_2 exchange then the noise level is lower, about 3%.

2.5.2.1 Precision of the EC Fluxes

The main sources of random and systematic errors in EC fluxes are related to sampling of stochastic turbulent records, which exhibit variations over several orders of magnitude in a frequency scale. Due to a finite frequency response of the EC system, high frequency fluctuations are not detected and the flux is systematically underestimated. However, if properly configured, the EC systems for the CO_2 and H_2O fluxes have sufficiently fast a frequency response time, and after application of spectral correction procedures the high-frequency flux attenuation can be considered negligible.

The main source of limited precision in short-term, i.e., half-hour to hourly average EC fluxes, is finite averaging over a turbulent record and resulting random error of the flux estimate, typically 10–20% of the flux value. With further aggregation to fluxes to daily or monthly level, random uncertainty becomes negligible.

2.5.2.2 Accuracy of the Fluxes

Similarly to a high frequency attenuation, the finite averaging time (work as a low pass filter for turbulent records) in the flux calculation effectively removes some fraction of the low frequency co-spectrum in the total covariance. To compensate for the flux underestimation at low frequencies, co-spectral corrections are made. The (co-)spectral transfer function of the high-pass filtering is well known and the main uncertainty of such correction is related to a turbulent spectrum at the low frequencies. Such a correction can be typically on the order of 10%. The accuracy of this correction is however expected to be better: For each 30 min average flux the correction can be relatively uncertain, but on average, over many half-hour periods the correction removes systematic bias introduced by the finite averaging.

The EC measurements are undertaken in order to evaluate the ecosystem-to-atmosphere exchanges, most often of CO_2, H_2O and heat. In practice, the EC system determines the vertical turbulent flux at the measurement level. The ecosystem exchange equals the sum of vertical turbulent flux and storage change of the compound below the observation level under horizontally homogeneous conditions.

Under horizontally inhomogeneous conditions, which are always the case in practice, the complete mass balance equation involves also the vertical and horizontal flux advection terms. Thus the main inaccuracy of short as well as long-term EC fluxes is related to violation of the main assumption behind the method, the assumption on the horizontal homogeneity.

The sources of uncertainty involve mainly determination of night-time fluxes, when emitted carbon dioxide is not captured by the EC system but escapes via other transport routes. To overcome this limitation, such night-time observations are frequently replaced by respiration estimates from a regression equation obtained from measurements under turbulent conditions (e.g., Aubinet et al., 2000) or from alternative methods. The poor accuracy of the night-time flux measurements and/or respiration modelling has been believed to be the main limitation of the over-all accuracy of yearly carbon balance of ecosystems. Also in day-time EC flux measurements uncertainty might exist due to an advective component of the total flux (e.g., Paw et al., 2000). The uncertainty of the night-time as well as day-time measurements results in the total uncertainty of the long-term EC measurements, estimated to be about 50 g C m^{-2} year^{-1} for the measurements at nearly ideal sites (Baldocchi, 2003). A majority of ecosystems, however, are located at non-flat terrain and the measurements are frequently performed at such non-ideal conditions because the ideal sites are difficult to find and there is also interest in the variety of the ecosystems located in prevailing landscapes.

2.5.3 Examples of a Measured Time Series

A metatheory and measurements at SMEAR II and I form a backbone of this book. The measurements run the year around producing time series of several aspects in the forest ecosystems. There is a break in chamber measurements during winter at SMEAR I because of strong ice formation on instrumentation. The measurements including the chambers at SMEAR II are running year around. Examples of the obtained time series are shown in Figs. 2.5.6 and 2.5.7.

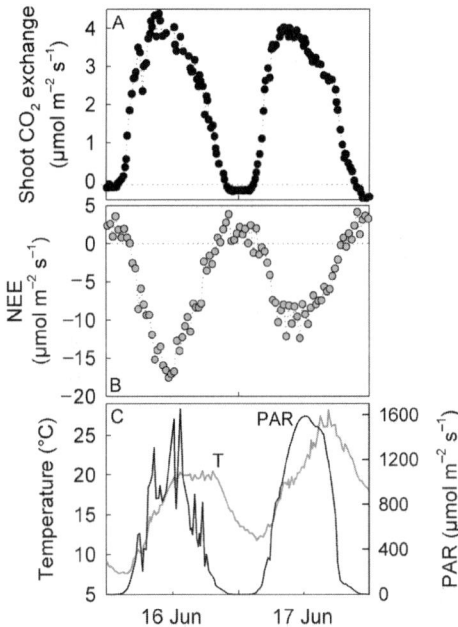

Fig. 2.5.6 Measurements of shoot exchange per leaf area (μmol m^{-2} s^{-1}, A) and net ecosystem exchange (NEE) per ground area (μmol m^{-2} s^{-1}, B) in 16–17 June 2006. We use a common, although not logical sign convection: Positive values denote uptake on the shoot scale (A) but release by the whole ecosystem (B). Contemporary temperature (°C) and PAR (μmol m^{-2} s^{-1}) measurements are shown in the panel lowest panel (C)

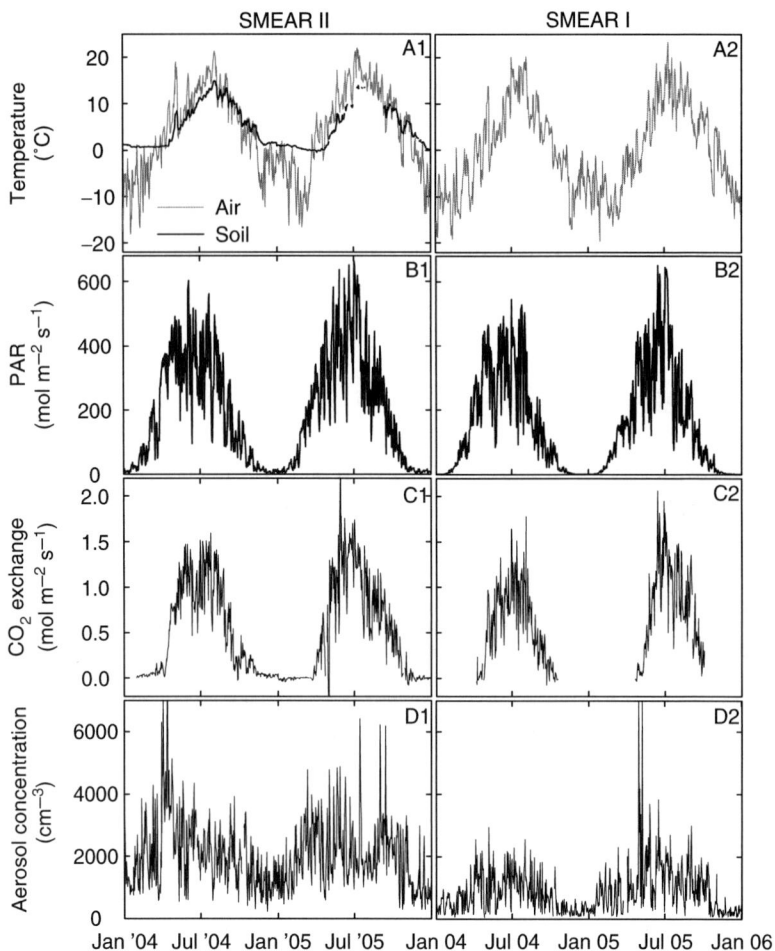

Fig. 2.5.7 Daily averages of measured temperature, PAR, shoot CO_2 exchange and daily median of aerosol particle concentration in 2004 and 2005 at SMEAR II (left panels) in southern Finland and SMEAR I (right panels) in northern Finland. Positive values denote uptake of CO_2 (C1-2)

Chapter 3
Environmental Factors

3.1 Annual Cycle of Environmental Factors

Üllar Rannik[1], Timo Vesala[1], Larisa Sogacheva[1], and Pertti Hari[2]

[1]University of Helsinki, Department of Physics, Finland
[2]University of Helsinki, Department of Forest Ecology, Finland

Our globe circles the Sun and one orbit takes a year. In addition, the Earth spins on its own axis once in 24 h and the rotation axis tilts $23.5°$ from the perpendicular to the Earth–Sun plane (Fig. 3.1.1). This geometry generates a strong annual cycle, especially at high latitudes. The distance between the globe and the Sun varies but is of lesser importance for the climate on the globe. On the other hand, the tilting of the rotation axis is the major factor in generating the annual cycle in the climate, a cycle that is particularly strong in the boreal zone. In summer, the day is very long; beyond the Arctic Circle we have polar day and the Sun does not set daily. The radiation energy input effects temperature variation. In winter, the days are short and beyond the polar circle we have polar night without daily sun rise.

The energy input at the top of the atmosphere depends on the variation of solar radiation and the position of the Earth in the solar system, as well as the angle of the Earth's axis with respect to the ecliptic plane. The factors related to position and orientation of the Earth affect the climate system also via gravitational influence by the Sun and other planets of the solar system.

The value of the solar energy reaching the top of the atmosphere is characterized by the solar constant (I_0). The solar constant is equal to $1.37 \, \mathrm{kW \, m^{-2}}$ (total solar output measurement varies by approximately 0.1% or about $1.3 \, \mathrm{W \, m^{-2}}$ peak-to-trough of the 11 year sunspot cycle). This is the flux of solar radiation through a tangential plane that represents the average distance between the Earth and the Sun. The distance varies during the year because the Earth's orbit is eccentric with an eccentricity value around $e = 0.0167$. The difference between the maximum and minimum distance during the year is equal to $2er_0$ which is about $5 \times 10^6 \, \mathrm{km}$ (the mean distance between the Earth and the Sun being $1.50 \times 10^8 \, \mathrm{km}$). As a result,

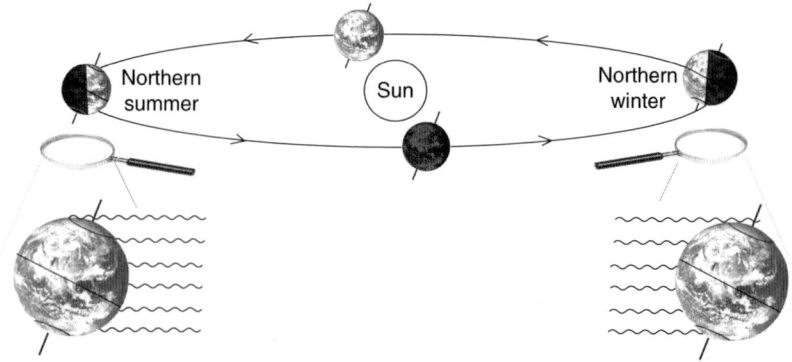

Fig. 3.1.1 The Earth revolves around the Sun and spins on its axis. Earth's axis is tilted 23.5° from the perpendicular to the Earth–Sun plane

the distance (r) of the Earth from the Sun depends on the time of year and solar radiation I input at the top of the atmosphere varies according to

$$I = I_0 \frac{r_0^2}{r^2} \tag{3.1.1}$$

where the subscript $_0$ determines the average values.

The amount of solar energy reaching the unit surface area parallel to Earth's surface depends in addition on the angle between the Sun and Earth's surface – called the solar height. The solar height is different for each geographic location on Earth and depends on the declination angle of the Sun, δ (the angle between the rays of the Sun and the plane of the Earth's equator), latitude of the location φ and time of day.

Thus the amount of solar flux through a unit area perpendicular to Earth's surface is given by

$$I = I_0 \frac{r_0^2}{r^2} \left(\sin \varphi \sin \delta + \cos \varphi \cos \delta \cos \psi \right) \tag{3.1.2}$$

where ψ is the hour angle (see Eq. 3.1.4).

The declination angle can be evaluated approximately as

$$\delta = \frac{23.45° \sin(2\pi(284 + DN))}{365.242} \tag{3.1.3}$$

where DN denotes day number since the beginning of the year. The declination angle varies between $-23.5°$ during Northern Winter Solstice and $23.5°$ in Northern Summer Solstice with small deviations attributable to the small wobble of the polar axis. The exact value of δ for any time can be found from astronomical tables.

The hour angle is determined as

$$\psi = \frac{2\pi \left(\dfrac{t - 12 + TE}{60} \right)}{24} \tag{3.1.4}$$

where t is the local time in hours and time equation (in minutes)

$$
\begin{aligned}
TE = {} & 0.017 + 0.4281\cos\left(\frac{2\pi DN}{365.242}\right) - 7.351\sin\left(\frac{2\pi DN}{365.242}\right) \\
& - 3.349\cos\left(\frac{4\pi DN}{365.242}\right) - 9.371\sin\left(\frac{4\pi DN}{365.242}\right)
\end{aligned}
\tag{3.1.5}
$$

The tilted positioning of the Earth's rotation axis is the primary cause for seasonality. Thus one can talk about strong seasonal variation of environmental factors such as radiation and consequently temperature, especially at high latitudes. Equation 3.1.2 gives the radiation above the atmosphere. It gives the same daily radiation pattern as in measurements of global radiation in clear sky conditions at SMEAR I (Fig. 3.1.2). The difference between radiation energy above the atmosphere and at the surface level is caused by absorption in the atmosphere. The annual cycle in the radiation energy input generates a corresponding annual cycle in temperature. However there is delay in the phase and temperature reaches maxima and minima about 1 month later than radiation (Fig. 3.1.3).

The parameters related to Earth positioning and orientation with respect to the Sun – eccentricity of the orbit and declination angle – vary themselves over different but very long time scales. These variations are caused mainly by gravitational interference of planets of the solar system. Time scales of variation are, however, from thousands to millions of years[1] and will not be discussed in the current context.

Figure 3.1.4 presents a daily sum of solar radiation at the top of the atmosphere (TOA) for three latitudes. It is noteworthy that the daily sum for 70° N exceeds the values for 50° N and also for the equator close to and around midsummer.

Figure 3.1.5 indicates that at the North Pole the solar insolation equals zero between autumn and spring solstice (Northern hemisphere) and that in the Polar region (from the Polar circle 66.5543° northward) the polar night gets shorter when

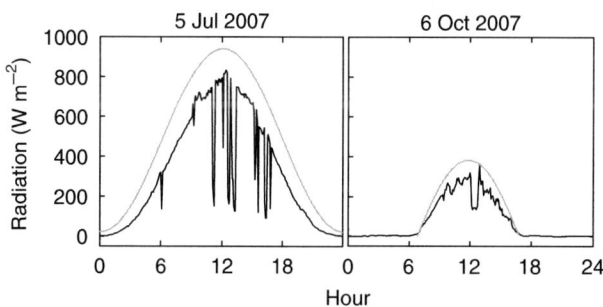

Fig. 3.1.2 Comparison of the radiation energy at the top of the atmosphere according to Eq. 3.1.2 (grey line) and measured values at SMEAR I (black line) on the dates 1.7.2007 and 6.10.2006

[1] Variations in eccentricity, axial tilt, and precession of the Earth's orbit have been related to influence on climate system and ice ages. These variations occur with cycles from 21,000 years to 41,000 years, resulting in 100,000 year ice age cycles.

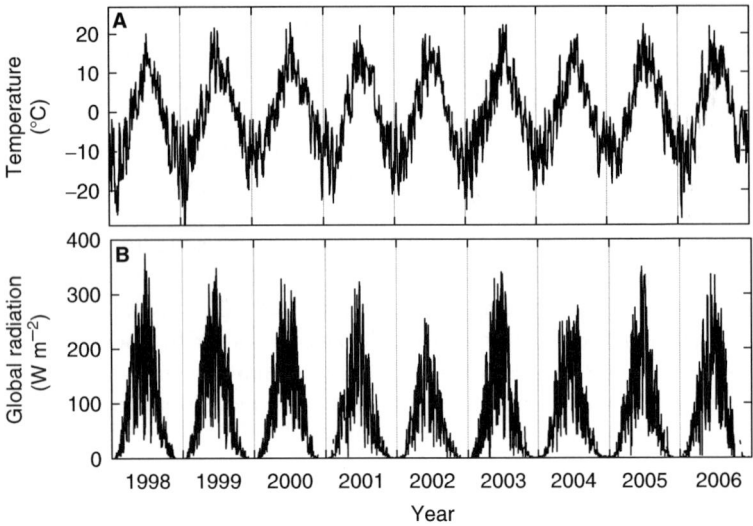

Fig. 3.1.3 Annual cycle of mean daily temperature (A, °C) and radiation (B, W m^{-2}) values measured at SMEAR I, Northern Finland, during 1998–2006

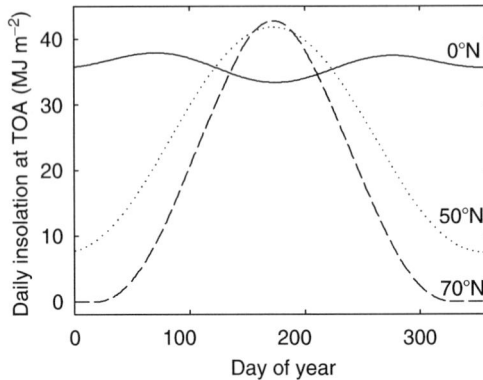

Fig. 3.1.4 Daily incoming solar radiation energy at the top of the atmosphere as a function of time of the year for 0, 50 and 70 North

moving southwards. In contrast, the northern areas get plenty of light in summer. In addition, the light can be better utilised in photosynthesis which saturates at high light intensities. Latitudinal variation of solar insolation at the top of the atmosphere is the primary cause for seasonal variation of environmental factors, which is strongest at higher latitudes.

The maximum of solar energy is located in visible light at about 500 nm. Altogether, about half of the radiation emitted by the Sun is in the short-wave part of the spectrum. As an example, the spectral irradiance at the Earth surface is presented for wavebands between 300 and 575 nm (Fig. 3.1.6). The atmosphere modifies the

Daily isolation at TOA (MJ m^{-2})

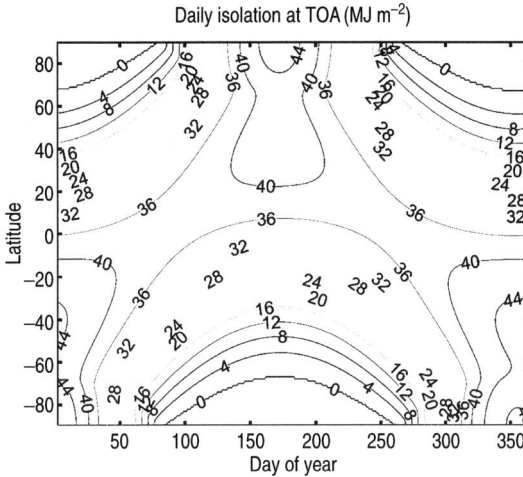

Fig. 3.1.5 Isobars of daily incoming solar radiation energy at the top of the atmosphere as a function of time of the year and latitude

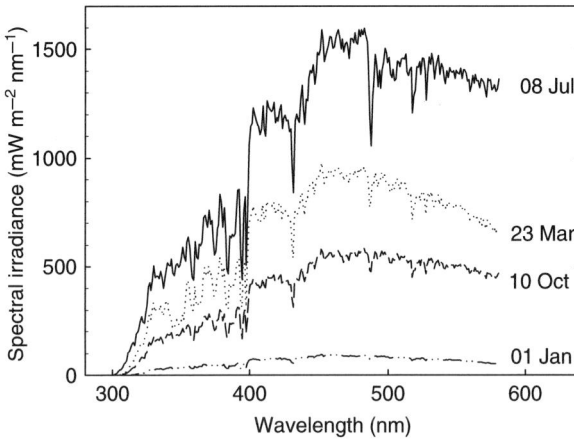

Fig. 3.1.6 Spectral irradiance at noon on 4 selected days for the SMEAR II in Hyytiälä, Finland, as an example for incoming solar radiation from clear sky in different seasons

solar spectrum and the spectral distribution reaching the Earth surface is different than that at the top of the atmosphere. The main absorbers of solar radiation are so-called greenhouse gases including water vapor, carbon dioxide, ozone and others, which each absorb Solar energy at distinct wavelengths.

The incoming solar energy is the main driving force of the climate system and thus the strongest influence on environmental factors. However, the environmental factors are also affected by the climate system itself. The influence ranges from global to regional scales.

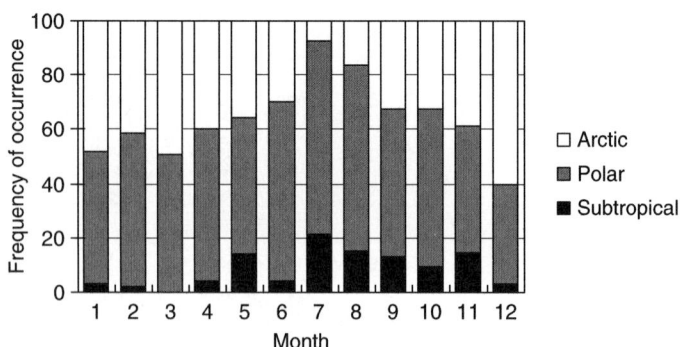

Fig. 3.1.7 Monthly distribution of different types of air masses at SMEAR II, Southern Finland, in 2003–2005

On a global scale, transport of heat and humidity by ocean and atmosphere forms an important mechanism of redistribution of heat reaching unevenly the Earth's surface. Particularly for the Scandinavian climate, the Gulf Stream provides an important energy source. According to climate model simulations the air temperatures over lands are due to oceanic heat transport by a few degrees higher. The influence to air temperature over ocean is even much larger (Seager et al., 2002).

The properties of air masses are strongly affected by the region of their formation; continental and maritime air masses differ strongly in humidity. Depending on the season and geographic location, certain types of air masses prevail and affect the characteristics of environmental factors. This influence is direct via transported air mass properties in terms of temperature and humidity. However clouds, whether transported or formed locally, influence in turn very strongly the resulting radiation and temperature conditions. Figure 3.1.7 present seasonal frequency of occurrence of different air mass types in Finland.

The main environmental factors exhibit strong seasonal variation. As an example, the variation of daily average (or daily sum for radiation) values are presented in Fig. 3.1.8 for Hyytiälä, Southern Finland. The seasonal variation in global radiation, i.e., the spectrally integrated radiation on horizontal surface area, is primarily driven by astronomic factors and strongly modified by cloudiness. The variation of air temperature and water content of the air is positively correlated, while the carbon dioxide concentration is lowest in summer.

Long-term time series of temperature and carbon dioxide concentration measurements show similar seasonal variation from year to year, with an increasing trend superimposed on the carbon dioxide record (Fig. 3.1.9). This is a manifestation of the well-known global increase of atmospheric carbon dioxide concentration. During the given period of about 8 years the concentration has increased more than 10 ppm. The seasonal variation in daily average concentration (i.e., the variation range does not include the diurnal cycle, which can be also significant depending on the location and height of the observation point) is about 25 ppm, which is the result of photosynthetic uptake of carbon dioxide during the summer season in the Northern hemisphere: average concentrations are lower in the lower troposphere.

Fig. 3.1.8 Annual cycle of
global radiation, temperature,
water vapour and carbon
dioxide measured at the height
of 33.6 m at SMEAR II station
in the year 2003

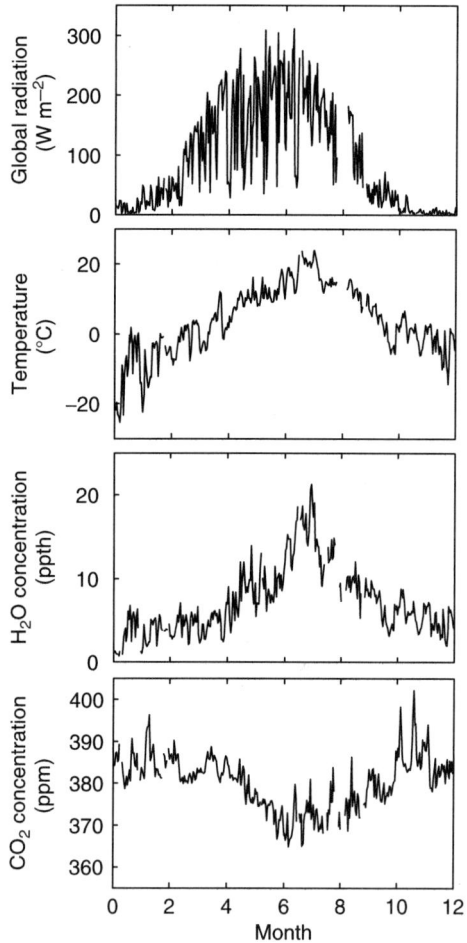

Stability of Atmosphere

When the air is warmer than its surrounding air it has lower density due to thermal
expansion and is affected by upward buoyancy forces. In turn, colder air is less dense
and experiences downward accelerating forces. In addition, the air water content
affects the air density so that the air parcel with higher water content has lower
density than the dry air at the same temperature and pressure.

In the atmosphere the pressure decreases with height. When an air parcel
moves upward without heat exchange with surroundings, it experiences temperature decrease due to adiabatic expansion and thus cooling. The temperature lapse
rate of such an isolated air parcel would be $9.78°\,C\,km^{-1}$. If an actual temperature
lapse rate of the atmosphere is lower, it is stably stratified (an elevated air parcel
remains cooler than its surrounding experiencing downward buoyancy forces). If

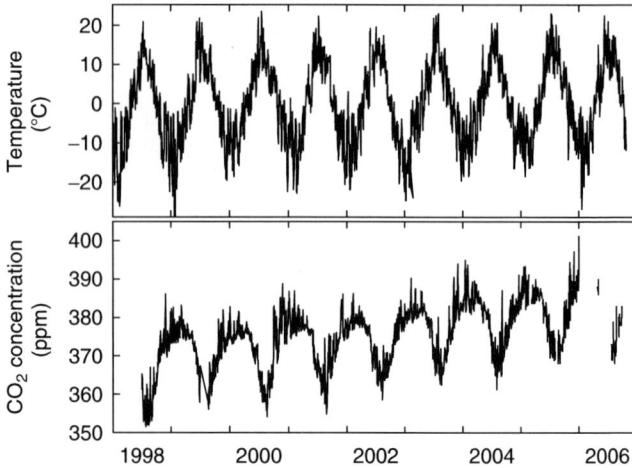

Fig. 3.1.9 Temperature and CO_2 concentration records from Northern Finland, 1998–2006. Temperatures were recorded in SMEAR I and carbon dioxide in Sammaltunturi (67°58' N, 24°07' E, elevation 565 m above mean sea-level, see Aalto et al., 2002b)

opposite occurs, i.e. when the temperature decreases more than $9.78°\,C\,km^{-1}$, it is unstably stratified: An air parcel with upward shift becomes warmer and thus with lower density than the surrounding air, experiencing further upward acceleration.

The influence of pressure decrease with height on air temperature is taken into account by potential temperature θ defined as $\theta = T \left(\frac{P_0}{P}\right)^{\frac{R}{c_p}}$, where P is the actual pressure and P_0 is the reference pressure usually taken as 10^5 Pa, R is the gas constant and c_P is the specific heat capacity at a constant pressure. In unsaturated, neutrally stratified atmosphere the vertical gradient of potential temperature is zero, in stable atmosphere it is positive and in unstable negative.[2]

The unstable stratification occurs typically during the day when ground surface is heated by solar radiation. The upward air currents can form resulting in vertically extensive motions up to several kilometres. The relatively moist air originating from surface becomes saturated and formation of occasional fair weather clouds occurs.

In Boreal summer, the stable stratification usually occurs at nights. In winter the stratification is typically stable, especially under cloud-free weather conditions. As an example, daily average temperature differs during the winter period up to $-15.6°\,C$ between closely located two sites differing in altitude by about 220 m. This is the effect of atmospheric stratification and not due to different air masses at different sites. The periods correspond to very cold weather conditions occurring during high-pressure weather systems (Fig. 3.1.10).

---

[2] The atmosphere contains water vapour which effect on air density and stability cannot be neglected. The influence of air water content on stability is taken into account by virtual potential temperature – the potential temperature of dry air that would have the same density as the actual moist air.

Fig. 3.1.10 Daily average
temperatures at SMEAR I
(67° 46′ N, 29° 35′ E, 400 m
a.s.l) and in Sodankylä
(67°22′ N, 26°38′ E, 179 m
a.s.l)

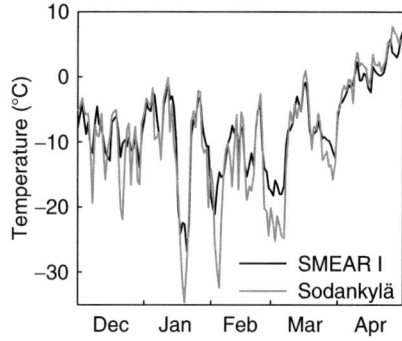

3.2 Temporal and Spatial Variation: Atmosphere

Üllar Rannik

University of Helsinki, Department of Physics, Finland

Scales of Variation

Apart from seasonal variation of environmental factors, typical shorter time scale variation exists in meteorological observations. One of the methods to identify periodicities in observations is the spectral (sometimes called Fourier[3]) analysis. The spectral analysis relies on finite Fourier series representation of a discrete time series $x(t)$ as a sum of cosine and sine terms

$$x(t) = \bar{x} + \sum_{p=1}^{N/2-1} \left[a_p \cos\left(\frac{2\pi pt}{N}\right) + b_p \cos\left(\frac{2\pi pt}{N}\right) \right] + a_{N/2} \cos(\pi t) \quad (3.2.1)$$

where the coefficients a_p and b_p are given by

$$a_p = \frac{2}{N} \sum_{t=1}^{N} x_t \cos\left(\frac{2\pi pt}{N}\right)$$

$$b_p = \frac{2}{N} \sum_{t=1}^{N} x_t \sin\left(\frac{2\pi pt}{N}\right) \quad (3.2.2)$$

---

[3] The Fourier or harmonic analysis dates back to 1822. The novel time-frequency analysis approach is the wavelet transform. The main difference between the wavelet transform and the standard Fourier transform is that wavelets are localized in both time and frequency whereas the Fourier transform is only localized in frequency. Notable contributions to wavelet theory and application can be attributed to works in the 1980s (e.g. Kumar and Foufoula-Georgiou, 1997).

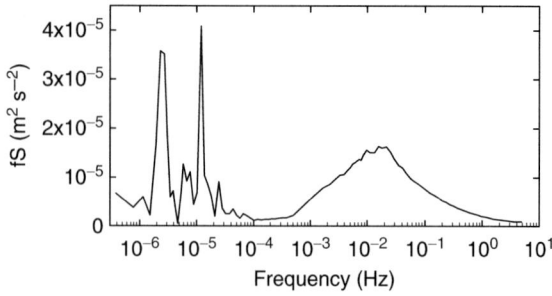

Fig. 3.2.1 Power spectrum of wind speed at SMEAR II observed during 1 month period in April 1999 at 23.3 m height, about 10 m above forest

for $p = 1, 2, \ldots, \frac{N}{2} - 1$ and

$$a_{N/2} = \frac{1}{N} \sum_{t=1}^{N} (-1)^t x_t \qquad (3.2.3)$$

(see, e.g., Jenkins and Watts, 1968). The coefficients a_p and b_p determinate the amplitude of variation at each frequency $\omega_p = 2\pi p/N$ and the sum of squares $a_p^2 + b_p^2$ is proportional to variance. This is the quantity presented in power spectra plots, normalized usually such that the integral over frequencies gives the total variance of the series. For a pure sign wave with fixed frequency the power spectrum consists of a single peak.[4] Geophysical or economic time series usually exhibit spectral energy over a wide range of frequencies.

Figure 3.2.1 presents the power spectrum of variation of wind speed in the lower part of the atmosphere, about 10 m above forest. The spectrum is weighted with frequency and presented on a logarithmic frequency scale such that the area below the power spectrum is proportional to variance. In the figure three distinct peaks can be identified, which correspond to the following approximate frequencies and corresponding time scales:

10^{-6} Hz – 5 to 10 days (synoptic scale)

10^{-5} Hz – 1 day (diurnal variation)

10^{-2} Hz – a few minutes (boundary layer, turbulent scale)

In addition, atmospheric variation occurs occasionally at about 10^{-3} Hz – a few tens of minutes – corresponding to mesoscale variation in the atmosphere.

Each of these time scales is related also to a spatial scale of variation. The synoptic scale is on the order of 1,000 km or more and is a typical horizontal distance between mid-latitude high or low pressure areas. The synoptic and mesoscale time scales result from horizontal advection of spatially different air mass properties. The diurnal time scale of variation is related to solar heating and locally driven flows (e.g., breezes and gravitationally driven flows). Turbulent scales are more related

---

[4] For a non-sinosoidal time series with a large spectral peak at some frequency ω; also related peaks may occur at multiple frequencies 2ω, 3ω.... These multiple frequencies of the main cyclical component are called harmonics.

to local phenomena like boundary layer and surface properties, thus time scales (or frequencies) are influenced also by the elevation of the observation point from the surface.

Further on, we will discuss the three distinct time scales of atmospheric variation in more detail.

Synoptic Scale

To illustrate the influence of synoptic scale weather systems on environmental conditions at a fixed location, Fig. 3.2.2 shows the latitudinal surface positions of fronts at a fixed longitude. During the observed period, arctic and polar air masses prevail at the latitude of the location of interest (Hyytiälä, Southern Finland), with a few exceptions when subtropical air masses reach the position.

The Arctic air masses are associated with low water content of the air and also low temperatures. Each front passage suggested by Fig. 3.2.2 corresponds to changes in temperature and humidity. Especially the sequences of Julian days 92–96 and 102–104 occurred in periods of cold air with low relative (and absolute) humidity (Fig. 3.2.3). The period following on Julian day 77 also had Arctic air conditions. Unlike the later cold periods, the relative humidity was high in the lower troposphere, and Stratus and Nimbostratus clouds caused overcast conditions that gave snow and rain, see the indication for presence of clouds the low radiation values during the period of days 77–83.

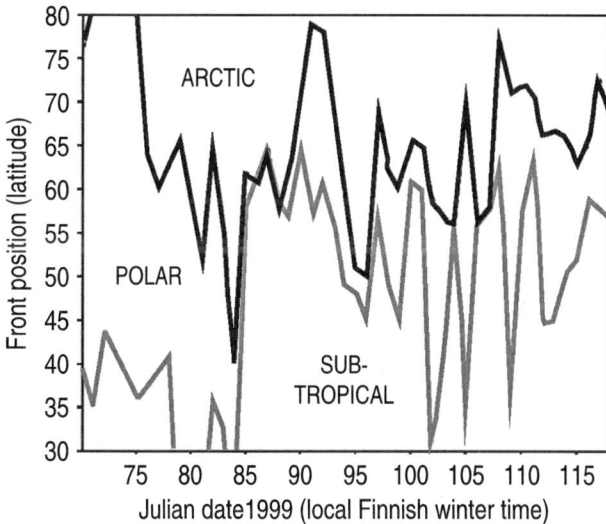

Fig. 3.2.2 The latitudinal surface position along the 25° E longitude of the Polar front (grey line) and the Arctic front (black line) based on the Berliner Wetterkarte in 1999

The weather conditions are closely related to synoptic scale weather systems, which according to surface pressure are separated to High (also Anticyclones) and Low Pressure Systems (Cyclones).

The Low Pressure Systems mean counterclockwise circulation in the Northern Hemisphere, low pressure in the middle of the weather system and flow convergence close to surface. The rising air in the middle of the system experiences expansion cooling and therefore supports cloud formation and subsequent precipitation. Due to significant impact of clouds on short-wave radiation field the amount of solar radiation at surface is decreased.

In turn, the High Pressure Systems are characterised by clockwise circulation in the Northern Hemisphere, with high pressure in the middle of the weather system, divergence of flow close to surface and corresponding vertical subsidence. Such a circulation pattern prevents formation of large-scale stratiform clouds and is therefore typically cloud-free or partly cloudy with more solar radiation available. During the High Pressure Systems the diurnal amplitude of temperature is larger because of relatively cold air associated with the weather system and intensive solar heating in cloud-free conditions during the day.

The surface pressure and radiation conditions appear in close accordance in Fig. 3.2.3. The periods of lower air pressure are also associated with cloudiness and decreased radiation values. The Arctic air conditions according to front positioning during the period of Julian days from 77 to 83 correspond actually to low surface pressure values, which is accompanied by cloudiness and reduced radiation.

The air water content and temperature determine the relative humidity of the air – the ratio of the actual water vapour pressure to saturation vapour pressure at given temperature, expressed usually in percents. The RH indicates how close is the air to saturation conditions: at low RH values the departure of vapour pressure from saturated conditions is larger and driving force for evaporation from surfaces or trough stomatal openings into air is larger. At RH value 100% the air is saturated and therefore evapotranspiration is suppressed, which is typical to night-time conditions. The pattern of RH is very well correlated with temperature and anticorrelated with radiation (Fig. 3.2.3). Since the saturation vapour pressure is a non-linear function of temperature, the diurnal variation of temperature causes strong diurnal variability of RH.

Diurnal Variation

Much of the diurnal variation in temperature and concentrations is related to the cycle of solar heating and Atmospheric Boundary Layer development (see Section 5.3). Temperature has an especially strong diurnal cycle, which is strongest during high-pressure clear-sky conditions (Fig. 3.2.3). Diurnal variation in solar heating also drives the airflow. The cloud cover inhibits heating by solar radiation. The daily temperature cycle is then strongly reduced and the temperature difference between daily maxima and minima is smaller than during clear sky conditions.

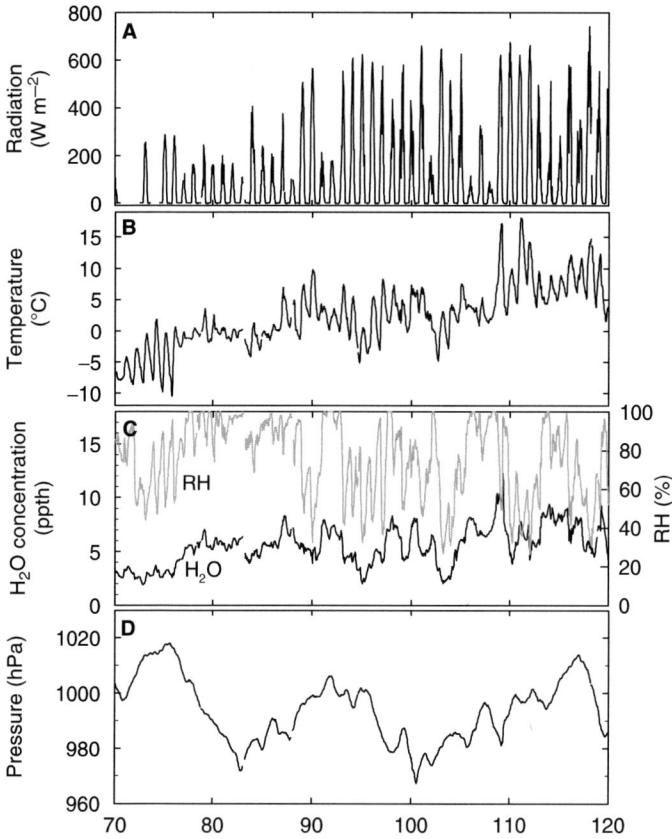

Fig. 3.2.3 (A) Global radiation, (B) temperature, (C) air water content with relative humidity and (D) atmospheric pressure at SMEAR II, 1999. Days are counted from the beginning of the year

   Diurnal variation of carbon dioxide and water vapor results from uptake/emission of these quantities as well as diurnal variation in atmospheric mixing intensity. This is because during a vegetative period carbon dioxide uptake occurs during the day and emission at night. (See Fig. 3.2.4 for the forest-atmosphere exchange determined the above forest areas). Together with inhibited atmospheric mixing at night, this leads to accumulation of carbon dioxide close to the surface.

Turbulence Scale

The turbulence time scale of variation ranges from tenths of seconds to approximately an hour. Such a wide variation range is related to the stochastic nature of turbulence, at high frequencies limited by the so-called Kolmogorov microscale

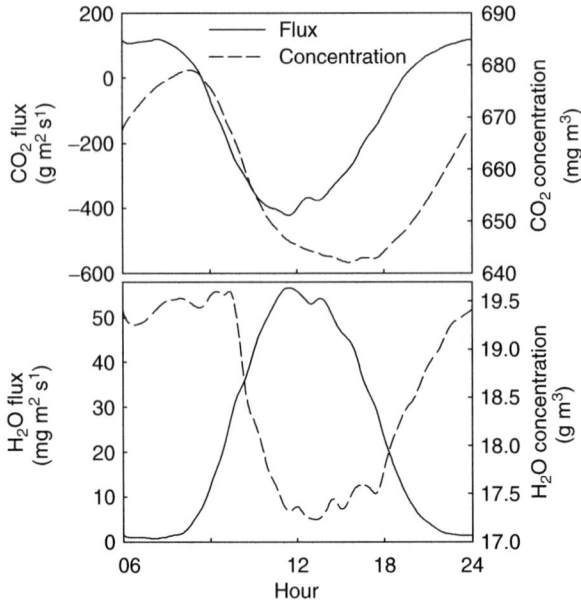

Fig. 3.2.4 Ensemble average diurnal curves of water vapor and CO_2 concentrations and fluxes above pine forest at SMEAR II during 3 months period from June to August 1996

being in the order of 1 mm (e.g., Kaimal and Finnigan, 1994) and at low frequencies roughly by the vertical transport time scale inside the Atmospheric Boundary Layer.

Variation of concentrations in turbulent flow results from spatial heterogeneities in a concentration field, which in turn is the result of the activity of sources and sinks. The Atmospheric Boundary Layer contains sources and sinks usually close to the surface due to biogenic activity (e.g., carbon dioxide uptake and emissions from respiration) or anthropogenic emissions. At its top the Atmospheric Boundary Layer is bounded by inversion and entrainment from the free atmosphere (day-time) or intermittent mixing of the Stable Boundary Layer causes concentration variations inside the Boundary Layer from above. Such concentration differences are reduced by turbulent mixing.

Figure 3.2.5 shows a time record of point measurements of vertical wind speed and temperature over a 12 min period. The measurements were performed about 10 m above a canopy with a fast-response anemometer so that the smallest variations in records are resolved. During daytime the average temperature gradient above a canopy is very small, on the order of 0.01 K m^{-1}. We see that the fluctuations are a few degrees. Heating of an upper canopy (mainly) serves as a temperature source, which is seen in air temperature fluctuations. The vertical wind speed is positively correlated with the air temperature on the average (lower panel), thus transporting warm air from a canopy to higher levels.

The spectral shape of a turbulence record in the Atmospheric Surface Layer has two distinct regions: the energy-containing range where energy input to turbulent

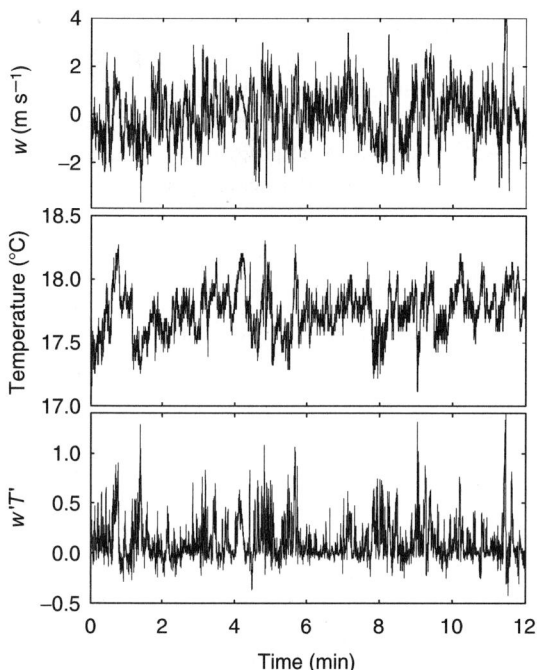

Fig. 3.2.5 Measurements of vertical wind (w) speed and temperature (T) with a fast-response ultrasonic anemometer above a pine forest at SMEAR II on 4 August 2003 starting at 13:00. The primed quantities denote deviations from means and $w'T'$ is instantaneous vertical turbulent flux of heat

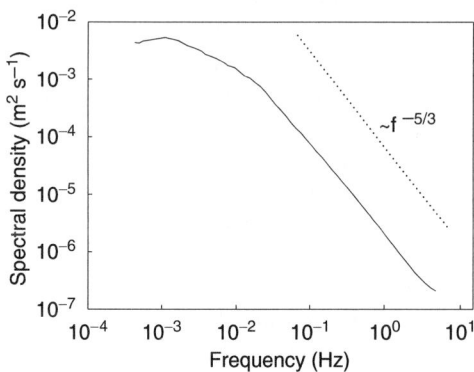

Fig. 3.2.6 Spectrum for turbulent kinetic energy measurements above pine forest at SMEAR II. The Kolmogorov power law (line with slope –5/3 on log-log plot) is also indicated

flow occurs; and the inertial subrange of the spectrum, where energy is transferred from lower frequencies to higher frequencies. Figure 3.2.6 shows the turbulent part of the spectrum, i.e. roughly frequencies higher than 10^{-3} Hz, presented in Fig. 3.2.1 on log-log scale. Such presentation enables to follow the inertial subrange of the spectrum where spectral energy decreases with frequency according to the Kolmogorov power law of $-5/3$ for the inertial subrange. The inertial subrange separates the energy containing and dissipation ranges and in this range no energy production or dissipation occurs but energy is transferred form larger to smaller turbulent scales, where it is finally converted into heat. The dissipation range cannot be seen in the figure, being much smaller than the measurement volume of the sonic anemometer and thus not measured.

3.3 Temporal and Spatial Variation: Soil

Jukka Pumpanen

University of Helsinki, Department of Forest Ecology, Finland

Temperature

Temperature is the most important environmental factor affecting the temporal variation in biological activity of the soil, because the underlying metabolic processes are temperature dependent. In a boreal region, the temperature has a strong seasonal variation that follows the variation in air temperature. The top soil (O-horizon) at SMEARII usually reaches a daily maximum of 15–17°C in July–August. Soil temperature decreases towards deeper soil layers (C-horizon), which reaches a maximum of 10–12°C in late July. On sunny days in June and July the diurnal amplitude of temperature in the humus is 5°C. Deeper in the soil, the daily temperature fluctuation is smaller (Fig. 3.3.1).

In winter, the deepest soil horizons are the warmest. At SMEAR II station, the soil temperature hardly ever drops below 0°C under the humus layer although the mean air temperature is mostly below zero degrees (Fig. 3.3.1). When permanent snow cover has been established, the diurnal variation in soil temperature is very small. The snow cover insulates the soil effectively, which prevents the penetration of frost to deeper soil layers. Frost formation depends on the insulation of snow, soil heat storage and thermal conductivity of the soil.

At SMEAR II station, the soil is very shallow, the soil depth varying between 0.05 and 1.6 m, the average depth being about 0.8 m. The soil layer is formed on top of a solid bedrock, which has relatively high heat conductivity compared to the porous soil matrix (Hillel, 1998). Thus the heat stored in the bedrock during the summer time maintains the deepest soil layers at a relatively warm level throughout the winter. Soil material and soil water content have a big influence on the heat conductivity and specific heat capacity of the soil. SMEAR II is very exceptional place; usually the top soil is frozen in boreal forests during winter.

The depth of insulating snow cover depends on the canopy structure. In forests with dense canopy, a substantial part of the snow is retained in the tree canopy and never reaches the soil. Under a dense canopy, the frost penetration may be much

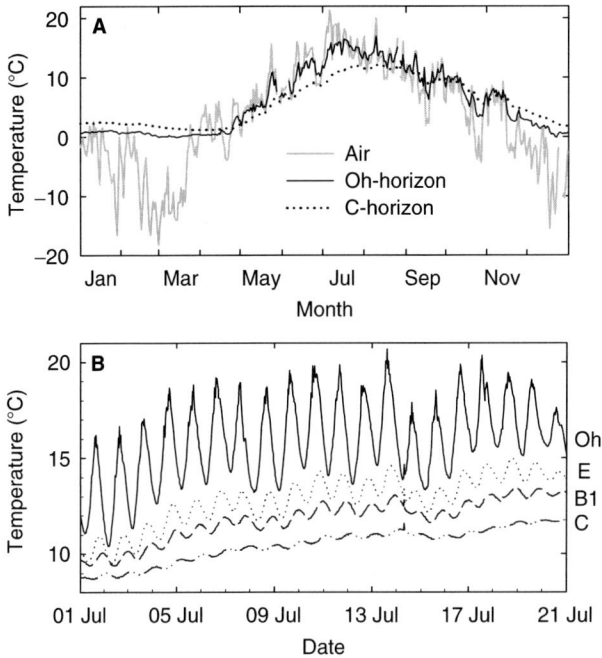

Fig. 3.3.1 (A) Seasonal pattern in air and soil temperature profiles at SMEAR II station in year 2005 and (B) a closer look on the temperatures in different soil profiles (Oh, E, B1 and C; see Section 7.4) in July 2006

deeper than on an open area. Also the interannual variation in snow cover affects soil temperatures during winter. There are sometimes years when permanent snow cover is not established until January. If there are periods with very low air temperatures before that, the frost penetrates much deeper than during average years.

Soil Water Storage

Soil water storage has a typical seasonal pattern in boreal forests. In Fig. 3.3.2 we discuss soil water storage, precipitation and runoff measured at SMEAR II station during year 2000, which represents a typical seasonal pattern in soil water storage. Soil water storage is at its highest during the snowmelt in spring and decreases towards the end of the summer because evapotranspiration exceeds rainfall. The maximum water storage at SMEAR II station is about 200 mm which equals about $0.40 \, m^3 \, m^{-3}$ volumetric water content. Water storage decreases during the summer by about 50% to 100 mm. During average years, lack of water does not limit photosynthesis and transpiration, but during the extremely dry period in July and August 2006 the water storage decreased down to 50 mm. This corresponds to volumetric water content of $0.10 \, m^3 \, m^{-3}$ in the surface soil and $0.15 \, m^3 \, m^{-3}$ in the deeper soil layers and soil water tension of –2 MPa in the surface soil and –0.7 MPa in the

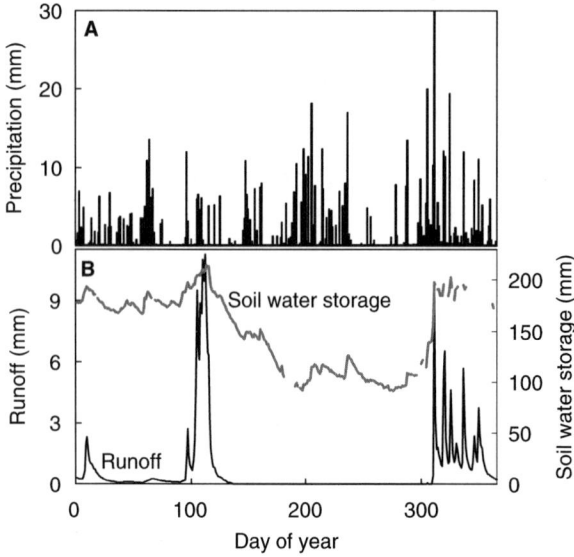

Fig. 3.3.2 (A) Precipitation, (B) runoff and soil water storage during a typical year 2001 at SMEAR II station, southern Finland

deep soil. When the autumn rains start, soil water storage recovers quite quickly to 150–170 mm. The water storage has recovered on average years by September and October.

The average annual runoff from the SMEAR II catchment was 308 mm between years 1998–2001, which was 40% of the average precipitation, 744 mm, during the respective years. According to the statistics of Finnish Meteorological Institute, the 30-year annual average precipitation was 713 mm between 1971–2000 (Drebs et al., 2002). The runoff takes place after the spring thaw in April and in most years during October and November when the soil is saturated with water.

Chapter 4
Transport

Timo Vesala, Üllar Rannik, and Pertti Hari

The system of the tree-atmosphere continuum is spatially large, which sets challenges for trees to be capable to transport various chemical compounds within them and to exchange compounds with the atmosphere. Transport phenomena and resulting mass flows are governed only by few fundamental physical laws giving rise to different transport mechanisms. The mutual effectiveness of different transport modes depends on the spatial scale. In a way, trees utilize effectively this frame of physical laws in their functioning and interactions with environment.

By transport we especially mean the movement a certain quantity of mass or energy between the storage locations of the quantity. Storages may be apparent as sources and sinks of the quantity and unequal amounts of any entity in the storage locations lead to the concept of a non-zero gradient of the quantity. In practical terms, ecosystems store materials as mass (such as carbon, nutrients, water) and heat, and any differences in the amounts of mass and heat tend to disappear by physical mechanisms called transport phenomena. The third fundamental quantity besides *mass* and *heat* is *momentum* which is defined as the product of the mass and velocity of a homogenous fluid (fluid means either gas or liquid) parcel. All of these can be transported by two physically different mechanisms, which are *molecular transport* and *bulk transport*, called also *convection*. Transport of heat may happen also by means of a third heat transport mechanism which is *electromagnetic radiation*. Next we consider three transport mechanisms starting with the molecular one.

4.1 Molecular Transport

Molecular transport refers to spontaneous and irreversible transport of momentum, energy (heat) and mass of various chemical species along the gradients (profiles) of

T. Vesala and Ü. Rannik
University of Helsinki, Department of Physics, Finland

P. Hari
University of Helsinki, Department of Forest Ecology, Finland

velocity, temperature and concentration, and from the higher velocity, temperature and concentration to the lower one. Molecular transport stems from random molecular motion. Besides velocity, temperature and concentration, pressure and density and their spatio-temporal behaviour are the elemental quantities governing physics of transport phenomena (Fig. 4.1.1).

Conduction and diffusion are examples of molecular transport related to the movement of heat and mass, respectively. They are easily recognized and they are familiar concepts, whereas the idea of momentum transport may not be so easily identified. However, it is the movement of fluids, which involves the transport of momentum, and the capability of gas or liquid to transport momentum that gives rise to the *viscosity* of a fluid, analogously with heat conductivity and diffusivity. If we consider a collision of two billiard balls, moving at different speeds, the faster one will have more momentum, part of which it will give off to the slower moving ball. Fluid is not a body, like a billiard ball, but it contains areas of higher and slower flow velocities and, similar to the transfer of momentum from a faster body to a slower one, momentum is transported from areas of higher flow velocities to those of slower movement.

Momentum associated with moving molecules is also the molecular cause for the macroscopic concept of *pressure*. Let us consider bombardment of gas molecules onto a surface. When a single molecule hits the surface and rebounds, an impulse is exchanged between the molecule and the surface and this produces a *force* on the surface during the collision. In normal conditions near the Earth's surface the bombardment rate is very high and the net result of many hits is the macroscopic force, which is defined as pressure when divided by the surface area. Pressure is a fundamental quantity for flows of fluids: pressure difference in the fluid produces flow and on the other hand, surfaces nearby the flowing fluid "absorb" momentum from the flow and give rise to a pressure drop in the fluid. For air, at standard conditions the number of molecules is $2.5 \times 10^{19}\,cm^{-3}$ and their average velocity is $460\,m\,s^{-1}$. The rate of molecular collisions with any surface can be further estimated to be $2.9 \times 10^{23}\,cm^{-2}\,s^{-1}$ (Hinds, 1982), at standard conditions.

Conductivity, diffusivity and viscosity are macroscopic concepts characterizing a fluid, but they naturally stem from molecular structure of the substance

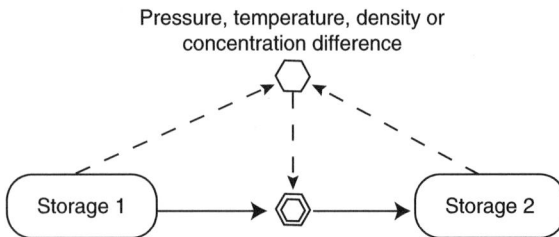

Fig. 4.1.1 Transport occurs between two storages creating sinks or sources. Transport is governed by pressure, temperature, density or/and concentration differences. The symbols are introduced in Fig. 1.2.1

and the associated intermolecular forces. The fundamental driver of conduction, diffusion and molecular momentum transport is the random thermal motion of atoms/molecules modified by interactions due to molecular collisions. If we simplify without losing much from the inherent physics, the collisions resemble elastic collisions between moving billiard balls. Irrespective of whether we consider billiard balls or molecules, if the objects are moving randomly the random motion tends to remove any inhomogeneities. Lack of homogeneity is caused by (macroscopic) sinks and sources. Note that a single molecule under thermal motion does not "know" in what direction it should move (it is determined by the last collision with its partner) in a mixture of two substances, but a net movement of a large group of molecules is along the gradient from the higher to lower concentration, and is predictable.

The measure of the transport is the *flux density* giving the amount of energy, mass (or moles) or momentum crossing a unit area per unit time. The flux density is naturally a *vector*, a quantity possessing both magnitude and direction (like velocity). In contrast to the vector, a quantity possessing only magnitude is called a *scalar* (such as speed). At the *equilibrium* all flux densities are zero and no transport occurs. In the case of small disturbances from the equilibrium,[1] the transport is measured by the gradients in temperature, concentration and velocity components. The flux density $\vec{I}$ due to the molecular transport is proportional to the gradient

$$\vec{I} \propto \nabla S \equiv \frac{\partial S}{\partial x}\vec{i} + \frac{\partial S}{\partial y}\vec{j} + \frac{\partial S}{\partial z}\vec{k} \qquad (4.1.1)$$

where S refers to the temperature (conduction), concentration (diffusion) or velocity component (viscous flow). The mathematical concept *gradient* means the rate of change with respect to distance of a quantity (scalar) in the direction of maximum change and results in a vector, the coordinates of which are the *partial derivatives* of the quantity. The partial derivatives in Eq. 4.1.1 are represented by the *x-y-z* co-ordinates and the corresponding *unit vectors* along *x-*, *y-* and *z*-axis are $\vec{i}$, $\vec{j}$ and $\vec{k}$, respectively. If the flux density is integrated over a certain area, the resulting surface-averaged flux density is called *flux* and its dimension is mass, energy or momentum per unit time. The concepts of flux density and flux are used loosely in the literature. Molecular transport refers to diffusive transfer described above. Note that "diffusion" commonly means diffusion of mass and diffusive transport of heat is called conduction. As was explained above, momentum can be transported also by diffusion.

The physical meaning of the diffusion law can be illustrated by a one-dimensional case starting from the given concentration difference $\Delta C = C(x + \Delta x) - C(x)$ over the distance Δx (see also e.g., Nobel, 2005 for similar reasoning). Let us set a plane perpendicular to the x-axis between the points x and $x + \Delta x$. Since the movement by diffusion is a statistical phenomenon it is reasonable to assume that the flux of molecules A, across the plane from two sides of it, is proportional

---

[1] For applications in this book the assumption of small disturbance is always valid enough.

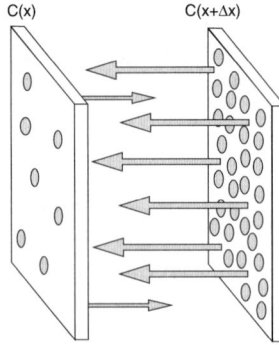

Fig. 4.1.2 Diffusive (molecular) transfer giving rise to the net transport from higher concentration $C(x+\Delta x)$ towards the lower one $C(x)$

to the concentration. Thus the flux from the direction of the point $x+\Delta x$ is proportional to $C(x+\Delta x)$ and the flux from the opposite direction to $C(x)$ (Fig. 4.1.2). Let as assume further that Δx is positive and $C(x+\Delta x) > C(x)$. There is now a greater probability for molecules to move from the concentrated to the dilute region and the flux g_A is

$$g_A \propto C(x+\Delta x) - C(x) \qquad (4.1.2)$$

However, the magnitude of the flux is naturally also affected by the distance over which the concentration difference occurs. Namely, if the given concentration difference exists over a long distance or the same ΔC exists over a shorter distance, the flux must be accordingly smaller in the former case. So we get

$$g_A \propto \frac{C(x+\Delta x) - C(x)}{\Delta x} \qquad (4.1.3)$$

At the limit $\Delta x \to 0$, the ratio

$$\frac{C(x+\Delta x) - C(x)}{\Delta x} = -\frac{C(x) - C(x+\Delta x)}{\Delta x} \to -\frac{dC}{dx} \qquad (4.1.4)$$

which is the x-derivative of C (one-dimensional gradient), and we obtain that $g_A \propto -\frac{dC}{dx}$.

The result is easy to interpret: the derivative tells the steepness of the concentration profile (the steeper the profile is the larger is the flux) and the minus sign says that the transport is negative along the x-axis, that is from larger x-values to lower ones. By convention, a net movement in the direction of increasing x is positive. The formula derived is still just the proportionality and the final form requires the coefficient, which is known to be the diffusivity D_{AB}.

The most rigorous and general transport equations are rather complex and we refer to the book by Bird et al. (2002). For applications considered in this book the approximated formulas of the mass flux density $\vec{g}$, heat flux density $\vec{q}$ and the momentum flux density τ_{yx} can be simplified. The mass flux density is given by

$$\vec{g}_A = -\rho D_{AB}\nabla\omega_A \text{ (Fick's first law)} \tag{4.1.5}$$

which states that the mass transport rate per unit area is proportional to the gradient of the mass fraction ω_A of substance A. The mass fraction characterizes the concentration of A. When the density ρ of the mixture of two compounds A and B is multiplied by the mass fraction of A, the mass concentration C_A is obtained.

The three-dimensional extension of the one-dimensional case is obtained when the x-axis is replaced with the gradient of the concentration, which tells the direction of steepest concentration drop, to which the diffusive transport is directed (in analogy, a ball would roll to a direction of the steepest slope characterized by ∇z where z is the elevation of the surface)

Similarly to diffusion the heat is transported by conduction according to

$$\vec{q} = -k\nabla T \text{ (Fourier's law for conduction)} \tag{4.1.6}$$

This is fully analogical to Fick's law of diffusion: heat is transferred from the higher temperature to a lower one and the magnitude is governed by the heat conductivity.

In the presence of diffusion in the multicomponent fluid, the heat transport by each of the diffusing species must be added. The heat carried is measured by *enthalpy*, describing the internal energy of the material, and consequently the generalized energy-flux expression for fluid consisting of N species is

$$\vec{q} = -k\nabla T + \sum_{i=1}^{N} \frac{H_i}{M_i}\vec{g}_i \text{ (conduction and mass transport)} \tag{4.1.7}$$

where H_i is the partial molar enthalpy and M_i is the molecular weight of species i. The inclusion of enthalpy carried by individual species is important for a fundamental treatment of phase transitions; see Section 6.2.1).

Finally, the momentum transport requires, even in its simplest form, two-dimensional analysis. Let us consider the flow and momentum transport by means of x and y axis, which are orthogonal to each other. The momentum transport is then given by

$$\tau_{yx} = -\mu\frac{dv_x}{dy} \text{ (Newton's law of viscosity)} \tag{4.1.8}$$

This form of Newton's viscosity law tells how much x-momentum associated to the bulk flow in x-direction (characterized by velocity component v_x) is transported along the y-axis. Again, the momentum transport is analogous to diffusion and conduction but one must notice that the quantity, momentum, to be transferred is generally a vector and not a scalar as are concentration and temperature. This is the reason that the formula above includes two dimensions, x and y, although it describes one-dimensional (along the y-axis) diffusive momentum transport.

Several notes for the above formulas should be addressed:

(i) The present form of Fourier's law describes the heat transport in isotropic media, i.e., heat is conducted with the same thermal conductivity in all directions; some solids (for example wood stem) may be anisotropic and k must be "replaced" by a symmetric second-order tensor; in addition, the so-called work flux occurring in flow systems is ignored in the present form.

(ii) Preconditions for diffusion is that the material is composed of at least two different compounds, but the concept of self-diffusion of chemically identical species exists for physically different species (for example radioactive isotopes); the present form is for a binary mixture and for multicomponent mixtures the formula must be replaced by the Maxwell-Stefan equations.

(iii) Besides concentration-gradient driven ordinary diffusion there are other kinds of diffusion mechanisms, such as forced diffusion caused by unequal external forces acting on the chemical species; in addition, the driving force of ordinary diffusion is strictly speaking the activity (defined by means of the partial molar Gibbs free energy), which must be taken into account for mechanisms such as osmosis discussed below.

(iv) As for the heat transport, for some anisotropic solids and structured fluids the diffusivity must be "replaced" by the diffusivity tensor.

(v) The mass flux density can be equivalently formulated as a mole flux density using a mole fraction gradient as the driving force.

(vi) The present form of Newton's law describes the simplest shearing flow in which v_x is a function of y alone and v_y and v_z are zero; the generalization for the viscous stress tensor is a much more complicated vector-tensor formula.

(vii) The momentum flux density can be equivalently interpreted as a shear stress, i.e., as a force exerted by the flowing fluid in the x direction on a unit area perpendicular to the y direction.

Osmosis is important in many biological systems. For example, the transport and cycling of water in a tree is partly driven by osmosis, which is the mechanism of the radial (in contrast to vertical water movement, sap flow) movement of water in Münch's cycle (Section 7.5.2). Osmosis can be considered as a special case of ordinary diffusion but its treatment requires the usage of activity rather than concentration (see e.g., Nobel, 2005). Molecules of solute species interact with each other as well as with other solute and solvent molecules, and this influences the behaviour of all compounds in the solution. The use of concentration for describing the thermodynamic properties of the solution is generally only an approximation and generally the concept of *activity* is needed. The activity can be regarded as a concentration corrected for the interactions between solution constituents. For a so-called ideal solute, the activities are equal to concentrations.

The classical example of osmosis is the transport of water through a membrane impermeable to the dissolved solutes but permeable to water (Fig. 4.1.3). Solutes in an aqueous solution decrease the activity of water, which leads to the concept of *osmotic pressure*. In the first approximation, for dilute solutions, the pressure is

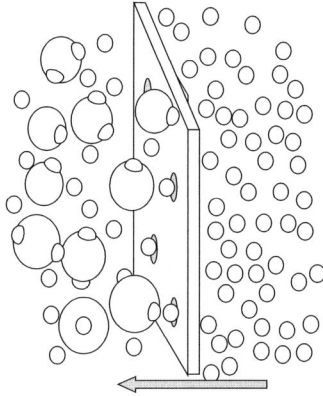

Fig. 4.1.3 Osmotic flux of water. Small molecules can penetrate the holes, but large molecules are unable to penetrate the holes. Small molecules also interact with larger ones

directly related to concentration of solutes and for the osmotic flux density g_{H2O} of water through the membrane (e.g., Nobel, 2005) we have the proportionality

$$g_{H2O} \propto -\sigma_j RT \Delta C_j \qquad (4.1.9)$$

where R is the gas constant and ΔC_j is the concentration difference of the dissolved species j across the membrane. T is the temperature in Kelvin. σ_j is the reflection coefficient of the species j, which is the phenomenological coefficient[2] describing the relative ease with which the solutes cross the membrane. It is unity when the membrane is completely impermeable to solutes and is zero when the membrane is nonselective.

When the membrane is not very permeable to water, rather high osmotic pressure difference can accumulate across the membrane. If the permeability increases (for example water conducting pores become more opened), large water flow is induced. The flow naturally dilutes the more concentrated solution and decreases the driving force and the flow, if no mechanism produces more solutes.

4.2 Convection

4.2.1 Laminar and Turbulent Flows

The spontaneous transport along the gradients by molecular movements and collisions is not the only transport. It is intuitively clear that any flow of the material has a capability to carry, and thus transport, its own mass and momentum and heat

---

[2] Linked to Onsager reciprocity relations of the general thermodynamic framework of transport phenomena (see Onsager, 1931a, b).

associated to it. Hence, besides the molecular transport of energy (conduction), mass (diffusion) and momentum (viscosity) the quantities can be transported by the bulk flow of the fluid, which is called convective transport. Note that in fields of physics and engineering sciences of fluid mechanics and of transport phenomena, convection refers generally to the bulk flow of the fluid, but in meteorology it refers also to the vertical flow of air resulting from density variations due to the temperature field.[3]

Generally, the flow appears in two "forms": *laminar* or *turbulent*. Laminar flow is orderly and turbulent flow is chaotic and dispersive with the quantities (velocity components, pressure, temperature, concentrations) violently fluctuating. Atmospheric flows are practically always turbulent, while for example the sap flow in a tree is laminar because the flow velocities are low. With respect to the causes for flows, they are classified into two types: *forced* and *free* (or natural) convection. Free convection refers to the flow determined by the buoyant forces arising from the fact that heated air rises because it is less dense.[4] Forced convection refers to all other situations where some external force generating pressure difference generates the flow. In some cases both effects are important and this situation is called mixed convection. Both forced and free convection can be either laminar or turbulent.

Again, we measure the amount of transport by flux densities. The convective mass flux density due to the flow velocity $\vec{v}$ is given by

$$\vec{g}_{A,con} = \rho\,\vec{v}\omega_A \text{ (convective mass flux)} \tag{4.2.1}$$

The interpretation of this formula is simple: ρ gives the amount of mass per unit volume and if the mass is transported by the velocity $\vec{v}$, $\rho\vec{v}$ gives the amount moved per unit time per unit area perpendicular to the velocity vector. Namely, the volume rate of flow across the surface element dS perpendicular to, for example, the x-axis is $v_x dS$ and the volume rate multiplied by the density gives the mass flow rate. Multiplication by the mass fraction ω_A gives the amount of the compound A carried by the bulk flow.

Similarly we get for the bulk transport of energy

$$\vec{q}_{con} = \rho\,\vec{v}\left(U + \frac{1}{2}v^2\right) \text{ (convective energy flux)} \tag{4.2.2}$$

where U consists of the sum of the kinetic energies of all atoms/molecules (relative to the flow velocity), the intramolecular potential energies and the intermolecular energies and is called the internal energy per unit mass (see e.g., Bird et al., 2002). This is the molecular scale basis for U, although U is a macroscopic concept. $\frac{1}{2}v^2$ is the kinetic energy per mass related to the flow velocity $v^2 = v_x^2 + v_y^2 + v_z^2$.

As for Newton's law of viscosity (Eq. 4.1.8), the convective momentum flux is formulated here only for x-momentum carried along the y-axis:

$$\tau_{yx,con} = \rho\,v_x v_y \text{ (convective momentum flux)} \tag{4.2.3}$$

---

[3] Which is only a special case of fluid physicists' various convective mechanisms.

[4] Note that the concept of convection in meteorology is free convection in terms of fluid physicists.

where v_x represents now the x-momentum per unit mass and the momentum transferred per unit time and unit area in the y direction is obtained by multiplying with ρv_y.

Together with molecular transport Eqs. 4.1.5–4.1.8 the convective equations provide the full description of possible transport forms, excluding only electromagnetic radiation as one transport mechanism for energy. The molecular and convective fluxes do not exclude each other but the total, combined flux is the sum of molecular and convective terms. However, in the direction of flow the molecular transport is typically rather slow compared with the convective one. The formulas simply say that flow, measured by the flow velocity $\vec{v}$ (or its component like v_x), carries with it some amount of energy, mass or momentum through a plane perpendicular to its direction. The flow velocity has replaced the gradients in Eqs. 4.1.5–4.1.8 as a driver of the transport. Thus flow seems to transport quantities independently of whether any gradients exist or not, and this turns out to be true.

The above formulas are valid both for laminar and turbulent flows. However, in the presence of turbulence the flow velocities fluctuate rapidly and the equations as such are not very viable. Let us consider first a simple case of the laminar steady (time independent) incompressible (constant density) flow through a circular tube, which is applicable and relevant also for many ecological topics, such as the sap flow in a tree stem. Based on the equation of motion (Navier-Stokes equation, see Eq. 7.2.8), one can show that the average velocity along the tube axis $\langle v_z \rangle$ (averaged over the tube cross section) is proportional to the profile of the pressure $\frac{dp}{dz}$ in the tube according to $\langle v_z \rangle = \frac{R^2}{8\mu} \frac{dp}{dz}$ where R is the radius of the tube. Furthermore it can be shown that the pressure profile has a constant value of $\frac{\Delta p}{L}$, where Δp is the pressure drop over the longitudinal distance L. Thus we get, for example, for the convective mass flux

$$\langle j_{A,tube} \rangle = C_{tube}\, \rho\, \omega_A \frac{\Delta p}{L} \qquad (4.2.4)$$

where the constant C_{tube} is a function of the tube radius and fluid viscosity. Note that the time-independence of this example means that there is no acceleration of the flow and thus no net forces affect the fluid. The pressure force driving the flow is in balance with the viscous friction force, which results from the radial molecular transport of momentum from the flowing fluid to the tube wall (see also the discussion on the concept of pressure in Section 4.1).

Next we consider the classical statistical way to analyze complicated turbulent transport. Turbulent flow is a very efficient transporter of energy, mass and momentum and for example the lowest most turbulent part of the atmosphere could be called a "blender". Turbulent flow can be thought of as being built up eddies (whirls which are a type of persistent motion structures) of different size. Eddies move along the main flow and they cause the fluctuations of velocity, temperature and concentrations in an observation point when passing it. Large eddies correspond to slow, and often high amplitude fluctuations while small eddies cause rapid random-like fluctuations. From the perspective of transport mechanisms, it is eddies which can be thought to carry quantities. For a mathematical analysis the time-smoothed (time-averaged) velocity $\bar{w}$ and temperature $\overline{T}$ are introduced so that $w = \bar{w} + w'$ and $T = \overline{T} + T'$ (see e.g., Bird et al., 2002), where w' and T' are the instantaneous turbulent fluctuations in the vertical velocity and temperature, respectively. Let us take an example to illustrate the idea. Let us consider an eddy which transports air momentarily from up to down while passing an observation point. This would be observed as a momentary decrease of the vertical wind velocity w, i.e., a negative fluctuation w'. If at the same time for example the average temperature of the air decreases upwards, we could observe also a momentary decrease of the temperature since the colder air transported down was replacing the warmer air (Fig. 4.2.1A). We see that fluctuations in the vertical motion and temperature are positively correlated, as would be also in the case of the positive w' (Fig. 4.2.1B). Over some time, eddies of different size would transfer as some net amount of heat. The averaging time must be chosen so that it is large with respect to the periods of fluctuations but shorter than possible slow changes in the flow and temperature. Now w' and T' averages to zero so that $\overline{w'} = 0$ and $\overline{T'} = 0$. However, quantities like $\overline{w'T'}$ will not be zero because of the correlation between the velocity and temperature fluctuations at any point. When w and T are replaced by $w = \bar{w} + w'$ and $T = \overline{T} + T'$ in governing equations (see Eq. 7.2.5) and equations are time-smoothed, the nonvanishing correlation terms appear describing turbulent phenomena. Consequently a turbulent heat transport appears as

$$q_{tur} = \rho C_p \overline{w'T'} \qquad (4.2.5)$$

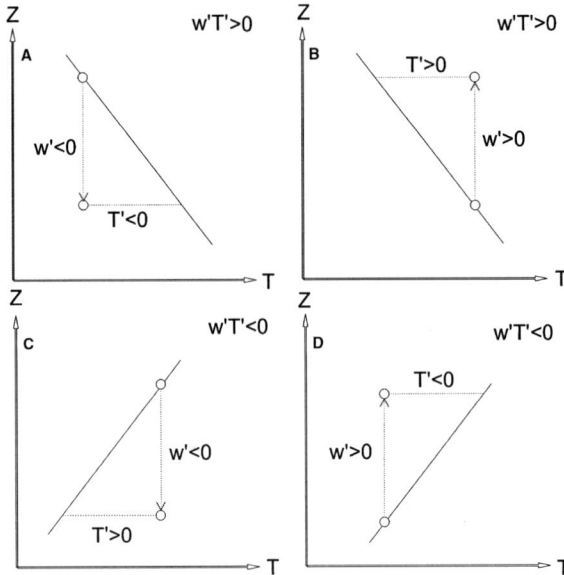

Fig. 4.2.1 Behaviour of the relationship between the turbulent fluctuations in vertical velocity (w') and air temperature fluctuation (T'). Four different cases are presented depending on the sign of w' and the average temperature profile. In Figs. A and B the heat transport is upwards and in Figs. C and D downwards. Z is the height.

If the temperature gradient was reversed in the previous example, the velocity and temperature fluctuations would be negatively correlated over a specified time and $q_{tur} < 0$ indicating downward transport of heat, as it should since now the temperature would increase upwards (Fig. 4.2.1C, D). Analogously for the turbulent mass transport

$$g_{A,tur} = \rho \overline{w' \omega'_A} \qquad (4.2.6)$$

In the same fashion, turbulence also transports momentum, leading to so-called Reynold's stress (see e.g., Bird et al., 2002).

The so-called eddy covariance method is based on formulas 4.2.5 and 4.2.6 and has been effectively utilized in recent micrometeorological biosphere-atmosphere heat and gas exchange studies (e.g., Aubinet et al., 2000). The technique is based on the high-response measurements of wind, temperature and concentration fluctuations and the flux is directly calculated according to formulas that are normally $1/2$ h averages.

For practical purposes, turbulent transport is often parameterized by semi-empirical formulas mimicking molecular gradient transport

$$q_{tur,K} = -K_T \frac{dT}{dy} \qquad (4.2.7)$$

$$g_{A,tur,K} = -K_\omega \frac{d\omega}{dy} \qquad (4.2.8)$$

where coefficients K are empirical turbulent heat and mass transport coefficients. Note, however, that the rigorous formulation of turbulent transport follows Eqs. 4.2.5–4.2.6, Eqs. 4.2.7–4.2.8 are good approximations only when even the largest eddies are much smaller than the scale where the temperature or concentration is significantly changed. In this case the eddies can be taken as "molecules" transporting quantities proportionally to gradients. This approximation is often called turbulent diffusion and for example the coefficient K_ω is called turbulent (or eddy) diffusivity. Since turbulent transport is not driven by diffusion, the concept of turbulent diffusion is misleading and although generally used, it should be avoided.

4.2.2 Eddy Covariance and Ecosystem Fluxes

One of the challenges of scientific community is global warming and the influence of greenhouse gases, carbon dioxide in particular. Atmospheric concentration of carbon dioxide has shown to increase continuously (Section 3.2). To date ecosystem-atmosphere exchange of carbon dioxide is monitored worldwide on a continuous basis (e.g. Baldocchi et al., 2001) by eddy covariance technique based on Eq. 4.2.6. Long-term *in situ* measurements of air turbulence and of the concentration of the scalar rely on the assumption of horizontal homogeneity. The conventional instrumental setups are unable to capture the horizontal transport. The

large, even undetermined uncertainty of ecosystem-atmosphere exchange derived
by single-point micrometeorological measurements has become one of the most
important topics of canopy micrometeorology (Finnigan et al., 2003).

The soil and canopy CO_2 exchange as denoted by Net Ecosystem Exchange
(NEE) in horizontally inhomogeneous conditions can be presented via scalar conservation equation in terms of vertical turbulent flux above canopy (measured by
micrometeorological techniques), the storage term below flux observation level, and
the vertical and horizontal advection terms (Feigenwinter et al., 2004; for details see
Finnigan et al., 2003),

$$NEE = \overline{w'C'_{CO_2}}_{z_r} + \int\limits_0^{z_r} \frac{\partial \overline{C_{CO_2}}}{\partial t}dz + \int\limits_0^{z_r} \overline{w}\frac{\partial \overline{C_{CO_2}}}{\partial z}dz + \int\limits_0^{z_r} \left(\overline{u}\frac{\partial \overline{C_{CO_2}}}{\partial x} + \overline{v}\frac{\partial \overline{C_{CO_2}}}{\partial y}\right)dz$$

$$(4.2.9)$$

where z_r refers to observation level, C_{CO_2} denotes absolute concentration ($C_{CO_2} = \rho\omega_{CO_2}$), u, v and w are the wind components in x, y and z directions, overbar denotes
time averaging and primed quantities deviations from the averages. The storage term
is typically important when turbulent mixing is relatively weak at night or in the
morning transition period. The vertical and horizontal advection terms are zero at
horizontally homogeneous conditions. Figure 4.2.2 shows diurnal variation of the
magnitude of the vertical turbulent flux, the storage and the vertical advection terms
for pine forest located in a moderately sloping terrain (see also Mammarella et al.,
2007). Deviation of the vertical turbulent flux as determined by the EC flux measurements from NEE is strongest during morning and night transition periods.

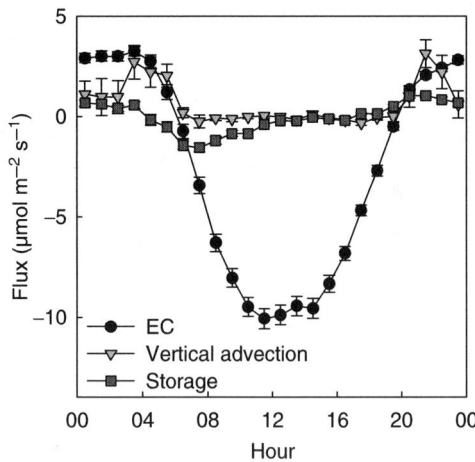

Fig. 4.2.2 The NEE equation terms (see Eq. 4.2.9) as determined for pine forest in Hyytiälä at
SMEAR II station

Fig. 4.3.1 Radiative heat flux.
$T_1 > T_2$. The intensity of
the radiation is proportional
to the fourth power of the
temperature, that is $\propto T^4$ (see
Eq. 6.2.3)

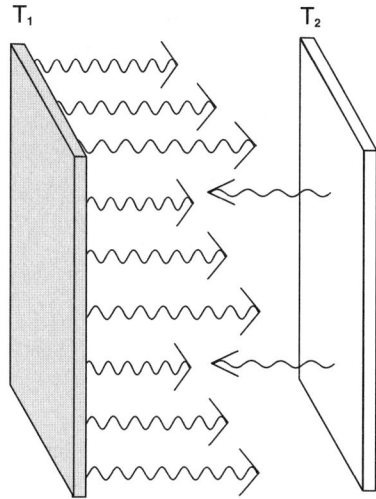

4.3 Radiative Transport

Finally, we consider the electro-magnetic radiation (Fig. 4.3.1) which is the third
mode of energy transport, for which there is no analogy in mass and momentum
transport. Radiation does not even require the existence of the material but it transports energy at the speed of the light in a vacuum. In practice, we can ignore radiation as a transport mechanism for a specific point in a transparent medium and it
should be taken into account only for opaque surfaces. It also differs drastically from
molecular modes of transport, which are proportional to gradients, while radiative
transfer between two bodies is proportional to the difference of the fourth powers
of their absolute temperatures. We discuss the radiative energy transport between
opaque bodies later in Section 6.2.1. Note that this is the reason that the discussion of
radiative transport here is so short, namely the transport between the bodies belongs
to the processes, as defined in this book, and they are discussed in Section 6.2.1.
The books by Thomas and Stamnes (1999) and Ross (1981) cover all elemental as
well as advanced aspects of radiative transfer in the atmosphere and plant stands.

4.4 Summary of Transport Phenomena

We summarize the main classification of the transport phenomena (Fig. 4.4.1). The
fundamental quantities of mass, heat and momentum are all carried by convection
(bulk flows) and molecular transport (diffusion, conduction and viscosity, respectively). Heat is carried also by radiation. Flows have two distinct modes: laminar
and turbulent one.

Mass Transport   **Heat Transport**   **Momentum Transport**

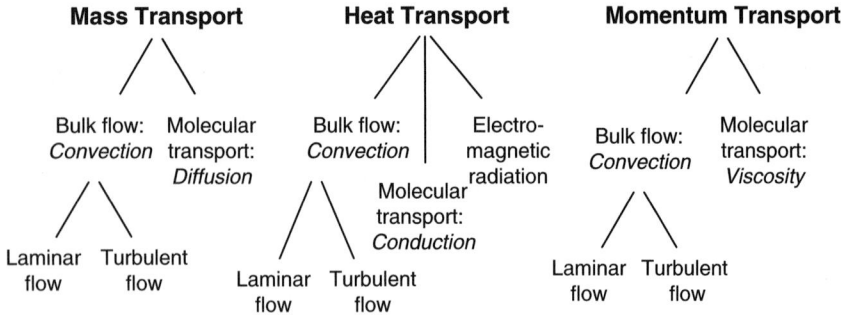

Bulk flow: Molecular   Bulk flow: | Electro-   Bulk flow: Molecular
Convection transport: *Convection* | magnetic   *Convection*   transport:
            Diffusion | radiation   *Viscosity*
                                Molecular
                                transport:
Laminar   Turbulent   *Conduction*   Laminar   Turbulent
flow   flow   Laminar   Turbulent   flow   flow
                      flow   flow

Fig. 4.4.1 General classification of transport phenomena of three basic quantities

Transport phenomena are fundamental drivers of many processes in the plant-atmosphere continuum. As an example, we decompose the tree water cycling and water exchange with the atmosphere in the terms of transport phenomena (Fig. 4.4.2). For the gas phase (air) there are three major components: diffusion inside a stoma, diffusion across the leaf boundary layer and turbulent transport in the canopy air. Inside a tree, water is transported in vertical and radial directions in the stem. Detailed description of transport in twigs and leaves is omitted here as they resemble those in the stem.

Transport in the air outside   **Transport through**
a leaf boundary layer: **a leaf boundary layer:**
Turbulent convection   *Diffusion*
Concentration difference due to   Concentration difference due to
higher water content close to   higher water vapour content
leaves/canopy   close to stomata

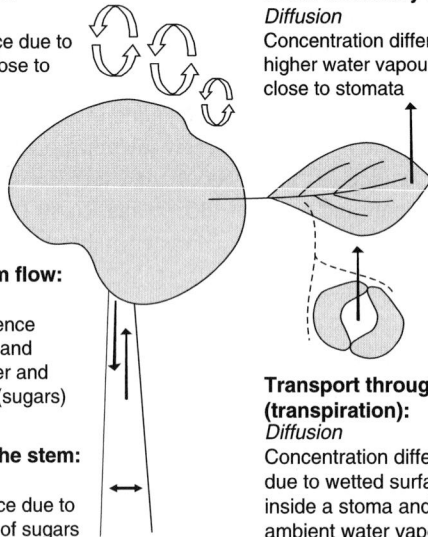

Sap flow and phloem flow:
Laminar convection
Driving pressure difference
from transpiration loss and
varying amount of water and
dissolved compounds (sugars)   **Transport through a stoma**
                                          (transpiration):
Radial transport in the stem: *Diffusion*
Diffusion by osmosis   Concentration difference
Concentration difference due to   due to wetted surfaces
loading and unloading of sugars   inside a stoma and lower
                                          ambient water vapour content

Fig. 4.4.2 Main water transport phenomena in the tree-atmosphere continuum

Chapter 5
Structure

5.1 Hierarchy of Structure

Pertti Hari, Jaana Bäck, and Eero Nikinmaa

University of Helsinki, Department of Forest Ecology, Finland

The scope of our considerations in the present book covers a wide area from subcellular plant parts to a global vegetation zone, in particular the boreal forest. There are, however, natural hierarchical levels in the structure which are necessary to understand in the study of natural phenomena in such a zone. The hierarchical structure we will describe here will be utilized later in our discussion in Chapters 7–10.

A central challenge in ecology and forest growth studies is to understand the emergence of ecosystem behavior from the interaction of its components. For example, a forest stand can be regarded as an assemblage consisting of living organisms, especially trees, their organs and ultimately molecules at different levels of structural hierarchy. Since evolutionary selection mechanisms act upon individual organisms, these organisms are the basic building blocks of any ecosystem. The success of an individual organism in dealing with its environment determines its capacity to survive and spread its off-spring.

Individuals have to accomplish multiple tasks such as photosynthesis, nutrient and water uptake, water transport, growth and mechanical support. Specialized organs (vegetative or reproductive) have developed to perform these tasks. For example, leaves perform photosynthesis, fine roots take up water and nutrients, and stems transport water and assimilates. All organs include several tissues, involved in specialized functions within the organ.

Cells are the basic building blocks of all living organisms. They often have tissue-specific features but within the plant kingdom, their basic structure is rather similar. Cells have their own metabolism, but are not independent in the sense that they need to take raw materials and information from other cells or from the environment, and in response they also may produce material for other cells. The inner, subcellular structure of the cell reflects the specific metabolic task of the tissue formed by similar cells.

An interesting feature of plant tissues is that living cells may be surrounded by a non-living region, the apoplast, that is never-the-less an important structure from the whole plant perspective. For example, a major part of the transport phenomena in plant (see Chapter 7) take place in the non-living tissue of tree woody axis.

Stands are formed by individuals, either trees or those of ground vegetation, interacting with each other. Trees modify environmental factors within the stand, resulting in alterations in metabolic processes, especially in the lower part of the canopy and in the ground vegetation. Furthermore, the senescing parts of living plants when falling to the ground feed the organic component in the soil.

Similarities in the physical environment of the plants generate vegetation zones on the globe. Between 50° and 70°N, the environment favors a mosaic of evergreen coniferous forests and peatlands depending on the water retention in the soil and the resulting influence on the rate of decomposition. The forested zones may be broken into forest stands of different size following the action of catastrophic disturbance events. Within these stands, trees grow and become mature and die forming more heterogeneous canopy structures over time. Silviculture has greatly changed the natural spatial dynamics of forest stands making them more homogeneous in size and internal canopy structure, and thus easier to harvest (Kuuluvainen, 2002).

5.2 Vegetation

5.2.1 Chemical Structure

Jaana Bäck and Pertti Hari

University of Helsinki, Department of Forest Ecology, Finland

Plant cells have hundreds of different biochemical components, but the cells and organelles share many similar features in respect to their macromolecular structure. The major components are large carbohydrate molecules, which can roughly be classified as structural, functional, regulatory, storage and secondary molecules. Some of them contain only carbon and hydrogen, others have significant quantities of other molecules such as nitrogen or sulphur, which are important as functional groups of molecules (Table 5.2.1). Here we describe the chemical structure of the most important macromolecule groups in mature gymnosperm cells. In addition to general reference to Buchanan et al. (2000) and Taiz and Zeiger (2006), specific references concerning boreal tree species are given.

Table 5.2.1 Element fractions in some cellular macromolecule groups

Fraction (%)	Carbon	Nitrogen	Hydrogen	Oxygen
Lipids	75–78	–	8–11	9–17
Sugars + starch	40	–	7	52
Hemicellulose + cellulose	43	–	7	49
Lignin	67	–	6	27
Proteins	40–50	10–25	5–10	22–40

5.2.1.1 Macromolecules

5.2.1.1.1 Functional Substances

Functional substances enable, by definition, processes in living organisms. All metabolic pathways consist of several interlinked processes, involving numerous functional and regulating substances.

Proteins

Most functional substances are proteins, some of which are composed of several chemical entities which are functional only when they are together. Proteins form about one tenth of the total dry matter in living plant cells, and they can be either soluble or bound to various membraneous structures. Proteins or their parts function as enzymes, signaling molecules, structural compounds and membrane pumps. Proteins have molecular masses ranging from 1 to 10^3 kDa (Da = Dalton, the unified atomic mass unit based on the mass of a ^{12}C atom; 1 kDa = 1.66 * 10^{-24} g), and they may be composed of several subunits. All proteins are formed of 20 different amino acids, which are organized into long chains called polypeptides (primary structure). There are generally 200–300 amino acids in one protein molecule, but some proteins can be considerably larger. The organization of polypeptides (secondary structure) determines the chemical and physical properties of the proteins. The three-dimensional conformation of protein molecules is determined by their folding and arrangement of atoms in the molecules, and the quaternary structure is formed when several subunits are linked together to form a functional protein.

The vast majority of proteins are **enzymes**, which act as biological catalysts in all metabolic processes that occur in living tissue. Enzymes are highly specific to their substrates and end products, which partially explains why there are so many different enzymes involved even in the simplest metabolic chains. They also have rather specific temperature and pH requirements where they can operate. Their catalytic capacity makes the enzymatically catalyzed reactions typically in the order of 10^8 to 10^{12} times faster than the corresponding uncatalyzed reactions. Isozymes are enzymes with a similar catalytic function but localized in different cellular compartments, such as, e.g., isoprene synthase which can be found both as soluble, stromal form and as bound in thylakoids, or nitrite reductase in plastids of either roots or leaves. The amounts of enzymes in the cell compartments are determined by the relative rates of their synthesis and degradation, i.e., turnover rates of the enzymes.

The enzyme molecule has an active site where the substrate and potential catalytic groups or cofactors are bound. In the process of catalysis, the properties of the enzyme itself are not changed, although reversible changes in the conformation of the active site may occur. Most enzymatic reactions show an exponential increase in rate with increasing temperature, up to the point where the enzymes are denatured and a rapid and irreversible decline in activity results.

Of the total soluble protein in green leaves, 40–50% is in the plastidial ribulose bisphosphate carboxylase/oxygenase (Rubisco), making it the most abundant single protein in plants. Rubisco is unique in that the same enzyme catalyzes two important reactions, namely carboxylation and oxygenation. Rubisco is a relatively large enzyme (560 kDa) with eight large, chloroplast-encoded subunits and eight small, nuclear-encoded subunits. Rubisco concentration in chloroplasts is ca. $8 \, mg \, m^{-2}$ (based on leaf area; Warren et al., 2003).

Transport Structures

The penetration of membranes by ions, metabolites and solutes is a prerequisite for most essential cell processes, and needs therefore to be carefully controlled (e.g., Flügge, 1999; Fig. 6.3.17). Details in membrane trafficking have been a subject for extensive research in recent years, but only the general structures and phenomena are described here and in Section 6.3.2.4. The penetration of membranes by molecules is performed with the help of a variety of intrinsic membrane proteins, which function either as active pumps for hydrophilic substances, or as passive channels or carriers. Hydrophobic molecules can passively move through membranes proportional to their lipid solubility.

Membrane pumps are proteins which are essential in the exchange of molecules and ions through the semipermeable membranes separating the compartments from each other, and in controlling cell and organelle pH and osmolality. They occur e.g. in mitochondrial membranes, plastid envelopes, tonoplasts surrounding the vacuoles, and in plasma membranes. Also the nuclear envelope possesses specialized transport structures. Nuclear pores are elaborated structures of hundreds of different proteins, involved in movement of macromolecules and ribosomal subunits both into and out of the nucleus.

Water transport through the lipid bilayer membranes occurs via aquaporins. Aquaporins are relatively small, integral membrane proteins, belonging to a larger Major Intrinsic Protein (MIP) family of transmembrane channels. They form a selective channel for water molecules across the plasma membrane or tonoplast, and their expression and functions are controlled in response to plant water availability (Maurel, 1997). They have been assumed to form narrow pores of 0.3–0.4 nm in diameter, excluding all larger molecules than water, which would flow molecule by molecule through the pore down to the pressure gradient.

Light-Absorbing Pigments

Light quanta are absorbed in the chloroplast thylakoid **pigments** by chlorophyll *a* and *b* and carotenoid molecules in photosynthetic light reactions (e.g., Bendall, 2006). Chlorophylls (Table 5.2.2) are lipid-like molecules which form a tetrapyrrole ring with a Mg atom in the centre. The molecule is nonpolar due to its long hydrocarbon tail. Small chemical changes render the chlorophyll *a* into chlorophyll *b*, which has a different action spectrum, i.e., absorbs different wavelengths of incoming radiation. Carotenoids and xanthophylls are products of the isoprenoid pathway and can be rapidly converted to each other by changes in light intensity (see Section 6.3.3.2.1).

Photosynthetic reaction centers PS I and PS II are integral membrane protein complexes, containing both chlorophyll molecules and several electron acceptor molecules acting as converters of light energy into chemical energy in a highly

Table 5.2.2 Chlorophyll *a* and Xanthophylls and their structural formulas

Name	Molecular formula
Chlorophyll *a*	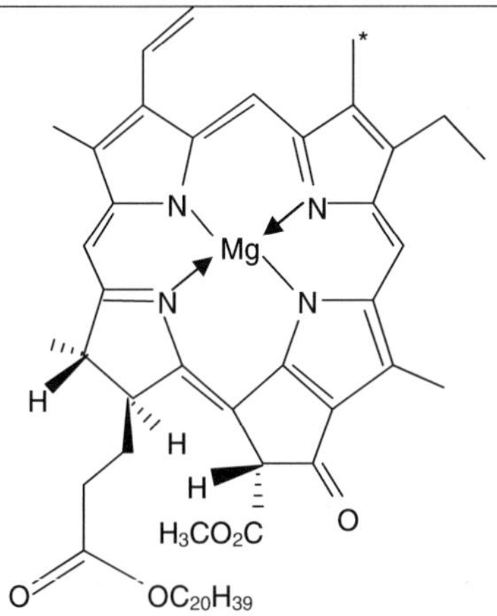
Xanthophylls

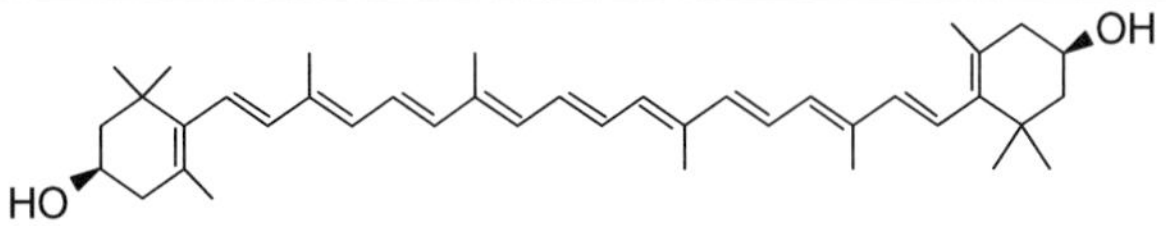

Zeaxanthin

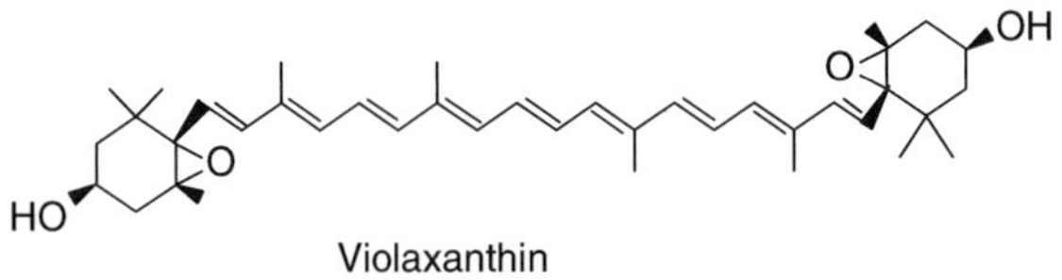

Violaxanthin |

organized manner. Reaction centers are localized in thylakoid membranes where they are distributed in a nonrandom fashion and interlinked by many other molecules.

In Scots pine needles the total chlorophyll content is 0.9–1.8 mmol m^{-2} (based on leaf area, Warren et al., 2003).

5.2.1.1.2 Regulating Substances

The functional substances are under strict control by the biochemical regulation system which provides homeostasis and enables the annual cycle of metabolism and acclimation to prevailing conditions. This control is achieved by regulating substances, which participate in synthesis, decomposition, activation or deactivation of the functional substances. The mechanisms to determine how regulation occurs in response to various stimuli are copious, and we present here only a rough schematic outline for the purpose of the following analyses.

We consider the regulating substances to be a very heterogeneous group of compounds, including e.g.:

- Many metabolites and proteins (for example fructose-2,6-bisphosphate, chaperones such as Hsp70, protein kinases, proteases and many other enzymes), taking part in the activation and turnover of enzymes or expression of genes
- Hormones and other signaling molecules (e.g., ethylene, polypeptides, polyamines, jasmonic acid), transmitting information between distant plant parts
- Inorganic molecules such as, e.g., NO_3^-, S, Ca or P

Regulation may take place at several different levels, from gene expression to activation of catalytic sites by cofactors. At a genetic level the amounts of enzymes are regulated by mechanisms that control expression of genomes in nucleus, chloroplast or mitochondria. Changes in protein biosynthesis also regulate enzyme concentrations, whereas post-translational modifications mostly contribute to regulation of enzyme activity. Some of the regulating substances are constitutively expressed (permanently turned on), others are inducible (involved in specific stages of development or specific environmental stimuli).

An enzyme molecule, once synthesized, has a finite lifetime in the cell, ranging from a few minutes to several days or weeks. Therefore, steady-state levels of cellular enzymes result from an equilibrium between enzyme synthesis and enzyme degradation, or turnover, which is achieved via the action of one or several regulating substances. Defective or abnormal proteins have to be disassembled to prevent the formation of large insoluble aggregates that would eventually inhibit the proper cellular functions. Protein degradation by protease enzymes also facilitates the recycling of amino acids.

One important example of post-translational regulation is phosphorylation and dephosphorylation of enzymes such as nitrate reductase enzyme in roots, which provides a far more rapid control over the nitrate assimilation process than could be achieved through synthesis and degradation of enzyme molecules (minutes *vs.*

hours) (e.g., Kaiser et al., 1999). The regulation of nitrate reductase depends, e.g., on root carbohydrate levels which stimulate the protein phosphatase activity, and consequently nitrate reductase is activated or deactivated.

Another example of metabolite-mediated regulation is the chain of reactions by which fructose-2,6-bisphosphate, a cytoplasmic sugar phosphate, regulates the carbon partitioning into starch and sucrose synthesis in green leaves (e.g., Stitt, 1990). The effect of fructose-2,6-bisphosphate is determined by the relative concentrations of activators and inhibitors, which include several Calvin cycle metabolites, inorganic phosphate and indirectly also light, through reactions associated with photosynthesis.

Signaling compounds are a heterogeneous group of molecules which are synthesized and degraded via a variety of metabolic pathways, and their pool sizes are rapidly changing (e.g., Davies, 1995; Babst et al., 2005). Most plant hormones and signaling compounds are present in tissues in micromolar concentrations. The signal is transmitted from the site of synthesis to distant parts where specific receptor molecules interact with the signaling molecule. Despite their small concentrations, they play vital roles in regulating the plant responses to environmental factors. Most importantly, they regulate cell division and extension, and more generally, growth and senescence of cells and organs. The signaling compounds include e.g. gibberellins, abscisic acid, auxin, cytokinins, ethylene, sterols and jasmonic acid.

5.2.1.1.3 Structural Compounds in Cell Walls and Membranes

All plant cells are separated from their environment by the cell wall consisting mainly of a network of cellulose and lignin (e.g. Sjöström, 1993; Morrell and Gartner, 1998). Primary walls are formed during the cell division, and the complex secondary walls during the cellular differentiation. Covalent and non-covalent cross-linking between molecules is a major contributor to the structural characteristics of secondary walls. The inner side of the primary wall is connected to the plasma membrane, consisting of a lipid bilayer and embedded proteins. Also subcellular organelles are surrounded by principally similar membraneous structures.

Cell Wall Polysaccharides

Cellulose (Table 5.2.3) molecules consist of numerous (from 2,000 to 20,000) D-glucose units arranged parallel to one another and hooked together with a β-linkage into glucan chains. The linkage is unique in the sense that it forms almost perfectly straight chains. Furthermore, the chains of glucose molecules bond to each other noncovalently with H – OH-bonds, which make cellulose microfibrils insoluble in water or organic solvents, and provide the molecule's high tensile strength. Cellulose microfibrils vary in length and width, and generally compose more than 30% of the dry weight of the wall. In Scots pine heartwood the cellulose concentration varies between 130–150 mg cm^{-3} (Harju et al., 2003). Cellulose is synthesized by an enzyme complex associated with the outer layer of the plasma

Table 5.2.3 Structural compounds in cell walls and membranes and their structural formulas

Cellulose	
Pectin	
Lignin	
Chitin	
Linoleic acid[a]	

[a] An example of a membrane fatty acid.

membrane, which means that following their synthesis, cellulose molecules are already positioned in their final location and don't need to be transported. The enzyme uses UDP-glucose as a substrate.

Pectins (Table 5.2.3) are cell wall matrix polysaccharides formed of galacturonate, rhamnose, galactose and arabinose. They comprise about 30% of the cell wall dry weight, and participate in, e.g., modulating wall pH and porosity. Pectins in the middle lamella bind the cells strongly together by branching, aggregating with calcium ions and cross-linkages between other cell wall components.

Hemicelluloses are a heterogeneous group of tightly bound polysaccharides, which do not assembly into microfibrils but are covalently linked to lignin molecules. They are rich in glucose, xylose or mannose. The most abundant hemicelluloses are xyloglucans which can form about 20% of dry weight of the wall. In Scots pine heartwood, the hemicellulose content is ca. $60\,\mathrm{mg\,cm}^{-3}$ (Harju et al., 2003).

Lignin (Table 5.2.3) is a complex mixture of several phenylpropanoid alcohols, and due to its mechanical rigidity, it is most common in the cell walls of supporting and conducting cells such as in vascular tissue. Lignin fills the spaces between cellulose, hemicellulose and pectin components in cell wall matrix. Deposition of lignins into cell wall implies specific irreversible changes in the cells that ultimately lead to cell death, and result in the formation of wood secondary xylem tissues. Lignin partially replaces the water in the wall structure, thus increasing the hydrophobicity of the wall structures. Lignin deposits may account for about 20–35% of the dry weight of wood, i.e. 80–$100\,\mathrm{mg\,cm}^{-3}$ (Harju et al., 2003). In conifer cell walls the most abundant lignin polymers are coniferyl alcohol and p-coumaryl alcohol. Particularly abundant lignin is in compression wood.

Chitin (Table 5.2.3) is the major component of cell walls in most fungi, and it occurs also in cell walls of many insect and crustacean species (Lezica and Quesada-Allue, 1990). It is a polysaccharide, formed of N-acetyl-D-glucos-2-amine molecules bound together with β-1,4 linkages, and resembles thus chemically the structure of cellulose molecule in plant cell walls. Chitin content in ectomycorrhizal roots can be used as an indicator of fungal biomass, integrated over the whole life span of the roots (Ekblad et al., 1998).

Lipids

Most lipids contain long-chain fatty acids which are esterified to glycerol, a 3-carbon alcohol. In plants, lipids are exclusively synthesized in plastids in the fatty acid pathway by cyclic condensation of two-carbon units. Their precursors can be either plastidial or cytoplasmic, and the central intermediate is acetyl-CoA, a 2-carbon acetylated compound which is one of the most central intermediates in all cellular metabolism and provides an important link between many biosynthesis pathways.

The composition of lipids varies between cellular organelles, and membrane and storage lipids often have quite distinct composition. Plasma membrane lipids are phospholipids, whereas in plastids the membrane lipids are almost entirely glycosylglycerides. Minor components are sphingolipids and sterols. In the membrane double bilayers, the lipid molecules are organized so that the polar heads in the outward facing region are hydrophilic, whereas the inner region (towards the center of the membrane) fatty acids are hydrophobic. These properties are extremely important for the membrane functions, as they prevent random diffusion of solutes

between cell compartments. The polar glycerolipids are 16- or 18-carbon straight
fatty acid chains with varying degree of unsaturation, and this regulates the fluidity
of the cellular membranes.

Many pigments, such as chlorophylls, plastoquinone, carotenoids and toco-
pherols are actually lipid-related substances, and they account for about 30% of
the lipids in plant leaves. The epidermal cells are covered with an epicuticular layer
(composed of a network of cutin and embedded polysaccharides, and overlaid by
wax crystals) to prevent uncontrolled flux of water from the leaves. Root endodermal cells also have a hydrophobic layer composed of suberin.

5.2.1.1.4 Storage Compounds

Starch and Sucrose

The main form of carbon storage in coniferous trees is starch (Table 5.2.4 and
Fig. 5.2.1), whereas the most important transportable carbohydrate is sucrose

Table 5.2.4 Glucose, sucrose and starch and their structural formulas

(Table 5.2.4). The syntheses of starch and sucrose are competing processes depending, e.g., on the inorganic phosphorus level in cytoplasm. Sucrose is a disaccharide and the major soluble end product of photosynthesis in green leaves.

Starch consists of two types of long carbohydrate molecules, unbranched amylose and branched amylopectin, which are formed of α-(1→4)glucose units (Table 5.2.4). Starch synthesis occurs in chloroplasts where it is also stored as large granules. Specialized storage plastids called amyloplasts serve as energy reserves in rapidly developing tissues such as seeds or buds. Starch grains vary in size from less than 1 μm in leaves to larger than 100 μm in seeds, and can grow in multiple layers. The amount of starch varies greatly within season and plant organ.

Energy Stored in ATP

ATP (adenosine triphosphate) (Table 5.2.5) is the universal energy carrier which is readily useful for fueling numerous cellular reactions. ATP is also a metabolic effector, regulating many reactions by its turnover rate. It is formed in both photosynthetic light reactions and in the respiratory pathways, and consumed, e.g., in

Table 5.2.5 ATP, NAD/NADH and NADP/NADPH and their structural formula

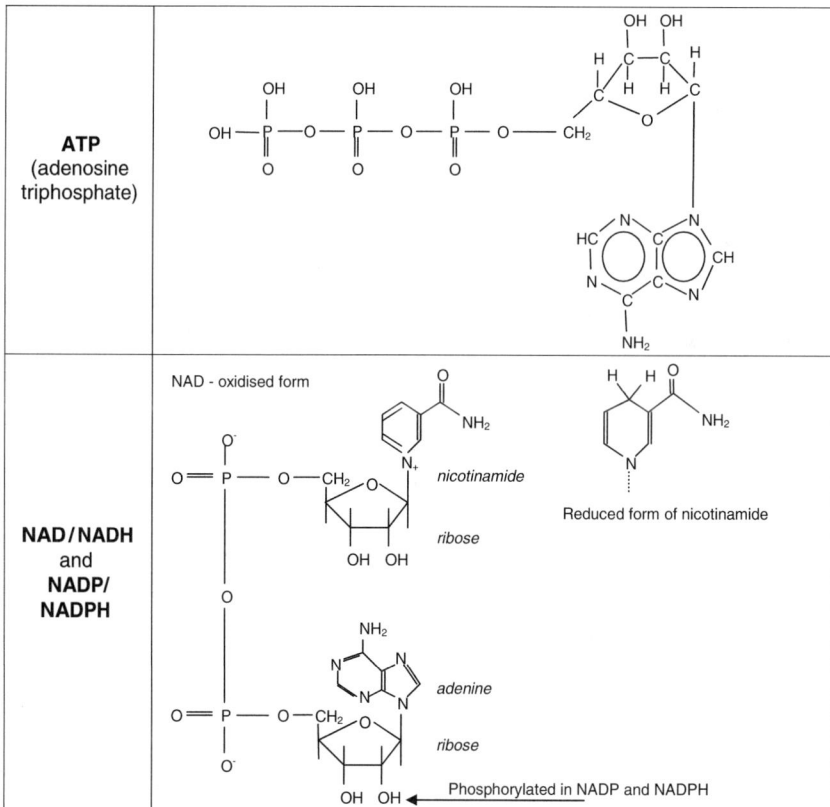

the Calvin cycle for reducing CO_2 into carbohydrates, and in almost all anabolic reaction chains thereafter. In the inner mitochondrial membrane and chloroplast thylakoids, the ATP synthesis involves an H^+ potential across the membrane, created by a transmembrane ATP synthase enzyme. ATP can be translocated between organelles, and especially the mitochondrial ATP-ADP translocator is an important component in cell energy balance (Hoefnagel et al., 1998).

ATP is hydrolyzed to liberate the energy and yield ADP and inorganic phosphate, which then can be recycled. Other forms of chemical energy (e.g., NAHD or $FADH_2$) need to be converted to ATP before they can be used as energy sources.

Reducing Power Stored in NADH, FADH and NADPH

All energy demanding or binding biochemical reactions involve the formation or participation of cofactors NAD^+ (nicotinamide adeninine dinucleotide) (Table 5.2.5), FAD^+ (flavin adeninine dinucleotide) or $NADP^+$ (nicotinamide dinucleotide phosphate). These cofactors are reduced by accepting electrons and one or two protons to produce the corresponding reduced compounds NADH, $FADH_2$ and NADPH. Glycolysis generates both ATP and NADH, whereas NADPH is supplied by the oxidative reactions of the pentose phosphate pathway and citric acid cycle. $FADH_2$ participates in the citric acid cycle and oxidative phosphorylation in mitochondria.

5.2.1.1.5 Secondary Compounds

In addition to the primary carbon metabolites, plant cells include numerous compounds performing specialized functions. Although they commonly are known as secondary metabolites, the group of compounds included in this category is very heterogeneous, and many of their functions are vital for plant survival and growth. These compounds are synthesized via, e.g., phenylpropanoid, isoprenoid or shikimic acid pathways, and can occasionally form a significant sink for assimilates. One important feature of secondary compounds is that the same compounds can be either constitutive (permanent storage pool in specialized cells) or induced (temporary pool which is produced after activation of metabolic pathways). The most important group of compounds in respect to the topic of this book is isoprenoids.

Isoprenoids are a large group of carbohydrates varying both in size and in chemical structure. The smallest molecule is isoprene (Table 5.2.6), which is the basic structural component in all larger isoprenoids. Their synthesis occurs in both plastids and cytoplasm, and two partially independent pathways are involved (e.g. Kesselmeier and Staudt, 1999; Pichersky and Gershenzon, 2002). Isoprene and many other small isoprenoids such as mono- and sesquiterpenes (Table 5.2.6) are volatile at ambient temperatures, and thus tend to evaporate. Many isoprenoids are specifically synthesized for, e.g., defense purposes and stored in specialized storage tissues, although not all of their precise functions are fully elucidated to date. Recently, a protective role against high temperatures and reactive oxygen species

Table 5.2.6 Secondary compounds and their structural formulas

Myrcene[a]	
Isoprene	

[a] An example of a monoterpene.

has also been postulated for many isoprenoids. Plants produce also copious numbers of non-isoprenoid volatile compounds, which are by-products of other major biological processes such as pectin biosynthesis (methanol), fermentation (acetaldehyde), or decarboxylation of carbon compounds in cells (acetone) (Fall, 2003).

5.2.1.2 Elements in Chemical Compounds

In addition to carbon, hydrogen and oxygen, plants are composed of 14 essential mineral nutrients. Many of these are minor constituents, yet their deficiencies can produce growth abnormalities or lead to plant death, and thus maintaining balanced uptake rates is essential for the plant survival. The tissue concentrations of macronutrients range from 10^3 to 10^4 µg (g dry weight)$^{-1}$, whereas micronutrients are found in ng (g dry weight)$^{-1}$ concentrations. Many mineral nutrients are assimilated as parts of organic compounds such as chlorophyll (Mg) or calmodulin (Ca), while others maintain their ionic identity within the plant (K).

Nitrogen

From plant dry matter, nitrogen forms about 1.5%, being the most abundant macronutrient after carbon. Since leaves contain most of the functional substances rich in nitrogen, the N content in leaves can be considerably higher than that, especially in some broad-leaved trees. In conifer needles, however, the N content is normally between 1–1.5% (Helmisaari, 1992; Pensa and Sellin, 2002). Nitrogen is the main element in amino acids and proteins, and an average of 15–17% of the molecular weight of proteins is nitrogen. Nitrogen is also a constituent in many other macromolecules such as nucleic acids, nucleotides, chlorophyll and many coenzymes. A sufficient supply of nitrogen is a major determinant for plant growth, because it is needed for functional substances.

Phosphorus

Phosphorus concentration in plant tissue is around 0.2% (dry weight) and its compounds are involved in the main energy storing and transport components, namely ATP, ADP, NAPD and NADPH. Additionally, a majority of biosynthetic pathways, such as starch synthesis and degradation, nitrogen and sulphur assimilation, and biosynthesis of cell wall components and isoprenoids, involve phosphorylated intermediates. Also nucleic acids, nucleotides, phospholipids and many coenzymes contain phosphorus. P is the macronutrient which is least available from the soil for root uptake, due to its strong chemical bonds to soil clay minerals. However, the uptake of phosphorus is greatly enhanced by mychorrizal associations, and also by root exudates that release organically bound P from the soil.

Sulphur

Sulphur plays an important role in many macromolecules, and regulates, e.g., protein activity by forming disulfide bridges in molecules and participating in catalytic site formation. Further, it is needed in electron transport chain molecules (e.g., ferredoxin and thioredoxin). An important sulphur containing a tripeptide molecule, glutathione, is involved in growth, redox control and responses to many environmental stresses. Sulphur content in conifer needles ranges between $0.7–1.7\,\mathrm{mg}$ (g dry weight)$^{-1}$ (e.g. Manninen et al., 1997; Rautio et al., 1998).

Other Mineral Nutrients

Magnesium (Mg) is the central atom in chlorophyll molecules and is required in many enzymes involved in phosphate transfer. **Calcium** (Ca) is a constituent in the middle lamella of cell walls, required as a cofactor in some enzymes and vitally important in signaling of a variety of physical and chemical factors through Ca-channels in many membranes and the cytoplasmic Ca-binding protein, calmodulin.

Potassium (K) is required by many enzymes as a cofactor and it is essential in establishing and maintaining cell turgor. It is the most abundant cation with millimolar concentrations, and needed for example in photosynthesis, oxidative metabolism and protein synthesis.

Also **boron** (B), **chlorine** (Cl), **zinc** (Zn), **manganese** (Mn), **copper** (Cu), **iron** (Fe), **molybdenum** (Mo), and **nickel** (Ni) are essential for normal plant growth and reproduction. Their physiological roles involve participation in, e.g., protein synthesis and electron transport reactions.

5.2.2 Cellular Level

Jaana Bäck

University of Helsinki, Department of Forest Ecology, Finland

The principal component of all living organisms is the cell, but there are several important features which characterize plant cells in particular. The general structure of plant cells and tissues in this chapter has been compiled from Taiz and Zeiger (2006), Evert (2006) and Buchanan et al. (2000). The following structural characterizations apply mostly for boreal gymnosperm tree species (and mainly the genus /Pinaceae/), although in some cases specific structural differences compared with angiosperm species are also indicated. Specific references are added in places where detailed and/or updated information is given.

The structural properties of the most important cell components are characterized in Fig. 5.2.1 and in the following chapters.

Cell Wall

The cell wall (Fig. 5.2.1) is the strong, outermost layer of each cell in plant tissues, governing cell shape and size, but also affecting cell expansion during growth (for a review see Fry, 2001). Cell walls differ between tissue types in their chemical composition, thickness, embedded substances, and fine structure. Generally, an elastic primary wall is formed during cell division and elongation. Primary walls of plant cells contain typically 70% water, and 90% of the dry matter is polysaccharides. Secondary cell walls are much thicker and account for most of the carbohydrates in plant biomass. Secondary walls are formed after the cells have reached their final size.

The majority of cell wall carbohydrates consist of complex polysaccharides, which are covalently linked with each other. The most abundant ones are cellulose

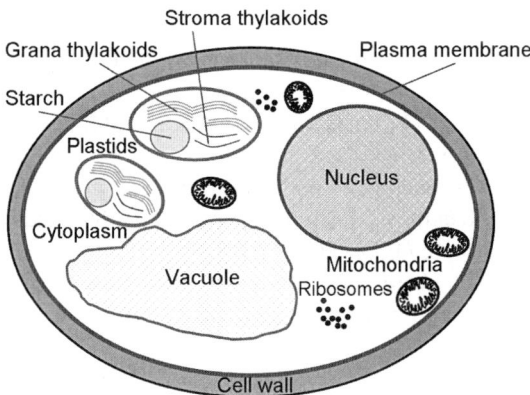

Fig. 5.2.1 A schematic presentation of the structural components typical of a plant cell. Details are given in the text

microfibrils (Table 5.2.3), forming long, chain-like layers which are embedded in hydrated, gel-like pectin (Table 5.2.3) and hemicellulose matrix. Other molecules such as suberin and cutin are deposited in those cell walls where prevention of water movement is especially important. In many secondary cell walls, especially in xylem vessels, lignin (Table 5.2.3) molecules are incorporated to improve the mechanical strength of the structure. A strong, pectinaceous middle lamella separates two adjoining cells. The inner face of the cell wall is in contact with plasma membrane, although not firmly attached to it. Meristematic and parenchymatous cells have thinner walls than collenchyma or sclerenchyma cells, and even within one cell the wall thickness can vary.

Mature cell wall generally does not participate in cell metabolic reactions. Walls allow free passage of most small, water soluble molecules and ions such as oxygen, CO_2, nitrate, phosphate, sugars and amino acids into and out of the cell, but prevent the permeation of larger molecules. Plasmodesmata (see Section 6.3.2.4 and Fig. 6.3.17) are small openings in the cell wall and plasma membrane, which form a continuum between the cells, and allow movement of water and solutes, ions and also many large molecules between the cells. The walls also have high affinity to some inorganic cations, e.g., Ca^{2+} or Cu^{2+}.

Membranes

Cells as well as most cellular organelles (e.g., nucleus, vacuole, mitochondria and plastids) are separated from their environment by a semipermeable membrane (Fig. 5.2.1). The most important functional properties of membranes follow from their biochemical nature as lipid bilayers, where large protein molecules are embedded. Proteins and lipids make roughly equal proportions of the membrane's mass. Membrane lipids (Table 5.2.3) belong to several chemical classes, including, e.g., phospholipids, galactosylglycerides and sterols, and their composition is highly variable between cell organelles and tissues, but it also varies with environmental factors such as growth temperature. Membrane proteins are either embedded into the lipid bilayer, such as proteins acting as ion pumps and receptors for signal transduction pathways, or bound to the membrane surface.

Membranes are selectively permeable to many molecules, which is reflected in their structure in many ways (see Sections 5.2.1.1.1 and 6.2.1.1). The membranes facilitate and actively control the continuous molecular trafficking as the cell/organelle takes up nutrients and metabolites, exports wastes and controls its turgor pressure. This is made possible with the specific, membrane-attached functional substances: membrane pumps, channels, antiporters and symporters (e.g. Véry and Sentenac, 2003; Weber et al., 2005; see Section 5.2.1.1.1). Membranes also form complicated internal structures, such as photosynthetic thylakoids in plastids and endoplasmic reticulum in cytoplasm, participating in many primary metabolic functions.

Plastids

Plastids are cylindrical, lense-shaped organelles which are one of the main characteristic features of plant cells. Each mesophyll cell in a mature leaf or needle contains from several tens to several hundreds of the chlorophyll-containing chloroplasts. They are arranged near the cell walls as a single-plastid layer (see Fig. 5.2.1). Chloroplasts are responsible for primary carbon assimilation, i.e., photosynthesis, and also for many fundamental intermediary metabolic reactions involved, e.g., in nitrogen metabolism. Other types of plastids are, e.g., amyloplasts (storage), leucoplasts (biosynthesis of secondary metabolites such as isoprenoids), and proplastids (the undifferentiated precursors from which all other plastids develop).

Plastids are separated from their cellular environment by a double layer membrane, a plastid envelope. The envelope, as all membranes, allows strict control of plastidial metabolism, independent from the rest of the cell. The outer and inner membranes in the chloroplast envelope differ in their composition and structure, and thus also in their functions. Surrounded by the envelope, the plastidial matrix called stroma contains a small amount of DNA, ribosomes and abundant functional substances, enzymes (e.g. ribulose bisphosphate carboxylase/oxygenase; Rubisco), as well as storage compounds such as starch and lipids. Starch grains can periodically make up the majority of the chloroplast volume.

Photosynthetic light capture occurs in chloroplast thylakoids (Fig. 5.2.1), which form a continuous network of membrane sacks, partially stacked in piles on top of each other (lamellae). Thylakoid membranes contain the photosynthetic pigments, mainly chlorophyll a (Table 5.2.2) and b and carotenoids, and their associated proteins, embedded in the glycosylglyceride bilayer. The integral membrane proteins include the photosynthetic reaction centres (PSI and PSII), ATPases and electron transport enzymes, positioned in highly specialized arrangements and unique orientation within the membrane (e.g., Chitnis, 2001; Allen and Forsberg, 2001). The majority of the pigment is clustered in antenna complexes capturing the energy of the light quanta into the reaction center complex, where the light energy is converted into chemical energy as ATP (adenosine triphosphate) and NADPH (nicotinamide adenine dinucleotide phosphate) (Table 5.2.5), simultaneously as an oxygen molecule is produced (see Section 6.3.2.3). The two photosystems and the reaction centre are closely linked with an electron transport chain which is formed of integral proteins.

Mitochondria

Mitochondria (Fig. 5.2.1) are spherical or rod-like organelles which are surrounded by a double membrane and range from 0.5 μm to 1.0 μm in diameter and up to 3 μm in length. Mitochondria are abundant in all plant cells, and they house the essential respiratory machinery which generates ATP by the citric acid cycle and electron transfer chain (see Section 6.3.2.1). Their number is closely correlated with the metabolic activity of the cells; guard cells and actively dividing meristematic cells for example are rich in mitochondria.

The smooth outer membrane completely surrounds the strongly invaginated inner membrane and the aqueous matrix located inside the double membrane. The membraneous formations of the inner membrane (cristae) contain high amounts of proteins. Small molecules can pass through the outer membrane rather freely, whereas the inner membrane has very low permeability. Mitochondria, as plastids, are semi-autonomous; they contain DNA, RNA and ribosomes and are thus able to carry out a part of their own protein synthesis.

Nucleus

Genetic codes determine all cellular functions and their responses to different regulatory agents. The nucleus (Fig. 5.2.1) is the organelle housing most of the plant genome, although plastids and mitochondria also contain a small genome of their own. The nucleic acids, **DNA** (deoxyribonucleic acid) and **RNA** (ribonucleic acid), are large polymers that store and transmit genetic information in cells. DNA constitutes the genome of the cell, and it forms the chromatin where the huge molecule is coiled in a double helix form. The three different RNA molecules contribute to the transcription, translation and processing of the encoded genetic information into protein structure. Both DNA and RNA are long, unbranched molecules, each composed of four different types of nucleotides. The nucleus is surrounded by a porous nuclear envelope, which is permeable to ribosomal subunits and some macromolecules through the nuclear pore complexes housing an active membrane transporter.

DNA comprises the chromosomal material, chromatin, which builds up genes when assembled in a highly organized manner. In chromatin, long DNA molecules are tightly packed together with proteins into a double helix structure where two DNA strands are associated with their complementary base pairs, linked with hydrogen bonds. The recently unveiled poplar (*Populus trichocarpa*) genome size is 480×10^6 base pairs (Tuskan et al., 2006); for the gymnosperm species genome, such a detailed analysis is not available at the moment.

RNA is the primary product of gene expression after DNA transcription by RNA polymerases, and it has an essential role in protein synthesis and many other cellular functions. Most of the RNA is ribosomal, i.e., bound into small, two-subunit structures which serve as platforms for reading of the information which translates into the amino acid sequence of each individual protein.

5.2.3 Tissue

Jaana Bäck and Eero Nikinmaa

University of Helsinki, Department of Forest Ecology, Finland

In general, the anatomy of any individual plant structure is highly specialized for some particular process or for transport. Therefore plant tissues entail specific

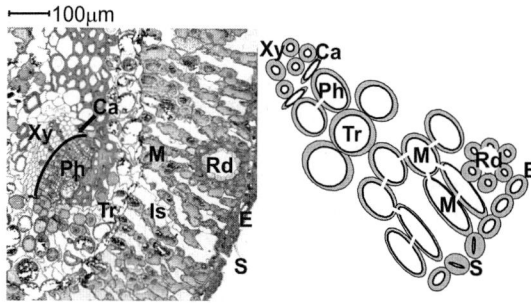

Fig. 5.2.2 Light micrograph (left) and schematic illustration (right) of epidermal, ground and vascular tissues of a Scots pine needle. Stoma (S), resin duct (Rd), phloem (Ph), xylem (Xy), cambium (Ca), transfusion parenchyma (Tr), epidermis (E), mesophyll (M), intercellular space (Is). Plasmodesmal connections are seen in (b) between mesophyll cells and between sieve elements and companion cells

anatomical and biochemical features to fulfill these functions. A tissue is considered to consist of a group of functionally similar cells, acting in concert in response to external or internal cues. The vascular plant tissues are either dermal, vascular or ground tissues, and these can be found in all organs, although in varying proportions.

Dermal tissues are the outermost layer of organs facing the atmosphere or soil. They protect the inner tissues from extreme environmental conditions, but also selectively exchange compounds between the plant and the environment. In needle-like leaves (Fig. 5.2.2), the dermal tissues consist of one or two layers of thick-walled cells (epidermis, hypodermis), which are covered with a thin cuticle and a complex, waxy epicuticular layer to prevent uncontrolled water loss and to act as protective cover. The thickness of the epicuticular layer in Scots pine needles is only few micrometers, and that of the dermal layer 30 μm (Bäck et al., 1994).

Gas phase diffusion of many substances occurs through stomata, which are a highly specialized type of dermal tissues in leaves (Fig. 5.2.3). Stomata actively control the exchange of water and gaseous compounds (CO_2, O_3, VOCs etc.) between the plant and environment. They are formed by pairs of specialized guard cells, which perceive environmental signals and control the opening of a small stomatal pore. The stomatal pore is approximately 5 μm wide when it is fully open. Stomatal density is sensitive to many environmental factors (e.g. Woodward, 1987; Woodward and Kelly, 1995; Brownlee, 2001), varies a lot between species, and in many species it also depends on the surface of the leaf. In the boreal region, the Scots pine needle stomatal density lies between 40–140 stoma mm^{-2} (Lin et al., 2001; Luomala et al., 2005; Turunen and Huttunen, 1996).

In the developing root tips the epidermis is specialized for absorption of mineral elements and water, and therefore the outer surface differs from that in leaves. Bark is the outermost layer in stems and roots of woody plants, and describes all tissues outside the vascular cambium. Outer bark of the stem, periderm, is a three-layered region of phellogen, phelloderm, and phellem (Fig. 5.2.8). Phellogen (i.e., cork cambium) produces phellem (cork) toward the outside and phelloderm (cortical

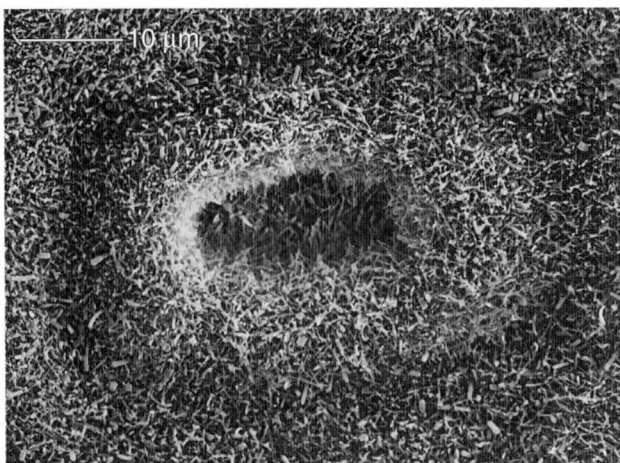

Fig. 5.2.3 A stoma of a Scots pine needle viewed under scanning electron microscope. A special feature of conifer stomata is that they are sunken deep in the epidermis, and the ledges are covered with a thick epicuticular wax layer preventing uncontrolled water penetration from guard cells (Courtesy of Satu Huttunen)

parenchyma) toward the inside of the phellogen. The cork layer is a strong tissue, and it efficiently protects the vulnerable transport tissue in stem from the external factors, and it can range from a few millimeters up to several centimeters in thickness.

Ground Tissues

Ground tissues are present throughout the plant body and formed of parenchyma, collenchyma (supporting) or sclerenchyma (protective and storage) cells. Ground parenchyma in leaves is formed from actively metabolizing cells involved in photosynthesis and storage. Most leaves have two or more layers of parenchyma cells called mesophyll cells (Fig. 5.2.4). The outermost mesophyll cell layer close to the epidermis is in many species formed by elongated, thin-walled and closely packed palisade parenchyma cells, which are rich in plastids and act efficiently in light capture and CO_2 assimilation. The palisade cell columns also channel light towards lower cell layers. In Scots pine needles, irregularly shaped and loosely arranged spongy mesophyll cells, which also photosynthesize, are located underneath the palisade tissue (Bäck et al., 1994). Light scatters in spongy mesophyll as it is reflected from cell-air interfaces, resulting in more uniform light absorption even in deeply laying cells.

In stems and roots the ground tissue is called pith or cortex, and composed of cells with thickened walls containing suberin and lignin. The detailed anatomy and location of ground tissues in stem vary a lot between tree species. In roots the central cortex usually consists of thin-walled parenchyma cells with numerous intercellular

Fig. 5.2.4 Light micrograph
of pine needle mesophyll
cells. Vacuole (V), plastids
with starch grains (P), cell
wall (Cw), nucleus (N), intercellular space (Is, apoplast)

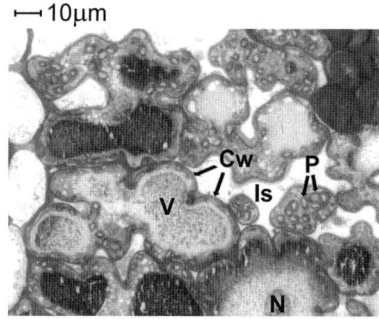

spaces. Cortical tissue allows the diffusion of water, nutrients and oxygen from the
root hairs inwards, and also serves as starch reserve for root metabolism. The endodermis forms the innermost layer of the cortex in roots. Endodermal cells are more
rectangular in shape, the side walls being thickened with suberin. These thicken-
ings are called Casparian strips, and they are involved in the transport of water from
cortex to the xylem.

Vascular Tissues: Cambium, Phloem and Xylem

Vascular tissues are vital for the long-distance longitudinal transport of substances
within the plant individual, and they form a continuous structure from roots to
leaves. In stems and needles of gymnosperms, they run parallel in a vascular bundle
(Figs. 5.2.2 and 5.2.7), and in roots the vascular tissues occupy a separate central
cylinder area, where xylem and phloem tissues are arranged in close connection
with each other. Both tissues have primary cells that originate from recently differentiated cambial cells, and secondary cells which are metabolically mature and can
form several cell layers.

The majority of the xylem in mature conifer tissues consists of thick-walled secondary xylem, where the regularly organized tracheids conduct water and solutes in
the system. In the xylem of broadleaved trees the tissue is more variable, consisting of vessels, tracheids and fibers, and also the proportion of living parenchyma in
wood is higher. Tracheids and vessels are dead when they reach maturity. They are
predominantly vertically oriented (with exception of rays, Fig. 5.2.6) very long and
narrow cells that are connected to neighboring elements by small pores (Fig. 5.2.5).
Normally these cells are tens to hundreds of micrometers in diameter but may be
several millimeters to centimeters long. In broadleaved trees, the vessels consist of
several vessel elements and they are generally much wider and longer than the tracheids that are normally only a few millimeters long. The diameters of connecting
pores are normally a few micrometers to 20 μm. In the pores there is a mesh-like
structure in the direction of cell walls in the middle of the pore called margo that
may have, especially in conifers, a thickening in the middle called the torus. Margo
and torus are important structures in preventing the spreading of the air bubbles in
the xylem vascular tissue.

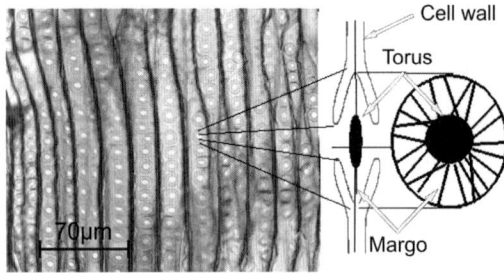

Fig. 5.2.5 A light microscope photo showing the interconduit pits in the tracheid wall of Scots pine and a schematic illustration of a pit structure with torus (black centre) and margo (surrounding net-like structure)

Fig. 5.2.6 A microscope photo of a tangential view of a Scots pine xylem. The groups of cells perpendicular to the longitudinal tracheids are rays with living cells (Courtesy of Tuuli Timonen)

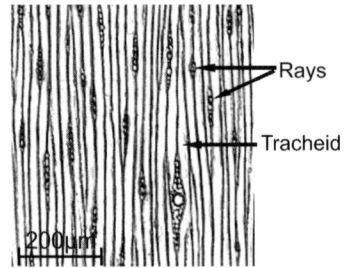

Apart form the vertically oriented vascular tissue, xylem also has radially oriented rays that occupy from a few percent (normally conifers) up to some tens of percent (broad leaved species) of the xylem volume. In contrast to the non-living, vertically oriented vascular tissue, in ray tissue there is also living vascular parenchyma that forms an important storage of energy-rich and phenolic compounds. Radially arranged ray tracheids and associated vascular parenchyma form continuous radial channels of transport within the woody tissue extending through the xylem all the way to the phloem tissue in the bark. Thus these channels provide a symplastic connection to otherwise apoplastic wood tissue. These radial channels may have an important role in maintaining the vertical vascular tissue functional, as there seems to be at least one radial ray channel associated with each vertical vascular cell even in a conifer that consists almost entirely of numerous vertical tracheids (see Fig. 5.2.5). In conifers there are also resin ducts in the wood that are also associated with living cells.

Translocation of photoassimilates occurs within highly specialized, living phloem sieve cells (gymnosperms) or sieve elements (angiosperms). The phloem functional conduit occurs via the cell-to-cell connections at their perforated end walls or through pore-like openings in the conifers. The sieve elements are living cells although they have lost their nucleus. However, there are a number of cell organelles left such as plastids and mitochondria in a thin cytoplasmic layer anchored in the plasma membrane, and the sieve elements are thus capable of individual metabolism, e.g. sealing damaged sieve elements from the vascular connection and thus preventing leakages from the phloem. The mature sieve cells are closely connected to large Strasburger cells, also called companion cells, which have numerous mitochondria and ribosomes indicative of high metabolic activity. The sieve cells and sieve elements and companion cells are connected to each other through a number of plasmodesmata and these companion cells have an important role in controlling the loading and unloading of sugars into the sieve cells. This is why the phloem transport tissue normally consists of sieve element-companion cell (SE-CC) complexes (Fig. 5.2.2). The plasmodesmata between the companion cells and the surrounding tissue (transfusion parenchyma) are far fewer, which is also assumed to be linked to loading and unloading processes (Fig. 6.3.17).

Meristematic Tissue

The new tree tissue is formed at specific meristematic tissues either at the ends of the growing axis (apical, primary meristem) in shoots and roots, or surrounding the xylem tissue in the growing axis (secondary meristem, or the cambium). The primary meristem gives rise to new tree organs, forming also primary wood in the shoot axis. After elongation growth in a region is complete, the secondary meristem is formed in the newly formed axis, and produces secondary phloem and xylem tissues, thus increasing the thickness of the stem or the root. Phellogen is the cork cambium which forms on the peripheral region of the expanding axis and forms periderm, a secondary protective structure to the stem. Meristematic activity is under a genetical regulation and forms a large sink for resources during the periods of active growth.

5.2.4 Organs

Jaana Bäck[1], Eero Nikinmaa[1], and Asko Simojoki[2]

[1] University of Helsinki, Department of Forest Ecology, Finland
[2] University of Helsinki, Department of Applied Chemistry and Microbiology, Finland

The traditional characterization of vascular plant body involves the differentiation of plant parts as organs that are more or less specialized to certain functions. The main organs are stem, leaf and root, and sometimes the flowers also are considered as separate organs. The stem and leaf (and flowers) can be regarded to form the shoot (above-ground organs) in correspondence to the root system (below-ground organs). Leaves are specialized for photosynthesis, roots for anchorage, nutrient and water uptake, and the stem tissues support the above ground plant parts and form a continuum acting as pipelines for transport of water and assimilates. Structurally these organs mainly differ in the relative distribution and organization of the vascular and ground tissues.

Fig. 5.2.7 A light micrograph
of a cross-section of a Scots
pine needle. Resin duct (Rd),
mesophyll (M), xylem (Xy),
phloem (Ph), stoma (S), epidermis (E)

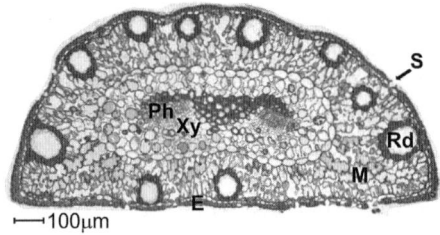

⊢—100μm

Leaves

Leaves are the organs where light is captured for providing the energy to drive the
chemical CO_2 fixation and other vital functions in plant. Leaves also participate in
gas exchange between the plant and atmosphere through the stomata (see Fig. 5.2.3).
Leaves can vary greatly in shape and size, but a general phenomenon is that they
have a flat and rather thin leaf blade (broad-leaved species) or a thicker, cylindrical
structure (needle- or scale-like leaves, Fig. 5.2.7). Usually the leaf is attached to
the shoot with a petiole, where the veins connect the vasculature of the leaf to that
of the shoot and further to other plant parts. The vasculature can be visible also
from outside of the leaf as lateral veins and midvein in flat leaf blades, whereas in
other species vascular tissues are fully embedded in the ground tissue (needle-like
leaves in conifers). Leaves may be possessing external structures such as scales,
hairs, glands or trichomes, mainly functioning in protection against external factors.
The spatial distribution of stomata and other epidermal structures is nonrandom and
a minimum spacing is existing between them. In conifer needles the stomata form
several rows, where the stomata are situated 20–50 μm apart.

The leaves are usually grouped together in a species-specific manner. In many
species (e.g. in the genus *Pinus*) the needles form a cluster (fascicle) consisting of
two or more needles which can be from 2 up to 15 cm in length. Each fascicle is
produced from a small bud on a dwarf shoot in the axil of a scale leaf. The fascicles
are combined to form a unit, which consists of a group of needles of same age, i.e.
they all are formed during the same growing season (or during one flush in those
species having several flushes per growing season). As the shoot grows, the leaves
are arranged in such a way that they tend to maximize the yield of light. Leaves
can be shed annually (in deciduous trees), or sustain and maintain their vitality for
2–10 years (even 40 years in some evergreen trees). Many leaves show dorsiventral
anatomy, where the leaf surfaces have somewhat different construction and may
serve different functions, for example the organization of the mesophyll tissue and
stomatal density may differ between the surfaces.

Leaf and shoot development is predetermined during vegetative bud development
in the previous growing season in many coniferous trees, and controlled by several
hormonal cues (Davies, 1995). Normally one apical and several lateral buds forming
the branch whorl are formed in e.g. Scots pine and Norway spruce. The meristematic apex of the buds is very small, only about 0.1 mm high. Downwards from the
apex, procambial strands, and shoot and needle primordia appear. Already the shoot

and needle primordia have highly developed vascular connections to the previous year's shoot, making carbon import possible for the needs of the developing tissues. Buds are covered by protective, waxy scales and often also by resinous excretion. The timing of shoot and needle tissue differentiation differs between boreal conifer species; e.g. in Norway spruce the protoxylem and other needle tissues are more or less fully differentiated already before bud burst, whereas in Scots pine buds and young shoots the needle tissues differentiate and their elongation growth continues longer.

Stem

The stem of a vascular plant is specialized in structural support and conducting functions between the above-ground parts (shoot) and below-ground parts (roots). Therefore the main proportion of a stem is formed of thick walled, water conducting tissue (xylem), participating in both of these functions (Fig. 5.2.8). Assimilates are

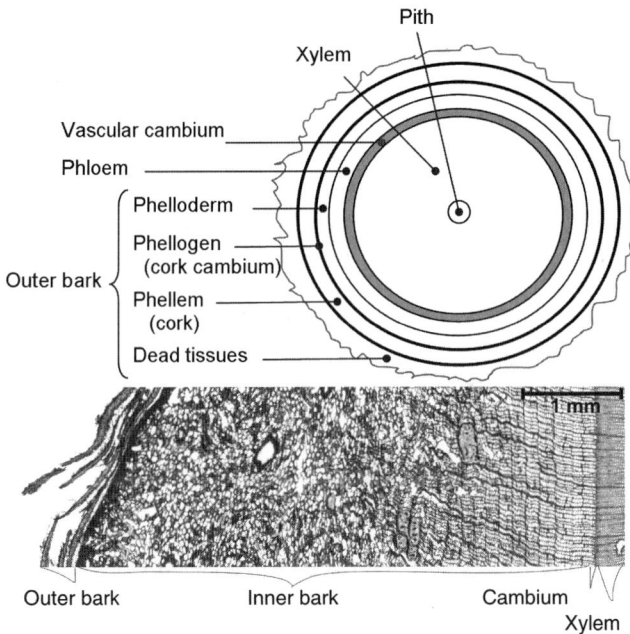

Fig. 5.2.8 Upper picture: Schematic illustration of radial components of a mature coniferous tree stem: pith, secondary xylem, vascular cambium, secondary phloem and outer bark including phelloderm, phellogen and phellem. Lower picture: The transverse, radial light micrograph is from a Norway spruce (*Picea abies* (*L.*)Karst.) Vascular cambium produces secondary xylem (only partly seen) internally and secondary phloem (i.e., inner bark) externally. Inner bark consists of the conducting, uncollapsed phloem that has functional sieve cells, and older, nonconducting, collapsed phloem. In the phloem, there are also radial rays, axial parenchyma, thick-walled, lignified sclereids, and resin canals with epithelial cells. Outer bark (also called rhytidome) consists of periderm and all tissues external to it. Periderm consists of phellogen (i.e., cork cambium) producing phellem externally and phelloderm internally (Courtesy of Tuula Jyske)

transported in the living cells of the phloem tissue, which form a minor proportion
of the cross section in a stem. Stem also forms a significant storage compartment
for water and reserve carbohydrates during dormancy period.

A particular feature of perennial tree structure is the capacity for growth year
after year, in some cases over thousands of years. In tree stems, a lateral meristem
called vascular cambium, separates the primary xylem and phloem cells of the vascular bundle. During secondary growth, the vascular cambium continuously divides
to form new phloem elements to the outside and xylem elements to the inside.
The cambium exhibits periodic activity forming wide tracheids with thin walls in
the rapid growth during spring and early summer (earlywood). Later in the growing season, relatively narrow tracheary elements with thick walls are formed (latewood). These alternating growth phases form growth ring patterns characteristic of
the woody tissues and the annual increment of the stem tissue can be measured from
these rings.

The growth of secondary xylem constitutes the major part of the diameter growth
in the stem, as well as in roots and branches. In conifers and in many hardwoods the
living parenchyma cells in the central part of the stem die and may release secondary
substances in the process of changing the chemical composition of the wood. This
part of the stem is called heartwood, whereas the water-conducting part of the stem
is called sapwood. The xylem cells in wood are arranged in cylinders parallel to the
length axis of the stem, and the growth rings form concentric layers with alternating
lumen and wall dimensions, further improving the strength of woody tissues.

The outermost layer of the stem, bark, consists all tissues out of the vascular
cambium. It is consisting of the secondary phloem, the periderm (including phelloderm, phellogen and phellem) and the dead tissues lying outside the periderm. A
distinction can be made between the living, inner bark, and nonliving, outer bark.
The inner bark consists of secondary phloem, which is composed both of the conducting, uncollapsed phloem with functional sieve cells (gymnosperms) or sieve
tube elements (angiosperms), and of older, nonconducting and collapsed phloem
tissue.

Roots

Roots anchor the tree to the soil, absorb water and nutrient ions from the soil, produce many functional substances such as growth regulators, and can also store sugars and starch. Efficient acquisition of soil resources depends on both the quantity
and functioning of the below-ground structures. Both non-woody and woody roots
participate in water and nutrient uptake.

Most boreal forest trees possess a tap root system, where the primary root grows
vertically down into the soil. Later growing lateral or secondary roots and other
branches form the branching root system capable of supporting large trunks and
acquiring water and nutrients from large area surrounding the tree trunk.

Roots are divided into both functionally and structurally different coarse and
fine roots. Generally, fine roots (diameter <0.5 mm) located in the uppermost soil

layers are the principal sites for nutrient and water absorption, and they can account for >90% of the whole root length (Finér et al., 1997). Under drought conditions, however, the tap root system penetrating deeper into soil layers will be of greater importance.

The roots are formed by an apical meristem at root tips. This meristem is protected by a root cap surrounding the root apex. On aging, the root tissues differentiate into three anatomically different longitudinal zones: the white zone of active cells nearest to the root apex, the brown-coloured condensed tannin zone with dead cortical cells, and the cork zone characterized by a secondary growth by vascular and cork cambia (Peterson et al., 1999; Hishi, 2007). The non-woody white root tips are the main part of the absorption of nutrients occurs in most species. Behind the apical meristem there is an elongation zone, where the root elongation growth occurs by both elongation and division of cells. The elongation zone is important as the main site for root exudate liberation to the surrounding soil. This is followed by the root hair zone, where fragile epidermal structures have developed to facilitate interaction between the root and soil surrounding it. In roots, there is no waxy cuticular layer like in above ground located plant organs, but instead abundant epidermal cell extensions, root hairs, which greatly increase the absorptive area and contact with soil particles and micro-organisms.

In mature root cells of the lateral (woody) root zone, the cellular differentiation has been completed, and this has great effects on the functions of root tissue. Endodermal cells divide the root into the cortex (outside) and stele (inside), which contains the phloem and xylem tissues (Fig. 5.2.9). The endodermal cell walls contain a thin suberin layer called a Casparian strip, which acts as a hydrophobic barrier preventing the uncontrolled apoplastic movement of water and solutes across the root. Suberization of entire cortical cells in endodermis or near the rhizoderm (exodermis) may act in a like manner. Nevertheless, the endodermis and external tissues of aging roots are generally sloughed off. At this stage, the cork layer becomes the outermost layer. It is produced inside the endodermis by a cell layer called pericycle, which has the capacity to produce lateral roots and cork. Newly formed cork layer is completely impermeable against water and nutrients; cracking, dead lateral roots, and lenticells may however allow some movement of water and solutes through older cork zones (Peterson et al., 1999).

Most of the boreal woody species, including Scots pine and Norway spruce, are associated with an ectomycorrhizal symbiont, which greatly increases the effective absorbing surface area of roots and improves water and nutrient (particularly phosphorus, zinc and copper) availability for the plant, especially in non-homogeneous environments such as forests. It is commonly accepted that there is mutual benefit for the mycorrhizal partners, due to the exchange of plant-derived soluble carbohydrates for amino acids and nutrients supplied by the fungal associate (review Nehls et al., 2001).

The ectotrophic mycorrhizal hyphae form a thick sheath around the root tips and partially penetrate between the root live cortical cells, forming a net-like structure called a Hartig net. Outside of the roots, the mycorrhizal hyphae (diameter few micrometers) extend to the surrounding soil layers and form large structures inside

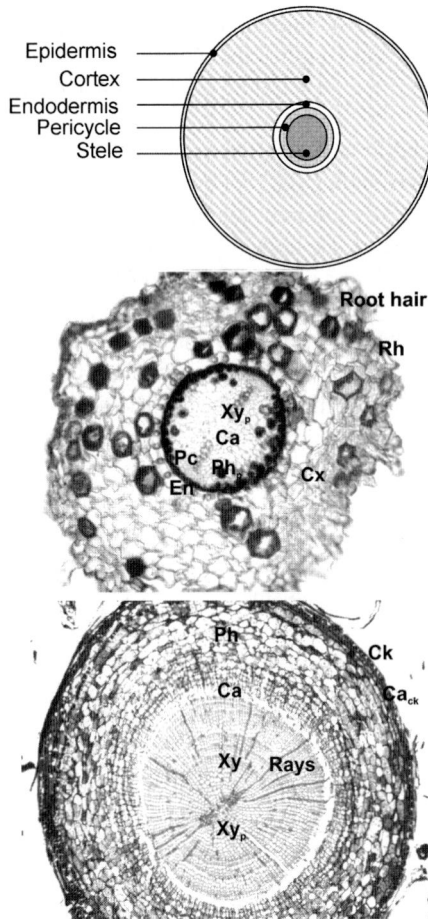

Fig. 5.2.9 Radial components of a conifer root (common yew Taxus baccata). Two upper pictures: A cross-section of a young root. Primary xylem (Xy_p), vascular cambium (Ca), primary phloem (Ph_p), pericycle (Pc), endodermis and a phi cell layer (En), primary cortex (Cx), rhizodermis (= root epidermis) with root hairs (Rh). Root diameter ca. 0.75 mm. Lower picture: A cross-section of a mature root. Secondary xylem (Xy), secondary phloem (Ph). Endodermis and external tissues have been sloughed off. Cork cambium (Cack) has arisen in a pericycle and produces a protective layer of cork (Ck). Rays extend radially across secondary vascular tissues. Root diameter ca. 2.0 mm (Courtesy of Arja Santanen)

the soil and fruiting bodies above the soil (Fig. 5.2.10). Woody plants are probably largely dependent on the symbiotic fungi to supply mineral nutrients, as there is little root surface area capable of nutrient absorption outside the fungal mantle (Taylor and Peterson, 2005).

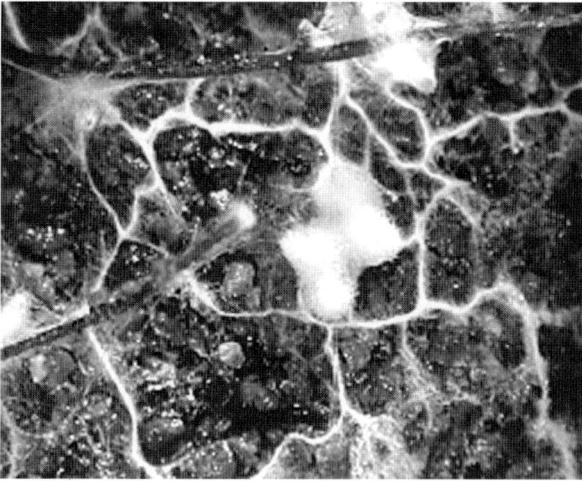

Fig. 5.2.10 A photo of roots and mycorrhiza (Courtesy of Jussi Heinonsalo)

5.2.5 *Individual*

Eero Nikinmaa and Jaana Bäck

University of Helsinki, Department of Forest Ecology, Finland

The individual organism is the central level of the hierarchy of biological organisms, since the evolutionary selection mechanisms act upon it. The success of the individual determines its capacity for producing off-spring. A critical component of tree success is its structure. Structure determines how well trees are able to reach favorable positions in the vegetation to capture the resources it needs, how well they are able to withstand high mechanical stresses that wind or heavy snow loads can cause, and how well they balance their allocation of resources between growth and reproduction over their life span.

The central driving force for the structural development of plants is photosynthetic production. Photosynthesis takes place predominantly by leaves using the solar radiation that they intercept. However, photosynthesis also requires a continuous supply of water from other parts of the tree to the leaves, as carbon dioxide intake is associated with loss of water vapour. The detailed biological machinery responsible for photosynthesis in leaves is built up of resources that originate from soil. Thus the vascular system that supplies the crown with water and mineral nutrients, and that supplying the root system with energy and structural carbohydrates have to be developed in concert.

The particular feature of trees that distinguishes them from other plants is their large crowns and stems. An important feature of tree structure is the above and below ground architecture as it determines the extent over which trees extract resources. Height and size are critical structural features as they determine the proportion that the individual captures from the total flux reaching the site. The architecture of the crown and root system determines the spatial distribution of the exchange sites. The closer the sites are the more they interact with each other. From the whole plant point of view this is unimportant since the resources captured are used by the same organism. However, different spatial arrangements of the exchange sites can affect the amount of resources used to bring about the uptake (e.g., Givnish, 1985; Küppers, 1989). Closely clumped structures require lower woody growth in branches and stem and even if they are not able to intercept as much light energy the saving in structural investments may make such architectural design more efficient than a more widely spread crown.

At crown level, the structure has a central role in determining the intercepted light at a certain point (e.g., Kellomäki and Oker-Blom, 1983; Oker-Blom and Smolander, 1988; Kuuluvainen and Pukkala, 1987, 1989). Closely clumped foliage will have strong internal shading while widely spread foliage does not interfere as much.

To characterize the relationship between the leaves and e.g. precipitation or light interception, a Leaf Area Index (LAI) has been widely used. LAI is generally defined as the ratio of total upper leaf surface of vegetation divided by the surface area of the ground. Usually LAI means one-sided leaf area but the meaning is not unequivocal with conifers: definitions of total, half or projected needle surface area are used as well. With needle leaves, the specific leaf area (SLA) is commonly used too. SLA is the ratio of projected leaf area to leaf dry mass.

Distribution of LAI can be derived from the height distribution of leaf mass and the leaf geometry, which usually vary in different parts of the crown: e.g. big needles have less surface area in relation to mass than small needles. LAI can be determined also indirectly by measuring the light extinction in the canopy (Chen et al., 1997).

LAI changes over the growing season as new leaves and needles emerge, and old ones are senescing and dropped down. In coniferous forests the seasonal changes in LAI are far less than in deciduous forests, due to the long-living, evergreen foliage of most conifers. Figure 5.2.11 demonstrates a needle mass distribution within the canopy in a 40-years-old Scots pine stand in Southern Finland. The average LAI in boreal needle-leaved forests is 3.5 ($\pm$3.3) $m^2\,m^{-2}$ (Scurlock et al., 2001; Global leaf area index[1]). Crown architecture is a major determinant for the radiation intensity distribution within the crown, and therefore LAI is a very commonly used characteristic to describe the light environment inside and below canopy. As the light response of photosynthesis is saturating, different shading patterns within the crown may bring about different productivity per leaf area. Narrow leaves far apart will have large penumbra area, i.e., shading only part of the sun disk while wide leaves close to each other will have very patchy light environment with bright and deeply shaded areas (Horn, 1971). In the former case the light is more evenly distributed within the crown, enhancing the production efficiency in comparison to the latter case.

---

[1] Global Leaf Area Index Data from Field Measurements, 1932–2000
http://daac.ornl.gov/VEGETATION/lai_des.html

Fig. 5.2.11 Average needle
mass distribution at SMEAR
II measuring station in a 40-
years-old Scots pine stand
(Redrawn from Ilvesniemi
and Liu, 2001)

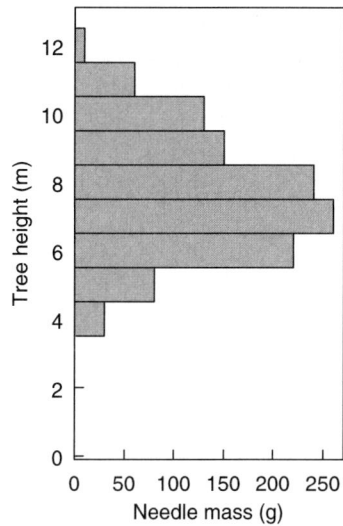

Tree and canopy structure can also be viewed from the point of view of the developmental processes that brings it about or from the functional significance that the structure has. From the developmental point of view a tree is simply a collection of elementary units that form larger structures (branches etc.) over passage of time (Prusinkiewitcz and Lindermaier, 1990). Analyses of plant architecture have suggested a number of possible elementary units, such as the metamer and growth unit (Caraglio and Barthélémy, 1997; Room et al., 1994). A metamer is defined as an internode with axillary bud(s) and leaf (leaves) in its upper end, but without any shoots resulting from growth of the axillary buds (Caraglio and Barthélémy, 1997). A growth unit, initially called a unit of extension by Hallé et al. (1978), is the part of the shoot resulting from uninterrupted extension growth. Room et al. (1994) describe it as "extension of the contents of a previously dormant apical bud followed by growth of neoformed leaves (if any) and formation of a new, dormant, apical bud".

The crown structure has a dual role in a tree's life. On the one hand it influences the energy capture for photosynthetic production and on the other hand it influences the way in which a tree can exploit space in the future. Tree crown is formed as a result of reiteration of bud formation, their activation and flushing, axis growth and new bud formation. Successful tree development seems to presuppose that only a few specific buds grow and the rest either become dormant or abort (Stafsform, 1995). Depending on the competitive environment, different height growth strategies and thus also bud formation and activation, are favoured (Givnish, 1995). Shoots grown in the upper and lower parts of the tree crown play different roles in crown dynamics. Because young, first-order shoots in the upper crown are unlikely to be immediately shaded, they play an important role in supporting tree structure for many years. They will be incorporated in the structures that will provide material for the active crown above for many years. On the other hand, shoots in the lower crown, even if at first in a strong light environment, become shaded and are less able to support other shoots and are more likely to be structures that a tree can shed.

Tree stems and crowns reflect their past developmental history, in that sense they are unique. However, a closer look at the stems reveals repeating patterns in different species. This is quite natural as tree stems have certain functional requirements that they need to fulfill. Already in his notes, Leonardo da Vinci (MacCurdy, 2002) put forward the concept that was later called the pipe model theory (Shinozaki, 1964a, b). This observation has shown that there is a constant relationship between the sapwood area and foliage area above in many species. Therefore, stem thickness increases from the tip of the tree crown downwards, in a saturating manner, as the foliage area above accumulates. Below the living crown, the sapwood area should not increase anymore according to the pipe model principle, but the whole stem thickness should increase. This is due to accumulation of disused pipes into the stem, in conifers very often into distinctly distinguishable heartwood.

This structural linking between the wood and foliage produces, at least qualitatively, the well known tapering of tree stems. However, quantitatively the assumptions of pipe model do not exactly hold below the crown or even within the crown. Below the living crown it has often been observed that the sapwood thickness, rather than area, seems to remain constant (Longeutaud et al., 2006) or even within the crown there seems to be variation in the relationship (Berninger and Nikinmaa, 1994). However, in many coniferous species, such as Scots pine, the relationship seems to be fairly constant when measured at the same reference point (e.g., below crown) in trees grown in very different growing conditions and positions within the same climatic zone.

A number of structural changes occur in the longitudinal direction in conifer stems and also from the core of the stem outwards. In the upper part of the stem the annual tree rings are wider, with considerably thinner sections of latewood than at the base of the stem. In wood technology, the wood within the crowns of fairly young trees is commonly referred as the juvenile wood. It has also been observed that the average size (length and thickness) of tracheids in the conifers become larger as the cambial age, i.e., the number of years when the cambium was initiated, increases. This observation is in agreement on theoretical derivations of structural scaling of all living creatures where the transport of substances equal to different tissues is fundamental to maintain metabolic activities (West et al., 1997).

Also the mechanical properties of xylem tissue change as a function of cambial age. The elasticity of wood rapidly decreases as the cambium ages until a rather constant value is reached (Mencuccini et al., 1997). The wood dimensions seem to follow also the requirement to withstand mechanical loads as they follow the pipe model principle. Critical factors are the diameter reference point and its distance from the centre of the loading point. A common feature in many species is the formation of so-called reaction wood at places which are under strong mechanical stress. In reaction wood the cell wall thickness normally increases vs. the lumen size in comparison to normal wood.

5.2.6 Stand

Eero Nikinmaa, Pertti Hari, and Liisa Kulmala

University of Helsinki, Department of Forest Ecology, Finland

Groups of rather even-aged trees growing on similar soil are known as stands, and they can be easily recognized in boreal forests. A stand is usually dominated by one species, although often other species are present in low numbers. During the stand development, the trees grow in height and diameter, but simultaneously the difference between big trees and small trees becomes larger due to the competition. Eventually the smallest trees start to die because of insufficient light for photosynthesis and lack of nutrients or water. This kind of orderly changes in the composition or structure of a stand is called succession.

Tree stands are structurally clearly distinguishable units in the landscape. The structural features that set the stands apart are the species composition and the tree size distribution. The reasons for the spatial differences of these features lie in the edaphic growing conditions of trees and the disturbance history of the region. A common feature of the boreal landscape is the alternation of lakes, wetlands and uplands that form a mosaic of site growing conditions. Superimposed on these site dependent differences is the impact of the disturbance history. Forest fires of different intensity, storms and large scale insect outbreaks create areas with different number and size of remaining trees of different species. These sites are gradually further stocked with newly-born and often fast-growing trees. More protected areas remain uninfluenced by the large scale disturbances and there only internal dynamics of gap establishment, growth, site dominance and tree mortality create vertically very heterogeneous stand structures with a small number of very large trees and increasing number of smaller trees. Due to the variation of factors causing the stand properties to vary, also their size variation in the landscape in the natural forests is large.

Most well-stocked stands, in terms of tree stem biomass, are the result of strong and large scale disturbance that has created fully-stocked rather even-aged stands where all the trees have established approximately at the same time. The proportion of these types of stands have increased due to silvicultural practices in the boreal region that try to mimic this type of stand development dynamics so that they are by far the most common stand structure presently, particularly in Fennoscandia (Kuuluvainen, 2002).

Within a tree stand, trees interact with each other via shading or via competition for nutrients and water in the soil. Competition for limited resources starts when the domains of resource uptake overlap (Walker et al., 1989). In a structurally heterogeneous stands there are clear differences between the trees both in terms of their height and biomass which quite clearly determines their competitive status. The development of a small understorey sapling into a canopy tree require several incidents of gap opening (Canham, 1985). In the case of even aged stand development, the trees in the stand develop together and the differentiation into size classes results from the competition. Simultaneously with the size increase the number of trees decrease due to self thinning that can be described with a power relationship between the tree number and their average size (Zeide, 1987).

The competition for light drives the differentiation between trees in closed canopies so that small differences in height are quickly reflected on the captured energy and growth that further increase the size differences leading into vicious circle and eventual death of the most suppressed individuals. Particularly light demanding tree species allocate more of their growth into height extension than canopy width, foliage biomass and stem thickness when shaded (Vanninen and Mäkelä, 2005; Ilomäki et al., 2003) that maintains the differences between trees in the stand always more prominent in leaf area, total biomass and stem thickness than in height. This competition is a continuous process during the first 100 years of stand development in boreal forests as the individual size of trees increase, however slowing down as the tree size increase. Because of the continued change, there is always a number of different sized trees present in the canopy layer that are the same age despite their size differences. These size differences have been used in silviculture to divide the canopy into dominance classes and as the basis for different silvicultural treatments.

As a typical example, an even-aged pine stand has grown around the SMEAR II measuring station (Section 2.5.1). Its age was estimated to be 40 years when it was measured in autumn of 2002. Within a circle of radius 100 m, five size classes of trees were determined according to their diameters, and five trees were chosen as samples from each of the five classes. The annual height and radial increments were measured back to the age of three years in each size class.

The height developments in each size class were very similar and the biggest trees at 40 years were also the biggest at the age of three years and retained their ratios respectively to other size classes (Fig. 5.2.12A). The diameter development was similar to that of height development. However, the differences between size classes were larger (Fig. 5.2.12B).

Ground vegetation typically covers the forest floor level. The species composition and density usually changes during the growth and environmental effects of surrounding trees. In the early phases of succession, the radiation environment is unshaded and fast-growing and opportunist species with rapidly reproducing new tissues are dominating. Such opportunist species are e.g. *Deschampsia flexuosa, Calamagrostis* spp., *Epilobium angustifolium* and *Rubus* spp. After 10–15 years, seedlings start to shade the ground level. Then evergreen and slowly growing species are more competitive than the earlier pioneer species. Such evergreen species are e.g. *Vaccinium* ssp. (Fig. 5.2.13), *Calluna vulgaris, Rhododendron tomentosum* and mosses (e.g. *Pleurozium schreberi, Polytrichum* spp., *Dicranum* ssp.).

The species composition and density of ground vegetation are spatially very heterogeneous in all phases of succession. Small disturbances like thinning, windfall or insect damage generate living space and locally improve living condition in the environment at ground level after the death of even a single tree. Stones and boulders are typical for areas which underwent glaciation. Mosses and lichens grow on stones,

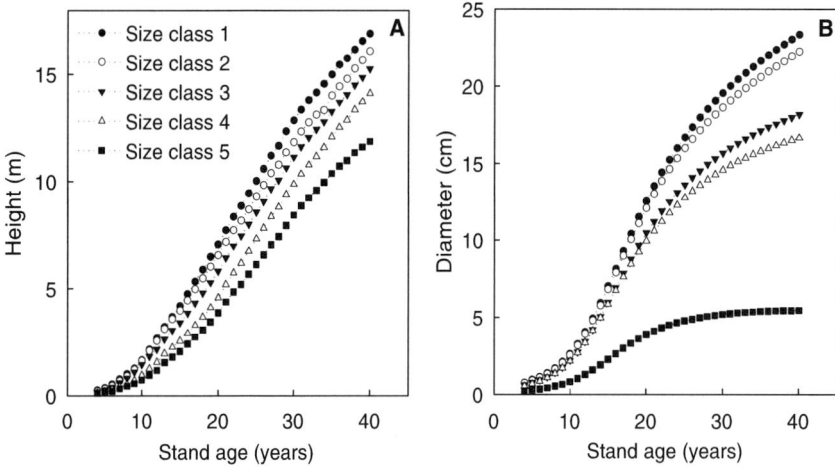

Fig. 5.2.12 The height (A) and diameter (B) development in five size classes in the stand around SMEAR II

Fig. 5.2.13 Ground vegetation at SMEAR II dominated by *Vaccinium vitis-idaea*

unlike vascular plants they do not require soil as a substrate. Small differences also in soil characters affect ground vegetation more notably than trees which are usually dominated by only one tree species and therefore have no interspecific competition. Under a homogenous tree canopy, ground vegetation type can vary and consist of small areas of diverse types of vegetation depending on small-scaled differences in water holding capacity and radiation environment.

5.2.7 Boreal Zone

Kalev Jögiste[1], Liisa Kulmala[2], Jaana Bäck[2], and Pertti Hari[2]

[1]Estonian University of Life Sciences, Institute of Forestry and Rural Engineering, Estonia
[2]University of Helsinki, Department of Forest Ecology, Finland

Boreal Forest Zone

By the boreal forest, we mean the set of coniferous forest ecosystems that can survive in northern, specifically subarctic, regions. Covering most of inland Alaska, Canada, Sweden, Finland, Norway and Russia, parts of the northern continental United States, northern Kazakhstan and Japan, the boreal forest is the world's largest terrestrial biome (Fig. 5.2.14). The boreal forest zone covers circa 15% of the Earth's land area and the zones of the boreal forest are the largest biome in the world.

The temperature regime in the boreal forest is variable across the latitudinal gradient. The mean annual temperature is low, and also the amount of precipitation and the climate is mostly continental. Winter temperatures fall to between $-50°C$ and $0°C$. The summer in Siberia is short, rather warm and sunny. The Atlantic cyclones cause an unstable climate with more rain in European boreal forest zones than in others (Hytteborn et al., 2005). The mean temperature in July varies between $10°C$ and $22°C$ in Siberian boreal forest zones. The annual precipitation amounts in general between 300 and 600 mm.

The boreal forest belt in Eurasia is over 1,000 km wide; in some areas reaching 3,000 km. Different classifications of zones try to describe regional variations of climate variables. The main principle is to divide boreal zones latitudinally into subzones and longitudinally into sections (Ahti et al., 1968). High topographic vari-

Fig. 5.2.14 The boreal forest zone
(Source Wikipedia: http://en.wikipedia.org/wiki/Image:Taiga.png)

ation creates problems with the zone classification approach (e.g., Norway is the area with the most complex topography and consequently is difficult to classify by zones).

In the southern parts of boreal forest we see transition between temperate and boreal forest called hemiboreal. Those forests are dominated by coniferous species, but the presence of hardwoods (angiosperms) changes the main characteristics of ecosystems. These temperate species diversify tree stand composition. This zone is called boreo-nemoral or hemiboreal (Ahti et al., 1968). In the north the boreal forest is smoothly changing into the tundra vegetation type, and the transition zone between boreal forest and treeless tundra is called forest-tundra.

Much of the boreal forest was glaciated over 10,000 years ago. Glaciers left depressions in the topography that have since filled with water, creating lakes and bogs, found throughout the boreal forest. Most of the Eurasian boreal forests are situated far from the sea and form a fine pattern of lowland forests and wetlands (Hytteborn et al., 2005). The positive precipitation balance and pure runoff from the wetlands results in humus accumulation and formation of raw humus and peat are characteristic of boreal forest zone wetlands.

Boreal forests are characterized by a low below-ground temperature. The permafrost forms as a maximum effect of a low mean annual temperature and very low winter temperatures. The permafrost stage coincides with the inception of palsas and peat plateaus (Bhiry et al., 2007). Low soil temperatures hinder the development of soil, as well as the ease with which plants can use its nutrients. Soils of boreal forest tend to be young and nutrient-poor; they lack the deep, organically-enriched profile present in temperate deciduous forests. Soil formation can be characterized by the podzolization process: the acidic reaction of decomposed litter causes leaching and moving of minerals at the top of the soils, resulting in a light horizon just below the litter (Kaurichev et al., 1989; Section 7.4).

Successional trends in boreal forest zones depend on disturbance regimes. Forest fires, windfalls and impact of humans and herbivores are main factors causing drastic changes in the forest community. Changing climate will affect the fire danger levels in boreal forests: more fire activity can be predicted (Stocks et al., 1998). Fire frequency has been determined as the main successional factor in boreal forest zones of North-America (Brassard and Chen, 2006).

Flora

The number of tree species forming the forest is rather low. Dominating species belong to the genera *Abies, Larix, Picea* and *Pinus*. Also deciduous species (*Alnus, Betula, Populus*) are represented in the composition of stands, particularly after disturbances like fires, windstorms or clear cuts where most if not all trees are removed from an area of forest. The volume of tree trunks can be as high as $400\,\mathrm{m}^3\,\mathrm{ha}^{-1}$ and needle mass can be $10,000\,\mathrm{kg}\,\mathrm{ha}^{-1}$.

The conical or spire-shaped dark coloured needle-leaf trees are adapted to the cold and the physiological drought of winter (when soil is frozen) and to the short growing season. The conical shape of trees help them shed snow and prevent loss of branches. Needles usually also have a thick waxy cuticle. Leaves may survive for several years and the perennial foliage allows plants to photosynthesize as soon as temperatures permit in spring.

Ground vegetation in boreal forests is usually species-poor. Most common species are evergreen: low shrubs, mosses and lichens. In nutrient poor sites, dwarf-shrubs dominate the field layer. The ground layer is often conspicuous and dominated by mosses and/or lichens and covers the ground completely as a dense carpet (Hytteborn et al., 2005). However, the ground vegetation can provide much more information and more effectively express biotope conditions, therefore it has been used in site type classifications (Frey, 1978). The life history characteristics determine the sequence of species in a particular site. In comparison with other biomes the boreal forest has a low biological diversity.

Fauna

The boreal forest is the home environment of large herbivorous mammals and smaller rodents which also have adapted to the harsh climate. Moose, caribou, hare, lemmings, and voles are common herbivores in the boreal forest. Fur-bearing predators like the lynx, bear, wolves and members of the weasel family are characteristic of the boreal forest. Insect-eaters birds are migratory and leave after the breeding season. Seed-eaters and omnivores tend to be year-round residents.

> The boreal forest zone in Eurasia stretches from Siberia to European North. Also the term "northern coniferous forest" refers to forest ecosystems in the north of Eurasia (Hytteborn et al., 2005). The Russian word "taiga" was first associated with coniferous forest by Siberians. Today the word has obtained a more general meaning, referring to different forests of the Northern Hemisphere.

5.2.8 Quantitative Description of Vegetation Structure

Pertti Hari and Annikki Mäkelä

University of Helsinki, Department of Forest Ecology, Finland

The basic quantitative characterisation of plant structures is mass, i.e., dry weight since water content is very variable and fresh weight does not properly describe the stable structure of cells, tissues and individuals. Dry weight can be a useful measurement for tissue, individual and stand level. Sometimes a chemical characterisation using proteins, cellulose, lipids, or lignin may provide additional information. Especially the proteins, which are functional substances, are useful to treat separately.

Quantitative descriptions at tissue element level require definition of a new concept, mass density which is denoted with ρ as in physics. Consider a small space element around a point x, denote its volume with ΔV and the mass of vegetation in the space element with $m(x, \Delta V)$. The mass density, ρ, is defined as the limit of the ratio $m(\Delta V)$ and ΔV, when the volume ΔV approaches zero, in mathematical terms

$$\rho = \lim_{\Delta V \to 0} \frac{m(x, \Delta V)}{\Delta V} \tag{5.2.1}$$

When the mass density is determined at points in the canopy, we get density distribution, i.e., the density as a function of space. This enables a very detailed description of the structure in the canopy.

It is often reasonable to assume that the canopy is horizontally homogenous, i.e., objects at the same height in the canopy are so similar with each other that we can pool them together. Then the mass height distribution, ρ_h, is introduced to describe the canopy structure. Let $m_h(h)$ denote the mass below the height h. The mass height distribution is defined as the height derivative of the mass below a certain height, with mathematical symbols

$$\rho_h(h) = \frac{dm_h(h)}{dh} = \lim_{\Delta h \to 0} \frac{m_h(h + \Delta h) - m_h(h)}{\Delta h} \tag{5.2.2}$$

When the mass height distribution is determined in practice, then the canopy is divided into thin layers having the thickness Δh and mass is measured in each layer. The mass height distribution is obtained with the right-hand side of the above equation.

In the rest of this book, the structural features are expressed using the notation introduced in this chapter. When the spatial distribution of the mass density of the vegetation elements are linked to the metabolic functions of these elements and transport between them the dynamic description of structural development can be obtained. Of course the metabolic activity of each vegetation element depends on their structure at different level of hierarchy. The challenge for the dynamic description of structural development is to select an appropriate detail of description that facilitates the consideration of most significant features.

5.3 Structure of the Atmosphere

Üllar Rannik[1], Douglas Nilsson[2], Pertti Hari[3], and Timo Vesala[1]

[1] University of Helsinki, Department of Physics, Finland
[2] Stockholm University, Department of Applied Environmental Science, Sweden
[3] University of Helsinki, Department of Forest Ecology, Finland

Earth's atmosphere contains roughly 78% nitrogen, 21% oxygen, 0.93% argon, 0.04% carbon dioxide, and trace amounts of other gases, in addition to about 3% water vapor. The mean molar mass of air is $28.97 \, g \, mol^{-1}$. The average atmospheric pressure at sea level, is about 101.3 kPa. Atmospheric pressure decreases with height, dropping by 50% at an altitude of about 5.6 km and 90% of the atmosphere by mass is below an altitude of 16 km. The pressure drop is approximately exponential, so that each doubling in altitude results in an approximate decrease in pressure by half.

The molecules/atoms that constitute the bulk of the atmosphere do not interact with infrared radiation significantly. The gases that effectively absorb the long-wave radiation emitted by the Earth's surface and emit themselves the long-wave radiation both upward and downward, contribute to greenhouse effect by the downward part of the emitted radiation, thus warming the Earth's surface. Without greenhouse gases the average temperature of the surface (15°C) would be 33°C lower. In the Earth's atmosphere, the dominant greenhouse gases are water vapor, carbon dioxide, and ozone (O_3). Water has multiple effects on infrared radiation, through its vapor phase and through its condensed phases in clouds. Other important greenhouse gases include methane, nitrous oxide and the chlorofluorocarbons.

Troposphere

The troposphere is the lowest layer of the atmosphere starting at the surface and going up to about 10 km at the poles and 15 km at the equator. The troposphere is vertically well mixed due to solar heating at the surface. It contains approximately 75% of the mass of the atmosphere and almost all the water vapor and aerosol.

In the troposphere, temperature decreases at an average rate of 6.5°C for every 1 km increase in height. This decrease in temperature is caused mainly by adiabatic cooling of air when an air parcel moves upward and pressure decreases. Dry adiabatic lapse of temperature occurs under unsaturated conditions at the rate of 9.78°C km^{-1} (Section 3.1). The average lapse rate is lower because of inversion layers occurring in the atmosphere and release of latent heat in saturated air conditions. Temperature decreases at middle latitudes from about +17°C at sea level to approximately −52°C at the beginning of the tropopause, separating the troposphere and stratosphere. At the poles, the troposphere is thinner and the temperature only decreases to −45°C, while at the equator the temperature at the top of the troposphere can reach −75°C.

The lower part of the troposphere, where friction on the Earth's surface influences the air flow, is the planetary boundary layer, also known as the atmospheric boundary layer (ABL). The behavior of ABL is directly influenced by its contact with the surface, and it responds to surface forcings such as surface heating or cooling in a timescale of an hour or less. In this layer physical quantities such as flow velocity, temperature, moisture etc., display rapid fluctuations (turbulence) and vertical mixing is strong. Above the ABL is the free atmosphere where the influence of surface on wind direction is negligible. The free atmosphere is usually non-turbulent, or only intermittently turbulent.

There are two principal types of atmospheric boundary layers: the convective planetary boundary layer (CBL) is the ABL where heating and evapotranspiration at the surface creates thermal instability and thus generates turbulence. The CBL is typical in summer hemisphere during daytime, when turbulence tends to mix air properties and quantities such as conservative concentrations and potential temperature are close to constant. An atmospheric layer with uniform air properties is called the mixed layer (ML). ML is frequently used as a synonym for CBL. The ABL is separated from the rest of the troposphere by the capping inversion layer – the stably stratified layer that limits exchange between the ABL and free atmosphere above. The CBL height varies from a few hundred meters during the early morning development stage up to 2–3 km in midafternoon. Occasionally solar heating assisted by the heat released from the water vapor condensation could create such strong convective turbulence that the CBL penetrates deep into the troposphere.

The stably stratified atmospheric boundary layer (SBL) is the ABL where negative buoyancy flux at the surface damps the turbulence. The height of the SBL is not well defined as the turbulence levels decrease gradually with height and the state of the SBL depends very much on the history of the ABL evolution. The SBL is solely driven by the wind shear turbulence and hence the SBL cannot exist without free atmospheric wind. The SBL is typical in nighttime at all locations and even in daytime in places where the earth's surface is colder than the air above. The SBL plays a particularly important role in high-latitudes where the long-lived (days to months) SBLs result in very cold air temperatures during winter.

In high-pressure regions, typical ABL diurnal development occurs according to Fig. 5.3.1. During the night, a stable stratification is formed due to radiative cooling. With sunrise, surface heating begins and a new shallow ML is formed close to the surface. The ML continues to grow via entrainment of air from above the ML. After the ML has eaten up the stable layer, fast growth of the ML through the residual layer occurs (which has near-neutral stratification and therefore little energy input is needed to mix the layer).

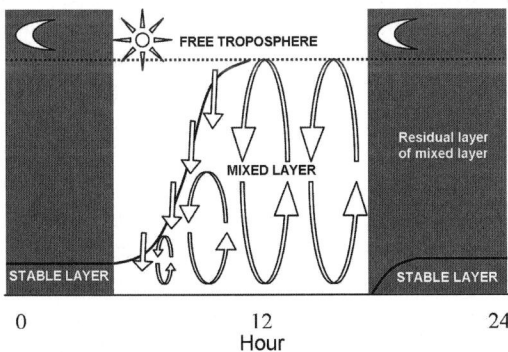

Fig. 5.3.1 Schematic picture of atmospheric boundary layer diurnal development during high-pressure conditions. The daytime convective boundary layer is separated from the free atmosphere by capping inversion layer

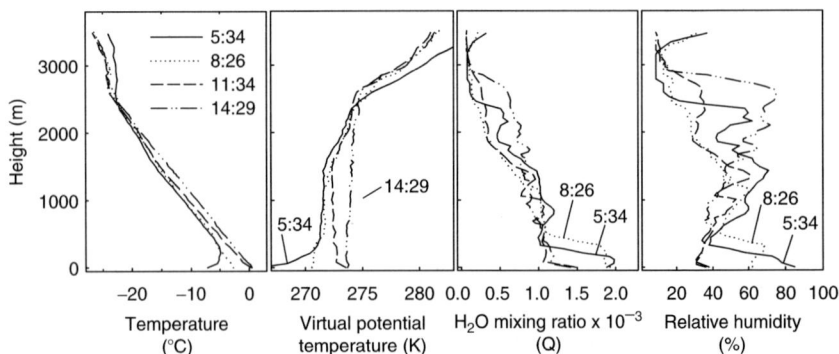

Fig. 5.3.2 The vertical profiles of temperature, virtual potential temperature, water vapor mixing ratio and relative humidity by height from four radiosondes launched at SMEAR II in Hyytiälä, Southern Finland, on April 4, 1999

With sunset, surface cooling by thermal radiation initiates formation of SBL close to the surface. Above the SBL, the residual layer of the previous day's ML persists.

Diurnal variation of the ABL height and mixing influence strongly the humidity and concentrations close to the surface. During the day ABL is deep and very well mixed, thus surface emissions are diluted and concentrations are low. At night, in turn, mixing is weak and high concentration levels occur for compounds with emission at the surface.

Figure 5.3.2 shows the vertical profiles of temperature and water concentration and development of these profiles during ABL growth periods. Except close to the surface in early morning, where the lower layer is very strongly stratified (the SBL), temperature decreases with height. Temperature decrease is due to pressure and corresponds to dry adiabatic cooling of air.

The virtual potential temperature, which takes into account the pressure reduction with height as well as the influence of air water content on the density of air, is a direct measure of stability of atmospheric layers. Around mid-day, the virtual potential temperature indicates mixing of BL up to about 1,500 m, which is the height of the ML. Also the water vapour is well mixed in this layer. Solar heating together with boundary layer development and stratification effects causes strong diurnal variation of temperature and concentrations in ABL, especially close to the surface.

At the top of the ABL, a capping inversion layer exists. In virtual potential temperature profiles, Fig. 5.3.2, the stable stratified layer above a ML rises together with ML growth until the ML depth is about 2,500 m in the afternoon. By that time the ML is covered by a strong inversion layer above.

The layer at the surface proximity where turbulence is produced by velocity gradient is conventionally called a surface layer – the lower part of the ABL. The surface layer constitutes about 10% of the total ABL depth. In the surface layer the wind profile is approximately logarithmic. However, stability affects strongly

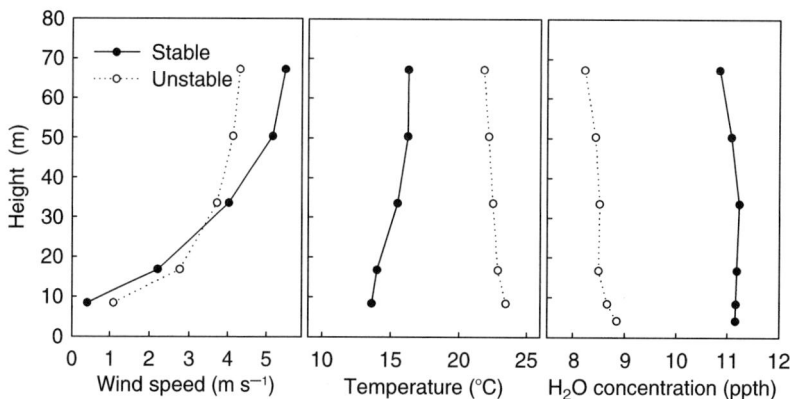

Fig. 5.3.3 Vertical profiles of average wind speed, temperature and water vapor concentration at day-time unstable (July 16, 2002, 14:30–15:00) and night-time stable (July 16, 2002, 02:00–02:30) conditions inside and above forest. Forest height is approximately 14 m

the wind profile (Fig. 5.3.3). In addition, in the vicinity of rough surfaces such as forests a roughness sublayer exists where wind gradient is reduced in comparison to the form of the profile above. The roughness sublayer thickness is between one to two characteristic heights of the rough surface, in case of forests this is the forest height.

During daytime typically unstable conditions prevail which promote turbulence and mixing becomes very efficient. Therefore turbulence tends efficiently to 'smooth' profiles, and gradients of concentrations (and wind speed) are smaller. At night the situation for the temperature is the opposite; radiative cooling at the surface drives the heat flux downward from the atmosphere above and negative temperature gradients occur.

Evapotranspiration of water occurs throughout the canopy during daytime, resulting in higher air water content inside canopy. At night the water transpiration is limited because of stomatal closure, also evaporation from surfaces is typically small, being limited by available energy. Therefore the concentration of water vapor in air is relatively constant with height at night (Fig. 5.3.3).

Clouds

Most clouds are located in the troposphere. Only a few clouds can be found above the troposphere; these include noctilucent and polar stratospheric clouds, which occur in the mesosphere and stratosphere respectively.

Clouds are divided into three general categories according to their outward characteristics: cirrus (denoting "fiber" or "hair"), cumulus (convective clouds) and stratus (layered) clouds. Three major types occur in combination with nimbus denoting the rainy combination of the cloud type. Cloud types are divided into four more groups that distinguish the cloud's altitude by the cloud base height: High, middle-level, low and in addition vertical clouds, which can have strong up-currents, rise far above their bases and form at many heights.

High clouds, denoted by cirrus, generally form above 5 km, in the cold region of the troposphere. However, in Polar regions, they may form as low as 3 km. At this altitude, water almost always freezes so clouds are composed of ice crystals. The clouds are often transparent. Cirrus clouds do not usually bring precipitation.

Middle-level clouds develop between 2 and 5 km and are denoted by the prefix alto-. They are made of water droplets and are frequently supercooled. The middle-level clouds can bring precipitation in the form of rain or snow; nimbostratus clouds tend to bring constant precipitation and low visibility.

Low clouds are found up to 2 km and include the stratus clouds (dense and grey). From low level clouds stratocumulus clouds often follow a cold front and can produce rain or drizzle. Stratus clouds are layer-like clouds associated with wide-spread precipitation or ocean air, and often produce drizzle. Cumulus clouds, the fair weather clouds, can also develop into more storm-condition clouds (cumulonimbus, for example), and produce showers, typically in the afternoon.

Clouds affect significantly the short-wave as well as long-wave radiation field in the atmosphere and at the surface. Clouds increase the albedo of the Earth-atmosphere system by scattering the solar radiation and thus have global average short-wave cooling effect of about 50 W m^{-2}. By reducing outgoing long-wave radiation by 30 W m^{-2}, clouds contribute to the greenhouse effect by about 20 W m^{-2} (Wielicki et al., 1995).

5.4 Soil

Asko Simojoki[1], Heli Garcia[1], Mari Pihlatie[2], Jukka Pumpanen[3], Jukka Kurola[4], Mirja Salkinoja-Salonen[1], and Pertti Hari[3]

[1] University of Helsinki, Department of Applied Chemistry and Microbiology, Finland
[2] University of Helsinki, Department of Physics, Finland
[3] University of Helsinki, Department of Forest Ecology, Finland
[4] University of Helsinki, Department of Biological and Environmental Sciences, Finland

5.4.1 Mineral Soil

Soil Textural Classes

Soil denotes the loose upper layer of the Earth's crust covering the solid bedrock. It is a heterogeneous mixture of mineral and organic matter. In the boreal zone, the bedrock is composed mainly of magmatic and metamorphic rocks, such as granite and gneiss. Soils are classified according to particle size into textural classes. Coarser mineral soil particles include silt and sand fractions that are 20–200 µm and 200–2,000 μm in diameter, respectively. Particles larger than 2,000 μm in diameter are classified as gravel or stones. Coarse particles are composed mainly of such primary silicates as quartz, feldspars and mica, reflecting the mineralogy of parent rocks. The smallest soil particles with <2 μm in diameter are called the clay fraction. Clay particles are dominated by secondary minerals formed by modification of original rock minerals or re-crystallization of their decomposition products during soil formation. These include the secondary layer-silicates called clay minerals as well as hydrous oxides (hydroxyoxides) of Al, Fe and Mn. The oxides are partly poorly crystalline. In contrast to the more or less spherical shape of coarser soil particles, the clay particles are plate-like, reflecting the shape of layer-silicate minerals.

Properties

Soil chemical properties are largely dominated by the reactions occurring at the surfaces of small particles. This is due to the large specific surface area and polar nature of clay minerals and oxides in the smallest particle size fractions. Most clay minerals have a permanent negative electrical charge due to isomorphic substitution of cations within the crystals. These substitutions occur during the crystallization of minerals and remain unchanged thereafter. On the other hand, the oxides may be negative, neutral or positive depending on soil acidity that determines the protonation and deprotonation of their hydroxyl groups. Soils generally have a net negative charge and thus adsorb cations on their surfaces by electrostatic forces and retain anions on the hydroxyl groups of clay mineral edges and oxide surfaces by specific mechanisms. These processes are discussed in more detail later (Section 6.2.4).

Larger Mineral Soil Fragments

Boreal forest soils often contain considerable amounts of gravel and stones. However, these larger fractions are chemically inert and mainly occupy space from more fertile and porous soil composed of finer fractions. In practice, soil texture refers to the size fractions finer than 2 mm in diameter. Moreover, the conventional pretreatments for the determination of particle size distribution include the removal of organic matter and oxides that bind the silicate particles together.

5.4.2 Soil Organic Matter

Definition and Origin of Soil Organic Matter

Soil organic matter denotes all living and dead biologically derived organic material at various stages of decomposition in the soil or on the soil surface, excluding the

above-ground portion of living plants (Baldock and Nelson, 2000). In practice, the
concept has often a narrower scope referring to materials passing through a 2-mm
sieve (Nelson and Sommers, 1996) excluding undecayed plant and animal residues.

Origin and Molecular Structures of Soil Organic Matter Compounds

Soil organic matter includes all organic compounds entering the soil through the
root growth and the litterfall from above-ground plant parts as well as the products of
their microbial decomposition and re-synthesis. Microbial decomposition of organic
macromolecules produces a range of small molecules as well as the decay-resistant,
dark, amorphous organic matter called humus.

The chemical composition of soil organic matter is similar to that of litter unless
not changed by microbial decomposition. Biological tissues are composed of various organic macromolecules and a lesser amount of small molecules. The chemical composition of organic residues depends on the tissue (reviewed by Baldock
and Nelson, 2000; Derenne and Largeau, 2001; Blume et al., 2002; Kögel-Knabner,
2002). Within the plasma membrane, the cells contain phospholipids, carbohydrates,
lignin, nucleic acids and proteins, as well as such smaller molecules as sugars, amino
acids and dissolved inorganic ions. The cell walls of plants are composed mainly
of cellulose, hemicellulose and lignin. Cellulose is a carbohydrate polymer of 1–4
β-linked glucose units, whereas hemicellulose contains other sugars as well. In contrast, lignin is not a carbohydrate and has a complex structure composed of linked
phenyl propane units. The nitrogenous compounds constitute a small but important
group, as nitrogen is a structural component of enzymes and other proteins as well
as nucleic acids, ATP and such pigments as heme and chlorophyll.

The chemical composition of microbial cells resembles plant cells. However, the cell walls of fungi contain mainly chitin, a polymer of 1–4 β-linked
N-acetyl-glucoseamine units, and those of bacteria peptidoglucanes and/or
lipopolysaccharides. As a consequence, the composition of soil microbes is dominated by phospholipids and protein-like chemical species. The chemical composition and functional role of biological macromolecules in plant and microbial tissues
is discussed in more detail in the preceding chapter.

Humus is traditionally considered to be composed of large molecules with a complex structure not amenable for detailed structural characterization. The classical
fractionation of humic substances is based on their acidity-dependent solubility
characteristics. The humic substances that are extracted by an alkaline solution and
remain soluble after subsequent acidification are fulvic acids, whereas those in the
acid-precipitated fraction are humic acids. Still, large amounts of humic substances
called humins may remain unextracted by the alkali. Humins are mostly humic acids
that have very high molecular weight or are protected by mineral soil particles. The
classical fractionation of humus has some practical merits, as it separates the fulvic
acids that are soluble and potentially mobile in a broad pH range in soil, in contrast
to the humic acids that are immobile in acid conditions. Nevertheless, the molecular
structure of humus has long been and still is a topic of active research and discussion

(Schnitzer, 2000; Burdon, 2001; Piccolo, 2001; Sutton and Sposito, 2005). Current views question the macromolecular nature of humus. Instead, the humic substances are best described as collections of diverse, relatively low molecular mass components that form supramolecular associations stabilized by hydrophobic interactions and hydrogen bonds (Sutton and Sposito, 2005).

Humus contains numerous acidic groups that can dissociate in a broad pH range which gives rise to pH-dependent negative charge of soil particles. For this reason, it is the dominant soil constituent responsible for the cation exchange capacity in coarse-textured soils, such as those in boreal forest.

Analysis of Soil Organic Matter

The amount and composition of soil organic matter can be analysed by various methods. The crudest approaches involve the determination of organic matter based on the loss-on-ignition or the organic carbon content. However, the treatment of soil organic matter according to its chemical structure allows efficient use of quantitative methods, since all soil components can, at least in principle, be directly measured. Quantitative data on the different chemical groups of biological macromolecules can be obtained by combinations of selective extractions, such as the proximate methods used in the food chemistry, or more generally, by thermochemolytic degradation of macromolecules followed by compound-specific analysis of the products (see Kögel-Knabner, 2000). Non-destructive spectroscopic methods allow the characterization of various groups of atoms, but the quantification of various molecular groups is not yet reliable (Kögel-Knabner, 2000), although, e.g., Nelson and Baldock (2005) made an attempt to apply a solid-state ^{13}C NMR technique towards this aim.

Traditionally, different empirical wet chemical methods have been used for the analysis of litter and humus. Nonetheless, the strict separation of soil organic matter into litter and humus fractions is not feasible with conventional extractions (Beyer, 1996). For this reason, Beyer et al. (1996) recommended applying litter compound analysis to soil as well. Moreover, application of the same analytical procedures for both plant and soil samples provides methodological advantages.

For the above reasons, we planned and carried out a uniform analytical procedure for the determination of chemical composition of plant tissues, litter and soil organic matter by combining dry and wet chemical methods of food and soil chemistry. The procedure is outlined in Figs. 5.4.1 and 5.4.2 and Box A. The analyses take advantage of the fact that biological macromolecules are composed of specific monomers. This allows the estimation of main organic macromolecules, including (1) lipids, (2) sugars and starch, (3) hemicellulose + cellulose, (4) lignin, and (5) proteins and other nitrogenous compounds. The lignin is determined as undissolved Klason lignin residue. In soil samples, this residue contains humus as well. Nitrogenous organic compounds fall into several fractions (Fig. 5.4.2). For simplicity, however, the protein content in this study was estimated according to the convention of food chemistry as a crude protein by multiplying the content of total hydrolysable N

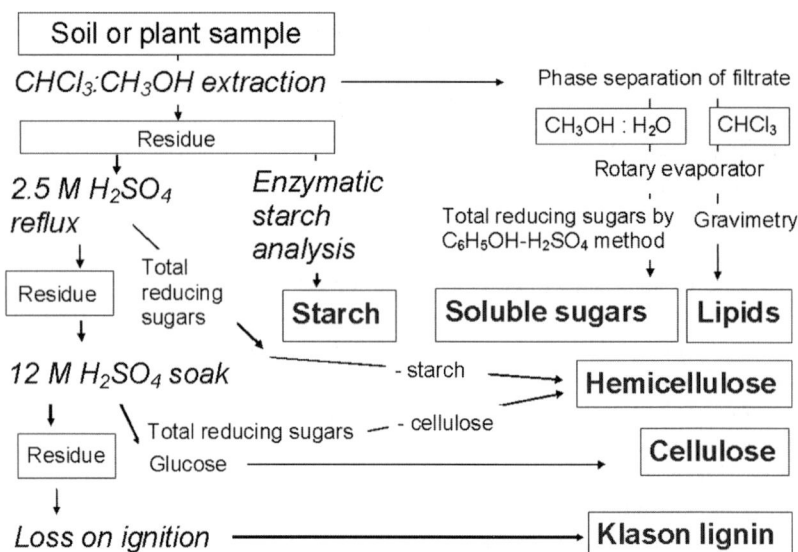

Fig. 5.4.1 Sequential extraction and analysis of various organic fractions of soil and plant by solvent, mild acid and strong acid. Lipid extraction with chloroform:ethanol solution (Bligh and Dyer, 1959; Kögel-Knabler, 1995). Double hydrolysis scheme for polysaccharides (Oades et al., 1970; Swift, 1996). Analysis of total reducing sugars by phenolic sulphuric acid method (Dubois et al., 1956; Wood and Bhat, 1988). Enzymatic analysis of starch by hot gelatinisation and amyloglucosidase (AG) incubation. Analysis of glucose by glucose oxidase peroxidase method (GOP; Karkalas, 1985)

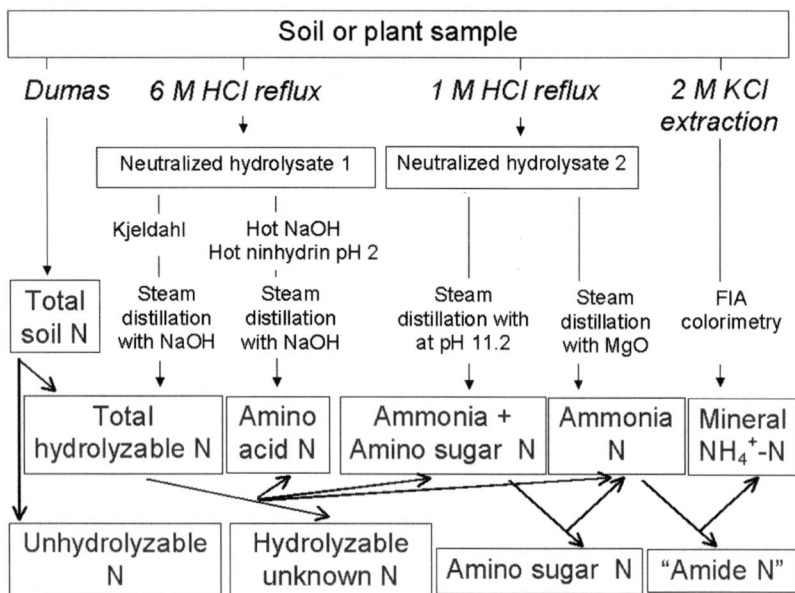

Fig. 5.4.2 Methods for determining the various forms of nitrogen in soil and plant (Bremner, 1965; Yonebashi and Hattori, 1980). Reflux in 70 ml screw-capped Pyrex bottles. Steam distillation with boric acid indicator in the collection flask. Flow injection analysis by QuikChem® 8000 (LACHAT Instruments, QuikChem® method 12-107-06-2-A)

by 6.25 (assuming 16% N in the proteins). "Humus N" was obtained as a difference between the total and hydrolysable N.

Box A

Air-dried and ground/sieved (<2 mm) materials were used for the analyses (soil from O and A horizons, green needles, litter, wood). Lipids, carbohydrates and lignin were determined by sequential extraction with a chloroform-methanol solution and sulphuric acid solutions. Nitrogenous compounds were determined by dry combustion and extraction with neutral salt and hydrochloric acid solutions. The amount of organic matter was determined as loss on ignition (within 2 h to 550°C plus 1 h burning).

Lipids and soluble sugars were extracted by a chloroform-methanol solution (N_2 bubbled samples, 16 h) (Bligh and Dyer, 1959; Kögel-Knabler, 1995). After the phase separation, the chloroform layer was evaporated to dryness by a rotary evaporator and the lipids were determined by gravimetry. The soluble sugars in the methanol-water layer were determined as total reducing sugars by the phenolic-sulphuric acid method (Dubois et al., 1956; Wood and Bhat, 1988). Polysaccharide and Klason lignin fractions were determined from the residue after the chloroform-methanol extraction. Starch was determined enzymatically by hot gelatinisation and amyloglucosidase (AG) incubation. Glucose was analysed by the glucose oxidase peroxidase method (GOP, Karkalas, 1985) and converted to starch (×0.9). A double-hydrolysis scheme with 2.5 M H_2SO_4 (reflux, 20 min) and 12 M H_2SO_4 (soak 20°C, 16 h) was used for polysaccharides (Oades et al., 1970; Swift, 1996). Cellulose was determined by the amount of glucose (×0.9) in the 12 M H_2SO_4 extract. Hemicellulose was determined by the sum of the amount of total reducing sugars (×0.9) in the 2.5 M H_2SO_4 and 12 M H_2SO_4 extracts (phenolic-sulphuric acid method) minus the contributions due to starch and cellulose in the extracts, respectively. The amount of Klason lignin was determined by the loss on ignition of the residue after 12 M H_2SO_4 extraction. This was done at 400°C for 16 h (Nelson and Sommers, 1996), as the glass fibre filters (Whatman GC/C) were not stable at higher temperatures. In soil samples, the Klason lignin fraction includes also humus compounds (humic acids and humin).

Total N content was determined by dry combustion (Leco® CSN-analyser). Fractions of organic nitrogen were determined by the extraction with 6 M and 1 M hydrochloric acid solutions and steam distillation techniques (Stevenson, 1996; Yonebashi and Hattori, 1980). Neutralised hydrolysates were steam distillated and titrated with Kjeltec Auto 1030 Analyzer (macro-Kjeldahl tubes, 40 ml hydrolysate, boric acid indicator, 0.025 M H_2SO_4). Total acid hydrolysable N was determined by the extraction with 6 M HCl (reflux in capped pyrex bottles, 110°C, 24 h, Salo-Väänänen, 1996). Half of the hydrolysate was analysed for the total hydrolysable N by Kjeldahl digestion and steam distillation with NaOH. For the determination of amino acid N in the other half of the hydrolysate, the hydrolysate was pretreated with hot

NaOH (100°C, 20 min) to remove amino sugars plus hot ninhydrin (pH 2, 100°C, 10 min) to convert α-amino-N to ammonium ions, and then steam distillated with NaOH. Amino sugars and ammonium N were determined by extraction with 1 M HCl (reflux, 3 h). Half of the hydrolysate was steam distillated with a phosphate-borate buffer (pH 11.2) to give the sum of amino sugars and ammonium N. The contribution of ammonium was estimated from the other half by steam distillating it with MgO. The amount of mineral ammonium was small, as estimated by the extraction of samples with 2 M KCl (10 g sample, 100 ml solution, 1 h) and colorimetric analysis of ammonium by flow injection analysis (Lachat Instruments QuikChem® 8000, QuikChem® method 12-107-056-2-A).

5.4.3 Soil Structure, Water and Air

Forest soil is a mixture consisting of soil particles, water and air. Soil structure generally denotes the spatial arrangement of soil particles and pore space between them. Natural structural development of soil involves spatial heterogenization of soil. In fine-textured soils that contain much clay, the individual mineral and organic particles are clustered together into larger units called aggregates or peds of various shapes and sizes. The aggregates in soils containing even modest amounts of organic matter show a hierarchical structure (Tisdall and Oades, 1982; Oades, 1993): the particles form microaggregates that are joined together into macroaggregates. Clay, humus and oxides often coat the coarse particles and aggregate surfaces. This kind of secondary soil structure is not well developed in boreal forest soils that are mostly coarse in texture. The structure of coarse-textured soils is predominantly single-grained, with roots, fungal hyphae, organic biopolymers and oxides providing only limited binding of particles together. However, the channels of old plant roots and earthworms or other burrowing animals often provide a stable system of continuous large pores in both fine and coarse textured soils (Lee and Foster, 1991; Oades, 1993; Angers and Caron, 1998).

The pore space between soil particles or aggregates is filled with water or air and is a site for chemical reactions and a habitat for soil organisms. Most reactions and microbial activity occur in the water-filled pore space. The facts that soil particles have charged surfaces and that water molecules are permanent dipoles, i.e. water molecules have electrically positive and negative areas, explain the behaviour of water in the soil (Hillel, 1998). Electric forces bind the water molecules to particle surfaces and with each other. Water forms layers on soil particles in which the first layer of water molecules are bound electrically to the particle surface (adhesion) and thereafter the water molecules are bound electrically with each other (cohesion).

At the border (interface) between water and air, the neighboring water molecules interact with each other unequally at different directions, as the molecules inside the water phase pull the molecules from the surface inwards, while the pull towards the air phase is negligible (Hillel, 1998). This net inward force causes a flow of water molecules from the surface deeper into the water until an equilibrium is reached at which the electric forces in the interface between water and air are strong enough to resist further flow and the surface area of water is minimized. As a result, the water molecules at the interface are bound electrically more strongly with each other than with the molecules of bulk water, and form a clear film (meniscus) separating the water and air phases. The film acts like an elastic membrane between water and air phases resisting any forces that tend to increase its surface area. The magnitude of phenomenon is quantified by the coefficient of surface tension (unit $N\,m^{-1}$), or equivalently by the coefficient of surface energy (unit $J\,m^{-2}$).

The shape of water-air interface at equilibrium with external forces neglecting gravity satisfies the Young-Laplace equation.

$$\Delta P = \gamma \left(\frac{1}{r_1} + \frac{1}{r_2} \right) \tag{5.4.1}$$

or

$$\Delta P = \gamma \frac{2}{r}, \text{with } r = \frac{2\,r_1\,r_2}{r_1 + r_2} \tag{5.4.2}$$

where $\Delta P\;(= P_{water} - P_{air})$ is the pressure difference across the water-air interfacial membrane, γ the coefficient of surface tension, r the equivalent radius of curvature of the surface, and r_1 and r_2 the radii of curvature lying in planes normal to each other at a point on the interface as shown in Fig. 5.4.3 (Adamson and Gast, 1997; Hillel, 1998). The radius of curvature r is taken as positive, if the radius lies within water (such as r_1 in Fig. 5.4.3). In words, these equations state that at equilibrium the soil water-air interfaces are characterized by a uniform curvature that may be produced by innumerous combinations of radii of curvature. The radius of curvature is determined by the boundary conditions determined by the contact angle (α) of water with the solid phase and the pore size (for a hemispherical surface $r = R/cos\ \alpha$, where R is the equivalent pore radius). The contact angle is commonly taken as zero for soil minerals, indicating a completely wetting surface, although in reality it depends on the relative magnitudes of interfacial surface energies between solid, liquid and gas phases. However, humic coatings on dry soil mineral particles may increase the contact angle up to hydrophobism ($>90°$). In wet soil, water-repellency is less pronounced. Bachmann and van der Ploeg (2002) reviewed the recent developments in interfacial tension and temperature effects on soil water retention.

According to the Young-Laplace equation, no pressure difference exists across the flat surface of free water. In unsaturated soils, however, the water molecules generally adhere to the solid particles. Owing to high surface tension of water, this produces concavely curved water-air interfaces with contact angles $<90°$ with the solid phase, which decrease the pressure in soil water below that in soil air. In such a case, soil water is under tension (underpressure, suction), the magnitude of which gives the strength of water retention in the capillary pores of soil (unit Pa or $J\,m^{-3}$).

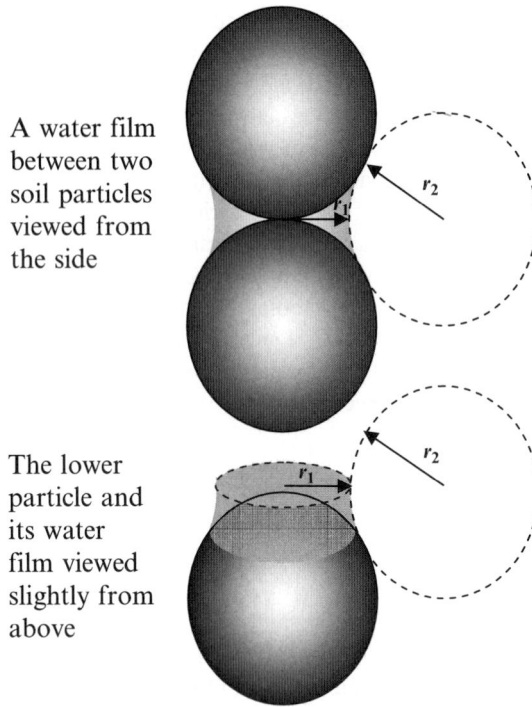

A water film between two soil particles viewed from the side

The lower particle and its water film viewed slightly from above

Fig. 5.4.3 Two radii of curvature (r_1 and r_2) in planes normal to each other in a pendular water between two spherical soil particles

The phenomenon of capillarity explains, why the largest pores empty first in a drying soil, while the smaller pores remain filled with water, because the curvature of water film and the resulting pressure difference increase (the radius of curvature decreases) with decreasing pore size. Small-sized humus is thus important in increasing the retention of water in coarse textured soils. Capillarity is also important, as the pressure difference between water and air phase drives water flux from lower to higher tension, and cause capillary rise from the free ground water table upwards into drier soil until the capillary forces are balanced by the gravity (the weight of water column).

Besides water, soil solution contains small amounts of organic and inorganic dissolved species. They typically decrease the surface tension of soil solution only slightly below that in pure water and thus have minor influence on the retention of water in soil.

In contrast to the small pores mostly filled with water, larger soil pores drain out quickly and are air-filled for most of the time. Soil air is in essence atmospheric air nearly saturated with water vapour and continuously modified by consumption and production of gases in soil. The amount, shape and distribution of air-filled pores is determined in a complementary manner by soil solid and liquid phases. Soil structure is thus an important determinant of soil functions, as it ultimately

Fig. 5.4.4 Soil water retention curves at various depths
of mineral soil horizons in a
podzol soil at the SMEAR II
station

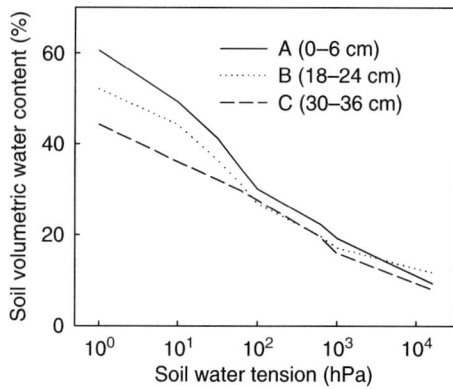

determines the retention (Fig. 5.4.4) and flux rates of water, nutrients and gases
supporting the processes that change material from one form to another in the soil.

5.4.4 Soil Organisms

Besides plant roots, numerous organisms live in the soil. The primary function of
soil organisms in the soil ecosystem is the processing and mixing of soil organic
matter (Killham, 1994). They can be classified according to size into microflora
(archae, bacteria, fungi), micro- and mesofauna (e.g., protozoa, rotifers, tardigrades)
and macrofauna (e.g., nematodes, earthworms, Enthytraeidae, arthropods) (see e.g.,
Blume et al., 2002). The microfauna (less than 100 μm in dimension), also called
protozoa, consists of flagellates, ciliates, amoebae and sporozoa (Paul and Clark,
1989; Killham, 1994). Protozoa live mainly by grazing soil microbes, bacteria and
fungi. The most important members of larger soil fauna, meso- (from 100 μm to
1–2 mm) and macrofauna (greater than 1–2 mm) are the earthworms, enchytraeid
worms, nematodes, arthropods (e.g. mites and Collembola), molluscs and mammals.
Macrofauna can move quite freely in soil pore space, and the largest ones can even
burrow their own channels through the soil. On the contrary, the smaller organisms
can not move around in the soil and are limited to the small water-filled pores and
films covering the soil particles to prevent dehydration by drying and predation by
soil animals.

The largest amounts of soil organisms live in the immediate vicinity of roots (rhizosphere) and in soil surface layers containing ample decomposable organic compounds. Some bacteria and fungi live even within plant roots, such as the symbiotic
nitrogen-fixing bacteria and mycorrhiza (Fig. 5.2.10). Moreover, a part of the soil
microbiota belongs to the phyla *Archaea*. It was earlier thought that *Archaea* can
only grow under very extreme conditions (in high salinity, or in high temperature,
or in anaerobic conditions), but it has been shown that *Archaea* exist also in great
numbers in acid boreal forest soils (Jurgens et al., 1997).

5.4.5 Soil Horizons and Distribution

The soil covering the Earth surface is not similar everywhere; its composition, structure and properties vary in space. Such differences appear on a global scale and in different vegetation zones as well as within landscapes and individual soil profiles. The differences originate partly from the transport and sedimentation of parent material, and partly from the gradual processes of local soil formation. As a consequence, dominant soil texture differs in different geographical areas, and the surface layers of a well-developed soil profile show distinct horizons differing in colour, chemical properties, or structure.

Boreal forest soils are mostly coarse-textured for several reasons. In the boreal zone, much of the soil parent material originates from the Holocene ice age. During and after deglaciation, much of the fine soil material was transported to more southern latitudes by melting waters, whereas the coarser material remained in place (see e.g. Salminen, 2005). Within the boreal zone, the finer soil fractions have been deposited at the bottom of lakes and seas, whereas the supra-aquatic soils at higher elevations are coarse-textured glacial till. As the land level rose after deglaciation, the lake and sea sediments partly mixed with coarser coastal sediments and became eventually dry land. Later, the fine-textured soils were preferred for agricultural land use rather than forestry due to their fertility.

Soil organic matter accumulates near the soil surface. This may even form an organic surface horizon (litter or humus layer) above the mineral soil. Deeper in the soil, humus covers larger mineral soil particles and forms complexes with clay particles.

In humid climates such as in the boreal zone, coarse-textured forest soils typically have a podsolic profile. It is characterized by an acidic litter and humus layer underlain by a dark-coloured mineral soil horizon, a greyish eluviation horizon and a reddish dark brown illuviation horizon. The bottom of the profile is practically unchanged parent material. Podsol soils are most common under boreal forests (see Section 7.4).

Chapter 6
Processes

6.1 Temporal and Spatial Scale of Processes and Fluxes

Pertti Hari and Annikki Mäkelä

University of Helsinki, Department of Forest Ecology, Finland

The spatial and temporal variation in environmental factors is strong (Section 1.2: Basic idea 1), especially in the light inside the canopy. Both environmental factors and active biochemical substances, that are synthesised and transported in the plant, affect processes such as growth and development (Section 1.2: Basic idea 2). Thus great variations in environmental factors generate corresponding variations in the processes. In order to fully understand these processes, we need mathematical tools that are able to cope with such variability. Thus, within this book, we introduce a new concept of a mass-specific process rate, defined below.

Consider a moment of time t and a point x in space. Let Δt be a short time interval after the moment t, ΔV a small space element centred at the point x and $Q_{PE}(t,x,\Delta t,\Delta V)$ the mass of products produced in the process during Δt in the volume ΔV. Consider the ratio of Q_{PE} to $\Delta t \Delta V$.

$$\frac{Q_{PE}(t,x,\Delta t,\Delta V)}{\Delta t \, \Delta V} = \frac{m(x,\Delta V)}{\Delta V} \frac{Q_P(t,x,\Delta t,\Delta V)}{m(x,\Delta V)\Delta t} \tag{6.1.1}$$

The term $m(x,\Delta x)$ is the mass of the active tissue in the volume ΔV.

Let the length of the time interval and the volume of the space element approach zero. Then the first term on the right-hand side approaches the mass density (Eq. 5.2.1). The limit of the second term on the right-hand side is defined as the mass specific process rate, $q_P(t,x)$. Thus

$$q_P(t,x) = \lim_{\substack{\Delta t \to 0 \\ \Delta V \to 0}} \frac{Q_{PE}(t,x,\Delta t,\Delta V)}{\Delta t \, m(x,\Delta V)} \tag{6.1.2}$$

The term "mass specific process rate" is a rather complicated expression and we shorten it to "process rate" in the following.

When the definitions of mass specific process rate and mass density are combined we get

$$\lim_{\substack{\Delta t \to 0 \\ \Delta V \to 0}} \frac{Q_{PE}(t,x,\Delta t,\Delta V)}{\Delta t \, \Delta V} = \rho_m(x) \, q_P(t,x) \qquad (6.1.3)$$

The fundamental relationships between processes and environment occur at the point level in time and space. The variation in driving environmental factors smears the relationship at more aggregated levels. Thus the great variation in environment (Section 1.2: Basic idea 1) forces us to use mathematical formalism that can cope with the variation. If we relate the processes to environmental factors on more aggregated levels than in a point or space element, then the nonlinear relationships, common in ecology, generate bias in the results.

6.2 Physical and Chemical Processes

6.2.1 Phase Transitions and Energy Exchange

Timo Vesala[1], Pasi Kolari[2], Tuukka Petäjä[1,3], Erkki Siivola[1], and Pertti Hari[2]

[1] University of Helsinki, Department of Physics, Finland
[2] University of Helsinki, Department of Forest Ecology, Finland
[3] Earth and Sun Systems Laboratory, Division of Atmospheric Chemistry, National Center for Atmospheric Research, Boulder, CO, USA

6.2.1.1 Phase Transitions

First-order phase transitions, like freezing-melting and condensation-evaporation of water/water vapour, are associated with heat of transition. The phase transitions are important for the behaviour of the atmosphere since they release and absorb energy affecting distribution and transport of it (Section 7.2). Physico-chemical analysis is based on the concept of thermodynamical potentials which describe energies associated to different thermodynamical states and systems (e.g., Mandl, 1971). One of the potentials is *enthalpy* which is conserved for a thermally isolated system held at a constant pressure. Enthalpy is lower for the more ordered phase (like liquid water) than for the less ordered one (water vapour). Thus when water vaporizes, the difference of energy states appears as a heat of vaporization. *Evaporation* needs external energy which is then transported from surroundings, in the absence of extra heat sources temperature with evaporating water is always lower than the ambient one. Correspondingly, in *condensation* heat is released and transported to surroundings, and, in the absence of extra heat sources, the surface temperature is elevated. When phase transitions occur, they must be taken into account in any considerations of energy balances and exchange. Transpiration is physical evaporation from wetted interfaces along the pores in the cell walls of mesophyll, epidermal and guard cells, via stomatal openings to the atmosphere, lowering the leaf temperature. Conversely, a leaf can gain latent heat if dew or frost condenses onto it (see Nobel, 2005). Note that evaporation and condensation are reverse processes and the amount of heat required to vaporize liquid water is the same as the released heat if the same amount of water vapour is condensed to liquid. The latent heat of vaporization of water is 2,260 kJ/kg and the same amount is released in condensation.

Freezing and melting are other relevant examples of phase transitions and, similarly to evaporation/condensation, external energy is needed for melting and the same amount is released in freezing. The latent heat of freezing of water is much smaller than the heat of vaporization, only 333 kJ kg^{-1}.

6.2.1.2 Energy Exchange

By energy exchange we mean the process occurring at the interface of two phases. Often energy exchange between bodies or between solid/liquid and gas phase includes both the interfacial process and transport of energy onto/away from the interface. We discuss transport phenomena in Chapter 4. At the interface, molecules of two phases interact and exchange energy in collisions. In non-metal materials the energy carried by molecules arises from their motion (translational movement, vibrations or rotations). More energetic molecules deliver energy for less energetic ones and the process may be coupled with phase transition. In metals, free electrons are the major heat carriers (see e.g., Bird et al., 2002). The material can absorb energy also from photons of electromagnetic radiation incident onto an interface.

6.2.1.3 Electromagnetic Radiation

Solar radiation energy is the source of energy for atmospheric processes and transport phenomena (Chapter 3). The atmosphere and surfaces of solid objects absorb and emit electromagnetic radiation. Conservation of energy means that absorption and emission are always connected with temperature changes, which has strong consequences to the behaviour of the atmosphere (Section 7.2).

All materials emit radiation, since there is a tendency for the atoms and molecules to return spontaneously to lower energy states (see Bird et al., 2002) and in this process energy is emitted. Because the emitted radiation results from changes in the electronic, vibrational and rotational states of the atoms and molecules, the radiation is distributed over a range of wavelengths.

Let us consider first a so-called *black body*, a perfectly efficient emitter, which has a surface emitting hemispherically isotropic radiation at all frequencies. Most natural objects are an approximation to this idealised body. Irradiance $I_{E,\lambda}$ as a

146 T. Vesala et al.

function of wavelength (λ) and surface temperature (T) can be approximated with Planck's law of black-body radiation (e.g., Bird et al., 2002).

$$I_{E,\lambda}(\lambda,T) = \frac{2hc^2}{\lambda^5} \cdot \frac{1}{e^{ch/k\lambda T} - 1}, \qquad (6.2.1)$$

where c, h and k are speed of light in vacuum ($2.9979 \cdot 10^8$ m s^{-1}), Planck's constant ($6.626 \cdot 10^{-34}$ J s) and Boltzmann's constant ($1.381 \cdot 10^{-23}$ J mol^{-1} K^{-1}), respectively. Planck's law is one of the triumphs of explaining observed phenomena by using quantum mechanics. It can be derived by applying quantum statistics to a photon gas (or generally to a system of elementary particles called bosons) in a cavity, the photons obeying Bose-Einstein statistics (Bird et al., 2002).

Theoretical spectral irradiance of two blackbodies of 6,000 and 300 K are depicted in Fig. 6.2.1. The selected temperatures are close to the surface temperatures of Sun and Earth, or any bodies on Earth. The Planck distribution predicts the entire energy versus wavelength curve and the shift of the maximum toward shorter wavelengths at higher temperatures. In fact, the emission spectrum of many bodies at room temperature is very close to 300 K blackbody spectrum in Fig. 6.2.1 and the spectrum of sun radiation (detected outside the atmosphere) is very close to 6,000 K blackbody spectrum (Figs. 6.2.1 and 6.2.2). The solar spectrum has a maximum in visible light wavelengths whereas terrestrial radiation emitted into space has a maximum approximately at 10 μm (infra-red).

One can easily find numerically the solution for the maximum in the Planck distribution. The maximum at a certain wavelength (λ_{max}) depends on the temperature according to the so-called Wien's displacement law

$$\lambda_{max} = \frac{hc}{4.9651\,kT} \qquad (6.2.2)$$

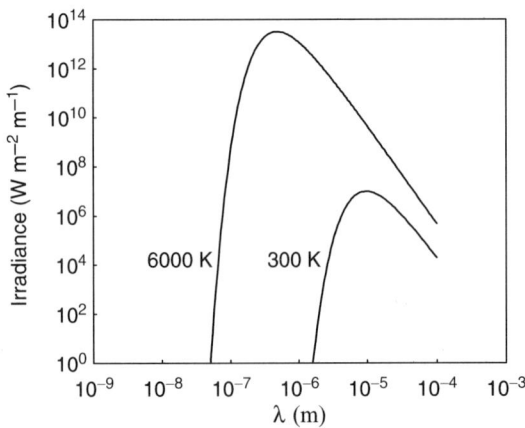

Fig. 6.2.1 Blackbody emission spectra for 6,000 and 300 K. The temperatures correspond to surface temperature of the Sun and the Earth, respectively

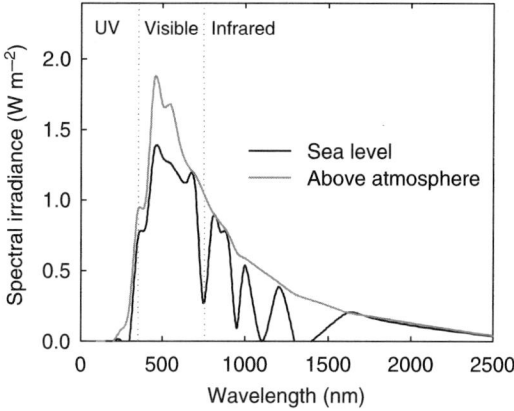

Fig. 6.2.2 A simplified wavelength distribution of radiative energy from the sun above the atmosphere and at sea level

The law predicts that the apparent colour of radiation shifts from long wavelengths (red) to short ones (blue) as the temperature increases. A more important result, the Stefan-Boltzmann law, is obtained if the Planck distribution is integrated over all wavelengths:

$$Q = \frac{2\pi^5 k^4 T^4}{15\,c^2 h^3} \equiv \sigma T^4 \qquad (6.2.3)$$

where σ the Stefan-Boltzmann constant, equal to 5.67×10^{-8} W m$^{-2}K^{-4}$. As Bird et al. (2002) notes, the formula is truly remarkable, interrelating the k from statistical mechanics, the c speed of light from electromagnetism and the h from quantum mechanics.

For non-black surfaces the emitted energy flux is

$$I_E = e\sigma\, T^4 \qquad (6.2.4)$$

where e is the *emissivity* and must be evaluated at temperature T. Unoxidized, clean, metallic surfaces have very low emissivities (like 0.02 for polished copper), whereas most nonmetals and metallic oxides have emissivities above 0.8 at room temperature. e increases with increasing T for nearly all materials. Note that the emissivity may depend on the frequency of radiation. If it is <1 but is the same over all frequencies, the body is called grey. If we consider the small opening in a blackbody cavity, with insulating walls completely isolating it from external influences, we obtain that the hole itself is very nearly a black surface.

Formula 6.2.4 has been effectively used in measurements of surface temperatures of bodies, by infrared sensors and cameras. If the surface emissivity is known, the measured energy flux can be readily inverted to the temperature by the formula.

Emissivity is closely linked to *absorptivity a*. Radiation hitting the surface of an opaque body is either absorbed or reflected. Absorption is the process by which the energy of a photon is taken up by atoms or molecules of the absorbing medium.

The photon is destroyed in the process. The absorbed energy of the photon can excite the atomic state (arrangement of electrons in atomic orbitals) or the molecular state (molecule's modes of vibration and rotation). The absorbed energy may be re-emitted as radiation at a longer wavelength or transformed into heat energy and dissipated by convection or conduction (known as sensible heat). Absorptivity normally depends strongly on the frequency of radiation. For a perfect black body the absorptivity is unity over all frequencies. When the radiation is in equilibrium with the solid surface, i.e., no radiative energy exchange occurs between the surface and its environment, the absorptivity and emissivity e are the same (Kirchhoff's law). This is assumed often to describe also non-equilibrium transport although it is not strictly true.

6.2.1.4 Interaction of Radiation and Matter

The functioning of all land ecosystems is driven by radiative energy from the sun. Light energy directly drives photosynthesis and transpiration in plants and also affects stomatal action and the rates of biological processes, such as respiration, through temperature by warming up the plants. Infra-red energy, which is about half of all the incident energy from the sun, has a significant warming effect, and thus increases the tendency of water to be evaporated from leaves in the process of transpiration.

Photons interacting with material are either absorbed, reflected or transmitted through the material. The fraction of incident radiation flux density that is absorbed by a surface is called absorptivity. Reflectivity is the fraction that is reflected by a surface, and transmissivity the fraction that is transmitted through a surface. The sum of absorptivity, reflectivity and transmissivity equals 1.

Scattering is a physical process where radiation is forced to deviate from a straight trajectory by the particles in the medium through which the radiation passes. In solid or liquid medium, scattering results in part of the incident radiation being reflected from the surface. Also photons absorbed by the medium or transmitted through it may have undergone scattering in the medium. Because scattering depends on wavelength, the spectral composition of scattered (diffuse) light is altered from that in the sun's direct beam. The spectrum of diffuse light in a clear sky is shifted towards shorter wavelengths and thus appears blue to the human eye. Under clouds, scattering causes the sky to be greyish. Correspondingly, the spectrum of light inside a tree canopy is different from the spectrum of incoming solar radiation: Light transmitted through or reflected from foliage is enriched in green because blue and red wavelengths are absorbed by photosynthetic pigments in the leaves.

The interactions of radiation and matter strongly depend on wavelength of the radiation (Ross, 1981). There are three wavebands that are of particular importance for living organisms (Fig. 6.2.2). These are the ultraviolet (UV), photosynthetically active and infrared bands. The most important of them is the photosynthetically active radiation (PAR), or visible light (380–750 nm), that provides the energy for photosynthesis. Depending on cloudiness and solar elevation angle, 20–55%, on average about 46%, of solar radiation on the earth surface falls in this band. Flux density in the shorter wavelengths is relatively low since most of the radiation is absorbed by gases in the atmosphere. Up to 4% of incoming solar radiation on the earth surface is in the UV band that consists of wavelengths below 380 nm. UV radiation is energetic and can break molecular bonds. The range of shortest wavelengths in the infrared band, called near infrared (NIR, 750–3,000 nm), is important in the heat budgets of living organisms and ecosystems, and it contributes to about half of the incoming solar radiation. Flux density of solar radiation at wavelengths of higher than 3 μm is negligible, but bodies on earth emit thermal radiation in wavelength range of 3–100 μm with flux density that depends on the surface temperature and emissivity of the body.

In plant leaves, light energy in the photosynthetically active waveband is primarily absorbed as excited states of pigment molecules. The predominant pigments in leaves are chlorophyll a and b. There are also additional pigments, such as carotenoids and xanthophylls that assist in light absorption and also serve to dissipate excess light energy. The maximum absorbance of plant leaves, about 90%, is at blue-violet (380–480 nm) and orange-red (620–680 nm) wavelengths (Fig. 6.2.3). The leaves appear green due to this selective absorption. At near-infrared range, absorbance is only 5–25% (Fig. 6.2.3). It increases towards the longer wavelengths and peaks at 1,500 and 2,000 nm at liquid water bands. Because the whole solar spectrum is not utilised in photosynthesis, photon flux at the photosynthetically active wavelengths is more convenient than energy flux when considering photosynthesis.

Optical properties of leaves (Table 6.2.1) are determined by surface roughness, reflectance of cuticle, composition, amount and distribution of pigments, internal leaf structure and distribution of water (Ross, 1981). The dimensions of cells in the mesophyll of leaves are large compared to the wavelength of light (Gates et al., 1965). Interaction of light in a leaf thus takes place at cell organell level. They are

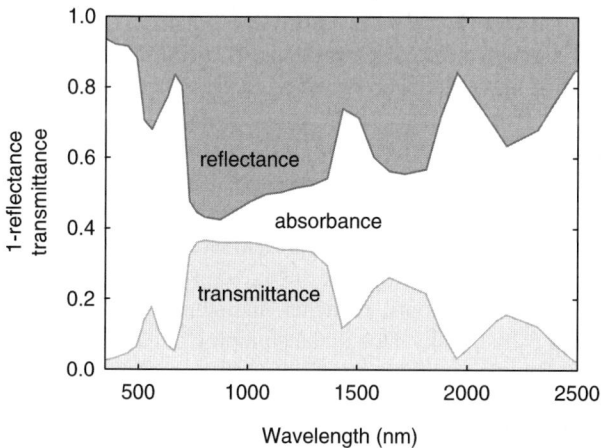

Fig. 6.2.3 A schematic picture of leaf reflectance and transmittance distribution at visible light and near-infrared wavelengths adopted from Monteith and Unsworth 1990. The area between the graphs of reflected and transmitted fractions indicates the fraction incident radiation absorbed by the leaf. Absorbance is highest at the visible (photosynthetically active) wavelength range

Table 6.2.1 Typical optical properties for green leaves

	PAR	NIR	Global shortwave
Reflectance	0.10	0.50	0.30
Transmittance	0.05	0.35	0.20
Absorbance	0.85	0.15	0.50

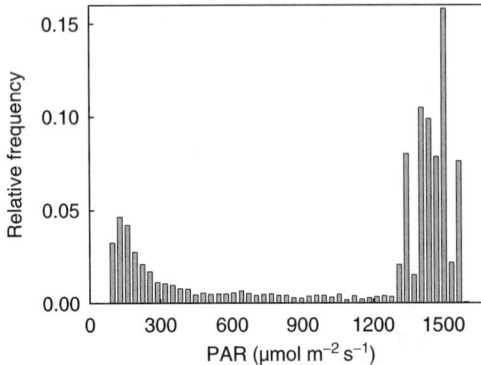

Fig. 6.2.4 Distribution of irradiance (PAR) inside a Scots pine canopy at SMEAR II on a clear summer day. Irradiance was measured at 48 locations in the lower part of the canopy in 30 s intervals for 1 h. The average incident PAR above the canopy was 1,560 μmol m^{-2} s^{-1}

small and can scatter light. Light is reflected back and forth between the surfaces inside the leaf, and cuticle can also reflect exiting photons back inside the leaf. Scattering of light inside a leaf increases its optical path length and thus the probability of a photon being absorbed. Only a fraction of the radiative energy absorbed by photosynthetic pigments is ultimately utilised in photosynthesis, the excess is released as fluorescent light and sensible heat (Section 6.3.3.2.1). The maximum instantaneous light-use efficiency of leaf photosynthesis, defined as the fraction of photosynthetically active light energy absorbed by the leaf that is converted to formation energy of sugars is about 10%. At the ecosystem level and over longer periods, chemical energy bound in biomass is typically in the order of 1% of the absorbed photosynthetically active light energy (Monteith, 1977).

The strong absorption of photosynthetically active radiation by green foliage as well as by other plant structures is an important issue also at plant and stand level. Irradiance at any given point in the tree foliage depends on the amount of shading canopy elements (leaves, shoots, branches and stems) that are intercepting solar radiation entering from the sky. Due to the clumping of canopy biomass into leaves, shoots and branches, the amount of light-absorbing material in the path of a solar beam to each observation point varies, which leads to a heterogeneous irradiance field (Fig. 6.2.4). Attenuation of light in the vegetation and its consequences on forest ecosystem gas exchange are discussed in more detail in Sections 7.3, 7.5.1 and 7.6.2.

6.2.1.5 Energy Balance

There are several simultaneous processes transporting bi-directionally energy into and out of a body. The tool to combine the different processes and transport phenomena is the concept of energy balance, based on conservation of energy. We consider the energy balance of bodies, which can be used for example as a tool to estimate body temperature. The energy balance is composed of energy exchange with the environment and storages. Following (Nobel, 2005) the balance can be formulated as

[change of energy storage by body] = [energy transport into body] – [energy transport out of body]

By body, we mean here all surface elements of ecosystems, like leaves, branches, stems etc. or bare soil. If we take a leaf as an example, the energy storage changes are related to leaf temperature changes, photosynthesis and other metabolism (respiration) being mostly marginal. The energy transport into the leaf is composed of absorbed solar irradiation and absorbed infrared (long wave) radiation from surroundings. The energy transport out of the leaf includes emitted infrared radiation, heat convection, heat conduction and heat loss accompanying transpiration with the contribution of reflected solar irradiation being mostly marginal.

Formally, the energy transport into and out of a body, that is the (net) energy exchange, can be solved as in Section 7.2. Although the machinery for full numerical tools would be available, the brute-force solution is seldom the way to do it. We present next some practical easy-to-use formulas assisting in mathematical formulations of energy balance.

In the vicinity of surfaces the flow velocities are suppressed by friction, and at the surface the velocity is equal to the velocity of the body. If the body stands still, the flow velocity is also zero at the surface. The region in the fluid where the flow velocity is significantly lower than the ambient (wind) velocity is called a boundary layer (Fig. 6.2.5). It is dominated by the shearing stresses originating at the surface

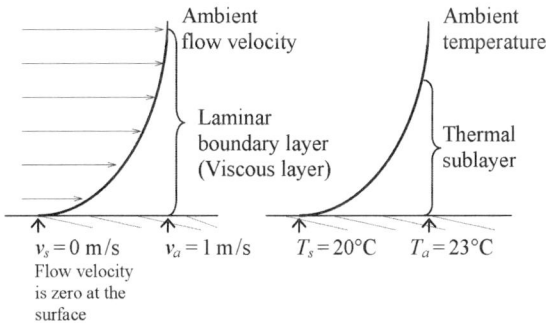

Ambient flow velocity

Ambient temperature

Laminar boundary layer (Viscous layer)

Thermal sublayer

$v_s = 0$ m/s
Flow velocity is zero at the surface

$v_a = 1$ m/s

$T_s = 20°C$

$T_a = 23°C$

Fig. 6.2.5 When a fluid flows over the surface the laminar boundary layer is formed. Across the boundary layer, the flow velocity decreases towards the surface. Similarly, if the surface temperature deviates from the ambient one, a thermal sublayer exists close to the surface of the solid (see e.g., Nobel, 2005), and such layers arise for any solid immersed in a fluid, such as a leaf in air. Although the wind away from the leaf is always turbulent, the flow in the boundary layer is laminar and is predominantly parallel to the leaf surface. Since there is no velocity component perpendicular to the surface, heat is first conducted across the boundary layer and then convected away in the more rapidly moving ambient airstream. Note also that every molecule entering (like CO_2) or leaving (like water vapour) a leaf must cross the boundary layer, that is they are carried perpendicularly, at macroscopic level although single molecules move randomly to all directions, away from the leaf by diffusion. Average boundary-layer thickness depends on the mean dimension of the body and the ambient flow speed and there exist various semi-empirical formulas for estimating the boundary layer depth for flat and bluff bodies (see e.g., Nobel, 2005).

Let us consider first heat conduction across the boundary layer. According to Eq. 4.1.6 and replacing the derivative $\frac{dT}{dx}$ (∇T is reduced to one-dimension since only one direction perpendicular to the surface is relevant) by the "macroscopic" ratio of the temperature difference ΔT over the boundary layer thickness δ_T, we can approximate the heat flux on the surface over the area A

$$Q = \frac{kA\Delta T}{\delta_T} = \frac{kA(T_\infty - T_s)}{\delta_T} \qquad (6.2.5)$$

where T_∞ is the ambient air temperature far from the surface, that is at the outer edge of the boundary layer and T_s is the surface temperature. δ_T is strictly speaking a thermal boundary layer (*thermal sublayer*) characterizing the layer where the temperature is effectively changing from the surface temperature to the ambient one, and it does not generally coincide with the boundary layer (*viscous sublayer*) reflecting the flow velocity change (Fig. 6.2.5). Now the numerator can be easily fixed, since the heat conductivity k is specified by the fluid, the area A depends on the size of the studied object and ΔT is the environmental input parameter, or at least T_∞ is known and T_s is in fact the unknown to be solved. How do we obtain the value for δ_T? The answer is that the heat loss from objects of different size and shape is tabulated utilizing so called non-dimensional groups of quantities of fluid mechanics, from data collected by engineers working with perfect geometrical shapes in a wind tunnel, or sometimes by others working with models of leaves. For conduction across the thermal sublayer the tabulated non-dimensional number is the Nusselt number, $Nu \equiv \frac{l}{\delta_T}$ where l is the characteristic length scale of the body, like the mean length of the plate (leaf) in the downwind direction, cylinder (needle) diameter or sphere diameter depending on the object concerned. Thus the heat flux can be written as

$$Q = Nu\,kA\frac{T_\infty - T_s}{l} \qquad (6.2.6)$$

(conduction and convection)

When Nu is now found from the tables and l is known for the object, we can calculate the heat flux or the equation can be used in estimating the surface temperature as a part of the heat balance equation (see below). The Nusselt number is further a function of the ambient flow velocity according to

$$\text{Nu} = a \left(\frac{\rho\, v_\infty\, l}{\mu} \right)^b \equiv a Re^b \tag{6.2.7}$$

(forced convection)

where ρ is the density of the fluid and v_∞ is the ambient flow velocity. Re is another non-dimensional number called the Reynolds number, which is one of the most important quantities in fluid mechanical analyses. Values of the constants a and b for different types of geometry are given in (e.g., Monteith and Unsworth, 1990). The factor b is generally positive and <1.

We can analyze qualitatively how the Eqs. 6.2.6 and 6.2.7 behave. If the wind is increased, Nu is increased and thus the heat flux is enhanced, which is a natural result. For example, the stronger the wind is, the higher is the cooling rate of a hot body. This can be interpreted also in the way that the larger Nu means a thinner thermal sublayer and thus the heat more easily escapes from the body through δ_T. Furthermore, $\frac{Q}{A} \propto l^{b-1}$ and since $b < 1$ the smaller the object, the larger is the heat flux per unit area. That is a smaller hot object is cooling faster than a larger one. Often the heat flux formula is written in the form $Q = h_T A \Delta T$ where $h_T \equiv \text{Nu} \frac{k}{l}$ is the heat transfer (or convection) coefficient.

The concepts presented above are valid for heat transported by forced convection, but not for the case of free convection (see convection types in Section 4.2), which may dominate at very low flow velocities and large temperature difference. For free convection, no ambient flow velocity scale can be defined, since the flow is created by the object itself, creating the density deviations in the fluid. The formula 6.2.6 is still valid but instead of the Reynolds number Nu is given by the Grashof number Gr.

$$\text{Nu} = a \left(\frac{g \rho^2 \beta \Delta T l^3}{\mu^2} \right) \equiv C Gr^D \tag{6.2.8}$$

(free convection)

where β is the coefficient of volumetric thermal expansion being equal to $\frac{1}{T}$ for the ideal gas (also for air). Values of the constants C and D for different types of geometry are given in (e.g., Monteith and Unsworth, 1990).

It can be shown that the ratio $\frac{Re^2}{Gr}$ describes the importance of forced convection over free convection (e.g., Nobel, 2005). Forced convection accounts for nearly all heat transfer when $\frac{Re^2}{Gr}$ is greater than 10 and forced convection can be neglected when the ratio is <0.1. The intervening region has mixed convection, which is more challenging to analyze. For leaves, needles and tree stems under normal conditions, forced convection is the dominating mode.

Let us next consider the effect of phase transition. In the presence of evaporation or condensation, heat transfer is coupled with mass transport via Eq. 6.2.9, where the mass flux density arises due to the net mass transport perpendicular to the surface because of phase transition. The latent heat vaporization is in fact the difference of the vapour phase and liquid phase enthalpies. Accordingly, the heat flux, Q, can be formulated as

$$Q = Nu \; k \, A \frac{(T_\infty - T_s)}{l} + L A \, g_m(T_s)$$ (6.2.9)

(conduction, convection and phase transition)

where L is the heat of evaporation and g_m is the mass flux density on the surface. The mass flux depends often strongly on the surface temperature via the exponential temperature dependence of the saturation vapour of the liquid phase (see Eq. 6.3.3 in Section 6.3.2). To stress that, we wrote $g_m(T_s)$ in the formula.

Finally we must add the effect of radiative energy transport. Many calculations of radiative energy exchange are prone to difficult geometrical analysis. We give here simplified cases to be adopted in the formulation of the energy balance. Let us consider first two black bodies which are in full view of each other. The heat flux density $I_{E,12}$ between them is given by

$$I_{E,12} = F_{12}\sigma(T_2^4 - T_1^4)$$ (6.2.10)

(two black body surfaces)

according to Eq. 6.2.3 and where T_1 and T_2 refer to respective surface temperatures (see Fig. 4.3.1). F_{12} is the so-called view factor (or angle factor) which represents the fraction of radiation leaving another body and directly intercepted by the other body, for example in the case of Fig. 4.3.1 all radiation emitted by another plate is not absorbed by the other one and hence the view factor would be less than 1. It takes into account the extensions and orientations of the surfaces and F_{12} values for basic geometries are tabulated. The second case concerns a small convex surface in a large nearly isothermal enclosure, but the surfaces are not necessarily black as was assumed in the first example. The net radiation exchange to the surface 1 from the surroundings is given by

$$I_{E,12} = \sigma(a_2 T_2^4 - e_1 T_1^4)$$ (6.2.11)

(convex surface in a large enclosure)

according to Eq. 6.2.4 and the discussion below it. For a heat balance of an ecosystem surface component, T_1 is to be replaced by the surface temperature and T_2 represents an effective radiative temperature of the surroundings. Often the radiative flux of the surroundings can be just replaced by the sum of the incoming and absorbed energy flux by short and longwave radiation $a_{short} I_{E,in}^{short} + a_{long} I_{E,in}^{long}$ where a_{short} and a_{long} refers to absorptivities of short and longwave radiation, respectively.

The full energy exchange formula would be then

$$Q = Nu \, kA \frac{(T_\infty - T_s)}{l} + LA \, g_m(T_s) + a_{short} I_{E\,in}^{short} + a_{long} I_{E\,in}^{long} - \sigma e_s T_s^4$$ (6.2.12)

(conduction, convection, phase transition and radiation)

If the change of the storage is zero, we set $Q = 0$. The above formula describes then a stationary energy balance corresponding to a situation so that T_s remains constant, if nothing in the surroundings changes. If all transport mechanisms are important, calculation of the surface temperature requires a numerical or iterative solution. If some transport mechanisms are minor, the corresponding terms can be dropped out.

If the heat storage must be taken into account, the energy exchange must be set equal to the heat storage rate of change:

$$\frac{\Delta Q_{stor}}{\Delta t} \equiv C_p V \frac{\Delta T_s}{\Delta t} = Q \qquad (6.2.13)$$

where V is the volume undergoing a change in temperature ΔT_s in the time interval Δt. Note that it is now assumed that the body has a homogeneous temperature field. If the internal temperature field is not constant, the heat balance must include also the term describing the internal heat transport.

Radiative transport facilitates the feeling of warmth of the high-temperature body without touching the object (Fig. 6.2.6). For example, you can feel the warmth of the fireplace from a distance of even several metres, although no convection would be present and even though the air is a very good heat insulator, that is its heat conductivity is relatively low and the heat transport rate over 1 m in the air by conduction is very small. The heat exchange occurs according to 6.2.10 where T_2 and T_1 refer to the fireplace and human body temperatures.

Fig. 6.2.6 The sense of heat away from a fire is mainly due to radiative transport, as here above the traditional Finnish "rakovalkea", that is log fire (literally "slit bonfire"), burning in the woods nearby SMEAR II station in Hyytiälä during the celebration of ICOS (Integrated Carbon Observation System) meeting on 10 May 2008 (The hand of L. Kulmala photographed by T. Vesala)

Let us assume that both the fireplace and the human face are the same size disks and they are opposed in parallel planes. Let the radii be 0.2 m and the separation 1 m. The view factor is then according to Bird et al. (2002) about 0.05. Let us further assume that the fireplace temperature is 773 K (500°C). Now, according to the formula 6.2.10 multiplied by the surface area, the heat flux on the human face is approximately

$$0.05\,\sigma\,773^4\,K^4\,\pi 0.2^2\,\text{m}^2 = 127\,\text{W} \tag{6.2.14}$$

The value is not overwhelmingly high but certainly notable (about 1000 Wm^{-2}), as everybody knows from experience. Note that the emission of radiation by the face was omitted here since it is very small compared to that of the fireplace.

6.2.2 Chemical Reactions in the Air

Michael Boy[1] and Boris Bonn[2]

[1]University of Helsinki, Department of Physics, Finland
[2]Frankfurt University, Institute of Atmospheric and Environmental Sciences, Germany

In this section we will focus on chemical reactions taking place in the planetary boundary layer and in particular reactions relevant for the chemistry in and above the boreal forest. Most of all trace gases emitted by the biosphere hold a low oxidation state (indicator of the degree of *oxidation* of *a chemical compound*, e.g., terpenes or methane). In contrast, most of substances removed from the atmosphere by dry and wet deposition are predominantly completely oxidized (e.g., H_2SO_4, HNO_3, CO_2). In this way we have to focus on chemical processes to reduce emitted gases to photochemical oxidized compounds. In the atmosphere the major oxidants are the hydroxyl and nitrate radicals plus ozone because of its high concentration. In this chapter we will discuss the sink and source processes of these molecules and their chemical reactions with biogenic hydrocarbons.

6.2.2.1 The Hydroxyl Radical Process (OH)

In the troposphere the hydroxyl radical is the most important oxidation agent in sunlight. Because of its radical character (one free electron) there is a tendency for it to obtain another electron by reaction with other molecules. In this way OH is a very effective oxidant for trace gases and organic species. The production of the hydroxyl radical starts by photolysis of ozone with UV-B light ($\lambda = 280$–320 nm) as illustrated in Fig. 6.2.7. The chemical reactions are

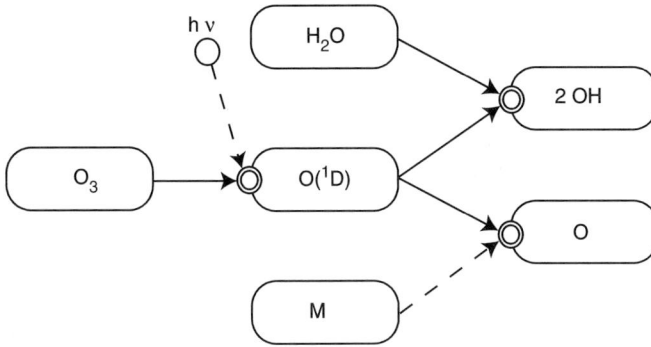

Fig. 6.2.7 Formation of hydroxyl radical from ozone. Boxes indicate amounts, arrows (solid line) flows, dubble circles processes, and arrows (dashed line) effects on process as indicated in the Fig. 1.2.1

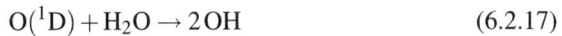

$$O_3 + hv \rightarrow O(^1D) + O_2 \qquad (6.2.15)$$

$$O(^1D) + M \rightarrow O + O_2 \qquad (6.2.16)$$

$$O(^1D) + H_2O \rightarrow 2\,OH \qquad (6.2.17)$$

The recombination of $O(^1D)$ (an excited oxygen atom compared to O which represents an oxygen atom in its ground state) with molecular oxygen or nitrogen (M) via Eq. 6.2.18 is much faster than the reaction with water vapour at atmospheric concentrations and therefore only a few percent of the produced excited oxygen atoms will form hydroxyl radicals (<10%). By using these equations we can estimate the amount of produced OH radicals from the photolysis of ozone and water vapour. However, to calculate OH concentrations additional processes have to be considered. Figure 6.2.8 gives a schematic overview of the most important chemical reactions for the hydroxyl radical cycle. Obviously, the hydroperoxy radical (HO_2) mainly formed by the reaction of OH with other gases plays a major role in this cycle and leads to HO_2 concentrations on average 10 to 100 times higher compared to the OH radical.

The relevance of the individual reactions above the boreal forest is demonstrated for the three most important sink and source terms for a six month period (April to September 2003) for SMEAR II, Finland in Fig. 6.2.9 (annual distribution of several trace gasses used in this calculation are given in Section 7.7). In the last few years, measurements of hydroxyl radicals conducted at several locations worldwide (Eisele and Tanner, 1993; Handisides et al., 2003) and computer simulations (Boy et al., 2005) showed that OH concentrations exhibit a diurnal profile with daytime maximum peaks of several 10^6 molecules cm^{-3}.

6.2.2.2 Ozone Process (O_3)

Tropospheric ozone is the precursor for the OH radical and it reacts with many volatile organic compounds (VOCs) emitted from the biosphere; therefore it plays

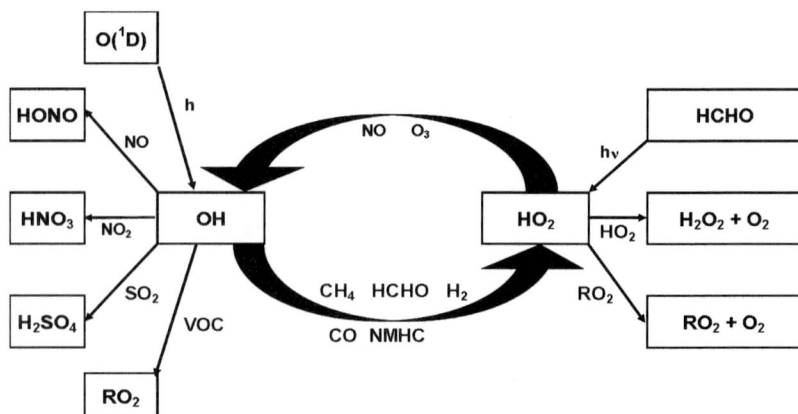

Fig. 6.2.8 Block diagram of the chemistry in the photo stationary state for the OH radical

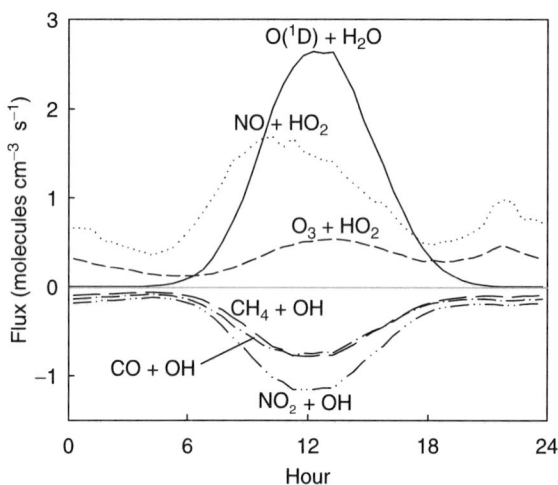

Fig. 6.2.9 Calculated contributions of various sink and source terms for the hydroxyl radical concentration for April to September 2003 (Boy et al., 2005)

a key role in the oxidizing power of the troposphere. In earlier days it was suggested that tropospheric ozone originated only by downward transport from the stratosphere. In the stratosphere oxygen atoms are produced by the photolysis of oxygen molecules (Fig. 6.2.10). The reaction with chemical notation is

$$O_2 + h\nu(\lambda < 242\,nm) \rightarrow O + O \qquad (6.2.18)$$

The oxygen atom reacts rapidly with O_2 in the presence of a third molecule M to form ozone (M is usually O_2 or N_2, Fig. 6.2.11). The reaction with chemical notation is

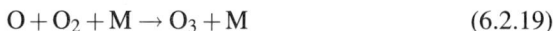

$$O + O_2 + M \rightarrow O_3 + M \qquad (6.2.19)$$

Fig. 6.2.10 Formation of
excited oxygen atoms

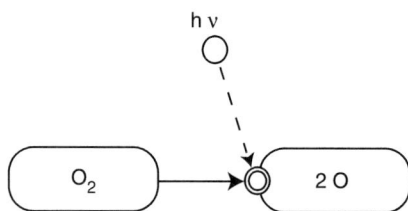

Fig. 6.2.11 Formation of
ozone molecules

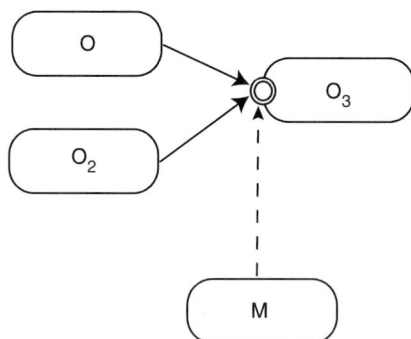

Nowadays we know that this source contributes maximally up to one third of
the ozone concentration in the planetary boundary layer. The major source for
ozone production here is by chemical processes depending on the NO/HO_x ratio
(Fig. 6.2.12). The two competing reactions are:

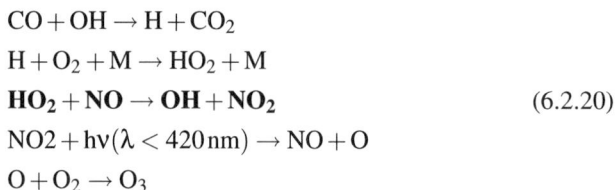

$$CO + OH \rightarrow H + CO_2$$
$$H + O_2 + M \rightarrow HO_2 + M$$
$$\mathbf{HO_2 + NO \rightarrow OH + NO_2}$$
$$NO2 + h\nu(\lambda < 420\,nm) \rightarrow NO + O$$
$$O + O_2 \rightarrow O_3$$

(6.2.20)

This process leads to ozone production and under low NO_x concentration ozone will
be destroyed as illustrated in Fig. 6.2.13. The chemical reactions are

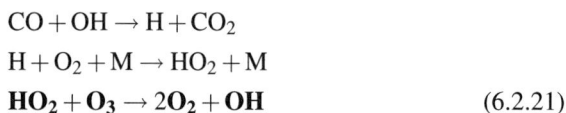

$$CO + OH \rightarrow H + CO_2$$
$$H + O_2 + M \rightarrow HO_2 + M$$
$$\mathbf{HO_2 + O_3 \rightarrow 2O_2 + OH}$$

(6.2.21)

The NO concentration at which point ozone is produced or destroyed depends on
the local ozone concentration. Measurements at the SMEAR II in Hyytiälä (see
also Section 7.7) show that the boreal forest over southern Finland during daytime
always acts as an ozone source with NO mixing ratios around 100 ppt. Besides the
loss of ozone by reaction with HO_2 via Eq. 6.2.22; Fig. 6.2.14, ozone photolysis
with radiation in the UV-B spectra is explained in Eq. 6.2.21.

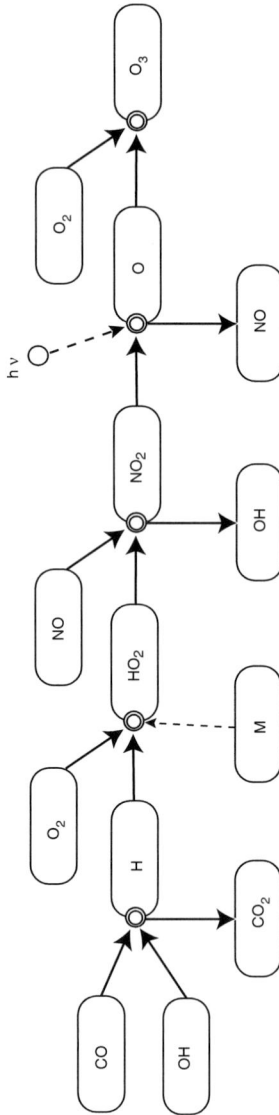

Fig. 6.2.12 Ozone production path in the troposphere

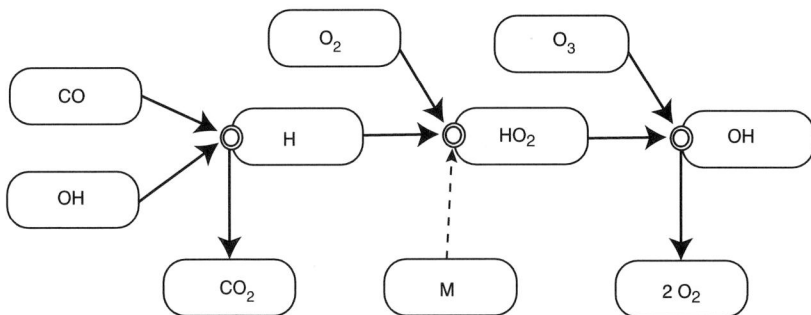

Fig. 6.2.13 Ozone degradation path in the troposphere

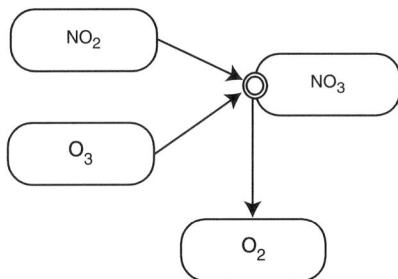

Fig. 6.2.14 Nitrate radical
formation path

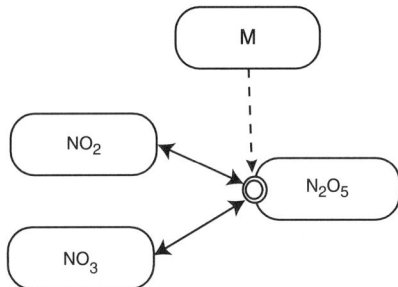

Fig. 6.2.15 Nitrate radical
reaction with NO_2

6.2.2.3 The Nitrate Radical Process (NO_3)

The third important oxidant is the nitrate radical. NO_3 is a strong oxidizing agent
and reacts with most emitted organic species. It is formed via the following reaction

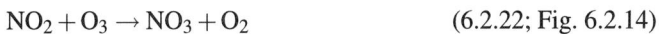

$$NO_2 + O_3 \rightarrow NO_3 + O_2 \qquad\qquad (6.2.22; \text{Fig. } 6.2.14)$$

and is in equilibrium with N_2O_5 according to

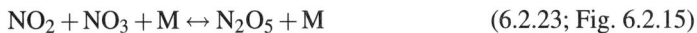

$$NO_2 + NO_3 + M \leftrightarrow N_2O_5 + M \qquad\qquad (6.2.23; \text{Fig. } 6.2.15)$$

During daytime it will photolyze rapidly by two reactions leading to an atmospheric
lifetime of a few seconds.

Fig. 6.2.16 Nitrate radical
photolysis, when wavelength
is larger than 700 nm

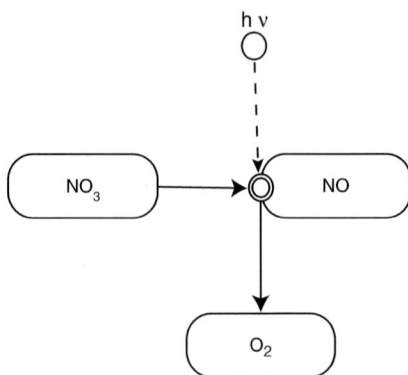

Fig. 6.2.17 Nitrate radical
photolysis, when wavelength
is smaller than 580 nm

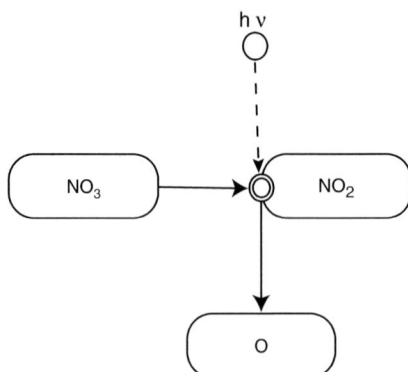

The photolysis processes for different wavelengths of incoming short wave radiations are

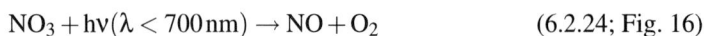

$$NO_3 + hv(\lambda < 700\,nm) \rightarrow NO + O_2 \qquad (6.2.24;\ Fig.\ 16)$$
$$NO_3 + hv(\lambda < 580\,nm) \rightarrow NO_2 + O \qquad (6.2.25;\ Fig.\ 17)$$

It also reacts with NO and forms two NO_2 molecules. Typically the NO_3 mixing ratios drop down during daytime to values of less than a few parts per ton. During the night when NO concentration drops to near zero because of the reaction with O_3, the nitrate mixing ratio can reach values up to several hundreds of parts per ton. An interesting feature about this molecule is its roughly equal concentration with N_2O_5, which can react with water molecules on the aerosol surface to form nitric acid in the particle phase.

6.2.2.4 Nitrogen Oxides Processes (NO_x)

In the last paragraph we discussed the importance of nitrogen dioxide for the formation of the nitrate radical, now we need to consider the sink and source terms of the

Fig. 6.2.18 Nitric acid production during day-time

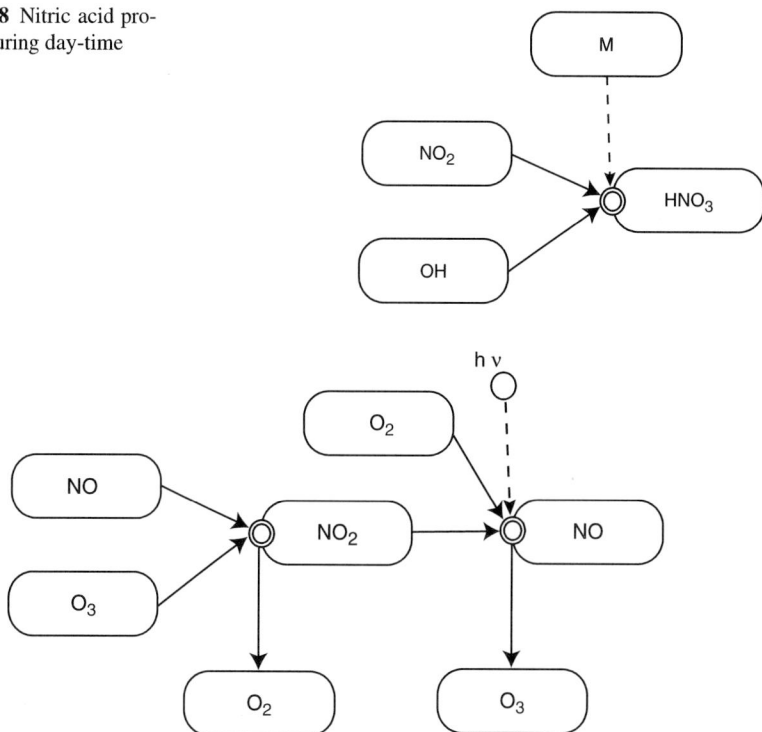

Fig. 6.2.19 NO-NO$_2$ reaction processes

nitrogen oxides. The main sources for these molecules are fossil fuel combustion, biomass burning and the emission from soils (see Section 6.4.2). Although NO$_x$ is mainly emitted as NO, cycling of NO and NO$_2$ takes place in the troposphere on a timescale of a minute during daytime. The reaction with chemical notation is

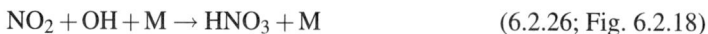

$$NO_2 + OH + M \rightarrow HNO_3 + M \qquad (6.2.26; \text{Fig. } 6.2.18)$$

At night ozone reacts with Nitrogen dioxide to form nitric acid as illustrated in Fig. 6.2.19. The chemical ratios are

$$NO + O_3 \rightarrow NO2 + O_2 \qquad (6.2.27)$$
$$NO_2 + h\nu + O_2 \rightarrow NO + O_3 \qquad (6.2.28)$$

At night NO$_x$ is exclusively present as NO$_2$ because of Eq. 6.2.28. The sink of NO$_x$ is the transformation into nitric acid (HNO$_3$) – following a daytime path according to Fig. 6.2.20. The chemicals reactions are

$$NO_2 + O_3 \rightarrow NO_3 + O_2 \qquad (6.2.29)$$
$$NO_3 + NO_2 \rightarrow N_2O_5 \qquad (6.2.30)$$
$$N_2O_5 + H_2O \rightarrow 2HNO_3 \text{ (aerosol)} \qquad (6.2.31)$$

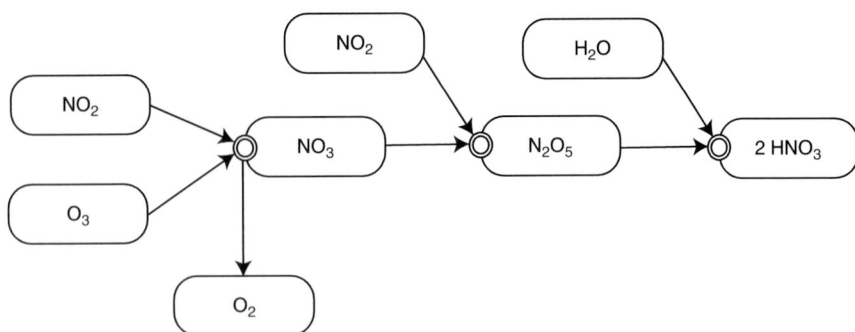

Fig. 6.2.20 Nitric acid production during night-time

In the troposphere HNO_3 is scavenged by precipitation because of its high water-solubility and is removed by deposition in this way. In general water-soluble species have a lifetime in the lower troposphere of a few days and we can conclude that HNO_3 is not an effective reservoir for NO_x. A more efficient mechanism for transporting NO_x from anthropogenic sources to rural areas is the transformation into peroxyacetylnitrate (better known as PAN. PAN is produced by photochemical oxidation of carbonyl compounds such as acetone in the presence of NO_x and in contrast to HNO_3 it is less soluble in water and therefore only to a small extent removed by wet deposition. Its principle sink is by thermal decomposition, regenerating NO_x with high temperature dependence. Although PAN is only one of many organic nitrates produced during the oxidation of carbonyl compounds under the presence of NO_x, it is the most important reservoir for NO_x in the troposphere.

6.2.2.5 Methane Processes (CH_4)

Methane is a long-lived compound (atmospheric lifetime ≈ 7 years) with about one third of its concentration having a natural source (wetlands and termites). Its ambient concentration is fairly stable throughout the year at roughly 1,800 ppm. The major sink term with over 85% is the reaction with OH (Fig. 6.2.21; additional sink terms: oxidation in the stratosphere and uptake by soil – see Section 6.4.3). The chemical reactions are

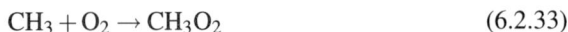

$$CH_4 + OH \rightarrow CH_3 + H_2O \qquad (6.2.32)$$
$$CH_3 + O_2 \rightarrow CH_3O_2 \qquad (6.2.33)$$

The produced methyl radical (CH_3) reacts immediately with O_2 to form a methyl peroxy radical (CH_3O_2 – also written as RO_2). This radical proceeds to react with NO, HO_2 or another peroxy to form, at the end of a chain of reactions, CO, CO_2 and H_2.

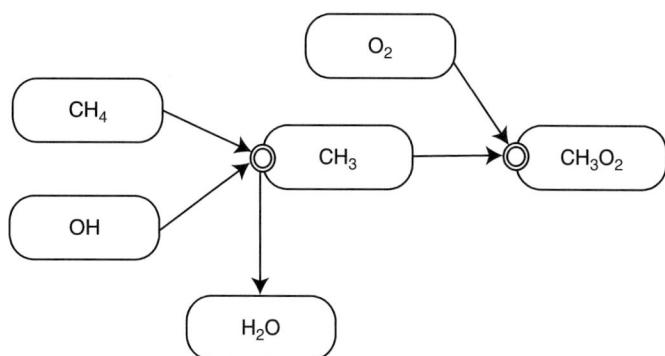

Fig. 6.2.21 Methane degradation processes

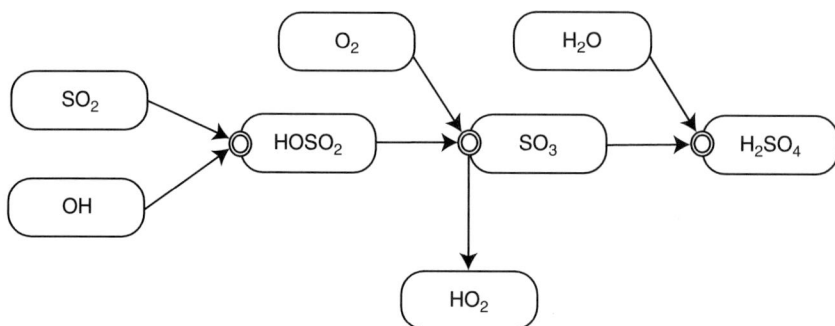

Fig. 6.2.22 Sulphuric acid production process

6.2.2.6 The Sulphuric Acid Processes (H₂SO₄)

Sulphuric acid is a key parameter for the formation and growth of atmospheric aerosols (see Section 6.2.3). In the gas-phase, H_2SO_4 is produced in a process including sulphur dioxide and the hydroxyl radical via Eq. 6.2.34 (Fig. 6.2.22). The main sink term of this acid is the conversion into the particle phase with following wet and dry deposition. In this way the lifetime and atmospheric gas concentration of sulphuric acid depends mainly on the amount of existing aerosols. The chemical reactions are

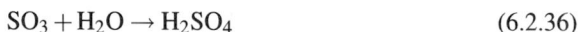

$$SO_2 + OH \rightarrow HOSO_2 \tag{6.2.34}$$

$$HOSO_2 + O_2 \rightarrow SO_3 + HO_2 \tag{6.2.35}$$

$$SO_3 + H_2O \rightarrow H_2SO_4 \tag{6.2.36}$$

6.2.2.7 Tropospheric Chemistry of Biogenic None-Methane-Hydrocarbons

The biosphere, and especially the boreal forest, emits a high amount of volatile organic compounds which even today are not all detected (Goldstein et al., 2004). Most of these compounds are highly reactive in the atmosphere because of their carbon double bounds and are susceptible to attack by O_3, OH and NO_3. By combining the rate constant data for the VOC with typically tropospheric concentrations of O_3, OH and NO_3, we can estimate their atmospheric lifetimes. Isoprene (C_5H_8 – most emitted VOC on a global scale) features a lifetime of less than 2 h through the reaction with OH and at night through the reaction with NO_3 less than 1 h. Most monoterpenes ($C_{10}H_{16}$) show a lifetime regarding OH of a few hours and more than one day concerning O_3. At night their reaction rate with NO_3 leads to lifetimes in the range of minutes to hours. A different situation is to be found for even higher terpenes like sesquiterpenes ($C_{15}H_{24}$). They react exclusively and very fast with ozone and stay in the atmosphere less than a few minutes (SMEAR II: 1–2 min, Lyubovtseva et al., 2005; Bonn et al., 2006).

Isoprene is the most important biogenic hydrocarbon in the atmosphere and reacts with all three oxidants (OH, O_3 and NO_3). The OH-isoprene reaction path is almost entirely understood and proceeds by OH addition to the carbon double bound. The major products of this reaction are formaldehyde (HCHO), methacrolein (MACR = C_4H_6O) and methyl vinyl ketone (MVK = C_4H_6O). The reaction of O_3 with isoprene continues by initial addition of O_3 to the carbon double bound to form in the end the same products compared to the reaction with OH. More complex are the end products of the processes with NO_3, which in the same way starts by adding a NO_3 radical to the C = C double bound. This reaction leads under the presence of NO_x to the formation of nitrato-substituted unsaturated C5-carbonyls.

The reaction schemes of the higher terpenes like mono- and sesquiterpenes are more complex and their reactions are only poor understood even today. The reaction rates of these biogenic hydrocarbons with OH, O_3 and NO_3 have been measured, but comparatively little is known about the distribution of the products. Strong efforts and time is needed to elucidate the atmospheric oxidation mechanisms of biogenic hydrocarbons especially because of their important role in the formation of secondary organic aerosols.

Abbreviations	
CO	Carbon monoxide
CO_2	Carbon dioxide
CH_4	Methane
CH_3	Methyl radical
CH_3O_2	Methyl peroxy radical
C_4H_6O	Methacrolein (MACR)
C_4H_6O	Methyl vinyl ketone (MVK)
O	Oxygen atom in its ground state
O_3	Ozone
$O(^1D)$	Excited oxygen atom

OH	Hydroxyl radical
HO_2	Hydroperoxy radical
HCHO	Formaldehyde
HNO_3	Nitric acid
H_2O_2	Hydrogen peroxide
H_2SO_4	Sulphuric acid
NO	Nitrate monoxide
NO_2	Nitrate dioxide
NO_3	Nitrate radical
RO_2	Organic peroxy radical
VOC	Volatile organic compound

6.2.3 Aerosol Formation and Growth

Miikka Dal Maso[1,2], Ilona Riipinen[2], Tuukka Petäjä[1,3], and Markku Kulmala[2]

[1] ICG-2: Troposphäre, Forschungszentrum Jülich, Germany
[2] University of Helsinki, Department of Physics, Finland
[3] Earth and Sun Systems Laboratory, Division of AtmosphericChemistry, National Center for Atmospheric Research, Boulder, CO, USA

Aerosol particles affect the absorption and reflection of radiation energy in the atmosphere. In this way, they play an important role in the energy balance of the atmosphere (Section 7.2) and in the climate change (Section 10.3.6). Atmospheric aerosol particles can be roughly divided to primary and secondary particles. Primary particles, such as dust, sea salt or pollen, have entered the atmosphere in the condensed phase, whereas secondary particles have been formed from condensable atmospheric vapours. The formation of stable nanosized clusters and particles by condensation of gaseous matter is thus an important source of atmospheric aerosol particles. Such formation often occurs as bursts, called nucleation events, in which new small particles appear in an ambient size distribution and grow to sizes large enough to have climatic importance. Nucleation events have been observed in several different environments around the world, and boreal forests have proven to be an important source of such secondary aerosol particles (see e.g., Mäkelä et al., 1997; Kulmala et al., 2001a, 2004a; Tunved et al., 2006).

An example of a particle formation event taking place at the SMEAR II station in Hyytiälä is shown in Fig. 6.2.23 (see also Section 2.4.4.7 and Mäkelä et al., 2000b). A basic feature of a particle formation event is an increase in particle number, especially in the smallest observable sizes (3–25 nm). The formed particles then grow for several hours until they have reached sizes of 50–100 nm, where they can activate as condensation nuclei for cloud droplets and therefore have a climatic effect. During growth the freshly formed particles coagulate with pre-existing particles, and their number is therefore reduced.

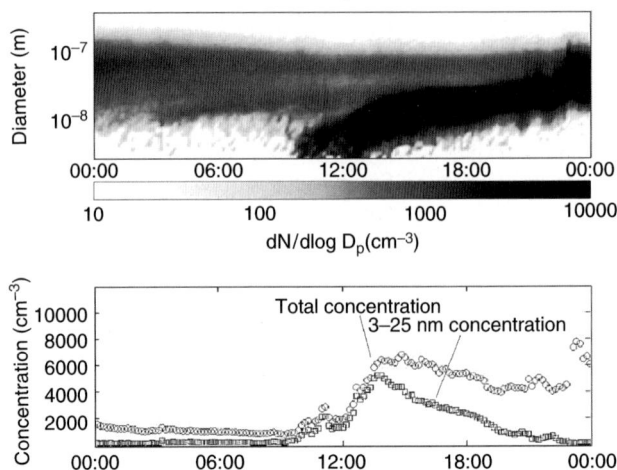

Fig. 6.2.23 Measured aerosol size distribution data during a particle formation event day. The upper panel shows the time evolution of the particle size distribution, with time on the horizontal axis and particle size on the vertical axis. The colour depicts particle concentrations, the darker the colour, the more particles of that size were observed. The lower panel shows the evolution of the total number concentration (circles) and the concentration of the smallest particles (squares). One can see that the increase of the particle number is almost exclusively due to formation of the very smallest particles; later, when the particles grow, the small particle number falls below the total number

If no combustion sources are nearby, the only possible source of such small particles is transformation of atmospheric vapours to the particle phase either by nucleation or condensation onto some very small and thus undetectable clusters. The growth of the particles is caused by condensation of one or several vapours onto the surface of the particles, while the decline in fresh particle number concentration is mostly due to coagulation or changes in meteorological conditions. Some of the particles are also lost by their deposition onto different surfaces (e.g., the ground and canopy), or washed away by precipitation.

The time evolution of the particle size distribution can be mathematically formulated by calculating the sources and sinks that different dynamical processes provide for the distribution (see e.g. Seinfeld and Pandis, 1998).

A schematic picture of the new particle formation and its effects on total aerosol mass and number is presented in Fig. 6.2.24. However, it should be pointed out that the effect of new particle formation on the total aerosol mass is in fact minuscule, as the smallest formed particles have masses that are negligible compared to the larger ones. If we flag each size section in the particle size distribution with a subscript i, the source provided by new particle formation from condensable vapours to the smallest section $i = g^*$ is

$$\frac{dN_{g*}}{dt} = J_0(t) \tag{6.2.37}$$

where N refers to the particle number concentration, t is time, J_0 is the particle formation rate and g^* is the size section to which the particles are formed.

Fig. 6.2.24 A schematic illustrating the new particle formation processes. The particles are formed via nucleation of a condensable vapour. The particle formation increases mainly the number concentration of particles – the increase in the particle mass is usually minuscule. The symbols are introduced in the Fig. 1.2.1

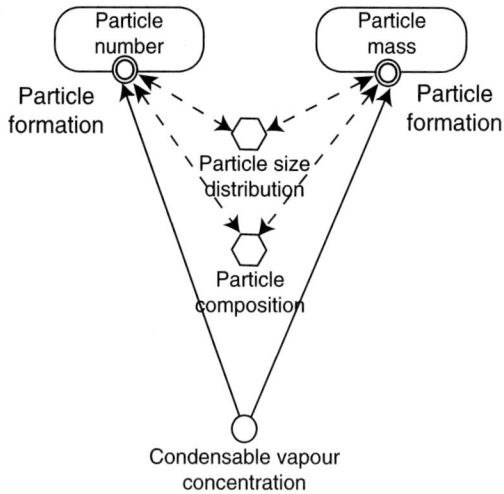

Fig. 6.2.25 A schematic illustrating condensation. In condensation, gas-phase species are transferred to the aerosol phase. The number of aerosol particles is not affected, but the size increases. Therefore, also the total aerosol mass increases

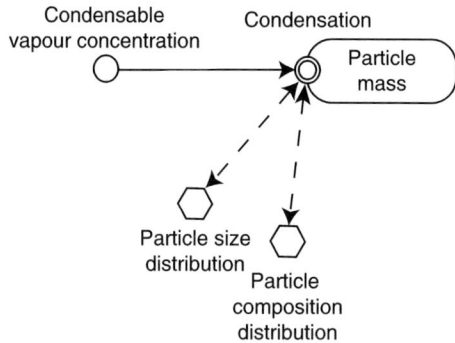

Figure 6.2.25 illustrates the effect of vapour condensation on the total particle mass – total number of particles is naturally not affected by condensation/evaporation as mass is transferred to already existing particles (or vice versa in the case of evaporation). Condensational growth and evaporation to and from the ith section can be described using the condensation and evaporation coefficients p and γ:

$$\frac{dN_i}{dt} = p_{i-1}N_{i-1} - (p_i + \gamma_i)N_i + \gamma_{i+1}N_{i+1} \tag{6.2.38}$$

The first term in the right-hand side of Eq. 6.2.38 describes the growth of particles smaller than i, the second term the condensation/evaporation to/from the i-sized particles and the third term the evaporation of particles larger than size i. The condensation coefficient depends mainly on the concentration and properties

Fig. 6.2.26 A schematic illustrating the coagulation losses of aerosol particles. As the particles collide with each other, the total number concentration is reduced. The total mass of the particles, however, stays constant

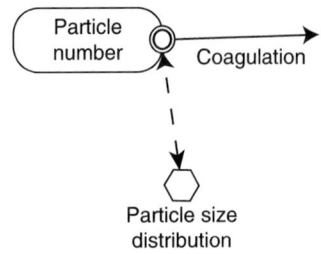

Fig. 6.2.27 The deposition losses of aerosol particles. Both particle number and mass are reduced as the particles are deposited onto existing surfaces or washed away with rain

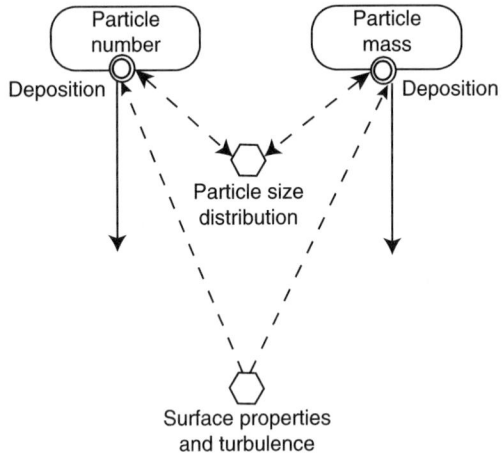

of the condensing vapour, whereas the evaporation coefficient is determined by the properties of particles in the ith section.

Coagulation modifies the number concentration in the section i, on one hand as the particles in the section are colliding with other particles and on the other hand as particles of size i are formed via collisions of smaller ones (see also Fig. 6.2.26):

$$\frac{dN_i}{dt} = C_i^+ - CoagS_iN_i \qquad (6.2.39)$$

In Eq. 6.2:41, C^+ now refers to the particle flux to section i resulting from coagulation, and $CoagS$ is the coagulation sink for the particles in the investigated size class. Coagulation reduces the total number of particles but keeps their total mass constant. The intensity of coagulation depends mainly on the properties of the particle size distribution: collisions between particles of different sizes are more likely to happen as compared with those between particles of similar size.

Particles are removed from the size section i also via their deposition onto existing surfaces (dry deposition, see Fig. 6.2.27) or as they are washed away with rain (wet deposition). Both deposition mechanisms are size specific (Seinfeld and Pandis, 1998). Dry deposition of aerosol particles depends principally on particle size, atmospheric turbulence and stability and the collecting properties of the surface. Wet deposition is affected by rain droplet size distribution, rain intensity and collision efficiency between particles and rain droplets.

6.2.4 Ion Exchange and Retention

Asko Simojoki

University of Helsinki, Department of Applied Chemistry and Microbiology, Finland

Nutrients in the soil solution originate from weathering of soil minerals, decomposition and mineralization of soil organic matter, as well as wet and dry deposition from the atmosphere.

Most nutrients in the soil water exist as ions, i.e., an atom or groups of atoms, which are either positively charged cations such as ammonium (NH_4^+), potassium (K^+), calcium (Ca^{2+}), or negatively charged anions such as nitrate (NO_3^-), nitrite (NO_2^-) and phosphates $(H_2PO_4^-, HPO_4^{2-})$. Soil solution is in a dynamic equilibrium with the contacting particles' surfaces. Soil particles usually have a slightly negative charge that attracts and holds positively charged ions by electrostatic forces on the particle surfaces. They thus form an electrical double layer, where the negative charge is exactly balanced by the positive charges of cations. The distribution of ions in the soil solution surrounding a negative surface is described by the diffuse double layer theory and other electric double layer theories (Singh and Uehara, 1999).

The cations adsorbed on the surfaces can be exchanged into the soil solution by other cations. The exchange occurs on an equivalent charge basis: monovalent cations (with one positive charge) occupy one negatively charged site, and divalent cations (with two positive charges) occupy two sites. The ease of exchange is determined by the charge density of exchange surfaces as well as the valence and size of cations. The higher the charge density of exchange surfaces is, the stronger are the cations adsorbed. Multivalent and small cations are adsorbed more strongly than monovalent and large cations, respectively.

The amount of net negative charge per unit mass of soil is termed the cation exchange capacity. It expresses the sum of positive charges of cations adsorbed on soil particle surfaces. The cation exchange capacity increases with increasing clay and organic matter contents. *Base saturation* is the proportion of cation exchange sites occupied by the so-called basic cations $(Na^+, K^+, Mg^{2+}, Ca^{2+})$ rather than acidic cations (H^+, Al^{3+}). The cations in the soil solution and those adsorbed on soil exchange surfaces are considered easily available to plants. Cation exchange capacity and base saturation thus describe the availability of plant nutrients in the soil.

Soil particle surfaces generally do not retain anions by electrostatic forces, as the anion exchange capacity of soil is usually negligible. However, the hydroxyl groups on the clay mineral edges and oxide surfaces retain anions by specific mechanisms, such as the ligand exchange of phosphate ions. Neutral molecules are adsorbed on the particle surfaces by weak molecular and hydrophobic interactions. In addition, some weakly soluble molecules may precipitate and become part of the solid phase depending on the ion composition of soil solution.

6.3 Vegetation Processes

6.3.1 General

Pertti Hari and Jaana Bäck

University of Helsinki, Department of Forest Ecology, Finland

Cell walls and membranes form a stable and inert structure in which the processes occur. Each process has its specific biologically determined functional substances, i.e., enzymes, membrane pumps and pigments. Usually, several interconnected steps form a process. Each step in the process is driven by one or several functional substances, thus the number of different functional substances needed for a process is large, and it can exceed 100. The functional substances are under the control of the biochemical regulation system and they act in a very coherent way.

Plant structure is formed by large organic molecules, which are not available in the surroundings of the plant. Plants synthesize small organic molecules, such as sugar and amino acids, from simple inorganic molecules and ions, by utilising solar energy. The small organic molecules are transported within a plant to the location of utilization where they are converted to larger molecules which then can be used further in, e.g., growth, protection or reproductive purposes. The biosynthetic processes of small and large organic molecules respond to the environment and to the amount of functional substances (Section 1.2: Basic idea 2) as demonstrated in the Fig. 6.3.1. The synthesis of new molecules is a fast process, often on a time scale of seconds.

The amount of the functional substances changes usually so slowly that we can assume that it is constant during a day. This means, for example, that a model linking photosynthesis and environment should be able to describe photosynthesis well during a day. Consider, as example, CO_2 exchange during spring. Figure 6.3.2

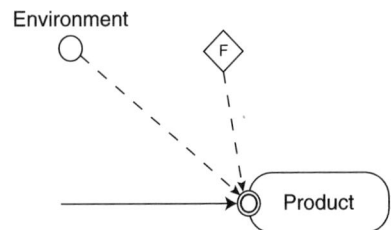

Fig. 6.3.1 The effect of environment and amount of functional substances (F) on a process at short time scale when the involved functional substances i.e. enzymes, membrane pumps or pigments do not change. The symbols are introduced in the Fig. 1.2.1

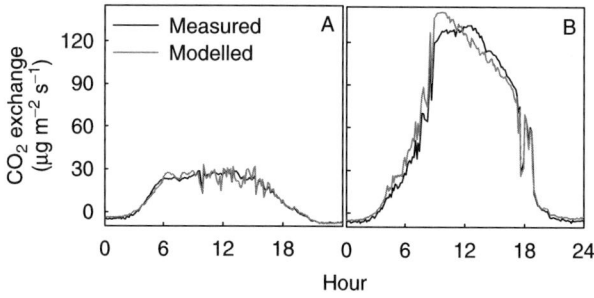

Fig. 6.3.2 Measured (black line) and modelled (gray line) CO_2 exchange of a Scots pine shoot on two clear days with similar temperature, 10 May 2003 (A) and 17 May 2003 (B)

Fig. 6.3.3 The effect of environment and amount of functional substances on process during a prolonged period when the regulation acts on the functional substances of the process as response to the environment. The symbols are introduced in the Fig. 1.2.1

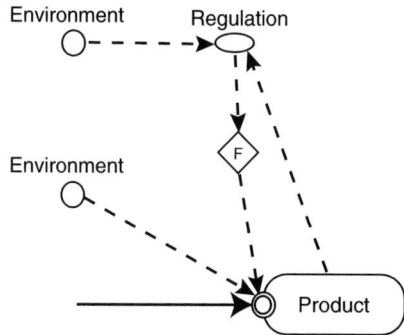

depicts measured and modelled CO_2 exchange of a Scots pine shoot at SMEAR I during two relatively clear days when the tree is recovering from the winter. Although the weather on those days was similar, the level of CO_2 exchange more than doubled during a period of one week. The model (see Section 6.3.3.1.3) is the same in both figures but the value of one parameter has been changed and it is able to capture the variation in CO_2 exchange rather well during both days. The increase in the level of CO_2 exchange reflects changes in the amounts of functional substances and we cannot anymore explain photosynthesis with models including only the effect of environment on the process.

Modern technical systems include nearly always regulation, which means that the properties of the system or its functioning are changed to achieve a predetermined goal. Complicated webs of enzymes and hormones act to regulate the functional substances of processes (Fig. 6.3.3) in plant cells and tissues. This structure is rather similar to the regulators in modern technical systems. This is why we analyse the changes in the amount of functional substances as regulation and use terminology derived from the construction of modern technical systems. Constructed systems including regulation are clearly more efficient than those without it. This explains why complicated webs of interconnected enzymes and hormones have emerged in evolution, which act on functional substances and are similar to regulation in

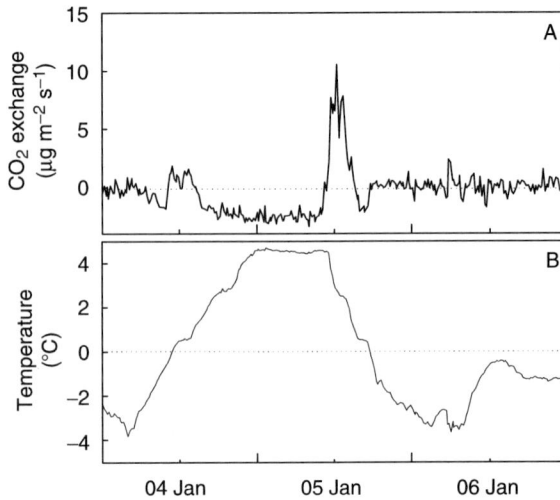

Fig. 6.3.4 Measured CO_2 exchange of a Scots pine shoot (A) and air temperature (B) over three days in January, indicating clear respiration and photosynthesis in winter

technical systems. We refer to these enzymes and hormones as regulation substances. Their general chemical and physiological characterization is presented in the Chapter 5.

The processes response to the prevailing environment and the amount of functional substances changes slowly, often on a time scale of days. The two time scales, simultaneous and delayed, open versatile possibilities for research. We can generate rapid changes in the environment and study the instantaneous responses caused by the action of the functional substances. Then we can assume that the amount of the functional substances in the plant is constant and we see only the simultaneous response. The same separation can also be done when field-monitoring data is analysed. This opportunity will be utilised later in this chapter.

Cold, dark winters and rather warm irradiation rich summers are characteristic for the climate of boreal forests. Regulation systems have developed in their evolution to cope with the very regular alternation of favourable and unfavourable periods during the year (Section 1.2: Basic idea 6). The plants have to be active during the favourable period and they have to tolerate hard winter conditions. Proper timing of active and inactive periods has been crucial for the survival and success of plats in the boreal zone (Section 1.2: Basic idea 7).

The annual cycle is manifested in very many ways in metabolism during a year. Dormant plants can tolerate very low temperatures, often as low as below −40°C or −50°C, and their metabolic activity is minimal. However, clear diurnal pattern of CO_2 exchange, respiration in the night and small photosynthetic CO_2 uptake in the day can be observed even in midwinter if temperature momentarily rises near or above zero (Fig. 6.3.4). Figure 6.3.5 shows that in early spring photosynthesis can be almost totally blocked. The daytime CO_2 exchange of a Scots pine shoot at

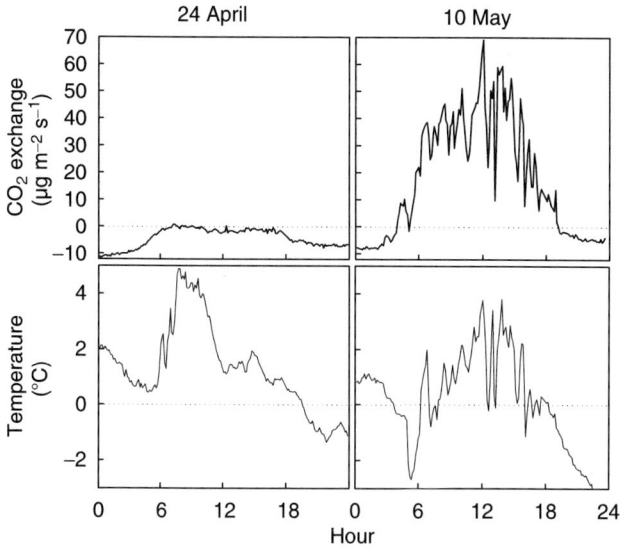

Fig. 6.3.5 Measured CO_2 exchange of a Scots pine shoot (upper panels) and air temperature at SMEAR I (lower panels) on 24 April and 10 May 2001. Photosynthesis is almost totally blocked in April: Photosynthetic CO_2 uptake is lower than release of CO_2 by respiration which results in negative daytime CO_2 exchange. At similar air temperature in May there is strong CO_2 uptake by the shoot

SMEAR I is near zero or even negative during a cool cloudy day in April. Evident photosynthesis can be seen later in May in similar conditions.

Growth reflects very clearly the regulation of the annual cycle. Buds and the meristems producing wood in branches, stems and coarse roots are inactive in winter. They activate simultaneously with the melting of snow and very fast shoot elongation, leaf and wood growth takes place. The growth stops or is strongly reduced soon after midsummer when the conditions are still very favourable. In addition, the fine structure of coniferous wood reflects the annual cycle. Large tracheids are formed in spring and early summer and small ones with very thick cell walls in summer. The cambium in branches, stems and coarse roots is rather inactive in late summer. The amounts of functional substances play a key role in the timing of growth phenomena.

Buds are formed in late summer. In several species, for example Scots pine, the next year's growth of branches and leaves is programmed into the buds. All leaves to be grown on the branch in the next year are present as primordia in the bud of the current year. The biochemical regulation system is evidently able to adjust growth in the coming summer with the resources available, although processes will produce the photosynthates and nutrients for the new tissues nearly a year later than the bud is formed.

The biochemical regulation system reacts also to several irregular changes in the environment, resulting in acclimations of the processes. Photosynthesis and

Fig. 6.3.6 Visualisation of
the interplay between genetic
code, regulation system and
functional substances on a
process. A new symbol for
genetic code is introduced i.e.
symbol of structure having
G in it. Other symbols are
introduced in Fig. 1.2.1

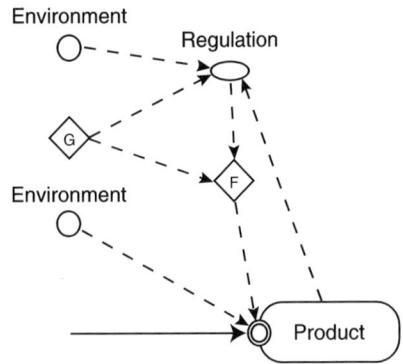

its stomatal regulation change during drought. Energy utilization in light reactions is reduced at high irradiance and the increase of atmospheric CO_2 concentration changes light and dark reactions of photosynthesis. Acclimations are common and they evidently strongly contribute to the success of plants because they have been thoroughly tested during evolution (Section 1.2: Basic idea 8).

The biochemical regulation system is formed by a complex web of regulation substances. The genetic code determines the structure of the web. The regulation system makes operational the information gained during evolution. For example, the annual cycle and acclimations are determined by the genetic code and regulation substances. The adaptation is an additional level in the temporal hierarchy (Fig. 6.3.6), it operates on a time scale of several generations. We use the genetic code as background in this book and we will not treat the control of gene expression in the analysis. For example, currently, more than 100 genes are known to be involved in changes in temperature responses occurring in plant cells (Thomashow, 2001).

6.3.2 Stable Functional Substances

6.3.2.1 Respiration

Jaana Bäck, Pasi Kolari, and Pertti Hari

University of Helsinki, Department of Forest Ecology, Finland

In living cells, there are several processes, such as synthesis of new molecules (Section 6.3.3.1.4) or transport through membranes (Section 6.3.2.8), which need energy. This energy is released from oxidation of energy-rich carbon molecules, such as sugars, and stored in intermediate compounds for use (Fig. 6.3.7). This process, called respiration, also produces many important carbon precursors for cellular metabolism, and releases CO_2 into mesophyll. In general, living

Fig. 6.3.7 Respiratory
process converts ADP to
ATP using carbon compounds
as their source of energy

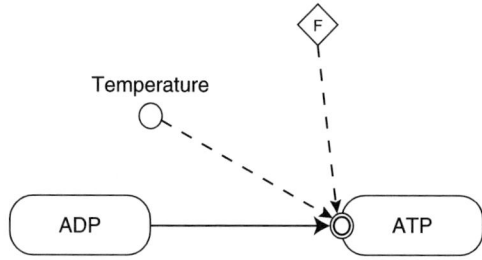

cells constantly need energy provided by ATP, and formation of ATP mainly occurs through respiration. Thus respiration is a characteristic feature of living organism.

Traditionally, a distinction is made between maintenance and growth components of respiration (McCree 1970; Amthor 1994; but see Cannell & Thornley 2000; Kruse & Adams 2008). However, the biochemical processes in both are similar. Growth respiration is associated with formation of new biomass while the maintenance respiration is associated with maintaining the vital functions of already existing biomass. Maintenance respiration in evergreen tissues includes processes related to e.g. protein turnover, seasonal adjustments of lipid and pigment molecules and active transport of molecules between organelles and tissues. Separating growth and maintenance respiration is, in practice, very difficult, since they both occur simultaneously, and all growing organs also contain some mature tissues which use energy only for maintenance. Respiration is the primary mechanism for energy supply in plant tissues (for a comprehensive description of plant respiration, see e.g. Taiz and Zeiger (2002, pp. 223–258) and Amthor (1991). In photosynthesizing tissues of C3 plants, there is also mechanism called photorespiration (reviewed by Ogren, 1984; Noctor and Foyer, 1998; Section 6.3.2.3). Although it eventually leads into CO_2 release analogously to respiration, photorespiration consumes energy from the light reactions of photosynthesis instead of producing energy, and should not be mixed with "true" respiration.

Mitochondria are the main sites for respiration in cells, but also plastids and cytoplasm house some important respiratory reactions. The respiratory processes are organized in a serial manner and can be grouped into glycolysis, citric acid cycle, pentose phosphate pathway (PP-pathway) and oxidative phosphorylation. Glycolysis occurs both in cytoplasm and plastids, as also the PP-pathway, whereas the citric acid cycle and oxidative phosphorylation are located in the mitochondria. Substrates are generated in various cellular processes such as, e.g., starch or sucrose breakdown, and they enter the respiratory pathways at various points. Glycolysis and PP-pathway convert sugars (fructose, glucose, hexose or triose phosphates) into organic acids such as pyruvate, and generate ATP (adenosine triphosphate) and NADH (NADPH) (nicotinamide adenine dinucleotide and the corresponding phosphate). The organic acids are in turn oxidized in the mitochondrial citric acid cycle, and the NADH and $FADH_2$ (reduced flavin adenine dinucleotide) produced are consumed in the oxidative phosphorylation to further generate ATP. Many respiratory intermediates are important substrates for cellular metabolism and maintenance, and they also stimulate or inhibit the respiratory reactions by changing the participating enzyme activities.

During aerobic respiration, the reduced carbon compounds are oxidized, whereby free energy is released and transiently stored in ATP, which then can be used for maintenance, development and growth of the cells. Respiration consumes carbon substrates (glucose, sucrose, hexose phosphates as well as lipids, organic acids or even proteins) to provide energy (ATP) and reducing power (NAD(P)H) for all the processes requiring energy in cells, and in the course of this, CO_2 is released. In general, ATP is needed for phosphorylation of the numerous compounds in the biosynthetic pathways of plant cells, and for signaling (protein kinases) to control the activity of many enzymes and transcription factors. ATP is used also in root N uptake and nitrate reduction, phloem loading, protein turnover, macromolecule biosynthesis and maintenance of ion concentration gradient (e.g., Cannell and Thornley, 2000).

Chemically, energy production and carbohydrate consumption during aerobic plant respiration is

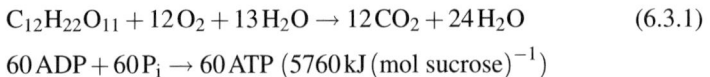

$$C_{12}H_{22}O_{11} + 12O_2 + 13H_2O \rightarrow 12CO_2 + 24H_2O \qquad (6.3.1)$$

$$60\,ADP + 60\,P_i \rightarrow 60\,ATP \; (5760\,kJ\,(mol\;sucrose)^{-1})$$

Respiration can be controlled by both energy demand of the tissues and supply of carbon substrates. Respiration is relatively more sensitive to temperature changes than photosynthesis (Amthor, 1994). In cold environments, the properties of respiratory enzymes are the most important factors limiting respiration, whereas under warmer conditions the substrate and adenylate concentrations control the respiration rate (Atkin and Tjoelker, 2003).

In daytime, cells in leaves are simultaneously photosynthesizing and respiring. The leaf respiration in light is, however, considerably smaller than in the dark, although exact measurements are impossible to obtain and controversy regarding the magnitude still exists (e.g. Hoefnagel et al., 1998; Pinelli and Loreto, 2003). Although the activity of, e.g., pyruvate dehydrogenase, one of the key enzymes in the citric acid cycle, decreases considerably in light, it is clear that even in illuminated leaves the mitochondrial respiration is a major supplier of ATP to the cytosolic processes.

In leaves, respiration is an additional source of CO_2 for photosynthesis. Its contribution to photosynthetic carbon fixation is, however, in many cases small when compared with the uptake of CO_2 through the stomata. If the stomatal pores are nearly closed and the temperature is high, the role of respiration may be important. Maintenance respiration rates of foliage, stem and roots correlate well with their N content, and thus also with their protein content (e.g. Ryan, 1991; Reich et al., 1998). As respiration uses up sugars produced in photosynthesis, the rates of these two processes are related to each other and respiration over longer periods of time is suggested to be proportional to photosynthetic production (Dewar et al., 1998). In

---

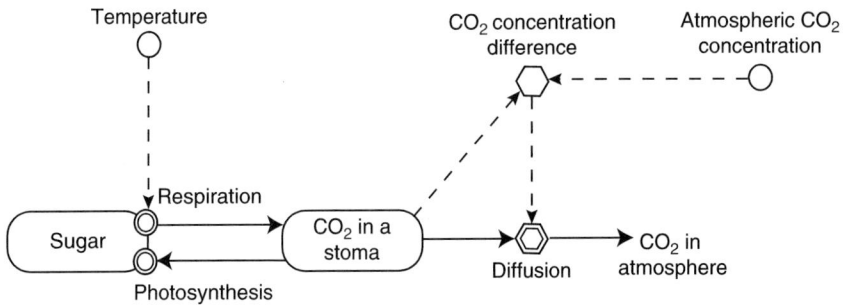

Fig. 6.3.8 The CO_2 fluxes within a stoma. The symbols are introduced in the Fig. 1.2.1

trees the proportion of photosynthates utilised for maintenance respiration locally in leaves is in the order of 20–30% annually (Ryan et al., 1997; see also carbon fluxes and budgets of SMEAR II stand in Sections 7.6.2 and 7.9).

Strong dependence on temperature is characteristic for enzymatic reactions, and rate of respiration in plant tissues is often described as an exponential function of temperature. Let r denote respiration rate and T temperature. Thus

$$r = r_o Q_{10}^{T/10} \qquad (6.3.2)$$

where r_o is respiration at 0°C and Q_{10} temperature sensitivity or the increase in respiration rate for every 10°C. In biological processes Q_{10} is often about 2 at ambient temperatures between 0°C and 30°C (reviewed by Atkin and Tjoelker, 2003).

The production of CO_2 is proportional to the ATP formation. Carbon dioxide produced by respiration diffuses out of the respiring tissue (Fig. 6.3.8). The rate of respiration can thus be determined from gas exchange measurements. If there are other CO_2 exchange processes, such as photosynthesis by the foliage, respiration must be extracted from the measured net CO_2 exchange. Night time CO_2 exchange directly gives the rate of respiration, but respiration in light must be determined indirectly. This is normally done by parameterising temperature regression of night time fluxes and extrapolating that regression to daytime.

Figure 6.3.9 illustrates the instantaneous temperature response of respiration of a Scots pine shoot over three nights. The measuring noise is superimposed on the exponential temperature response. The large noise dominates the residuals and hinders seeing small systematic differences between measured and modelled values.

Conclusions

ATP produced in respiration is the source of energy in the cells of living organisms. The rate of respiration follows temperature approximately exponentially. The noise in our measurements hampers seeing small discrepancies between the theoretical exponential temperature relationship and measured respiratory CO_2 efflux.

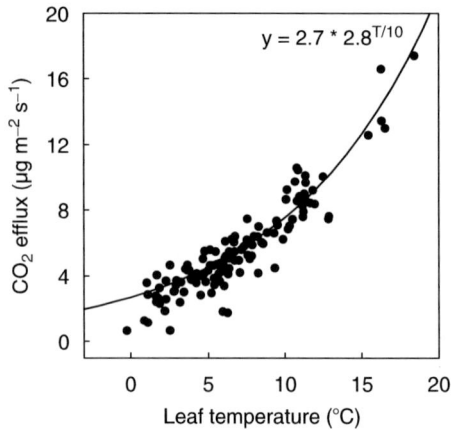

Fig. 6.3.9 Temperature response of respiration (night time CO_2 exchange) in a Scots pine shoot. The data was measured during three nights in May 2007 at SMEAR II

6.3.2.2 Transpiration

Pasi Kolari, Jaana Bäck, Martti Perämäki, and Pertti Hari

University of Helsinki, Department of Forest Ecology, Finland

Photosynthesis requires water in formation of sugars, and in general all biochemical processes in the plant cells take place in a liquid phase. Hydrostatic pressure of water in the cells, called turgor pressure, is vital to the cell function. In many plants turgor pressure is also the means of retaining the leaves in shape ("hydro-skeleton"). Water is also the medium used for transporting essential nutrients from soil up to the canopy and the assimilates from source organs to the sinks. The flow of water from the soil into the leaves is driven by evaporation from the leaf. The surface tension of the water film between parenchyma cells inside the leaf draws upwards the water columns that extend from the leaves through the water conducting tissues into the soil (Section 7.5.2.2).

Mesophyll cells in leaves are arranged loosely, so that they form air spaces underneath the leaf surface. There is a thin water film on the mesophyll cell surfaces (apoplastic liquid), and water evaporates from these surfaces to the intercellular space. Water vapour diffuses out of the leaf through stomatal pores that cover about 1% of leaf area when fully open (Wilmer and Fricker, 1996; Morison and Lawson, 2007). This diffusive flux of water vapour from the intercellular space through the stomata is called transpiration (Fig. 6.3.10).

The stomatal pores enable diffusion of water vapour into the atmosphere and simultaneous diffusion of CO_2 into the mesophyll. CO_2 molecules diffused into the stomatal cavity dissolve into the water laying on the cell surfaces and further continue their journey deeper into the cells. Since the diffusion of carbon dioxide and water share the same route through the stomatal aperture, transpiration and photosynthesis are strongly coupled with each other.

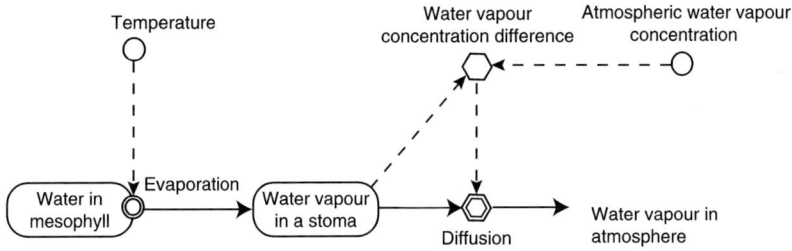

Fig. 6.3.10 Transpiration as combination of evaporation and diffusive transport. The symbols are introduced in the Fig. 1.2.1

Water vapour in the substomatal cavity is saturated since nearly all cell walls are covered with a thin water layer and the stomatal pore is small compared with the surface area of mesophyll cell walls. The water vapour concentration in the ambient air is usually clearly lower than that in the stomatal cavity. The concentration difference generates diffusive flow out from the leaf. The concentration of saturated water vapour in the air depends on air temperature approximately exponentially. Several different forms of exponential or logarithmic equations have been derived to describe the relationship between temperature T and the saturation concentration of water vapour C^s_{H2O} over a surface of liquid. For example, Preining et al. (1981) derived temperature relationship of saturation pressure which can be converted to concentration by multiplying with molecular mass M_{H2O} and dividing with the universal gas constant R and temperature in Kelvin

$$C^s_{H2O} = \frac{M_{H2O}e^{77.34 - \frac{7235}{T} - 8.2\ln T + 0.005711T}}{RT} \qquad (6.3.3)$$

Diffusion transports water vapour out from the substomatal space. Let g_{H2O} denote transpiration rate and D_{H2O} water vapour concentration difference between stomatal cavity (C^s_{H2O}) and ambient air (C^a_{H2O}). Thus

$$D_{H2O} = C^s_{H2O} - C^a_{H2O}.$$

The diffusive flux is proportional to the concentration difference, thus

$$g_{H2O} = c_{H2O}\, D_{H2O} \qquad (6.3.4)$$

where c_{H2O} is a parameter called stomatal conductance. Stomatal conductance is related to how open the stomata are, its maximum value, c_{max}, corresponding to fully open stomata. The value of c_{max} in turn depends on the size of the stomatal pores and on stomatal density. These can vary greatly among different plant species, among individuals of the same species growing in different conditions, and even in different parts of one individual (e.g., sun and shade leaves of a tree).

Transpiration can be measured following the same principles as in measuring CO_2 exchange, by detecting the change in water vapour concentration inside a chamber enclosing a shoot or leaves (Section 2.4.3.8). The measuring accuracy is clearly lower though because water tends to adsorb on surfaces. The thickness of the adsorbed water film depends on relative humidity. At high relative humidity (typically above 70%) the amount of water on the inner surfaces of the measurement chamber and the sample tubing surface increases very steeply when RH increases. Eventually most of the water transpired by the plant is adsorbed by the chamber and the tubes and does not reach the gas analyser at all. The adsorption can be overcome to some extent by heating the analysers, tubing and chamber walls to decrease relative humidity over the surfaces, but the natural conditions are lost with climate control. When calibrating the measuring system at SMEAR II, we have noticed that the absolute level of the systematic underestimation in the observed transpiration at RH range of 30–70% varies between 15% and 40% depending on chamber configuration, dirt on the chamber walls etc., however on a time scale of a few days the temporal patterns are fairly reliable. At high RH ($>75\%$) the data has to be interpreted cautiously and may even need to be rejected.

The measured daily transpiration pattern shows that, with moderate water vapour concentration differences between the stomatal cavity and ambient air, the model (6.3.4) gives a similar diurnal pattern of transpiration as the measurements (Fig. 6.3.11). If the concentration difference is high, the model assuming constant stomatal conductance overestimates the measured transpiration. Evidently, the stomatal action reduces transpiration at high water vapour saturation deficit. When the optimal stomatal regulation (Section 6.3.2.3) is considered, the agreement between observed and modelled transpiration is fairly good.

When determining the value of stomatal conductance from transpiration, the uncertainty in leaf temperature also decreases the accuracy because water vapour deficit depends on temperature as can be seen from Eq. 6.3.3. Convective heat transfer into the ambient air tends to keep the leaf temperature near the air temperature, but solar radiation heats up the leaf and transpiration cools it down (Section 6.2.1). Most of the time the leaf temperature is different from ambient air temperature.

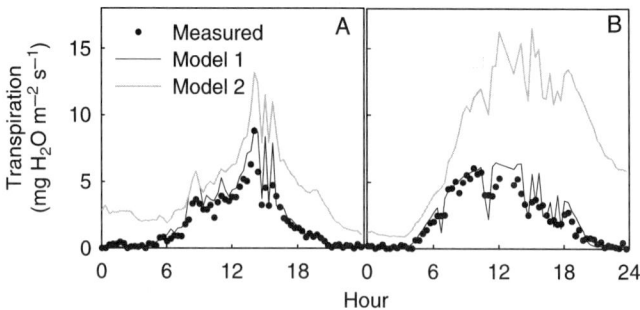

Fig. 6.3.11 Diurnal course of measured (dots) and modelled transpiration (lines) in Scots pine shoot on a cloudy day (A) and on a sunny day (B). Model 1 (black line) is calculated transpiration under optimal stomatal regulation (Section 6.3.2.3). Model 2 (grey line) is transpiration calculated with Eq. 6.3.4 assuming constant stomatal conductance (equal to conductance when stomata are fully open)

At high solar irradiances the leaf temperature can be a couple of degrees above the air temperature, whereas at low irradiances and low wind speed the leaf can cool down below the air temperature. Thus, at a fixed value of stomatal conductance, transpiration is higher in high light than in low light due to higher water vapour concentration difference between the substomatal cavity and free air.

Conclusions

Transpiration is driven by saturation deficit of water vapour concentration. At high saturation deficit the stomata restrict the rate of transpiration in a systematic fashion, the observed stomatal behaviour agrees with the theory of optimal stomatal control. Systematic errors in transpiration measurements disturb determination of transpiration rate and evaluation of the theory.

6.3.2.3 Photosynthesis

Pertti Hari, Pasi Kolari, Jaana Bäck, Annikki Mäkelä, and Eero Nikinmaa

University of Helsinki, Department of Forest Ecology, Finland

Carbon is the dominating element in living organisms: about 50% of plant dry weight is carbon. Photosynthesis is the most important single process in a plant, and the main process governing the plant carbon content. In photosynthesis, radiation energy is converted into chemical form as sugars using atmospheric CO_2 and water originating from soil. Sugars formed in photosynthesis are the dominating source of raw material for synthesis of macromolecules (Section 5.2.1), as well as the source of energy for metabolism, obtained via respiration (Section 6.3.2.1). Thus understanding the effect of environment on the photosynthetic process is the key to linking climate in the boreal zone with the growth and development of ecosystems as well as to understand the effects of climate change on boreal forests (Section 10.2.3).

In C-3 plants dominating the northern ecosystems, photosynthesis occurs in chloroplasts in mesophyll cells (Fig. 5.2.2). The process itself is a complicated chain of separate reactions each of them having its specific functional substances. There are two main reaction complexes in photosynthesis. The light reactions occurring in thylakoids (Fig. 5.2.1) of chloroplasts convert solar radiation energy into chemical form as ATP and NADPH; in addition, water is split and oxygen is released. Carbon reactions (Calvin cycle) in chloroplast stroma join a carbon atom to five-carbon sugar utilising energy released from ATP and NADPH. They were traditionally called 'dark reactions' to distinguish them from the photosynthetic light capture, although the reactions actually are occurring in light and are directly affected by light.

Photosynthetic light reactions comprise a series of reactions starting from excitation of chlorophyll molecules by light energy. Almost all reactions take place in four integral thylakoid membrane protein complexes; photosystem II, cytochrome B_6F complex, photosystem I and ATP synthase. Photosystem I (PS I) prefers far-red light (>680 nm), whereas Photosystem II (PS II) preferentially absorbs red light of 680 nm. Another difference is that PS I produces a strong reductant in the stroma, capable of reducing $NADP^+$, and a weak oxidant, whereas PS II produces a very strong oxidant, capable of oxidising water in the thylakoid lumen, and a weaker reductant, re-reducing the oxidant produced by PS I. The end products of photosynthetic light reactions are ATP and NADPH, which are subsequently used in the Calvin cycle, and O_2 which is liberated from the chloroplast into the atmosphere.

The two photosystems are chemically and physically distinct, having their own antenna pigment complexes and photochemical reaction centers. Light harvesting complexes surround the photosystem I and II reaction centers and serve as an antennae (e.g., Malkin and Fork, 1981), gathering the light quanta into the reaction center. The antenna complexes contain chlorophyl a and b molecules and carotenoids, and they aggregate around the periphery of the photosystem reaction center complexes. The membrane-bound pump, ATP synthase, converts ADP to ATP by moving protons from the lumen back into the stromal side.

After the light energy has been converted into chemical energy, it can be used in Calvin cycle reactions in plastid stroma to reduce carbon dioxide into small carbohydrate molecules. These reactions are rather independent of incident light intensity. In the Calvin cycle, a five-carbon molecule (RuBP, ribulose 1,5-bisphosphate) is bound with CO_2 in a reaction catalyzed by Rubisco enzyme (ribulose bisphosphate carboxylase/oxygenase; see Section 5.2.1), to form two molecules of three-carbon intermediates (3-phosphoglycerate, 3-PGA). The 3-PGA molecules are phosphorylated by ATP and the resulting diphosphoglycerate (DPGA) is reduced by NADPH to glyceraldehyde 3-phosphate (G3P). After several enzymatically catalyzed reactions, the resulting pentose phosphates are finally converted to ribulose 5-phosphate (Ru5P), and phosphorylated to RuBP. The whole cycle consumes nine molecules of ATP and six molecules of NADPH in the formation of one triose phosphate molecule (see chemical equation below). Triose phosphates are either used for storage starch biosynthesis in chloroplasts, or exported into cytoplasm in exchange to inorganic phosphorus by a triose phosphate pump, and converted into sucrose, which is the main soluble sugar compound in photosynthetizing cells. Starch is the main carbohydrate storage compound in most plants, whereas sucrose is the compound commonly translocated in phloem to be used in various organs for metabolic processes. The synthesis of sucrose and starch draws one carbon molecule out of six CO_2 molecules that are fixed under conditions of continuous CO_2 uptake, thus maintaining a constant flow of carbon through a Calvin cycle.

In chemical notation we get (Lawlor, 1993):

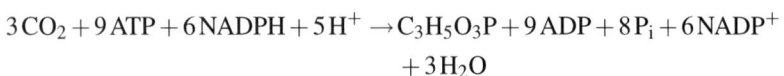

$$3\,CO_2 + 9\,ATP + 6\,NADPH + 5\,H^+ \rightarrow C_3H_5O_3P + 9\,ADP + 8\,P_i + 6\,NADP^+ \\ + 3\,H_2O$$

Rubisco is a non-specific enzyme catalysing also O_2 binding in the same catalytic site of the enzyme as the CO_2 molecule. In low mesophyll CO_2, the enzyme prefers O_2 as a substrate instead of CO_2. The oxygenation reaction (start of the reaction chain called photorespiration) of RuBP with O_2 produces one three-carbon and one two-carbon intermediate. Photorespiration is a complicated pathway which involves synchronized cooperation and close interaction between plastids, peroxisomes and mitochondria, and eventually results in the net loss of 25% of fixed carbon in a decarboxylation reaction in mitochondrion at ambient concentrations and at 25°C. The balance between carboxylation and oxygenation reactions in plastids depends on the relative concentrations of CO_2 and O_2 at the Rubisco catalytic site, prevailing temperature and the kinetic properties of Rubisco.

The availability of light energy often limits the photosynthetic process. At low light, the photosynthetic rate increases almost linearly with increasing photon flux density. In addition, the low CO_2 concentration in the mesophyll reduces photosynthesis, especially at high photon fluxes. Diffusion transports CO_2 molecules from the atmosphere into the sub-stomatal cavity through the stomatal pores. The carbon dioxide molecules dissolve into the water on the mesophyll cell walls, forming bicarbonate ions. Thereafter the bicarbonate ions are transported into the plastids of mesophyll cells. As a result the relationship between light and photosynthetic rate is saturating due to low availability of CO_2 at high light intensities.

Photosynthesis and transport of photosynthates require flow of water from soil into leaves. This water flow is driven by transpiration (see Section 6.3.2.2). Uncontrolled transpiration would be, however, often so large that the water transport system would be unable to provide enough water to maintain the leaf water status needed for cell function and the maintenance of turgor. Therefore, a control system of the opening of stomatal pores has developed through evolution. At high water vapour deficit it is beneficial for the plant to restrict water loss at the cost of reduced carbon gain, whereas at low water vapour deficit the evaporative demand is low and stomata can be kept open to enhance CO_2 uptake. Besides controlling water loss, the stomata allow diffusion of CO_2 into the mesophyll in a way that maintains mesophyll CO_2 concentration at a level needed for photosynthetic CO_2 fixation. At low light the internal CO_2 concentration tends to increase due to low consumption of CO_2 by photosynthesis. In such conditions keeping the stomata fully open does not give any increase in CO_2 fixation because the availability of energy from the sun is the limiting factor. Stomata tend to close, at least partially, in darkness. This can be seen in the diurnal course of measured transpiration (Fig. 6.3.11). Thus stomata are operating both to limit water loss and maintain internal CO_2 concentration.

Partial closure of the stoma decreases CO_2 concentration in the stomatal cavity, because inward diffusion is restricted. The increased concentration difference enhances the driving force of diffusion through the stomatal aperture and compensates to some extent for the reduction in diffusivity through stomatal pores due to closure. When the photosynthetic and respiratory processes and transport by diffusion is combined, we get the flow of carbon from atmosphere to sugars (Fig. 6.3.12). The next step in the analysis is to derive exact descriptions for the processes and transport described in the figure.

Diffusion transports CO_2 into and out from stomatal cavity depending on whether concentration is lower or higher than in the outside air. Let g_{CO2} denote the diffusive flow through a stomatal pore, C_{CO2}^s the CO_2 concentration in the stomatal cavity and

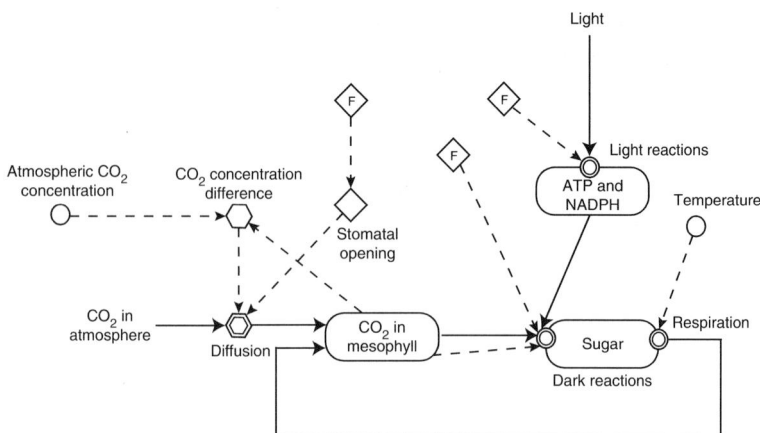

Fig. 6.3.12 The role of biochemical processes and transport in photosynthesis: ATP is produced in light reactions and utilised in dark reactions to form sugar utilising atmospheric CO_2. Diffusion transports CO_2 into mesophyll. The symbols are introduced in the Fig. 1.2.1

C_{CO2}^a ambient CO_2 concentration. The diffusive flow is proportional to the concentration difference, thus

$$g_{CO2} = c_{CO2}(C_{CO2}^a - C_{CO2}^s) \tag{6.3.5}$$

where c_{CO2} is a parameter, frequently called stomatal conductance.

The energy captured in light reactions is used to form sugars utilising CO_2 from the stomatal cavity as raw material. Photosynthesis involves two phases; (i) low light phase when photosynthetically active radiation determines photosynthesis and (ii) high light phase when the low CO_2 concentration in mesophyll reduces photosynthesis. Let p denote the photosynthetic rate and I_{PAR} the photosynthetically active radiation flux. Assume that the photosynthetic rate is proportional to the product of CO_2 in the stomatal cavity and to a function of photosynthetically active radiation, f(I). Thus

$$p = \beta \, f(I_{PAR})C_{CO2}^s \tag{6.3.6}$$

where β is a parameter, called efficiency of functional substances. The function f is

$$f(I_{PAR}) = \frac{I_{PAR}}{I_{PAR} + \gamma} \tag{6.3.7}$$

where the parameter γ introduces saturation of light reactions. This kind of saturating light response function is frequently used for describing the relationship between photosynthetically active radiation and photosynthetic rate.

Diffusive flux between the stomatal cavity and ambient air, photosynthesis and respiration change the CO_2 concentration inside the leaf. Now we can apply the general principle of conservation of mass to construct a dynamic model of a mass in a container, as treated in Chapter 2, Eq. 2.2.3. The amount of CO_2 in the stomatal cavity is concentration multiplied by the volume of the cavity, V_i. The differential equation for the CO_2 in the cavity is

$$\frac{d(V_i C_{CO2}^s)}{dt} = A_s \left(c_{CO2}(C_{CO2}^a - C_{CO2}^s) - \beta \ f(I_{PAR})C_{CO2}^s + r \right) \qquad (6.3.8)$$

where A_s is the leaf area which a stoma feeds with CO_2.

Assume that the CO_2 concentration in the cavity is in steady state, which means that the concentration does not change. Then

$$\frac{d(V_i C_{CO2}^s)}{dt} = 0 \qquad (6.3.9)$$

Thus

$$A_s \left(c_{CO2} (C_{CO2}^a - C_{CO2}^s) - \beta \ f(I_{PAR})C_{CO2}^s + r \right) = 0 \qquad (6.3.10)$$

When the above equation is solved for C_{CO2}^s, we get

$$C_{CO2}^s = \frac{c_{CO2} \ C_{CO2}^a + r}{c_{CO2} + \beta \ f(I_{PAR})} \qquad (6.3.11)$$

When the obtained dependence of C_{CO2}^s, on light is introduced in Eq. 6.3.6 we get a model for photosynthesis:

$$p = \beta \ f(I_{PAR}) \frac{c_{CO2} C_{CO2}^a + r}{c_{CO2} + \beta \ f(I_{PAR})} \qquad (6.3.12)$$

The above model assumes that diffusivity from the atmosphere through stomatal pores into the stomatal cavity does not change. The stomatal action that reduces diffusion is, however, evident during high evaporative demand.

Cowan (1977) and Cowan and Farquhar (1977) introduced the idea of optimal stomatal control. They were, however, unable to solve the proposed optimisation problem. We introduced the above described additional assumptions to derive a simple model for photosynthesis and this enabled solution of the optimisation problem with a pen and a piece of paper. Our formulation (Hari et al., 1986) is as follows.

Stomatal closure reduces the diffusion of CO_2 and water vapour into and out from a leaf. The parameter c_{CO2}, called stomatal conductance for CO_2, takes this reducing effect in consideration in our model. We introduce the concept "degree of stomatal opening", u, to describe the reduction in diffusion. We assume

$$c_{CO2} = u c_{max} \qquad (6.3.13)$$

where c_{max} is stomatal conductance for CO_2 when the stomata are fully open. The degree of stomatal opening can take values only between 0 and 1. When u is introduced into Eq. 6.3.13 we get

$$p = \beta \ f(I_{PAR}) \frac{u c_{max} C_{CO2}^a + r}{u c_{max} + \beta \ f(I_{PAR})} \qquad (6.3.14)$$

Light intensity and the degree of stomatal opening vary during a day, and photosynthesis reflects this variation. We get

$$p(t) = \beta \, f(I_{PAR}(t)) \frac{u(t)c_{\max} C_{CO2}^a + r}{u(t)c_{\max} + \beta \, f(I_{PAR}(t))} \tag{6.3.15}$$

CO_2 is dissolved into the water on mesophyll cells in the stomatal cavity. This generates a close coupling between photosynthesis and transpiration, since water molecules evaporate, the stomatal cavity is saturated with water vapour, and consequently the vapour concentration is higher in the cavity than in the surrounding air. Thus photosynthesis is impossible without transpiration.

The stomatal action has two conflicting goals, (i) to maximise photosynthetic production and (ii) minimise amount of transpiration. These goals can be combined by assuming that transpiration of 1 kg of water requires λ kg CO_2 for formation and maintenance of water uptake and transport capacity. The optimal stomatal control can now be formulated as maximisation of photosynthetic production minus transpiration costs in CO_2.

Let g_{H2O} denote transpiration rate and D_{H2O} the water vapour concentration difference between stomatal cavity and ambient air (Section 6.3.2.2). The diffusion out from the leaf is analogous to the diffusion of CO_2 into the cavity. As water molecules are lighter than CO_2 molecules, diffusivity of water vapour is higher than diffusivity of CO_2. Stomatal conductances for H_2O and CO_2 are related to each other by factor $a = 1.6$. Thus

$$g_{H2O} = ac_{CO2}D_{H2O} = auc_{\max}D_{H2O} \tag{6.3.16}$$

When we introduce the changing stomatal regulation and concentration difference we get

$$g_{H2O}(t) = au(t) \, c_{\max} D_{H2O}(t) \tag{6.3.17}$$

The optimisation problem is to find the best possible pattern of the degree of stomatal opening u(t), that maximises the photosynthetic production minus transpiration costs during a prolonged period from t_1 to t_2. The solution can be found with calculus of variations using rather old mathematical tools.

Let $P_P(t_1, t_2)$ denote photosynthetic production during the interval from t_1 to t_2 and $G_{H2O}(t_1, t_2)$ the amount of transpiration during the same interval. They are obtained by integration (Section 7.1)

$$P_P(t_1, t_2) = \int_{t_1}^{t_2} p(t)dt \tag{6.3.18}$$

and

$$G_{H2O}(t_1, t_2) = \int_{t_1}^{t_2} g_{H2O}(t)dt \tag{6.3.19}$$

The optimisation problem can now be formulated in an exact form:

$$\underset{u}{Max} \left\{ P_P(t_1, t_2) - \lambda G_{H20}(t_1, t_2) \right\} \tag{6.3.20}$$

The above optimisation problem can be solved with the Lagrange method, which is rather commonly used in systems analysis and economics. As a first step the Lagrange function, L, is formed:

$$L = \beta \, f(I) \frac{uc_{max} C_a + r}{uc_{max} + \beta \, f(I)} - \lambda \, uc_{max} D_{H2O} \qquad (6.3.21)$$

The next step is to differentiate the Lagrange function with respect to u. The optimal stomatal opening during the interval from t_1 to t_2 is obtained as a zero point of the derivative of the Lagrange function.

The derivative of the Lagrange function with respect to u includes u as first and second power. Thus it can be solved resulting in

$$u^* = \left(\sqrt{\frac{C^a_{CO2} - r/(\beta \, f(I))}{\lambda \, a \, D_{H2O}}} - 1 \right) \frac{\beta \, f(I)}{c_{max}} \qquad (6.3.22)$$

The above solution yields values that are greater than 1 or negative which are outside the range of degree of a stomatal opening. In these cases, u takes value 1 or 0. The final solution is

$$u = \begin{cases} 0, & if \, u^* \leq 0 \\ u^*, & if \, 0 < u^* < 1 \\ 1, & if \, u^* \geq 1 \end{cases} \qquad (6.3.23)$$

Figure 6.3.13 illustrates the modelled behaviour of stomatal conductance and CO_2 exchange in more detail.

The light intensity and water vapour concentration difference varies during a day and the degree of stomatal opening responds to this variation. Since the water vapour concentration difference is low in the mornings, the stomata are usually fully open after sunrise and they partially close if the day is sunny and the water vapour concentration difference is large. These variations in the light and water vapour concentration difference generate strong daily patterns in photosynthesis.

We face the fundamental problem of scales when arranging measurements to test a model with field data. The model deals with photosynthesis in a small space element during a short interval, or at a point in space and time. The present field instrumentation is able to measure photosynthesis at shoot or leaf level during a prolonged time. If the environmental factors inside the measuring chamber remain constant during one chamber closure period, we can consider to obtain information on the shoot or leaf gas exchange at space and time element level. We installed the chambers on top branches in the stand at SMEAR I in order to eliminate the spatial variation in photosynthetically active radiation caused by shading. In addition, the variation in PAR falling on needle elements was reduced by bending the needles in a plane. The light sensor was installed parallel to the needle plane. These practical arrangements reduced considerably the problem of environmental factors changing during one measurement and within the object being measured.

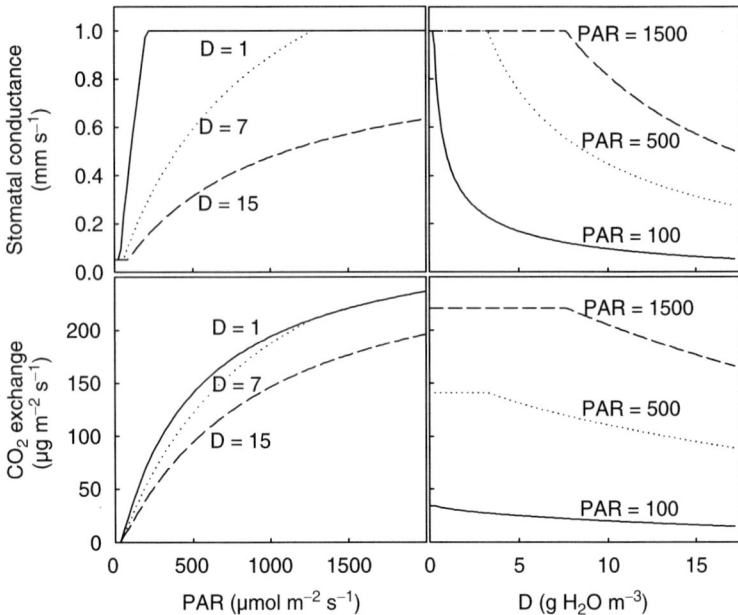

Fig. 6.3.13 Modelled stomatal conductance and CO_2 exchange at different combinations of light (PAR, μmol m^{-2} s^{-1}) and water vapour concentration deficit (D, g H$_2$O m^{-3}). Note how the large differences in stomatal conductance at different values of D (top left) result in relatively small changes in photosynthetic light response (bottom left)

The monitoring of CO_2 exchange, transpiration, photosynthetically active radiation, temperature and water vapour concentration at SMEAR stations provide informative data to test the optimal stomatal control model of photosynthesis and to determine its prediction power. The data set of over one decade is unique worldwide and enables a thorough analysis of photosynthetic CO_2 fixation over wide variety of different environmental conditions, including wintertime and occasional drought periods. Also the quality of the data has been analysed, i.e., the measuring noise level has been determined and also systematic measuring errors of the whole gas-exchange measuring setup have been evaluated for CO_2 (Hari et al., 1999) and H$_2$O (Kolari et al., 2004a).

We use the normal statistical least squares method for estimating the values of the model parameters. Thus we determine those parameter values that give the smallest sum of the second power of residuals. If the explaining factors are independent of each other and if the parameters in the model do not compensate each other, then the estimation of the parameter values is a straightforward task. However, in the case of our model the explaining variables (solar radiation, leaf temperature and water vapour deficit) are causally connected with each other, and the structure of the model is such that the parameters also compensate each other. This means that there is an infinite number of different combinations of estimated parameter values that all give the sum of squared residuals close to the global minimum. This leads to instability in numerical estimation and indicates that the model may be overparameterised. The parameter estimation procedure, especially finding the value of λ, can be made more robust by using also transpiration measurements because they give more direct information on stomatal conductance than CO_2 exchange. The data must be weighted so that the sum of squared residuals is roughly equal for CO_2 exchange and transpiration. Obtaining accurate and precise parameter values would ultimately require breaking the correlations between the explaining variables. In field measurements this is unfortunately not possible.

We selected gas exchange measurements on one Scots pine shoot at SMEAR I in the summer of 2004 for the first test of the optimal stomatal control model. First we selected a period of five days from the middle of July for parameter estimation. The estimation procedure was somewhat unstable indicating correlated parameters. A more stable parameter set could be found when transpiration measurements were also introduced. The problem in those is the systematic error in the measurement and that the data must be rejected when air humidity is high. The fit of the model for the test period was good: the proportion of explained variance in CO_2 exchange over the five-day period was 95.1%. The model could reproduce well the daily measured pattern as well as the short term variation in CO_2 exchange caused by movement of clouds.

When judging the model's performance, we must be aware that the parameter estimation improves the model fit because good fit itself is the estimation criteria. For understanding behaviour of a system, it is more important how well the model can predict it in the future. The proportion of explained variance is too large, i.e. biased, when the parameters are estimated with the same dataset. This problem is evident in regression analysis when the number of explaining factors is large when compared with that of measurements. This bias in proportion of explained variance may disturb the conclusions. A more appropriate criterion for the model's goodness is the proportion of explained variance when the model parameters are not estimated by fitting from the same data set. If we take a new dataset, which is not utilized in the parameter estimation, and predict the system behaviour, we get a better idea of the model's performance. To stress the difference with the normal statistical practice, we speak about prediction power instead of proportion of explained variance.

We predicted the CO_2 exchange for the period for all of July using the parameter values estimated previously from the measurements in five days of July. The five-day period was excluded in the analysis. The fit of the model is demonstrated for sunny and cloudy days in Fig. 6.3.14. The model is able to predict properly the daily pattern of photosynthesis, without exception. There is, however, a slight tendency for the prediction to be too low on sunny days and too high on cloudy days. The predictive power of the model is 94.6%, compared to the proportion of explained variance of 95.1% in the fitted five-day data and 97.9% when the model was fitted daily over the whole July (Fig. 6.3.14).

We then expanded the analysis by introducing more shoots and years. Once the shoot-specific parameters β, λ, γ and c_{max} were estimated for a five-day period, the model was able to predict CO_2 exchange over July with the proportion of explained variance of 87–97% (Fig. 6.3.15).

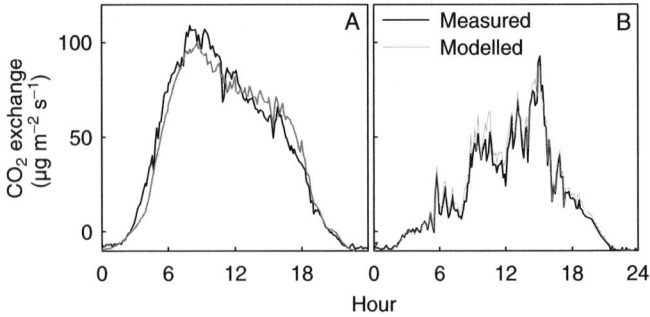

Fig. 6.3.14 Measured and predicted CO_2 exchange of a Scots pine shoot during a sunny day (27 July, panel A) and a cloudy day (28 July, panel B) in 2004 at SMEAR I. The model parameter values were estimated from data measured during the period of 19–23 July

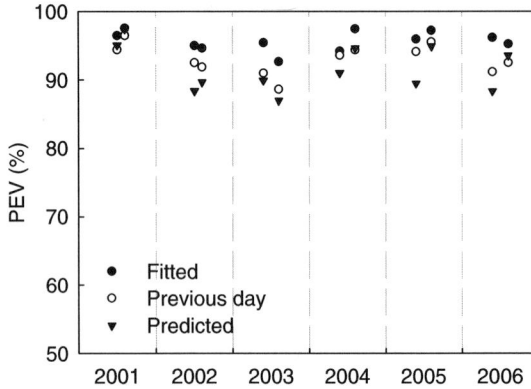

Fig. 6.3.15 Proportion of explained variance (PEV) in shoot CO_2 exchange at SMEAR I over July in different years. "Fitted" represents the case when the efficiency of functional substances (β) was estimated daily, "Previous day" refers to CO_2 exchange predicted with β estimated for the previous day, and "Predicted" CO_2 exchange predicted using shoot-specific fixed value of β estimated from a five-day period of data that was excluded in the calculation of PEV. The other parameters in the optimal stomatal control model were shoot-specific and fixed over the whole summer. One dot represents one shoot

We split the observed variance of CO_2 exchange into six components. The most important is the prediction power of the theory, i.e., the proportion of the observed variance that the theory explains when no fitting improves the agreement between the model and the measurements. The measuring noise is treated as four components, each of them connected with measuring noise of CO_2 exchange or explaining factors, light, temperature and water vapour concentration. The last component is associated with shortcomings in the theory and systematic measuring errors.

The above ideas can be formulated as follows. We use f_{opt} as short notation for the optimal stomatal control model and m_{CO2} as measured CO_2 exchange. Let $\varepsilon_{m,g}$ denote the measuring noise of CO_2 exchange, $\varepsilon_{m,f_{opt}(I)}$ the random component in the modelled CO_2 exchange caused by measuring noise of the light, $\varepsilon_{m,f_{opt}(T)}$ of temperature, $\varepsilon_{m,f_{opt}(w)}$ of water vapour concentration and ε_{th} the random component caused by shortcomings of the theory and systematic measuring errors.

The difference between measured and predicted CO_2 exchange is considered to be a random number consisting of five random components:

$$m_{CO2}^i - f_{opt}(I^i, T^i, W^i) = \varepsilon_{m,g} + \varepsilon_{m,f_{opt}(I)} + \varepsilon_{m,f_{opt}(T)} + \varepsilon_{m,f_{opt}(w)} + \varepsilon_{th} \quad (6.3.24)$$

The index i refers to the number of measurements.

The noise components are assumed to be independent stochastic variables. The above equation enables decomposition of the variance of the measured CO_2 into six components; explained (i) by the theory, (ii) by measuring noise of CO_2 exchange, (iii–v) by measuring noise of light, temperature and water vapour deficit and (vi) by shortcomings in the theory and systematic measuring errors.

Noise in the measurements of the explaining factors was first determined from consecutive readings of PAR, air temperature, and H_2O concentration measured at 10-s intervals during chamber closure (see Section 2.4.1). The variance component generated by the measuring noise was then obtained with simulations where normally distributed random component, equal to the observed noise, was added to each measurement of the explaining factor. The proportion of the variance of the modelled CO_2 exchange, generated by the measuring noise in the explaining factors, was extremely small: 0.005% for PAR, 0.03% for temperature, and 0.02% for water vapour concentration.

On average, the theory predicted 91.6% of the variance of measured shoot CO_2 exchange over July; 0.6% originates in the measuring noise of CO_2 exchange, 0.06% in the measuring noise of explaining factors and the rest, 7.7%, in the shortcomings of the theory and systematic measuring errors. The theory had a dominating role in explaining the variation in the CO_2 exchange. The reduction of measuring noise has only marginal importance in improving the measurements since the contribution of noise to the residual variation is small.

Systematic errors are unavoidable in field measurements. The movement of the sun on the sky generates occasional disturbing shading and inconsistency between the light environment measured outside the chamber and that experienced by the shoot. The systematic errors in light measurements are large at low solar angle and the sensor is sensitive to temperature changes and dirt. Evidently, the biggest problems are in temperature measurements. We have used thermocouples to measure air temperature inside the chamber. This is, however, not satisfactory since the needle temperature in the closed chamber is evidently higher because convective cooling on the upper side of the shoot is weak (wind speed is less than $1\ m\ s^{-1}$). The large temperature variation also affects the electrical properties of the measuring systems causing temperature-dependent systematic errors.

Although the residual component caused by shortcomings in optimal stomatal control theory and systematic measuring errors is small, we tried to detect any systematic features in the total residuals of the model. There is only a small systematic deviation at high temperature and irradiance and at low irradiances (Fig. 6.3.16). The former may be due to the lack of direct temperature response of photosynthesis in the model. At low irradiances, when the sun is near the horizon, the shoot is illuminated horizontally but the PAR sensor cannot detect the incoming radiation accurately, which can explain the observed systematic deviation with respect to irradiance.

Statistical methods have commonly been used in the analysis of photosynthesis in field conditions. The relationship between light and photosynthesis has been described using simple functions, such as the Michaelis-Menten-type saturating response. The fit of the models has often been rather poor. This may be caused by the low quality of data or pooling measurements on several objects for the analysis. This introduces large additional variation as is discussed in Section 2.4.1.

Purely empirical statistical models give little aid in understanding the mechanisms behind the system behaviour. Farquhar et al. (1980) presented a mechanistic model of the biochemistry of photosynthesis, which can be combined with statistical models of stomatal regulation. Such a pair of models is commonly used in the analysis of field measurements. The model is more detailed than the optimal stomatal control model, and includes more parameters. The estimation of the values of all the parameters is problematic with field data only, and may require additional laboratory measurements or taking some of the parameter values from literature. Comparison of the Farquhar model and the optimal stomatal control model showed

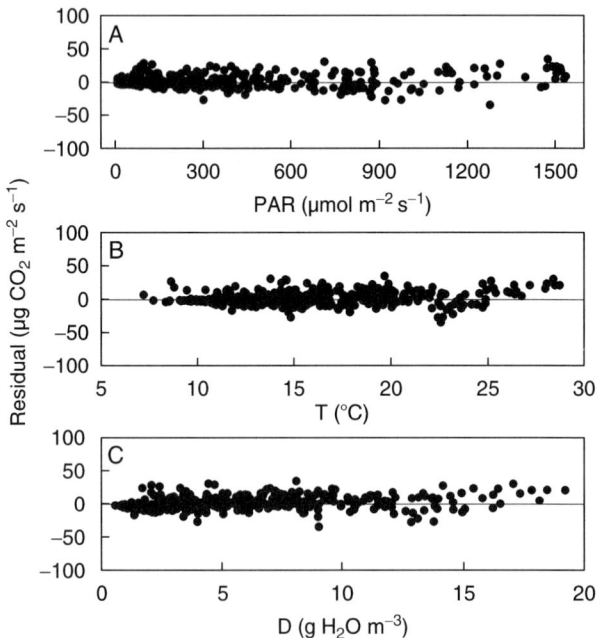

Fig. 6.3.16 Residuals of CO_2 exchange (measured flux minus modelled flux) of one shoot as a function of PAR (A), air temperature (B), and water vapour concentration difference between stomatal cavity and ambient air (B) at SMEAR I in July 2004. One tenth of the data were randomly chosen for the graphs, the five-day period used for parameter estimation was excluded

rather similar performance with shoot CO_2 exchange data measured at SMEAR I (Aalto et al., 2002a). The difference in the number of parameters to be estimated hampered the comparison. Although the theoretical basis of the optimal stomatal control model is completely different from the biochemical Farquhar model, the behaviour of those two models is remarkably similar.

The biochemistry of photosynthesis is complicated by the fact that both light reactions and carbon fixation include many steps, and each of them has its own set of functional substances. In the literature, Rubisco has received attention as the most important enzyme in modelling of carbon reactions of photosynthesis, although it is only one among a large group of functional substances.

The derivation of the optimal stomatal control model includes several oversimplifications and approximations. When the theory was formulated into operational form, only the main steps, i.e., light and carbon reactions, were introduced in a very rough form. The physiology of stomata or empirical relationships between stomatal function and environmental driving factors were omitted totally, instead the optimality principle was used to connect stomatal action to photosynthesis. Thus it is really surprising that the prediction power of the optimal stomatal control model is about 92% in field conditions. The model is able to properly describe essential features in photosynthesis and transpiration in field conditions. This result will largely be utilised in the remaining chapters of our book.

Conclusions

Photosynthetic CO_2 uptake and water loss by transpiration are connected by stomatal control. The theory of optimal stomatal control suggests that the stomata act so as to optimise photosynthetic production minus transpiration costs over a prolonged period. A mathematical formulation of this theory, combined with idealising assumptions, resulted in the model of optimal stomatal control. The prediction power of the model was high for an extensive set of field measurements, thus the model is able to catch permanent and general regularities in photosynthesis.

6.3.2.4 Carbon and Nitrogen Metabolism and Senescence

Jaana Bäck[1], Eero Nikinmaa[1], and Asko Simojoki[2]

[1]University of Helsinki, Department of Forest Ecology, Finland
[2]University of Helsinki, Department of Applied Chemistry and Microbiology, Finland

The primary metabolites in plant tissues are small compounds containing, in addition to hydrogen and oxygen, only a few carbon atoms (e.g., sucrose), or in addition also a few nitrogen atoms (e.g., amino acids). All large macromolecules containing hundreds or thousands of carbon atoms, such as proteins, lignin, starch or cellulose, are synthesized from these primary metabolites. Nitrogen in the plant, on the other hand, can either be inorganic as nitrate or ammonium ions, or bound as organic molecules such as amino acids and proteins.

Primary products of photosynthesis (Calvin cycle) are three-carbon compounds (DHAP, 3-PGA) which are subsequently processed in either chloroplasts or cytoplasm. The main end products are either starch (inert storage) or sucrose (mobile), but in addition the 3-carbon skeletons can also be used in a variety of biosynthetic pathways, such as amino acid, nucleic acid or triglyceride synthesis, and in production of NAPDH via the pentose phosphate pathway. Exchange of 3-carbon compounds between chloroplast and cytoplasm occurs via an active triose phosphate translocator, placed in the inner membrane of a plastid envelope (Flügge, 1999). Starch and sucrose are enzymatically degraded to yield hexose phosphates, which in turn are substrates for, e.g., cell wall biosynthesis, repairing damage and building up defence chemicals.

Carbon and nitrogen metabolism are intimately linked in the amino acid biosynthesis, which is seen in the close connections with the conditions stimulating or inhibiting both pathways. The actual site where this linkage occurs is in the chloroplasts, where ammonium and carbon compounds are combined into glutamine and glutamate following photorespiration. The majority of amino acids (and thus proteins) and also other macromolecules are synthesized in cells from intermediates of the main metabolic pathways. For the operation of these closely linked pathways, efficient transport of precursors and raw materials over short (e.g. from cellular organelle to another) or long (e.g. from roots to shoot) distances is necessary (e.g. Lalonde et al., 2004; Weber et al., 2005).

Most macromolecules are rather inert and contained within the organelle where they were synthesized. Due to the limited permeability of membranes, only the small, primary metabolites or inorganic molecules are capable of moving between organelles or cells. Active penetration of many molecules (both inorganic and organic) through cellular membranes is possible with the ATP-dependent **membrane pumps** (see Section 5.2.2) which relocate specific solutes across the membranes against their electrochemical potential. This means that active solute movement is closely connected to the availability of energy for the pumps. A group of the most important membrane pumps are the vacuolar and plasma membrane H^+-ATPases, which create gradients of pH and electric potential across the membrane and thus drive the penetration of several molecules and ions through their specific active-penetration proteins (Taiz and Zeiger, 2006). This is called the primary active transport, in contrast to the secondary active transport by carriers, and passive transport through channels.

Carrier proteins perform facilitated diffusion which includes conformational changes in the protein involved, and therefore is much slower than the direct molecular diffusion through a passive channel. Carriers do not have pores crossing the whole membrane. Instead, they bind the molecule on one side and release it on another. Ions (such as phosphate and sulphate) and some solutes can be actively moved together with protons by carrier proteins such as symporters or antiporters, driven by the proton motive force rather than energy from ATP hydrolysis. The proton motive force represents stored chemical energy in the form of an H^+ gradient.

Some small ions use protein channels to passively diffuse according to a concentration or chemical gradient. Channels are essentially transmembrane selective pores that open and close in response to cellular signals ("gating"). The substances diffuse very rapidly through channels, if they are open (10^8 ions/s per channel). Anion channels allow anions to diffuse out of cells, whereas Ca channels only release calcium into the cytoplasm, and if an opposite movement is needed, then active means must be used. K channels are the only channels working both directions, depending on the potassium equilibrium potential on each side of the membrane (Véry and Sentenac, 2003).

Channels specific for water molecules are called aquaporins (Luu and Maurel, 2005). Transmembrane water flow through **aquaporins** accompanies many physiological processes in plants, including phloem loading, osmotic adjustments between the vacuole and the cytoplasm, stomatal movement, and cell expansion. The presence of aquaporins in the cell membranes seems to facilitate the transcellular symplastic pathway for water transport. Recent evidence suggests that aquaporins are also directly relevant in the transport of CO_2 through mesophyll plasma membrane (Kaldenhoff et al., 2008). Many factors that influence plant water relations, such as plant hormone ABA, also influence aquaporin pore opening. For example, Tournair-Roux et al. (2003) showed how the changes in acidity within root cells efficiently closed the aquaporins, reducing considerably the plant water uptake. Aquaporins have also been considered to have an important role in recovery of embolised water transport tissue in xylem (Vesala et al., 2003; Sakr et al., 2003).

Solutes can move within the plant individual either in the **symplast**, i.e., between the cytoplasms of neighbouring cells, passing via plasmodesmata from one cell to another, or in the **apoplast**, i.e., in the cell walls and extracellular spaces, without passing the plasma membrane (Fig. 6.3.17). Through plasmodesmata, solutes can

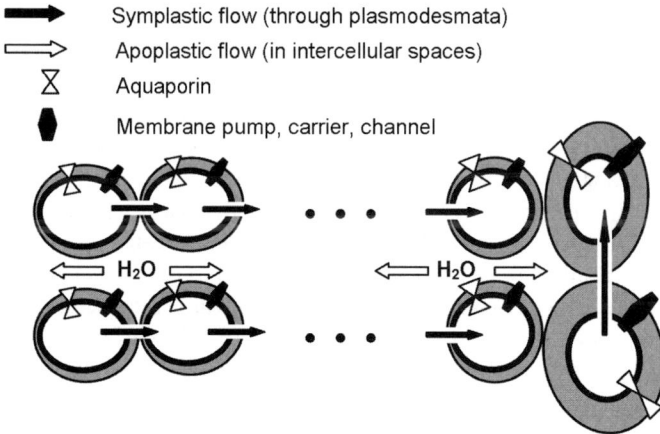

Fig. 6.3.17 Schematic illustration of symplastic (black arrows) and apoplastic (white arrows) transport pathways within and between tissues. Symplastic flow connects the adjacent cells within a tissue through plasmodesmata, and allows flow within the symplast according to osmotic pressure difference. Water passes membranes via aquaporins, and solutes and ions through membrane pumps, carriers and channels. Apoplastic flow of water and water-soluble compounds occurs on the cell surfaces and intercellular space, and is driven by the pressure difference

move according to osmotic pressure gradient. Symplastic flow is used in particular in root tissues to bring in nutrients from soil. Apoplast is formed by the continuum of extracellular spaces and cell walls of adjacent cells, but interrupted by the Casparian strips in roots and the cuticle on the outer surface of the plant. Apoplast is an important scene of plant's interaction with its environment: all gaseous components (CO_2, O_2, hormones, volatile organic compounds) entering or leaving the cell are passing through the apoplast, where dissolving or evaporation of molecules occurs. Ion uptake in roots (see Section 6.3.2.8) also involves an apoplastic diffusion step, before they are taken up into the symplast by specific ion channels. Further, the transpiration stream pulling water from soil up to the leaves is driven by apoplastic water flow. Apoplast is also an important scene for many enzymatic reactions such as detoxification of reactive oxygen species by apoplastic peroxidase enzymes.

In cool environments, such as in boreal forests, the tree species are mainly using an apoplastic loading of sucrose to phloem sieve element-companion cell complexes and further to be translocated towards the various sinks. This pathway involves symplastic passage of sucrose molecules through the mesophyll cells, active loading into apoplast and crossing of the relatively impermeable transfusion parenchyma cells, and an active sucrose-H^+ pump in loading the sucrose molecules into sieve elements against the concentration gradient (Fig. 6.3.17). Unloading of sucrose in the meristematic and elongating root tips occurs primarily via symplastic pathway, although in some other sinks sucrose unloading may involve also partial apoplastic steps (Patrick, 1997).

Nitrogen is a mobile element within a plant individual, i.e., it can rather freely be translocated as organic (amino acids, peptides) or inorganic (ammonium) molecules from one organ to another during the annual phases of development, meeting the needs for active protein synthesis in, e.g., meristematic tissues. Nitrogen is taken up from the soil as ammonium or nitrate (see Section 6.3.2.8). Subsequently, soil-derived nitrate is enzymatically converted in the cytoplasm of root or shoot cells into nitrite, which normally is immediately further reduced to ammonium. Plastids in both shoots and roots house the enzyme nitrite reductase which reduces nitrite into ammonium (NH_4^+) ion with the help of electrons either from photosynthesis (shoots) or the oxidative pentose phosphate pathway (roots). If soil nitrate supply is low, the primary site for its reduction is roots, whereas with increasing nitrate availability from the soil, the balance is shifted into assimilation in shoots.

Transport of nitrogen from roots to shoots occurs via the transpiration stream mainly as ions. The generated or soil-derived ammonium is rapidly converted into amino acids, which are the primary building blocks for protein synthesis. This conversion is enzymatic, glutamine synthetase and glutamate synthase being the first functional substances involved. Once assimilated into glutamine and glutamate, nitrogen will be incorporated into all other amino acids by the action of aminotransferases. The nitrogen assimilation is a dynamic process, which is strongly affected by external factors and by an internal balance between nitrogen and carbon pools. Some amino acids (glutamate, glutamine, aspartate, and asparagine) can also be transported in phloem for subsequent use in growth, defence and reproductive processes in other organs.

Once nitrogen is assimilated into protein molecules, they will operate a finite time. Turnover times of enzymes can range from few minutes up to few weeks in actively metabolizing cells (Vierstra, 1993). The damaged or malfunctioning proteins are disassembled either back into amino acids, which in their turn are recycled to form new protein molecules, or in the case of senescing organs, are dropped into the soil where micro-organisms will decompose the organic matter and ammonium ions are liberated into the humus layer (see Section 6.4).

Living plant parts exhibit typical metabolic activity and processes, but several characteristic features can also be related to senescing plant organs or tissues. Senescence of foliage can be initiated in response to a variety of environmental cues, such as changes in day length, temperature and water availability. Regardless of the initial stimulus, the plants' own genetic programme regulates the cascade of metabolic processes, which lead to recovery of many valuable resources such as some mineral elements, sugars, nucleic acids and amino acids (e.g. Näsholm (1994); Buchanan-Wollaston (1997); Keskitalo et al. (2005)). The hydrolytic reactions involved in senescence break down the macromolecules in a coordinated manner, and many resulting compounds and mineral elements are recycled into new growth in active meristems or during the next growth period. Chloroplasts are dismantled first, whereas the nucleus remains intact until the late phases of senescence. At cellular level, also the differentiation of tracheary elements in xylem follows a similar sequence of senescence as in the organ-level senescence.

6.3.2.5 BVOC Emissions

Jaana Bäck and Pertti Hari

University of Helsinki, Department of Forest Ecology, Finland

The volatile organic compounds are reactive in the atmosphere, thus their lifetime is short and they do not accumulate. Although the concentrations are low, the volatile organic compounds emitted by trees are important in the air chemical reactions (Sections 6.2.2 and 7.7.3) and in the formation and growth of aerosol particles (Section 7.8). The aerosols effect on the behaviour of radiation in the atmosphere and they play an important role in the climate change (Section 10.3.3).

The group of the biogenic volatile organic compounds (BVOC's) in the atmosphere-plant interface is very heterogeneous, both in respect to their chemical properties such as volatility and hydrophobicity, and in respect to their biogenic origin. The BVOC's are emitted by boreal trees as mixtures of a variety of individual compounds, and these mixtures change in time both seasonally and diurnally. BVOC's can be synthesized in both aerial and below-ground plant parts, in floral, vegetative and structural tissues, and they are stored in significant quantities in many plant genera. Their emissions show a conspicuous and compound-specific seasonal pattern, for most compounds the emission rates are highest in summer, and lower, although still significant in spring and autumn. Even during wintertime, emissions of some BVOC's can be measured (Tarvainen et al., 2005; Hakola et al., 2006).

BVOC's are not found uniformly within the plant. The constituent, defence-related compounds are often located in secretory tissues, such as resin ducts in conifer needles and resin canals in their trunks. From there they can be liberated if necessary for defence purposes, for example after a pathogen or herbivore attack. Another, temporally more fluctuating group of BVOC's is located in air spaces and apoplastic compartments, and is readily diffused to the boundary layer and further to the atmosphere. The turnover rates of these non-permanent BVOC pools depend both on prevailing precursor levels, and on factors controlling their synthesis and evaporation. The biological functions of some inducible compounds have been hypothesised to be related with heat or oxidative stress tolerance (Sharkey, 2005; Loreto and Velikova, 2001; Peñuelas and Munne-Bosch, 2005).

Here we restrict our model approach only to the volatile monoterpenes which are located in the non-permanent pools, and form the majority of emissions from boreal tree species such as Scots pine. The biosynthesis of monoterpenes occurs by two spatially distinguishable pathways, located either in cytoplasm or in plastids (for details see Lichtenthaler, 1999). The non-permanent pool is primarily synthesized in plastids. In a sequence of enzymatic reactions in the chloroplast stroma, cytoplasm-derived pyruvate reacts with plastidic triose phosphate (TP) to produce DOXP (1-deoxy-D-xylulose-5-phosphate), and finally IPP and DMAPP (isopentenyl pyrophosphate and dimethylallyl pyrophosphate, respectively) (Lichtenthaler, 1999). These are the common precursors for a plethora of terpenoids, including, e.g., isoprene, mono- and sesquiterpenes, xanthophylls and the phytol side-chain of chlorophyll. Monoterpene biosynthesis can be controlled either by the supply of substrates and availability of energy, or by enzyme activities in the metabolic branching points of the DOXP pathway (e.g., deoxyxylulose-5-P synthase or monoterpene synthase) (Bohlmann et al, 1998; Fischbach et al., 2002; Dudareva et al., 2004). Enzyme activities correlate well with measured emissions over the year, indicating that the level of monoterpene synthase activities is one of the factors controlling monoterpene emission rates in perennial species (Fischbach et al., 2002).

A mechanistic model was constructed (Fig. 6.3.18), based on the current physiological information on monoterpene biosynthesis and its relation to photosynthesis (Bäck et al., 2005). In the model, the synthesis of monoterpenes occurs in mesophyll chloroplasts, and the recently synthesized monoterpenes are temporarily located in liquid phase in the apoplast. From the liquid phase they diffuse to the leaf surface and are emitted into the atmosphere. There are two alternative processes for keeping the monoterpene precursor biosynthesis ongoing, either photosynthesis or photorespiration. Both depend on Rubisco activity, i.e., either carboxylation or oxygenation reaction.

It is assumed that two independent processes influence emissions from needles to the atmosphere: one is biosynthesis, related to the instantaneous irradiation and mesophyll CO_2 concentration, and the other is temperature-dependent emission from pools. The quantitative analysis of monoterpene emissions from Scots pine needles includes separate sub-models for substrate production, monoterpene biosynthesis, storage, transport within the leaf, and finally emission.

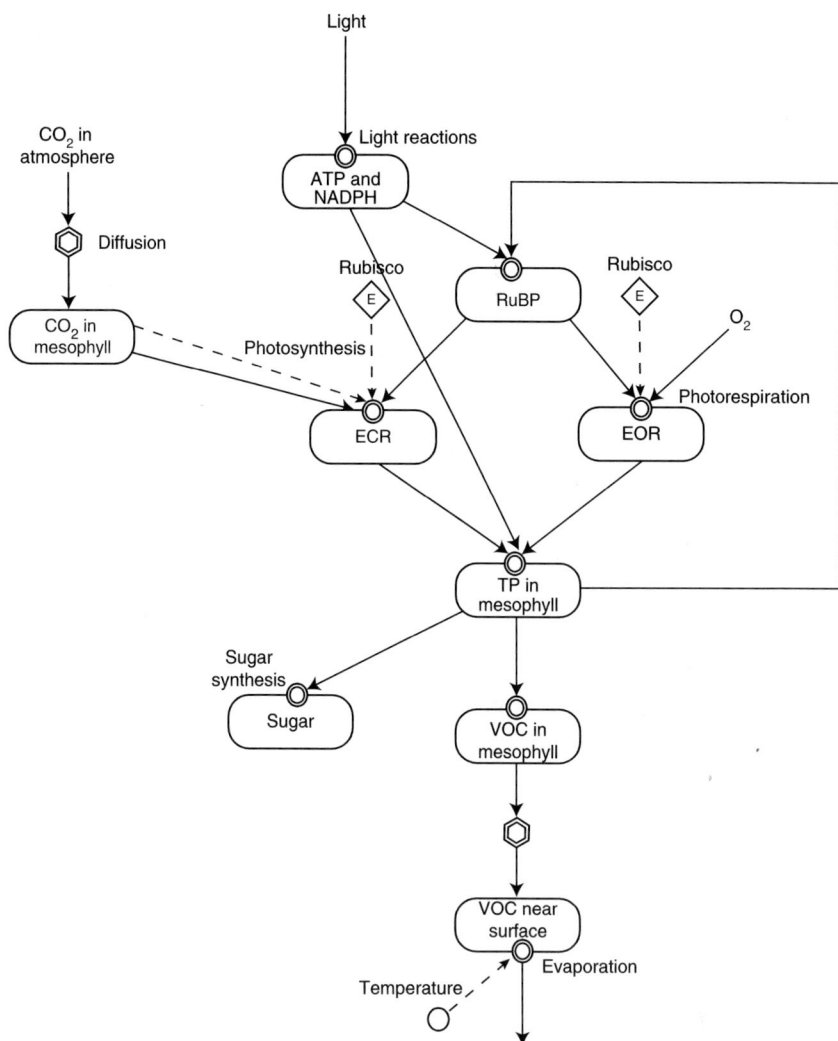

Fig. 6.3.18 Schematic illustration on connections between BVOC synthesis and photosynthesis. ECR = carbon reduction, EOR = carbon oxygenation, RuBP = ribulose-1,5-bisphosphate, TP = triose phosphate, BVOC = biogenic volatile organic compound, ATP = adenosine triphosphate, NADPH = nicotinamide adenine dinucleotide phosphate

Substrate Production

The photosynthetic sub-model is the 'Optimum stomatal control model for photosynthesis' (Section 6.3.2.3), which allows us to use detailed gas exchange measurements in the natural boreal forest conditions. The model calculates the photosynthetic rate, $p(g(CO_2)$ m^{-2} s$^{-1})$, and the CO_2 concentration inside stoma,

$C^s_{CO2}(\text{g}(CO_2)\text{m}^{-3})$, as a function of irradiance, I_{PAR} (mol m^{-2} s^{-1}), ambient temperature, $T(^\circ C)$, and water vapor pressure deficit, VPD (g (H$_2$O) m^{-3}):

$$p = p(I_{PAR}, T, VPD) \tag{6.3.25}$$
$$C^s_{CO2} = C^s_{CO2}(I_{PAR}, T, VPD) \tag{6.3.26}$$

Terpene Biosynthesis

The sub-model for plastidial monoterpene formation is based on simplified assumptions on the biosynthetic pathway. The availability of triose phosphate (TP) for monoterpene biosynthesis is controlled by the CO_2 concentration in mesophyll. The most important functional substance is Rubisco, which keeps the Calvin cycle running to produce the primary substrate for monoterpene biosynthesis.

Light availability influences both the regeneration of RuBP in the Calvin cycle, and monoterpene biosynthesis by providing ATP and NADPH from light reactions. However, energy availability does not usually restrict the synthesis. RuBP is carboxylated by Rubisco when the CO_2 concentration in the mesophyll is high and oxygenated when the availability of CO_2 is limited, giving rise to photorespiration in this case. The availability of TP is the limiting factor, whereas the availability of the other precursor, pyruvate, is considered to be non-limiting for monoterpene biosynthesis. We analysed two alternatives; either the monoterpene biosynthesis is linked with photosynthesis or with photorespiration. Let r_p and r_r denote the monoterpene synthesis rate (g m^{-2} s^{-1}), dependent on photosynthesis and photorespiration, respectively.

Assume

$$\text{CASE I} : r_P = \alpha_1^{VOC} \, p(I_{PAR}, T, VPD) \tag{6.3.27}$$
$$\text{CASE II} : r_r = \alpha_2^{VOC} \, (C^a_{CO2} - C^s_{CO2}(I_{PAR}, T, VPD)) \tag{6.3.28}$$

where α_1^{VOC} equals 0.5 mg(monoterpenes) g(CO_2)$^{-1}$ and α_2^{VOC} 140 ng(monoterpenes) mg (CO_2)$^{-1}$ s^{-1}, describing the efficiency of monoterpene biosynthesis, and C_a is the CO_2 concentration of air (g m^{-3}).

Storage Pools

For our approach, we consider the temporary monoterpene storage to consist of two spatially separated sub-pools: a mesophyll pool and a surface pool, both of which are physically located in the liquid phase, either in mesophyll apoplast or at the stomatal cavity. The presence and physicochemical characteristics of these multiple temporary pools produce delays on the order of minutes or hours in emission responses to external stimuli.

The synthesis and transport of monoterpenes change their mass in the mesophyll pool, thus

$$\frac{dM_m^{VOC}}{dt} = r_* - l_{voc} \tag{6.3.29}$$

where l_{VOC}(g m^{-2} s^{-1}) is the flux of monoterpenes inside the leaf, M_m^{voc}(g m^{-2}) is the monoterpene mass in the mesophyll pool in a square meter area of mesophyll tissue, and r_* is either r_p or r_r.

Transport Within a Leaf

The transport of monoterpenes from the mesophyll pool (i.e., the site of synthesis) out to the surface pool (i.e., the site of emission) is based on diffusion, which is proportional to the concentration differences between the components of the pathway.

The concentrations of monoterpenes in mesophyll and surface pools (C_m^{VOC}, C_s^{VOC}) (g m^{-3}) are derived from the masses of monoterpenes (M_m^{VOC}, M_s^{VOC}) and volumes of the pools (V_m^{VOC}, V_s^{VOC}) (m^3), which were calculated from experimentally determined anatomical dimensions for Scots pine needles (average mesophyll thickness 150 μm, average epidermal thickness 30 μm; Bäck et al., 1994). The flux l_{VOC} (g m^{-2} s^{-1}) from mesophyll pool to surface pool is expressed as

$$l_{VOC} = c_{voc}\left(C_m^{VOC} - C_s^{VOC}\right) \tag{6.3.30}$$

where c_{VOC} describes the monoterpene conductance in the mesophyll liquid phase (m s^{-1}), and was calculated using the concept presented by Niinemets and Reichstein (2002). The synthesized monoterpenes were assumed to possess equal conductances in the mesophyll, due to their equal molecular weights (136.24 g mol^{-1}).

Emission

Emissions of highly volatile compounds especially exhibit only limited stomatal control as shown in many studies for, e.g., isoprene and α-pinene (Loreto et al., 1996; Shao et al., 2001). Also the lipophilic nature of monoterpenes suggests possible liberation routes directly through the cuticle and implies a negligible stomatal control for these compounds (Guenther et al., 1991). Thus we chose to omit the stomatal conductance from the model, and suggest that monoterpene emission may occur homogeneously from the leaf surface.

Monoterpenes are freely emitted into the atmosphere from the leaf surface. Let f_h (g m^{-2} s^{-1}) denote the temperature dependence of monoterpene emission from the liquid phase. We assume that the temperature dependence of emission can be approximated with

$$f_h(T_a) = \exp(a + b/(T_a + 273)) \tag{6.3.31}$$

where T_a is the air temperature (°C) and the empirical parameter values $a = 24.5$ (dimensionless) and $b = -5{,}474$ K (for α-pinene, CRC (2003)).

Let M_s^{VOC} (g m^{-2}) denote the monoterpene mass in 1 m^2 area of leaf surface (surface pool size). The emission rate, g_{VOC} (g m^{-2} s^{-1}), is determined by the mass of monoterpenes in the leaf surface pool and temperature, thus

$$g_{VOC} = \beta M_s^{VOC} f_h(T_a) \qquad (6.3.32)$$

where β equals 0.005 m^2 g^{-1}.

The fluxes, i.e., flow inside the leaf (l_{VOC}) and emission (g_{VOC}), change the mass of monoterpenes in the surface pool as

$$\frac{dM_s^{VOC}}{dt} = l_{VOC} - g_{VOC} \qquad (6.3.33)$$

The values of the parameters α_1, α_2 and β in the synthesis and emission submodels were estimated using the data on Δ^3-carene emissions (the dominant component in Scots pine shoot emissions) measured between May 20 and 21, 2003 at SMEAR II (Tarvainen et al., 2005). The parameter estimation was done by minimizing the residual sum of squares.

The presented monoterpene emission model is composed of the most important processes and transport phenomena involved in monoterpene synthesis and emissions, and can be used for calculating the emissions, after the prevailing environmental conditions in the field and incident photosynthetic rate are known.

The model was tested with an independent dataset created from Scots pine monoterpene emissions in 11–18 September, 2005. The monoterpene emissions were measured with a dynamic shoot cuvette from a branch at the top whorl of a 40 year-old tree using a proton transfer reaction – mass spectrometer (PTR-MS, Ionicon, Austria). The branch was inserted into a transparent, Teflon-coated acrylic box cuvette, and the outgoing air flow was connected to monoterpene measurements. The method is described in detail in Ruuskanen et al. (2005).

Monoterpene emissions showed a typical diurnal variation, which was partially captured by the model (Fig. 6.3.19a). The emissions were in maximum during early afternoon and in minimum during nighttime. Fluctuations within and between days were caused by changes in both monoterpene synthesis and pool sizes (Fig. 6.3.19b). Due to the large internal storage pools, the emissions were not totally depleted during night although synthesis was inhibited due to lack of light. The model showed good agreement with the measurements in particular during high light intensities, such as in September 11 and 15, but failed in following the emission pattern during lower light levels (e.g. in September 12). This was probably due to the close relation between monoterpene synthesis rate and monoterpene emissions in the model, which obviously did not properly deal with the constitutive emissions from the large storage pools in Scots pine. Also the necessary biosynthesis parameter values may be different depending on the plant phenological status. The improvement of cuvette measurements will provide necessary information for development of our approach.

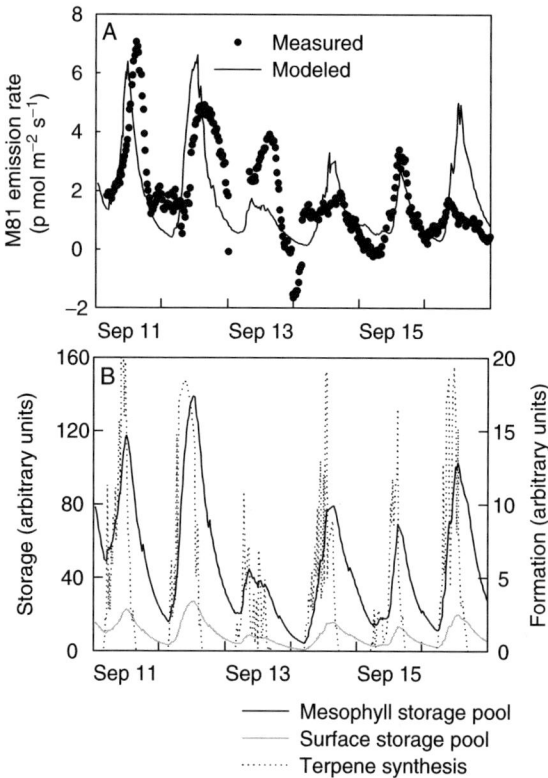

Fig. 6.3.19 Measured and modeled monoterpene emissions from a Scots pine shoot (A), temporal development of the modeled monoterpene synthesis and storage pools (B) during one week in September 2004

Conclusions

The monoterpene emissions from Scots pine fluctuate considerably on a diurnal and seasonal basis. These fluctuations are due to changes in biosynthetic processes caused by short-term variations in light and temperature levels, and in longer term, also due to changes in the plant phenological status. Since the instantaneous emissions are at least partially related to the photosynthetic CO_2 exchange, the monoterpene emissions can be modelled using a process-based model with a photosynthesis submodel. The process-based monoterpene emission model with some further developments, taking into account the seasonal emission variability and storage pool development, can also be used in estimating the emission development under changing climate.

6.3.2.6 Ozone Deposition

Nuria Altimir and Pertti Hari

University of Helsinki, Department of Forest Ecology, Finland

Gaseous ozone (O_3) is a normal constituent in the Earth's atmosphere, where it is photochemically formed (see Sections 6.2.2 and 7.7 for more on atmospheric chemistry). O_3 is an unstable gas, a highly reactive species and a strong oxidizing agent. The reduction of O_3 carries a standard reduction potential of 2.07 V, a value larger than for most materials. This means that the oxidation of most species by O_3 is thermodynamically favourable, and thus O_3 is capable of reacting with many species. In particular, O_3 readily reacts with compounds containing multiple bonds and simple oxidizing ions, since the kinetics of such reactions are markedly fast. Organic compounds, which often contain double bonds, are easily oxidised by O_3.

In the lower troposphere, where it comes in contact with vegetation, O_3 is not only destroyed in air chemistry reactions but also by encountering bare ground, snow cover, water bodies, and vegetation. The interaction with vegetation results in O_3 being removed from the atmosphere, ultimately through oxidation of plant-related compounds. The overall removal from the atmosphere into the Earth surface is known as deposition. This chapter looks into the details of this process at the leaf-air interface.

Convection transports O_3 near leaf surfaces. At the leaf-air interface it might react with components on the outer leaf surface or be transported by diffusion through the stomata pore and into the substomatal space in the mesophyll where it reacts further. Ozone is thought to react with the leaf surfaces via photochemical reactions (Rondón et al., 1993; Coe et al., 1995; Coyle, 2005) or thermal decomposition (Fowler et al., 2001) mediated by the foliage surface. In addition, the presence of wetness on the epicuticule strongly controls the intensity of the deposition, partly by enhancing heterogeneous reactions and partly by preventing possible dry reactions. These phenomena can be related to the presence of liquid film in the leaves surfaces, which are formed by absorption, a process related to ambient relative humidity (RH) (Altimir et al., 2006). Most chemical reactions of toxicological consequence to the plant occur on the stomatal and mesophyll cell surfaces (Sandermann, et al., 1997). O_3 reacts mostly with the antioxidant ascorbic acid (Chen and Galli, 2004), but also with other apoplastic (Castillo and Greppin, 1988; Langebartels et al., 1991) and cell wall components (Wiese and Pell, 2003).

There appear to be many possible reactions that are able to remove O_3, but the spectra of all the compounds involved in these reactions is not fully known. It is customarily assumed that they are to follow a first-order reaction scheme and further, that the O_3 concentration at the surface is negligible. This means the reaction rate is simply dependent on the O_3 concentration.

All in all, the process can be described in terms of ambient concentration and the main controlling factors. These are the RH for the outer deposition and the stomatal aperture for the inner deposition (Fig. 6.3.20).

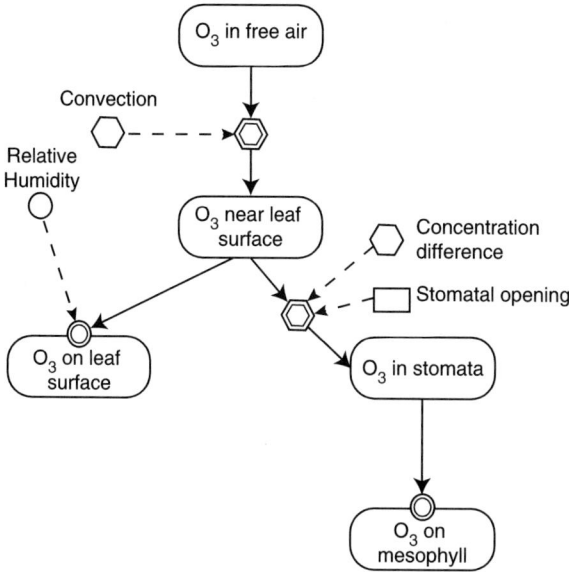

Fig. 6.3.20 The visualisation of ozone exchange at the air-leaf interface. Note concentration difference is effectively the ambient concentration since the concentration at the surfaces is negligible. The symbols are introduced in the Fig. 1.2.1

Let g_{O3} denote the O_3 flux between air and leaf. It is formed by two components, O_3 reactions with the leaf surface dependent on RH, and on mesophyll cells dependent on stomatal conductance (c_{O3})

$$g_{O3} = C_{O_3} f(c_{O3}) + C_{O_3} f(RH) \tag{6.3.34}$$

We can define concentration-normalised flux as

$$\frac{g_{O3}}{C_{O3}} = c_{O3}^* \tag{6.3.35}$$

where c_{O3}^* is the proportionality constant. This is analogous to the conductance for transpiration (see 6.3.2.3). It is a bulk parameter that may aggregate physical, chemical and biological processes affecting the flux. We have identified two components (Fig. 6.3.20):

$$c_{O3}^* = f(c_{O3}) + f(RH) \tag{6.3.36}$$

Let us assume, for lack of a better formulation, that the diffusive transit of O_3 through the stomatal is equivalent to the passage of water vapour, only in the opposite direction. Then we get

$$f(c_{O_3}) = c_{H2O} \frac{D_{H2O}}{D_{O_3}} \approx c_{CO_2} \frac{D_{CO_2}}{D_{O_3}} \tag{6.3.37}$$

where c_{H2O} is the stomatal conductance and D_{H2O}, D_{O3}, D_{CO2} are the stomatal conductance for of CO_2, water vapour, O_3 and CO_2, respectively.

As to the relation with RH we describe it as an absorption process

$$f(RH) = \varphi \cdot \Phi \qquad (6.3.38)$$

where φ refers to the relative amount of water on the surface and Φ relates to the chemical rate of O_3 decomposition. φ can be described with existing absorption isotherms, e.g., the BET isotherm.

The quantification of Φ, however, runs into the lack of information mentioned above on the spectra of compounds involved. In view of this shortcoming, f(RH) can be simply described empirically, that is, with a function that fits the data but has no mechanistic meaning. In that case, we find that:

$$f(RH) = \frac{a}{b - RH} \qquad (6.3.39)$$

where a and b are two parameters to be estimated. They roughly correspond, respectively, to the minimum O_3 deposition and the maximum RH (i.e., 100%).

At SMEAR II, O_3 shoot-scale fluxes were measured by a gas-exchange enclosure technique. The general performance of the chambers has been evaluated in Altimir et al. (2002, 2004) and Kulmala et al. (1999) with respect to O_3. The methodology is akin to the gas exchange measuring system used for, e.g., CO_2 and water vapour. However, the reactive nature of O_3 results in significant losses into the chamber material, especially the chamber walls. The systematic measuring error caused by ozone reactions with the instrumentation has to be reduced with corrections, e.g., with the aid of simultaneous measures of the O_3 flux produced by an empty equivalent chamber.

The mass balance equation for the change in O_3 concentration inside the chamber is:

$$\frac{V \cdot dC_{O_3}^c(t)}{dt} = q \cdot (C_{O_3}^a - C_{O_3}^c(t)) - V \cdot K \cdot C_{O_3}^c(t) - A \cdot c_{O_3} \cdot C_{O_3}^c(t) \qquad (6.3.40)$$

where the second and last terms on the left-hand side are the O_3 mass flux produced by the chamber walls and the shoot respectively. $C_{CO_3}^a$ is the ambient O_3 concentration, C_{O3}^c the concentration inside the chamber, $K(s^{-1})$ the rate constant of O_3 loss to the chamber walls, and c_{O3} (m s^{-1}) is total shoot conductance. K was fitted on measurements from the empty chamber (omitting $Ac_{O3}C_{O3}(t)$) and its value was used when fitting c_{O3} to measurements with a shoot. The fitted parameters, K and c_{O3}, are thus not a flux but a proportionality constant to be multiplied by ambient concentration in order to obtain an expression of flux.

Thus, once the bias due to the chamber wall loss is corrected from the measurements, we obtain the value of c_{O3}, which contains both the removal in the outer and inner surfaces, that is, the stomatal and non-stomatal deposition (Eq. 6.3.37). The stomatal part (Eq. 6.3.38) is estimated though the stomatal action and the reduction

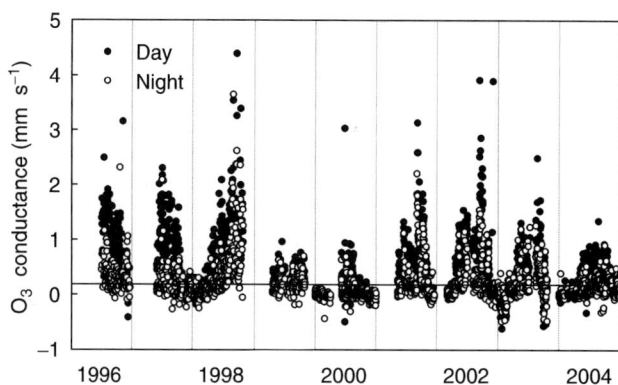

Fig. 6.3.21 Multi-year time series of O_3 removal on Scots pine shoots as measured in SMEAR II by the enclosure chamber technique. Data shows the daily averages for the daylight and night-time values. Removal is here expressed as O_3 conductance, g_{O3} in mms^{-1}

in diffusion can be obtained from photosynthesis or water vapour exchange with the formulation introduced in Sections 6.3.2.2–6.3.2.3.

At SMEAR II, a long time series of O_3 flux measurements at the shoot scale shows that the interannual variation in the pattern of the measured fluxes is large (Fig. 6.3.21). This reflects not only the peculiarities on different years but also the difference between measured shoots and, in case of O_3 fluxes, the differences between instrumentation performance. Nonetheless, an initial look at the measured values is very informative. The flux of O_3 is mainly positive, that is, O_3 essentially deposited from the atmosphere to the forest. Deposition is predominant throughout the years only to fade to small levels during the winter. A close inspection reveals that winter O_3 fluxes fluctuate slightly above, below or just around zero depending on the year. This behaviour can be interpreted as white noise or measurements under the detection limit of the system. It is also clearly shown that O_3 removal also happens during night and that the levels of nocturnal removal are often half as big as diurnal levels.

A typical monthy-scale time-series of O_3 deposition shows a daily pattern of removal with nocturnal minimums, and superimposed to it, a fluctuation in the general level of removal that relates to the general level of ambient RH so that the more humid the air the more intense the O_3 removal (Fig. 6.3.22). There remains aspects of the deposition patterns that are not explained by these estimations – as can be clearly seen in the detailed days. But the consideration of non-stomatal routes and the addition of f(RH) is a notable improvement compared with only the calculation based on stomatal action.

Analysis of the relative contribution of the different processes to the generation of the O_3 flux has shown that stomatal and non-stomatal routes share the importance, being each of them responsible for approximately 50% of the total yearly flux (Altimir et al., 2004, 2006). The value of f(RH) represented a variable percentage of the calculated total ozone conductance. This relative contribution was more

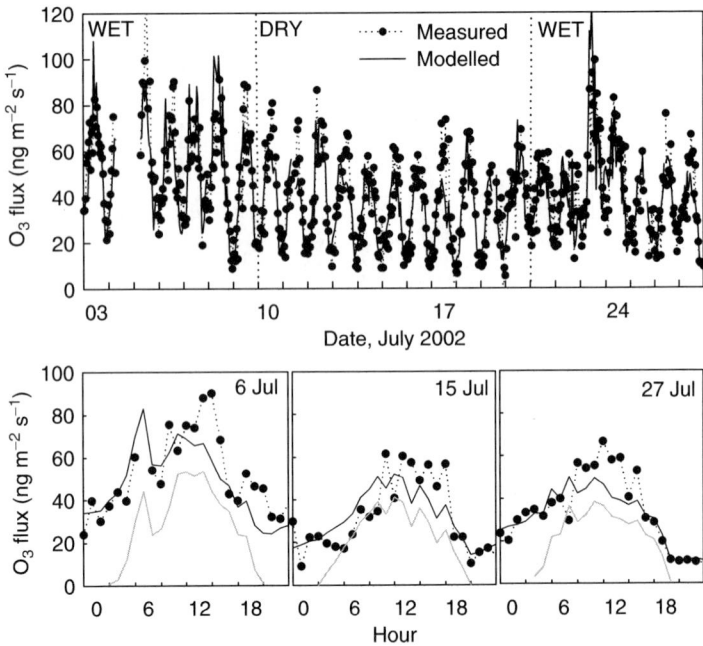

Fig. 6.3.22 An exemplary time series of measured (spheres) and modelled (lines) O_3 flux at the shoot scale. The series runs through a period that starts and ends with wet conditions and in between experiences a dry spell when there was no rain and the ambient RH stayed under 70%. The lower panel shows three days within this series and compares the modelled flux with only stomatal action (grey line) and with added RH enhancement (black line)

or less exponentially proportional to RH typically ranging from 2% at RH = 30% to 95% at RH = 100%. On average, during moist conditions the contribution of the non-stomatal component is around 50% and slightly lower and more variable under dry conditions.

6.3.2.7 NO_x Exchange of Needles

Maarit Raivonen and Pertti Hari

University of Helsinki, Department of Forest Ecology, Finland

Nitrogen oxides are atmospheric trace gases that play an important role in the tropospheric chemistry (Section 6.2.2). The major sources of the primary oxidized nitrogen species, NO (nitric oxide) and NO_2 (nitrogen dioxide), are fuel combustion, biomass burning, lightning and some microbial processes in the soil. Usual NO_x ($NO + NO_2$) concentrations in rural areas are only a few parts per billion (ppb), being well below 1 ppb in the most remote areas. In urban regions, the concentrations are generally at tens of ppb and the peak values can approach 1,000 ppb (Seinfeld and Pandis, 1998).

Vegetation plays a role also in the atmospheric balance of NO_x. It has been shown that plants absorb NO_2, and to a lesser extent NO, from the air, when the ambient concentration of these gases is high enough (Rondón et al., 1993; Sparks et al., 2001). Plants can even utilize the nitrogen of absorbed NO_x in their metabolism. Some observations exist also on NO_x emission from plants at ambient concentrations close to zero (Wildt et al., 1997; Sparks et al., 2001). The critical concentration at which plants start to absorb NO_2, also called the compensation point, has usually been below 2 ppb. This means that if plants were able to emit NO_x, vegetation would more likely be a source than a sink of NO_x in rural and remote areas. However, the general capability of plants to emit NO_x is controversial.

The absorption rate of NO_2 has been observed to depend on the degree of stomatal opening and on the concentration difference between plant interior and ambient air (e.g., Rondón and Granat, 1994; Geßler et al., 2002). It is still somewhat unclear whether NO_2 dissolves in the cell wall water so fast that the internal concentration could be assumed zero (e.g., Rondón et al., 1993), or is the dissolution slower and a non-zero internal concentration should be taken into account (e.g., Thoene et al., 1996). Atmospheric NO_2 deposits also to plant surfaces, but this non-stomatal absorption has been found to be small compared to the stomatal absorption (Geßler et al., 2002).

Emission of NO_x seems to be connected with nitrate metabolism of plant leaves. It has been shown that NO can be produced from nitrite and NADH by nitrate reductase (Yamasaki and Sakihama, 2000). High internal concentrations of nitrite are necessary for significant NO emission to occur, and this apparently requires that nitrate, not ammonium, is the nitrogen source for the plant (Wildt et al., 1997). Also NO_2 emission rates were found to correlate positively with leaf nitrogen content, which suggested a positive correlation between nitrate reductase activity and NO_2 emission (Sparks et al., 2001).

When the measuring station SMEAR II was planned and implemented, we established a programme of measurements of NO_y flux from the atmosphere into Scots pine needles, to quantify the nitrogen flux and to evaluate its importance. Our system detects the flux of total reactive nitrogen NO_y, which besides NO_x includes, e.g., nitrous acid (HONO), nitric acid (HNO_3) and peroxy acetyl nitrates (PAN).

When the measurements were started, it turned out that there was a NO_y flux *from* the needles *to* the atmosphere. In addition, the flux seemed to be connected with UV radiation (Hari et al., 2003). The UV-driven NO_y emission most likely originates from the surface of the needles. However, the mechanism is not yet fully understood. We have suggested that the NO_y emissions consist mainly of NO_2, and that they originate from photolysis of nitric acid that has deposited on the needle surface from the air (Raivonen et al., 2006). A similar phenomenon has earlier been observed to produce NO_2, NO and HONO on snow and glass surfaces (Dibb et al., 2002; Zhou et al., 2002). The NO_x concentrations in air are very low at SMEAR II (about $1–3 \mu g\,m^{-3}$), which explains why we have not constantly seen deposition on needles. However, there are spells when the air mass is coming from urban areas and

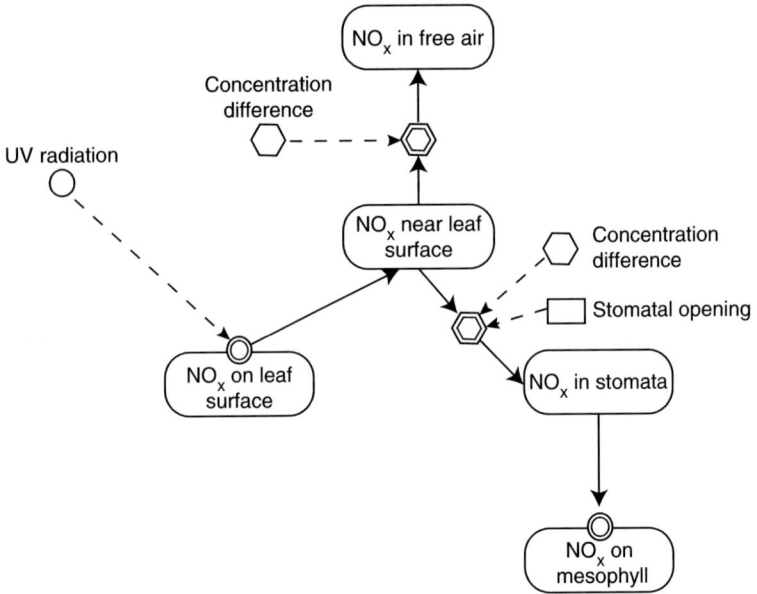

Fig. 6.3.23 Subprocesses and transport generating leaf-level NO$_x$ fluxes. The symbols are introduced in the Fig. 1.2.1

its NO$_x$ concentration is high (about $10\,\mu\mathrm{g\,m}^{-3}$). During these, we have observed also some deposition. Figure 6.3.23 illustrates our ideas on how these two processes determine the NO$_x$ fluxes on pine shoots.

Let g_{NOx} denote the NO$_x$ exchange, e_{NOx} emission and u_{NOx} uptake, respectively. Assume that emission depends linearly on UV-irradiance, thus

$$e_{NOx} = a_{NOx}I_{UV} \qquad (6.3.41)$$

where I_{UV} is UV-irradiance and a_{NOx} a parameter.

Assume further, that the uptake of NO$_x$ depends linearly on the NO$_x$ concentration in the air, C_{NOx}

$$u_{NOx} = b_{NOx}C_{NOx} \qquad (6.3.42)$$

where b_{NOx} is a parameter.

When the above equations are combined, we get

$$g_{NOx} = e_{NOx} + u_{NOx} = a_{NOx}I_{UV} + b_{NOx}C_{NOx} \qquad (6.3.43)$$

We constructed the measurements of NO$_y$ exchange with the chamber technique to test the above model. Measuring NO$_y$ fluxes is problematic, since the reactive NO$_y$ interacts with the chamber and slightly also with the tubing needed for the transport of gas (Raivonen et al., 2003). Figure 6.3.24 shows an example of the chamber effects. On a sunny day, the NO$_y$ emissions in a branch chamber followed the

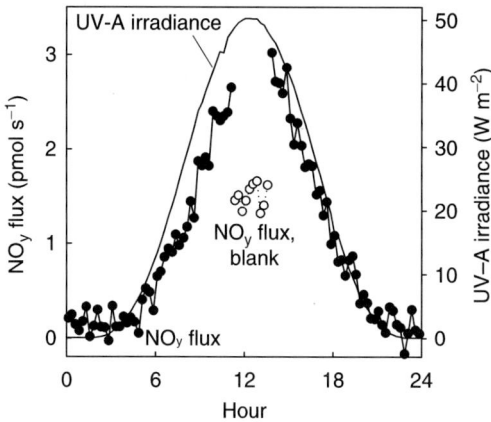

Fig. 6.3.24 Measured NO_y fluxes of one clear-sky day in a branch chamber. The branch was removed from the chamber for a few hours in the middle of the day

solar irradiance. When the branch was removed, the emission level did not drop to zero but to around half: the chamber was emitting NO_y as much as the pine branch did. We reduced the systematic measuring error caused by interactions of NO_y with the instrumentation by a correction method. Alongside the chambers enclosing pine branches, we monitored simultaneously an empty chamber. We considered the chamber blank of the branch chambers to be similar to the empty one.

The emission part of the model could be tested on a beautiful sunny day when the ambient NO_y concentrations were low (Fig. 6.3.25). The NO_y emissions of the branch had a linear relationship with UV-A irradiance (as similarly with the total solar irradiance, too), as suggested by the model. Uptake of NO_y was clear on a day when the airmass contained unusually high concentrations of NO_y (Fig. 6.3.26). The degree of stomatal opening apparently affects the uptake rate, too, but it is not included in this simple model. Because of this, we excluded measurements with low PAR (below $100 \mu mol s^{-1} m^{-2}$), suggesting small stomatal aperture, from the analysis. The relationship between ambient NO_y concentrations and the uptake rates with open stomata was roughly linear, again in agreement with the model.

The absorption of NO_x has been demonstrated frequently: one can relatively safely assume that vegetation acts as a NO_x sink at high ambient concentrations. However, measurements of small fluxes at near-zero concentrations are demanding, because reactions on the chamber surfaces disturb the analysis. So, the universality and the underlying processes of NO_x emissions remain unsolved. Emission from metabolism may be specific to certain plant species or habitats, or it may be possible everywhere, depending on the prevailing nitrate status. In case there really is a UV-induced surface process of nitric acid that generates NO_x, it could occur always when nitrate/nitric acid and sunlight are present, recycling deposited nitrate back to the atmosphere. Further studies are still needed, to clarify what happens with the NO_x fluxes below the compensation point.

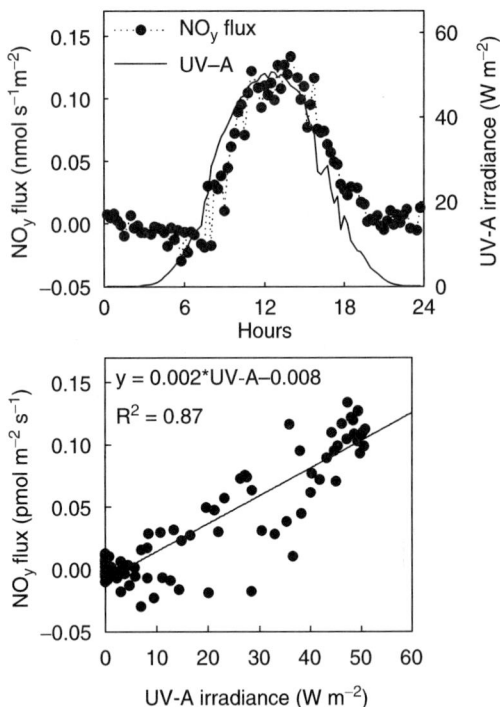

Fig. 6.3.25 Measured NO_y fluxes on a sunny day with low ambient NO_y concentrations. Upper panel: The daily patterns of UV-A irradiance and the NO_y flux. Lower panel: Relationship between the blank corrected NO_y flux and the UV-A irradiance. The line is fitted using the least squares method

Conclusions

Flux of nitrogen oxides in vegetation has been suggested to be bi-directional: plants can both absorb and emit NO_x. Emission of NO_x is still somewhat controversial. Our results indicate that at low ambient NO_x concentration and high UV-A irradiance emissions dominate and at high ambient NO_x concentrations deposition occurs.

6.3.2.8 Uptake of Water and Nutrients by Roots

Asko Simojoki

University of Helsinki, Department of Applied Chemistry and Microbiology, Finland

Most of the water and elements required for plant growth are taken up from the soil by roots. The main exceptions are carbon and part of oxygen taken up as carbon dioxide by plant leaves from the atmosphere for the photosynthesis (Section 6.3.2.3). Water flows from the soil through the plant to the atmosphere and

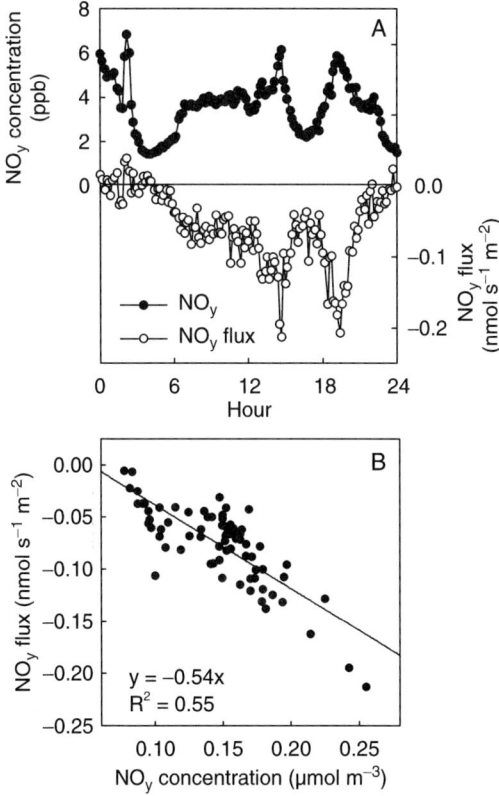

Fig. 6.3.26 Measured NO_y fluxes on a cloudy day with high ambient NO_y concentrations. (A) The daily patterns of NO_y concentration and the NO_y flux. (B) Relationship between the blank corrected NO_y flux and the NO_y concentration. The line is fitted using the least squares method

carries solutes along (Section 7.5.2). However, the uptake of water and nutrients by roots are not mere passive transport phenomena, but active processes continuously controlled in a selective manner by root metabolism. Roots apply this control mainly as water and solutes enter and exit the cell membrane through various substrate-specific membrane transport proteins. Similar transport proteins operate across other biological membranes as well, such as in tonoplast and mitochondrial membranes. Nitrogen is needed for synthesis of functional substances (Section 5.2.1), several other elements, like potassium, phosphorus, magnesium and calcium, are necessary for proper functioning of plants.

Water and nutrients are transported in the soil-plant-atmosphere continuum by convection and diffusion mechanisms (see Chapter 4). Spontaneous transport occurs passively along a decreasing chemical potential gradient so that the potential of the system to perform work (the free energy of the system) decreases in accordance with second law of thermodynamics. On the other hand, external work is required for transporting the species actively against their potential gradients. For details on how the concentrations, temperature, pressure, electrical forces and gravitation affect the energy status of solutes and water, and how the free energy of whole system is quantified by the sum of chemical potentials of all species present, we refer to standard text books in soil physics and plant physiology (Nobel, 1991; Hillel, 1998; Taiz and Zeiger, 2006 for more details). The practical outcome of this discussion is that water tends to flow spontaneously towards lower elevations, temperatures, pressures (or higher tensions) and higher solute concentrations, whereas the solutes carried by water tend to distribute themselves towards lower temperatures and concentrations (with all other things being equal). Electric forces influence the movement of solute ions but not the movement of such uncharged species as water. The movement of water and nutrients against these directions of spontaneous transport within plant requires metabolic energy.

In contrast to the non-living world, the water and nutrient fluxes in living organisms would frequently fail to support the organisms, unless their spontaneous directions were redirected by the work provided by cellular metabolism. In particular, plant roots must be able to maintain cellular solute concentrations far from equilibrium with the surrounding soil solution. Membrane transport proteins capable of pumping ions actively against their potential gradients are the most important in this respect (see Section 5.2.1: 'Transport Structures'). The energy for maintaining the concentration gradients and the transport protein system is provided by the ATP molecules produced by cellular respiration (see Section 6.3.2.1). In this way, root metabolism controls both the driving force (concentration gradients) and material properties (transport proteins) determining the flux of water and nutrients across the plasma membrane.

We notice that the uptake of water and nutrients by plants involves several passive and active transport mechanisms and gradients that often influence and oppose each other in a coupled way. For this reason, the full description of material fluxes in the systems with multiple constituents requires explicit accounting for all transport mechanisms for the species involved (see e.g. Corey and Logsdon, 2005). Below we outline the mechanisms involved in the uptake of water and nutrients by roots.

The gradients of both gravitational, pressure and osmotic components of water potential affect the water flow (see the box in Section 7.5.2; Nobel, 1991; Hillel, 1998), but in practice some of them are often small enough to be left out. In bulk soil, only the convective flow driven by the gradients in pressure (or tension) and gravity is relevant, whereas at the cellular level of water uptake by roots the gravity can be neglected, because the main driving forces for the uptake of water by root cells are the differences in the solute concentration and water tension between soil water and root cell.

Water and nutrients move in the plant roots through (1) the transmembrane pathways through plasma membranes and tonoplasts, (2) the network of cytoplasm interconnected by plasmodesmata, and (3) the continuum of extra-cellular space (Figs. 6.3.27 and 6.3.28; see Section 5.2.1 'Transport Structures'). The hydrostatic pressure (or tension) gradient is the main driving force for the convective water flow through the apoplastic and symplastic pathways that involve no crossing of semipermeable membranes. In the transmembrane pathway, however, the solute

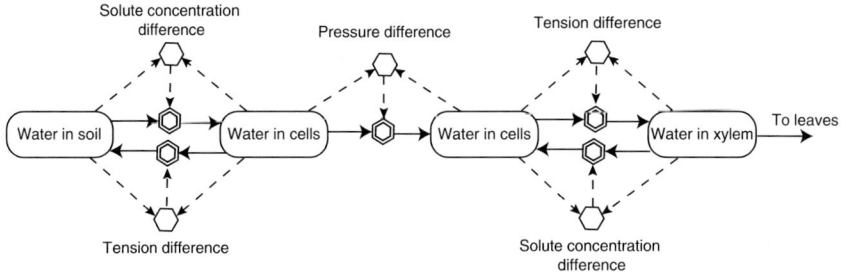

Fig. 6.3.27 Water flow from soil through roots to leaves via transmembrane and symplastic pathways. Solute concentration and pressure (or tension) gradients drive the water flow via a transmembrane pathway of water from soil into the cell and out of the cell into xylem across the plasma membranes. Pressure gradients drive the water flow via a symplastic pathway through the plasmodesmata between neighboring root cells (Section 6.3.2.4). The symbols are introduced in the Fig. 1.2.1

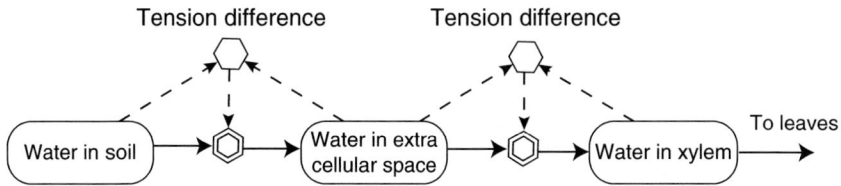

Fig. 6.3.28 Water flow from soil through roots to xylem via an apoplastic pathway. Tension gradients drive the water flow through the extra cellular space

concentration gradients (osmotic potential gradients) driving the diffusive water flow across the membrane by osmosis are the most relevant (Taiz and Zeiger, 2006). According to Fig. 6.3.27, water flows from soil to plant leaves only if the water tension in xylem is higher than in soil. Small positive pressures (typically <0.1 MPa) may build up in xylem due to accumulation of solutes, if soil water tensions and transpiration rates are low, but this mechanism is inadequate to move water up to actively transpiring (Fig. 6.3.28) tall trees (Taiz and Zeiger, 2006). The relative importance of the apoplastic pathway decreases with root age as the radial cell walls in the endodermis become impregnated by the wax-like substance suberin to form Casparian strips that force most water and solutes to cross the endodermis by passing through the plasma membrane (Mengel and Kirkby, 2001; Taiz and Zeiger, 2006).

Root metabolism continuously controls the concentrations of sugars, ions and other solutes within cells. Even if the cellular concentrations of individual species may be higher or lower compared with soil solution, the overall solute concentrations in plant cells are usually in the range of 0.2–0.8 kmol m^{-3} (Mengel and Kirkby, 2001), which means the cells are one or two orders of magnitude more concentrated than the typical soil solutions with solute concentrations of about 0.01 kmol m^{-3}. The high cellular concentrations of sugars and other solutes (decreased water potential) drives water from soil through cell walls and plasma membrane into root cells (see Section 4.1 'Conduction and Diffusion'). This causes the buildup of hydrostatic pressure (turgor) within cells at a rate depending on the rate of water flow and the mechanical properties of cell wall (Mengel and Kirkby, 2001; Taiz and Zeiger, 2006). The buildup of turgor, in turn, drives the diffusive and convective flow of water out of cells until an equilibrium or steady state is reached at which the inflow and outflow of water balance each other. Water crosses the membranes mainly by diffusion through small protein channels called aquaporins that are opened or closed by cellular metabolism (Fig. 6.3.27).

Besides its role in the control of water uptake, positive turgor is important for root growth because it provides the physical pressure for the expansion of growing root cells (Greacen and Oh, 1972; Bengough et al., 2006). Root growth in turn largely determines the accessibility of soil water and nutrients to plants. The theoretical maximum equilibrium turgor pressure is never reached in practice, because as the water flows continuously from the roots to upper plant parts and the atmosphere, water potentials are higher in soil than in the roots. If there were no cellular activities maintaining positive turgor in a transpiring plant growing in unsaturated soil, the cells would be under tension. In that case water could still flow via the apoplastic pathway to the atmosphere (Fig. 6.3.28), but the root cells would not grow (expand) any more and large tension would cause partial collapse of root cells.

Flowing water carries along solutes (Figs. 6.3.29, and 6.3.30) as dissolved inorganic ions, neutral species, and inorganic or organic complexes. The amounts and chemical forms of solutes depend on their reactions in soil, most notably on the exchange and retention by the soil solid phase as well as on the acid-base and

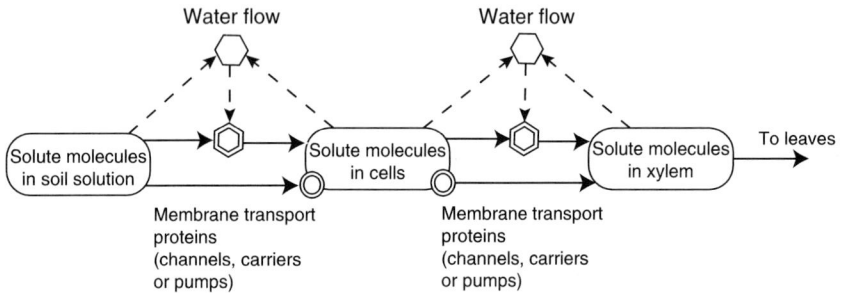

Fig. 6.3.29 The flux of solutes from soil into root cells and xylem to leaves through root cells (transmembrane pathway with different membrane transport proteins). The symbols are introduced in the Fig. 1.2.1

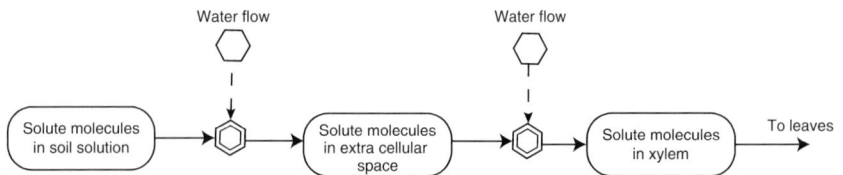

Fig. 6.3.30 The flux of solutes from soil through roots to leaves through extra-cellular space (apoplastic pathway)

oxidation-reduction reactions in soil. However, plant roots are remarkably selective in their uptake of different elements from soil solution. The relative uptake rates and cytosolic concentrations of different nutrients deviate much from their respective soil solution concentrations. The high selectivity is attributed to highly substrate-specific membrane transport proteins, including channels, carriers and pumps that facilitate most of the transport of water and solutes through biological membranes (Fig. 6.3.29; see Section 5.2.1: 'Transport Structures'). Only non-polar and small uncharged molecules permeate rapidly through the phospholipid part of the membranes.

The membrane transport proteins may increase the permeability of ions and water one or more orders of magnitude compared with the transport through a pure phospholipid bilayer, but have no effect on nonpolar gases and small uncharged molecules (Taiz and Zeiger, 2006). Due to the transport proteins, practically all substrates have their own specific uptake mechanism, although the specificity may not be absolute. A nice animation on the functioning of transport proteins is available via internet at www.teachersdomain.org/resources/tdc02/sci/life/cell/membraneweb (WGBH, 2003). The molecular biology of different transport proteins is currently an active field of research (e.g. Miller and Cramer, 2004; Luu and Maurel, 2005; Raghothama and Karthikeyan, 2005). The membrane transporters are commonly classified as channels, carriers and pumps according to their type of functioning.

Channels are membrane transport proteins that facilitate passive transport along the decreasing electrochemical gradient (Taiz and Zeiger, 2006). Channels are essentially transmembrane selective pores that open and close in response to cellular signals ("gating"). The substances diffuse very rapidly through channels, if they are open (10^8 ions/s per channel). Specific channels exist for numerous cations and anions. Those specific for water molecules are called aquaporins (Luu and Maurel, 2005). Aquaporins also allow transport of small neutral solutes and/or gases.

Carrier proteins mediate both passive and active transport (Taiz and Zeiger, 2006). Carriers do not have pores crossing the whole membrane. Instead, they bind the molecule on one side and release it on another. The slow rate of required conformational changes in the protein make the carrier transport 10^6 times slower compared with the channel transport. Active carrier-mediated transport uses energy stored in the electrochemical potential gradient of protons (higher concentration of protons outside than inside the cells) to drive the transport of other ions in the same direction (symport, coport) or the opposite direction (antiport) compared with that of protons.

Pumps are membrane transport proteins that use energy obtained from direct ATP hydrolysis to carry out active energetically uphill transport of substances, such as that of protons or calcium ions out of cytosol (Taiz and Zeiger, 2006). This is called the primary active transport, in contrast to the secondary active transport by carriers for which the energy is provided by the proton motive force rather than by direct ATP hydrolysis. Most pumps are ion pumps. Proton is the most important ion transported by this way in the membranes of plants.

Living cells exhibit a net negative charge as shown by a voltage difference in a range between -60 and -240 mV across the cell membrane (Taiz and Zeiger, 2006;

Ritchie, 2006). This unequal ionic charge distribution inside and outside of cells is caused by the unequal flux rates and active transport of ions into and out of the cells (Taiz and Zeiger, 2006). The net negative charge impedes the uptake of anions and enhances the uptake of cations by roots. Accordingly, the comparison of measured cytosolic concentrations to theoretical predictions shows that the uptake of anions is active, whereas the uptake of most cations is passive, with only potassium taken up actively at low external concentrations. On the other hand, such cations as sodium, calcium and protons are pumped actively out of cytosol into extracellular space and vacuole to maintain low cytosolic concentrations (Taiz and Zeiger, 2006).

In contrast, although the concentrations of nitrate and ammonium in soil solution of natural ecosystems vary widely ranging from several tens or hundreds of millimoles per cubic metre up to several moles per cubic metre (Britto et al., 2002; Miller and Cramer, 2004), these values are typically much less than the estimates for their cytosolic concentrations ranging from several moles per cubic metre to several tens of moles per cubic metre, with the values in the lower range being more probable (Mengel and Kirkby, 2001; Loqué and Wirén, 2004; Ritchie, 2006). For this reason, active uptake of nitrate anions by plant roots is required almost in the whole range of concentrations encountered in the soil (Miller and Cramer, 2004), whereas active uptake of ammonium cations is required at low concentrations lower as the uptake becomes passive only at concentrations higher than about 500 mmol m^{-3} (Britto et al., 2001).

The uptake of amino acids by roots is an active energy-requiring process as well. Many plant species are commonly able to take up amino acids by roots, but the importance of this as a major pathway for nitrogen acquisition in the field has not been confirmed (Jones et al., 2005; Persson et al., 2006). Ammonium and nitrate are thus the major nitrogen forms taken up by roots. In boreal forest soils, the nitrate concentrations are low due to acidity that impedes nitrification. Moreover, many plant species including conifers take up ammonium preferentially over nitrate (Kronzucker et al., 1996; Mengel and Kirkby, 2001; Persson et al., 2006). For these reasons, ammonium ion may be considered the major chemical form of nitrogen taken up by plants in boreal forests.

The mechanisms involved in the uptake of nitrate and ammonium by plant roots have been recently reviewed by Glass et al. (2002), Loqué and Wirén (2004), and Miller and Cramer (2004). Several high and low affinity transport systems have been indicated for both nitrate and ammonium. The cytosolic concentrations are determined by the net effect of influx and efflux, as well as the internal compartmentation and assimilation of ions within the cell, all of which are controlled largely by plant metabolism at the transcriptional, protein, cellular and whole plant levels. Nitrate is generally considered to be taken up by a secondary active $2\,H^+/NO_3^-$ coport or by a primary active ATP-driven pump (Miller and Cramer, 2004; Ritchie, 2006), whereas anion channels are the most obvious route for nitrate efflux (Miller and Cramer, 2004). Ammonium ions enter the cells by an active H^+/NH_4^+ coport, or through cation selective channels or carriers for NH_4^+ and K^+ as chiefly driven by the negative membrane potential of plant cell (Mengel and Kirkby, 2001; Miller and Cramer, 2004).

Gaseous ammonia (NH_3) is a weak base ($pK_a = 9.3$) that accompanies ammonium ions (NH_4^+) in aqueous solutions at high pH. The concentrations of gaseous ammonia increase with increasing pH, but this does not influence the uptake of ammonium by plant roots at the normal soil pH range. Nevertheless, gaseous ammonia does contribute to the overall ammonium efflux (Loqué and Wirén, 2004; Britto et al., 2001). The high cytosolic pH favours the passive efflux of gaseous ammonia. It may permeate the cell membranes directly through the phospholipidic bilayer, or through aquaporin-like transport proteins. The extent to which plants control the transmembrane ammonia fluxes by gating of such channels is not known (Loqué and Wirén, 2004). In addition, the energy-demanding active efflux of ammonium ions has been implicated as a possible cause for ammonium toxicity at high concentrations (Britto et al., 2002).

The secretion of protons and other ions, along with that of sugars, polysaccharides and enzymes (see Sections 6.4, 7.6 and 7.9) by roots into the surrounding soil contributes to the modified physico-chemical and microbiological properties of rhizosphere soil compared with bulk soil (recently reviewed by Hinsinger et al., 2005; Gregory, 2006). The capability of roots for nutrient uptake varies according to the anatomy of roots and architecture of the root system (Peterson et al., 1999; Hishi, 2007). Most of the nutrients are taken up by the active white zones near the tips of apical first order roots, and some through the pigmented zones of condensed tannin, whereas the pigmented cork zones forms a barrier for nutrient uptake (Hishi, 2007). The nutrient uptake of trees is additionally influenced by the symbiosis between roots and mycorrhizal fungi (Graham and Miller, 2005). They are important for the nutrient uptake of woody plants, as they may cover most of the root surface area capable of efficient nutrient absorption (Taylor and Peterson, 2005).

The external hyphae of mycorrhizas are considered to enhance the uptake of ammonium ions by increasing the effective absorbing surface of roots as well as by hydrolysing N-containing macromolecules, whereas the direct uptake of amino acids by mycorrhiza seem inefficient (Chalot et al., 2002; Miller and Cramer, 2004). The energy requirements and mechanisms of ion transport across biological membranes through transport proteins are broadly similar in mycorrhizal symbiosis and in plant roots, with the notable difference that now the ions first enter the fungal cells through transport proteins in the fungal plasmalemma. Ammonium ions may then be assimilated into amino acids, and amino acids and ammonium ions be transported to the host plant through the fungal plasmalemma, the interfacial apoplastic matric, and the root plasmalemma. The energy for fungal metabolism is provided by the flow of sugars from the host plant to the fungus. Ammonium ions, amino acids and sugars pass the fungal and plant membranes through specific transport proteins. Chalot et al. (2002) summarized the membrane transport proteins involved, part of which are fully characterized, whereas some remain putative or hypothetic. Recent papers by Govindarajulu et al. (2005), Chalot et al. (2006), Schüßler et al. (2006), Martin et al. (2007) and Müller et al. (2007) provide more detailed discussion of the molecular mechanisms of the transfer of nitrogen and carbohydrates in the mycorrhizal symbiosis.

6.3.3 Changing Functional Substances

6.3.3.1 Annual Cycle of Processes

Heikki Hänninen[1], Pertti Hari[2], Risto Häkkinen[3], and Tapio Linkosalo[2]

[1] University of Helsinki, Department of Biological and Environmental Sciences, Finland
[2] University of Helsinki, Department of Forest Ecology, Finland
[3] Finnish Forest Research Institute, Finland

The climate in the boreal zone is characterized by strong seasonal variation (Chapter 3). In the most continental parts of the boreal zone the minimum air temperatures may drop below $-70°C$ during winter, whereas the summertime maxima are often above $+30°C$. In these extreme conditions the annual amplitude of air temperature thus exceeds $100°C$. In less continental areas of the boreal zone the amplitude is smaller, but still well above $50°C$. Unlike ground vegetation overwintering below a sheltering snow cover, boreal trees experience – with the exception of roots – all these temperature extremes (Sakai and Larcher, 1987). Boreal trees are adapted to this extreme climatic seasonality with their regulation system of the annual cycle so that the alternation of the frost hardy dormant state and the susceptible growth state are synchronized with the annual climatic cycle (Sarvas, 1972, 1974; Fuchigami et al., 1982; Koski and Sievänen, 1985; Hänninen and Kramer, 2007).

The regulation system of the annual cycle takes its signal from the annual cycle of environmental factors. Subsequently the amounts and activities of several functional substances, such as enzymes, are regulated so that the timing of the growth and dormancy of the trees is synchronized with the climatic cycle prevailing at the growing site (Fig. 6.3.31). This requires a proper orchestration of the functional substances involved. The functional substances in turn affect the synthesis of several other macromolecules, such as proteins, cellulose, and lignin. Finally, all of these

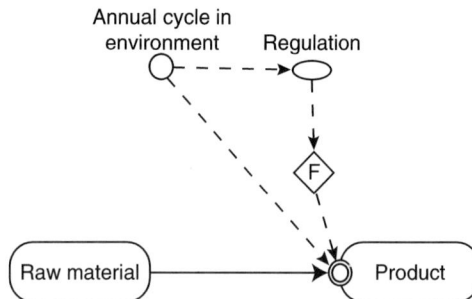

Fig. 6.3.31 Schematic presentation of the regulation of the annual cycle of boreal trees. "F" denotes functional substances such as enzymes; "Raw material" denotes carbon dioxide, water, carbohydrates and nutrients; and "Product" various organic substances and tree tissues formed on the basis of them. The regulation system is based on the genetic properties of the trees and it responses to the annual cycle of environmental factors

complicated physiological and morphological phenomena taking place at molecular, cellular and tissue levels are manifested in the annual cycle of phenological events which is readily visible to the naked eye at the whole tree level. This phenological cycle consists of such events as bud burst and onset of elongation in the spring and growth cessation and bud set during early autumn (Sarvas, 1972, 1974; Fuchigami et al., 1982; Koski and Sievänen, 1985; Hänninen and Kramer, 2007).

The activation of photosynthesis of evergreen coniferous trees is initiated during early spring, mainly as a result of rising air temperature. During this time of the year a lot of incoming solar radiation is available, but its utilization is impaired due to the low photosynthetic capacity of the needles (Pelkonen, 1980; Pelkonen and Hari, 1980; Hari and Mäkelä, 2003; Mäkelä et al., 2004). Impaired water uptake by frozen soil may also slow down photosynthesis. The photosynthetic capacity is recovered due to prolonged exposure to relatively high air temperatures (see Section 6.3.3.1.3). A simultaneous gradual dehardening, i.e. loss of frost hardiness, of the needles takes place (Repo et al., 2006). These physiological changes are to some extent reversible, so that the needles may reharden and their photosynthetic capacity may decrease during a cold spell occurring after the dehardening and activation of the needles during earlier periods of relatively high air temperatures (fluctuating development; Hänninen and Kramer, 2007). Later during spring, formation of new tissues starts, i.e., cambial and elongation growth are initiated (fixed sequence development, Hänninen and Kramer, 2007).

Early summer is the season of high metabolic activity in trees. At this time resource acquisition and use for tissue formation take place at a high rate. Towards the latter part of the summer the metabolic rates are slowed down. Generally, the formation of new tissues stops earlier than the processes of resource acquisition, i.e., photosynthesis and uptake of nutrients. For instance in central Finnish conditions the elongation growth of mature Scots pine trees stops in late June–early July, and the growth of needles and cambium in late July–early August (Kanninen et al., 1982), whereas photosynthesis continues until late autumn (Hari and Mäkelä, 2003; Mäkelä et al., 2004; Kolari et al., 2007). Thus the latter part of the growing season is characterized by resource acquisition and collecting carbohydrate reserves for the next growing season. During this time of the year also the overwintering buds are formed. The environmental conditions of the growing season affect bud formation. In the case of species with the so-called fixed growth habit, the number of needle primordia in the bud, and thus the number of needles per shoot during the next summer, are determined on the basis of the environmental conditions prevailing during the summer when the bud is formed (Garrett and Zahner, 1973).

Intracellular ice formation is always lethal to plant cells, and the boreal trees are no exception to this rule. Frost survival of boreal trees is based on two mechanisms: freezing avoidance and freezing tolerance (Sakai and Larcher, 1987). *Freezing avoidance* is based on the phenomenon of supercooling where the temperature of the cell sap drops below the freezing point without formation of any ice crystals. With supercooling the tree tissues can tolerate slight short-term frosts even during the active growing season. Depending on the species and tissue, supercooling is an important survival mechanism also during overwintering, but it is not alone sufficient. Rather, during winter the mechanism of *freezing tolerance* is also required. Freezing tolerance has two components: Firstly, tolerance to extracellular ice formation and secondly, tolerance to a drastic dehydration of the cell sap. As active growth processes are not possible in severely dehydrated tissues, the freezing tolerance is accompanied by a deep dormancy of the meristematic tissues (Sakai and Larcher, 1987).

During late summer, approximately at the time of cessation of tissue formation, the process of frost hardening is initiated, i.e., the two components of freezing tolerance start to develop in the tree tissues. Several functional substances, such as dehydrins, are involved in this process (Sutinen et al., 2001). Sugars are important cryoprotectants in plant tissues so transforming starch into soluble sugars is an essential part of the frost hardening process. For this reason the collection of carbohydrate reserves during the later part of the growing season is essential also with respect to the frost survival of the trees. When air temperature drops sufficiently below zero, ice crystals start to be formed in the extracellular spaces where the ion concentrations are quite low. This leads to a flow of water molecules from the cells to the extracellular spaces. In this way the size of the ice crystals in the extracellular spaces is increased and the cell sap is gradually dehydrated. As a result of the dehydration the freezing point of the cell sap is lowered so that no intracellular freezing takes place even during the lowest temperatures prevailing during winter (Sutinen et al., 2001).

According to the classical theory the frost hardening takes place in two phases so that increase of night length is essential during the first and declining air temperatures during the second phase (Weiser, 1970). According to a contrasting theory, the effect of these two is additive during the entire hardening phase (Chen and Li, 1978; Greer, 1983; Leinonen, 1996). During spring, dehardening of the tree tissues is driven by rising air temperatures (Repo and Pelkonen, 1986). During the dehardening the cells are gradually rehydrated, so the process of dehardening is closely related to recovery of the metabolic activity of the tissues. Thus in general, the development taking place during spring is opposite to that taking place during autumn. Repo et al. (2006) found, however, that the relationship between frost hardiness and photosynthetic capacity is not identical during autumn and spring. This finding suggests that partially different physiological mechanisms with their respectively different functional substances take place during these two phases of the annual cycle.

Overall, the regulation of various aspects of the annual cycle of development takes place as an interaction of genetic and environmental factors (Fig. 6.3.31). Production of functional substances is regulated by genes whose functioning responds to environmental factors. During the course of adaptation different genetic properties have been selected for under different climatic conditions. For this reason there is much genetic variation among tree species and provenances in their response to the environmental factors regulating the annual cycle (Aitken and Hannerz, 2001; Clapham et al., 2001), i.e., the amount and activity of functional substances respond to environmental factors in a different way in different provenances.

6.3.3.1.1 Bud Burst Phenology

Heikki Hänninen[1], Tapio Linkosalo[2], Risto Häkkinen[3], and Pertti Hari[2]

[1] University of Helsinki, Department of Biological and Environmental Sciences, Finland
[2] University of Helsinki, Department of Forest Ecology, Finland
[3] Finnish Forest Research Institute, Finland

Despite the great variation among species and provenances in the interaction of environmental and genetic factors in regulation of the annual ontogenetic cycle, some generalizations can be made. This is because variations exist mainly in the details of the regulation system, while its main principles appear to be quite common among the boreal tree species. Thus for instance when modelling the regulation of the annual cycle, similar equations can often be applied with different species, while different parameter values should be used for different species, provenances, and even individuals (Linkosalo, 2000; Hänninen and Kramer, 2007).

Bud development during early spring starts with formation of macromolecules, cellulose, lignin, etc., to construct new tissues. Each macromolecule has its specific enzymes for the biochemical steps in the synthesis. There is a clear order in the construction of new structures, for example, synthesis of cellulose in the primary cell walls must take place before formation of lignin. Thus the regulation system synthesizes, activates, decomposes and deactivates the enzymes involved in the bud developed in a very synchronised way. Functional substances synthesize new macromolecules, such as proteins, cellulose and lignin. The state of functional substances determines the new macromolecules to be synthesised.

Formation of new cells and their growth lead into irreversible ontogenetic development, i.e., to the microscopic anatomical changes within the bud which finally lead to the visible bud burst. The anatomical changes are tightly synchronised with the changes in the functional substances, since construction of the new structures in the buds has very specific requirements for raw material, i.e. for macro molecules. Thus there are regularities in the functional substances, we introduce the concept state of functional substances to describe these regularities.

The regulation system of the annual cycle can change the functional substances towards summer state or backwards towards winter. The formation of new cells can not go backwards since existing cell structures can not be removed (apart from senescence where whole organs are inactivated). Thus there are two different aspects in the bud development; the state of the functional substances and the state of bud development.

Assume that the state of the functional substances can be described with one scalar valued variable, S_F. The state of functional substances changes during the annual cycle and the synthesis of new molecules and cell formation respond to the changes in S_F. The rate of change of functional substances, s_F, is defined as the time derivative of the state of the functional substances, i.e.,

$$s_F = \frac{dS_F}{dt} \tag{6.3.44}$$

The regulation system of the annual cycle takes its signal from temperatures in the spring (Fig. 6.3.31) and it changes the functional substances towards summer or winter state according temperature. This can be introduced into the analysis by assuming that the rate of change of functional substances depends on temperature,

$$s_F = s_F(T) \tag{6.3.45}$$

The ontogenetic development within the bud is described quantitatively by the concept of *state of bud development*, $S_B(t)$ (Hari et al., 1970; Hari, 1972; Hari and Häkkinen, 1991; Hänninen, 1995; Hänninen and Kramer, 2007). This variable describes in quantitative terms the phase of the ontogenetic development of the bud, i.e., the anatomic changes taking place within it. The scale of $S_B(t)$ is arbitrary, depending on the specific model used (see next paragraph). The time derivative of the state of bud development, $dS_B(t)/dt$ is called *rate of bud development*, $s_B(t)$. It depends on the state of functional substances and environmental conditions, first of all on air temperature. If we assume that the role of the state of functional substances in the ontogenetic development is to start the growth in proper time early in the spring and that the states of functional substances and bud development are synchronised, then we can assume that after onset of ontogenetic development the rate of bud development depends only on temperature, thus

$$s_B = s_B(T) \tag{6.3.46}$$

When the temperature conditions are known, then the value of the rate of development can be calculated with the specific model used. Subsequently the state of bud development is by definition obtained as an integral of the rate of bud development:

$$S_B(t) = \int_{t_0}^{t} s_B(T(t)) \, dt \tag{6.3.47}$$

where t_0 is the beginning instant of the bud development, being determined by the state of functional substances.

Bud burst is predicted to occur when the state of bud development attains a critical value, i.e., the *high temperature requirement of bud burst (or flowering)*, H_{crit}:

$$S_B = H_{crit} \tag{6.3.48}$$

Ontogenetic development towards vegetative bud burst and growth onset is driven by exposure to high "forcing" temperatures (Sarvas, 1972, 1974). The ontogenetic development of generative buds towards flowering is similarly affected by air temperature. Within a certain range of air temperature, the rate of bud development, $s_B(t)$, increases with rising temperature. Traditionally this regulation of the ontogenetic development is approximated by linear day degree models (Wang, 1960), so in this case the dimension used for state of bud development, $S_B(t)$, and for high temperature requirement of bud burst, H_{crit}, is day degree. However, the real air

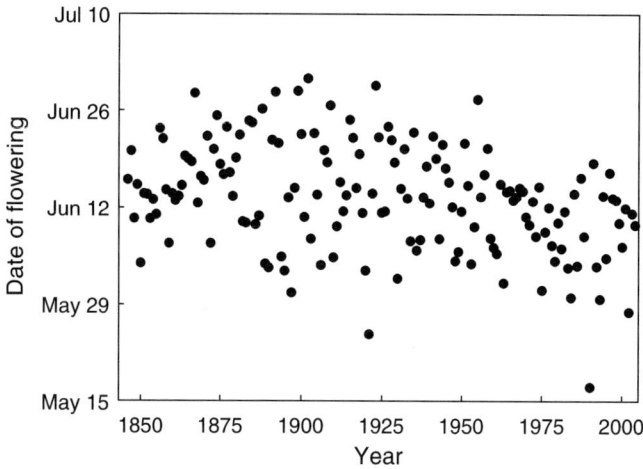

Fig. 6.3.32 Annual means of observed bud burst of Rowan (*Sorbus aucuparia*) in Central and Southern Finland, during the years 1848 to 2004. The phenological observation data was collected by the Finnish Society of Sciences and Letters, more recently in co-operation with the Finnish Museum of Natural History

temperature responses are nonlinear. The air temperature responses can be readily determined experimentally, but rather few responses are found in the literature (Sarvas, 1972, 1974; Campbell and Sugano, 1975, 1979). This is unfortunate since in order to describe the ontogenetic development in a realistic way, the linear day degree models should be replaced by the real nonlinear air temperature responses. This is especially the case when the models are applied in studies assessing the effects of the predicted climate change. It was shown recently that the predictions concerning the effects of climate change are sensitive to the air temperature response applied (Hänninen et al., 2005; Hänninen, 2006).

The spring development of boreal trees can be studied e.g. utilising historical phenological time series, that indicate the observed bud burst and flowering dates over a longer time period. Long time series are most useful, as the year-to-year variation of these events is large, often more than a month (Fig. 6.3.32). The date of flowering has become earlier during the last decades, evidently a sign of climate change. Only rather long data series, used with corresponding temperature data, reveal the dependency of the events on environmental cues. The modelling approach indicated by Eq. 6.3.47 is illustrated in Fig. 6.3.33, where the timing of flowering of rowan (*Sorbus aucuparia*) is predicted with a simple linear day degree model. Despite its simplicity this model can predict the timing of rowan flowering quite accurately.

Differences among species in their spring phenology are usually caused by corresponding differences in the high temperature requirement of bud burst, H_{crit}: bud burst occurs the later the greater H_{crit} is (Sarvas, 1972; Linkosalo, 2000). There are also corresponding differences in the spring phenology among different provenances

Fig. 6.3.33 Prediction of timing of flowering of rowan (*Sorbus aucuparia*) in Jyväskylä during 1953–2005. The modelling approach outlined with Eqs. 6.3.47–6.3.49 was implemented using a linear day degree model for modelling the air temperature response of rate of bud development, $s_B(t)$

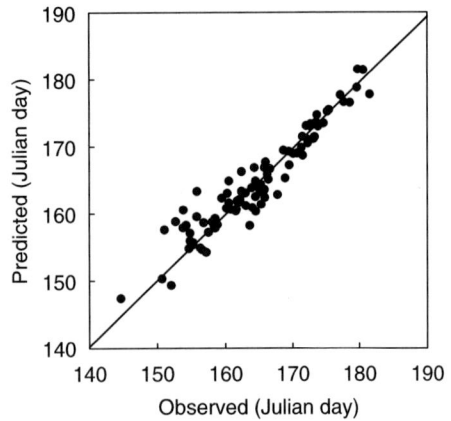

of a given species. The high temperature requirement of bud burst, H_{crit}, is generally the smaller the shorter and colder the growing season is in the climate prevailing at the natural growing site of the provenance (Sarvas, 1967). Thus, northern or high altitude provenances have usually lower H_{crit} than do southern and low altitude provenances, so when grown in a common garden the former burst buds earlier than the latter.

6.3.3.1.2 Seasonality of Respiration

Pasi Kolari, Jaana Bäck, and Pertti Hari

University of Helsinki, Department of Forest Ecology, Finland

Growth and maintenance of plant tissues continuously require chemical energy. Energy storages are built up in photosynthesis and during the subsequent partitioning of energy-rich carbon skeletons between starch and sucrose, which are the main forms of chemical energy. When energy is needed for growth or maintenance, these compounds are broken down in respiratory processes converting ADP to ATP, and simultaneously releasing CO_2. In growth respiration the reduced, energy-rich carbon compounds are used for building up new tissues, whereas the maintenance respiration keeps the basic cellular metabolism in operation. Maintenance respiration can account for more than 50% of the respiratory flux under normal conditions. Respiratory processes provide the plant significant metabolic flexibility which facilitates the survival and development also under less favourable conditions. The processes that depend on respiratory ATP as the source of energy, however, vary e.g. in different phases of the annual cycle of plants. In spring there is additional need for energy in leaves and other meristems, when the functional substances needed in photosynthesis and for cellular growth processes are recovered. Synthesis of macromolecules for formation of new tissues also requires lots of energy, and therefore

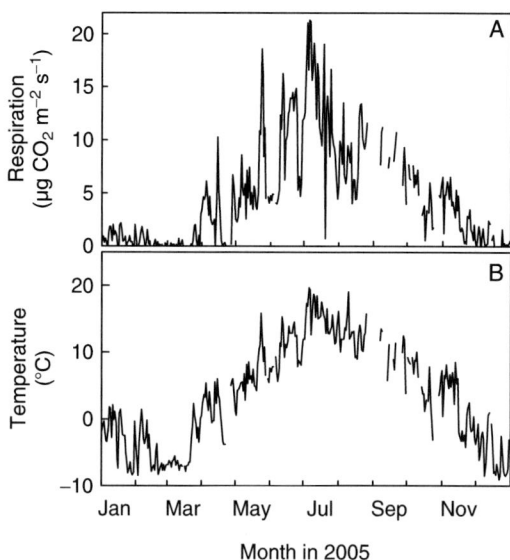

Fig. 6.3.34 Seasonal pattern of shoot respiration (daily mean night-time CO_2 exchange) (A), and night-time temperature (B) over one year. The fluxes and temperature were averaged between 22:00 and 2:00 solar time

respiration is high during the active growth (growth respiration). During the stable functional substance phase the need of energy for metabolism should be relatively stable (maintenance respiration), whereas in autumn the changes in functional substances for winter conditions again require some extra energy.

The respiration of the whole-plant or a plant organ can be measured as the CO_2 efflux during the dark period, which eliminates the photorespiration driven CO_2 production and gives a reliable estimate on the base respiratory metabolism. As the rate of ATP formation, i.e. respiration, is related to temperature (Section 6.3.2.1), respiration over a year in general follows the seasonal course of temperature (Fig. 6.3.34). The base level of respiration, however, varies. In spring and early summer the shoot respiration is higher than in the same temperature range in autumn (Fig. 6.3.35). The changes in photosynthetic functional substances and in the activity of tissue meristems can be seen in the enhanced respiration although the buds were removed in previous winter to prevent shoot elongation. Thus this seasonality cannot be strictly speaking considered as growth respiration.

We studied the seasonal pattern of pine shoot respiration by utilising night-time CO_2 exchange measured with chambers at SMEAR II to parameterise a simple empirical respiration model (see Section 6.3.2.1)

$$r(T) = c_r(r_0 Q_{10}^{T/10} + c_0) \qquad (6.3.49)$$

where T is leaf temperature, Q_{10} the temperature sensitivity of respiration, and r_0 describes the base level of respiration, i.e. respiration in a standard temperature (0°C). A parameter c_0 is introduced to force respiration to disappear at a point

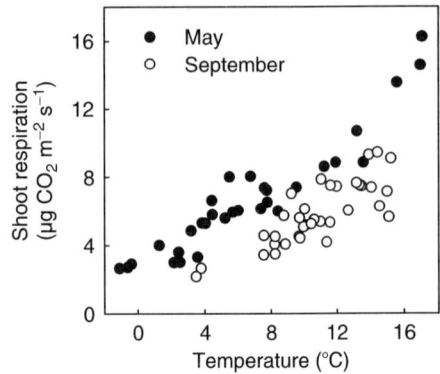

Fig. 6.3.35 Relationship between pine shoot respiration (daily mean night-time flux) and mean night-time temperature in May and in September 2004. Night is defined as the time when photosynthetically active radiation is below 3 μmol m^{-2} s^{-1}

where we have observed shoot CO_2 exchange to diminish below detection limit. This modification helps overcoming the non-constant Q_{10} issue reported in several sources (e.g. Atkin and Tjoelker, 2003). Temperature sensitivity of respiration is known to increase at low temperatures, although this is merely a consequence of using exponential equation that cannot properly describe the temperature response of respiratory CO_2 efflux at low temperatures (Davidson et al., 2006).

We first estimated the temperature sensitivity Q_{10} and the base level r_0 using chamber data from July when the functional substances are expected to be fairly stable and the base level of respiration constant. Parameter r_0 was then re-estimated daily in a moving time window of seven days. CO_2 exchange measurements indicate that respiration diminishes below detection limit at very low temperatures (for Scots pine shoots at SMEAR II this temperature is about –5°C). Thus we only analyse nights when there was a clear CO_2 exchange signal from the chamber and exclude the measurements made in midwinter.

The daily estimation of r_0 reveals a systematic seasonal cycle in the relationship between temperature and respiratory CO_2 efflux, and the instantaneous temperature response is superimposed on this cycle. The value of the parameter r_0 is high in spring and stays high until mid-June when it fairly quickly decreases to about 60–70% of the springtime level (Fig. 6.3.36). Later in the summer and in the autumn there is some variation in r_0 but it is less consistent from year to year and we cannot always say whether it is real or a measuring artifact. The sharp fluctuation in late September and October is coinciding with freezing temperatures, so it might be linked with development of frost hardiness. In autumn trees undergo cellular and whole-plant regulation of metabolism, whereby storage compounds are formed e.g. in stems and roots, and resources allocated to provide increased cold tolerance in overwintering plant organs. It is probable that some of the observed seasonal variation in the estimated base level is due to shortcomings in the model, despite the addition of a constant term to the model to increase temperature sensitivity at low temperatures. Values of r_0 estimated at the same temperature range, however, are directly comparable regardless of the model used, as the spring-autumn comparison in Fig. 6.3.35 indicates.

Fig. 6.3.36 Seasonal course
of the base level of respiration
r_0 in Scots pine shoots (mean
of three shoots) at SMEAR
II in 2004. Measurements
when daily mean temperature
was below $0°C$ were omitted
because shoot CO_2 exchange
at freezing temperatures is too
low for accurate determination of respiration

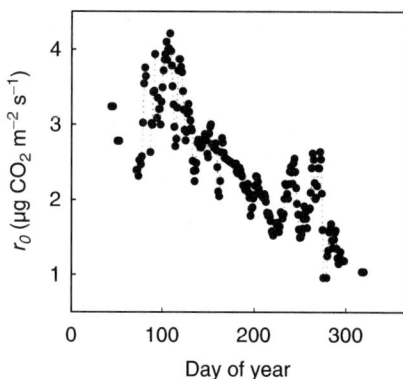

Similar behaviour of shoot respiration can be observed at SMEAR I, 200 km
beyond the Polar Circle, but the base level of respiration cannot be determined accurately from field measurements in the midsummer because the sun stays constantly
above the horizon for a couple of weeks. Measuring respiration accurately would
require covering the chamber to obtain darkness.

Conclusions

The annual pattern in respiration is rather weak, but evident. The clear peak in the
spring, as well as the stable phase in summer, is as expected from the changing
energy needs of cells. The variation in the base level of respiration in autumn might
be linked with regulation of metabolism to increase cold tolerance.

6.3.3.1.3 Photosynthesis

Pertti Hari, Pasi Kolari, Jaana Bäck, Annikki Mäkelä, and Eero Nikinmaa

University of Helsinki, Department of Forest Ecology, Finland

The annual cycle in photosynthesis is strong in the boreal forests; there is intensive
sugar formation in summer and very small, if any, photosynthetic activity in winter.
This annual pattern is partly directly related to changes in the environment since in
winter there is little sunlight available and beyond The Polar Circle, as at SMEAR I,
the sun never rises in the middle of the winter.

The annual cycle is also strong in the regulation of functional substances involved
in photosynthesis. Photosynthesis is based on a complicated chain of reactions, both
biochemical and physical. Each step in the biochemical chain requires its own functional substances: pigments, enzymes or membrane pumps. The number of functional substances involved in photosynthesis is large, evidently over 100. These

functional substances, such as chlorophyll a and b and Rubisco, work in a well balanced concert and are controlled by the genetically determined regulation system. The regulation system responds to the annual cycle of environmental factors such as light and temperature, and changes the concentrations and activities of photosynthetic functional substances, which further influence the efficiency of the photosynthesis process itself.

In the boreal zone, photosynthetic rate is very low in winter and early spring. Evidently an actively operating photosynthetic system would be unable to tolerate the hard conditions in early spring with low temperatures and high light intensity (see 'VOC Emissions Under Changing Climate'). Therefore, the functional substances need to be regulated in a sophisticated way to cope with the conditions. In a dormant plant, cells are adjusting their metabolism in order to avoid both intracellular freezing and photodamage. Intracellular freezing is avoided by changes in e.g. osmotic status of cells, mediated by changes in concentrations of sucrose and other small, cryoprotective molecules (e.g. dehydrins, Kontunen-Soppela and Laine, 2001; Öquist et al., 2001; 'Bud Burst Phenology'). The increases in concentrations of solutes inhibit freezing-induced water movement from cells into the extracellular spaces and stabilize the lipid molecules in light-harvesting complexes and in membrane bilayers (Thomashow, 1998; Öquist and Huner, 2003). In the extracellular spaces ice crystals can be formed rather harmlessly, but if freezing periods are prolonged and/or temperatures low enough, then water is drawn from cells and they become dehydrated. Experiments showing that ABA treatment improved the freezing tolerance (Mäntylä et al., 1995; Wilen et al., 1996) suggest that there may be many similarities in plant acclimation to drought and in the annual cycle of photosynthesis.

Photosynthetic rate is at its maximum in summer when there is little need to maintain freezing tolerance or downregulate the light harvesting system in leaves. The functional substances involved in photosynthesis can be considered fairly stable after the peak of new tissue formation that takes place in late spring and early summer. The optimal stomatal control model (Eq. 6.3.23) predicted CO_2 exchange rather well during this period in the midsummer (Section 6.3.2.3). Variation of irradiance was the most important factor behind the short-time fluctuations in photosynthetic rate. If the model parameter values obtained in mid-July are used for predicting CO_2 exchange over the whole year, the discrepancy between the predicted and the observed CO_2 exchange is evident, especially in spring (Fig. 6.3.37). The daily patterns seem to be rather similar but the level of photosynthesis varies greatly over the year (Fig. 6.3.38).

Previously it has been shown that although the absolute level of CO_2 exchange of Scots pine shoots varies during the year, the diurnal patterns and the relationships of CO_2 exchange with environmental driving factors remain similar qualitatively (Pelkonen and Hari, 1980; Hari and Mäkelä, 2003; Mäkelä et al., 2004; Kolari et al., 2007). The concentrations of functional substances change considerably during a year, which is reflected in the magnitude of CO_2 exchange. The similarity of the diurnal patterns and environmental responses, on the other hand, indicates that the regulation system changes the functional substances in a balanced way, so

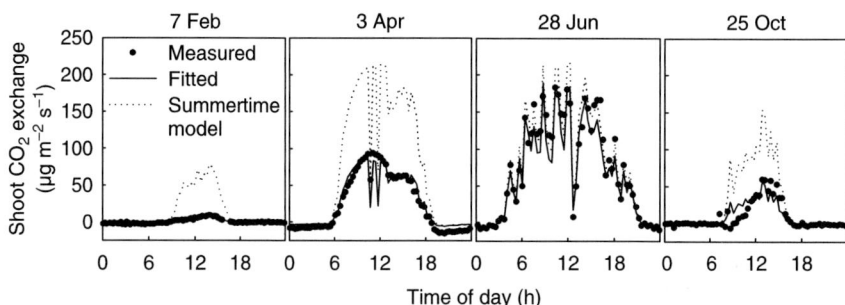

Fig. 6.3.37 Examples of diurnal courses of Scots pine shoot CO_2 exchange in winter, spring, summer and autumn at SMEAR II in Hyytiälä. The dots represent measured CO_2 exchange, dotted lines CO_2 exchange predicted with the optimum model and midsummer parameters, and solid lines the calibrated model when photosynthetic efficiency β is estimated daily

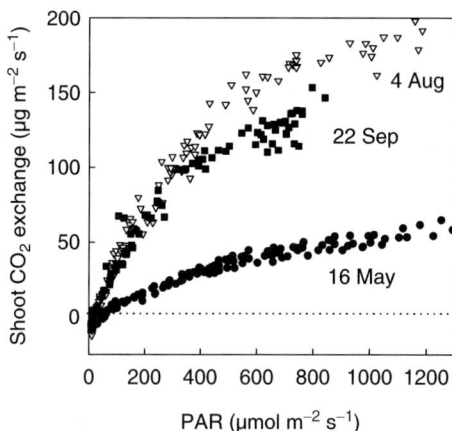

Fig. 6.3.38 Relationship between light (photosynthetically active radiation) and shoot CO_2 exchange at SMEAR I on three days in spring, summer and autumn (Redrawn from Kolari et al., 2007)

that the relations between different subprocesses, such as light and 'dark' reactions, remain unaltered. The resulting regularities in the functional substances form the state of the functional substances. This is analogous to the analysis of the annual cycle ('Bud Burst Phenology').

The derivation of the optimal stomatal control model is valid whenever the functional substances do not change during the period under consideration. The regulation system, however, changes the concentrations of functional substances during recovery of photosynthesis in spring (Fig. 6.3.39), but we can assume that these changes are so slow that the state of functional substances is constant during one day. Thus we can analyse within-day data with the optimal stomatal regulation model and study day-to-day changes in the parameter values.

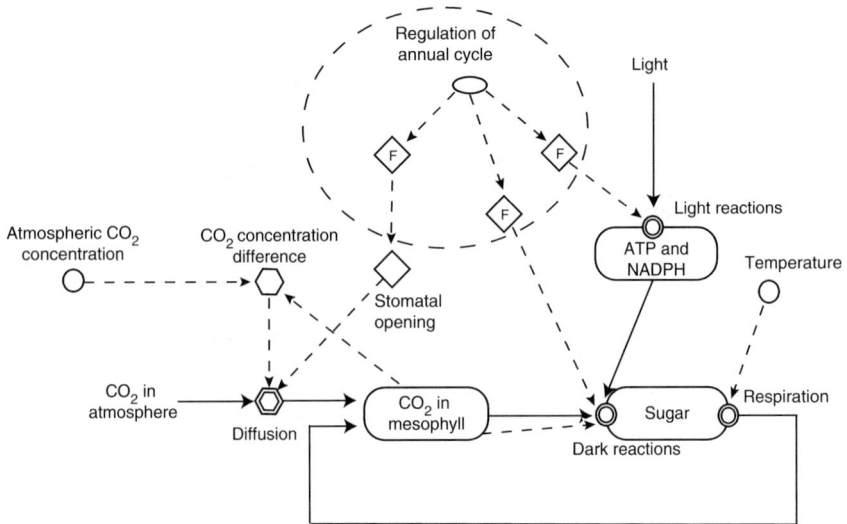

Fig. 6.3.39 Visualisation of the action of the regulation system of annual cycle on photosynthesis. The photosynthetic process is as in Fig. 6.3.12 but the regulation system changes the functional substances of stomatal action, light and dark reactions. The dash line circle stresses the role of regulation in the figure. The symbols are introduced in the Fig. 1.2.1

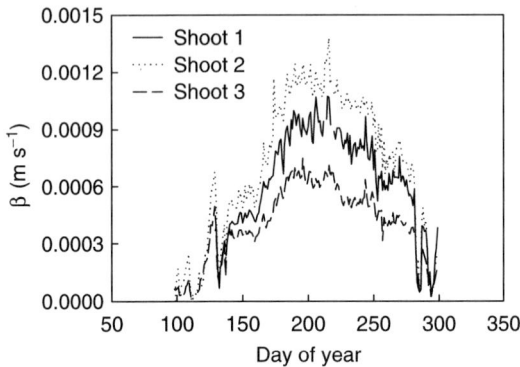

Fig. 6.3.40 Seasonal patterns of daily photosynthetic efficiency β in three individual shoots at SMEAR I in year 2001

Assuming that the functional substances change in a coherent way, the value of the parameter β, efficiency of functional substances, is the most obvious candidate to reflect the changes caused by the annual cycle of trees. The other key parameters in the optimal stomatal control model are λ, called cost of water, and γ introducing saturation of light reactions, and they are less likely to change along seasons. We choose measurements taken in the year 2004 at SMEAR I for testing the hypothesis of functional substances changing slowly. We first estimated values of the parameter β for each day (Fig. 6.3.40). Then we used the estimated parameter values

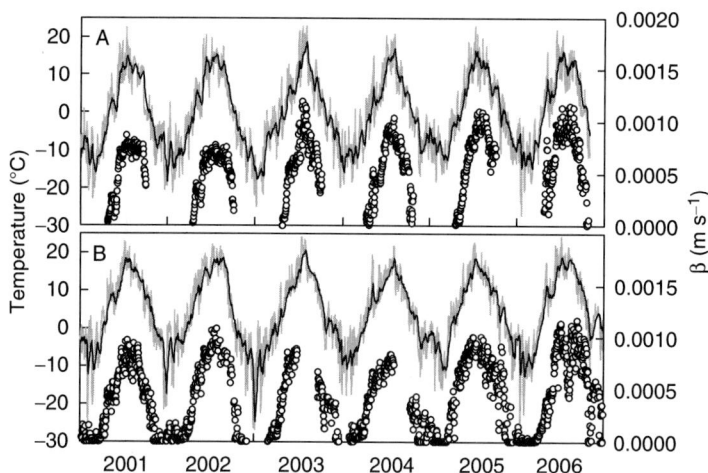

Fig. 6.3.41 Daily mean air temperature (grey line), temperature history S_P (black line) and the daily estimated photosynthetic efficiency β (mean of observed shoots, circles) in Scots pine at SMEAR I (A) and SMEAR II (B) over years 2001–2006

to predict the CO_2 exchange of the next day to determine the predictive power of the optimal stomatal control model without fitting the model. The model was able to explain the daily patterns of CO_2 exchange during the whole active period as well as the within-day rapid variation caused by movements of clouds (Fig. 6.3.37). There are, however, two exceptions. After freezing nights, CO_2 uptake is strongly reduced before noon (see the autumn day in Fig. 6.3.37). This effect seems to be related to daily minimum temperature (Kolari et al., 2007) and it's normally most obvious in spring. On sunny days in spring, stomatal regulation is sometimes stronger than the model predicts, and using the summertime value of the parameter λ overestimates CO_2 exchange and transpiration. We omit these small effects in our analysis because they do not disturb the analysis of photosynthetic efficiency.

The daily values of the parameter β showed a clear pattern, a rapid increase in the spring, fairly stable value in the midsummer and finally a slow declining trend in autumn (Figs. 6.3.40 and 6.3.41). When testing the predictive power of the optimal stomatal control model with previous-day parameter values, the average proportion of explained variance in the CO_2 exchange data was 93.9%, only slightly poorer than PEV when β was estimated daily (97.1%, Fig. 6.3.42). The explaining power of the optimal stomal control model was also in this case close to that obtained in predicting midsummer CO_2 exchange, and the model could produce the daily patterns as well as the short-term variation caused by clouds.

The good agreement between the measured fluxes and CO_2 exchange predicted with previous-day parameters supports the assumption that the state of functional substances changes slowly. Over the year, however, the changes are substantial and a fixed set of summertime parameters cannot predict the observed seasonal pattern in CO_2 exchange. The efficiency of functional substances is low early in the spring,

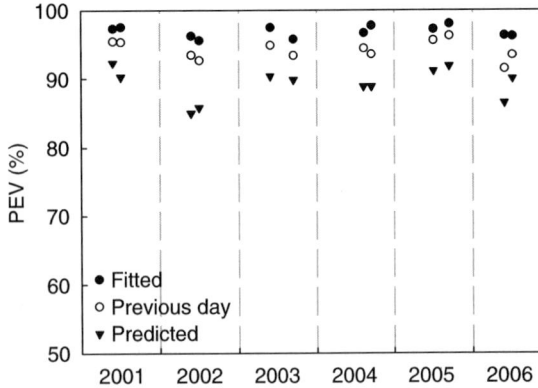

Fig. 6.3.42 Percentage of explained variance (PEV) in shoot CO_2 exchange at SMEAR I by the optimal stomatal control model. "Fitted" refers to PEV when photosynthetic efficiency β was estimated daily, "Previous day" is PEV when CO_2 exchange is predicted with the value of β estimated for the previous day, and "Predicted" is prediction power of modelled CO_2 exchange with the daily values of β calculated from temperature history (Eq. 6.3.52). Each dot represents data from one shoot and year

increases towards the summer, levels off for the midsummer and starts to slowly decline in the autumn, thus following a similar pattern as the daily mean temperature. The annual cycle of phenology is temperature driven (Section 6.3.3.1.1) and the phases of rapid increase in the efficiency of the functional substances of photosynthesis in spring seem to coincide with warm spells. Evidently, also the regulation of the functional substances involved in photosynthesis employs temperature as a signal for the phase of the annual cycle.

Our starting point is, as in the case of bud development (Section 6.3.3.1.1), that the regulation system slowly changes the state of the functional substances of photosynthesis to reflect the annual cycle of environmental factors, especially temperature (Fig. 6.3.43). Since the mechanism of the regulation is currently poorly known, we must use simple models to introduce a temperature driven change in the state of the functional substances. Assume that the state of photosynthetic functional substances can be described with one scalar variable, S_P, and that it follows temperature T with a delay determined by time constant τ.

$$\frac{dS_P}{dt} = \frac{T - S_P}{\tau} \tag{6.3.50}$$

In this formulation S_P will always move towards the prevailing temperature. The time constant τ describes the slowness of the changes in the state of photosynthetic functional substances, large values of τ indicating slower change than small values.

The efficiency of the functional substances, hereafter called photosynthetic efficiency, reflects the state of photosynthetic functional substances. Thus, assume

$$\beta = \beta(S_P) \tag{6.3.51}$$

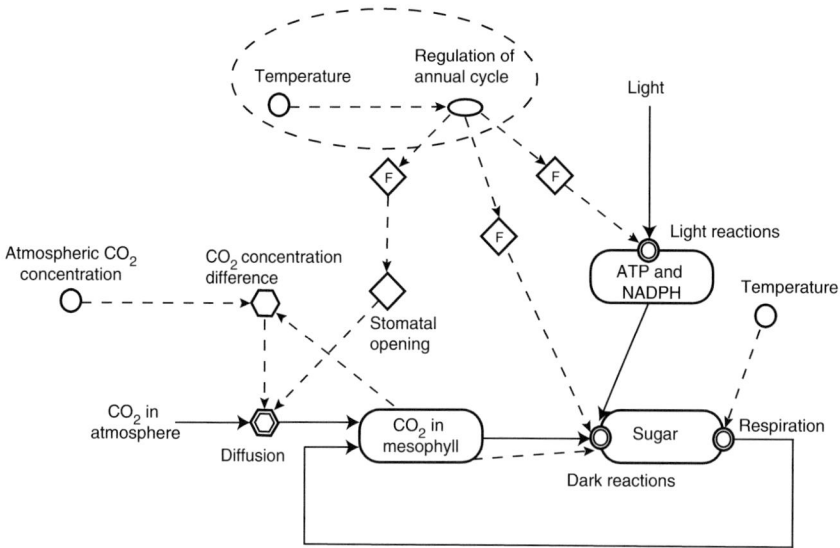

Fig. 6.3.43 The effect of temperature on the regulation of the annual cycle in photosynthesis. The figure is the same as that in 6.3.39 but the effect of temperature on regulation of the annual cycle is included. The dash line ellipse indicates the change. The symbols are introduced in the Fig. 1.2.1

There is no prior information on the physiological mechanism behind the dependence $\beta(S_P)$, thus we have to use the simplicity criterion and start with the linear relationship proposed by Mäkelä et al. (2004). Assume

$$\beta(S_P) = Max\{0, a_S(S_P - S_0)\} \qquad (6.3.52)$$

where a_S and S_0 are parameters. The threshold state of functional substances S_0 describes the onset of photosynthesis in the spring.

We selected data measured during the year 2004 for estimation of the relationship between photosynthetic efficiency and S_0 (Fig. 6.3.44). The value of time constant τ that gave the best match between the modelled and estimated seasonal courses of β was about 200 h. The spring recovery of photosynthesis is so slow that it would take several days or even weeks for photosynthetic efficiency to fully stabilise after a sudden stepwise change in temperature. Experiments in controlled environment have shown that a close-to-full photosynthetic capacity will not be attained until after a few days upon exposure to optimal environmental conditions (Ottander and Öquist, 1991; Ottander et al., 1995). This slowness confirms that the annual cycle of photosynthesis in boreal conifers is largely attributed to the slow changes in the concentrations and activities of pigments, enzymes and other functional substances involved in photosynthesis rather than just a direct response of biochemical process rates to temperature.

The optimal stomatal control model of photosynthesis describes phenomena in the time scale of a day and the phenological model introduces the slow changes in

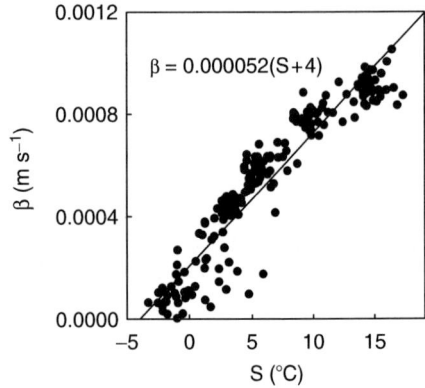

Fig. 6.3.44 The relationship between β and S_P in Scots pine (mean of three shoots) in 2001 at SMEAR I. Time constant in the calculation of S_P was 200 h. The proportion of variance in β explained by the fitted model was 91%

the values of the parameter β. When these models are combined we get a model, called PhenPhoto, covering the whole photosynthetically active season.

$$P = \beta f(I) \frac{u\, c_{max}\, C^a_{CO2} + r}{u\, c_{max} + \beta\, f(I)} \qquad (6.3.53)$$

$$f(I) = \frac{I}{I + \gamma} \qquad (6.3.54)$$

$$u^* = \left(\sqrt{\frac{C^a_{CO2} - (r/(\beta\, f(I)))}{\lambda\, a D}} - 1 \right) \frac{\beta\, f(I)}{c_{max}} \qquad (6.3.55)$$

$$u = \begin{cases} 0, & if\ u^* \leq 0 \\ u^*, & if\ 0 < u^* < 1 \\ 1, & if\ u^* \geq 1 \end{cases} \qquad (6.3.56)$$

$$\frac{dS_P}{dt} = \frac{T - S_P}{\tau} \qquad (6.3.57)$$

$$\beta(S_P) = Max\{0, a_S(S_P - S_0)\} \qquad (6.3.58)$$

We tested the above model with shoot CO_2 exchange measurements at SMEAR I from years 2001–2006. The seasonal patterns of shoot photosynthesis agreed well with the model, but there was a lot of variation in the absolute level of CO_2 exchange among the shoots that were monitored during these years. This is partly due to the uncertainties in determining the needle area or variation in the degree of self-shading by overlapping needles, but there also seems to be real shoot-to-shoot variation in the photosynthetic performance. The temporal patterns and the responses of shoot CO_2 exchange to the environmental driving factors were consistent and it is sufficient to calibrate the model for each shoot by just adjusting the parameter a_S in Eq. 6.3.58, which determines the slope of the relationship between the state of functional substances S_P and photosynthetic efficiency, β. Once the parameter a_S was calibrated for each shoot with a five-day data from July that was later excluded from the analysis, the combination of the photosynthesis model and the annual cycle model was on average able to predict 90% of the variance in the CO_2 exchange data. The range of prediction power among the individual shoots was 86–93% (Fig. 6.3.42). Figure 6.3.45 shows and example of predicted and observed CO_2 exchange in one shoot during one week in spring. When the efficiency of functional substances (parameter β) is calculated from temperature history the model can predict the increase in photosynthetic rate whereas using model parameters estimated in July greatly overestimates daytime CO_2 uptake. Measured and predicted daily mean midday CO_2 exchanges over several years are shown in Fig. 6.3.46.

The other parameters of the photosynthesis model seem to change less during the year. There is some seasonality in the base level of respiration r_0 (see Section 6.3.3.1.2) but it only plays a minor role compared with the seasonal variation in photosynthetic CO_2 uptake because photosynthetic CO_2 fixation is normally much higher than respiration. Cost of transpiration (λ) typically also shows a slightly decreasing course from the early spring to the beginning of summer and again an increase in the autumn (see Section 6.3.3.2.2). The higher spring and late autumn values of λ correspond to lower values of stomatal conductance and a higher sensitivity of the stomata in response to air humidity. Using fixed values for these parameters instead of separately estimating their seasonal courses does not affect the estimated annual pattern of β significantly but it greatly increases the stability if we estimate the daily values of β. The value of the light-saturation parameter γ did not show systematic changes over the year. This indicates that the regulatory processes in light reactions and carbon fixation follow similar seasonal patterns. The relatively temperature-insensitive light reactions need to acclimate to the energy consumption by the temperature-sensitive carbon fixation in order to maintain balance between the absorbed light energy and CO_2 fixation and to avoid deleterious

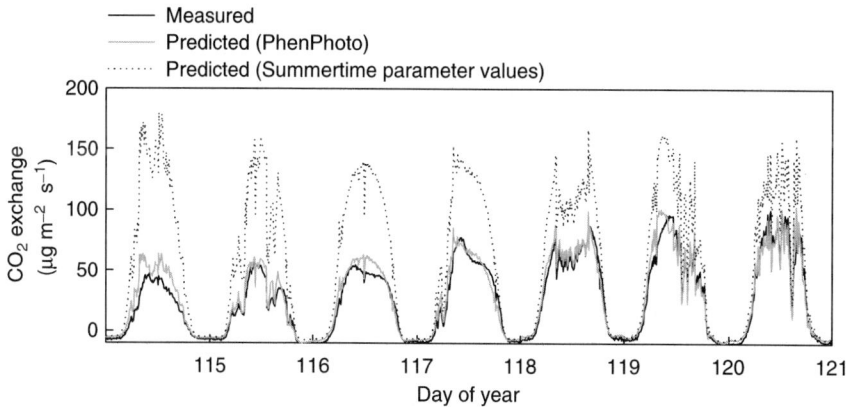

Fig. 6.3.45 Measured (black line) and predicted CO_2 exchange of a Scots pine shoot at SMEAR I during one week in spring 2002. Prediction with the optimal stomatal control model and a fixed summertime parameter set (dash line) results in clearly higher CO_2 exchange than the measurement, whereas PhenPhoto (grey line) that incorporates the temperature-driven annual cycle can accurately predict the spring recovery of photosynthesis

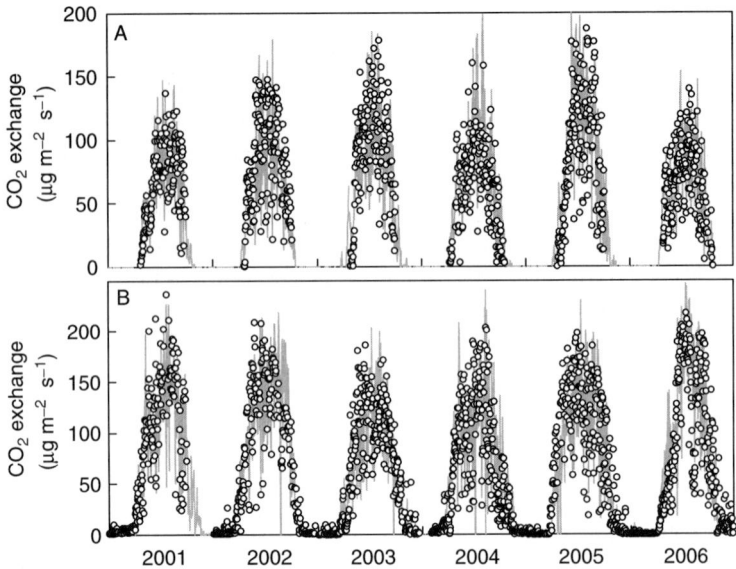

Fig. 6.3.46 Predicted (grey line) and measured (circles) seasonal courses of daily mean midday CO_2 exchange of Scots pine shoots over several years at SMEAR I (A) and SMEAR II (B). The CO_2 fluxes were averaged daily over all shoots that were under monitoring simultaneously (2–4 shoots) and between 11:00–13:00 solar time. The year-to-year variation in the absolute level of the CO_2 fluxes is mainly reflecting the natural variation of physiology and morphology among the sample shoots, and to lesser extent the variation of summertime weather conditions from year to year. At SMEAR I a set of fresh shoots is taken under monitoring each spring whereas at SMEAR II the measurements are going on year round and each shoot is kept in the chamber for about two years

effects on photosystems (Öquist and Huner, 2003, see also 'Short-Term Acclimation of Photosynthetic Light Reactions to Light'). Thus one parameter, photosynthetic efficiency, β, is sufficient in characterising the state of the functional substances involved in photosynthesis.

In the midsummer the model of stable functional substances (i.e. constant β) performed equally well compared to linear $\beta(S_P)$ relationship. When studying the relationship between β and S_P we have found that at high values of S_P, the photosynthetic efficiency seems to saturate. The $\beta(S_P)$ relationship can be modified such that there is an upper limit for β (Mäkelä et al., 2006), or that the relationship can be described by a sigmoid function (Kolari et al., 2007). Due to the simplifications of the model structure, e.g. omission of instantaneous temperature response of photosynthesis, and the systematic errors in the needle temperature measurements, the determination of the saturation of the relationship between β and temperature history is problematic with our field data. In warming climate the instantaneous and delayed responses of photosynthesis to high temperatures, however, become more important.

The PhenPhoto model was able to explain about 90% of the variation in the measured CO_2 exchange in a very large data set shown here, consisting of about 600,000 measurements during six years. The model can explain the seasonal patterns of shoot CO_2 exchange for the other years as well (Mäkelä et al., 2004), and also at SMEAR II in Hyytiälä, located 800 km south of Värriö, SMEAR I (Kolari et al., 2007). Figure 6.3.46 illustrates seasonal courses of CO_2 exchange over several years in Värriö and in Hyytiälä. Although there is small systematic difference between Värriö and Hyytiälä in the summertime values of the parameters of the optimal stomatal control model, the shape of the $\beta(S_P)$ relationship as well as the slowness of the change in photosynthetic efficiency is almost equal at both sites (Kolari et al., 2007). Thermal growing season begins about four weeks earlier in Hyytiälä than in Värriö, and almost as big year-to-year variation in the timing of spring can be observed at each site. The spring recovery of photosynthesis systematically seems to follow temperature and the temperature-driven model is able to predict the increase of photosynthetic efficiency in early and late springs equally (Fig. 6.3.46).

The unexplained variance component of CO_2 exchange, about 10%, is caused by measuring errors and shortcomings in the theory. The measuring noise plays only a minor role (Section 6.3.2.3), but systematic errors, most importantly those involved in determining irradiance and leaf temperature, contribute more to the unexplained variance. The evident decrease in the proportion of explained variance in the prediction compared to daily estimation is mainly generated by short-term variation in the value of the parameter β. The submodel linking temperature history with the photosynthetic efficiency may be oversimplified and could be improved by, for example, adding the effect of freezing nights on β. There are evidently functional substances that change at different rates and a model with one time constant of several days cannot explain those processes that change substantially within one day. The shortcomings in the theory generate clearly less than 10% of the unexplained variance.

We have not found in the literature any similar analysis of predicting the annual cycle of leaf or shoot photosynthesis over several active periods using temporally independent datasets for estimating the model parameter values and testing the model, thus we can not directly compare our results with previously obtained ones. Several empirical and modelling studies, however, have indicated strong relationship between temperature and the seasonal changes in light-saturated photosynthesis and in the efficiency of light reactions (e.g., Leverenz and Öquist, 1987; Lundmark et al., 1998; Bergh et al., 1998). The seasonal changes in photosynthesis are also linked with frost tolerance (e.g., Vogg et al., 1998; Repo et al., 2006).

We have also tested the concept of temperature-driven annual cycle with a larger dataset consisting of photosynthesis of several coniferous forests stands in northern and central Europe (Mäkelä et al., 2007; Section 7.6.2). We noticed that at the northern sites, the seasonal pattern of photosynthesis was well explained with the temperature-driven annual cycle. In warmer climatic zones with milder winters the role of temperature in the annual cycle of photosynthesis is not as dominating as in the boreal zone, whereas the dry periods occurring more or less regularly every year are more important.

The model PhenPhoto combines two traditions in our research group. The analysis of phenology started already in the 1970s and the optimal stomatal control approach in the 1980s. The result means that the theory is able to capture the essential regularities in the nature and we can predict photosynthesis using light and temperature history very accurately, if no water deficit exists. This result will have a key role in the remaining chapters of the book.

Conclusions

The regulation system changes the concentrations and activities of functional substances in coherent way during annual cycle. The resulting regularities in the functional substances form the state of functional substances. Photosynthesis during a day is determined by the state of the functional substances, light and temperature. The slow changes in the state of functional substances are driven by temperature. A test with extensive field data strongly corroborates the theory of temperature-driven seasonal cycle of photosynthesis.

6.3.3.1.4 Shoot Elongation

Pertti Hari, Päivi Väänänen, and Eero Nikinmaa

University of Helsinki, Department of Forest Ecology, Finland

Shoot elongation is a vital component of the annual cycle of trees and other vegetation. Then new shoots and needles (leaves) in them are formed using sugars from photosynthesis (Section 6.3.2.3) for construction of macromolecules and nitrogen from nutrient uptake by roots (Section 6.3.2.8) and from reuse in internal circulation (Section 6.3.2.4) as raw material. The regulation system of the annual cycle takes care of the timing of different phenomena in proper order in respect to the annual environmental cycle.

The meristems in buds and cambium become active in the spring when the conditions for the synthesis of new macromolecules and formation of cells become favourable and these processes begin. As was shown in Section 6.3.3.1.1, temperature plays an important role in the bud burst in the spring. The critical role of proper timing is particularly important in the growth processes in the seasonal boreal environment. Delayed start of growth will translate into smaller size which leads into lower competitive ability. However, too early growth means exposure to frost damage which will be even more detrimental for the tree development. In the early summer after winter, soil moisture conditions are favourable and once the temperature levels rise the conditions for growth are very good. The growth phenomena are very

Fig. 6.3.47 Growing shoots of Scots pine early in the summer

intensive in early summer and they gradually slow down during summer and end well before autumn. In trees with determinate growth pattern, such as Scots pine, the cessation of growth is associated with fulfilment of the maximum size of the organ. During favourable year this takes place earlier while during e.g. cold summers the development may still be incomplete when the growth stops due to the starting of the hardening processes before winter. In trees following indeterminate growth pattern, the growth stops only due to the latter reason when conditions turn unfavourable. All the growth processes, be it shoot elongation (Fig. 6.3.47), growth of leaves or wood in stems follow rather similar patterns, but their timing during summer may vary. For example in pine, shoot extension growth gets ready first, followed by the needle growth. The stem thickness growth extends well into late summer and the root growth in the soil may continue as long as the soil temperature allows.

Due to regular pattern of development, the timing of the growth processes can be described with rather simple temperature dependent model. Small molecules or ions, such as sugars and amino acids, are used as raw material for growth processes forming new cells with complicated chemical structure. The compounds needed for growth, such as cellulose, lignin and proteins, are synthesized and placed in proper locations in the cell structure. This requires activity of a large number of functional substances, such as enzymes. The composition and amounts of the functional substances change under the control of regulation system during the growing season since the needs of new structures develop along the season. We assume as previously that the regularities in the functional substances can be described with one number, called the state of the functional substances. The regulation system changes for

Fig. 6.3.48 Visualization
of the two time scales of
growth process; immediate
response to temperature and
slow response via regulation
of functional substances. The
symbols are introduced in the
Fig. 1.2.1

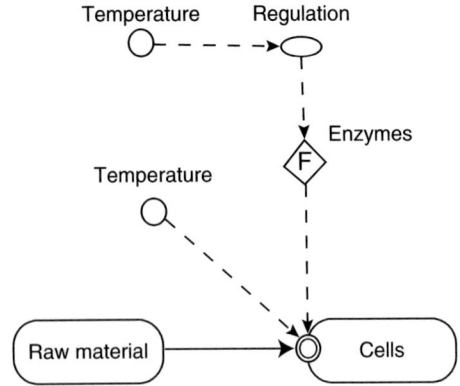

example the concentrations of enzymes and thus the state of functional substances
slowly. The regulation seems to be linked to temperature (e.g. Pietarinen et al.,
1982). On the other hand, the new macromolecules are synthesized in biochemical
reactions that are catalysed by enzymes, thus their synthesis depends on temperature. Thus the growth responds to temperature also without time lag (Fig. 6.3.48).

Based on the above argumentation, we should be able to describe growth with a
model that has two distinct response time scales to temperature. Let $g_{shoot}(\mathrm{mm\,s}^{-1})$
denote the growth rate. It depends on temperature T and the state of functional
substances S_F:

$$g_{Shoot}(t) = g(T(t), S_F(t)) \tag{6.3.59}$$

where t denotes time.

Assume that the effects of temperature and state of functional substances on
growth are multiplicative,

$$g_{Shoot}(t) = g_T(T(t)) g_S(S_F(t)) \tag{6.3.60}$$

where g_T describes the effect of temperature and g_S the effect of the state of functional substances, respectively.

The measurements of continuous growth rate are difficult to do both due to
small changes of growing dimensions in time, especially in the boreal environment,
and also since changes in the water status of plant reflect also in the dimensional
changes. However, daily shoot elongations can be determined rather easily. The
transition from process rate to daily values is done by integration (Section 7.1).
We get

$$G_{Shoot}(t_{i+1}) - G_{Shoot}(t_i) = \int_{t_i}^{t_{i+1}} g_{Shoot}(t)dt = \int_{t_i}^{t_{i+1}} g_T(T(t))g_S(S(t))dt \tag{6.3.61}$$

where $G_{shoot}(t_i)$ is the length of the shoot at the moment t_i.

The state of development changes slowly, so lets assume that it remains constant during a day. We get

$$G_{Shoot}(t_{i+1}) - G_{Shoot}(t_i) = g_S(S(t_i)) \int_{t_i}^{t_{i+1}} g_T(T(t))dt \qquad (6.3.62)$$

The identification of this type of a model is feasible since the daily shoot elongations and diameters are rather non-problematic to measure as well as temperature.

To operationalise the model, we need the dependence of growth on temperature which, in principle, can be obtained with measurements of shoot elongation at short interval, less than 1 h. This measurement is rather problematic to arrange and the accuracy and precision of the result is poor to reasons mentioned above. We need to find temperature dependence of some quantity which would reflect the effect of temperature in a similar way than growth and we use the obtained relationship as proxy. The synthesis of new macromolecules requires energy in the form of ATP, which is formed in respiration. The respiratory release of CO_2 is easy to measure and it enables approximation of the temperature dependence of growth. Let $r(T)$ denote the relationship between respiration and temperature, thus

$$g_T = a r(T) \qquad (6.3.63)$$

where a is a parameter.

As mentioned previously the instantaneous growth and its response to temperature is changing during the ontogenetic development (Eq. 6.3.60). This regulation of growth reacts to temperature as visualised in Fig. 6.3.48. This is introduced into the analysis with the change of the state of functional substances, analogously to the treatment of bud burst. The rate of change of the state of the functional substances, s_F, is by definition (as in Eq. 6.3.44)

$$s_F = \frac{dS_F}{dt} \qquad (6.3.64)$$

The rate of change of the state of the functional substances is temperature driven as shown in the Fig. 6.3.49. Assume further

$$s_F(T) = r(T) \qquad (6.3.65)$$

Finally, the function g_s has to be specified. Assume that it is formed from three linear sections, for increasing, stable, and declining growth as shown in Fig. 6.3.49.

We measured 35 years ago daily shoot elongations and temperature in Hyytiälä, rather close to the present SMEAR II (Pietarinen et al., 1982). The direct temperature responses were derived from published respiration measurements (Dahl and Mork, 1959). The parameters in the model were estimated annually numerically with the normal least squares principle. The fit of the model with the measurements was evidently within measuring precision Fig. 6.3.50. The model was applied to the analysis of height increments of several boreal species from fern to herbs. The performance of the model was within measuring precision (Vuokko et al., 1977).

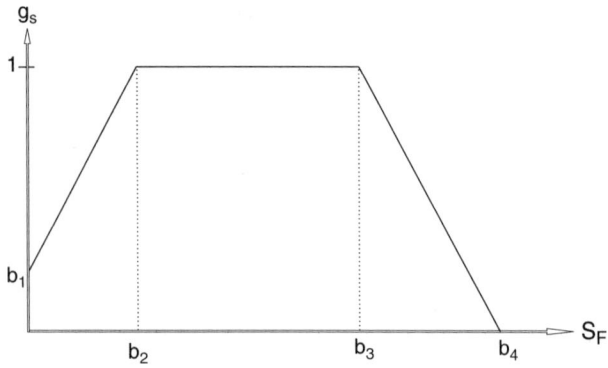

Fig. 6.3.49 The relationship between the effect of the state of the functional substances on growth, g_S and S_F

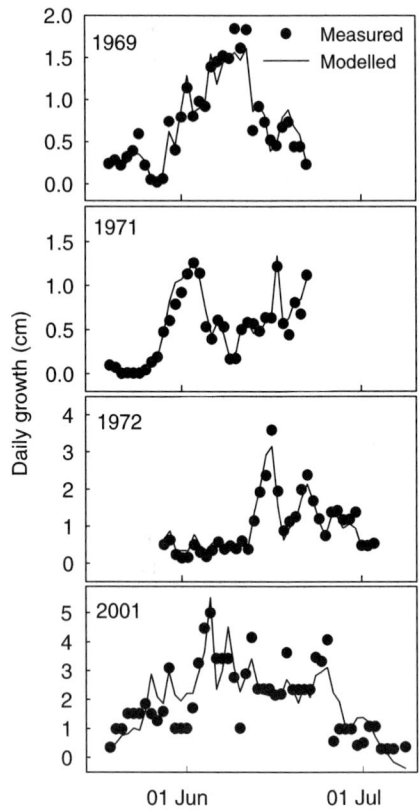

Fig. 6.3.50 Measured and modelled daily shoot elongation of Scots pine in Hyytiälä, near SMEAR II station in summers 1969, 1971, and 1972 and at SMEAR II station in 2001

We compared the model behaviour against measured daily shoot elongations also at SMEAR II in 2001 without any changes in the model. The fit of the model is clearly weaker in these new measurements although the main pattern of growth is still predicted. The observed trees were from the same stand. During the first measurements they were still rather small sized saplings with very large height growth while in the latter case the canopy had fully closed already about 15 years ago and considerable differentiation in the stand into canopy classes had taken place. One practical consequence of the development is that the access to the tree crowns had become more difficult and thus the extension growth of the measured shoots was smaller and proportion of measurement error had become bigger. This may partially describe the differences.

In the model parameterisation we used the same as that instantaneous temperature response (Eq. 6.3.65) as was used in early 1970s. The respiration measurements behind the temperature response (Dahl and Mork, 1959) showed similar pattern in temperature response as observed at SMEAR II station and fitted within the observed variation we did not change the instantaneous temperature response in the model. It is possible that some changes had taken place between the two occasions in the response for example to increasing tree size as e.g. the shoot water relations change as a function of tree size. Water status influence the transport processes (see Section 7.5.2) but it also determines the cell expansion as this is accomplished by trugor pressure of the growing cells in plants. The major obstacle in growth modelling has been the proper description of the transport processes. To improve the realism of the growth modelling with source-sink relationship (Thornley, 1972; 1991; Berninger et al., 2000), the transport should be considered. The approach that we describe in Section 7.5.2 ('Phloem Transport') is promising as it is build on simple physical factors and can be parameterised. While that is still missing the approach presented here offers possibility to describe the day to day variation in growth over the growing season.

Conclusions

The main idea with the model was to study whether similar principles as applied with bud burst could be easily applied to growth processes and predict the rather regular timing of seasonal growth with a simple temperature dependant model. Growth processes consist of differenciation of new cells, their extension and formation of new cell walls. High rate of biosynthesis is involved in all of these processes. In boreal environments, temperature often limits the rate of biosynthesis. However, temperature is also correlated with carbon and nutrient uptake which ultimately influence growth (see Section 6.3.3.1.1). Therefore it could be expected, that simple models using temperature as a driving variable would be able to predict the seasonal variation in growth. The overall good behaviour of the model shows that this type of approach can be used as a first approximation. It is likely that very good description of the timing of a single shoot growth can be achieved with appropriate parameterisation (e.g. Fig. 6.3.50).

6.3.3.2 Process Acclimation

6.3.3.2.1 Short-Term Acclimation of Photosynthetic Light Reactions to Light

Albert Porcar-Castell, Jaana Bäck, Eija Juurola, and Pertti Hari

University of Helsinki, Department of Forest Ecology, Finland

Light Fluctuations: Time-Scale and Response of the Leaf

Imbalances between energy supply by the light reactions of photosynthesis and energy consumption by the dark reactions take place almost constantly in plant leaves. For example, low CO_2 concentration in the mesophyll during partial closure of stomata reduces photosynthesis (Section 6.3.2.3), yet light absorption continues, increasing the imbalance between energy capture and its conversion into stable form as sugars. These imbalances eventually increase the risk of photo-oxidative damage to the thylakoid membrane. Acclimation of the light reactions is therefore needed to protect the leaf from excess light.

Leaves of plants are constantly exposed to fluctuations in the light environment. Long-term fluctuations (days-months) are periodical due to the diurnal and annual movement of the planet. In contrast, short-term fluctuations in the light intensity (seconds-minutes) have a more random nature caused by the passing of clouds, movement of sunflecks, or fluttering of the leaf. Upon light absorption by the leaf, the lifetime of excitation in the antenna of the photosystem is on the order of nanoseconds, i.e., the time that the exciton resulting from the absorption of a photon remains as such before being dissipated as heat, emitted as radiation, or used in photosynthesis. The capture of electrons by a single chlorophyll molecule, even under large photon flux density, is rare and in the order of two to three photons per second. This was shown by Emerson and Arnold in the 1930s (Lawlor, 2001).

Thus, under natural conditions light absorption will never saturate due to over-excitation of chlorophyll molecules and if the photosynthetic photon flux density (PPFD) to which a leaf is exposed increases two-fold, the rate of light capture by the leaf pigments will immediately double. In contrast, the rate at which the absorbed energy is chemically bound does not directly depend on the light intensity and varies in relation to the functional substances involved in the dark reactions. Subsequently, when the rate of light capture exceeds the photochemical capacity for energy sequestration by the light reactions, the lifetime of the exciton will tend to increase, causing increased formation of chlorophyll triplet-states, and the associated risk of photo-oxidative damage to the thylakoid membrane (Demmig-Adams and Adams, 1996; Gilmore, 1997; Eskling et al., 2001). To cope with these fluctuations in the light environment, light reactions undergo acclimation in order to maintain the equilibrium between energy absorbed by light capture and energy utilized in photochemistry.

In evergreen species, there are several long-term acclimation mechanisms of the light reactions (days, months), e.g., adjustments in the chlorophyll and carotenoid contents (Öquist et al., 1978; Öquist and Huner, 2003), or structural rearrangements in the thylakoid membrane (Ottander et al., 1995; Gilmore and Ball, 2000; Busch et al., 2007). Long-term acclimation of light reactions responds to long-term changes in the performance of the dark-reactions, induced by persistent changes in water status and nutrient status. In contrast, short-term acclimation processes (seconds, minutes) include structural and biochemical changes in the relative absorption cross section of the photosystems known also as state-transitions (Haldrup et al., 2001), or changes in the concentration of specific xanthophyll pigments (e.g., xanthophyll-cycle pigments), respectively (Demmig-Adams and Adams, 1996; Nixon and Mullineaux, 2001). In this chapter we focus on the short-time acclimation processes of the light reactions, and describe the acclimation of light reactions to short-term fluctuations in light intensity with the help of a dynamic model (Porcar-Castell et al., 2006). Finally, long-term acclimation processes are shortly discussed in connection with the model.

Short-Term Modulation of Energy Partitioning in PSII

When a photon is captured by a chlorophyll molecule, and in a dark-acclimated and non-photoinhibited leaf, the absorbed energy (exciton) may be consumed in different processes: (i) used to reduce the plastoquinone pool, i.e., electron transport or photochemistry (p), (ii) emitted as fluorescence (f), (iii) dissipated as heat by internal conversion or constitutive heat dissipation (d), or (iv) experience intersystem crossing and form a chlorophyll triplet-state (Parson and Nagarajan, 2003). Later, when the same leaf is illuminated the rate of ATP and NADPH formation by the light reactions will partly decrease due to their buildup in the chloroplast stroma, causing an accumulation of protons in the thylakoid lumen (decrease in lumen pH), and an increase in the redox state of the plastoquinone pool (Nixon and Mullineaux, 2001). As a result, the rate of electron transport through PSII decreases, increasing the excitation lifetime and the number of excitons present in the antennae (Fig. 6.3.51), and the yields of the alternative energy-consuming processes increase, including the formation of hazardous chlorophyll triplet-states.

To control the formation of triplet-states and counteract the decrease in the photochemical rate, PSII has an efficient mechanism that uses the lumen pH as a clue to adjust the rate of regulative heat dissipation (n) in PSII. When lumen pH decreases due to saturation of the dark-reactions, some PSII proteins are protonated and violaxanthin molecules are released and de-epoxidized to antheraxanthin and zeaxanthin, as part of the xanthophyll-cycle (Demmig-Adams and Adams, 1996; Eskling et al., 2001). It has been suggested that zeaxanthin combined with the protonation of specific PSII proteins triggers the conformational change needed for the formation of a quenching complex that favours the dissipation of excitation energy as heat (Horton et al., 1996; Müller et al., 2001), represented in Fig. 6.3.51 as a quenching site ON. In Fig. 6.3.51, lumen pH is not represented for the sake of simplicity; however the number of excitons in the antenna and electrons in the quinone pool can be equally used as clues for the acclimation of the number of quenching sites and the rate of regulative heat dissipation. An increase in lumen pH will be correlated with an increase in the redox state of the plastoquinone pool (i.e., number

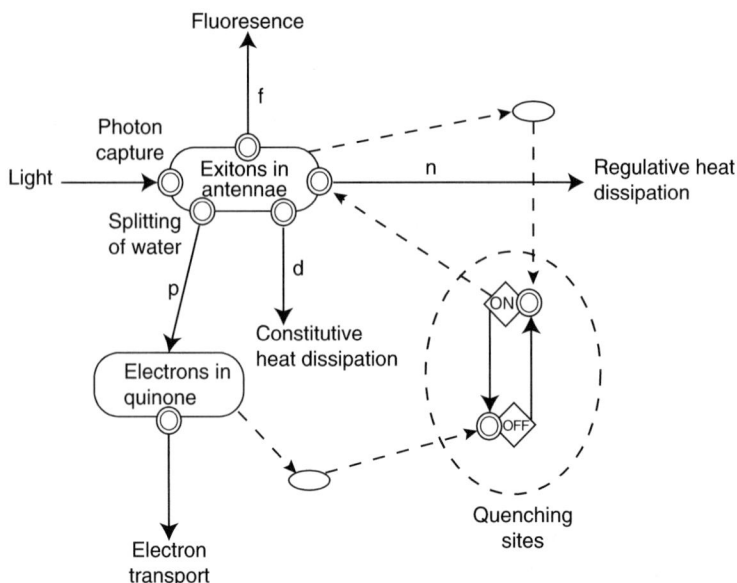

Fig. 6.3.51 Scheme of the processes and short-term acclimation of the light reactions of photosynthesis to light; f stands for fluorescence, n for regulative heat dissipation or NPQ, d for constitutive heat dissipation and p for photochemistry. The symbols are introduced in the Fig. 1.2.1

of electrons in the quinone), as well as with the level of excitation present in the antenna (i.e., number of excitons in the antenna). This acclimation process is reversible and responds on the time-scale of seconds to minutes to changes in light intensity (Krause and Weis, 1991; Müller et al., 2001).

A Tool to Follow the Energy Partitioning in PSII

Chlorophyll fluorescence techniques can be used to study the dynamic changes in the partitioning of absorbed energy in light reactions. In the standard fluorescence yield equation, the rate constant associated to fluorescence (k_f) is divided by the sum of the rate constants of each of the energy-consuming processes in PSII (Kitajima and Butler, 1975; Krause and Weis, 1991). Using this model it is possible to derive the maximum quantum yield of photochemistry (Kitajima and Butler, 1975), the rate of electron transport (Genty et al., 1989), the overall yield of thermal dissipation (Demmig-Adams and Adams, 1996), or the energy fluxes into the main energy-consuming processes in PSII (Hendrickson et al., 2004). Following the same fundamental assumptions, Porcar-Castell et al. (2006) derived a dynamic model that includes adjustments in the pH-dependent mechanisms for the continuous estimation of yields and energy fluxes of each of the energy-consuming processes in PSII.

The study of the dynamics of acclimation of light reactions and energy partitioning is dominated by a series of fluorescence-centered parameters: the rates of each of the energy-consuming processes are commonly expressed in relation to the degree of quenching they produce to the measured chlorophyll fluorescence signal. Two types of quenching are differentiated: the quenching produced by the photochemical process, denoted as photochemical quenching, and other sources of quenching or non-photochemical quenching (NPQ). Subsequently, we will refer to the pH-dependent or regulative heat dissipation as pH-dependent NPQ or, in the absence of any other source of quenching, simply as NPQ.

The Model at the Short-Term Scale (Seconds-Minutes)

The model by Porcar-Castell et al. (2006) considers a population of N PSII units where each PSII has a given amount of Chla, Chlb and xanthophyll molecules, as well as a pool of quinone-equivalent molecules capable of accepting one electron each. The system of N PSII units is assumed to follow a lake-type organisation model (Kitajima and Butler, 1975; Dau, 1994) where excitons can move freely between PSII units. Pigment molecules and quinone-equivalents change from an OFF to an ON state once they accept a photon, an exciton, or an electron, and return to the OFF state after losing it. Once a Chla molecule has absorbed a photon, it can transfer the exciton to an oxidized quinone-equivalent, it can reemit a photon of fluorescence, dissipate all or part of its energy as constitutive heat, or dissipate it as a result of the non-photochemical quenching processes (NPQ). The faith of the exciton will depend on the prevailing rate constant for each process. Finally, reduced (ON) quinone-equivalents are re-oxidized by the downstream electron transport, and return to the OFF state.

The flow of energy and the processes included in the model are depicted in Fig. 6.3.51. Next, each of the different process rates are derived from chlorophyll fluorescence measurements.

The rate of light capture (c) is determined by the light intensity (I), and the number of chlorophyll molecules in the ground state:

$$c = \alpha I (Chla^{OFF} + Chlb^{OFF}) \tag{6.3.66}$$

where I is the PPFD at the leaf surface ($\mu\,mol\,m^{-2}\,s^{-1}$), $Chla^{OFF}$ and $Chlb^{OFF}$ are amount of Chla and Chlb molecules in the ground state, and α is an efficiency parameter [m^2 (chlorophyll molecule)$^{-1}$].

The rate of constitutive heat dissipation (d) (excitons s^{-1}) is linearly proportional to the number of excitons in the antennae ($Chla^{ON}$):

$$d = k_d\,Chla^{ON} \tag{6.3.67}$$

where k_d is a parameter identified as the rate constant of heat dissipation. The rate of fluorescence (f) is:

$$f = k_f\,Chla^{ON} \tag{6.3.68}$$

where k_f is a parameter identified as the rate constant of fluorescence.

Changes in the efficiency of NPQ are introduced by the model with the variable NPQ-efficiency, denoted with E, where E can range from 0 (no NPQ) to 1 (maximum NPQ efficiency). The rate of NPQ processes (n) is calculated as:

$$n = k_n E Chla^{ON} \tag{6.3.69}$$

where k_n is a parameter that corresponds to the rate constant associated to NPQ processes.

The efficiency of photochemistry varies depending on the redox state of the plastoquinone pool, i.e., when the pool of plastoquinone is totally reduced the rate of photochemistry is 0, and the rate is maximum when the plastoquinone pool is fully oxidized. These variations are represented in the model as the oxidized fraction of the quinone-equivalent pool, denoted with Q. Where Q can range from 0 (all quinone-equivalents reduced) to 1 (all quinone-equivalents oxidized).

Thus, the rate of photochemistry (p) is:

$$p = k_p Q Chla^{ON} \tag{6.3.70}$$

where k_p is a parameter that corresponds to the rate constant associated to the photochemical process.

Combining Eqs. 6.3.66–6.3.70, we obtain the differential equation for the number of excitons ($Chla^{ON}$):

$$\frac{dChla^{ON}}{dt} = \alpha I \left(Chla^{OFF} + Chlb^{OFF} \right) - k_f Chla^{ON} - k_d Chla^{ON} - k_n E Chla^{ON} - k_p Q Chla^{ON} \tag{6.3.71}$$

We set Eq. 6.3.71 equal to 0, and assume that the sum of $Chla^{OFF}$ and $Chlb^{OFF}$ remains constant over time and equal to the total chlorophyll content (Chl_T) under consideration, which holds true under natural illumination since under natural illumination there is no exciton accumulation in PSII (Dau, 1994). The number of excitons in the system can be calculated as:

$$Chla^{ON} = \frac{\alpha I Chl_T}{k_f + k_d + k_n E + k_p Q} \tag{6.3.72}$$

Conventional modulated fluorometers combine an actinic light source and a modulated beam of constant light intensity (I_{MB}); the latter is used to record the variations in the fluorescence yield induced by the actinic light. Then, combining Eqs. 6.3.38 and 6.3.72, the rate of chlorophyll fluorescence emission (f) is:

$$f = \frac{k_f \alpha I_{MB} Chl_T}{k_f + k_d + k_n E + k_p Q} \tag{6.3.73}$$

Next, after darkening a leaf for a long enough time we can assume that all the pH-dependent NPQ has relaxed and $E = 0$. In addition, the quinone-equivalents will be fully oxidized giving $Q = 1$. Accordingly, the minimum fluorescence rate (f_o) under beam light, can be expressed:

$$f_o = \frac{k_f \alpha I_{MB} Chl_T}{k_f + k_d + k_p} \qquad (6.3.74)$$

and after applying a saturating light pulse to the leaf, which completely reduces the quinone-equivalent pool $(Q = 0)$, the maximum fluorescence rate (f_m) can be expressed:

$$f_m = \frac{k_f \alpha I_{MB} Chl_T}{k_f + k_d} \qquad (6.3.75)$$

Conversely, after exposing a leaf to light in excess (i.e., light that saturates the Calvin cycle) during several minutes at ambient temperatures, we assume that the maximum efficiency of pH-dependent NPQ processes is attained (E_{max}), and after applying a saturating light pulse to the leaf the maximum fluorescence rate in the light (f'_m) can be expressed:

$$f'_m = \frac{k_f \alpha I_{MB} Chl_T}{k_f + k_d + E_{max} k_n} \qquad (6.3.76)$$

Subsequently, by substituting the previous fluorescence rates (f) with their proportional measured modulated fluorescence signal (F), and combining Eqs. 6.3.74 and 6.3.75, we obtain the rate constant for photochemistry (k_p):

$$k_p = \frac{F_m}{F_o}(k_d k_f) - (k_d + k_f) \qquad (6.3.77)$$

and combining Eqs. 6.3.75 and 6.3.76, we obtain the rate constant for non-photochemical quenching processes (k_n):

$$k_n = \frac{\frac{F_m}{F'_m}(k_d + k_f) - (k_d - k_f)}{E_{max}} \qquad (6.3.78)$$

The values for k_d and k_f can be obtained and derived (see Porcar-Castell et al., 2006 for details) from the literature (Barber et al., 1989).

The rate of electron transport to PSI determines the rate of reoxidation of the plastoquinone pool, while the rate of photochemistry (p) determines the rate of oxidation of the plastoquinone pool. The number of electrons in the quinone equivalent pool is represented by Q^{ON}, whereas the number of oxidized quinone equivalents is Q^{OFF}. In order to calculate the changes in the oxidized fraction of the quinone-equivalent pool, we denote by Q the proportion of oxidized quinone equivalents, i.e., $Q = \frac{Q^{OFF}}{Q^{ON} + Q^{OFF}}$. And assuming that the reoxidation process is proportional to Q^{ON}, and that the reduction process is equivalent to p (Eq. 6.3.70), changes in the number of oxidised plastoquinone equivalents can be estimated as:

$$\frac{dQ^{OFF}}{dt} = \gamma Q^{ON} - k_p Q Chla^{ON} \qquad (6.3.79)$$

where the first term of the equation represents the reoxidation of the plastoquinone pool, the second the reduction of the plastoquinone pool, and γ is a parameter indicative of the rate constant of reoxidation of the quinone-equivalent pool.

Finally, the short-term acclimation in the fraction of heat dissipation is modelled through the proportion of active quenching sites (ON). In the model, a quenching site is defined as a site where NPQ heat dissipation takes place. These sites can be in an active state, when they are engaged in heat dissipation, or in an inactive state. The number of active quenching sites is denoted with S^{ON} and with S^{OFF} the number of inactive sites.

Let E denote the proportion of active quenching sites, i.e., $E = \frac{S^{ON}}{S^{ON}+S^{OFF}}$ as an equivalent of the efficiency of NPQ. The rate of violaxanthin/zeaxanthin interconversion and the protonation of PSII proteins, are processes controlled by the thylakoid lumen pH (Demmig-Adams and Adams, 1996; Demmig-Adams et al., 1996; Eskling et al., 2001; Müller et al., 2001). Under physiological conditions, variations in the lumen pH will be correlated to variations in $Chla^{ON}$, and with the oxidized fraction of the quinone-equivalent pool (Q), the building process of NPQ (activation of inactive sites) is assumed to be proportional to the product of $Chla^{ON}$ and S^{OFF}, and the relaxation process (inactivation of active sites) proportional to the product of Q and S^{ON}. Thus variations in the number of active sites will correspond to:

$$\frac{dS^{ON}}{dt} = \lambda_b \, Chla^{ON} \, S^{OFF} - \lambda_r \, Q S^{ON} \qquad (6.3.80)$$

where the first term of the equation represents the building of NPQ and the second term its relaxation, and λ_b and λ_r are experimentally obtained parameters.

Model Results

The model gives us the rates of each of the energy consuming processes in the light reactions and their changes upon changing the illumination conditions. This can be used to follow the constant adjustments in energy partitioning at the light reactions under fluctuating light. In Fig. 6.3.52, we show an example of model output modified from Porcar-Castell et al. (2006), where a dark-acclimated leaf was supplied with light for 400 s and then placed for another 400 s under very low light. One can clearly see how the yield of photochemistry (ΦP), the yield of pH-dependent non-photochemical processes (ΦN), and the yields of fluorescence (ΦF) and constitutive heat dissipation (ΦD) acclimate to the new illumination conditions. With the model it is possible to visualize, e.g., the time that it takes before a new steady-state is attained, or how the photochemical yield adjusts quickly to increasing light but more slowly to decreasing light, due to the slower dynamics of the pH-dependent NPQ.

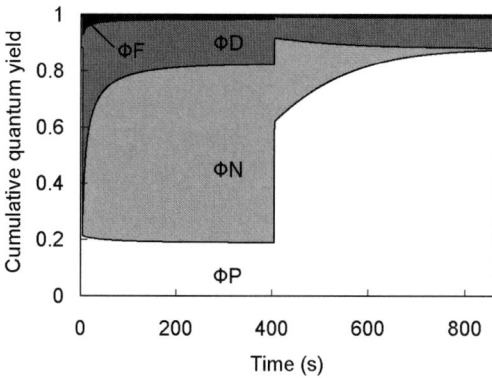

Fig. 6.3.52 Modelled development of the yields of fluorescence (ΦF), constitutive heat dissipation (ΦD), non-photochemical quenching processes (ΦN), and photochemistry (ΦP) in previously dark-acclimated alder leaves. The first 400 s the illumination was $1,200\,\mu mol\,m^{-2}\,s^{-1}$, and the remaining 400 s $0.001\,\mu mol\,m^{-2}\,s^{-1}$

Model Considerations at the Long-Term Scale (Days-Months)

At the longer time-scale of days to months, a whole new set of considerations needs to be taken into account when using chlorophyll fluorescence to monitor changes in the energy partitioning at PSII. These considerations can be updated into the model presented here in order to follow and study the acclimation of light reactions at the seasonal time-scale. At a short-time scale, it is feasible to assume that pigment contents remain constant. Chlorophyll content, for instance, will determine the amount of light absorbed by the leaf (Eq. 6.3.66), and by extension, the intensity of the chlorophyll fluorescence recorded by the fluorometer, however the extent to how much light absorption will increase with increasing chlorophyll contents will depend also on the light extinction coefficient inside the leaf. Therefore, seasonal changes in absorptance need to be considered when comparing fluorescence values at the former scale. Another factor that would need consideration is the total pool of xanthophyll-cycle pigments and how this pool influences the kinetics and capacity of NPQ.

Conclusions

Acclimation of photosynthetic light reactions to light is necessary to avoid damage, since the light energy capture depends linearly on light intensity and the performance of dark reactions saturate. The measurements support the analysis, but more versatile data is needed.

6.3.3.2.2 Photosynthesis and Drought

Pasi Kolari, Eero Nikinmaa, and Pertti Hari

University of Helsinki, Department of Forest Ecology, Finland

Plants rely on water in their metabolism. Water is used with CO_2 for formation of sugars in photosynthesis, and supply of water from the soil is required for transporting nutrients from the soil up to the leaves as well as assimilates from leaves to other parts of the plant (Section 7.5.2.2). Hydrostatic pressure of water in the cells, called turgor pressure, is also essential in maintaining cell function and retaining the leaves in shape. Loss of water is an unavoidable side effect of photosynthesis because uptake of CO_2 into the leaf by diffusion simultaneously allows water vapour diffuse out of the leaf (Sections 6.3.2.2 and 6.3.2.3). As water vapour concentration difference between the substomatal cavity and the ambient air is normally one or two orders of magnitude larger than CO_2 concentration difference, diffusive flux of water from the leaf correspondingly exceeds CO_2 uptake by a magnitude or two.

Most of the area on the globe is more arid than the humid northern forests and the availability of water is crucial for vegetation. As soil water tension increases in drying soil, water transport in the soil-plant-atmosphere system becomes increasingly difficult (Sections 6.3.2.8 and 7.6.3), and a higher osmotic pressure in leaf cells is required to maintain turgor pressure. At some point the leaves cannot maintain turgor any more and desiccation occurs. Cavitation, i.e. breaking of the water column in xylem, also limits water transport rate at high xylem water tensions (Section 7.5.2.3). Adaptation and acclimation to drought by reducing transpiration has thus been very important in the evolution of plants.

Physiologically, drought is understood as conditions where soil water content is so low that it becomes a major limiting factor for plant metabolism. The availability of water from the soil is more strongly related to water tension than to water content, thus, it is more relevant to describe soil water status in pressure units. A frequently used threshold value of soil water status is wilting point that refers to soil water tension at which plants loose their turgor pressure and cannot recover without addition of water into the soil. Traditionally its value is -1.5 MPa. There is no unique definition for the water tension where drought begins, however, as the effect of soil moisture on plant function is gradual and related to soil properties and especially to the type of vegetation. Hydromorphic plants are adapted to moist conditions and cannot tolerate dry conditions, whereas xeromorphic plants, like pines, can withstand prolonged drought and have wilting point lower than -1.5 MPa.

Availability of water can also be reduced when the water in the soil or in the water-conducting tissues of the plant itself is frozen. In addition of blocking the water conduits, ice tends to very strongly attract liquid water from surrounding tissues (Section 6.3.3.1.1). The evergreen boreal conifers typically are xeromorphic to withstand this kind of drought stress related to freezing.

Changes in soil water storage are determined by precipitation, runoff and evaporation (Section 7.9). Evaporation is driven by the energy of solar radiation and water vapour concentration deficit in air (Section 7.2). The big seasonal variation in the energy input in the boreal zone affects the annual evaporation pattern: in winter evaporation is negligible and in summer considerable. Evaporation through plant leaves, transpiration, is a major component of the forest ecosystem water balance during the growing season. In addition to the seasonal variation in the driving factors, the plants themselves modulate the annual patterns of ecosystem evapotranspiration and soil water conditions by controlling transpiration. Also rainfall has a conspicuous annual pattern, the maximum typically occurring in late summer and the minimum in late winter. These annual patterns of climatic factors and functioning of vegetation are reflected in the water pool in the soil. Water begins to accumulate in the soil in autumn and continues accumulation as snow on the ground. The water storage in the soil after melting of snow cover in spring is large, up to 300 mm H_2O per square metre ground in the uppermost 1 m layer of soil, depending on soil type. The storage decreases in spring and summer when cumulative evaporation and transpiration exceed precipitation (Section 3.3).

Normally the large initial soil water storage in the spring and regular replenishment of the storage by rain showers is able to provide water supply for the plants over the whole growing season in the boreal zone. During dry periods, however, strict control of water loss may be needed. The plants have evolved means to avoid excessive loss of water and the decline of photosynthetic capacity and physical damage due to water loss. Reduction of transpiration by stomatal closure allows a plant to maintain necessary water pool in the cells. When soil water storage decreases, the regulation system changes the functional substances involved in stomatal action in a way that the stomata become more sensitive to close to prevent transpiration rates that would exceed water uptake capacity and eventually lead to desiccation. Closure of the stomata also restricts diffusive flow of CO_2 into the mesophyll and thus reduces the rate of photosynthesis (Figs. 6.3.53 and 6.3.54).

Stomatal action is affected by different cues from the environment and by the physiological state of the plant itself. The regulation system changes concentrations of functional substances in the leaves, and the response of photosynthesis and transpiration is determined by these functional substances. The changes in the functional substances are slow but during longer periods they are reflected in plant metabolism and gas exchange. The actual regulation mechanisms and functional substances behind the stomatal responses are not yet fully understood but we can point out the most important relationships between stomatal action, environment, and the state of the plant. First, stomatal control is linked with photosynthetic rate, indirectly or via mesophyll CO_2 concentration. This mechanism allows for increasing photosynthetic CO_2 uptake in favorable conditions while keeping the stomata less open when the rate of photosynthesis is limited by light or low temperature instead of availability of CO_2. Stomata also respond to air humidity, either by directly sensing the rate or evaporation from guard cell surfaces or through change in osmotic pressure in the guard cells due to evaporative water loss. This air-humidity-induced feedforward adjustment of stomatal aperture serves for drought avoidance also when there is plenty of water available. Finally, the water deficit, that develops in the plant when the soil dries out, can directly close stomata because the control of stomatal aperture is based on the turgor pressure of the guard cells. During prolonged drought

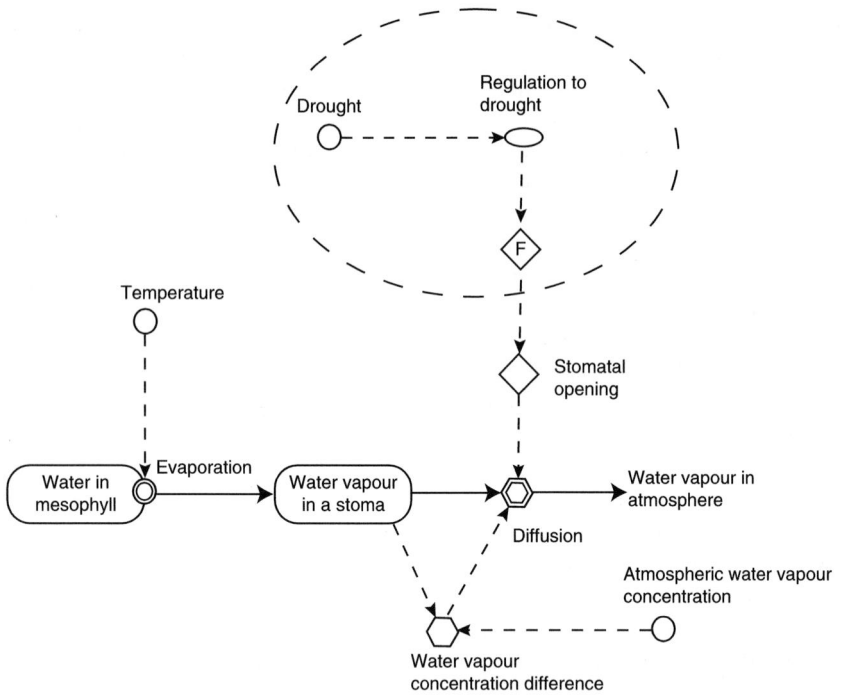

Fig. 6.3.53 The effect of drought on transpiration via regulation of functional substances. The regulation responds to drought, otherwise the mechanism is similar with Fig. 6.3.10

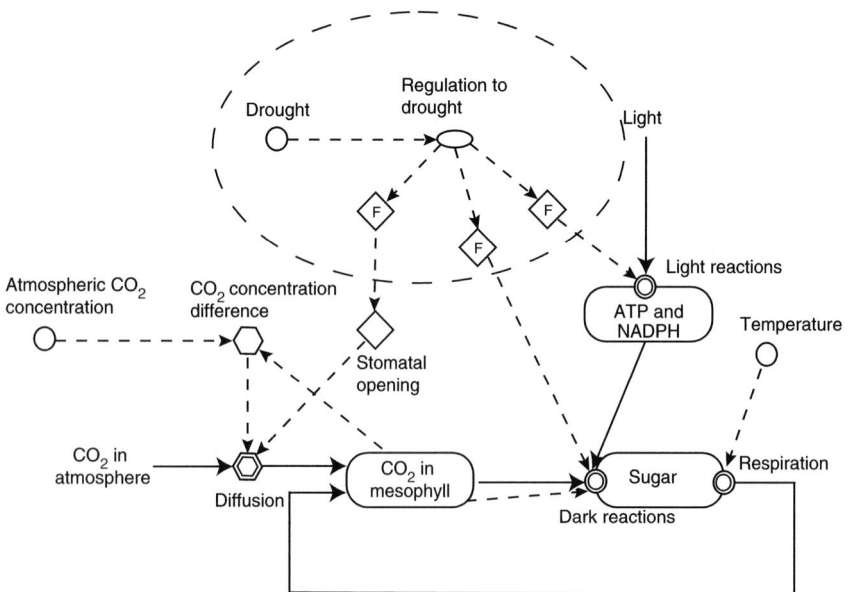

Fig. 6.3.54 The effect of drought on photosynthesis via regulation of functional substances. The regulation responces to drought, otherwise the figure is rather similar with Fig. 6.3.39

also downregulation of photosynthesis occurs in a way similar to downregulation in cold-stressed leaves, simply to avoid light damage (Flexas and Medrano, 2002). Eventually the pools of functional substances needed in photosynthesis and maintenance of cells are depleted which can lead to premature senescence of leaves. In extremely severe drought, separation of plant cell cytoplasm from the cell wall, plasmolysis, can also result from water loss.

In normal conditions the boreal conifers seldom need to modify the behaviour of stomatal action to prevent excessive water loss. Photosynthetically active radiation, temperature, water vapour deficit and annual cycle can explain stomatal movements and photosynthetic rate. Most of the time, the previously used optimal stomatal control model is able to predict well the daily behaviour of gas exchange over the summer with a fixed parameter set, as seen in Section 6.3.2.3.

In the optimal stomatal control model, parameter λ, cost of transpiration, describes the sensitivity of stomatal action, i.e. the state of functional substances involved in stomatal control. The value of λ is reflected in the model behaviour in such a way that the stomata are less open at high values of λ. Normally the value of λ does not vary much over the growing season (Section 6.3.2.3; Fig. 6.3.57a) but we can expect it to increase when the plant needs considerably reduce its water loss, that is, during drought.

The summer of 2006 was extremely dry in southern Finland. There was very little rain from early July to mid-August. During late July and the first half of August there was a clear reduction in CO_2 uptake and transpiration in Scots pine shoots at SMEAR II (Fig. 6.3.55). The decline could be first seen in the afternoon and later, when the soil kept drying out, over the whole day. Eventually transpiration and CO_2 exchange dropped to about 10% of their pre-drought levels. The rate of respiration (night-time CO_2 exchange in Fig. 6.3.55) also declined with photosynthetic CO_2 uptake. This can be attributed to downregulation of photosynthesis or simply to shortage of sugars for use in ATP synthesis. The role of stomatal action in the reduction of gas exchange can be concluded from the measured transpiration that indicates almost complete closure of stomata in the last few days of the drought period. When the soil water storage was replenished in mid-August, gas exchange recovered to nearly normal level in few days. Evidently the decrease in CO_2 uptake during the drought was mainly due to stomata being closed, not persistent damage to cell metabolism.

During the drought in the summer of 2006, prediction of daily CO_2 exchange pattern of the Scots pine shoots at SMEAR II failed if fixed photosynthetic parameter values were used (Fig. 6.3.56). With fixed λ, the predicted CO_2 exchange is strongly overestimated during the dry period. Parameter λ was then estimated daily using both CO_2 exchange and transpiration. Photosynthetic efficiency β was obtained from the annual cycle model (Eqs. 6.3.53–6.3.58; Section 6.3.2.3) and calibrated for each shoot using values estimated for the period of 15 June–15 July when there was still ample water in the soil. The other photosynthetic parameters were fixed. With λ estimated daily, the optimal stomatal control model was able to capture the reduction in photosynthesis and transpiration rather satisfactorily (fitted lines in Fig. 6.3.56).

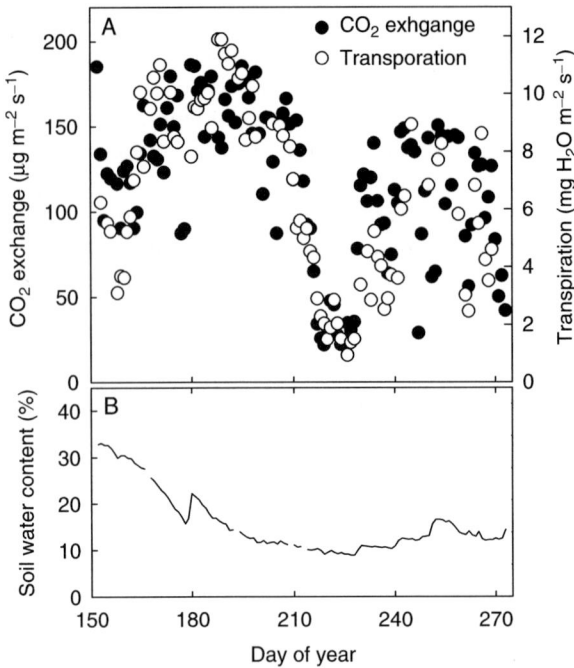

Fig. 6.3.55 CO_2 exchange (dots) and transpiration (circles) of Scots pine shoots (A) and soil water content (B) at SMEAR II in the summer of 2006. The fluxes are means of three shoots, averaged daily between 11:00 and 13:00 solar time

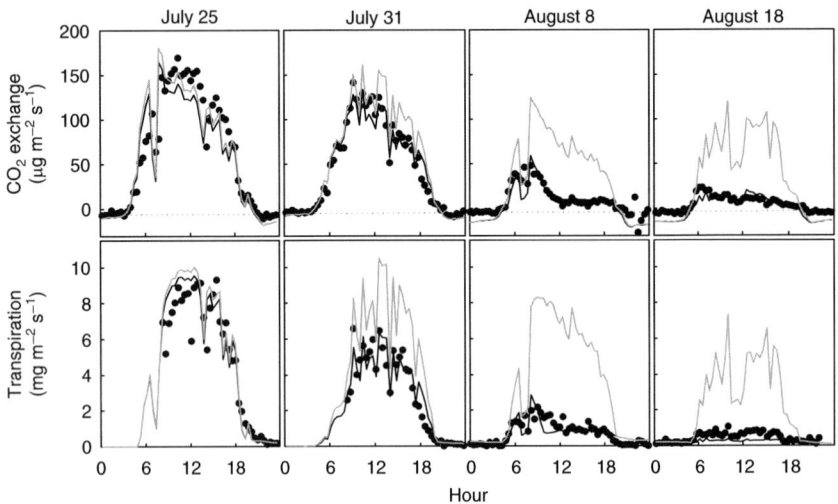

Fig. 6.3.56 Modelled (black and grey lines) and measured (dots) diurnal course of CO_2 exchange (upper panels) and transpiration (lower panels) of a Scots pine shoot at SMEAR II on four selected days during progressive drought in the summer of 2006. The grey line indicates the gas exchange predicted with the optimal stomatal control model and constant cost of water (λ), and the black line modelled gas exchange when λ was estimated daily

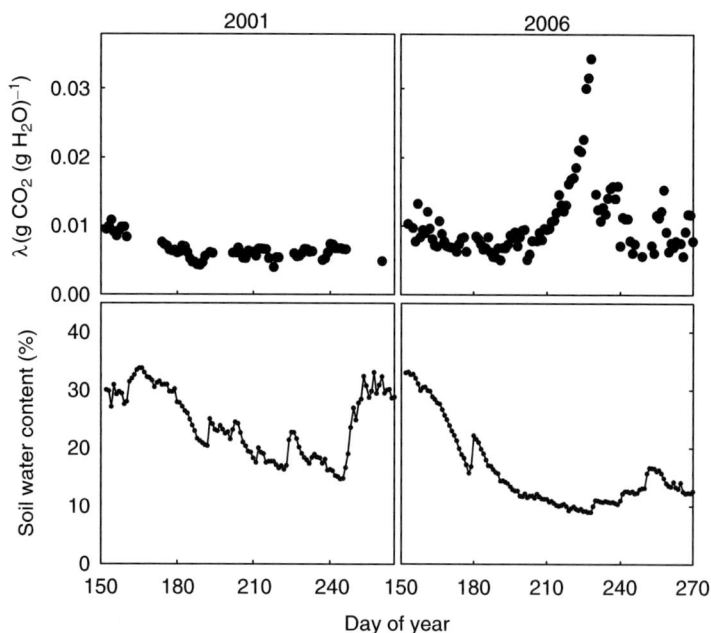

Fig. 6.3.57 Seasonal course of cost of water (λ) estimated daily (mean of three Scots pine shoots) (upper panel) and volumetric water content in the uppermost 5-cm layer of mineral soil in a typical summer (2001) and in a summer when there was severe drought (lower panel) at SMEAR II (2006)

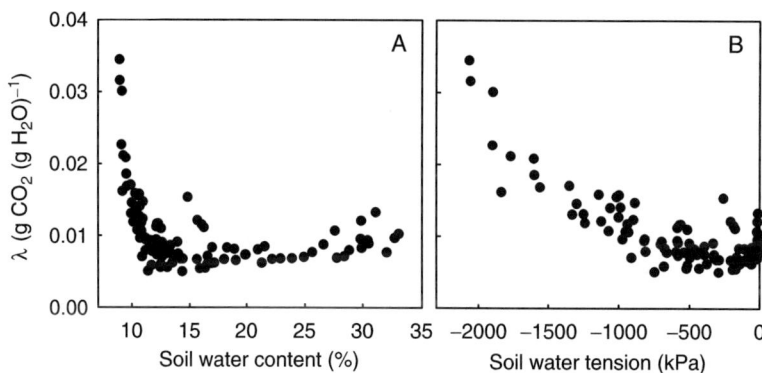

Fig. 6.3.58 The relationship of estimated daily values of cost of water (λ) with soil volumetric water content (A) and with water tension in the uppermost 10 cm of mineral soil (B) at SMEAR II in the summer of 2006

The daily values of λ show a strong rise along with declining soil water content and a rapid decrease after a rain shower (Fig. 6.3.57). When compared against soil water storage (Fig. 6.3.58A), the increase in λ is very steep when soil water content drops below 10%. This is because soil water tension follows water storage in approximately negative exponential fashion. Because trees actually respond to soil water tension rather than to absolute volumetric water content, it is more convenient to analyse the relationship using water tension as the explaining factor (see Section 6.3.2.8). Figure 6.3.58b shows the relationship between the estimated daily values of λ and soil water tension at SMEAR II. Water tension in the uppermost mineral soil layer actually dropped below wilting point (−1.5 MPa) but the deeper soil layers provided just enough water for the trees to survive over the drought.

The occurrence of severe drought periods varies in the boreal zone. In the fairly moist conditions of the northern boreal zone, there are seldom warm and dry periods long enough to dry out the soil. We haven't been able to observe clearly drought-induced decline in shoot CO_2 uptake at SMEAR I near the arctic-alpine timberline. The forest ecosystems in the continental and the southern regions of the boreal zone are more prone to suffer from soil water deficit. In southern boreal conditions of SMEAR II severe drought is still rare but in few years there have been short periods in late summer when the effect of soil water deficit has been obvious in the gas exchange data. However, conditions like in the summer of 2006 described above are exceptional in the present climate. In the warming climate the frequency of drought periods and the ability of forest ecosystems to withstand drought may become a more important question. Therefore, it is important that we are also able to mechanistically address the effects of dry soil in predicting photosynthesis (Section 7.6.3).

Conclusions

Photosynthesis and transpiration of trees declines in drying soil because stomatal regulation restricts gas exchange. The optimal stomatal control model of photosynthesis and transpiration is able to capture rather well the essential regularities in the response of transpiration and photosynthesis to drought. However, our humid climate has hindered sufficient testing of the approach.

6.3.3.2.3 Increasing CO_2 and Photosynthesis

Eija Juurola and Pertti Hari

University of Helsinki, Department of Forest Ecology, Finland

Background

Atmospheric CO_2 concentration has varied considerably in the time scale of million or billion years. During the last 600,000 years, it has fluctuated between 200–300 ppm (Chapter 1) and affected the fine structures of leaves as well as the functional substances that control photosynthesis. Still during the lifetime of an individual tree, atmospheric CO_2 concentration has been rather stable due to the large atmospheric carbon pool. The present large emissions of CO_2 have, however, caused very rapid change that has never occurred before – at least not in millions of years. Hence, the availability of CO_2 changes even during the lifetime of a single tree.

As CO_2 is a key substrate for photosynthesis, its availability largely determines the speed of the process. Initially, with increasing CO_2 concentration, photosynthesis is accelerated due to the increased availability of main substrate and the biochemical properties of CO_2 fixation (see Section 6.3.2.3). Thus the primary acclimation is in the CO_2 fixing enzyme Rubisco and all the other acclimations originate from this primary effect. As a consequence the water use efficiency (WUE, CO_2 fixed per water lost) increases. In the long term, if the CO_2 concentration stays high, more complicated acclimations arise leading to alterations in concentrations or activities of functional substances involved in photosynthesis and finally to reallocation of resources within the photosynthetic apparatus (e.g., Woodrow, 1994; Drake et al., 1997; Luo et al., 1998).

During the last 250 years the atmospheric CO_2 concentration has increased from near 280 ppm in the pre-industrial era (i.e., before 1750, Houghton et al., 2001; Keeling and Whorf, 2004), to 380 ppm, and it is expected to reach 700–1,000 ppm by the end of the 21st century (Houghton et al.,2001). Such a rapid increase in atmospheric CO_2 concentration is unprecedented in the genetic history of tree species. The acclimation of structure and function of trees to such a rapid change in CO_2 concentration was not tested during their evolutionary history, and even unfavourable acclimations may occur. Indeed, despite extensive research on long-term effects of elevated CO_2 concentration on plants, it has remained unclear why large differences in photosynthetic response to elevated CO_2 concentration exist among (Tjoelker et al., 1998) and within species (Pettersson et al., 1993; Gunderson and Wullschleger, 1994; Rey and Jarvis, 1998).

Recent results show that coniferous and broadleaved tree species have different strategies in investing nitrogen to CO_2 fixing enzyme Rubisco (Warren and Adams, 2004), which in turn may lead to different acclimation patterns. On top of all this, the Rubisco reaction is a temperature dependent process and increasing CO_2 concentration may affect the temperature dependence of CO_2 assimilation that is shown to vary between species and according to growth conditions (Björkman, 1981a, b; Bernacchi et al., 2003).

Biochemical Properties

In acclimation of C3 plants to elevated atmospheric CO_2 concentration, Rubisco plays a key role (e.g., Eamus and Jarvis, 1989; Stitt, 1991). In the long term, increasing CO_2 concentration may lead to reduction in Rubisco activity or concentration because the amount of Rubisco required for maintaining the same assimilation rate decreases (Woodrow, 1994; Drake et al., 1997). Despite this decrease, the increase in CO_2 concentration may still allow for a higher CO_2 assimilation rate (e.g., Hikosaka and Hirose, 1998). Furthermore, because of the general kinetics of

Fig. 6.3.59 The temperature
dependence of CO_2 exchange
in three silver birch seedlings
grown in different CO_2 concentrations (spheres), and for
comparison a temperature
dependence of birch seedling
grown at 360 ppm but mea-
sured at 2,000 ppm of CO_2
(crosses)

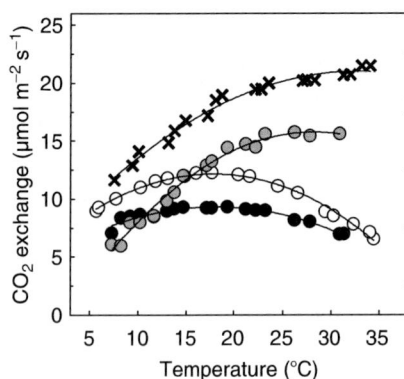

the Rubisco-catalyzed enzymatic reactions, i.e., saturation at high substrate concentrations and two competitive substrates, the acclimations mediated through Rubisco should be nonlinear in response to increasing CO_2 concentration (Woodrow, 1994; Luo et al., 1998). Finally, due to the general kinetics of Rubisco, the prolonged effect of increased CO_2 concentration on photosynthesis should be temperature dependent leading to altered optimum temperature (e.g., Long, 1991; Drake et al., 1997, Fig. 6.3.59).

The acclimation of Rubisco is not, however, the only changes in the photosynthetic functional substances we should expect since there are strong connections in the regulation of functional substances. The increased performance of dark reactions is reflected also in light reactions and stomatal action. Thus we can assume that concentrations of all functional substances involved in photosynthesis react to the increasing CO_2 concentration as visualised in Fig. 6.3.60.

When the effects of increasing CO_2 concentration are studied in plants grown at one or two elevated CO_2 concentrations, it is implicitly assumed that the responses are linear, although it was suggested already in 1990s by Bowes (1991) and Woodrow (1994) that this was unlikely. Thus, to reveal the pattern of acclimation and establish whether there are truly species-specific differences, it is essential to study changes in plant physiology in response to a wide range of treatments. We studied the acclimation of young silver birch and Scots pine seedlings to a series of increased CO_2 concentrations from 350 to 2,000 ppm for two growing seasons (Juurola, 2003). We measured chlorophyll fluorescence, Rubisco and chlorophyll contents, C/N ratios and contents, and the gas exchange properties at the end of both growing seasons. All gas exchange measurements were performed in the laboratory with a dynamic system for measuring gas exchange (described in detail in Aalto and Juurola, 2001).

Both an unchanged and decreased chlorophyll concentration as well as PSII photochemistry has been observed (Wilkins et al., 1994; Lawlor et al., 1995; Lewis et al., 1996; Scarascia-Mugnozza et al., 1996; Rey and Jarvis, 1998; Jach and Ceulemans, 2000; Gielen et al., 2000; Riikonen et al., 2005). In Juurola (2003),

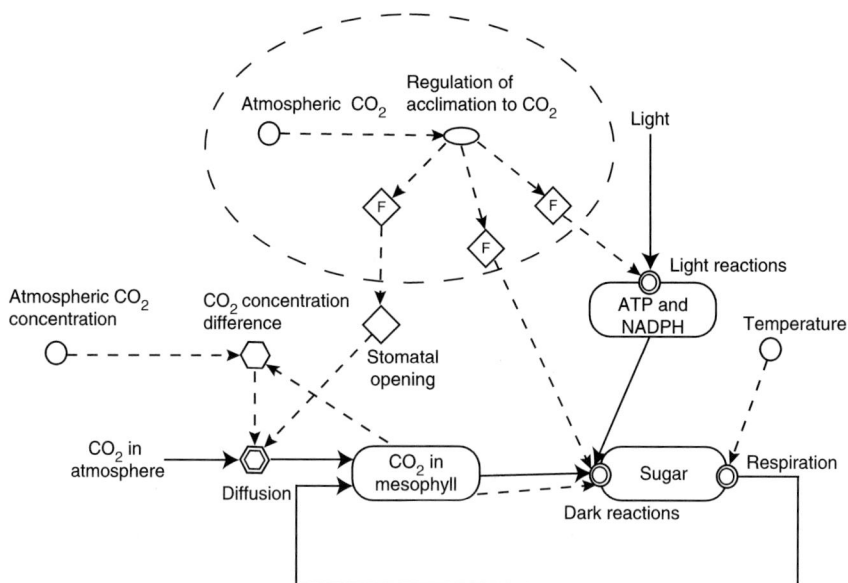

Fig. 6.3.60 Visualisation of the acclimation of photosynthesis to increasing atmospheric CO_2 concentration. The regulation react to the atmospheric CO_2 concentration. The symbols are introduced in the Fig. 1.2.1

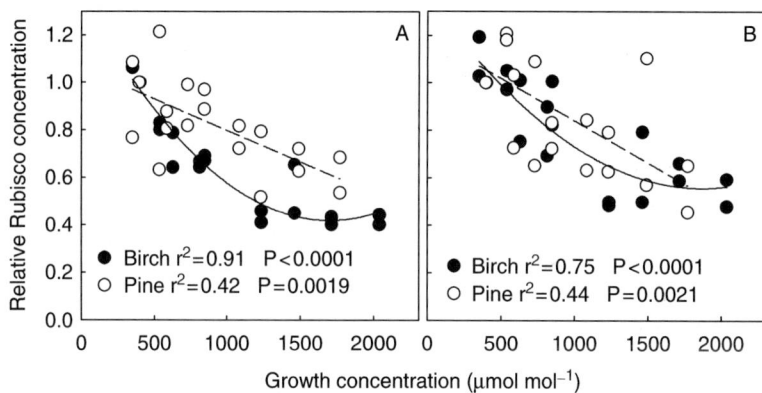

Fig. 6.3.61 Patterns of relative Rubisco concentration (A) and relative chlorophyll concentration (B) in response to growth CO_2 concentration in silver birch (dots) and Scots pine (circles) when data from two years were pooled and expressed as relative values with respect to seedlings grown at 400 μmol mol^{-1}. Nonlinear or linear regressions were applied separately for both species. Second order regressions were applied only when the nonlinear term in the model was significant

however, a clear decreasing trend was also observed in chlorophyll concentrations in silver birch and Scots pine leading to unchanged balance between light and dark reactions in seedlings grown at elevated CO_2 concentrations (Fig. 6.3.61). This implies also that the acclimation is a complicated phenomenon.

There exist several possible reasons for diverse results on gas exchange in different species. First, due to the higher availability of CO_2, plants can more efficiently control the transpiration stream by changing stomatal aperture. The response may be acclamatory, or a direct effect of CO_2 on stomatal control. In fact, a doubling of CO_2 concentration has often been reported to have no effect on stomatal conductance in conifers in contrast to deciduous species, reflecting perhaps the differences in leaf structure and in the strategy for water use (Eamus and Jarvis, 1989; Roberntz and Stockfors, 1998; Juurola, 2003).

The temperature dependence of CO_2 assimilation is also an important attribute in acclimation to increasing CO_2 concentration. Therefore, the acclimation patterns observed in CO_2 exchange may also reflect changes in the temperature response of photosynthesis originating from the combined effect of the CO_2/O_2 specificity of Rubisco (Ghashghaie and Cornic, 1994; Hikosaka and Hirose, 1998) and structural differences between species. Indeed, it is often found that the temperature optimum increases at elevated CO_2 concentration (e.g., Juurola, 2005).

Changes in the temperature response of CO_2 assimilation reflect properties of the Rubisco-catalysed reaction, but also acclimation of the photosynthetic functional substances to increasing CO_2 concentration (e.g., Bunce, 2000). This acclamatory effect is highlighted when the traditional Farquhar–model (Farquhar et al., 1980) was compared to the measured temperature responses of birch seedlings acclimated to different CO_2 concentrations (Fig. 6.3.62). The ratio of CO_2 exchange rate at growth concentration to that at current atmospheric CO_2 concentration (A_{growth}/A_{350}) was calculated both for the birch seedlings acclimated to elevated CO_2 concentrations and by utilising the temperature response from the Farquhar model that was parameterised for silver birch (Aalto and Juurola, 2001). In the model analysis the measured respiration rates and intercellular CO_2 concentrations

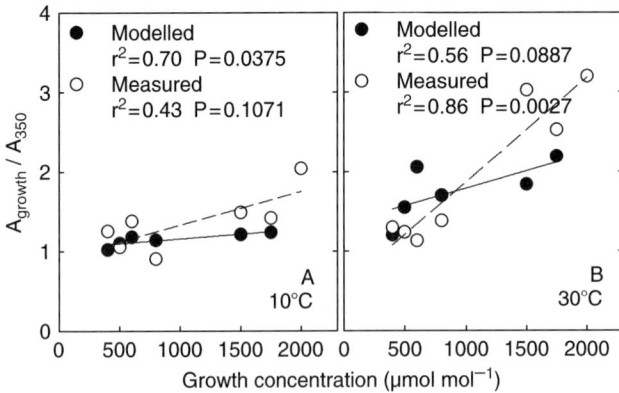

Fig. 6.3.62 The modelled (black dots) and measured (circles) ratios of CO_2 exchange at growth CO_2 concentration to that at current ambient CO_2 concentration (A_{growth}/A_{350}) in silver birch leaves at 10°C (A) and at 30°C (B). Linear regressions were applied separately for both modelled and measured A_{growth}/A_{350}

were taken in account to ensure that the possible differences would not originate from differences in the respiration or stomatal conductance. The results at different temperatures show that the observed changes in temperature response are not only due to properties of Rubisco. The changes are also due to the acclamatory effect on other functional substances, which is shown as a discrepancy in the calculated ratio especially at high temperatures.

Although there is an increasing amount of information on the response of photo-synthetic functional substances to increasing CO_2 concentration, the picture is not clear. There are observed differences between species and even within species in the acclimations to elevated CO_2 concentrations. Some of the results may be caused by experimental technique, since proper experiments are difficult to implement. How-ever, these differences can be expected, since acclimation to present atmospheric concentrations has not been tested in evolution, at least during the last one million years in which the atmospheric CO_2 concentration has been below 300 ppm.

The controlled environment studies give some idea about the acclamatory responses but it makes a big leap of fate to generalise them to field conditions with a gradually changing climate (Körner, 2006). Recent results from long term FACE experiments, as well as chamber experiments show that the response to increasing CO_2 at natural ecosystems and in mature trees may be temporal and less than expected (Körner, 2006; Morgan et al., 2005). Also, it seems that higher rates of photosynthesis will not produce 1:1 increase in growth (Körner, 2006). However, Kimball et al. (2007) showed a clear increase in growth with a decreasing effect on photosynthesis. It seems that the soil properties and underground processes play an important role in photosynthetic responses to increasing CO_2 in the long run.

Conclusions

Photosynthetic functional substances acclimate evidently to increasing atmospheric CO_2 concentration, but the acclimations vary. This is to be expected since the acclimations have not been tested in evolution during the last one million years.

6.3.3.2.4 VOC Emissions Under Changing Climate

Jaana Bäck

University of Helsinki, Department of Forest Ecology, Finland

Effects of climate change on ecosystem-level BVOC emissions are produced by many, partially opposite, responses to changing factors. General and consistent response patterns to climate change, e.g., increasing CO_2 concentration, periodic drought and warming, among plant species, seasons and years are difficult to detect, which makes evaluating the net effects at the ecosystem level very difficult. Positive effects on emissions can be caused by drought, increased leaf area and biomass, increased availability of nitrogen and phosphorus, and increased temperature. However, elevated CO_2 effects on emissions are extremely variable, and the mechanisms of these responses are still largely unveiled.

Short-term exposures to elevated CO_2 are often shown to rapidly reduce leaf-level isoprenoid emissions (Monson and Fall, 1989; Loreto and Sharkey, 1990; Loreto et al., 1996), whereas growth in subambient CO_2 levels tends to increase emissions (Possell et al., 2005). Also increases in monoterpene emissions in, e.g., oak species growing at elevated CO_2 have been reported (Tognetti et al., 1998; Staudt et al., 2001). Emission reductions due to elevated CO_2 have been connected with down-regulation at biochemical and/or enzymatic levels (e.g., Loreto et al., 2001; Rosenstiel et al., 2003). Clear and indisputable evidence for a direct action of CO_2 on the functional substances involved in isoprenoid biosynthesis has not been found so far.

However, the functional substances involved in BVOC emissions evidently can be modified by elevated CO_2 concentrations. In a longer perspective (several growing seasons), adjustments may be visible in the variable CO_2 responses with species, environmental conditions and exposure time (e.g., Sharkey et al., 1991; Constable et al., 1999; Staudt et al., 2001; Loreto et al., 2001; Baraldi et al., 2004; Rapparini et al., 2004). Elevated CO_2 may inhibit the expression of isoprenoid synthesis genes, change protein turnover or affect the phosphorylation and dephosphorylation processes of the functional substances in plants growing at elevated CO_2 levels (Loreto et al., 2001; Scholefield et al., 2004). Despite many conjectures, the information dealing with acclimation of perennial plants to elevated CO_2 levels is far from sufficient for any clear conclusions regarding the effects on BVOC emissions at the ecosystem level.

In general, BVOC biosynthesis is strongly limited by carbon availability (see Section 6.3.2.5), and good correlations with photosynthesis have been found in many species. However, under elevated CO_2, the correlations between the measured emissions and both leaf temperatures and photosynthetic rates become more vague (Loreto et al., 2001; Rapparini et al., 2004; Scholefield et al., 2004). Since the volatile isoprenoids can also be derived from a number of alternative C sources, and their biosynthesis requires a large energetic input, it is possible that the biosynthesis would be controlled by photosynthetic electron transport rates (i.e., incident light availability) rather than by carboxylation, especially if the carbon limitation would diminish under elevated CO_2 (Delwiche and Sharkey, 1993; Karl et al., 2002; Kreuzwieser et al., 2002; Possell et al., 2004).

According to our model, the BVOC emissions will clearly respond to elevated CO_2 and temperature. Increasing temperature and atmospheric CO_2 concentration accelerates photosynthesis, thus producing more substrates for isoprenoid biosynthesis, and therefore the BVOC emissions increase proportionally to the increase in photosynthesis. Also direct effects of temperature rise to the volatilization of BVOCs increase their emissions. This model behaviour is not in conflict, although not strongly supported by the present experimental results either, since the mechanisms of CO_2 influence on BVOC biosynthesis are still ambiguous.

Conclusions

The BVOC emissions in elevated CO_2 concentrations are rather problematic to predict due to many interactions and interlinked processes. However, the assumption that BVOC emissions follow photosynthesis at least on a longer time scale seems to be the most reasonable one.

6.4 Soil Processes

6.4.1 Decomposition of Soil Organic Matter

Asko Simojoki[1], Jukka Kurola[2,1], Mari Pihlatie[3], Jukka Pumpanen[4], Mika Kähkönen[1], and Mirja Salkinoja-Salonen[1]

[1] University of Helsinki, Department of Applied Chemistry and Microbiology, Finland
[2] University of Helsinki, Department of Ecological and Environmental Sciences, Lahti, Finland
[3] University of Helsinki, Department of Physics, Finland
[4] University of Helsinki, Department of Forest Ecology, Finland

Plant cells have limited lifetime and they die and fall on the ground as litter (Section 6.3.2.4). Microbes start to decompose the macromolecules (Section 5.2.1) in the litter and utilise them as a source of energy and raw material. Litter input and rain generate a typical structure in the top layer of the soil (Section 7.4). The decomposition has a key role in releasing the nutrients in the litter for reuse by vegetation (Section 6.3.2.8) and in the formation of soil structure.

The decomposition of organic macromolecules in soil proceeds in two main chemical steps (Fig. 6.4.1): (1) the breakdown of macromolecules to small molecules – such as the enzymatic hydrolysis of biopolymers to their monomers – and (2) the subsequent uptake and oxidation of the monomers by soil organisms (see reviews by Warren, 1996; Tate III, 2002; Sinsabaugh et al., 2002; Schimel and Bennett, 2004). The small molecules are converted by microbial catabolic reactions

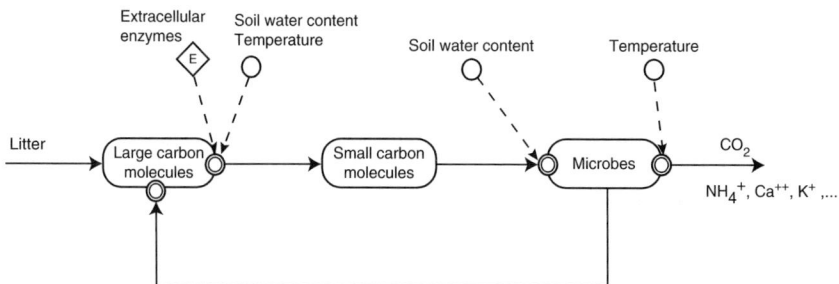

Fig. 6.4.1 Microbial decomposition of large carbon molecules with extracellular enzymes. The symbols are introduced in the Fig. 1.2.1

partly to inorganic forms, such as carbon dioxide, water and ammonium ions, while
the rest is re-synthesized by the growth of microbial cells. The release of inorganic species from organic compounds to soil is called mineralization. The organic
decomposition products too large to be taken up by microbes may contribute to
chemical reactions producing the decay-resistant organic matter called humus.

Soil microbes deal with the structural complexity of decomposable compounds
by producing complex enzyme systems, and by growing as communities that produce a large variety of different enzymes (Warren, 1996; Sinsaugh et al., 2002).
The enzymatic character of soil organic matter decomposition has several consequences. First, the reaction kinetics is affected not only by the availability of substrates and products but additionally by the activity level of enzymes. The activity
of enzymes depends on soil properties and conditions such as soil temperate, moisture content and acidity, which all have direct impact on the functioning of enzymes
in soil. Secondly, the enzymatic reactions are highly compound-specific: each compound and type of bond discussed in Section 5.2.2 requires a specific enzyme of
its own (Sinsaugh et al., 2002). For example, the hydrolysis of starch, cellulose
and hemicellulose in soil are catalysed by glucosidases, the hydrolysis of proteins
by peptidases, and those of lipids and chitin by esterases and chitinases, respectively. Thirdly, the physical structure of soil and decomposing organic residues adds
a spatial component to the factors determining the reaction kinetics. They affect the
soil volume available for the reactions, the binding of enzymes with soil, and the
microscale mass transfer of reacting species (Tate III, 2002).

The ease of decomposition of biological tissues is largely determined by their
chemical composition (as reviewed by Derenne and Largeau, 2001; Blume et al.,
2002; Kögel-Knabner, 2002). For example, the cell contents surrounded by the
plasma membrane are rapidly decomposed by the numerous active enzymes within
the cell. This occurs in part already during the senescence of living cells.

The lipids found in biological membranes and bark are relatively short-chained
molecules not bonded chemically to each other. They have both hydrophilic and
hydrophobic parts. As a consequence, their decomposition proceeds relatively
rapidly in soil. In contrast, the macromolecules in cell walls and extra-cellular
enzymes in the soil are too large and insoluble to enter microbial cells. They must
first be hydrolysed to their monomers by the enzymes excreted by microbes. The
monomers can then enter the microbial cell and be metabolized for energy needs
and cell construction.

Cellulose, a major component of plant cell walls, decomposes slowly in soil,
notwithstanding its simple chemical structure of straight chains of 1–4 β-linked glucose units. This is because the linear cellulose chains pack close together and form
a crystal-like structure stabilized by hydrogen bonding between the molecules. This
gives cellulose mechanical strength and effectively prevents the action of hydrolyz-
ing enzymes. Chitin, the 1–4 β-linked polymer of N-acetyl-glucoseamine units in
fungal cell walls, resembles cellulose in structure but is mechanically even stronger.
In the other hand, hemicellulose and other cell wall carbohydrates composed of different sugar units have more branched and open structure, which facilitates more
rapid hydrolysis by the enzymes compared with that of cellulose.

Lignin, another major component of plant cell wall, is a high-molecular weight substance composed three-dimensionally of phenyl propane units, which makes it particularly recalcitrant to microbial decomposition. White rot and brown rot fungi belong to the few organisms that can decompose lignin, but even they cannot utilize lignin as the only carbon or energy source for growth. Lignin is decomposed by an extra-cellular, co-metabolic radical mechanism that partly breaks the aromatic rings and side-chains. This direct oxidation releases small molecules (e.g. organic acids) that may be taken up and metabolized by the microbes, whereas other parts of lignin only change in structure. The chemically modified lignin may react with other soil compounds, such as amino acids and polyphenols, to form humus in a process of humification.

To assess the microbial induced decomposition of soil organic matter at SMEAR II measuring station, hydrolytic activities of ten enzymes related to the decomposition of organic matter were measured with fluorometric microtiter plate method described by Wittmann et al. (2004). Seven of them are related directly to C and N cycles (beta-cellobiosidase, alfa- and beta-glucosidases, N-acetyl-glucosaminidase, beta-xylosidase, asetate esterase, and aminopeptidase). For the measurements 5–7 soil cores (to 25 cm depth) were taken nine times at two to three month interval during two years at SMEAR II (Fig. 6.4.2). The activities of enzymes catalyzing the decomposition of soil organic matter generally decline with soil depth during warm seasons. In contrast, the activities during cold seasons the activities are lower and less variable with soil depth in comparison with warm season.

6.4.2 Nitrogen Processes in Soil

Mari Pihlatie[1], Asko Simojoki[2], Jukka Kurola[3,2], Jukka Pumpanen[4], and Mirja Salkinoja-Salonen[2]

[1] University of Helsinki, Department of Physics, Finland
[2] University of Helsinki, Department of Applied Chemistry and Microbiology, Finland
[3] University of Helsinki, Department of Ecological and Environmental Sciences, Lahti, Finland
[4] University of Helsinki, Department of Forest Ecology, Finland

Functional substances enable processes, such as photosynthesis and the formation of ATP (Section 6.3.1), to take place. These functional substances are large protein molecules formed by amino acids (Section 5.2.1). Amino acids contain about 15% nitrogen, which determines the importance of nitrogen as a nutrient. The release of nitrogen from functional substances in forest soil is very important for the cycling of nitrogen in a forest ecosystem.

6.4.2.1 Mineralization

Boreal forest ecosystems are starved of nitrogen. The only external nitrogen input, atmospheric nitrogen deposition, is very low (Section 7.9). This gives the important

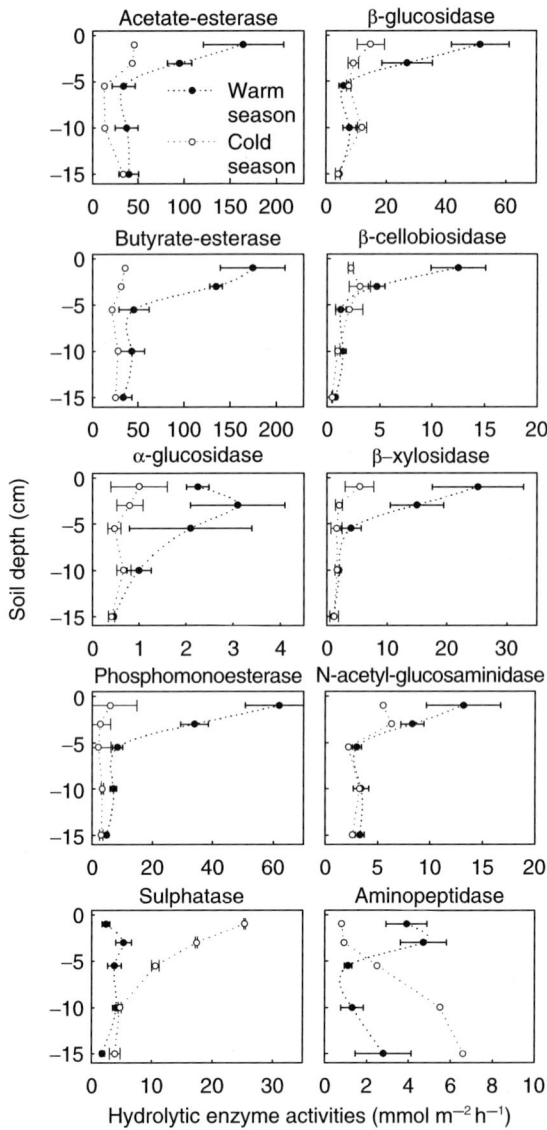

Fig. 6.4.2 Profiles of hydrolytic activities of enzymes involved in the decomposition of organic matter in soil during warm and cold seasons as measured with fluorometric microtiter plate method described by Wittmann et al. (2004)

role for microbial decomposition of proteins in the soil. Internal nitrogen cycling, the uptake of nitrogen by trees and microbes, is largely driven by mineralization of organic matter in the soil. Mineralization is a process in which organic molecules in soil are broken down and inorganic forms of nitrogen, such as ammonium (NH_4^+) or nitrate (NO_3^-), are formed. Once released in the soil solution ammonium and

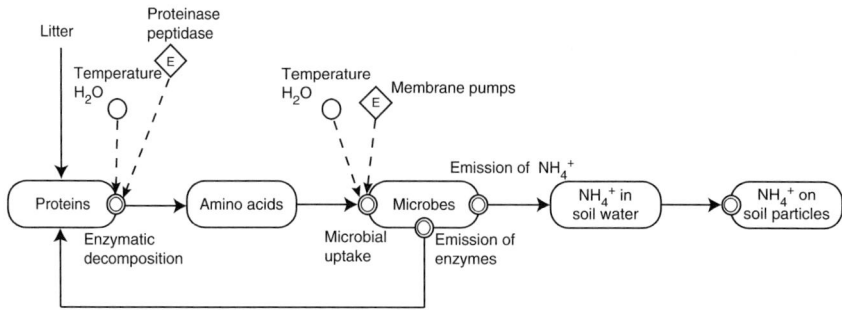

Fig. 6.4.3 Decomposition chain of organic matter into mineral forms of nitrogen in the soil. The symbols are introduced in the Fig. 1.2.1

nitrate ions can be taken up by plants and microbes and utilized in the build-up of new cellular material. Most of the nitrogen containing organic matter in the soil is in non-soluble complex forms that are too large to pass through microbial membranes. Therefore, microbes in the soil produce extracellular enzymes such as proteinases and peptidases that can break down proteins into smaller, water soluble compounds such as amino acids (see Fig. 6.4.3). These amino acids are then usable for microbes as they can pass through the microbial membranes via membrane pumps (see Fig. 6.4.3). The rate of protein decomposition in the soil is driven by soil temperature and the availability of water, as well as the input of new organic matter and the activity of extracellular enzymes in the soil.

Further release of inorganic forms of nitrogen from the bacterial cells into the soil depends on the rate of nitrogen cycling in the soil. Microbes release ammonium (NH_4^+) into the soil as soon as the need for the microbial cell growth is fulfilled. Ammonium-ions in the soil solution can be attached on to the negatively charged clay particles or taken up by plants or microbes. Microbial uptake of inorganic nitrogen from the soil solution hence reduces the availability of nitrogen to the growing plants. This process is called immobilization and it usually takes place in nitrogen limited ecosystems where the dissolved organic carbon is not sufficient to meet microbial N need.

6.4.2.2 Nitrogen Fixation

Nitrogen fixation is a biological process in which nitrogen from the atmosphere can be brought into a soil ecosystem. Many bacteria and algae living in soil or aquatic environments are able to reduce molecular nitrogen (N_2) from the atmosphere into NH_3, and incorporate it into amino acids for further protein synthesis (see Fig. 6.4.4). The reaction chain from the molecular nitrogen present in soil air into amino acids involves the nitrogenase enzyme to fix the molecular nitrogen from the atmosphere. Further reactions and enzymes involved in the biological nitrogen fixation are summarized in Fig. 6.4.4. The nitrogen fixing microbes are either

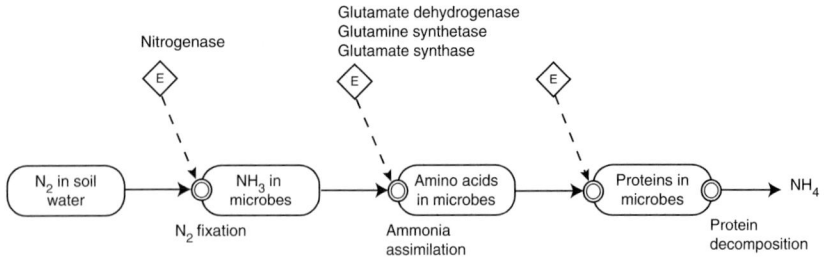

Fig. 6.4.4 Biological nitrogen fixation

free-living or they live in symbiotic association with plants. Nitrogen fixation by free-living microbes occurs for example during decomposition of litter and soil organic matter (Vitousek and Hobbie, 2000). The symbiotic microbes form nodules in the roots of the parent plant, and hence receive carbohydrates from the plant to fulfill their energy needs and in return supply the plant with ammonium ions. In agricultural systems, legumes are well known for their association with bacteria of the genus *Rhizobium*, whereas in natural ecosystems alder trees are associated with the actinomycete *Frankia*.

Recent findings indicate that some cyanobacteria may contribute significantly to the N fixation in nitrogen-poor ecosystems such as in boreal forests. DeLuca and colleagues (2002) found that a cyanobacterium (*Nostoc* sp.) and the common feather moss *Pleurozium schreberi* can fix large amounts of nitrogen and hence act as a major contributor to N accumulation.

6.4.2.3 Nitrification

Nitrification is oxidation of ammonium (NH_4^+), ammonia (NH_3) or organic N compounds to nitrite (NO_2^-) and nitrate (NO_3^-) (Groffman, 1991; Bremner, 1997; Wrage et al., 2001) (see Fig. 6.4.5). A range of soil microorganisms are capable of biological nitrification, but the autotrophic ammonia and nitrite oxidizing bacteria are probably one of the most important groups (Killham, 1990; Bothe et al., 2000). Bacteria of the genus *Nitrosomonas* and *Nitrosospira* carry out the initial oxidation of NH_4^+ to NO_2^- and those of the genus *Nitrobacter* oxidize NO_2^- further to NO_3^-. These bacteria obtain their energy for growth from the oxidation of ammonia or nitrite and assimilate carbon as carbon dioxide. In contrast, the heterotrophic microbial oxidation of ammonia is not linked to cellular growth (Killham, 1990; De Boer and Kowalchuk, 2001).

While the general biogeochemistry of bacterial ammonia oxidizing is well understood, there have been several recent discoveries, which suggest a potential capacity for ammonia oxidation by *Archaea* in soil (Francis et al., 2007). The ammonia-oxidising archaeota seem to be globally distributed and abundant in many terrestrial environments, including forest, agricultural, grassland and alpine soils (Treusch et al., 2005; Nicol and Schleper, 2006; Leininger et al., 2006; He et al., 2007). Still,

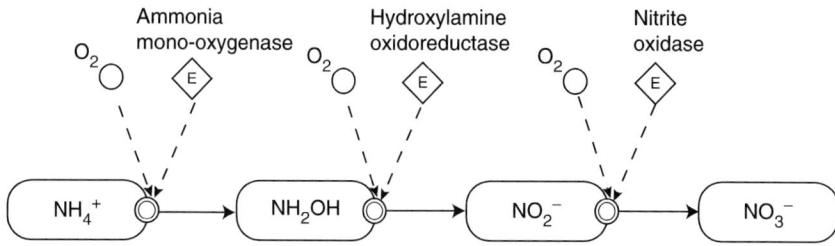

Fig. 6.4.5 Nitrification of ammonium ions

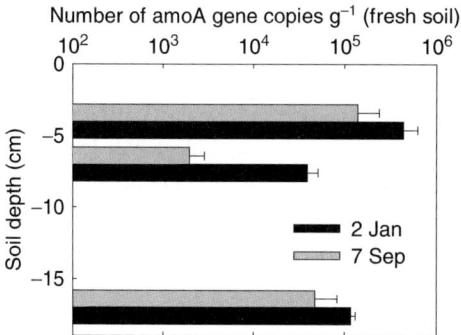

Fig. 6.4.6 The copy numbers of A-subunit of bacterial ammonia mono-oxygenase gene during warm (September) and cold season (January) at SMEAR II soil measured in 2007. Bars show standard error of the mean

the suggestion of Nicol and Schleper (2006) that the *Archaea* are also the major contributors to the biogeochemical transformations of nitrogen remains to be confirmed in acid boreal forest soils.

The stepwise oxidation of ammonium (NH_4^+) to nitrate (NO_3^-) is catalyzed by several enzymes (see Fig. 6.4.5). The initial ammonium oxidation to hydroxylamine (NH_2OH) is catalyzed by ammonia mono-oxygenase, the oxidation of NH_2OH to NO_2^- by hydroxylamine oxidoreductase, and the oxidation of NO_2^- to NO_3^- by nitrite oxidoreductase (McCarty, 1999; Wrage et al., 2001). The oxidation of ammonia is usually the rate-limiting step for nitrification as nitrite is rarely found to accumulate in soils (Kowalchuk and Stephen, 2001). In nitrification, availability of oxygen is essential, since each step of the oxidation reaction requires O_2. Gaseous nitrogen oxides (NO and N_2O) can be formed through the chemical decomposition of NH_2OH or NO_2^- (Poth and Focht, 1985).

The genes encoding the key enzyme in nitrification, ammonia mono-oxygenase, can be used for determining population size of soil ammonia oxidizing microorganisms in combination with quantitative real-time polymerase chain reaction (PCR) (Francis et al., 2007). In the soil at the Smear II station the population size of ammonia oxidizing bacteria was highest in the top-soil humus layer (Fig. 6.4.6). The number of ammonia monooxygenase genes in the soil decreased order of magnitude from the humus to the elluvial layer, but increased again deeper in the soil. This trend was similar in the soils sampled in winter (January) than soils from the early autumn (September).

Factors controlling nitrification in soil are the availability of carbon dioxide, NH_4^+ ions, oxygen, moisture and pH (Simek, 2000). Carbon dioxide is always present in biologically active soils due to microbial and root respiration, whereas the same factors may lead to localized depletion of soil molecular oxygen. The concentrations of carbon dioxide and molecular oxygen vary depending on the rates of respiration activity and soil aeration, which is affected by the balance between soil air filled porosity and soil moisture (Simek, 2000; Simojoki, 2001). In well-aerated soils, the availability of NH_4^+ is the limiting factor in autotrophic nitrification. Environments with very small soil nitrogen content often have low nitrification activity, and consequently low N_2O emissions (Martikainen, 1984; Martikainen, 1985; Priha et al., 1999; Priha and Smolander, 1999). It has been shown that high atmospheric load of ammonium stimulates nitrification in boreal (Martikainen et al., 1993) and forest soils (Van Breemen et al., 1987; Tietema et al., 1992, 1993). Especially, in relation to nitrification activities, additions of urea and liming increase nitrification activities in acidic forest soils (Martikainen, 1984). This effect is partly due to the increase in available mineral nitrogen but also due to increased soil pH, which favours nitrifying microorganisms (Priha and Smolander, 1999; Paavolainen et al., 2000).

6.4.2.4 Ammonia-Oxidation by Planctomycetes

Recent findings on the ability of microbial communities to oxidize ammonium anaerobically have indicated that the global nitrogen cycling may be much more diverse than previously considered (Francis et al., 2007). For example, a group of planctomycete-like bacteria can oxidize ammonia with nitrite in strictly anoxic conditions in a process called anaerobic ammonium oxidation (anammox) (Schmidt et al., 2002; Strous and Jetten, 2004; den Camp et al., 2006). The anammox bacteria have been found in many marine and freshwater ecosystems (Lam et al., 2007; Schmid et al., 2007; Woebken et al., 2007) however, their contribution to ammonium oxidation in terrestrial ecosystems remains largely unknown.

6.4.2.5 Denitrification

Denitrification is defined as the respiratory bacterial reduction of nitrate (NO_3^-) or nitrite (NO_2^-) to nitrogen oxides or molecular nitrogen (Knowles, 1982; Bremner, 1997; Einsle and Kroneck, 2004) (see Fig. 6.4.7). This chain of reduction reactions is catalyzed by the enzymes nitrate reductase, nitrite reductase, nitric oxide reductase and nitrous oxide reductase (Einsle and Kroneck, 2004) (see Fig. 6.4.7). Nitrogen monoxide (NO) and nitrous oxide (N_2O) are obligatory intermediates of the reduction process. Denitrifying micro-organisms are often facultative aerobic bacteria that are able to reduce nitrogen oxides when O_2 becomes limiting. Denitrifying micro-organisms derive energy from organic substrates, and

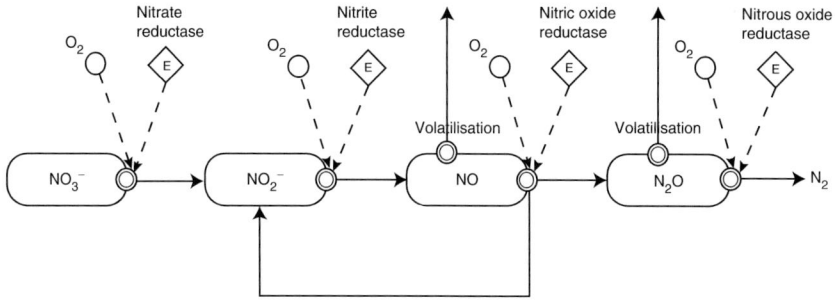

Fig. 6.4.7 Chain of denitrification steps

hence denitrification is limited by the amount of readily-decomposable organic matter and mineral substrates (NO_3^-, NO_2^-) in the soil.

Factors controlling the rate and end-products of denitrification are the O_2 concentration, the availability of readily-decomposable organic matter, and the amount of NO_3^- and NO_2^- (Knowles, 1982). The proportion of N_2O in the end product is higher if the soil pH is low, as the N_2O reductase is inhibited at low pH (Knowles, 1982). In general, an increase in soil water content and temperature or addition of fertilizer N, plant residues or animal manures increase denitrification activity and consequently increase N_2O emissions from soils (e.g. Davidson, 1991, 1993; Kaiser et al., 1998; Baggs et al., 2003; Schindlbacher et al., 2004).

6.4.3 Methane Processes in Soil

Mari Pihlatie[1], Jukka Kurola[2,3], Asko Simojoki[4], Jukka Pumpanen[5], and Mirja Salkinoja-Salonen[4]

[1] University of Helsinki, Department of Physics, Finland
[2] University of Helsinki, Department of Ecological and Environmental Sciences, Lahti, Finland
[3] University of Helsinki, Department of Biological and Environmental Sciences, Finland
[4] University of Helsinki, Department of Applied Chemistry and Microbiology, Finland
[5] University of Helsinki, Department of Forest Ecology, Finland

Methene is an important greenhouse gas in the atmosphere (Section 10.1.1). Boreal forest soils may act as sinks and sources of methane due to microbial activity in the soil. Thus the methane processes in boreal forest soils are relevant in order to understand the role of boreal forests in the present climate change.

6.4.3.1 Methane Production

Methane is produced in soils as an end product of anaerobic decomposition of organic matter. This anaerobic decomposition occurs through methanogenic fermentation, which produces both methane (CH_4) and carbon dioxide (CO_2)

(Le Mer and Roger, 2001). The methane production includes four major steps and involves four major populations of micro-organisms (Le Mer and Roger, 2001):

1. Hydrolysis: Large carbon molecules in organic matter are solubilized and converted into smaller monomers, such as fatty acids and amino acids.
2. Fermentation (Acidogenesis): Small monomers are fermented into simple carbon molecules such as volatile fatty acids, organic acids, alcohols, and carbon dioxide (CO_2).
3. Acetogenesis: Simple carbon molecules produced during primary fermentation are converted into acetic acid, CO_2, and H_2.
4. Methanogenesis: Acetate or CO_2 are converted into CH_4 and CO_2, while hydrogen is consumed.

The steps of CH_4 synthesis in the soil are shown in the Fig. 6.4.8. The large group of micro-organisms involved in the decomposition of carbon molecules can be aerobic, facultatively anaerobic, or strictly anaerobic. The last step of methane synthesis involves methanogens, which belong to the domain *Archaea* (Woese et al., 1978; Valentine, 2007). Methanogens are strictly anaerobic microorganisms and highly sensitive to oxygen. Hence, methane production in soils only occurs under highly reducing conditions in the absence of such potential electron acceptors as nitrate, sulfate or ferric iron (Topp and Pattey, 1997).

Methanogens can be divided into five groups: hydrogenotrophs, formatotrophs, acetotrophs, methylotrophs, and alcoholotrophs according to the substrates that they utilize (Le Mer and Roger, 2001). Hydrogenotrophy, the reduction of CO_2 by H_2, and acetotrophy, the reduction of acetate to CH_4 and CO_2, are the two most common pathways for CH_4 production (Le Mer and Roger, 2001). When we follow the anaerobic decomposition of a glucose molecule, we get that one mole of glucose produces anaerobically two moles of acetate, and four moles of H_2. The production of one CH_4 requires four H_2, but only one acetate. Consequently, during the degradation of polysaccharides, acetate contributes more to the CH_4 formation than H_2/CO_2. In fact, Conrad (1999) estimated that the contribution of acetate and H_2/CO_2 to the total CH_4 formation is $>67\%$ and $<33\%$, respectively.

The key enzyme in methanogenesis is methyl-coenzyme M reductase (MCR), which catalyses the last step in methane formation and is present in all methanogens (Friedrich, 2005). All methanogenic pathways have in common the conversion of a methyl group to methane; however, the origin of the methyl group can vary (Ferry,

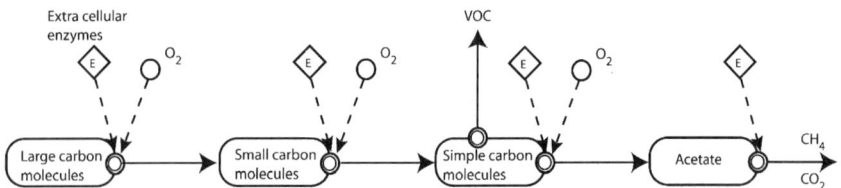

Fig. 6.4.8 Visualization of methane synthesis from acetate. The symbols are introduced in the Fig. 1.2.1

2002). Coenzyme M (CoM) is the smallest cofactor known in nature. The cofactor is methylated on the sulfhydryl group, forming CH_3-S-CoM, the substrate for the methyl-coenzyme M reductase (MCR). Coenzyme B is the second substrate for MCR and, as a consequence of the reaction, forms the heterodisulfide complex with CoM (CoB-S-S-CoM). The major steps in methanogenesis can be classified into three general categories based on methyl group acquisition (Ferry, 2002).

1. Reduction of CO_2: CO_2 is reduced to a methyl group with electrons derived from the oxidation of electron donors (primarily H_2 or formate), which are also the source of electrons for the reduction of CoM-S-S-CoB.
2. Fermentation of acetate: Acetate is cleaved to provide the methyl group and a carbonyl group for oxidation to CO_2, providing electrons for reduction of CoM-S-SCoB.
3. Dismutation of methanol or methylamines: A molecule of substrate is demethylated to provide the methyl group and another molecule is oxidized to provide electrons for reduction of CoM-S-S-CoB.

6.4.3.2 Methane Oxidation

Methane (CH_4) oxidation takes place in a vast range of soil environments, such as dry savannas, landfill cover soils, rice fields, boreal peatlands and forest soils. Methane oxidizing bacteria can be divided into methane-assimilating bacteria (methanotrophs), which use CH_4 as their sole source of carbon and energy, and autotrophic ammonia-oxidizing bacteria (nitrifiers), which co-oxidize CH_4 as they derive energy from the oxidation of ammonia (Bédard and Knowles, 1989). Although nitrifiers have the capacity to oxidize CH_4, they do not significantly contribute to the CH_4 oxidation in forest soils (Jiang and Bakken, 1999; Klemedtsson et al., 1999; Saari and Martikainen, 2001). Hence, readers interested in nitrifier CH_4 oxidation are referred to, for instance, the review by Bedard and Knowles (1989).

Methanotrophs are a phylogenetically and physiologically diverse group of microorganisms. They can be divided into two physiologically distinct communities: high affinity oxidizers and low affinity oxidizers, reviewed for instance by Segers (1998) and Le Mer and Rogers (2001). The high affinity oxidation occurs at close to ambient atmospheric CH_4 concentrations (below 12 ppm), and the low

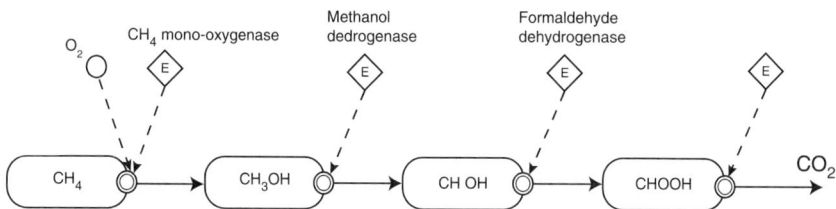

Fig. 6.4.9 Visualization of the steps in the decomposition of methane

affinity oxidation takes place at CH_4 concentrations higher than 40 ppm (Le Mer and Rogers, 2001). High affinity CH_4 oxidizers are largely unknown due to difficulties in bacterial cultivation, whereas low CH_4 affinity oxidation is often associated with common methanotrophs that have been studied in pure cultures for decades (Bender and Conrad, 1992; Conrad, 1996).

Methane is oxidized to carbon dioxide (CO_2) through a sequential reaction chain, which is catalyzed by several enzymes (Fig. 6.4.9). Methanotrophs derive energy from the oxidation reactions and utilize formaldehyde, the intermediate product of the reactions, as their carbon source for growth (Topp and Pattey, 1997). The first step in the oxidation reaction of methane to methanol (CH_3OH) is catalyzed by methane mono-oxygenase (MMO). Further, the break-down of methanol to formaldehyde ($CH \cdot OH$) is catalyzed by methanol dehydrogenase, and the breakdown of formaldehyde to formic acid (HCOOH) by formaldehyde dehydrogenase (see Fig. 6.4.9). Availability of oxygen is crucial in the first step of CH_4 oxidation as one atom of oxygen is needed to form methanol and one atom of oxygen is used to form water (H_2O). The energy for the reaction is obtained from $NADH_2$ as it is converted to NAD.

Chapter 7
From Processes and Transport to Trees, Ecosystems and Atmosphere

7.1 Mathematical Tools

Pertti Hari and Annikki Mäkelä

University of Helsinki, Department of Forest Ecology, Finland

The circulation of the globe around the sun generates the annual cycle and rotation of the globe the diurnal cycle in the environmental factors (Chapter 3). In addition, there are more rapid variations in the atmospheric environmental factors (Section 3.2) affecting considerably the biological processes both in space and time (Section 6.3.1). The attenuation of light (Sections 6.2.1 and 7.3) is an additional and strong source of variation in photon flux density. Forest soils are spatially very inhomogeneous (Section 7.4). The great and often irregular variation is a challenge for the methodology to be applied.

Boreal forest ecosystems are hierarchical in space and time. For our studies, the spatial levels of hierarchy are space/point element, individual tree, ecosystem and boreal forest; the temporal levels are time element, year and rotation period. The space and time elements have to be so small that environmental factors can be considered as constants within the elements. The proper utilization of information gained on a more detailed level is the key for transition between the levels (Section 1.2: Basic idea 5).

The amounts of produced organic material during a period are needed to understand the metabolism of a forest ecosystem. The fundamental phenomenon processes are driven by environment at any point due to the inherent variation in environment. Thus we need tools to utilise the results obtained at point level to get amounts of process products at individual and ecosystem level.

The different levels in an ecosystem form six combinations (cf. Fig. 7.1.1), each of them having its own character. During prolonged periods, amounts of material are formed in processes i.e. material is synthesized or transported through a membrane. The time derivatives of the produced amounts are rates. The spatial derivatives of produced amounts are specific rates or amounts. The hierarchy of the concepts is demonstrated in the Fig. 7.1.1.

Time Space	Moment	Period from t_1 to t_2
Point	Specific process rate $q_m(t, x)$	Specific produced amount $Q_m(t_1, t_2, x)$
Invidual	Process rate $q_I(t, V_I)$	Produced amount $Q_I(t_1, t_2, V_I)$
Ecosystem	Process rate $q_E(t, V_E)$	Produced amount $Q_E(t_1, t_2, V_E)$

Fig. 7.1.1 The temporal and spatial hierarchy of processes in a forest ecosystem

Consider a process in a small volume ΔV during a short time interval Δt. Let x be the centre point of the volume ΔV and t the beginning moment of the interval Δt. Let $Q_p(t, x, \Delta t, \Delta V)$ denote the production in the process during Δt in the volume ΔV. The amount produced in the volume ΔV during Δt, $Q_{PE}(t, x, \Delta t, \Delta V)$ is according to Eq. 6.1.3,

$$Q_{PE}(t, x, \Delta t, \Delta V) \approx \rho_m(x) q_P(t, x) \Delta V \Delta t \qquad (7.1.1)$$

Consider the volume V_I within an individual in the ecosystem. A process in the individual has mass specific rate q_P and mass density ρ_M. The problem is to determine the link between the environment and the amount of material produced by the process, $Q_P(t_1, t_2, V_I)$, in the volume V_I during the period from t_1 to t_2.

The amount $Q_P(t_1, t_2, V_I)$ can be approximated by dividing the volume V_I into n subvolumes ΔV_i, i = 1, 2, 3, ...,n and the time interval from t_1 to t_2 into m subintervals of equal length. Let t_j denote the beginning moment of the jth subinterval. We can approximate the amount by summing the subvolumes and intervals

$$Q_P(t_1, t_2, V_I) \approx \sum_{i=1}^{n} \sum_{j=1}^{m} Q_{PE}(t_j, x_i, \Delta t, \Delta V) \qquad (7.1.2)$$

where x_i is the centre point of the subvolume ΔV_i.

When the approximation in Eq. 7.1.1 is introduced into Eq. 7.1.2 we get

$$Q_P(t_1, t_2, V_I) \approx \sum_{i=1}^{n} \sum_{j=1}^{m} \rho_m(x_i) q_p(t_j, x_i) \Delta V_i \Delta t_j \qquad (7.1.3)$$

In the above formula, the smaller the volume and time element, the better is the approximation. Finally, when the volume and lengths of the intervals approach zero we get

$$Q_P(t_1,t_2,V_I) = \int_{V_I} \int_{t_1}^{t_2} \rho_m(x)q_p(t,x)dtdV \qquad (7.1.4)$$

The above expression is an exact mathematical formulation of the relationship between specific process rates at points in time and space and the produced amount by an individual during a prolonged period. The environmental factors and the active substances in the tissues determine the process rate. The environmental factors used in calculations are, nearly always, measured values, not simple mathematical functions; this is why we have to use numerical methods for integration. The numerical calculations are based on the approximation in Eq. 7.1.3. The specific process rate describes metabolism at a point in space. We need also the corresponding concept on the individual level, i.e., process rate of an individual. Let $Q_I(t_1,t_2,V_I)$ denote the amount of material produced in a metabolic process by an individual in the volume V_I, during the time interval from t_1 to t_2. The process rate of an individual $q_I(t,V_I)$ is defined as the time derivative of $Q_I(t_1,t_2,V_I)$

$$q_I(t,V_I) = \lim_{\Delta t \to 0} \frac{Q_I(t,t+\Delta t,V_I) - Q_I(t,t,V_I)}{\Delta t} \qquad (7.1.5)$$

(Note $Q_I(t,t,V_I) = 0$).

The transition from process rate of an individual to produced amounts by the individual is integration over time

$$Q_I(t_1,t_2,V_I) = \int_{t_1}^{t_2} q_I(t,V_I)dt \qquad (7.1.6)$$

Consider n subvolumes of V_I for transition from rates at point level to rates at individual level. The process rate of an individual can be approximated utilising the subvolumes

$$q_I(t,V_I) \approx \lim_{\Delta t \to 0} \sum_{i=1}^{n} \frac{Q_{PE}(t,x_i,\Delta t,\Delta V_i)}{\Delta t \Delta V}\Delta V \qquad (7.1.7)$$

where $Q_{PE}(t,x_i,\Delta t,\Delta V_i)$ is as in Section 6.1 the amount produced in the subvolume ΔV_i during Δt, x_i is the centre of the subvolume ΔV_i. The order of the limit and sum operations can be changed

$$q_I(t,V_I) \approx \sum_{i=1}^{n} \lim_{\Delta t \to 0} \frac{Q_{PE}(t,x_i,\Delta t,\Delta V_i)}{\Delta t \Delta V}\Delta V \qquad (7.1.8)$$

The production term in the above formula can be approximated with the derivatives (Eq. 6.1.3)

$$q_I(t,V_I) \approx \sum_{i=1}^{n} \rho_m(x_i)q_P(t,x_i)\Delta V \qquad (7.1.9)$$

Thus the process rate of an individual $q_I(t, V_I)$ is obtained with integration over space

$$q_I(t,V_I) = \int_{V_I} \rho_m(x)q_p(t,x)dV \qquad (7.1.10)$$

The specific production amounts, $Q_m(t_1,t_2,x)$, are obtained by integrating specific process rate over time

$$Q_m(t_1,t_2,x) = \int_{t_1}^{t_2} q_p(t,x)dt \qquad (7.1.11)$$

The transition from individual level to ecosystems is summed over the individuals in the system. Let $Q_P(t_1,t_2,V_E)$ be the amount produced in the process by the ecosystem during the interval from t_1 to t_2 and $Q_p^i(t_1,t_2,V_I)$ the corresponding amount produced by the individual i. If the ecosystem includes N individuals, then

$$Q_E(t_1,t_2,V_I) = \sum_{i=1}^{N} Q_p^i(t_1,t_2,V_I) \qquad (7.1.12)$$

Conclusion

Integration over space and time, by summation of produced amounts in small volume elements during short intervals, is an efficient means to utilise process information at individual and ecosystem levels during prolonged periods.

7.2 Atmospheric Processes and Transport

Timo Vesala[1], Pertti Hari[2], and Jouni Räisänen[1]

[1] University of Helsinki, Department of Physics, Finland
[2] University of Helsinki, Department of Forest Ecology, Finland

The solar radiation energy input on the globe is $1.37\,\text{kW}\,\text{m}^{-2}$ through a tangential (perpendicular to radiation flux) plane (Section 3.1) outside the atmosphere. Several processes convert the solar energy in the atmosphere into other forms. Absorption in the atmosphere and on the earth's surface transform this radiation energy into heat warming the atmosphere. Emission of thermal radiation and transpiration and evaporation processes cool and condensation of water vapour releases heat. Heat exchange tends to level temperature differences (Section 6.2.1). Pressure and temperature differences generate vertical and horizontal flows in the atmosphere. Conservation of mass, energy and momentum forms the basis for the quantitative analysis of the behaviour of the atmosphere (see Section 1.2: Basic idea 9).

7.2.1 Conservation Principles

The *conservation principles of mass, energy and momentum* are fundamental theoretical principles in classical physics as well as in quantum mechanics and the theory of relativity. For phenomena considered in this book, the theory of relativity and quantum mechanics per se can be ignored, although they are needed for example in fundamental understanding of electro-magnetic radiation. The conservation principles in classical physics have been tested since the Newtonian era and they have passed the tests very successfully. In fact, modern technology can be seen in general as rigorous tests of conservation principles.

The conservation of basic entities sets strict frame of analysis, which any solution of a problem and a model must follow. An excellent and thorough discussion on conservation principles in the context of transport phenomena is given by Bird et al. (2002). The important feature of conservation equation is that they follow the form

$$[\textit{rate of increase of quantity}] = [\textit{net rate of addition of quantity}]$$
$$+ [\textit{rate of production of quantity}]$$

where quantity stands for mass, momentum or energy. The net rate of the addition of the quantity is related to the net transport of the quantity, which is the flux of it.[1] Mathematically, the net rate of addition per unit volume is the negative of *the divergence* of the flux. The divergence is the vector operation (vector possesses both magnitude and direction while scalar only magnitude as was discussed in Section 4.1.), which is the dot product of the given vector $(\vec{a})$ and the vector whose components are the partial derivatives with respect to each coordinate $(\nabla \bullet \vec{a})$, i.e. $\nabla \bullet \vec{a} \equiv \frac{\partial a_x}{\partial x} + \frac{\partial a_y}{\partial y} + \frac{\partial a_z}{\partial z}$. It gives the limiting ratio of "the flux passing through the area surrounding a point" to "the area" as it decreases to zero, for a given vector at the point in a vector field.

In derivations of conservation equations one considers the volume element where the amounts of momentum, heat and mass may change and those changes are balanced by transport of the quantities through the volume boundaries and the rate of production within the volume. The equations are thus derived from a fundamental concept of conservation where the phenomenological transport mechanisms are embedded and the fluxes are replaced by expressions (Eqs. 4.1.5–4.1.8 and 4.2.1–4.2.3) involving transport properties and gradients of concentration, velocity and temperature. This enables the derivation of formulas that form a closed set of equations for velocity components, temperature, concentration, pressure and density.

---

[1] The rates of production arise from chemical reactions in the case of mass conservation, from external force fields in the case of momentum and from photon emission and absorption in the case of energy conservation. Obviously, the sum of net rate of addition and rate of production must be the same as the rate of increase, for the conserved entity.

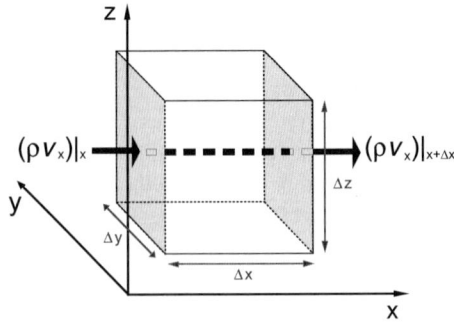

Fig. 7.2.1 Fixed volume through which a fluid is flowing. The arrows indicate the mass flux in and out at the two shaded faces located at x and $x + \Delta x$ (See also Bird et al., 2002)

7.2.2 Continuity Equation

We start from the conservation of the total mass and follow the derivation presented by Bird et al. (2002). Let us consider a mass balance over a (cube-shaped) volume $\Delta x \Delta y \Delta z$ (Fig. 7.2.1), which is fixed in space and through which a fluid is flowing. The principle of mass conservation states now that

$$[rate\ of\ increase\ of\ mass] = [rate\ of\ mass\ in] - [rate\ of\ mass\ out]$$

The rate of mass *entering* in the direction of x-axis $(\rho v_x)|_x \Delta y \Delta z$, where $|_x$ refers to the value of ρv_x at the location x and the area of the face is $\Delta y \Delta z$ (compare with the similarities between Eq. 4.2.1.) Let the other face perpendicular to x-axis be located at $x + \Delta x$ in which case the rate of mass *leaving* in the direction of x-axis is $(\rho v_x)|_{x+\Delta x} \Delta y \Delta z$. Thus the flow in the direction of x-axis through the volume contributes to the net rate of addition of mass by the amount of $\lfloor (\rho v_x)|_x - (\rho v_x)|_{x+\Delta x} \rfloor \Delta y \Delta z$. Similar expressions can be written for the other two pairs of faces perpendicular to y- and z-axis. Now the rate of increase of mass within the volume is the rate of density change multiplied by the volume, that is $\Delta x \Delta y \Delta z \partial \rho \partial t$.

The mass balance equation is thus

$$\Delta x \Delta y \Delta z \partial \rho \partial t = \left[(\rho v_x)|_x - (\rho v_x)|_{x+\Delta x} \right] \Delta y \Delta z + \left\lfloor (\rho v_y)|_y - (\rho v_y)|_{y+\Delta y} \right\rfloor \Delta x \Delta z$$
$$+ \left[(\rho v_z)|_z - (\rho v_z)|_{z+\Delta z} \right] \Delta x \Delta y$$

So far the derivation has been simple physical statement of the conservation of mass.

The both sides of the above equation can be divided by $\Delta x \Delta y \Delta z$ and then we obtain

$$\partial \rho / \partial t = \frac{\left[(\rho v_x)|_x - (\rho v_x)|_{x+\Delta x} \right]}{\Delta x} + \frac{\left\lfloor (\rho v_y)|_y - (\rho v_y)|_{y+\Delta y} \right\rfloor}{\Delta y} + \frac{\left[(\rho v_z)|_z - (\rho v_z)|_{z+\Delta z} \right]}{\Delta z}$$

Now at the limit as Δx, Δy and Δz go to zero, we get the general mathematical form from the fundamental definition of the derivative

$$\frac{\partial \rho}{\partial t} = -\left(\frac{\partial}{\partial x}\rho v_x + \frac{\partial}{\partial y}\rho v_y + \frac{\partial}{\partial z}\rho v_z\right) = -\nabla \bullet \rho \vec{v} \qquad (7.2.1)$$

which is the *equation of continuity*, stating that the rate of increase of mass per unit volume equals to the net rate of mass addition per unit volume by convection. $\nabla \bullet \rho \vec{v}$ is the divergence of the mass flux density vector $\rho \vec{v}$ and physically means the net rate of mass efflux per unit volume. If the divergence is positive, the rate of density change is negative, that is the positive efflux is balanced by the decreasing density. Mass cannot appear as if nowhere but only by diminishing density. The negative divergence is called convergence (especially in meteorology meaning a net flow of air into a given region) and then the density increases in time.

This derivation of the continuity equation was based on a (Cartesian) coordinate frame fixed in space and it is called *Eulerian* derivation. We next follow partly the illustrative discussion on the continuity equation presented by Holton (2004). It can be shown that Eq. 7.2.1 is mathematically equivalent with

$$\frac{1}{\rho}\frac{D\rho}{Dt} = -\left(\frac{\partial}{\partial x}v_x + \frac{\partial}{\partial y}v_y + \frac{\partial}{\partial z}v_z\right) = -\nabla \bullet \vec{v} \qquad (7.2.2)$$

where

$$\frac{D}{Dt} \equiv \frac{\partial}{\partial t} + \vec{v} \bullet \nabla \qquad (7.2.3)$$

and it is called *substantial (or total or material or hydrodynamic) derivative* (derivative following the motion) referring to the *Lagrangian* frame of reference of the formulation. It describes the changes which occur in the fluid parcel containing the same material but moving with the flow whereas the partial time derivative $\frac{\partial}{\partial t}$ refers to changes occurring in the fixed place. This alternative, but mathematically equivalent, form of the continuity equation is the velocity divergence form and it states that the fractional rate of increase of the density following the motion of a fluid parcel is equal to minus the velocity divergence. The motion of the parcel is described by the term $\vec{v} \bullet \nabla$ in the derivative 7.2.3. In physical terms, the Lagrangian form should be clearly distinguished from the Eulerian form (7.2.1), which states that the local rate of change of density is equal to minus the mass divergence, but in mathematical terms the two formulas are simply identical.

Physically the Lagrangian form means the consideration of the fluid parcel with the fixed mass that moves with the fluid, in contrast to Eulerian derivation where fluid was flowing through an immobile volume element with a fixed volume. Equation 7.2.2 implies that the density is only changing if there exists divergence in the velocity field. Because the mass of the parcel is fixed, the change of the density must mean the change of the parcel volume and this is exactly the case. Namely the velocity divergence means that the velocity is spatially changing and then the different faces of the parcel are moving with the different velocity, which modifies the parcel volume. In the case of the positive divergence the velocity is increasing across the parcel, which involves that the faces are drawn away and the volume is increased and the density diminished. In terms of Eq. 7.2.3, when $\nabla \bullet \vec{v} > 0$ then $\frac{D\rho}{Dt} < 0$.

In the case of *incompressible* flow the continuity equation is much simplified. Incompressibility means that the density is constant and Eqs. 7.2.2 and 7.2.3 both imply that then

$$\nabla \bullet \vec{v} = 0 \qquad (7.2.4)$$

No fluid is truly incompressible but frequently in ecophysiological and many atmospheric applications, the assumption results in very little error. The criterion is fulfilled when flow velocities are low compared with the speed of sound in the fluid (see e.g. Tritton, 1988), so that $\frac{U^2}{a^2} \ll 1$, where U is the characteristic flow velocity (typical value of velocity for a studied case) and a is the speed of sound in the studied fluid.

7.2.3 Energy Equation

Similarly to the derivation of the continuity equation, the equation of change for energy is obtained by applying the principle of conservation of energy to a small element of volume and then allowing the dimensions of the volume to become vanishingly small. The derivation is somewhat more complicated than for the mass and we present here only essential features following the discussion in Bird et al. (2002). The principle of conservation of energy is based on the thermodynamical notion that the internal energy is changed by adding heat to the system or doing work on it. In the presence of energy transport, both kinetic and internal energy may be entering and leaving the fluid parcel by convection and by conduction. The energy balance can be formulated by the statement

[*rate of increase of in internal energy*] = [*net rate of addition of internal energy by convection*] + [*rate of internal energy addition by heat conduction*] + [*reversible rate of internal energy increase by compression*]

This formulation omits the viscous dissipation term always heating the flowing fluid but being important only in flows with enormous velocity gradients. We also omit, for the present, effects of radiative energy transport and phase transition and discuss on them later.

Similarly to the derivation of Eq. 7.2.1, the heat balance equation can be written as (Bird et al., 2002)

$$\Delta x \Delta y \Delta z \partial E_{tot}/\partial t = \left[(e_x)|_x - (e_x)|_{x+\Delta x} \right] \Delta y \Delta z + \left[(e_y)|_y - (e_y)|_{y+\Delta y} \right] \Delta x \Delta z$$
$$+ \left[(e_z)|_z - (e_z)|_{z+\Delta z} \right] \Delta x \Delta y$$

where E_{tot} is the sum of the internal and kinetic energy of the unit volume and e includes the convective transport of kinetic and internal energy, the heat conduction and the work associated with molecular processes. At the limit as Δx, Δy and Δz go to zero, the equation of energy is obtained in the general vector form. However, the general form is quite cumbersome for many practical purposes and often the most useful form is one in which the temperature appears. We do not present the general form and skip over the tedious derivations and refer to the book by Bird et al. (2002). For many problems in atmospheric sciences and ecophysiology, two forms of the energy equation are very relevant (in both of them the heat conductivity k is assumed to be constant)

$$\rho C_p \frac{DT}{Dt} = k\nabla^2 T + \frac{Dp}{Dt} \qquad (7.2.5)$$

which is valid for ideal gases. C_p is the specific heat capacity at constant pressure. The time derivative of the pressure describes the effect of compression.

For fluids flowing in a constant pressure system or with constant density the equation is simply

$$\rho C_p \frac{DT}{Dt} = k\nabla^2 T \qquad (7.2.6)$$

This is the form which is applicable for incompressible flows obeying Eq. 7.2.4. If $\vec{v} = 0$ the derivative $\frac{D}{Dt}$ in Eq. 7.2.3 is reduced to $\frac{\partial}{\partial t}$ and the formula 7.2.6 is reduced to the famous heat conduction equation for solids.

7.2.4 Continuity Equation for Each Species in a Multicomponent Mixture

Similarly to the derivation of the continuity equation for the total mass, the equation of change for mass of individual species in the fluid mixture is obtained by applying the principle of conservation of mass to a small element of volume and then allowing the dimensions of the volume to become vanishingly small. Note that while in the case of general continuity equation diffusion and chemical reactions were ignored (diffusion requires at least two different species and chemical reactions do not alter the total mass), they must be taken into account in the mass balance of individual constituents. The derivation is somewhat more complicated than for the total mass and we present here only essential features following the discussion in Bird et al. (2002). Especially, some problems can be conveniently formulated using mass and using moles. The mass- and mole-based formulas are written in different set of notations and the rigorous forms may look quite different although they express exactly the same physical content.

The mass balance can be formulated by the statement

[*rate of increase of mass of A*] = [*net rate of addition of mass of A by convection*] + [*net rate of addition of mass of A by diffusion*] + [*rate of production of mass of A by reactions*]

Similarly to the derivation of Eq. 7.2.1, the mass balance equation can be written as (Bird et al., 2002)

$$\Delta x \Delta y \Delta z \partial C_A / \partial t = \left[(g_x - (g_{Ax})|_{x+\Delta x} \right] \Delta y \Delta z + \left[(g_{Ay})|_y - (g_{Ay})|_{y+\Delta y} \right] \Delta x \Delta z$$
$$+ \left[(g_{Az})|_z - (g_{Az})|_{z+\Delta z} \right] \Delta x \Delta y + r_A \Delta x \Delta y \Delta z$$

where C_A is the mass concentration of species A and its mass flux g_A includes both the molecular flux and the convective flux. r_A is the rate of production of mass per unit volume by chemical reactions. At the limit as Δx, Δy and Δz go to zero, the equation of continuity for species A is obtained in the general vector form. Often the most useful form is one in which the concentration-related variable (either mass concentration, mass density, mass fraction, molar concentration, molar density or molar fraction) appears. In practise, the choice of the form can be based on the constancy of the variables in the specific problem, for example whether the mass density or molar density of a mixture is constant. For more details, we refer to the book by Bird et al. (2002).

Assuming the binary system of A and B species with constant ρD_{AB} the mass equation is

$$\rho \frac{D\omega_A}{Dt} = \rho D_{AB} \nabla^2 \omega_A + r_A \tag{7.2.7}$$

This form is the most appropriate for describing the diffusion in dilute liquid solutions at constant temperature and pressure. Note the analogy with energy equations, especially with Eq. 7.2.6 if no chemical reactions occur. If $\vec{v} = 0$ the derivative $\frac{D}{Dt}$ is reduced to $\frac{\partial}{\partial t}$ and the formula 7.2.7 is reduced to the form called Fick's second law of diffusion.

7.2.5 Equation of Motion

After considerations of conservation of total mass, energy and individual species we finally formulate the conservation equation for momentum, which is in fact the equation of motion for fluids, analogical to the Newton's second law for motion of rigid bodies.

Again, a momentum balance over the volume element is considered

[rate of increase of momentum] = [rate of momentum addition by convection]

+[rate of momentum addition by molecular transport] + [external force on fluid]

Similarly to the derivation of Eq. 7.2.1 the balance equation for the x-component of momentum can be written as (Bird et al., 2002).

$$\Delta x \Delta y \Delta z \partial \rho v_x / \partial t = \left[(\phi_{xx})|_x - (\phi_{xx})|_{x+\Delta x} \right] \Delta y \Delta z + \left[(\phi_{yx})|_y - (\phi_{yx})|_{y+\Delta y} \right] \Delta x \Delta z$$
$$+ \left[(\phi_{zx})|_z - (\phi_{zx})|_{z+\Delta z} \right] \Delta x \Delta y + \rho g_x \Delta x \Delta y \Delta z$$

where ϕ_x is includes both convective and molecular transport of x-component momentum and the last term takes into account the external (gravitational) force acting on the fluid in the volume element. Similar equations can be written for y and z- components. Adding all component equations and at the limit as Δx, Δy and Δz go to zero, the general equation of motion is obtained. However, the general form is quite cumbersome for many practical purposes and the one given below is often the most useful form.

For applications considered in this book the equation of motion with the assumption of incompressible flow and constant viscosity μ is adequate, and in this case the momentum balance leads to the famous Navier-Stokes equation (NS)

$$\rho \frac{D}{Dt}\vec{v} = -\nabla p + \mu \nabla^2 \vec{v} + \rho \vec{g} \qquad (7.2.8)$$

where $\vec{g}$ is the acceleration due to gravitational force (it is assumed that the only external force present is gravity). The NS equation is a kind of generalization of Newton's second law stating that the resultant of forces affecting a body equals the product of the body mass and its acceleration. The left side of NS corresponds to mass times acceleration; it consists of density multiplied by the derivative of velocity and the right side is composed of three forces: pressure gradient, viscosity (friction) and gravity force. Related to the momentum balance formulated above, the terms $\rho \frac{D}{Dt}\vec{v}$ and ∇p include the rate of increase of momentum and rate of momentum addition by convection, $\mu \nabla^2 \vec{v}$ describes the rate of addition by molecular transport and $\rho \vec{g}$ represents the external gravitational force on fluid.

NS is well known for its complexity and even for simple systems its analytical solution may not be feasible. However, NS inherently governs both laminar and turbulent behaviour and it is turbulence that poses special difficulties in mathematical analyses. The complete solution of viscous flow problems is a formidable task (Bird et al., 2002). The beginner may feel inadequate when faced with such equations. The Clay Mathematics Institute[2] has promised to pay US $1,000,000, if anyone (a) proves existence of smooth solutions or (b) gives an example of non-existence for incompressible NS equations: (a) Existence either (i) in whole space R^3 with initial conditions an arbitrary smooth velocity field vanishing outside some finite ball and zero external force or (ii) initial conditions an arbitrary smooth velocity field, periodic in all spatial directions, and zero external force (b) Non-existence (i) in whole space R^3 shows there exist some smooth initial velocity field and some smooth external force field so that no (smooth) solutions exist, (ii) the same for spatially periodic initial conditions and forcing. Remark: Weak (i.e., only square integrable) solutions always exist. If they are unique then they are smooth and conversely smooth solutions are always unique.[3]

---

[2] http://www.claymath.org/

[3] For more details see the official formulation http://www.claymath.org/millennium/Navier-Stokes_Equations/.

In many cases it is wise to begin by making some postulates (i.e., guesses) about
the form for the pressure and velocity distributions using intuition, which is based
on our daily experience with flow phenomena.[4]

7.2.6 Use of the Equations of Change to Solve Transport Problems

Principles of conservation together with phenomenological transport equations lead
to the equations of change 7.2.5 (energy), 7.2.7 (mass of individual species) and
7.2.8 (momentum).

Let us set now the general problem to solve seven unknowns: three velocity components, pressure, density, temperature and concentration. Thus seven independent
equations are needed. Six of them are NS 7.2.8 (includes three equations, one for
each velocity component), energy Eqs. 7.2.5 or 7.2.6, concentration Eq. 7.2.7 and
the continuity Eqs. 7.2.1 or 7.2.2 or 7.2.4. One missing is the *equation of state*, like
the ideal gas law for the air in atmospheric conditions

$$pV = nRT \qquad (7.2.9)$$

where V is the volume, p is the gas pressure, n is the amount of moles, R is the
universal gas constant being $8.31\,\mathrm{J\,K^{-1}\,mol^{-1}}$ and T is the temperature.

With appropriate initial and boundary conditions the unknowns can be thus
solved. Furthermore, the heat, mass and momentum fluxes can be calculated by
means of molecular and convective flux formulas using the known temperature, concentration and velocity fields. Note that heat conductivity, diffusion coefficients and
viscosity depend on temperature, pressure and composition, which may need to be
taken into account if large variations in these quantities occur. If no changes in time
occur, the system is in steady state, also called *stationary state*, and all derivatives in
respect of time disappear. The concept of stationarity should not be confused with
the *equilibrium* where all fluxes are zero, that is no temperature or concentration
profiles exist and the fluid is still.

> **Ideal gas law:**
> The ideal gas law links together four independent variables: pressure, vol-
> ume, number of molecules (moles) and temperature. It is a very accurate
> description of the state of a gas at normal conditions and at lowered pressures.
> Especially, it provides the molar density n/V by means of the pressure and tem-
> perature. If the molar density is multiplied by the molecular weight, one gets
> the density. Note that T must be obviously given in Kelvin units. The ideal gas
> law is often written in various forms. In fact, $R = N_o k$, where N_o is Avogadro's
> number and k is the Boltzmann constant and the alternative form is $pV = NkT$,

---

[4] A book by Darrigol (2005) presents a history of hydrodynamics and big challenges in the practical
work of hydraulics, navigation, blood circulation, meteorology and aeronautics.

where N is the number of the molecules. In the other form R is replaced by species specific gas constant R_M and then $pV = MR_MT$, where M is the mass. For dry air $R_M = 287$ J/kg K. The density is then readily given by $\rho = p/R_mT$. The inverse of the density $\alpha = R_mT/p$ is called the specific volume.

For gas mixtures under normal conditions or at lowered pressures, the ideal gas law holds for each individual compound and then the pressure is replaced by the corresponding partial pressures and the amount of moles, molecules or mass refers to the species considered. The total pressure is the sum of the partial pressures, which is called Dalton's law.

7.2.7 Heat Transport and Exchange in the Atmosphere and Between Atmosphere and Surface

We discuss on a general level the conservation and transport of energy in the atmosphere. Energy input by solar radiation is the "engine" driving the processes and the associated transport phenomena in the atmosphere and biosphere. The numbers in the following discussion refer to Fig. 7.2.2. A fraction of the energy carried by solar radiation penetrates the atmosphere and is absorbed at the surface (1). The heated surface warms the adjacent air layers composed of nitrogen, oxygen and, in lesser extent, greenhouse gases (GHG) (2 and 3). Conduction and convective transport are coupled with evapotranspiration (4), which transforms a fraction of heat from the absorbed radiation to the latent heat associated with vaporized water vapour. The surface emits thermal (longwave) radiation (5). A part of the incident solar radiation is reflected, depending on surface albedo. A large fraction of the thermal radiation is absorbed by greenhouse gases (6) and clouds (7) (in Fig. 7.2.2 aerosols include also clouds). Absorbed energy is distributed to the main materials of the air (8 and 9), nitrogen and oxygen, and further transported by thermal radiation (10 and 11) and convection. Not all solar radiation penetrates to the surface; some is reflected (12) mainly by clouds and absorbed (13) mainly by gases such as water vapour and ozone. The absorbed energy is redistributed among gases and radiated away (14). The overall energy balance is maintained by the outward thermal radiation of the atmosphere (11 and 14). This is the key issue in understanding the greenhouse effect. The atmosphere of the increased concentration of GHGs is more opaque for thermal radiation and thus the loss of heat by outward thermal radiation occurs effectively from upper atmospheric layers. Upper layers are however colder and thus the intensity of radiation tends to be reduced. For maintaining the (radiative) energy balance, the temperature is consequently increased at the lower atmosphere.

Figure 7.2.2 represents the day-time situation when insolation is at high levels. At night and occasionally also on winter days at high latitudes, the solar radiation component is naturally missing or is very low, and thermal radiation (5) may effectively cool the ground surface and surface layer.

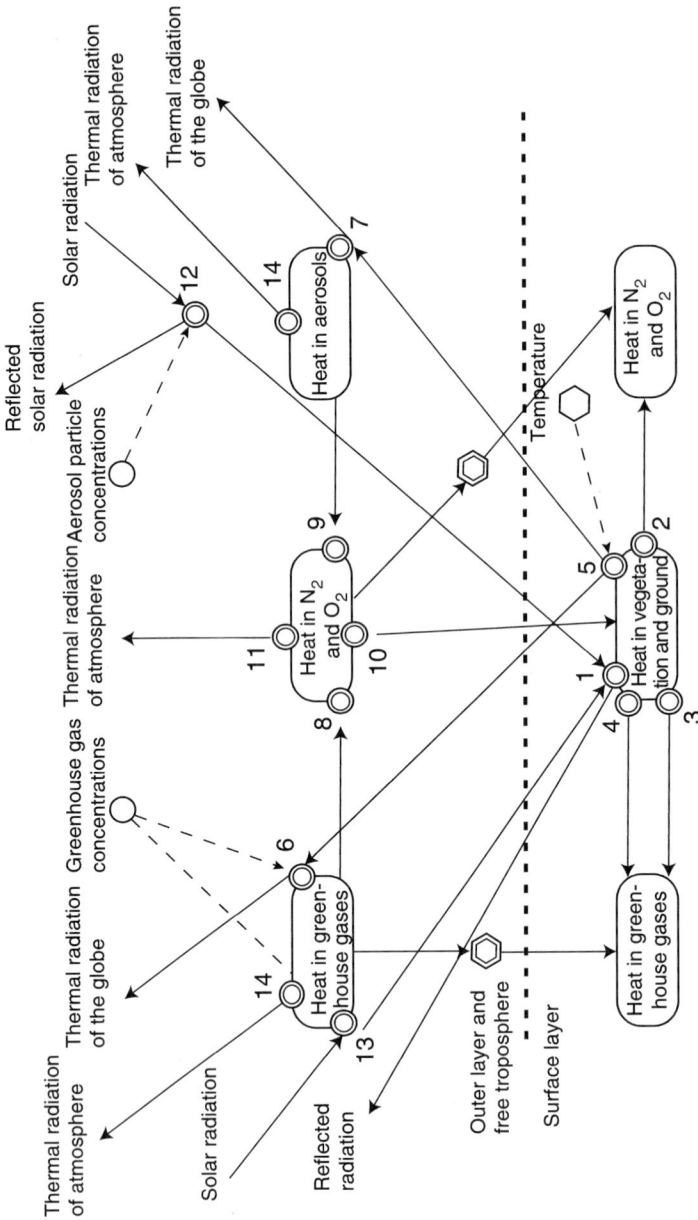

Fig. 7.2.2 Energy exchange, transport and storages in the atmosphere-biosphere continuum day-time. The atmosphere can be divided into several layers, but here only the surface layer is indicated since ecosystems are located there. The surface layer is located within the atmospheric boundary layer and the boundary layer above the surface layer is called the outer layer (see e.g., Arya, 2001). The free troposphere extends above the boundary layer up to the tropopause. Heat in N_2 and O_2 includes also a contribution from non-greenhouse gases. Note that aerosols include also clouds since physically cloud is aerosol. The box "Heat in vegetation and ground" is discussed in detail in Section 6.2.1 on "Energy balance". The symbols are introduced in the Fig. 1.2.1

7.2.8 Basic Equations of Atmospheric Models

General Circulation Models (GCM) are invaluable tools in analysis of past, present and future climate and are thus linked with many themes presented in this book. There is debate on how much we can trust complicated numerical models. However, it is demonstrated below that in fact the basic formulae are very simple and they are rigorously based only on conservation principles and transport equations and do not contain any esoteric components. The analysis of atmospheric flows normally utilizes some coordinate transforms, but they are just straightforward mathematical treatments.

Present GCMs are based on so-called *hydrostatic primitive equations*, which are normally formulated in Cartesian *pressure co-ordinates*. At synoptic scale ($\sim$1,000 km), the atmospheric flows are very close to incompressible flows, and the condition of hydrostatic balance provides an excellent approximation for the vertical dependence of the pressure field in the real atmosphere (Holton, 2004). Note that incompressibility does not mean a fixed constant density, since it is allowed to change due to changing temperature, but refers to incompressibility regarding the density variations induced by the flow itself, that is the criterion discussed below Eq. 7.2.4 is fulfilled for atmospheric flows. This can be summarized by saying that, from a thermodynamic point of view, whole flow systems are only small perturbations around the equilibrium (see Tritton, 1988). That remark may sound odd for someone standing in a wind of 20 m/s, but it is thermodynamical fact. Under hydrostatic balance the gravity force is exactly balanced by the vertical component of the pressure gradient force leading to

$$\frac{dp}{dz} = -\rho g \qquad (7.2.10)$$

Thus a single-valued monotonic relationship exists between pressure and height and pressure can be used as the independent vertical coordinate and height as a dependent variable to be derived from pressure. The use of pressure as the vertical coordinate simplifies the mathematical analysis of equations. In addition, it is convenient to formulate the flow equations in the coordinate system which rotates with the Earth. Thus the flow velocities represent flow velocities of air relative to the Earth, and not in an inertial reference frame. The coordinate transform leads to additional apparent forces, appearing as the Coriolis acceleration and centripetal acceleration, to be included in the equation of motion (see e.g., Holton, 2004).

The atmospheric equations in the rotating pressure coordinates are

$$\frac{\partial u}{\partial t} = -u\frac{\partial u}{\partial x} - v\frac{\partial u}{\partial y} - \omega\frac{\partial u}{\partial p} - \frac{\partial \Phi}{\partial x} + fv + F_x \qquad (7.2.11)$$

(equation for west-east wind u)

$$\frac{\partial v}{\partial t} = -u\frac{\partial v}{\partial x} - v\frac{\partial v}{\partial y} - \omega\frac{\partial v}{\partial p} - \frac{\partial \Phi}{\partial y} - fu + F_y \qquad (7.2.12)$$

(equation for south-north wind v)

$$\frac{\partial T}{\partial t} = -u\frac{\partial T}{\partial x} - v\frac{\partial T}{\partial y} - \omega\frac{\partial T}{\partial p} + \omega\frac{RT}{C_p p} + \frac{Q}{C_p} \qquad (7.2.13)$$

(thermodynamic equation (energy conservation))

$$\frac{\partial \Phi}{\partial p} = -\frac{RT}{p} \qquad (7.2.14)$$

(hydrostatic balance equation)

$$\frac{\partial u}{\partial x} + \frac{\partial v}{\partial y} + \frac{\partial \omega}{\partial p} = 0 \qquad (7.2.15)$$

(equation of continuity (mass conservation))

$$\frac{\partial C_{H2O}}{\partial t} = -u\frac{\partial C_{H2O}}{\partial x} - v\frac{\partial C_{H2O}}{\partial y} - \omega\frac{\partial C_{H2O}}{\partial p} + S_{H2O} \qquad (7.2.16)$$

(continuity equation for water vapour concentration C_{H2O})

$$\frac{\partial C_A}{\partial t} = -u\frac{\partial C_A}{\partial x} - v\frac{\partial C_A}{\partial y} - \omega\frac{\partial C_A}{\partial p} + S_A \qquad (7.2.17)$$

(continuity equations for other species A)

where $\omega = \frac{dp}{dt}$ is usually called the "omega" vertical motion, the pressure change following the motion. It plays the role of the vertical wind in the pressure co-ordinate system. Φ is the geopotential equal to gz where g is the gravity including also the apparent centrifugal acceleration due to rotation of the Earth. The geopotential at the height of z is the work required to raise a unit mass to height z from mean sea level, that is the potential energy per unit mass. f is the Coriolis parameter equal to $2\Omega \sin\phi$ where Ω is earth's angular speed of rotation and ϕ is latitude. F_x and F_y describe the turbulent friction in the atmospheric boundary layer (ABL). R is the gas constant of air. Q involves the diabatic heating, including most importantly radiation and phase transitions of water. S_{H2O} describes effects of evaporation and condensation to water vapour concentration.

We next compare these equations with those in 7.2.1–7.2.8, but note that the above equations are divided by the density so that it does not appear on the left-hand side as it does in Eqs. 7.2.5, 7.2.7 and 7.2.8.

- 1st equation (7.2.11) for wind component u is based on Navier-Stokes equations (7.2.8) where the substantial derivative is written according to 7.2.3 in the pressure coordinates. In pressure coordinates the x-component of $\frac{1}{\rho}\nabla p$ is expressed as $\frac{\partial \Phi}{\partial x}$ (see e.g., Holton, 2004). The viscous friction term in 7.2.8 is negligible for atmospheric flows but the effect of vertical turbulent momentum transport in ABL is taken into account by F_x and can be parameterized for example using the

K-theory approach (analogously to Eqs. 4.2.7 and 4.2.8). The effect of gravitation on the x-component is zero in 7.2.8. fv arises from the rotating coordinate system.

- The 2nd equation (7.2.12) for a wind component v is analogical to the 1st equation.
- The 3rd equation (7.2.13) is a direct consequence of Eq. 7.2.5. The substantial derivative is written according to 7.2.3 in the pressure coordinates. Effects of diabatic heating (emission and absorption of radiation and phase transitions) are added by the term Q.
- The 4th equation (7.2.14) is the hydrostatic balance equation 7.2.10 written using the geopotential.
- The 5th equation (7.2.15) is the continuity equation and identical with incompressible form (Eq. 7.2.4) in the pressure coordinates.
- The 6th equation (7.2.16) is the mass balance equation for water vapour and follows from 7.2.7. Again, the substantial derivative is written according to 7.2.3 in the pressure coordinates. The diffusive transport term in 7.2.7 is negligible for the atmospheric flows. Water vapour is chemically inert but r_a is replaced by S_{H2O} accounting for the rate of production of water vapor mass due to phase transitions. Note that the balance equation here is not formulated for the total mass of water (liquid + vapour) and thus the rate of production appears in the formula. The 7th equation (7.2.17) is the analogical formula for any other compound or phase of the compound.

Equations of change form a full set with as many equations as there are unknowns and in principle they describe the behaviour of the atmosphere very accurately. The atmosphere is divided into grid boxes, and conservation laws are applied for each box. However, as computing resources make it necessary to use a rather coarse grid, small-scale and heterogeneous processes, like turbulence and convective clouds, cannot be solved directly. Even their numerical treatment is difficult. The challenge is to calculate "the source terms" of turbulent friction, diabatic heating and phase transitions accurately enough, since they describe the processes occurring in spatial scales smaller than the typical grid size of the numerical model. However, the source terms are very important in simulations of climate and its changes.

Conclusions

Conservation equations and transport laws provide a full theoretical framework for understanding and predicting atmospheric processes and atmosphere-biosphere interaction processes. Effects of radiation absorption and emission, phase transitions and chemical reactions are also included in the formulas. They form the basis for atmospheric models and according to the present knowledge they describe rather well the behaviour of the real atmosphere.

7.3 Environmental Factors in the Canopy

Üllar Rannik[1], Pasi Kolari[2], and SamuliLauniainen[1]

[1] University of Helsinki, Department of Physics, Finland
[2] University of Helsinki, Department of Forest Ecology, Finland

The properties of the atmosphere are determined by processes and energy, material and momentum fluxes combined with conservation of mass, energy and momentum (Section 7.2). Similarly also near earth's surface and in vegetation, the same principles determine the spatial properties of environmental factors (photon flux, temperature and concentrations). Processes related to energy transformation inside a canopy are absorption of direct and diffuse solar radiation, absorption and emission of thermal (infrared) radiation and evapotranspiration or condensation of water vapour (Section 6.2.1). Temperature differences generate more (or decrease) turbulence via thermal stability, the resulting convective flows (Section 4.2) level concentration differences.

7.3.1 Short-Wave Solar Radiation

The incoming solar radiation, called as short-wave radiation in contrast to terrestrial or long-wave radiation, contains most of the energy at waveband from 0.1 to 5.0 μm, which contains ultraviolet (0.1–0.4 μm), visible (0.4–0.7 μm) and near-infrared (0.4–0.7 μm) spectral regions. Solar energy enters the canopy in the form of direct and diffuse radiation, summed together as global radiation. The amount of global radiation available at the top of the canopy depends strongly on elevation of the sun as well as atmospheric conditions, especially cloudiness. At Boreal latitudes the global radiation values at noon in clear-sky conditions are approximately half of the solar radiation reaching the top of the atmosphere (Solar constant), i.e., in summer about 700–800 W m^{-2}. Note that the Solar constant corresponds to short-wave radiation flux perpendicular to a solar beam at an average sun-earth distance, but global radiation is the measure of solar energy through the plane parallel to the Earth's surface, which is strongly affected by the elevation of the sun. Below cloud cover the values can be only fractions of the clear-sky conditions. In case of broken cloudiness the global radiation can occasionally even exceed the clear-sky value due to reflection from clouds. Figure 7.3.1A presents a daily course of global radiation in Hyytiälä. This is an example of a day with clear sky before noon and some cloudiness in the afternoon, with a few observations exceeding the 'clear-sky' upper bound of the global radiation pattern.

From a biological point of view the important part of the solar spectrum is visible radiation in the waveband from 400 to 700 nm. This range of the solar energy spectrum, photosynthetically active radiation (PAR), contains approximately 40% of total energy of the spectrum emitted by the sun. Above the canopy, global PAR consists of two components, direct PAR and diffuse PAR. Partitioning of global

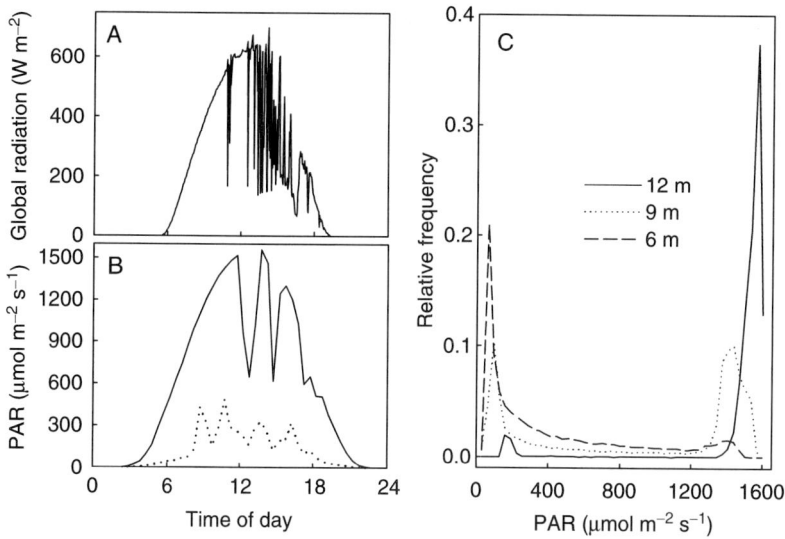

Fig. 7.3.1 Solar radiation on some summer days at SMEAR II (A) Daily course of global radiation above the canopy; (B) Daily course of PAR measured above (solid line) and below (dashed line) the canopy at the forest floor; (C) Distribution of photosynthetically active radiation at different heights inside a Scots pine canopy. The height of the canopy was 14 m

radiation into direct and diffuse components depends on atmospheric conditions, solar elevation and very much on cloudiness. The angular distribution of global radiation is important for penetration of radiation into the canopy.

Solar radiation interferes with canopy elements and is partly reflected and partly absorbed, the latter being a process of radiation transformation into heat (see Section 6.2.1) and into chemical energy in photosynthesis. The transport of solar radiation occurs differentially throughout the canopy down to the forest floor, where the global radiation is much reduced compared to above-canopy values (Fig. 7.3.1B).

The global PAR inside a canopy consists of the following components: direct PAR that penetrates through the canopy without scattering by phytoelements, diffuse PAR that penetrates through the canopy without scattering by phytoelements, additional diffuse PAR due to scattering of direct PAR by phytoelements, and additional diffuse PAR due to scattering of diffuse sky radiation by phytoelements. The PAR value observed inside canopy is the sum of these components and varies strongly in time and space, depending on spatial averaging by radiation sensor.

We can determine a unique value for irradiance only for an infinitesimally small point. When moving towards a larger scale, the irradiances over the space in consideration form a distribution of different values. The point values of global PAR inside a canopy vary to a great extent due to movement of canopy elements and random shading of the sun's disc by canopy elements. Figure 7.3.1C shows the frequency distribution of global PAR measured inside the canopy. The distribution is divided into three 'areas': sunfleck area, in which the sun's disk is not shaded by any phytoelements (highest PAR values); penumbra is an area in which the sun's disk is partly covered by phytoelements; umbra is an area inside the canopy in which the sun's disk is fully covered by canopy elements and the direct part of the global PAR is zero. The probability distribution of PAR inside a canopy depends on the level inside the canopy; at lower levels inside the canopy the sunfleck vales are less frequent and the probability of umbra values increases (Fig. 7.3.1C). Figure 7.3.2 illustrates the relative areas of sunflecks, umbra, and penumbra at SMEAR II as a function of optical path length, i.e., the shading foliage area along the direct solar beam (Vesala et al., 2000; Rannik et al., 2002).

The spatial distribution of light also varies in time. The sunflecks and shaded patches continuously change their location, size and shape when the position of the sun changes over the day, the variation being superimposed by relatively fast waving of shading elements due to turbulent air flow. If light intensity is measured over an extended period at one point inside the canopy, combinations of different solar elevation angles and azimuths will be represented in the measured time series of irradiance. The spatial distribution of light can also be considered equivalent to

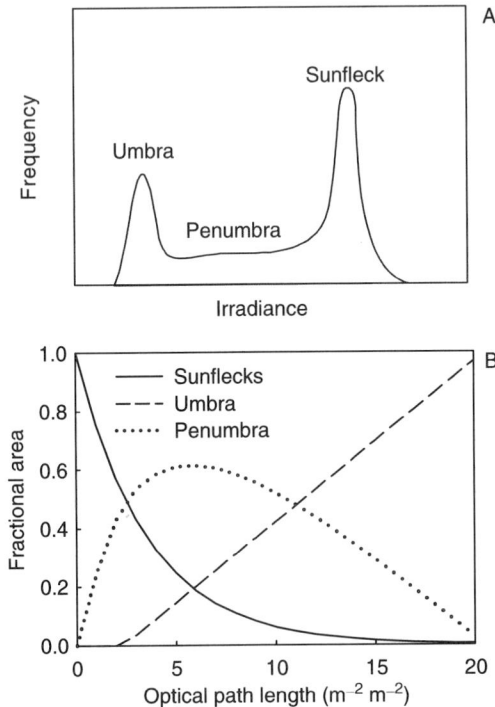

Fig. 7.3.2 Schematic presentation of (A) irradiance distribution inside a tree canopy in clear weather and (B) the fractional areas of sunflecks, umbra and penumbra inside a canopy as a function of optical path length. The optical path length is the all-sided foliage area (m^2 per m^2 ground) above the observation level, divided by the sine of the solar elevation angle

distribution over time at one point under the restriction that radiation conditions above canopy do not change significantly. In the following paragraphs we concentrate on discussing spatial patterns of irradiance over relatively short time period when solar angle and cloudiness can be considered constant.

The interception of light inside the canopy can be described with different degrees of complexity but in general there are two basic approaches: in a spatially explicit approach the canopy structure, i.e., the arrangement of individual trees and canopy elements in space, is known. A spatially inexplicit approach assumes that the stand foliage consists of a homogeneous light-absorbing medium without structure. In this simple approach, light intensity at different depths inside the canopy can be approximated using the Beer-Lambert law originally derived for calculating attenuation of radiation in a homogeneous medium. With this law, the reduction of global radiation is described as an exponential function of cumulative leaf area density above the level of interest.

Attenuation of light within the canopy depends on the optical path length, i.e., the distance travelled by solar rays inside the canopy, and on the density of canopy elements along the path. In the case of homogeneously distributed canopy biomass, the change in light intensity, ΔI, is proportional to the distance ΔL traveled down in the canopy:

$$\frac{\Delta I}{I} = k\Delta L \tag{7.3.1}$$

Parameter k is called the extinction coefficient, which can be determined empirically or mechanistically based on the dimensions and density of phytoelements inside the canopy space (Oker-Blom and Smolander, 1988; Stenberg, 1996). Equation 7.3.1 results in exponential decrease of light inside the canopy

$$I(L) = I_o e^{-kL} \tag{7.3.2}$$

where I_0 is the incident light above the canopy and L depth from the top of the canopy.

In reality the structure of the canopy is more complex. Leaves are grouped in shoots and branches. This creates gaps in the canopy and enhances penetration of light deeper down into the canopy. Clumping of foliage in the canopy reduces light interception and photosynthetic rate, compared with a homogeneous canopy with the same leaf area (Oker-Blom et al., 1989). The distribution of leaf surface orientations also varies in different parts of the canopy. Spatially explicit calculation of light environment involves description of canopy structure, i.e., locations of leaves, shoots and branches in space. At the ultimate end of complexity are models where the canopy structure is defined down to the dimensions and orientations of individual leaves (Perttunen et al., 1995). Irradiance is calculated for each point (leaf, shoot) in space by ray-tracing: collisions of incident rays with other canopy elements are detected. Typically the sky hemisphere is divided into sectors of equal solid angle. The sectors have their own radiation intensities that are reduced to "rays" that either penetrate the canopy and reach the observation point, or are intercepted somewhere else in the canopy. Also the intensity and the angular distribution of the light reflected from the canopy and the forest floor can be taken into account. The sun can be treated as a point source, or the solar disk can be further divided into subsectors to allow calculation of light coming from partly obscured sun (penumbra). Irradiance at a given observation point is the sum of all "rays", weighted by the cosine of the angle of incidence, reaching that point. When the irradiance calculation procedure is applied for several points inside the canopy, we obtain a distribution of different irradiances.

Spatially explicit methods require very detailed information on canopy structure, often more detailed than is normally obtained from routine measurements of stand structure. They also require a vast number of calculations of interactions between the beams of light and the canopy elements, thus consuming a lot of computing time. Therefore, their usefulness for practical applications is somewhat limited.

7.3.2 Radiation Balance

In the long-wave part of the spectrum, the canopy elements as well as the forest floor emit thermal radiation. Most of the natural surfaces are very close to 'full' radiators.[5] Terrestrial radiation, i.e., long-wave radiation emitted by the Earth's surface, is mostly absorbed by the atmospheric greenhouse gases. These gases in turn emit long-wave radiation, part of which is lost to outer space and part is downward propagating. Large fraction of the downward flux reaching the Earth surface originates from the lower atmosphere of about 1 km. The difference between upward and downward long-wave radiation in clear sky conditions is typically in the order of $100\,W\,m^{-2}$, see Fig. 7.3.3 at night hours for one month average conditions in summer.

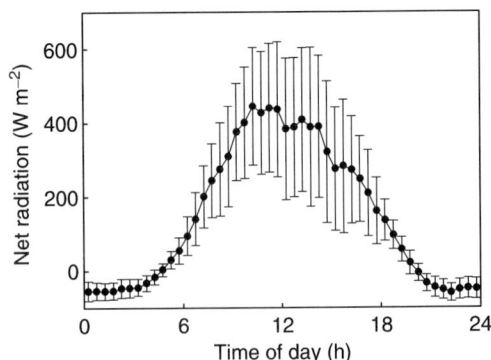

Fig. 7.3.3 Diurnal course of net radiation over a pine forest in Hyytiälä obtained by averaging data from July 2005. The error bars denote standard deviation

---

[5] A thermal radiator capable to provide a spectrum dependent on the temperature alone according to Planck's law is called a full radiator, see Section 6.2.4

The downward thermal radiation of the atmosphere is strongly dependent on cloudiness. The lower surface of the clouds emits approximately as a full radiator and the total radiation reaching the surface is supplemented by emission from clouds in wavebands that the gaseous emissions lack.

The sum of the downward and upward solar and thermal radiation forms the net radiation, the radiation balance. Net radiation at surface is typically positive in day-time (surface gains energy) and negative at night (loss of energy). Net radiation is very sensitive to both short-wave and long-wave components, both being sensitive to cloudiness. Net radiation above the canopy top is typically in day-time somewhat less than net global (short-wave) radiation in day-time, while at night it is between 0 and $-100 \, \text{W} \, \text{m}^{-2}$ (thermal cooling). Figure 7.3.3 is an example of a daily course of net radiation measured in Hyytiälä above the forest canopy. This is an average of about one month period, thus being an average of clear-sky and cloudy conditions at the particular site. The net radiation is the energy which is used (mainly) for heating up canopy elements and air being in contact with canopy elements. The absorption and heating occurs differentially with height inside the canopy, most of the radiation being absorbed at the upper part of the canopy.

7.3.3 Vertical Profiles and Fluxes

The absorption of solar radiation, net emission of thermal radiation and evapotranspiration or condensation of water inside the canopy leads to heating or cooling of the canopy elements, giving rise to vertical as well as horizontal temperature differences inside canopy. Energy exchange between canopy surfaces and air tends to reduce the spatial differences in temperature. Transport of energy in air occurs via advection of air parcels being first in contact with surfaces. The energy consumed in evapotranspiration or released in condensation is called latent heat. Thus the energy has two forms; sensible and latent.

The motion of air inside and above canopy is usually turbulent. Unstable thermal stratification typically generates more turbulence during day and stable stratification acts to reduce turbulence at night. The strength of energy sources and sinks and transport (turbulence) intensity determine vertical temperature profiles. Since the solar radiation is strong during day time and missing at night, it is useful to consider daytime and night-time separately: in following the average profiles of wind speed, temperature, water vapour and carbon dioxide and corresponding fluxes are demonstrated for different stability conditions for pine forest in SMEAR II (Figs. 7.3.4 and 7.3.5).

During day-time typically unstable conditions prevail which promote turbulence and therefore mixing is very efficient. Turbulence tends efficiently to 'smooth' profiles and gradients of concentrations (and wind speed) are smaller above canopy (see Section 5.3). At night for temperature the situation is opposite, the negative net radiation inside canopy drives also downward heat flux from the atmosphere above

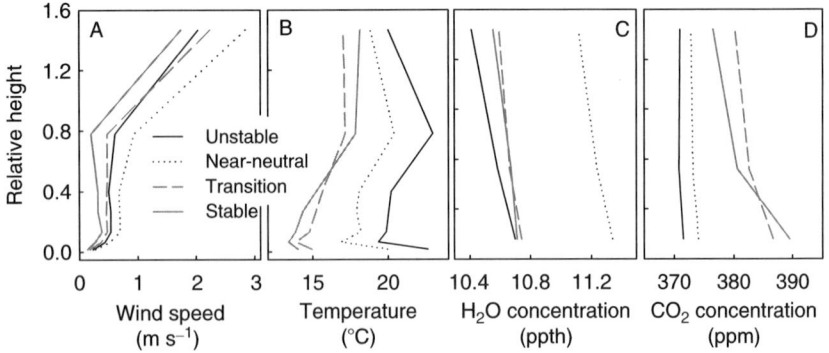

Fig. 7.3.4 Vertical profiles of horizontal wind speed (A), temperature (B), water vapor (C) and carbon dioxide concentrations (D) inside and above forest. Forest height (h) is approximately 15 m. Period from 27.06–15.07.2005 was considered with classification of measurements into four stability classes according to following criteria: unstable, $h/L < -0.05$; near-neutral, $|h/L| \leq 0.05$; transition stable, $0.05 < h/L < 0.8$; strongly stable: $h/L \geq 0.8$. L is the Monin-Obukhov stability length

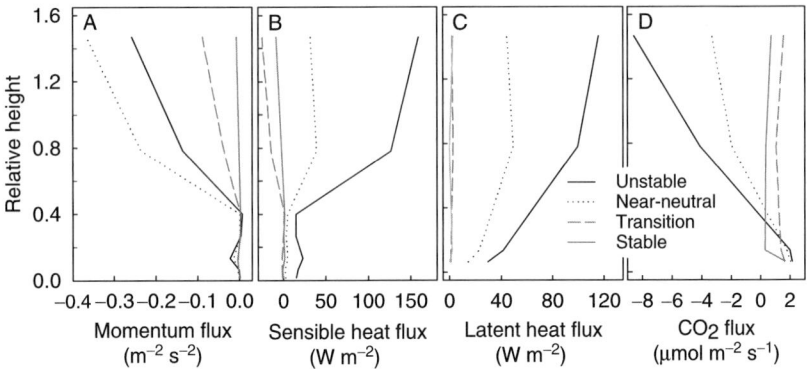

Fig. 7.3.5 Momentum flux (A), sensible heat flux (B), latent heat flux, LE (C) and CO_2-flux (D) profiles inside and above canopy in Hyytiälä. Forest height is approximately 15 m. Period from 27.06–15.07.2005 was considered with classification of measurements into four stability classes according to following criteria: unstable, $h/L < -0.05$; near-neutral, $|h/L| \leq 0.05$; transition stable, $0.05 < h/L < 0.8$; strongly stable: $h/L \geq 0.8$

and opposite temperature gradients occur above canopy. Inside canopy, however, temperature maximum exists in the middle of foliage also during day-time due to heating of upper canopy (Fig. 7.3.4B).

The wind speed decreases very sharply in the upper part of the canopy and is relatively constant in open trunk space. The wind profile inside canopy depends very much on the vertical leaf area distribution and in magnitude depends on the wind speed above canopy. Such a profile corresponds to momentum absorption in the upper part of the canopy and almost no momentum flux below canopy foliage

(Fig. 7.3.5A). On the average, the wind speed is highest during near-neutral conditions and lowest during stable conditions (Fig. 7.3.4A).

Water has an important role in the energy balance of forest canopy. Water is available at the surfaces or it diffuses via stomata into air. In both cases, energy is needed for phase transition from liquid water into vapor; condensation of water releases the same amount of energy. Since the energy consumption by evaporation is large (about five times more energy is needed for evaporation of water than for heating it by $100°$), the latent heat transport can be as significant as sensible heat transport in energy balance of forest. For day-time conditions, both of the heat fluxes can be around 300–400 W m^{-2} when observed above forest.

During day time evapotranspiration of water occurs throughout the canopy. At night the water transpiration throughout stomata is limited because of stomatal closure and high humidity (Fig. 7.3.5B). Even, condensation of water can occur resulting in dew formation and small downward water flux. Therefore the water vapor concentration profile in air is relatively constant with height at night (Fig. 7.3.4C).

The carbon dioxide has more complicated vertical source and sink distribution inside canopy, especially during day-time (Fig. 7.3.5D). Forest floor is a source, where belowground respiration leads to CO_2 flux via soil surface interface. Ground vegetation respires as well as participates in photosynthetic activity and the net result of production and consumption processes depends on light and temperature conditions. The stems respire CO_2, but the canopy is a major sink in day-light conditions during photosynthetically active period. Therefore positive and negative CO_2 exchange occurs at different parts of the forest during day-time, which shows up as a concentration maximum below trunk space. At the forest floor, the soil efflux and ground vegetation CO_2 exchange combined with the lower transport efficiency of turbulence lead to higher concentrations. Especially at night strong concentration gradient close to surface can build up Fig. 7.3.4D). Figure 7.3.6 summarises the processes responsible for carbon dioxide production or consumption as well as transport routes between pools or sources and sinks.

The fluxes and concentrations have usually strong diurnal variation. For example, the sum of sensible and latent heat fluxes corresponds approximately to net radiation, which is driving the diurnal variation of these fluxes. The carbon dioxide flux above forest is the measure of the carbon exchange of forest and is also strongly dependent on diurnal variation of solar and on annual cycle of metabolism. However, the transport can be limited under stable conditions, which leads to an accumulation of emitted gases and shows up as an increase in concentrations at night (as CO_2 and H_2O). Therefore the diurnal variation of concentrations indicates the influence of sources and sinks as well as transport efficiency of turbulence (cf. Fig. 3.2.4). Note that the absolute water content of the air is limited by 100% relative humidity, but there is no such a limitation for carbon dioxide. Therefore the carbon dioxide concentrations can have extremely large values close to soil surface at very calm nights.

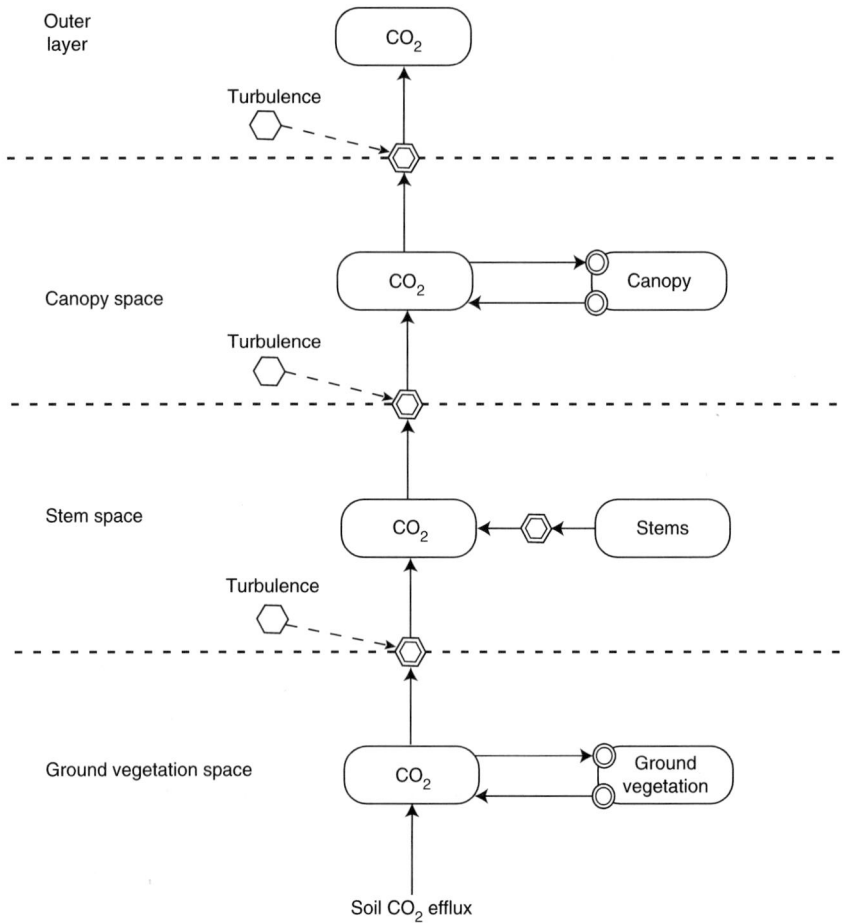

Fig. 7.3.6 Schematic visualisation of processes generating or consuming CO_2 in a forest canopy and transport within canopy and between outer layer and canopy air space. The symbols are introduced in the Fig. 1.2.1

7.4 Environmental Factors in Soil

Asko Simojoki[1], Heli Garcia[1], Mari Pihlatie[2], Jukka Pumpanen[3], Jukka Kurola[1], Mirja Salkinoja-Salonen[1], and Pertti Hari[3]

[1] University of Helsinki, Department of Applied Chemistry and Microbiology, Finland
[2] University of Helsinki, Department of Physics, Finland
[3] University of Helsinki, Department of Forest Ecology, Finland

Litter input provides new material for the soil organic matter (Section 6.3.2.4). Decomposition of the organic compounds release nutrients and CO_2 into the soil 6.4.1) and ion exchange stores cations on the surfaces of soil particles (Section 6.2.4). Transport by water flow and diffusion is ineffective in the soil and the processes generate strong spatial variation in the properties of the topsoil.

7.4.1 Classification and Distribution of Boreal Forest Soils

Soils in the boreal climatic zone are predominantly Histosols (peat soils) and Podsols, or soils developing towards Podsols (Mazhitova, 2006; Van Rees, 2006; USDA-NRCS 2005). Young and weakly developed soils (Regosols) are common as well. In addition, some Gelisols characterized by permanent frost (freezing temperatures) are found in northern-most parts of the boreal zone. Histosols and Gelisols are not discussed in more detail here due to their minor importance as forest soils. In contrast, Podsol soils are discussed in some detail as they are generally considered the predominant class of boreal mineral soils without permafrost.

7.4.2 Description of a Podsolic Soil Profile and Podsol Formation

In a podsolic soil profile, the uppermost humus layer of partly decomposed organic material (O horizon) and the humus-rich mineral soil horizon (A horizon) are underlain by the ash-grey eluvial horizon (albic E horizon), and the dark or reddish brown illuvial horizon (spodic B horizon) enriched with humus or sesquioxides (Al and Fe oxides), respectively. The deepest soil layers consist of practically unchanged parent material (C horizon). Podzols are identified based the occurrence of spodic materials and spodic horizons (Buol et al., 1997). The development of different soil horizons varies in response to soil and climatic conditions. In most extreme cases, either the A horizon or more seldom the E horizon may be lacking entirely (Buol et al., 1997).

The prevalence of podsolic soil formation, or podsolisation, in the Boreal zone is a sum of many processes and transport in soil. Podsols develop typically in coarse-textured, permeable mineral soil parent material under coniferous forest vegetation in a cool and humid climate, although they also occur outside the boreal zone on quartz-rich sands (Van Breemen and Buurman, 2002; Mokma and Evans, 2000). Podsolisation is favoured by poor nutrient status of soil, poor decomposability of litter, low temperature, high rainfall, poor drainage, and impeded biological mixing of soil (Van Breemen and Buurman, 2002). Podsols develop in soils with widely varying water table depths, excluding the soils with permanent water saturation to near soil surface and the soils of arid climate (Mokma and Evans, 2000).

The main processes and transport phenomena in the podsolic soil profile are schematically depicted in the Figs. 7.4.2 and in 7.4.1. Podsol formation includes the mobilization of sesquioxides and organic compounds in the upper soil horizons, and their translocation and subsequent immobilization deeper in the soil. The main theories explaining the podsol formation are (1) the metal-complexing organic acids

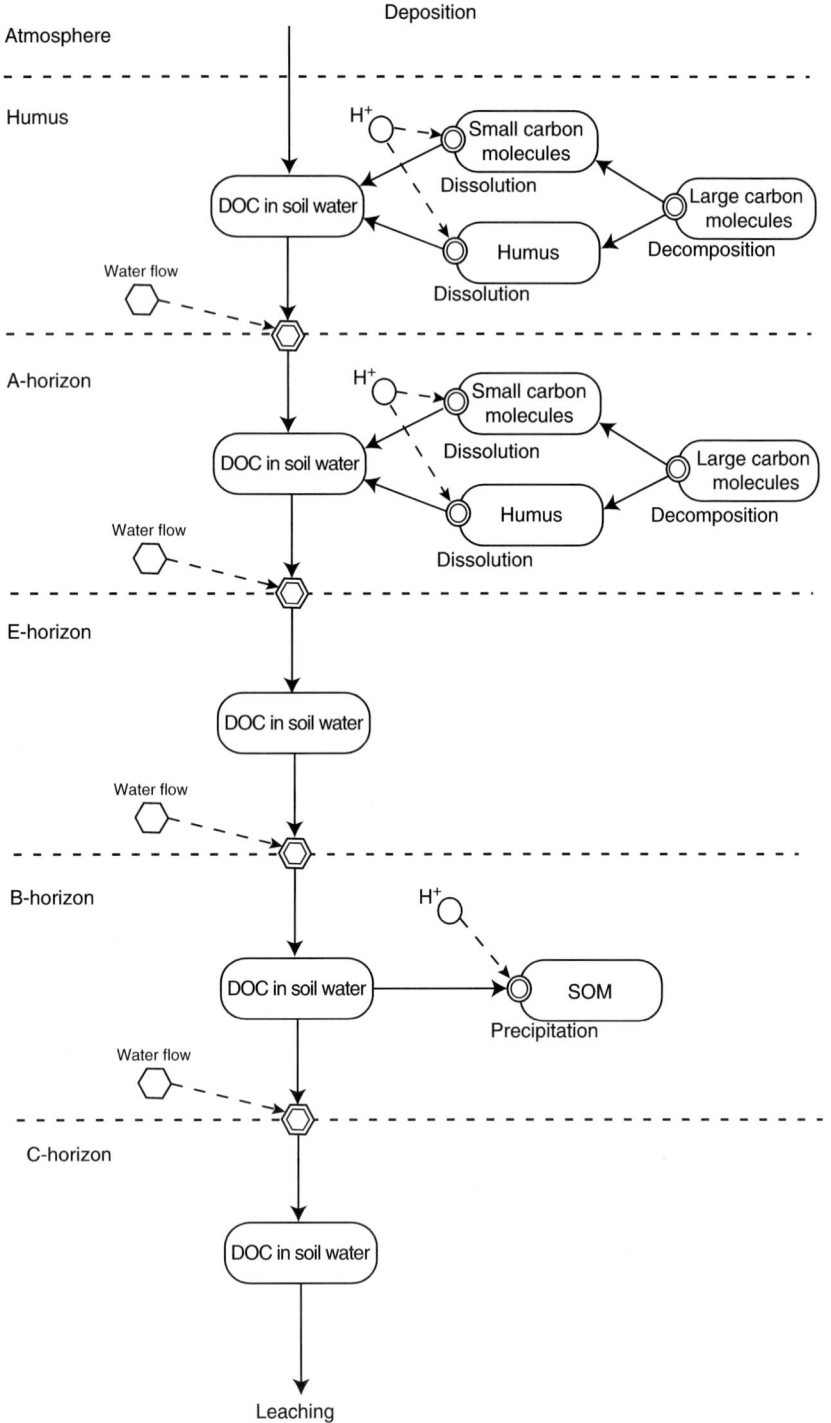

Fig. 7.4.1 Movement of dissolved organic compounds in a podsolic soil profile

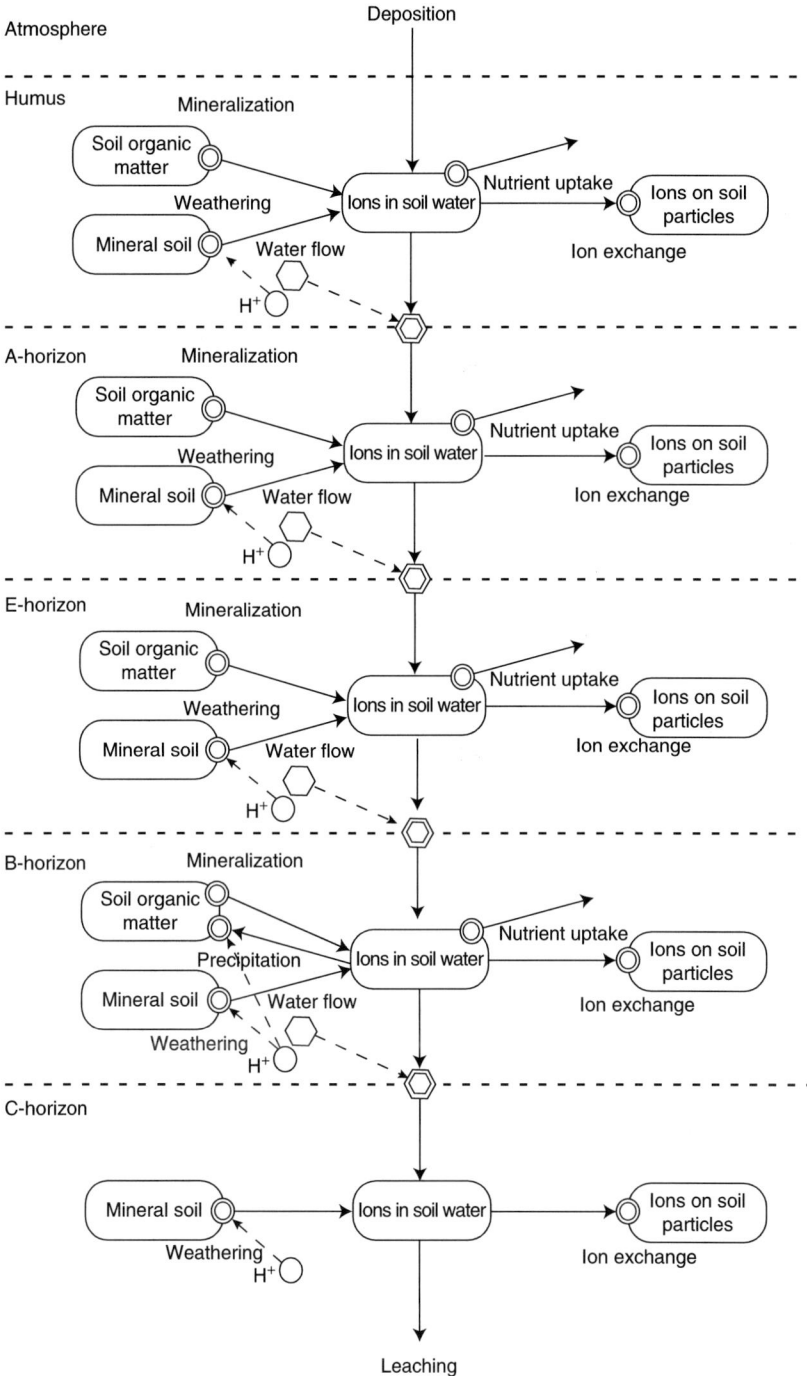

Fig. 7.4.2 Movement of dissolved ions in a podsolic soil profile. The symbols are introduced in the Fig. 1.2.1

theories (MCOA), (2) the proto-imogolite theory (PI) and (3) the root-decay theory of organic matter accumulation. The mechanisms involved in podsolization are a topic of on-going research. The following discussion is based mainly on results reviewed recently by Lundström et al. (2000a, b), Van Breemen and Buurman (2002), Buurman and Jongmans (2005), Jansen et al. (2005), Mokma (2006) and Tanskanen (2006).

According to MCOA theories, the decomposition process of litter in the surface layers produces organic acids that enhance the weathering of soil minerals and dissolution of iron and aluminium ions into the soil solution as well as their leaching deeper into the soil as soluble organometallic complexes. Other cations released by the weathering leach down mostly as simple ions. MCOAs include e.g. high molecular weight organic acids (HWOA), such as highly soluble fulvic acids and less mobile humus acids, as well as low molecular weight organic acids (LWOA), such as citric and oxalic acids. Both these groups have functional groups showing high acid strength (low pK_a values) and contributing to the strong acidity in the upper horizons of Podsol soils. As a consequence, all but the most resistant minerals leach out of the E horizon that will eventually become practically pure quartz sand. In the B horizon, the organometallic compounds of iron and aluminium are precipitated as a consequence of changes in pH and redox potential, by microbial decomposition of the complexing organic acids, or due to restricted water flow (Van Breemen and Buurman, 2002; Blume et al., 2002). The classical fulvate theory assumes that the metal-complexing organic acids are high-molecular weight fulvic acids that precipitate on saturation by metal-complexation on their way deeper into the soil (e.g., De Coninck, 1980).

In contrast, newer MCOA theories emphasize the different roles of small and large organic acids. According to Lundström et al. (2000b), low weight organic acids (LWOA) exuded by plants and mycorrhizal fungi are mainly responsible for the weathering of minerals and the initial complexation and mobilization of dissolved metals. LWOA are easily decomposed by soil microbes, especially in the surface organic horizon and at low metal/C ratios (Lundström et al., 2000b; Van Hees et al., 2002). Lundström et al. (2000b) emphasized the role of microbial degradation of LWOAs in the formation of spodic B horizons, as the release of metals from the complexes allows their subsequent precipitation as inorganic metal hydroxy polymers, proto-imogolite type materials, or organic complexes with high molecular weigh organic acids (HWOA). According to this view, HWOAs have a minor role in determining the mobility of metals.

However, microbial degradation is not always required for podsolisation, as the mobility of iron and aluminium may be determined by the solubility control, e.g., through the precipitation of metal hydroxides and organic complexes (Gustafsson et al., 2001; Jansen et al., 2005). Moreover, the decomposition at greater soil depths proceeds at a much reduced rate due to organo-metallic complexation and the adsorption of complexes to the solid phase (Van Hees et al., 2002; Buurman and Jongmans, 2005; Scheel et al., 2007).

Podsolization proceeds slowly: the first signs may appear in less than 100 years, but the formation of mature podsols takes typically centuries or millennia

(Lundström, et al. 2000a, Mokma et al., 2004). During the development, the E horizon becomes deeper with time. In addition, the weatherable minerals in the E horizon become typically criss-crossed by 5-μm tunnels that are probably produced by ectomycorrhizal "rock-eating" fungi (Jongmans et al., 1997; Lundström, 2000b; Van Breemen and Buurman, 2002). The contribution of this process to weathering increases with soil age but may nevertheless remain as low as 1% (Van Breemen et al., 2000; Smits et al., 2005).

According to a current view, the deepening of E horizon on podsolization is caused by the acidification of soil profile to ever greater depths with time. This causes the redissolution of organo-metallic complexes in the upper B horizon and their reprecipitation at greater soil depths (Buurman and Jongmans, 2005). Sometimes this may even lead to the separation of the B horizon into the humus-containing upper layer overlying the sesquioxide layer. Occasionally, the B horizon may develop into an Ortstein horizon where the sand grains may be cemented by organometallic complexes so strongly as to inhibit root growth (Mokma, 2006).

According to the proto-imogolite theory of podsolization, the transport of metals by MCOAs has no key role for podsol formation (Farmer and Lumsdon, 2001). Instead, inorganic and low weight organic acids dissolve metals from weatherable minerals, the metals move deeper into soil as positively charged silicate sols, and eventually precipitate in the subsoil as amorphous imogolite or allophone due to an increase in pH (Farmer and Lumsdon, 2001). Organic acids then precipitate on these allophonic deposits. The role of fulvates migrating through the E horizon is seen merely as recycling of the Al and Fe transported by biological processes from the B horizon to the O horizon again back to the B horizon. The absence of imogolite in some soils and the instability of Fe-Al-silicate sols in the presence of complexing organic acids (Van Breemen and Buurman, 2002; Jansen et al., 2005) undermine the general importance of this theory of podsolation. However, different podsolization mechanisms may dominate in different geographical regions, as e.g., the occurrence of imogolite becomes less common from north to south in Scandinavia (Jansen et al., 2005).

Buurman and Jongmans (2005) acknowledged the weaknesses of MCOA and proto-imogolite theories in explaining the morphology and chemical characteristics of organic compounds accumulated in the B horizon. Currently, podsolization is mostly viewed as a combination of fulvate and LWOA processes, although root-derived accumulation of organic, polymorphic pellets in the B horizon is important in podsolization of rich parent materials in a boreal climate (Van Breemen and Buurman, 2002; Buurman and Jongmans, 2005). In contrast, the root-derived organic matter contributes less to the development of the podsols on poor parent materials outside the boreal zone, and in the hydromorphic (very wet, poorly drained) podzols of boreal soils the composition of organic matter immobilized in the B horizon is monomorphic and similar in composition to that of the organic matter dissolved in soil water (Van Breemen and Buurman, 2002).

The processes and transport phenomena involved in podsolization are reflected in the chemical properties of the podsolic soil profile (Figs. 7.4.3–7.4.5). As a consequence of the podsolization, the upper parts of the soil profile become acidic and depleted of base cations that leach deeper into the soil and eventually out of the

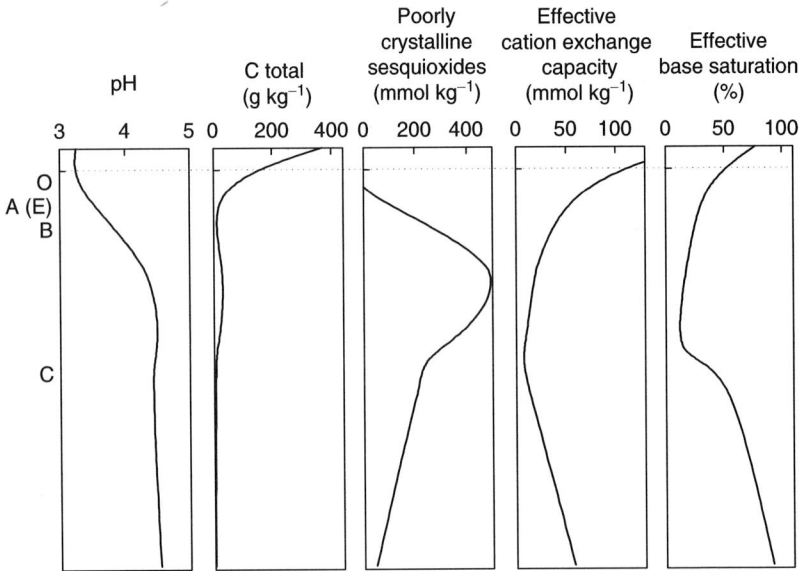

Fig. 7.4.3 Schematic presentation of typical chemical properties of a podsolic soil profile. Soil surface is illustrated with a dash line

soil. The organic matter decomposes slowly and accumulates mainly in the upper soil horizons. The Al and Fe ions weathered from soil minerals are complexed by organic acids, move down the soil profile with water flow, and precipitate in the B horizon as organomineral complexes or oxides due to pH rise or microbial decomposition of the complexing organic acids, respectively. The accumulation of poorly crystalline sesquioxides in the B horizon provides efficient adsorption of anions with a ligand exchange mechanism, such as phosphates, whereas the other anions, such as nitrate and chloride, leach easily down and out of the soil. The accumulation of organic matter in the B horizon of weakly developed podsols is less pronounced compared with sesquioxides (Fig. 7.4.3).

Most of the soil organic matter is located in the humus layer (Figs. 7.4.3–7.4.5). The amount and chemical composition of soil organic matter change during the decomposition of needles, wood and litter at different depths in the soil profile. In general, the contents of hemicellulose and cellulose decrease, while lignin and/or humus as well as proteins seem to become enriched during the decomposition. These results are in good qualitative agreement with results from decomposition studies carried out using long litterbag incubations and slightly different analytical procedures compared with ours (Berg et al., 1991; Berg, 2000; McTiernan et al., 2003; Berg and Dise, 2004). In addition, the proportion of protein N to total N seems to increase with increasing soil depth (Fig. 7.4.5). The same observation was confirmed under different forest types at several sites in Hyytiälä, Finland. Nevertheless, the extent to which this applies more generally to Podsols and other boreal forest soils remains to be shown by future research.

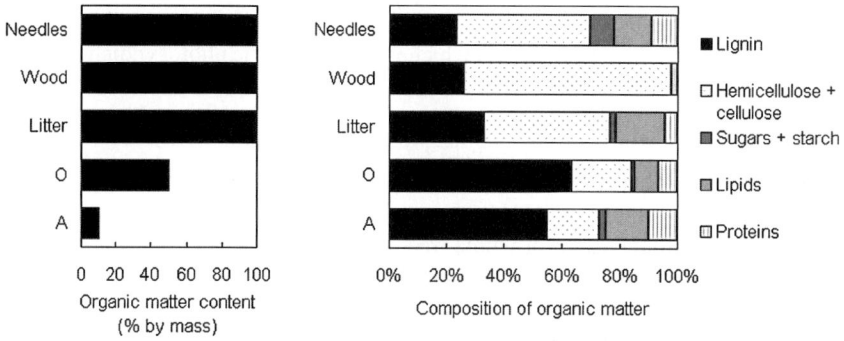

Fig. 7.4.4 The content and composition of organic matter in plant (green needles, wood, litter) and soil samples (soil horizons O and A) taken from a 40-year-old pine forest growing at the Smear II Forestry Field Station in Hyytiälä, Finland, in 2002

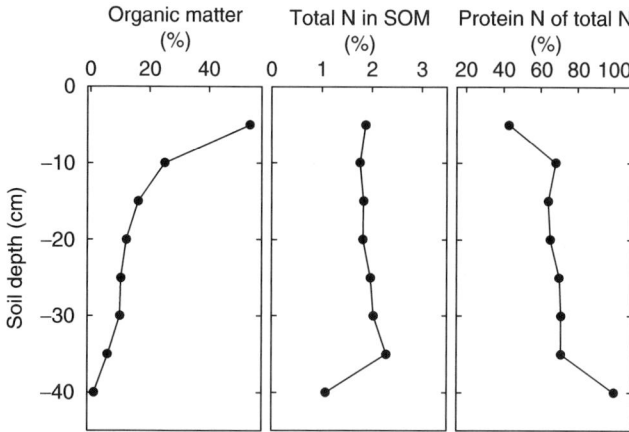

Fig. 7.4.5 The gravimetric percentages of soil organic matter (SOM) content, the content of total N in SOM, and the proportion of protein N (6 M HCl hydrolysable N) to total N at 5-cm intervals in a podsolic soil profile under pine forest near Smear II Forestry Field Station in Hyytiälä, Finland (Drawn after Rajasekar, 2007, unpublished)

7.5 Tree Level

7.5.1 Photosynthetic Production of a Tree

Pasi Kolari and Pertti Hari

University of Helsinki, Department of Forest Ecology, Finland

Photosynthetic production of a single tree is the key to understand tree growth since photosynthesis provides sugars for the synthesis of new macromolecules in growing tissues (Sections 6.3.3.1.4 and 9.1). The environmental factors vary strongly within canopy (Section 7.3) which is reflected in the photosynthetic process (Section 6.3.2.3).

Integration over Space and Time

Photosynthetic production of a whole tree or a stand is the sum of photosynthetic production of all individual leaf elements. The leaves located in different parts of the canopy experience different conditions, each leaf having its own microclimate and thus different photosynthetic rate (Fig. 7.5.1). Light has the largest spatial variation

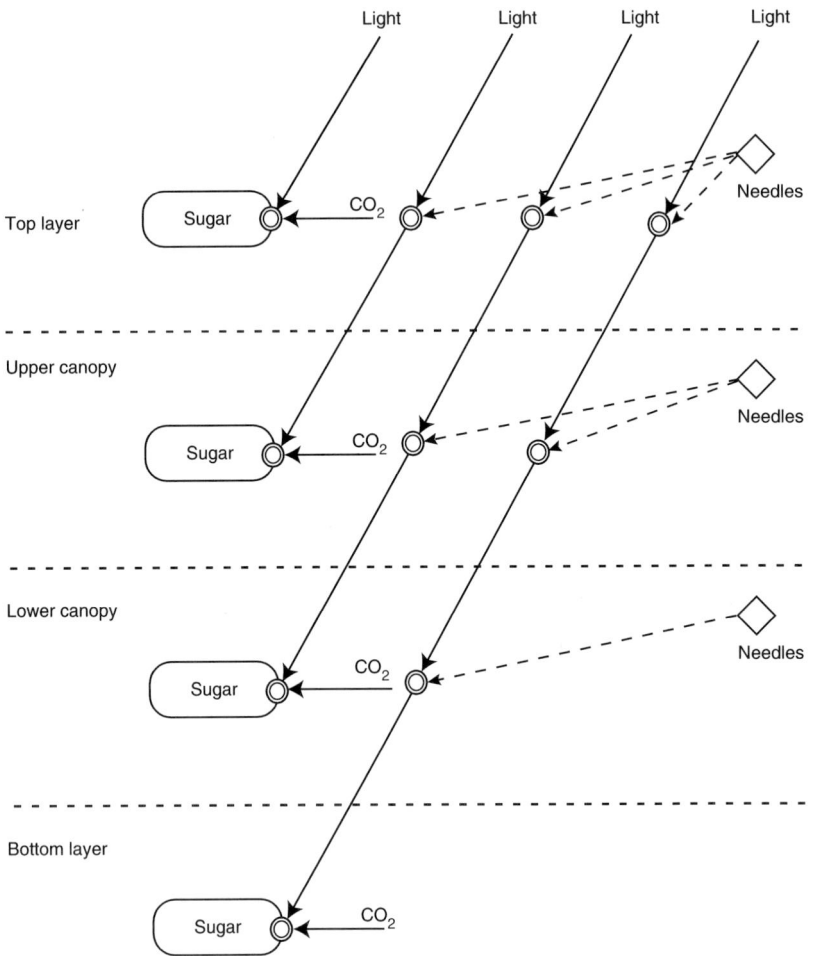

Fig. 7.5.1 Schematic presentation of the interactions between foliage, light and photosynthetic rate in a tree canopy. Boxes indicate amounts, arrows (solid line) flows, double circles processes, and arrows (dash line) effects on processes

among all the environmental factors affecting tree photosynthesis. Within-canopy variation of other driving factors (air or leaf temperature, VPD and CO_2 concentration) is smaller although they may still have important roles especially in very dense canopies.

Canopy structure modifies the light environment within the canopy but the structure itself also depends on the light environment. At the top of the canopy, irradiances are high and solar rays can reach the foliage with little obstruction regardless of the angular distribution of radiation. In the lower parts of the canopy the availability of light is much lower. The rays entering at high zenith angles (originating near the horizon) travel a longer distance inside the light-absorbing foliage than light entering from the zenith. Thus, when travelling deeper down the canopy the angular distribution of incident light also becomes more concentrated around the zenith. In shaded conditions, effective utilisation of the low irradiances is essential, whereas for the leaves at the top of the canopy light interception is less crucial.

Spatial variation in photosynthetic properties and environmental driving factors increases when moving from a small element of leaf surface area towards a larger spatial scale. This additional variation must be taken into account when determining the photosynthetic production of a whole tree or a stand. The photosynthetic rate at a point in space and time is determined by environmental factors and the physiology and structure of the leaf. The results obtained in Sections 6.3.2.3 and 6.3.3.1.3 in Chapter 6 can be utilised in the determination of photosynthetic production of a tree with integration over space and time as shown in Section 7.1.

The analysis in Section 6.3.2.3 resulted in a model for the dependence of photosynthetic rate on light, I, and water vapour concentration difference between mesophyll space and ambient air, D. In addition the model includes parameters β, γ, λ, and c_{max}. At this point we assume that the physiology and structure of the leaves do not vary in time or within the canopy and the values of the corresponding model parameters are constant. Thus

$$p(t,x) = p(I(t,x), D(t,x), \beta, \gamma, \lambda, c_{max}) \qquad (7.5.1)$$

Photosynthetic production by a tree, $P(t_1, t_2)$, is the substrate for maintenance and growth. Integration over space and time (see Section 7.1) is the bridge between process rates and produced amounts of metabolites. Thus

$$P(t_1, t_2) = \int_V \int_{t_1}^{t_2} \rho_m(x) p(I(t,x), D(t,x), \beta, \gamma, \lambda, c_{max}) dt \, dV \qquad (7.5.2)$$

where ρ_m is the spatial distribution of needle mass in the canopy (Section 5.2.5).

In practice, numerical methods have to be used in the integration of photosynthetic production over space and time (see Section 7.1, Eq. 7.1.4). The canopy is split into smaller elements where environmental factors can be assumed to be sufficiently constant; the photosynthetic rate is calculated for each element and finally integrated over the canopy. The time interval from t_1 to t_2 is divided into N subintervals of equal length and the volume of tree canopy to M subvolumes of equal size

$$P(t_1,t_2) \approx \sum_{j=1}^{M} \sum_{i=1}^{N} \rho_m(x_j) p(I(t_i,x_j), D(t_i,x_j), \beta, \gamma, \lambda, c_{\max}) \frac{V}{M} \frac{t_2 - t_1}{N} \qquad (7.5.3)$$

This kind of numerical integration method is often called upscaling. Ideally, the basic element should be so small that the spatial variation in driving factors within one element can be ignored, but in practice this is not possible, especially when light in the canopy is considered. Calculating photosynthesis thus requires several simplifying assumptions, and the integration method is always a compromise between the accuracy of the information on the canopy structure, the variation of photosynthetic properties, and the environmental driving factors within the canopy.

Determination of photosynthetic production of a tree can be done in several ways. In the simplest case, the environmental gradients in foliage are ignored and the whole stand is considered as a giant leaf without any physical structure (Fig. 7.5.2A). This approach, so-called big-leaf modelling, omits relationships between canopy structure, within-canopy variation of the driving factors, and CO_2 exchange. Irradiance and other environmental variables measured at one point, typically above the canopy, are used as the drivers for photosynthesis. The effect of canopy structure on light environment is only implicitly embedded in photosynthetic light response parameters. Thus it cannot be considered as an integration method. Big-leaf models are typically used in deriving simple relationships between, e.g., irradiance and stand CO_2 exchange to fill gaps in eddy-covariance measurements or making estimates of stand photosynthesis from satellite or airborne measurements of foliage area and absorbed light. The ecosystem light-use efficiency study reported in Section 7.6.2.4 also employs a big-leaf model of stand photosynthetic production.

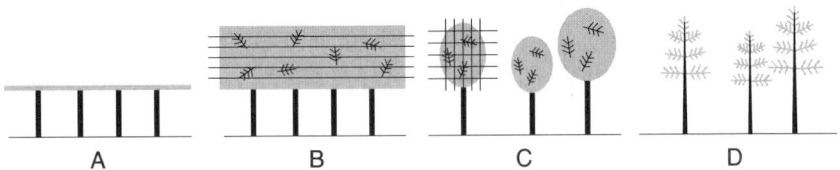

Fig. 7.5.2 Different ways to describe canopy structure in calculating the distributions of the environmental driving factors in the canopy and the rate of tree or stand photosynthesis. Foliage is indicated with grey colour. (A) Big-leaf approach considers the whole forest stand as a giant leaf. Incident light is used as the driving factor for photosynthesis, there is neither vertical nor horizontal variation in light (B) Homogeneous canopy without individual trees but vertical gradient in light. The canopy can be further stratified vertically into layers that each receive different amount of light depending on how much there is shading foliage area above the layer (C) Individual trees with crowns consisting of homogeneous matter, vertical and horizontal gradient in light. The crowns can be divided into volume units. Light in each volume unit is calculated from shading by the other volume units and shading by neighbouring trees (D) Individual trees with explicitly defined three-dimensional branch and shoot structure

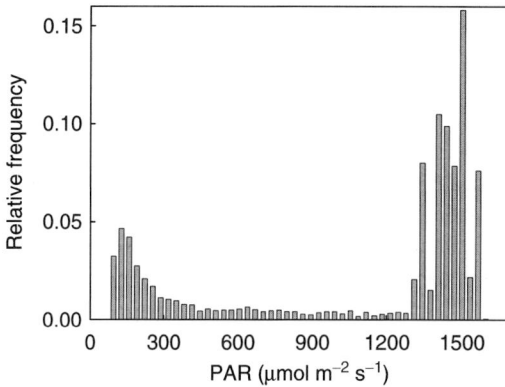

Fig. 7.5.3 Distribution of irradiance (PAR) inside a Scots pine canopy on a clear summer day. Irradiance was measured at 48 locations in the lower part of the canopy at 30 s intervals for 1 h. The average incident PAR was 1,560 μmol m^{-2} s^{-1}

In a typical application for integrating tree photosynthesis, the canopy is assumed to consist of a homogeneous medium. Let $D_{PAR}(h, t, I)$ denote the distribution of PAR at height h in the canopy at the moment t. In case of a horizontally homogeneous canopy with a vertical leaf mass distribution ρ_m, photosynthetic production is calculated as

$$P(t_1,t_2) = \int_0^{h_{max}} \int_{t_1}^{t_2} \rho_m(h) \int_0^{I_{max}} D_{PAR}(h,t,I)p(I,D(t,h),\beta,\gamma,\lambda,c_{max})dI\,dt\,dh \quad (7.5.4)$$

The integral above is usually calculated numerically. The canopy is stratified to horizontal layers h_i (Fig. 7.5.2B). In each canopy layer, irradiance forms a discrete distribution $D_{PAR}(h_i, t, I_j)$ that gives the probability of a leaf surface element in the layer i to fall into irradiance category I_j (Fig. 7.5.3). Instantaneous photosynthetic rates corresponding to each irradiance category are summed up, weighted by the probability of that category, and multiplied by the leaf mass or area in the layer to obtain the photosynthetic production of the layer. Photosynthetic production of the whole canopy is the sum of photosynthetic production in all layers.

Light Climate Inside the Canopy

The main differences between the various integration approaches in determining canopy gas exchange are related to the determination of light distributions in different parts of the canopy and in the description of the structure. In the tree canopy, there are points that are fully illuminated by direct solar radiation (sunflecks), points that only receive diffuse radiation, and areas of intermediate irradiance (penumbra) where part but not the entire direct beam enters the observation point. The irradiances over all foliage surface elements form a continuous distribution (see Section 7.3).

As a first approximation, the irradiance distribution can be reduced to one mean value of irradiance. Irradiance at any given point is calculated from above-canopy radiation as a function of shading canopy elements above the observation level, using the Lambert-Beer law of extinction (Eq. 7.3.2). The averaging tends to overestimate photosynthesis because of nonlinearity of the photosynthetic light response.

Attenuation of a beam of light depends heavily on the angle of incidence: for light coming near the horizon at low elevation angles, the distance that the rays travel in the canopy is longer than at high elevation angles. Therefore the irradiance at any given point inside the canopy also depends on the angular distribution of incident light. The accuracy of light environment calculations can be improved if the angular distribution of incident light is considered. The first step here is separation of total irradiance to direct and diffuse components. The direct beam has direction and intensity whereas the diffuse radiation field can be considered evenly distributed over the sky hemisphere (uniform overcast) or it can have non-even angular distribution. Attenuation of the radiation components is calculated separately and finally the component irradiances at the observation point are summed up. This approach better takes into account the different optical path lengths for direct and diffuse radiation. The averaged irradiance at the observation point still leads to overestimation in photosynthesis.

A simple and often used method of working around the problem of averaging irradiance is division of foliage area into leaves or shoots that are illuminated by both diffuse light and direct beam (sun leaves or shoots), and shade leaves or shoots that are only receiving diffuse light. The fractional area of sunflecks decreases and the shaded area increases when moving deeper down the canopy. If the sun is treated as a point source of light, the transmission probability of the direct beam can be satisfactorily approximated with the Lambert-Beer equation and the fractional area of sunflecks equals the probability for a direct beam to reach the observation level. The intensity of diffuse radiation also decreases following the Lambert-Beer equation.

The accuracy of determining the light environment can be further improved by treating the irradiance at each level as a distribution instead of a single value or a pair of shade and sunfleck irradiances. The distributions can be determined empirically from measurements at several locations inside the canopy. For example, Ross et al. (1998) developed empirical relationships between shading foliage area and irradiance distributions in a willow coppice. From the measured distributions, the intensities of direct and diffuse components as well as penumbra were derived empirically assuming normal distribution for sunflecks and shade. Irradiances in penumbra were distributed fairly evenly between the irradiance values in sunfleck and shade regions and approximated with beta distribution. Inside the Scots pine canopy at SMEAR II the measured distributions were similar (Vesala et al., 2000). An example of distributions measured at different heights in the canopy is shown in Section 7.3. Irradiances at each height form a bimodal distribution that becomes dominated by low irradiance values when moving down the canopy.

The complicated empirical distributions were further simplified to functions for fractional areas of sunflecks, penumbra and shade, and mean irradiances within these light regimes as functions of shading needle area above the observation level

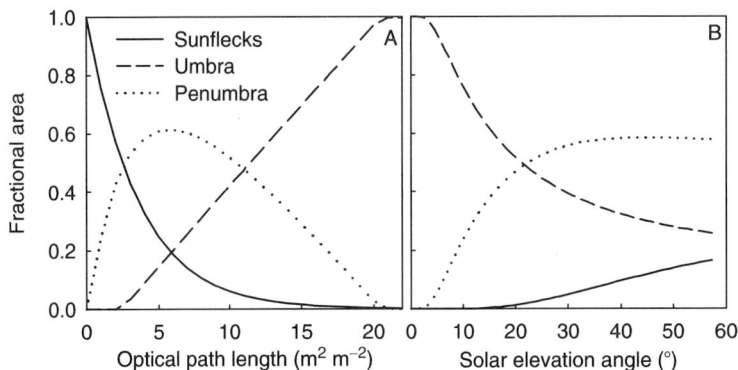

Fig. 7.5.4 The empirical irradiance model based on irradiance distributions measured inside the Scots pine canopy and at the forest floor at SMEAR II (Vesala et al., 2000; Kolari et al., 2006). (A) Fractional areas of sunflecks, umbra and penumbra in the canopy as a function of optical path length. The optical path length equals the all-sided needle area above the observation level (m^2 per m^2 ground) divided by the sine of solar elevation angle. (B) Fractional areas of sunflecks, umbra and penumbra below the tree canopy as a function of solar elevation angle, when the all-sided leaf area (optical path length in vertical direction) of the canopy is 8 m^2 per m^2 ground

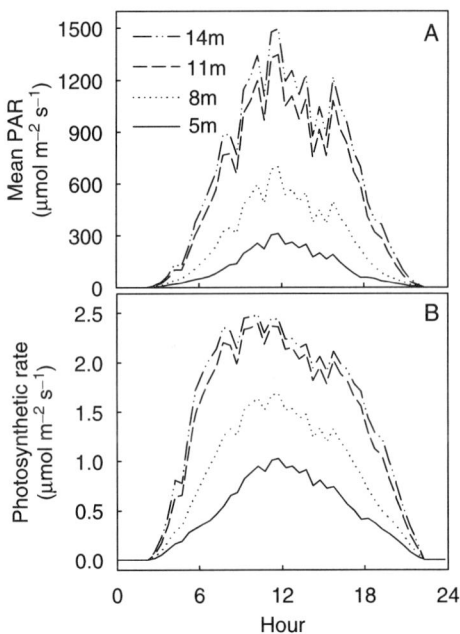

Fig. 7.5.5 Diurnal cycles of mean irradiance (A) and photosynthetic rate per unit all-sided needle area (B) at different heights in the canopy (determined with sun-shade-penumbra model) in 30 min time steps on a summer day (2 July 2004) in SMEAR II pine stand

(Fig. 7.5.4). Different variants of this model have been utilised for calculating light environment and photosynthesis of the Scots pine canopy (Rannik et al., 2002) and forest floor vegetation in the SMEAR II stand (Kolari et al., 2006). Mean irradiances and photosynthetic rates calculated with the empirical model for different heights in the Scots pine canopy at the SMEAR II stand are shown in Fig. 7.5.5.

The mean light in the lowest canopy layer is about one third of the light measured at the top of the canopy. Due to the saturating light response, the photosynthetic rate in the bottom is relatively higher, about half of the photosynthetic rate of the top shoots.

The problem in this kind of empirical distributions is that the shapes of the distributions cannot be bound mechanistically to canopy structure, so the results apply for similar stands only. A more precise determination of irradiance distributions would require spatially explicit calculation of light environment, based on detailed information on canopy structure, clumping and orientation of leaves, shoots and branches (Fig. 7.5.2D). Due to the complexity of such calculations and the structural information needed, there are only a few operational spatially explicit models for canopy light environment and photosynthesis (e.g., Perttunen et al., 1995). In order to avoid time-consuming numerical ray-tracing algorithms, the canopy structure can be idealised to mathematical functions that define the crown shape and also take into account the asymmetry of crowns and the irregular distribution (clumping) of foliage mass inside the crowns (e.g., Cescatti, 1997).

Usually the distribution of irradiances over all foliage surface elements, be it represented by a single intensity value or discrete or continuous distribution, is calculated for a horizontal plane or for a plane perpendicular to the direct beam. In tree canopies the leaves are arranged at different angles. At any moment of time, some of the leaves are parallel to the direct beam and virtually receive diffuse light only whereas some leaves are facing the direct beam. Therefore, the actual irradiance distribution over the surfaces of all leaves with more or less random orientation within the canopy is different from the irradiance distribution on a two-dimensional plane. This is especially important in conifers that have complex three-dimensional shoot architecture.

The irradiance distribution over the foliage surface also depends on the angular distribution of light. In natural conditions when light intensity changes, the angular distribution of light also changes. At high irradiances most of the radiation energy is in the direct beam whereas at low irradiances diffuse radiation dominates. Diffuse radiation is more efficient because it illuminates a larger proportion of the needle surface area. Due to the saturating light response of photosynthesis, a shoot with large leaf surface area receiving low light has a higher photosynthetic rate than a shoot with small leaf surface area receiving high light (Allen et al., 1974).

In forest stands, foliage is grouped not only in shoots and branches, but also in tree crowns. Crown dimensions depend on species, within a species the dimensions vary according to stand density. In dense stands individual crowns are narrower and canopy height to tree height ratio is lower than in sparse stands. When the leaf area of the crown is fixed, in a larger crown the canopy elements are more sparsely arranged which decreases self-shading and increases average light intensity on leaves and eventually leads to higher photosynthetic production (Duursma and Mäkelä, 2007).

Conversion of the values of photosynthetic parameters from leaf or shoot to the canopy is an often overlooked part of the integration procedure. The experimental shoots are illuminated perpendicularly by an artificial light unit, or they can be placed in fixed position where they receive natural light. Irradiance is measured with a single or a few sensors placed horizontally next to the leaf or shoot. The changes in shoot and irradiance geometry when moving from experimental leaves or shoots to the free shoots in the crown should be taken into account when considering canopy photosynthesis.

An approach that avoids dealing with the complex angular distributions and geometrical conversion of photosynthetic parameters is considering the amount of light intercepted by the leaf or shoot instead of irradiance (e.g., Stenberg, 1995; Thornley, 2002). The measured irradiance is converted to intercepted radiation and averaged over the all-sided leaf area. The intercepted radiation can be estimated using the concept of STAR (shoot silhouette area by total needle area) by Oker-Blom et al. (1989). The silhouette area is the projection of the shoot on a plane perpendicular to the beam of light. It depends on the angle between the shoot and the beam. Therefore, the distribution of different shoot orientations in the canopy must be estimated to calculate average STAR. The concept of intercepted radiation is used in the canopy light model of Stenberg (1996). In this model the canopy foliage mass is distributed evenly inside the individual crowns, i.e., there is no explicitly defined structure inside the crowns (Fig. 7.5.2C). The internal structure of the crown is condensed into an aggregated parameter k, which is equivalent to the Lambert-Beer light extinction coefficient. The tree crown is divided into volume elements and the light intercepted by the shoots in each volume element is calculated. We will utilise this kind of light calculation combined with shoot photosynthesis model (PhenPhoto, Section 6.3.2.3) in Section 7.6.2 for estimating the photosynthetic production of SMEAR II stand.

Because of the nonlinear response of photosynthesis to irradiance, the concept of intercepted light averaged over the needle surface is not entirely satisfactory, especially in conifers that have complex shoot structure. Computing irradiance distributions over all infinitesimally small surface elements is, however, not trivial due to within-shoot shading (Stenberg, 1995). Therefore, it is often more convenient to use the shoot as the basic functional unit for gas exchange measurements and integration of photosynthesis over the canopy (Gower and Norman, 1991) and either ignore the geometrical conversion or use the intercepted radiation instead of irradiance over the leaf surfaces.

The overestimation of photosynthesis resulting from the averaging of irradiance decreases with the complexity of determining the irradiance distribution. Apparent light response of canopy photosynthesis also becomes more clearly saturating when moving from horizontal irradiance to irradiance distribution over the three-dimensional leaf surface area. In Fig. 7.5.6 the apparent light response of the canopy photosynthesis is modelled with three different methods to calculate light distribution: averaged Lambert-Beer light on a horizontal plane, an empirical irradiance distribution model (Kolari et al., 2006) that calculates fractional areas and corresponding mean irradiances in sunflecks, shade and penumbra on a horizontal plane, and a modified version of the model that takes into account the directional distribution of light inside the canopy and calculates the irradiance distribution over the surfaces of imaginary spherical shoots. The canopy was divided into layers of 1 m height, irradiances or irradiance distributions were then calculated separately

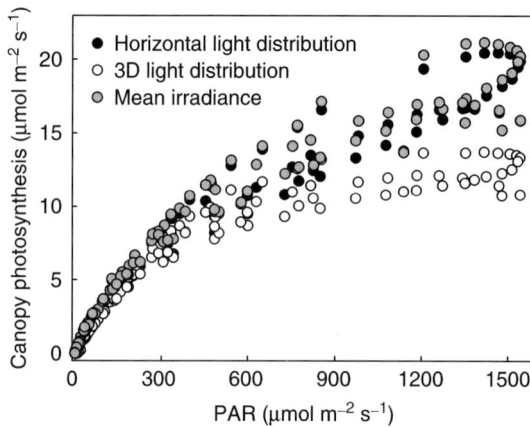

Fig. 7.5.6 The relationship between total photosynthetically active radiation above the canopy and canopy photosynthesis modelled with different light calculation methods: (1) average light intensity on a horizontal surface, calculated with Lambert-Beer law of light extinction, (2) sun-shade-penumbra differentiation, light calculated for a horizontal surface, (3) three-dimensional irradiance distribution over spherically distributed foliage area. The canopy was divided into layers, irradiances or irradiance distributions were calculated separately for each layer, and photosynthetic rate was calculated for each layer with the optimal stomatal control model (Section 6.3.2.3) using the calculated irradiance as the driving factor. The radiation data consists of half-hourly averages of actual measurements on a sunny midsummer day at SMEAR II. The proportions of direct and diffuse light at each irradiance level were measured and correspond to natural radiation environment, i.e., proportion of direct radiation increases with irradiance. Photosynthetic parameters were common to all models

for each layer. The photosynthesis model and its parameters were common to all cases (the optimal stomatal control model, Section 6.3.3.1.3), so the variation in the model output originates only in the different treatment of light environment. The differences of the graphs show that determining the light environment is at least as crucial as the accuracy of the photosynthetic parameters in integrating canopy photosynthesis.

Acclimation

Leaves acclimate to the prevailing environment, especially to the light climate. At shoot level, acclimation to light environment can be achieved in two different ways, via changes in leaf and shoot structure or via changes in functional substances. The shoot-level acclimation to the prevailing light conditions involves changes in shoot structure and orientation. In Scots pine, sun shoots typically are more vertically inclined and the angle between the twig and the needles is smaller than in shade shoots, whereas shade leaves are nearly horizontally arranged and more efficiently intercept light entering vertically through the canopy. Needles acclimated to high light are thicker, i.e., needle mass to surface area ratio is higher than in shade needles. Stenberg et al. (2001) found that in Scots pine the specific needle area varies from roughly $10\,m^2\,kg^{-1}$ in sun shoots up to $20\,m^2\,kg^{-1}$ in shade shoots. They also showed that shoots located lower down in the canopy have clearly higher projected to all-sided needle surface area ratio, which means less within-shoot shading than in sun shoots. This enhances light interception and allows a more efficient light utilization to compensate for the low availability of light.

Shade leaves typically have high photosynthetic efficiency at low light levels but more pronounced saturation of photosynthesis at high irradiances than in leaves higher up. Palmroth and Hari (2001) tested this on Scots pine at SMEAR II by measuring CO_2 exchange of pine shoots in the laboratory. The shoots were detached and were immersed to provide sufficient water. The differences in photosynthetic behaviour were relatively small. Photosynthetic light responses in sun and shade shoots were similar. CO_2 uptake was higher in sun shoots when based on needle surface area but there was no difference when the photosynthetic rate was expressed per unit needle mass. Similar conclusions can be drawn from field measurements of CO_2 exchange at SMEAR II with chambers located in the lower parts of the canopy (Palva et al., 2001). These measurements indicate that photosynthesis in shade shoots indeed saturates at lower irradiances than in sun shoots (Fig. 7.5.7). The photosynthetic rates in the graph are expressed on surface area basis. Specific leaf area in the shade shoot was about 50% higher than in the sun shoot, so photosynthetic rates per unit needle mass were similar (Fig. 7.5.7B). The difference in photosynthetic rate between sun and shade shoots seems to be fairly small in the relatively open canopy of Scots pine, however, in denser broadleaved canopies the difference between sun and shade leaves may be considerable (e.g., Kull and Niinemets, 1998).

The concentrations of functional substances and the structure of leaves and shoots vary within the canopy due to acclimation. However, the physical basis of CO_2 transport into and within leaves by diffusion and the biochemical reactions are the same for all leaves. This means that the theory linking environmental factors and photosynthesis is valid everywhere in the canopy, but the parameters may have location specific values. Thus acclimation can be introduced into the analysis as follows

$$p(t,x) = p(I(t,x),D(t,x),\beta(x),\gamma(x),\lambda(x),c_{max}(x)) \qquad (7.5.5)$$

The photosynthetic production by a tree can be obtained with integration over space and time, analogously to the previous case without acclimation of photosynthesis. The only change is that the parameters in the model are no longer constants, but they are location specific. This can be formalised as follows:

$$P(t_1,t_2) = \int_V \int_{t_1}^{t_2} \rho_m(x)p(I(t,x),D(t,x),\beta(x),\gamma(x),\lambda(x),c_{max}(x))dt\,dV \qquad (7.5.6)$$

The value of the integral is obtained with sums, analogously to the Eq. 7.5.3

$$P(t_1,t_2) \approx \sum_{j=1}^{M}\sum_{i=1}^{N} \rho_m(x_j)p(I(t_i,x_j),D(t_i,x_j),\beta(x_j),\gamma(x_j),\lambda(x_j),c_{max}(x_j))\frac{V}{M}\frac{t_2-t_1}{N}$$
$$(7.5.7)$$

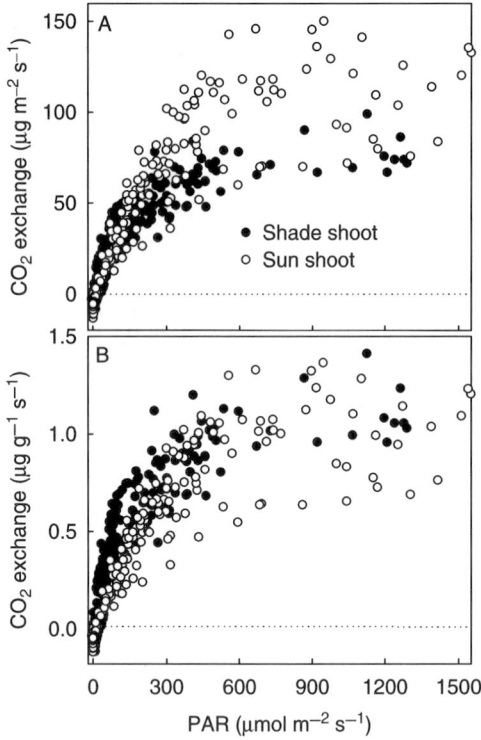

Fig. 7.5.7 CO_2 exchange vs. irradiance in a Scots pine shoot located at the top of the canopy and in another shoot in the lower canopy at SMEAR II during three days in June. The upper panel (A) illustrates CO_2 exchange per unit needle surface area and the lower panel (B) CO_2 exchange per unit dry needle biomass. Gas exchange rate per unit needle mass is relatively higher in the shade shoot because the needles are thinner and the ratio of needle mass to surface area correspondingly smaller. In the shade shoot graphs the irradiance is the average of about 200 PAR sensors inside the shoot silhouette area

Based on the chamber measurements at lower heights in the canopy, the integration of canopy photosynthetic production was carried out with two parameter sets: one with constant photosynthetic parameters and another with a stronger light saturation ($\gamma = 600\mu mol\ m^{-2}\ s^{-1}$) and correspondingly lower photosynthetic efficiency β and maximum stomatal conductance c_{max} (60% of the values obtained from the top cuvettes) in the lowest shoots. The parameters were assumed to decrease linearly as a function of shading needle area. Canopy photosynthesis was then calculated in time steps of 30 min. The effect of stronger light saturation and lower photosynthetic efficiency and maximum stomatal conductance on daily photosynthetic production in the midsummer was about 8%. In autumn the difference was only about 4%.

Stand photosynthesis only changes a little when the curvature of photosynthetic light response is considered as a function of light environment. Light in the lower parts of the canopy is most of the time below saturating intensity (see Fig. 7.5.4), i.e., the initial slope of the light response (quantum yield) largely determines photosynthetic production. The acclimation of light response can thus be thought of as avoiding the costs of maintaining high photosynthetic capacity rather than just maximising photosynthetic production.

When calculating tree or stand photosynthesis, an accurate determination of light interception and within-canopy variation of light intensities (see comparison in Fig. 7.5.6) is in general more crucial than taking into account small variations of light-saturated photosynthesis in the canopy. In practise the selection of light calculation method is determined by the available information on crown structure. In most cases sufficient information for properly applying Eq. 7.5.7 is lacking, the distribution of foliage in tree crowns is assumed homogeneous and the within-canopy variation in the response of photosynthesis to the driving factors ignored. Such an approach will be adopted in Section 7.6.2 for estimating the photosynthetic production of SMEAR II stand.

Conclusion

The photosynthetic production of a tree can be obtained by integration, the main problems being involved in the proper determination of the light climate inside the canopy.

7.5.2 Long-Distance Transport of Water and Solutes in Trees

Eero Nikinmaa[1], Martti Perämäki[1], Teemu Hölttä[1], and TimoVesala[2]

[1] University of Helsinki, Department of Forest Ecology, Finland
[2] University of Helsinki, Department of Physics, Finland

In trees, the active organs are located far apart from each other and still they need to be able to maintain sufficient supply of water, nutrients and assimilates between them. The maximum distance between locations of water transpiration and uptake can be several tens of meters and the amounts of transpired water are large (Sections 6.3.2.2 and 6.3.2.8). Thus efficient water transport within a tree is a very important determinant of tree performance. Also the maximum distance between locations of sugar production and consumption is of the same magnitude than that of water, although the amounts of sugars to be transported are one or two magnitudes smaller than those of water (Sections 6.3.2.2, 6.3.2.3 and 6.3.3.1.4). Nevertheless, an efficient sugar transport system is similarly needed for proper functioning of a tree.

7.5.2.1 General Mechanisms

Water and solute transport in trees may occur over distances of several tens of meters to more than 100 m in the largest trees. It is clear that over such large distances transport needs to rely on bulk flow of material. Two parallel conduit systems are responsible for water, nutrient, and assimilate product transport in plants (see Section 5.2.3 for detailed description). Transpiration-driven water flow occurs through the non-living tissue in xylem from soil to leaves. This bulk flow of water also transports with it dissolved solutes from roots to leaves. Transport of assimilation products and solutes from mature leaves to the stem, roots, immature leaves, and other organs occurs via the phloem. They are transported in the phloem along a continuous pathway of living sieve space made out of sieve elements (sieve cells in conifers) that are connected to each other either by sieve plates or by small pores.

According to the cohesion theory, evaporation at the leaf surfaces introduces a water tension gradient through a continuous pathway of water conduits extending from leaves to the roots due to cohesive forces between water molecules (Fig. 7.5.8). As the water is under tension in the xylem it has been said to be (Tyree and Zimmermann, 2002) in a metastable state. On the other hand, the driving force and mechanisms for phloem transport is an osmotically created turgor pressure gradient as first suggested by Ernst Münch in the 1930s. Accordingly to the nature of the driving pressure gradient, the major threat for xylem flow is a leakage (or spontaneous formation) of air bubbles into the xylem elements that breaks the connections between the water molecules while in the phloem flow, ruptures in the sieve elements allow the sap to flow out similarly to, e.g., the blood circulation system in humans.

Transpiration, photosynthesis, carbon allocation and growth, tree structure, and water flow are all linked to each other. Transpiration induces the sap flow where liquid water is pulled from soil through the stem and branches up to the leaves where it evaporates in the air. Roots uptake water and dissolved nutrients from soil and the wooden structure serves as a pathway for water flow from roots to shoots. Soil water availability, hydraulic conductance of the transport pathway and the evaporative demand of ambient air, influences the water status of the leaves, which, in turn, can impose a limitation to gas exchange by controlling the stomata (Hari et al., 1999). Stomatal control has an effect on the carbon gain and thus on the growth of the tree. The hydraulic pathway has to be constructed in such a way that it can feed the necessary amount of water to the leaves. However, stomatal control influences also the phloem transport of assimilates from the leaves to the sites of active growth. In leaves, the hydrostatic pressure due to transpiration in xylem needs to be in balance with the osmotically created sum of the hydrostatic and osmotic pressures in the symplast. Stomatal opening controls both the hydrostatic pressure in leaves and the potential for osmotic adjustment by influencing the carbon gain. Furthermore at sites where carbon is used in growth, water from phloem recirculates back to xylem and may have a role in refilling the possible air-filled xylem elements (Höltta et al., 2006). Thus the functioning of, and linkages between, the xylem and phloem are of central importance in the success of trees.

The term "water potential"

Water potential is defined as the difference in free energy or chemical potential per unit molal volume relative to pure bulk water at atmospheric pressure. Water potential is divided into the following components: the hydrostatic pressure (ψ_p) and the osmotic pressure (ψ_o) arising from dissolved solutes which lower the free energy of water:

$$\psi = \psi_p + \psi_o$$

Water potential may also contain other components such as the matric and electric components, but they are not relevant here. The unit of water potential is that of energy per volume [$J\ m^{-3}$], i.e., not energy as the term potential might indicate. The unit is the same as that of pressure [$Pa = F\ m^{-2} = J\ m^{-1}m^{-2} = J\ m^{-3}$]. In this book we refrain from using the term water potential as all of the phenomena can be readily explained in terms of pressure. In axial xylem and phloem sap flow the direction and magnitude of the flow follows the hydrostatic pressure gradient. Osmotic effects do not contribute to the driving gradient due to the absence of semipermeable membranes even when solutes are present. When semipermeable membranes are encountered, as is the case for example with radial water exchange between the xylem and phloem, the direction and magnitude of the water flow follows the gradient of sum of the hydrostatic and osmotic pressure.

7.5.2.2 The Xylem Flow

Transpiration is an inevitable consequence of photosynthesis as has been explained. When the stomata in the leaves are open for carbon dioxide uptake, water molecules diffuse from intercellular air spaces of leaf (RH 100%) through the open stomata to ambient air. The driving concentration difference can be up to a magnitude of 1 mol m^{-3}. Taking into account the maximum concentration difference of carbon dioxide ($0.0045\ mol\ m^{-3}$) and also the ratio of diffusitivies of these gases in air (D_w/D_c ≈ 1.6, Lawlor, 1993) we get that for each carbon dioxide molecule harvested from air up to 400 molecules of water may be lost. The quite substantial transpiration per unit leaf area together with total leaf area up to hundreds of square meters produces a significant flux of water at tree level. The diurnal water use of a single tree can exceed 1,000 kg (For a review, see Wullscheleger et al., 1998).

The physical basis for water ascent in plants derives from the structure of a water molecule, where an oxygen atom and two hydrogen atoms form an asymmetrical composition (Fig. 7.5.9). This asymmetrical composition with opposite electrical charges of oxygen and hydrogen ions leads to a dipolar structure, which enables water molecules to attach to each other (cohesion) and form clusters of molecules.

Fig. 7.5.8 Schematic presentation of water flow in the stem. The symbols are introduced in the Fig. 1.2.1

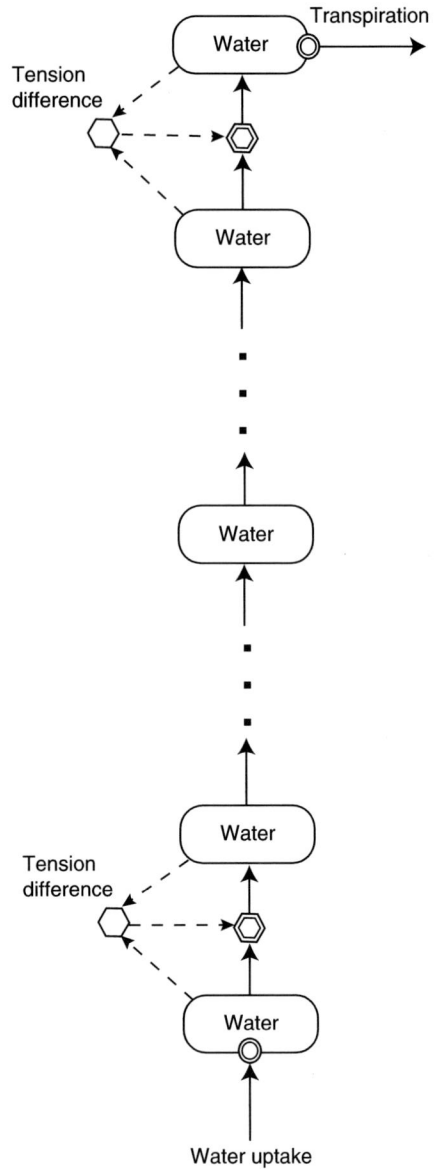

When water molecules are attracted to charged surfaces (wettable) of other substances, it is called adhesion.

Water flow through sapwood of a tree is initiated in intercellular air spaces of leaves from where water molecules diffuse to ambient air (Fig. 7.5.10). This causes a net evaporation of water molecules from water surfaces on parenchyma cells and

Fig. 7.5.9 A schematic illustration of the composition of a water molecule

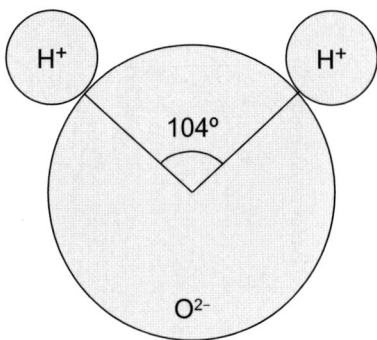

Fig. 7.5.10 Schematic illustration of a curved surface of water film on parenchyma cells inside a leaf

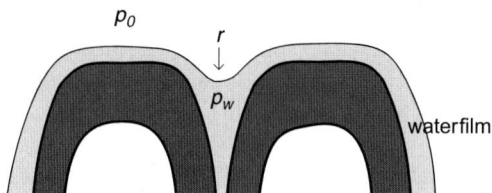

increases the concavity of the surface of the water film between cells (Fig. 7.5.10) or into pores in cell wall. Surface tension (relatively high on water, 0.073 N m^{-1}) which results from asymmetric forces between water molecules at the surface, tends to minimize the area of the surface and pulls the water molecules towards the surface. This results in pressure drop in water below the curved surface. The pressure drop is a function of the radius of the curvature and surface tension of water (Laplace equation (Pickard, 1981)):

$$p_W = p_0 - \frac{2T}{r} \qquad (7.5.8)$$

where p_W is the pressure under a curved surface of water, p_0 is the ambient pressure, T is the surface tension of water, r is the radius of the curved surface. The pull, i.e., tension between water molecules propagates along the continuous water chain down to the soil and cause them to rise through the tracheary elements of the stem. The hydrogen bonding between adjacent water molecules keeps the molecules together as a continuous chain. This is the cohesion – tension theory of sap flow in trees, which was presented already in 1894 by H.H Dixon and J. Joly (Pickard, 1981).

The cohesion theory implies that a tension gradient of $0.02\text{--}0.03 \text{ MPa m}^{-1}$ should exist inside xylem when transpiration is present (0.01 MPa m^{-1} due to gravity and $0.01\text{--}0.02 \text{ MPa m}^{-1}$ because of frictional pressure losses in flow). This leads to tension of a couple of megapascals at the top of tallest trees. This predicted tension is the reason why the cohesion – tension theory remained controversial for decades: nobody had convincingly demonstrated the existence of negative pressure of water in plants and how it is possible that water columns can remain continuous under high tension for long periods of time without breaking into separate

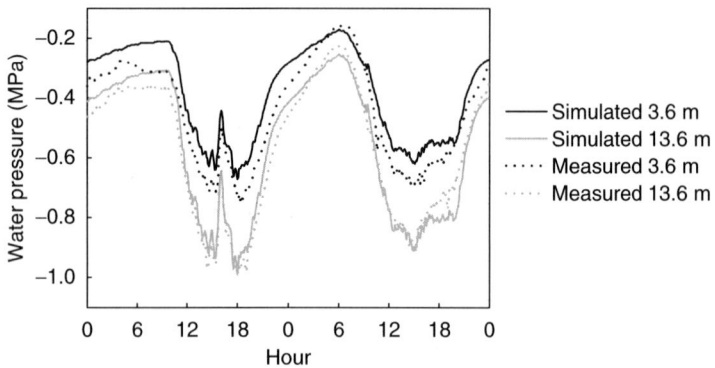

Fig. 7.5.11 Measured (dots) and simulated (lines) variation in stem water pressure at the heights of 3.6 (black) and 13.6 m (grey). The water pressure fluctuation is derived from diurnal diameter variation according to the Hooke's law. The modeled water pressure is be derived from pressure propagation from leaf transpiration according to Perämäki et al. (2005)

molecules. This intuitively unavoidable phase change should lead to an embolism when a conduit is filled with gas, and water flow would then be impossible.

While connected to each other with cohesive forces, the molecules in a stretched water column are also attached to the walls of tracheary elements with adhesive forces. With this mechanism the tension between water molecules is transferred to the elastic cell walls, they bend inwards and the diameter of the cell shrinks. On the scale of a stem (a bundle of individual cells) this adds up in sapwood to a diameter change of a detectable order of magnitude (Irvine and Grace, 1997; Perämäki et al., 2001). Figure 7.5.11 presents the time course of a sapwood diameter of a Scots pine in Hyytiälä during two days in August, 2000. Note that the shrinking and swelling of external elastic tissue is not included in the measurement. The magnitude of the changes in diameter and elastic modulus are consistent with the high pressures predicted by the cohesion theory and thus shows direct support for it.

The cohesion forces keep the water column continuous, and adhesion binds it to the cell walls of the conducting elements. Transpiration in the foliage creates a water tension gradient that initiates sap flow. Because of wood elasticity, wood volume decreases under tension. The water tension can be derived from the diameter decrease in the stem and branches by using Hooke's law (Irvine and Grace, 1997). Ignoring the effect of cavitation of water columns within the xylem, the change in wood volume equals the net change in the amount of water.

Let us consider a tree stem as a tapering pipe of elastic porous material with radial elastic modulus $E = E(h)$, where h is stem height, and permeability $k = k(h)$ and sapwood area $A_{sw}(h) = \pi(r^2 - r_{hw}^2)$, where $r = r(h)$ is stem radius and $r_{hw} = r_{hw}(h)$ is the radius of the heartwood. No changes in water saturation are assumed, i.e., embolism is not considered.

Sap flow g_{H20} is driven by the pressure gradient (Darcy's law)

$$g_{H20} = -\frac{k}{\eta}\left(\frac{dP_W}{dh} + \rho g\right)A \qquad (7.5.9)$$

where dP_W/dh is pressure gradient over distance h and ρg is the effect of gravity. The radius of the stem changes with change in pressure (Hooke's law),

$$\frac{dr}{dP_W} = \frac{(r - r_{hw})}{E} \qquad (7.5.10)$$

The water mass balance equation of a stem segment:

$$\frac{dm^i_{H20}}{dt} = g^{in,i}_{H20} - g^{out,i}_{H20} - g^b_{H20} \qquad (7.5.11)$$

where m^i_{H20} (kg) is the mass of water of an element i, $g^{in,i}_{H20}$ (kg s^{-1}) is the inward mass flow rate and $g^{out,i}_{H20}$, i is the outward mass flow rate, g^b_{H20} is the water sink caused by water flow to branches.

The above equations can be applied to a simplified presentation of a tree in which the stems and branches are divided into a chain of small elements of constant length.

The above equations can also be reduced to a single partial differential equation of water pressure propagation inside a tapering tree stem (Perämäki et al., 2005).

$$\frac{\partial p_W}{\partial t} = \left(\frac{Ek}{2\eta}\left(1 + \frac{r_{hw}}{r}\right)\right)\frac{\partial^2 p_W}{\partial h^2} + \left(\frac{E}{2\eta}\left(1 + \frac{r_{hw}}{r}\right)\frac{\partial k}{\partial h} + \frac{Ek}{\eta\,(r - r_{hw})}\left.\frac{\partial r}{\partial h}\right|_P\right.$$
$$+ \frac{k}{\eta}\frac{\partial p_W}{\partial h} - \frac{r_{hw}Ek}{r(r - r_{hw})\eta}\frac{\partial r_{hw}}{\partial h} + \rho g\frac{k}{\eta}\left)\frac{\partial p_W}{\partial h}\right.$$
$$+ \frac{E\rho g}{2\eta}\left(1 + \frac{r_{hw}}{r}\right)\frac{\partial k}{\partial h} + \frac{E\rho g k}{(r - r_{hw})\eta}\left.\frac{\partial r}{\partial h}\right|_P - \frac{E\rho g k r_{hw}}{r(r - r_{hw})\eta}\frac{\partial r_{hw}}{\partial h} - \frac{E}{2\pi r(r - r_{hw})}g^b_{H20}$$

The above equation, which is numerically solved, denotes that pressure propagates by 'diffusion' (the diffusion coefficient corresponds to the multiplier of the second spatial derivative of pressure) and 'advection' (advection velocity is the multiplier of the first spatial derivative of pressure) along the stem and is depleted according to the sink term (last line in the equation; usually negative). The diffusion term equals $Ek/2\eta$ in a stem without heartwood and approaches Ek/η with increasing heartwood percentage. The advection term introduces effects of change in permeability $(\partial k/\partial h)$, stem taper $(\partial r/\partial h)$ and $(\partial r_{hw}/\partial h)$ and gravity $(\rho\,gk/\eta)$. The sink term originates from transpiration and changes in water conduction capability of the stem $(\partial k/\partial h, \partial r/\partial h$ and $\partial r_{hw}/\partial h)$.

As transpiration causes the curvature of the water film at the transpiring surfaces to change (7.5.8), it propagates through the stem according to the above equation, causing changes in the fluxes and tree dimensions. How high the tension needs to rise at the top of the tree depends on the flow of water from the root system into the tree (see Section 6.3.2.8). Inflow to the base of the tree is caused by the difference between the soil water tension and the water tension at the base of the tree, and is directly proportional to the root surface area. If soil water tension is very low, the leaf water tension needs to drop down in order to maintain the flow of water at the same level (see Fig. 7.5.12).

With Scots pine the pressure propagation through a tree is very fast. In close to 30 m tall trees, changes in the transpiration rate are visible in elastic diameter changes at the base of the tree within 10 min (Fig. 7.5.13).

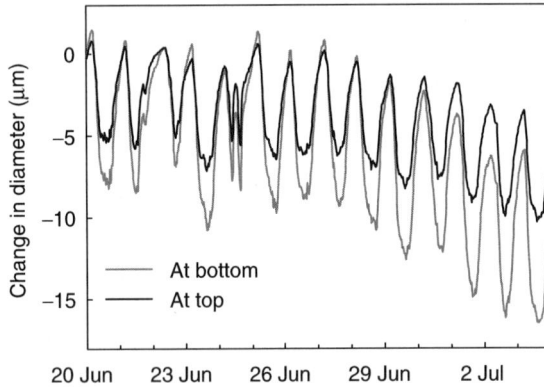

Fig. 7.5.12 Change in diameter relative to starting reference value at the measuring point during two weeks at the top and at the bottom of a Scots pine stem. Lower value signifies higher water tension. During the observation period the xylem tension decreased both at the top and at the bottom. The lower diameter decreased relatively more indicating increase in the tension difference between the measuring points

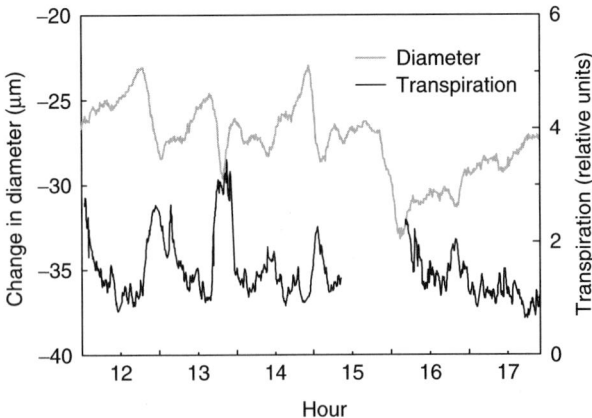

Fig. 7.5.13 Variation of xylem diameter at the base of 28 m tall Scots pine tree (black line) and changes in driving transpiration rate (grey line) measured at the top of the tree with portable gas exchange analyser CIRAS2

7.5.2.3 Embolism – Xylem Malfunctioning

Introduction of air bubbles into the water transporting xylem tissue is the major threat for functioning of the water transport system. Embolism causes a decrease in hydraulic conductivity by blocking the conduit with gas bubbles and therefore restricts sap flow, which, in turn, increases the water tension. If transpiration is not controlled, the tension in the water column may rise very high (as soil dries out, or when air is very dry) which may eventually lead to 'runaway cavitation' (Tyree and Sperry, 1988). This term describes a situation when an embolism starts to form in the stem but transpirational demand does not decrease sufficiently. The loss of conductivity due to embolism tends to increase the pressure gradient which leads to a further embolism, eventually leading to complete loss of hydraulic conductivity in the stem. In the longer run, it is crucial that excessive embolism be avoided in order to maintain the water transport capacity. As trees grow taller it is even more important because a longer transport pathway increases the water tension and makes the embolism more apparent.

An embolism may form either through the cavitation processes or due to so-called air seeding in which air leaks into the functional xylem elements from already air-filled parts of the stem. The saturated vapor pressure of water is the partial pressure of water vapor that is in equilibrium with liquid water at a particular temperature. When saturated vapor pressure equals ambient pressure, bubbles of water vapor are *usually* formed in water. If this is achieved by increasing temperature at constant pressure, the process is called *boiling*; if by decreasing pressure at constant temperature, it is called *cavitation* (Brennen, 1995). If phase-change has not occurred when pressure-temperature combination has gone outside the stability region in its phase diagram, water is in a metastable state (superheated or supercooled). Thus, liquid water, which is pulled upwards through xylem according to the cohesion tension theory, and in which pressure is lower than 0.023 MPa (at 20°C), is in a metastable state.

According to classical nucleation theory (CNT) (Brennen, 1995) cavitation is a special case of *nucleation,* which is the process of spontaneous formation of a new phase within the volume of a pre-existing phase, by random fluctuation. In cavitation gas bubbles are formed in liquid in regions of low pressure with constant temperature. Once the initial cavitation (nucleation) step has occurred, the tiny nucleus of the new phase tends to grow rapidly if the size of the nucleus exceeds a certain critical size. The probability of forming a void of critical size is dependent on tension and temperature of the subject water. If the size of the nucleus is below the critical size, surface tension will dissolve it. The CNT distinguishes two types of nucleation: homogeneous and heterogeneous, but only heterogeneous nucleation is relevant for plant processes. In heterogeneous nucleation, interactions between water molecules and impurities or non-wettable (or hydrophobic) surfaces of other substances increase the probability of the formation of nuclei with critical size, even in moderate tensions. Wettability of a surface is expressed as contact angle θ, which a water droplet forms with a surface. If $\theta < 90°$ the surface is wettable, i.e., hydrophilic, the adhesive forces are dominating; $\theta > 90°$ the surface is non-wettable, i.e., hydrophobic, the cohesive forces between water molecules are dominating (Pickard, 1981). Cavitation is a stochastic process and the probability of a cavitation event becomes larger with time and increases in hydrophobicity of the surface the hydrophobic surface area. A single successful cavitation event with bubble growth is enough to embolize the whole conduit.

In air-seeding (Zimmermann, 1983), tension draws an air bubble through a pore or pit in a cell wall into the cell lumen where water evaporates into the bubble which then enlarges and fills the whole conduit. The threshold value of pressure which leads to air-seeding is a function of pore size and surface tension of water (Laplace's equation, Eq. x1.). The third mechanism leading to embolism is the release of existing bubbles from cracks or crevices of the cell walls (Pickard, 1981). Vapor bubbles most efficiently stabilize by surface tension forces in steep non-wettable cracks or crevices of the cell wall. The bubble is released into liquid when the tension of the liquid exceeds the surface tension forces holding the bubble in the crevice. Air seeding and release of existing bubbles are deterministic processes: when tension has reached some threshold limit the initial nucleus starts to grow and the conduit is filled with gas. They are also currently thought as the most probable reasons for emboli formation in trees (see Fig. 7.5.14).

During embolization plants produce detectable acoustics (Milburn and Johnsson, 1966) and ultra-acoustic (Tyree and Dixon, 1983) emissions, which are thought to result from shock waves following the cavitation or air seeding event (Jackson and Grace, 1996). These emissions can be recorded with ultra-sonic acoustic sensors (Höltä et al., 2005). Figure 7.5.14 presents the time course of ultra-acoustic emissions and diameter change of a Scots pine stem in Hyytiälä during two days in August 2000.

Release of acoustic emissions (AE) coincide with the emboli formation in xylem. The figure shows that acoustic emissions are numerous as the diameter drops (tension increases) to lower values than previously observed. If the diameter remains at a higher level than previously, no AE are observed (16th to 19th of August). Only after the tension drops to a new low value (20th of August) that emissions

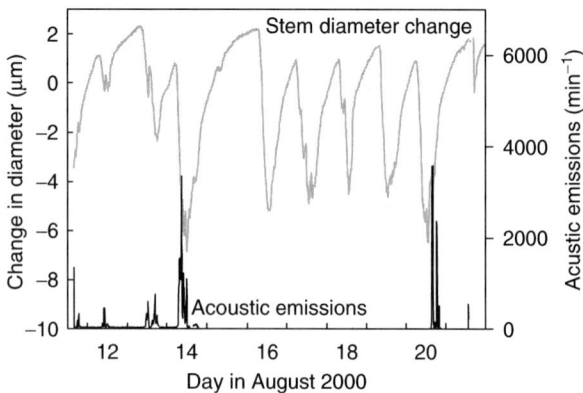

Fig. 7.5.14 Relationship between the stem diameter change (grey line, low value, large tension; high value, low tension) and acoustic emissions (black line) in Scots pine trees in field conditions in the August 2000

are observed. This behaviour would suggest that emboli are mainly formed with deterministic processes (air seeding). However, the behaviour on the 21st of August would lean toward cavitation. Note that after a prolonged exposure to low tension (large diameter) the acoustic emissions start again with lower tension than observed in the 20th of August, possibly suggesting that refilling of embolised tracheids did take place.

The nucleation processes described by the CNT set limits to strength of water to cope with tension. Theoretical *tensile strength* of pure water is from −140 to −230 MPa (Pickard, 1981) but as mentioned above the existence of heterogeneous nucleation lowers substantially the maximum sustained tension. In addition, according to the equations above, the two other mechanisms leading to embolism take place under considerably lower tensions.

Prevention of Embolism

During evolution, xylem of higher plants has undergone structural adaptation to avoid excess emboli formation (Sperry, 2003). The cell lumens are small (cavitation probability increases with cell surface area). Xylem walls are air free (Zimmermann, 1983) or air bubbles are unlikely in newly produced conduits (Sperry, 2003) and water uptake by roots excludes air bubbles (Zimmermann, 1983). Lignified and thickened cell walls are able to avoid wall collapse under tension (Hacke et al., 2001). Cell wall pores are small (1.2–3.3 nm) enough to prevent air seeding under tension lower than −15 MPa (Sperry, 2003). Most of the conduit walls are highly wettable; wettability of cellulose and hemicellulose, which are the main constituents of the conduit wall, is high (Pickard, 1981). The walls contain also hydrophobic lignin, which may destabilize xylem water under tension, but its concentration in cells of conifer xylem is highest in the middle lamella and drops to much lower values closer to the lumen (Donaldson, 2001). The interconduit pits are able to block the expansion of a gas bubble from one conduit to another by aspirating the torus against the pit aperture (Sperry, 2003).

There are several structural adaptations to avoid hazardous water tensions. The decrease of the leaf area:sapwood area ratio with increase of stem height (Vanninen et al., 1996) increases the leaf specific hydraulic conductance of the tree and decreases the maximum water tensions at treetop. The production of xylem vessels with increased permeability (Pothier et al., 1989), an increase in the fine root foliage ratio (Sperry et al., 1998; Magnani et al., 2000) and increased water storage in the stem (see next section; Phillips et al., 2003) also lead to lower water tensions.

The functional way to avoid excess embolism is to regulate the transpiration induced water tension by partial closure of the stomata. This leads to reduced CO_2 uptake and lower photosynthesis. This is typical in sunny afternoons when evaporative demand exceeds water uptake and transport capacity of the tree leading to reduced gas exchange (See Section 6.3.2.2). This can be seen in asymmetric CO_2 exchange, although PAR, which is the driving force of photosynthesis, is symmetric around noon.

Repair of Embolized Conduits

Embolism seems to be an usual occurrence in tree stems during periods of increasing transpiration. Because gas-filled cells cannot transmit tensions, embolized conduits are lost from the water transport system. To maintain hydraulic capacity, plants must replace embolized cells, maintain a highly redundant transport system, or repair embolized conduits (Holbrook and Zwieniecki, 1999). Pickard (1981) has suggested that embolized conduits might be restored to their functional state but this restoration has usually been connected to situations where the whole plant has been pressurized by root pressure (Tyree et al., 1986; Cochard et al., 1994; Fisher et al., 1997). In recent studies, it has been suggested that embolized conduits may be repaired also when the water in neighboring conduits is under tension (McCully et al., 1998; Zwieniecki and Holbrook, 1998; Tyree et al., 1999; Melcher et al., 2001). Holbrook and Zwieniecki (1999) suggested that living cells in xylem parenchyma provide the driving force for refilling the embolized conduits, when they are hydraulically isolated. Vesala et al. (2003) showed that spontaneous refilling of embolised vessels is physically feasible under such a scheme. If the properties of the semipermeable membrane between the living cells of the xylem and the embolised vessels (tracheids) would change, facilitating easier leakage, then the living cells could act as channels for the water to flow from the functional xylem into the embolised xylem cells. Höltää et al. (2006) also showed that the Münch circulation of water (see below) could under certain conditions provide for refilling through the connections of ray cells in the xylem to the bark tissue. The important role of bark tissue in emboli refilling has also been demonstrated experimentally; phloem girdling and metabolic poisons prevented embolism refilling (Salleo et al., 1996, 2004, 2006; Zwieniecki et al., 2000).

Large variation in stem water content has been reported, indicating that stems have an important role as water storage (Warind and Running, 1978; Waring et al., 1979). One possible source for this large variation in the stem water content is the emboli formation in the stem. As xylem cells cavitate or become air filled through air seeding the water tension is temporarily lowered as more water becomes available for the transpiration. This process rises the leaf water potential and helps to avoid conditions that would result into runaway cavitation (Hölttä et al., in press). On the long run the beneficial role of the cavitation processes depends on the size of the water volume in the stem relative to the transpiration rate. The bigger the volume the more important is the benefit. Another requirement is that large scale embolism need to be able to recover within reasonable time frame. As this has been seen difficult to be achieved, at least during the active periods, before the work by Holbrook and Zwieniecki (1999), and Vesala et al. (2003), emboli formation has not been often considered as an important water storage mechanism. However, the recent work with the recovery from embolism suggest that it is a common place phenomena and also physically feasible. This would then also suggest that the emboli formation may have a key role in stem water storage as well.

7.5.2.4 Phloem Transport

Transport of assimilation products from mature leaves to the stem, roots, immature leaves, and other organs occurs via phloem. Sugars are transported in the phloem along a continuous pathway of sieve space made out of sieve elements (sieve cells) that are connected to each other either by sieve plates or by small pores. Each sieve element is also connected to a so-called companion cell (see Section 5.2.3) that has an important functional role in maintaining the flow. The driving force and mechanisms for phloem transport have been under dispute. The pressure-flow model, as proposed by Ernst Münch in the 1930s, is now quite commonly agreed to explain phloem translocation and has also gained experimental support (Knoblauch and van Bel, 1998). According to the pressure-flow model, the active loading of solutes onto mature leaves, "sources", lowers the osmotic pressure of the semi-permeable sieve elements and draws in water from the surrounding tissue. The loading may take place either through symplastic connections between the surrounding cells and the companion cells or it may be apoplastic meaning that the sugars are taken in to the companion cells from the spaces between the cells. Phloem transport is presented schematically in Fig. 7.5.15.

The loading of the sieve elements causes the hydrostatic water pressure, i.e., turgor pressure, to rise. Unloading of solutes at the sink areas raises the osmotic pressure of the sieve elements and water is pushed out. This pressure gradient drives the bulk water flow in the phloem from the solute sources to sinks, and solutes move by convection along with the water stream. No membranes are crossed in the axial direction, and thus osmotic pressure differences do not contribute to axial water movement in the phloem. The majority of the axial resistance to water flow is caused by pore plates or pores between the sieve elements (Thompson and Holbrook, 2003).

When calculating the phloem flows one needs to consider also the xylem, as the water flow into and from the phloem depends on the xylem water pressure. For modelling purposes we can discretize a tree in the radial direction into different functional components: xylem, cambium, phloem sieve cells, and other living bark tissue. For the sake of simplicity, we, treat each component with a homogeneous radial distribution of water pressure and omit the influence of ray cells from the analysis. Each component is separated from adjacent components by a semi-permeable membrane, i.e., the reflection coefficient of the membrane is 1. Solutes cannot move across membranes, so osmotic pressure differences contribute to water pressure differences driving the radial water flow. Axially the model tree is divided into N equally long elements in each radial component. Both in the phloem sieve cells and in the xylem, axial water flow is caused by a hydrostatic pressure gradient. There is no axial water flow in the cambium and living bark tissue. In the xylem, the water pressure gradient is induced by water removal at the topmost element of the transport matrix, which propagates tension in the water column that drives water along the pressure gradient. At the bottom-most xylem element water is drawn from the soil (Eqs. 7.5.2–7.5.11). In the calculation we assume that sugars are loaded into phloem sieve cells at the top and unloaded at the bottom of the stem, and the consequent changes in hydrostatic, i.e., turgor, pressure, induces the water pressure

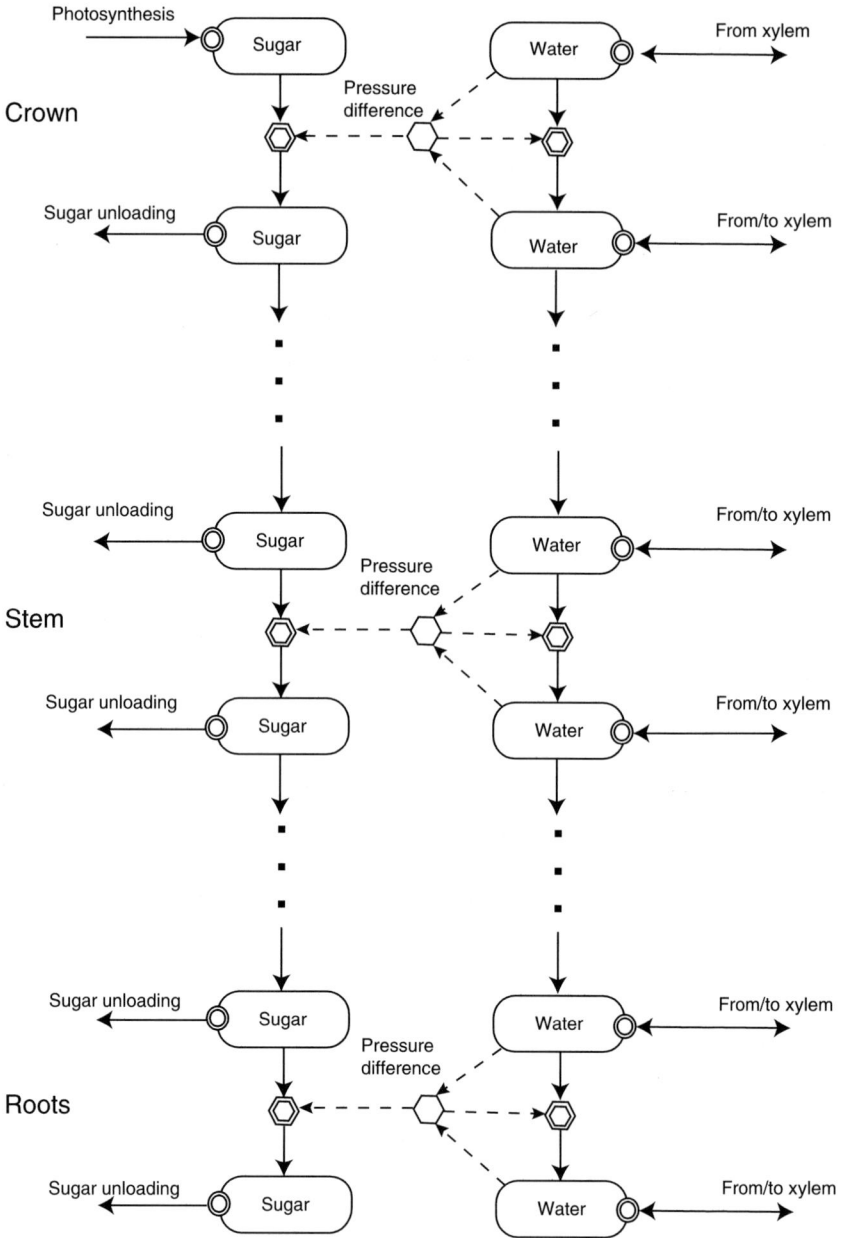

Fig. 7.5.15 A schematic presentation of phloem solute transport. The symbols are introduced in the Fig. 1.2.1

gradient. The solutes flow in the phloem sieve cell elements by convection at the same speed as water.

Now the mass balance Eq. 7.5.2 is extended to consider the radial fluxes

$$\frac{dm_{H2O}^{i,j}}{dt} = ga_{H2O}^{in,i,j} - ga_{H2O}^{out,i,j} + gr_{H2O}^{in,i,j} - gr_{H2O}^{out,i,j} \qquad (7.5.12)$$

and is solved for each axial element in each radial component. The subscript i refers to axial elements from $i = 1$ to $i = N$ and the lower index j refers to radial components: xylem, cambium, phloem sieve cells, living bark tissue (see Section 5.2.4, Fig. 5.2.8). $m_{H2O}^{i,j}$ is the mass of water in element i in component j, $ga_{H2O}^{in,i,j}$ and $ga_{H2O}^{out,i,j}$ are the corresponding values of the axial inflow and outflow rates ($m^3 s^{-1}$), and $gr_{H2O}^{in,i,j}$ and $gr_{H2O}^{out,i,j}$ are the radial inflow and outflow rates.

The axial water inflow rate for the xylem and phloem elements is calculated using Darcy's law (Eq. 7.5.3) which combines water flow with the pressure gradient in a segment of the component (Perämäki et al., 2001). The axial outflow rate $Q_{ax,out,i,j}$ for each element is equal to the inflow rate of the element above it. In the xylem, outflow from the topmost element $Q_{ax,out,N,j}$ equals transpiration, and water pressure under the bottom-most element equals the soil water pressure (see Fig. 7.5.8). In the phloem the outflow from the top-most element and inflow to the bottom-most element equals zero.

The radial flow rate of water ($m^3 s^{-1}$) between adjacent components becomes (Nobel, 1991)

$$gr_{H2O}^{in,i,j} = L_r \left[P_{i,j-1} - P_{i,j} - \sigma(C_{COOH}^{i,j-1} - C_{COOH}^{i,j})RT \right] A_{rad,i,j} \qquad (7.5.13)$$

where $A_{rad,i,j}$ is the surface area of the radial interface between the components j and $j-1$, and L_r ($mPa^{-1}s^{-1}$) is the radial hydraulic conductivity of the interface surface. $P_{i,j}$ is the hydrostatic pressure in element i in component j, $C_{i,j}$ is the solute (pure sucrose) concentration in element i in component j,

$$C_{COOH}^{i,j} = \frac{n_{COOH}^{i,j}}{V_{i,j}} \qquad (7.5.14)$$

where $n_{COOH}^{i,j}$ is the molar amount of solute in the volume $V_{i,j}$, σ is the reflection coefficient between the components, which identically equals 1 (i.e., semi-permeability) between radial components, R is the universal gas constant ($8.314 J mol^{-1} K^{-1}$), and T is the absolute temperature. Note that Eq. 7.5.3 holds in cylindrical coordinates as well as Cartesian coordinates when the surface area term $A_{rad,i,j}$ is calculated accordingly. Here van't Hoff approximation is used to calculate the osmotic pressure of the phloem solution.

If we assume no phase transitions in the wood, Hooke's law then gives the relationship between the change in water mass and pressure $P_{i,j}$ for each element in each component (Perämäki et al., 2001; Eq. 7.5.8).

If we assume constant loading at the sources and constant sugar concentration at the sinks, the equation describing the change in the amount of solutes becomes

$$\frac{dn_{COOH}^{i,j}}{dt} = (ga_{H2O}^{in,i,j} - ga_{H2O}^{out,i,j})C_{COOH}^{i,j}\rho + L_i - U_i \qquad (7.5.15)$$

where L_i is the constant loading rate of sugar in element i (and is zero outside the sugar loading zone), and U_i is the sugar unloading rate in element i. The unloading rate at the sugar sink zone is made solute concentration dependent and is explicitly written as

$$U_i = \left(C_{COOH}^i - C_{COOH}^0\right)\alpha \qquad (7.5.16)$$

where C_{COOH}^0 is a reference concentration and α is a constant. U_i is zero outside the sugar sink zone.

Using the Eqs. 7.5.2–7.5.16 we can calculate the long distance transport and variation in it depending on both the structural and functional parameters and boundary conditions (dimensions, permeabilities, soil water pressure, transpiration and sugar loading rates) and this way analyse the feasible range of tree functions from both the structural and functional point of view.

Phloem transport is very sensitive to changes in the amount of sugar loading. If the sugar-loading rate is too small, sap flow in the phloem is hindered by decreased turgor pressure at the top of the sieve elements and dilution of the phloem sap. At high loading rates the sap flow is slowed down due to accumulation of sugar and a consequent rise of viscosity at the top of the phloem. Viscosity of the phloem sap is a steeply growing exponential function of sugar concentration and varies also substantially with height. The model simulations by Höltta et al. (2005) showed that there is "an optimal value" for the sugar-loading rate for maximum sugar translocation to balance between the two scenarios of a drop in turgor pressure and a dilution of the sap solution, and a rise in viscosity. This "optimal value" is evidently dependent on the structure of the transport system and relationships between water demand of transpiration and photosynthate availability.

A critical place for development of excess viscosity is close to the loading sites. There the phloem cross-sectional area is about as large as that of xylem. However, when going down towards the base of the tree, from numerous twigs to just one stem, the phloem area becomes much smaller. The simulations of Höltta et al. (2005) showed clearly how this type of structural arrangements allowed a much higher sugar loading rate than would be the case if such a taper was not present. With this ad hoc structural variation more sugar loading is permitted without an increase in the solute concentration, but sugar transport is inefficient at smaller solute loading rates, as a larger volume of sieve elements keeps the turgor pressure and the axial pressure gradient in the sieve elements lower.

7.5.2.5 Linkages Between the Xylem and Phloem Transport

The two conduit pathways, the xylem and phloem, have mainly been studied as separate systems. However, they are next to each other in vascular bundles in tree leaves and in small roots. Also, in tree organs with secondary thickening they are connected through a narrow band of cambial cells and a number of intrusions of living and dead cells penetrating the xylem tissue in radial directions in the wood rays. As xylem and phloem are hydraulically interconnected, there should be water exchange between them in the direction of the gradient of the sum of the hydrostatic and osmotic pressures. Models by, e.g., Tyree et al. (1974), Ferrier et al. (1975), Smith et al. (1980), Phillips and Dungan (1993), Thompson and Holbrook (2003) have demonstrated that the pressure-flow hypothesis can produce a net transport of photosynthates in the phloem, but in their analysis discounted the influence of the transient transpiration stream in the xylem. The long distance transport of assimilate products in plants results from interplay between transpiration driven xylem flow and osmotically driven phloem flow. That such interaction should function imposes several requirements and limitations, not only for both the regulation of transpiration and physiology of sugar loading and unloading, but also on the tree structure.

Phloem translocation does not require transpiration to function, and the daily amount of sugar transport is only slightly reduced in the case of moderate transpiration increase. However, transpiration has a noticeable impact on the amount of sugar loading that can be sustained. Water tension build-up in xylem due to transpiration may even reverse the direction of the phloem flow during the daily peak transpiration if stomatal control does not regulate the transpiration rate. Too high a transpiration rate is liable to drop the turgor pressure below zero during the daily transpiration peak as water is "sucked" into the xylem. This suggests that the water tension in leaves might be controlled by stomatal conductivity, which would also facilitate Münch flow. The turgor drop and phloem sap flow direction reversal are faster and stronger if the radial permeability is high and if the sugarloading rate is not sufficient to resist the "pull" from the xylem. The different permeability components, radial permeability between the phloem and xylem, and the axial permeability of the phloem sieve cells and the xylem have to be high enough for efficient phloem translocation as the radial and axial water flows are coupled in water circulation.

An interesting and potentially very important linking between the xylem and phloem is in the role that the recycled "Münch water" passing from the phloem to the xylem can have in xylem embolism refilling when certain structural and environmental conditions are met. The recycled "Münch water" passes from the phloem to the xylem at sites of sugar unloading where the sum of the hydrostatic and osmotic pressure in the phloem water rises above the xylem water pressure. Thompson and Holbrook (2004) suggested that efficient phloem transport and signalling would require that phloem transport would occur in relays, i.e. that solutes would be unloaded and reloaded several times along the transport pathway. Apart from the signal speed this would also enhance the refilling process. Refilling may occur if there is a sufficient radial gradient in the sum of the hydrostatic and osmotic water pressure from the rays to the xylem conduit. If this gradient in the sum of the hydrostatic and osmotic water pressure between the rays and the xylem conduits is large enough, water is directed from rays also to embolised xylem water conduits in addition to non-embolised conduits. Whether this internal imbalance in the sum of the hydrostatic and osmotic water pressure is sufficient to push water into embolised conduits depends on many factors. At least the magnitude of the water flow passing from the phloem to xylem, bulk xylem water tension, and the structure of the pathway connecting the phloem and xylem, determine whether water will flow to embolised conduits. Important details within the structure of the pathway are the size of the rays and the hydraulic conductivities between the rays and xylem conduits and within the rays.

A necessary and sufficient requirement for water flow to embolised conduits to occur is that the sum of the hydrostatic and osmotic water pressure in the ray is higher than the sum of the hydrostatic and osmotic pressure in the embolised conduit. This requirement is the same as the difference in the sums of hydrostatic and osmotic pressures between the ray and non-embolised xylem conduits being larger than bulk xylem tension. Refilling as a consequence of the "Münch water circulation" is typically slow as the contact area of a xylem conduit to a ray is only a fraction of a xylem conduit's total surface area. Also the difference in the sum of the hydrostatic and osmotic pressure between the ray and embolised conduit is relatively small compared to the gradient over the ray to the functioning conduit interface.

According to Hölttä et al. (2006) the properties of interfacial area between a ray and xylem conduits are very central for determining whether embolism refilling due to "Münch water circulation" can occur. Low hydraulic conductivity and low interfacial area of the interface between the ray and the xylem conduits tend to increase the difference in the sums of the hydrostatic and osmotic pressure between the ray and xylem conduits larger at a fixed amount of radial water flux, so embolism refilling can occur at larger xylem water tensions. Direct measurements of the conductivity between the rays and xylem conduits have not been made, but there is evidence in the literature that this conductivity is relatively low. De Boer and Volkov (2003) state that the half-bordered pits between xylem parenchyma cells and xylem vessels are on the one hand the gates to the xylem conduits for water and ion transport, but also a serious 'bottle-neck' from a hydraulic perspective. Also (Siau, 1984) found simple and half-bordered pits revealed to be relatively impermeable. Zwieniecki et al. (2001) found the gradient in the sums of hydrostatic and osmotic pressure in the radial direction to be many orders of magnitude greater than in the axial direction, as the radial specific conductance (conductance per unit area) is approximately six orders of magnitude lower than the longitudinal conductance.

Trees may be able to enhance the refilling by increasing the loading and unloading rates. The driving force for Münch circulation drivenembolism refilling is the positive water potential gradient in between the phloem and embolised xylem element which is created by the unloading of sugars. Higher loading and unloading rates would also facilitate higher gambial growth directly or through accumulated stores. Analyses have shown that hydraulic conductance per unit leaf area remains remarkably constant (Magnani et al., 2000). Magnani et al. (2000) proposed that trees would tune the hydraulic resistance of a plant to match the tension created by transpiration. Higher refilling capacity with Münch circulation, and also lowering of hydraulic resistance in xylem in the long run through enhanced growth, go hand in hand and would be beneficial acclimations to conditions, provoking very negative water pressures in the transporting pathway. Such an interaction between the parallel conductive systems brings the notion of maintenance of xylem functionality under much stronger whole tree control, as previously thought. As mentioned earlier, stomatal conductance plays a very important role in controlling simultaneously the transpiration and the rate of sugar transport away from the assimilating surfaces. The linkages between the sugar transport and maintaining xylem functionality links the short term transport processes with the long term longevity of the functional xylem, but also with its new growth.

The long distance transport has a particularly significant role in trees that reach magnificent sizes of all land plants. The present understanding of the transport mechanisms show that trees rely heavily on the tensile properties of water and can utilise quite efficiently energy from evaporation to lift the water up from the soil. This reliance however, exposes the trees to serious risk of malfunction as the transport mechanism relies on the metastable state of water. Water transport in trees has been quite extensively studied and its implications on tree structure have received a lot of attention. Much less attention has been given to the existence of two parallel transport systems and what implications their interaction may have for tree performance although Ernst Münch put forward the idea of the internal circulation of water in trees already almost a century ago. When the parallel transport systems are treated simultaneously it becomes obvious that the state of water in xylem has strong influence on the phloem transport. The influence that phloem transport and the so called Münch flow may have on the embolism recovery in xylem suggests much more integrated picture of the long-distance transport in trees that has often been considered. The strength of the sources and sinks are no more merely driving forces of bulk flow of assimilates, but they may have significant role in maintaining the whole tree functions intact. The results of the significance of the bark tissue tapering on the assimilate transport also shows that maintaining the long distance transport mechanism require quite specific structure not only in the wood tissue but also in the living bark tissue. As the long distance transport system is constrained by the physical laws of fluid dynamics the feasible structural combinations are limited and they need to be linked with the leaf level control of both water loss and the carbon intake that are the two principle components transported in a tree.

7.6 Stand Level

7.6.1 CO_2 in Soil

Jukka Pumpanen[1], Hannu Ilvesniemi[2], and Pertti Hari[1]

[1] University of Helsinki, Department of Forest Ecology, Finland
[2] Finnish Forest Research Institute, Finland

The CO_2 production of soil plays an important role in the carbon balance of ecosystems. In mature forest soil, CO_2 emissions are about 50% of the canopy photosynthesis and about twice as much as the respiration of foliage (Kolari et al., 2004). A substantial part of the soil CO_2 production originates from root respiration and from the rapid decomposition of carbohydrates emitted from the root system, i.e., root exudates. These processes are described in more detail in Section 6.4. The CO_2 produced in the soil is transported mainly by diffusion towards the soil surface. In this section, we analyse the dynamics of CO_2 in soil.

Boreal forest soil consists of horizontal layers (Section 7.4) which have been formed in podzolization processes during thousands of years. For details of the podzolization process see Section 5.4 on soil structure. A layered structure can be used as a basis when describing the processes underlying soil CO_2 efflux such as CO_2 production in different soil layers and the CO_2 transport within the soil as shown in Fig. 7.6.1.

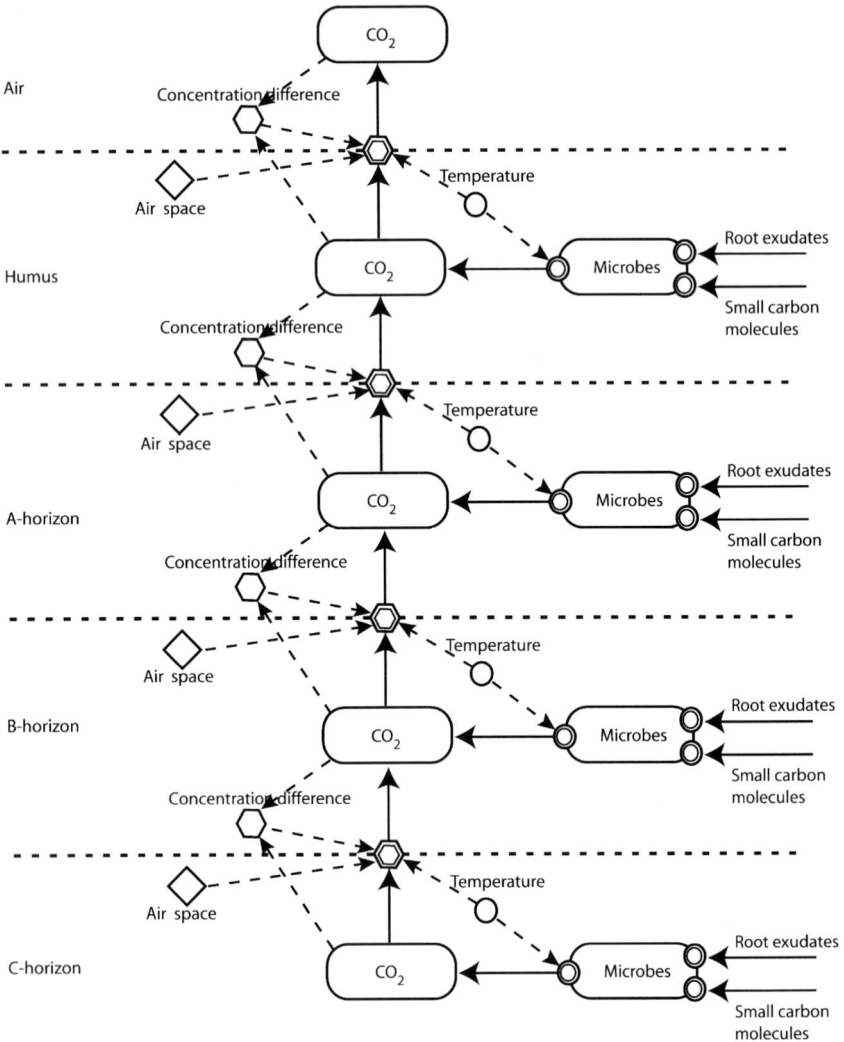

Fig. 7.6.1 Schematic presentation of the processes involved in soil CO_2 efflux. The symbols are introduced in the Fig. 1.2.1

7.6.1.1 CO_2 Production

CO_2 is produced in the soil by heterotrophic microbial respiration and by autotrophic root respiration. Soil microorganisms release CO_2 by oxidizing organic debris and return the carbon assimilated by the plants back to the atmosphere. Major factors affecting microbial respiration are the amount and quality of organic carbon in the soil, soil temperature and soil moisture (Kirschbaum, 1995; Davidson et al., 1998; Prescott et al., 2000). These factors are highly variable, depending on the geographical location of the site, the physical and chemical properties of the soil, and the age and species composition of the forest. A more detailed description of decomposition of soil organic matter is given in Section 6.4.1. Root and rhizosphere respiration is the second major component of soil CO_2 efflux. The contribution of root and rhizosphere respiration is strongly linked to the seasonal cycle in root exudates and phenology of trees. Here respiration in soil (r) is formed as an outcome of microbial respiration (r_m) and root respiration (r_r).

$$r = r_m + r_r \tag{7.6.1}$$

Both of these processes are temperature dependent since they are enzymatic processes. There are many possible expressions to relate the dependence of respiration in soil on temperature. Here we use an exponential function for the temperature response of respiration $f_T(T)$, an approach used by Boone et al. (1998), Buchmann (2000), Widén and Majdi (2001):

$$f_T(T) = \alpha e^{\beta T} l \rho \tag{7.6.2}$$

where T is the temperature (°C), α and β are fitted coefficients, which have been obtained by incubating soil samples collected from individual soil layers. The respiration values measured from the soil samples were scaled to $1\,m^{-2}$ surface area using the thickness l (m) and the bulk density $\rho(g m^{-3})$ of the soil layer. We use similar temperature responses for both microbial respiration and root respiration within each soil horizon. This is justified by studies showing similar temperature responses for these components of soil respiration. The temperature response of respiration is often described with Q_{10} value, which can be derived from parameter β in Eq. 7.6.2. as follows $Q_{10} = e^{10\beta}$. Widén and Majdi (2001) and Buchmann (2000) determined Q_{10} values ranging from 2.1 to 3.2 for coniferous forest soil. Grogan and Chapin (1999) estimated a Q_{10} value of 3.3 for the soil respiration from a range of arctic vegetation types. These are rather similar to temperature responses determined for root respiration.

Respiration depends also on soil moisture, and merely on the availability of substrates for the microbes. Microbes use extracellular enzymes to decompose soil organic matter, and they function in soil solution. This is partly a diffusion-based process where the enzymes and microbes work inside a water film. When the soil dries, the water film around soil particles disappears, microbes no longer have access to soil organic matter and the decomposition process gets slower. During extreme drought, even the microbes themselves are affected by the drought through desiccation.

Numerous studies have shown the relationship between soil moisture and microbial activity (Greaves and Carter, 1920; Linn and Doran, 1984; Davidson et al., 1998). Skopp et al. (1990) presented an equation taking into account both the effects of drought and anoxic conditions in wet soils approaching water saturation:

$$f_\theta(\theta_v) = Min\left[a\theta_v{}^d, \, b \, (E_o - \theta_v)^g, 1\right] \tag{7.6.3}$$

where $f(\theta_v)$ represents the CO_2 efflux evolved from soil (relative value where 1 represents the maximum respiration and 0 the minimum), θ_v is the volumetric water content ($m^3 \, m^{-3}$) and E_o is the total porosity ($m^3 \, m^{-3}$). Parameters a, b, d and g are empirical constants that are fixed for a given soil (Skopp et al., 1990). At low water contents, water availability limits respiration activity in soil. This aerobic microbial activity increases with soil water content until a point is reached where water starts to restrict the diffusion and availability of oxygen. According to Linn and Doran (1984) many studies involving a wide range of soil types indicate that a soil water content equivalent to 60% of a soil's water-filled pore space achieves maximum aerobic microbial activity. In the study of Doran et al. (1988), the maximum aerobic microbial respiration for 16 soils of varying texture occurred at volumetric water content equal to 0.55–0.61 times the value of total porosity.

Temperature and moisture both affect microbial activity in the soil and it is rather difficult to separate the effect of soil moisture and temperature explicitly. We have assumed that the temperature and moisture effects are multiplicative.

$$r(T, \theta) = f_T(T) \, f_\theta(\theta_v) \tag{7.6.4}$$

In the equation, $f_T(T)$ is the dependence of soil respiration on temperature only, and $f_\theta(\theta)$ is a scaling factor for the dependence of soil respiration on soil water content. The same kind of multiplicative approach has previously been used in several studies (e.g., Schlentner and van Cleve, 1985; Davidson et al., 1998; Fang and Moncrieff, 1999; Moncrieff and Fang, 1999).

There are also some abiotic processes that may contribute to soil CO_2 efflux, such as carbonate dte dissolution and chemical oxidation (Burton and Beauchamp, 1994). This is however a minor source of CO_2 in boreal forests in Scandinavia due to the mineral composition of soil, mainly acidic minerals such as granodiorite and gneiss and almost complete lack of lime stone (Wahlström et al., 1992). Therefore, we have ignored these processes here. However, if this model were applied for calcareous soils, these processes should be taken into consideration.

7.6.1.2 CO_2 Transport

CO_2 concentration in the soil air space between soil particles is often an order of magnitude higher than in the atmosphere (Fernandez and Kosian, 1987; Suarez and Šimunek, 1993) resulting in a large concentration gradient between the soil and the atmosphere. The primary mechanism for transporting CO_2 from the soil to the atmosphere is molecular diffusion (Freijer and Leffelaar, 1996). According to Fick's first law, the gas flux is dependent on the concentration gradient and the diffusivity of the soil. Thus the CO_2 flux in the soil is usually upwards, resulting in a CO_2 efflux out of the soil (Fig. 7.6.2). The diffusion rate is dependent on the total porosity of subsequent soil layers, soil water content, the distance and the concentration gradient between the layers.

CO_2 can also move between the soil layers as it dissolves in water (Šimůnek and Suarez, 1993). Also mass flow of CO_2 by convection caused by wind or atmospheric pressure fluctuations may affect the gas movement in soil especially in deep soils. However, the contribution of convection to the transport of CO_2 in shallow soils such as at SMEARII station is small. Diffusion is the dominating mechanism in

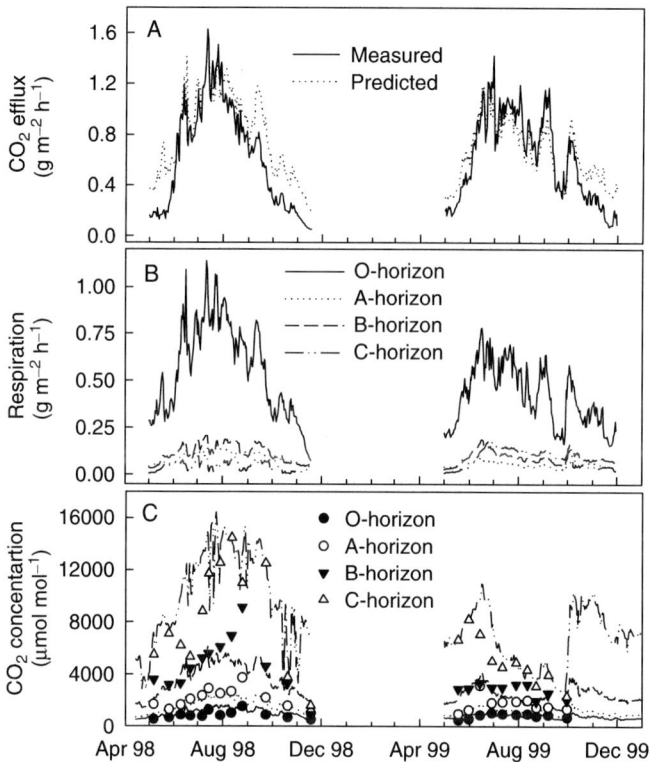

Fig. 7.6.2 (A) Simulated and measured soil CO_2 efflux, (B) CO_2 production in different soil layers and (C) simulated (lines) and measured (symbols) CO_2 concentrations in different soil layers (Redrawn from Pumpanen et al., 2003)

the CO_2 transport within soil. Other mechanisms of gas movement than concentration controlled diffusion have been shown to account for less than 10% of the CO_2 lost from the upper soil and even less for the deeper unsaturated zone (Wood and Petraitis, 1984).

A layered structure is very characteristic for podzolic soils. This is why we have treated the soil as a structure consisting of distinctive layers and formulated our flux equations in discrete variables. As an example, we have presented here the equations for O- and A-horizons. The soil layers are specified and denoted with capital letters referring to the horizons O, A, B and C. Other horizons can be obtained by changing the indexes referring to respective layers. The CO_2 flux between an A- and an O-horizon is:

$$g_{CO2}^{AO} = -D_{AO} \frac{C_{CO2}^{O} - C_{CO2}^{A}}{(l_O + l_A)/2} \tag{7.6.5}$$

where g_{CO2}^{AO} is the flux (g CO_2 m^{-2} s^{-1}), D_{AO} is the average diffusion coefficient of CO_2 in the O- and A-horizons (m^{2}s^{-1}), C_{CO2}^{A}, C_{CO}^{A}, l_O and l_A is the CO_2 concentration (g CO_2 m^{-3}) and thickness (m) of the O- and A-horizons, respectively.

The diffusion coefficient of CO_2 (D) in a soil layer is a fraction of the diffusion coefficient of CO_2 in air D_o(m^{2}s^{-1}) according to a model developed by Troeh et al. (1982):

$$\frac{D}{D_o} = \left(\frac{E_g - u}{1 - u} \right)^h \tag{7.6.6}$$

where E_g is the air-filled porosity of soil (m^3 m^{-3}) and u and h are empirical parameters obtained from the literature (Glinski and Stepniewski, 1985). D was determined separately for each layer.

We also assume that CO_2 is pushed between the layers by water replacing air in the soil pore space. The CO_2 flux from the A- to the O-horizon caused by a change in the air-filled porosity of A-, B- and C-horizons (q_{CO2}^{AO}) is:

$$q_{CO2}^{AO} = \frac{\{[V_{H2O}^{C}(t_i) - V_{H2O}^{C}(t_{i+1})] l_C + [V_{H2O}^{B}(t_i) - V_{H2O}^{B}(t_{i+1})] l_B + [V_{H2O}^{A}(t_i) - V_{H2O}^{A}(t_{i+1})] l_A \} C_{CO2}^{A}(t_i)}{t_{i+1} - t_i}$$

$$\tag{7.6.7}$$

where $V_{H2O}^{*}(t_i) - V_{H2O}^{*}(t_{i+1})$ is the change in the water volume in the layer *, where * denotes C, B or A layer in the soil horizon and $C_{CO2}^{A}(t_i)$ is the CO_2 concentration of the soil horizon A (g CO_2 m^{-3}) at moment t_i.

The amount of CO_2 in each soil layer for an area of $1 m^2$ is obtained using a CO_2 mass balance equation, which is expressed here for the A-horizon using time discrete formalism:

$$V_{Air} C_{CO2}^{A}(t_{i+1}) = V_{Air} C_{CO2}^{A}(t_i) + (r_A + g_{CO2}^{BA} + q_{CO2}^{BA} - g_{CO2}^{AO} - q_{CO2}^{AO})(t_{i+1} - t_i) \tag{7.6.8}$$

where V_{Air} is the volume of air in the A-horizon (m^3), C_{CO2}^{A} is the CO_2 concentration in the A-horizon (g CO_2 m^{-3}), r_A is the soil respiration rate in the A-horizon (g CO_2 m^{-2} h^{-1}), g_{CO2}^{BA}, q_{CO2}^{BA}, g_{CO2}^{AO} and q_{CO2}^{AO} are CO_2 fluxes from the B-horizon to the A-horizon and from the A-horizon to the O-horizon (g CO_2 m^{-2} h^{-1}), respectively.

7.6.1.3 Parameterization of the CO_2 Production Model

Values for parameters can be determined by estimation from measured fluxes, or by measuring the processes involved. If the parameters were estimated from the measured fluxes, it would be difficult to evaluate the performance of the model because the predicted values would depend on the measured fluxes. Because of this we have avoided this estimation of the values of the parameters from the measured fluxes and based the values of parameters on process measurements and published data, whenever possible.

Parameters a, b, d and g in Eq. 7.6.3 were determined by Skopp et al. (1990), for a soil of similar texture as that in this study (fine silty, mixed). With given parameters $f_\theta(\theta_v)$ values can be >1, when $\varepsilon > 0.50 \, m^3 \, m^{-3}$. This has been taken into account in the model by limiting the maximum $f_\theta(\theta_v)$-value to 1. In the O-, A- and B-horizons this results in a wider range for maximum microbial respiration than in the C-horizon. According to Howard and Howard (1993) the optimal moisture range for respiration is wider in organic soil than in mineral soil. Vanhala (unpublished data) measured maximum soil respiration in the humus layer at volumetric water content of $0.35 \, m^3 \, m^{-3}$. In this study the optimal water content in the O-horizon ranges from 0.36 to $0.50 \, m^3 \, m^{-3}$.

Values for parameters α, β and ρ for the O-horizon in Eq. 7.6.2 were obtained from laboratory measured CO_2 efflux and soil temperature response curves based on humus samples collected from the measurement site in July 1998 (Kähkönen et al., 2001). The respiration rate of the field moist humus samples was measured in the head space of 120 ml incubation bottles at four temperature levels ranging from 2°C to 17°C using the GC – TC method. An exponential curve of the form $r_m = \alpha \, e^{\beta T}$ where $e^{10\beta} = Q_{10}$, was fitted on the data. For the A-, B- and C-horizons, parameters α, β and ρ were obtained from soil core samples taken from the SMEAR II station and from the studies by Kähkönen et al. (2001) and Pietikäinen et al. (1999) accomplished for forest soils similar to that at SMEAR II station.

A value for the total porosity of the soil (E_o) was obtained from soil water retention curves determined separately for each soil layer (Mecke and Ilvesniemi, 1999). The soil water retention curves for each soil horizon were measured from volumetric samples collected from the walls of five pits excavated at the site. The thickness of the soil layers was measured at the pits and used as parameter l in Eqs. 7.6.2, 7.6.5, and 7.6.7. Values for parameters u and h in Eq. 7.6.6 were obtained from Glinski and Stepniewski (1985). Selected parameter values represent texture similar to the soil at SMEAR II station. The parameter values are summarised in Table 7.6.1.

7.6.1.4 Agreement Between Modeled Fluxes and Field Measurements

We ran the model with hourly measured soil temperature and soil moisture data during a very wet year (1998) and a very dry year (1999) for studying the effect of drought and extreme moisture on soil CO_2 efflux. In addition, we studied the vertical distribution of soil CO_2 production.

Table 7.6.1 Parameters for respiration and transport functions

Soil horizon	$\acute{\alpha}$	β	U	h	ρ	l	E_o
					$(Mg\ m^{-3})$	(m)	$(m^3\ m^{-3})$
O	7.613	0.1167	0	1.1	0.50	0.050	0.70
A	0.430	0.1026	0	1.4	1.68	0.054	0.61
B	0.237	0.1026	0	1.4	1.47	0.174	0.58
C	0.0800	0.1026	0	1.4	1.63	0.543	0.50

Soil CO_2 concentration showed a seasonal pattern that followed the soil temperature. The highest concentrations in the soil profile were measured in the summer ranging from 580–780 μmol mol^{-1} in the O-horizon to 14,470 μmol mol^{-1} in the C-horizon. High concentrations were also measured occasionally in the O-horizon in April because of the formation of ice crust on the soil surface. In winter, the concentrations were much lower, ranging from 498 μmol mol^{-1} in the O-horizon to 1,213–4,325 μmol mol^{-1} in the C-horizon (Fig. 7.6.2).

CO_2 concentrations predicted by the model agreed quite well with measured values, especially in the A-, B- and C-horizons. The coefficient of determination (r^2) for predicted CO_2 concentrations ranged from 67% in the A-horizon to 82% in the C-horizon. There was a gradient in CO_2 concentration, the concentrations being highest in deeper soil layers throughout the year, indicating that there was biological activity in the soil profile all year round. According to model simulations, most of the CO_2 production occurred in the humus layer throughout the year. However, the relative contribution of deeper layers to total respiration was at its highest in late autumn, because of low temperature at the soil surface (Fig. 7.6.2).

The two years studied represented extreme variation concerning soil water content. The summer of 1999 was very dry whereas the summer of 1998 was exceptionally wet (Fig. 7.6.2). In 1998 the soil was near field capacity most of the summer; in the humus layer, soil water content varied between 0.35 and 0.4 m^3 m^{-3} (from –0.02 to –0.08 MPa) and in the B- and C-horizons between 0.4 and 0.5 m^3 m^{-3} (from –0.004 to –0.001 MPa). Soil water content affected substantially CO_2 concentration in the soil. Soil air CO_2 concentration was much lower during the drought than when the soil was wet. The concentration increased in the spring, partly because of increased soil moisture by thawing and formation of ice crust on the soil surface.

Variation in soil temperature induced changes in soil CO_2 efflux both on daily and annual time scales. Soil surface temperature explained 69–85% of the temporal variation in soil CO_2 efflux. Highest CO_2 effluxes were measured in July and August, the daily average effluxes measured by automated chambers being 1.23 and 0.98 g CO_2 m^{-2} h^{-1} in 1998 and 1999, respectively. In winter the average effluxes were much lower, only between 0.0 and 0.1 g CO_2 m^{-2} h^{-1}. The spatial variation of CO_2 efflux was high; the coefficient of variation varied between 18% and 45%, being highest during high efflux.

Simulations with the process model showed that the model could quite accurately predict soil CO_2 effluxes measured by automated chambers. The coefficient

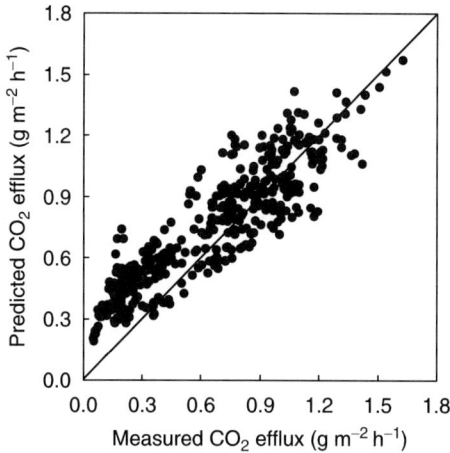

Fig. 7.6.3 Relationship between measured and predicted CO_2 efflux (Redrawn from Pumpanen et al., 2003)

of determination (r^2) for the regression of modeled vs. predicted CO_2 efflux was 82–86% (Fig. 7.6.3). However, the model slightly overestimated low effluxes and underestimated high effluxes.

7.6.2 Forest Ecosystem CO_2 Exchange

Pasi Kolari[1], Minna Pulkkinen[1], Liisa Kulmala[1], Jukka Pumpanen[1], Eero Nikinmaa[1], Annikki Mäkelä[1], Timo Vesala[2], and Pertti Hari[1]

[1]University of Helsinki, Department of Forest Ecology, Finland
[2]University of Helsinki, Department of Physics, Finland

Carbon exchange of a forest ecosystem is generated by processes binding CO_2 from the atmosphere and processes that release CO_2. Photosynthesis (Section 6.3.2.3) is the fundamental carbon-binding process and the origin of all organic carbon accumulated in land ecosystems. Photosynthetic products are either used directly for the metabolism of photosynthesizing tissues or transported to other parts of the plant where they are used for development and growth of new tissues or the chemical energy bound in photosynthates is utilised for maintenance of cell metabolism. CO_2 is released in the respiratory processes involved in maintenance (Section 6.3.2.1) and growth (Section 6.3.3.1.4). Eventually also the carbon bound in biomass will be released as CO_2 through microbial decomposition of dead biomass (Section 6.4.1).

Exchange of carbon between a forest ecosystem and the atmosphere can be determined directly by micrometeorological methods. Recent empirical research on forest ecosystem carbon balance has usually been based on measurements of net CO_2 exchange of the ecosystem by eddy covariance (Baldocchi, 2003). Stand photosynthesis and total ecosystem respiration can be extracted from the measured fluxes, but the measurements give no deeper understanding on partitioning of CO_2 uptake between the trees and the understorey vegetation, or on the relative magnitudes of above- and below-ground respiration components or root and microbial respiration. To better understand the functioning of the ecosystem, it is more useful to analyse the component fluxes, i.e., photosynthetic CO_2 fixation and release of CO_2 in respiration.

Here we compare the measured and modelled component fluxes to direct measurements of ecosystem CO_2 exchange at SMEAR II, and give an overview of the component CO_2 fluxes and their dynamics in boreal forests in general. Finally we study how the regularities in the environmental responses of stand photosynthesis change over a wide range of climatic variation (Section 7.6.2.4).

7.6.2.1 Measured Ecosystem CO_2 Exchange, Photosynthesis and Respiration

Eddy covariance is a method to measure turbulent transport of energy and matter between the land ecosystems and the atmosphere (Chapter 5). It directly gives the net gas exchange of the whole ecosystem (Net Ecosystem Exchange, NEE). The measuring system consists of an ultrasonic anemometer that measures the 3-dimensional wind speed components, and an infrared gas analyser that monitors gas concentrations in the air parcels moving through the anemometer. At SMEAR II the anemometer and sample tube intake are installed above the stand at a height of 23 m. The sampling frequency in the raw data is about 10 Hz. In the post-processing of the data the concentrations, fluxes and turbulence variables are normally averaged to half-hourly values. The change in storage of CO_2 below the measurement level (Section 7.3) must be added to the observed turbulent flux to obtain ecosystem CO_2 exchange. Eddy covariance fails to detect the actual ecosystem exchange under stable atmospheric stratification when there is little turbulent vertical movement of air, especially at night. Therefore, measurements that are expected to be biased, are rejected and replaced with calculated values that are based on the accepted data. This procedure is called gapfilling.

The net CO_2 exchange detected by eddy covariance is a sum of two main component fluxes of opposite direction, i.e., photosynthetic CO_2 uptake and CO_2 release as a result of all respiratory processes in the ecosystem. These major component fluxes can be determined directly from eddy covariance without tedious integration of CO_2 fluxes measured or modelled in a smaller scale. This basic knowledge is also utilised in replacing the missing or rejected measurements. A common gap-filling method is calculating net ecosystem CO_2 exchange with simple empirical models, for example with a combination of a big-leaf photosynthesis model and a relationship between temperature and the sum of all respiration components (Falge et al., 2001). Different subcomponents of CO_2 uptake and respiration are, however,

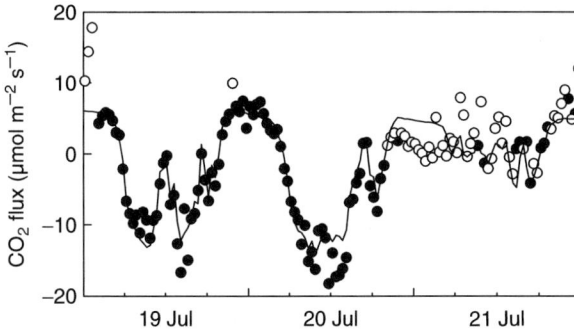

Fig. 7.6.4 Ecosystem CO_2 exchange measured by eddy covariance (dots) and modelled with a simple gapfilling model (solid line; Eqs. 7.6.10 and 7.6.11) over three days in July 2001. The gapfilling model was fitted to accepted turbulent flux data (indicated by closed circles). Open circles represent measurements rejected due to low turbulence; they are replaced with modelled CO_2 exchange when calculating daily and annual carbon budgets. The graph follows so called atmospheric sign convention: Negative values indicate CO_2 uptake by the forest (the atmospheric CO_2 storage decreases), positive values CO_2 efflux from the forest

hard to separate. The noise in half-hourly flux measurements (Fig. 7.6.4) as well as the uncertainty of determining day-time respiration limit the complexity of the component flux models and the number of model parameters. It is practical to only determine the combined photosynthetic production of all vegetation layers, often referred as Gross Primary Production (GPP), and the sum of all respiratory fluxes, often called Total Ecosystem Respiration (TER).

The standard way to extract the component fluxes is to use night-time data to estimate ecosystem respiration and extrapolate the obtained temperature relationship to day-time. Photosynthetic production is then calculated as the difference between net ecosystem exchange and ecosystem respiration. Typically ecosystem respiration is modelled with one exponential temperature regression and the whole stand photosynthesis with a simple big-leaf model as a function of incident solar radiation (Falge et al., 2001).

We derive stand photosynthesis and ecosystem respiration following the procedure described in Mäkelä et al. (2006). Ecosystem respiration, R_e, was modelled using a modified Arrhenius type exponential equation (Lloyd and Taylor, 1994):

$$R_e = r_0 e^{E\left(1-\frac{T_0}{T_s}\right)} \tag{7.6.9}$$

where T_s is temperature at a depth of 2 cm in the soil organic layer, r_0 the average night-time turbulent flux at soil temperature T_0, and E a temperature sensitivity parameter.

Only fluxes measured in turbulent atmospheric conditions (see Sections 2.4.4.9 and 2.5.2) can be used for deriving stand photosynthetic production directly from the measured NEE. During periods of weak turbulence, photosynthesis must be estimated with an empirical big-leaf model parameterised with the photosynthesis obtained directly from measured data. Stand photosynthesis P is modelled as a saturating function of light with a nonrectangular hyperbola

$$P = \frac{1}{2\theta}\left[\alpha I_{PAR} + P_{max} - \sqrt{(\alpha I_{PAR} + P_{max})^2 - 4\theta\,\alpha I_{PAR}\,P_{max}}\right] \qquad (7.6.10)$$

where I_{PAR} is the intensity of photosynthetically active radiation (PAR), P_{max} the rate of saturated photosynthesis, θ a parameter defining the convexity of the light response curve, and α the initial slope of the curve.

The temperature sensitivity of respiration was derived from regressions of accepted night-time turbulent fluxes on temperature in the soil organic layer over the growing season. To take into account the interannual and seasonal dynamics of photosynthesis and respiration, the base level of respiration and the parameters α and P_{max} in the photosynthesis model were estimated for each day of the year using an 11-day moving window of accepted flux data. The parameters were estimated simultaneously using both night-time and daytime measurements within the time window. An example of measured and gapfilled NEE above the SMEAR II pine stand is presented in Fig. 7.6.4.

Stand photosynthesis extracted as described above is further used in comparison with integrated CO_2 fluxes at SMEAR II and in a model analysis of daily stand GPP in Section 7.6.2.4. In the following subsections we refer to net CO_2 exchange, stand GPP and ecosystem respiration derived from eddy covariance as "measured" fluxes. The component CO_2 fluxes determined using chamber measurements and modelling are called "integrated" fluxes.

7.6.2.2 Ecosystem Photosynthesis

Photosynthesis of Trees

In forested ecosystems, photosynthesis by trees is the major carbon-binding process. Stand photosynthetic production is determined similarly to the tree photosynthesis described in Section 7.5.1, by integrating the instantaneous photosynthetic rate at shoot level over the tree and eventually over the whole stand. Light is the environmental driving factor that has the strongest variation within the canopy, therefore the integration of photosynthetic production over space essentially involves determining the light environment at different locations inside the canopy and calculating the respective photosynthetic rates.

Different approaches to calculate the canopy light environment (Sections 7.3 and 7.5.1) can be mixed to optimise modelling accuracy while keeping computing time acceptable and the amount of needed information on the canopy structure limited so that overly detailed biometric measurements are not needed. The canopy light model of Stenberg (1996) is of intermediate complexity. In the model, canopy foliage mass is distributed evenly inside the individual crowns, i.e., there is no explicitly defined structure inside the crowns. This simplifies the calculation of light attenuation inside the crown. The internal structure of the canopy is condensed into an aggregated parameter k, which is equivalent to Beer-Lambert light extinction coefficient. The light extinction coefficient depends on density of canopy elements inside the crown, thus it is a species-specific parameter. The shoot orientation is assumed to be spherical, i.e., the shoots are randomly pointing in all directions.

SPP (Stand Photosynthesis Program, Mäkelä et al., 2006) combines the model of shoot photosynthetic production (Section 6.3.2.3) with the canopy light model (Stenberg, 1996; Section 7.3). Photosynthetic production is modelled at tree level, but the individual crowns consist of a homogeneous medium. Trees of different species, size, leaf area density or physiology are represented as size classes. In each size class, the trees can have different photosynthetic parameters, canopy shape and dimensions. The trees are assumed to be randomly distributed (Poisson distribution) in the stand. When calculating the light environment inside the crowns, shading by the neighboring trees is taken into account in addition to within-crown shading. Tree crowns are described as ellipsoids or cones filled with shoots randomly distributed within the crown volume.

Irradiance at a given point in the canopy is calculated as a function of the distance travelled by the beam inside the absorbing medium, i.e., the distance through the neighbouring crowns intersected by the beam, and the distance from the crown surface to the point of observation. Incident radiation is divided into direct and diffuse components. Direct radiation arrives at a given elevation angle, while diffuse radiation is assumed to enter equally from all directions of the sky. Attenuation of the radiation components is calculated separately. In the version of SPP used here, the canopy is further divided into sun shoots illuminated by both direct beam and diffuse radiation, and shade shoots that are only receiving diffuse light. The calculation of light environment and photosynthesis is performed for volume units of the crown.

The photosynthesis component of SPP consists of the optimal stomatal control model combined with the annual cycle model (PhenPhoto, Section 6.3.3.1.3). The shoot is the basic unit. We first derived shoot-specific values for the optimal stomatal control model and the annual cycle model using shoot gas exchange measurements at SMEAR II during years 2000–2002 (Sections 6.3.2.3 and 6.3.3.1.3; Mäkelä et al., 2006; Kolari et al., 2007). The shoot-specific parameter values were then averaged into a common parameter set that was used in determining canopy photosynthesis.

The experimental shoots used in the model parameter estimation were located at the top of the canopy. Due to acclimation to prevailing conditions, response of photosynthetic rate to environmental driving factors can vary within the canopy, as was discussed in Section 7.5.1. This variation was, however, omitted. In the relatively sparse canopy of Scots pine there is also very little difference in the annual course of leaf temperature between the upper and the lower parts of the canopy. Therefore, we assume that all shoots in the canopy share the values of the annual cycle parameters of the top shoots, time constant of the change in the state of functional substances (Eq. 6.3.50, Section 6.3.3.1.3) being 180 h.

As the photosynthetic parameters were derived from flattened shoots (Section 6.3.2.3), light interception by the experimental shoots and by the free shoots in the canopy is different: a horizontal shoot catches solar radiation more efficiently than an average free shoot in the canopy (Stenberg, 1996). The inter-cepted radiation can be estimated using the concept of a STAR (shoot silhouette area by total needle area) by Oker-Blom et al. (1989). Assuming cylindrical needles, the STAR of an ideal flattened shoot in a horizontal position is about $0.32(1/\pi)$, whereas in free shoots with random orientation it averages about 0.14 (Stenberg et al., 2001). There was overlap of needles in the measurement chamber so that the ratio of horizontally projected to all-sided needle area was about 0.22, clearly lower than in the ideal flat shoot. The empirical light extinction coefficient at SMEAR II stand was 0.18. The larger value compared to randomly oriented shoots can originate in additional interception of light by branches and stems and in a possible deviation of the shoots from random orientation distribution.

Photosynthetic production was integrated over several years in half-hourly time steps, using incident PAR measured above the canopy and temperature and gas con-centrations measured inside the canopy at 8 m height as the driving factors for pho-tosynthesis. Seasonal variation in the stand foliage area (leaf area index, LAI) was taken into account in a coarse way: Foliage area is at its minimum in winter and spring, starts growing linearly in the beginning of June, stabilises to its maximum value for July and August and declines again in autumn when the oldest age class of needles is shed. Maximum all-sided LAI before the thinning of the stand in early 2002 was 8; after the thinning it was reduced to 6. The annual turnover of foliage was assumed to be 25% of the foliage, corresponding to needle longevity of three years. The seasonal minimum of LAI was thus 6 before the thinning and 4.5 thereafter.

Photosynthesis of Forest Floor Vegetation

Research on the photosynthetic production of boreal forest ecosystems has tradition-ally concentrated on studying the function of trees, and the trees indeed contribute to the majority of the forest ecosystem CO_2 uptake. However, to estimate the carbon balance of a whole ecosystem, we need to take into account the other photosyn-thesising component of a forest: the ground vegetation. In several coniferous boreal forest types, for example black spruce forests of North America, Scots pine forests in Northern Europe, and larch forests in Siberia, the canopy is relatively open. Present silviculture also supports a low stand density which enables sufficient light for pho-tosynthesis also on the ground level. In addition, small disturbances like thinning, windfall or insect damage generate gaps in the canopy, and locally improve greatly the living conditions for ground vegetation. This leads to potentially considerable contribution of the forest floor vegetation to stand CO_2 uptake.

The forest floor vegetation at SMEAR II stand is typical of boreal Scots pine forests. The field layer is dominated by dwarf shrubs including blueberry (*Vaccinium myrtillus*) and lingonberry (*Vaccinium vitis-idaea*) and the ground is covered by feather moss (*Pleurozium schreberi*) and other bryophytes.

The essential environmental factors behind photosynthetic production are the same for trees and ground vegetation, but the actual environment and the physiology of the plants is different. If we do not consider the periods when the stands are tem-

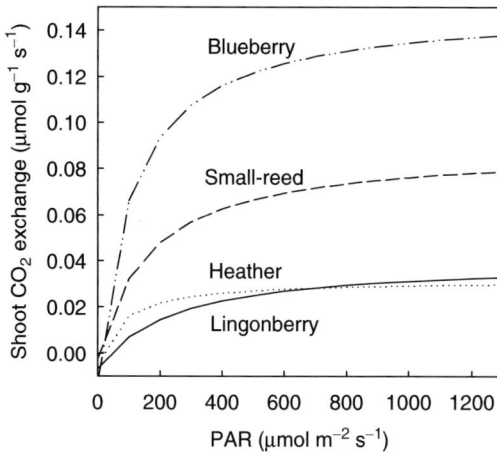

Fig. 7.6.5 Leaf mass based light response curves of CO_2 exchange for blueberry (*Vaccinium myrtillus*), heather (*Calluna vulgaris*), wood small-reed (*Calamagrostis epigejos*) and lingonberry (*Vaccinium vitis-idaea*) at SMEAR II on 20 August in 2006

porarily treeless, forest floor is continuously shaded by trees, thus we can expect that the light response of photosynthesis is such that low light levels are efficiently utilised but photosynthesis saturates at fairly low irradiances. Indeed, photosynthesis in forest floor species saturates at lower irradiances than in Scots pine (Fig. 7.6.5, Widén, 2002; Goulden and Crill, 1997; Skre, 1975; Kolari et al., 2006; Kulmala et al., 2008). Plants with annual leaves have higher mass-specific values of photosynthesis. For example, Kulmala et al. (2008) found that blueberry, with its thin annual leaves, has in the midsummer two to three times as high leaf mass specific rate of photosynthesis as heather or lingonberry, which have perennial leaves (Fig. 7.6.5).

In the boreal zone, forest floor is at least partly covered by snow for a considerable time over the year. The phenology of the forest floor vegetation may thus differ from that of the trees. However, Kolari et al. (2006) found that the annual cycle model originally parameterised for Scots pine (Section 6.3.3.1.3) could predict the seasonal course of photosynthesis in dwarf shrub forest floor vegetation as well. Measured annual pattern of light-saturated photosynthesis rate (P_{max}) for Blueberry, Lingonberry, Wood small-reed and Heather are shown in Fig. 7.6.6. The species with annual leaves (Wood small-reed and blueberry) have higher values and more obvious annual pattern of P_{max} than the species with perennial leaves (heather and lingonberry).

Another typical feature of boreal coniferous forests is the abundance of mosses or lichens on the forest floor. Their different physiology and water relations compared to vascular plants may complicate determining the seasonal pattern of photosynthesis in different forest floor vegetation types. The mosses lack roots and vascular system and largely rely on water supply from rain and dew. Therefore, they are very sensitive to variation in moisture.

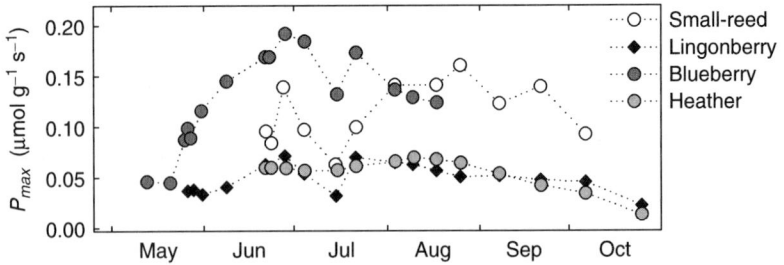

Fig. 7.6.6 Measured photosynthetic activity (P_{max} values) of common vascular plant species at SMEAR II in the year 2005

Roots of vascular ground vegetation species reach only the upper parts of mineral soil where soil moisture is usually lower and more dynamic than in the deeper soil layers that the trees utilise in their water uptake (Duursma et al., 2008). Kulmala et al. (2008) found that photosynthetic activity of heather seems to have no apparent connection to soil moisture in the prevailing conditions unlike the other measured species which were more sensitive to drought. For example, there was a short dry period in July which decreased maximum rates of photosynthesis of all the other ground vegetation species except for heather (see Fig. 7.6.6). In those soil moisture conditions, Scots pine did not show decrease in photosynthetic activity, presumably due to more extensive roots. In the changing climate, we can assume that increasing risk of drought (see Section 10.1.2) first affect the vegetation with roots confined in the upper soil but the evaluation of the effects need still more careful study.

Photosynthetic production of the whole forest floor vegetation can be determined using the same principles of integrating point measurements of momentary photosynthetic rate over space and time up to the stand level. In the integration procedure, we used the photosynthetic light response functions for the different vegetation types, the measured biomass distribution of the forest floor species, and modelled spatial distribution of irradiance at the forest floor. Photosynthetic light responses were determined from measurements of irradiance at the forest floor and forest floor net CO_2 exchange with opaque and transparent chamber. The measurements were taken at about two-week intervals at the study site during two growing seasons with a manual, cylindrical chamber based on the closed dynamic chamber technique. One set of measurements consisted of four to six repetitions with different light intensities and one dark measurement. The highest light intensity was direct sunlight and the other four to five light intensities were created by shadowing the chamber with layers of netted fabric. The difference between the dark and the transparent chamber fluxes directly gives photosynthetic rate of the patch being measured. Photosynthetic parameters were determined separately for blueberry, lingonberry, mosses, heather and wood small-reed.

The irradiance pattern at the forest floor was determined with an empirical model of light attenuation within the canopy (Kolari et al., 2006; Section 7.3). The model was parameterised using measurements at a height of 0.5 m with a multipoint light measuring system (Vesala et al., 2000; Palva et al., 2001; Section 7.3). Measured above-canopy photosynthetically active radiation is divided into direct and diffuse components. Attenuation of the component irradiances and probability of a given point at the forest floor to fall into sunfleck, penumbra or shade category were calculated as a function of the optical path length to give the momentary fractional areas of sunflecks, shade and penumbra and the corresponding irradiances. The upscaled momentary photosynthetic rates were integrated over time in steps of 30 min using half-hourly averaged irradiance data and a modelled seasonal pattern of maximum photosynthesis. The seasonal patterns of light-saturated rate of photosynthesis (P_{max}) were assumed to be similar to the efficiency of functional substances of Scots pine (Section 6.3.3.1.3). Daily values of light-saturated photosynthesis were thus calculated using the same annual cycle model as was used for the Scots pine canopy, the values measured in July representing the annual maxima of P_{max}.

Integrated and Measured Stand Photosynthetic Production

Integrated canopy photosynthesis shows similar diurnal patterns as shoot photosynthesis. It also agrees well with photosynthesis derived from eddy covariance (Fig. 7.6.7). The diurnal variation in stand photosynthesis is mainly driven by changes in solar radiation and cloudiness and modified by the stomata that also respond to water vapour concentration difference between leaf mesophyll and ambient air (Section 6.3.2.3). Photosynthetic rate follows the changes in irradiance in a time scale of seconds (Section 6.3.3.2.1). The half-hourly averaging of the driving factors means that the rapid variations of photosynthetic rate due to intermittent cloudiness are evidently lost.

Different methods for calculating the light environment in the canopy result in different apparent light responses of modelled canopy photosynthesis as shown in

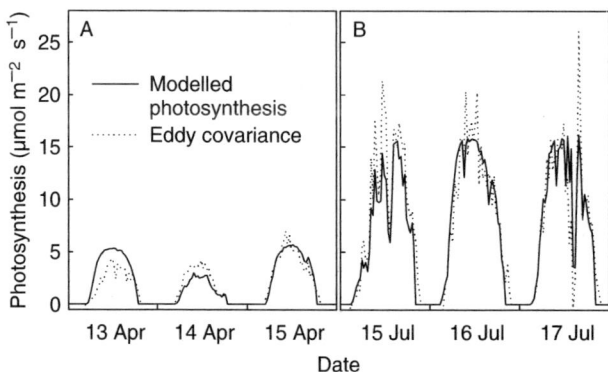

Fig. 7.6.7 Diurnal patterns of modelled stand photosynthesis (trees and forest floor vegetation combined) and photosynthesis extracted from eddy covariance during two three-day periods in spring (A) and in summer (B)

Fig. 7.5.6 (Section 7.5.1). Over one day this is mainly reflected in the absolute level of photosynthesis, not so much in the shape of the diurnal course. From day to day, however, systematically different deviations from measured stand photosynthesis on cloudy and sunny days will result if there is large systematic error in the calculated light environment. No such systematic pattern can be seen in the residuals of our data, so we can conclude that the light distribution model is relatively accurate.

The diurnal patterns of photosynthesis are superimposed on the temperature-driven annual cycle that determines the level of light-saturated photosynthesis. Photosynthetic production in winter is very low due to low temperatures and low light and there is notable photosynthetic activity only during warm spells. Photosynthetic CO_2 uptake increases in spring, levels off for summer and declines again in autumn (Fig. 7.6.8). The regular seasonal cycles of irradiance are similar from year to year. Temperature is related to irradiance but their seasonal courses are different enough to reveal that the spring recovery and the whole annual cycle of photosynthesis is primarily driven by temperature (Section 6.3.3.1.3).

Comparison of integrated and measured stand photosynthesis indicates that in general the stand photosynthesis models can capture the seasonal pattern of photosynthesis very well (Fig. 7.6.8) The models were not calibrated with the measured stand photosynthesis, so the modelled photosynthesis can be considered as true prediction. The regularities observed at the shoot level remain essentially the same when integrated over the canopy and over several years. There is, however, a small systematic difference in the absolute levels of the fluxes: the modelled photosynthesis is slightly lower than measured. The disagreement mainly originates in underestimation of canopy photosynthesis. In early spring the modelled stand photosynthesis is slightly overestimated because the simple annual cycle model cannot predict the decline of photosynthesis after freezing nights. When comparing the absolute level of the fluxes, we must realize that the "measured" stand photosynthesis is not a direct measurement but derived from the measured net CO_2 exchange by subtracting modelled ecosystem respiration. Using different models and explanatory factors for calculating daytime ecosystem respiration can affect the outcome of GPP estimation by as much as 10% (Mäkelä et al., 2006).

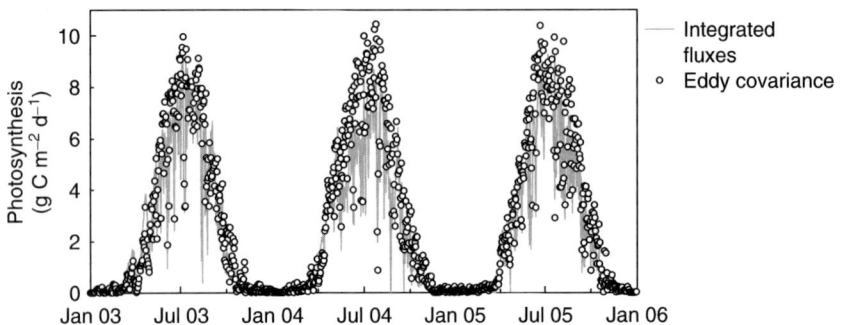

Fig. 7.6.8 Annual patterns of modelled stand photosynthesis (trees and forest floor vegetation combined; grey line) and photosynthesis extracted from eddy covariance (circles) over the years 2003–2005

The integrated photosynthetic production of trees at the SMEAR II stand varied roughly between 900 and 1,000 g (C) m^{-2} a^{-1} (Mäkelä et al., 2006). Photosynthetic production of the dwarf shrub and moss vegetation was 10–15% of the photosynthesis of the Scots pine canopy. Annually this means 100–150 g (C) m^{-2} (Kolari et al., 2006). The mosses contributed to about 40% of the cumulative forest floor photosynthetic production. Moss photosynthesis depends a lot on moisture (not taken into account here), so the actual proportion may vary considerably within just few days. The role of mosses in the whole-stand photosynthesis is so small, however, that we cannot distinguish moisture-related variation in moss photosynthesis from the measured stand photosynthesis.

Photosynthetic production at the forest floor is mainly light limited (Bisbee et al., 2001) and the potential productivity of the forest floor vegetation in unshaded conditions is much higher than under a closed tree canopy. A partial thinning performed in early 2002 at SMEAR II resulted in a notable change in the average proportion of incident light reaching the forest floor: before the thinning it was less than 20% and it increased to almost 30% after the thinning. The partitioning of the photosynthetic CO_2 uptake changed correspondingly: Tree photosynthesis declined by about 8% (80 g (C) m^{-2} a^{-1}) but the decline was partly compensated by an increase of about 20% (20–30 g (C) m^{-2} a^{-1}) in forest floor photosynthesis (Vesala et al., 2005). The total aboveground biomass of the forest floor vegetation at the SMEAR II stand is fairly low, less than biomasses found in the literature. In stands with a denser dwarf shrub and moss cover the forest floor photosynthesis may thus reach a more significant proportion of the whole ecosystem photosynthesis: Goulden and Crill (1997) estimated that the moss-dominated understorey of black spruce forests can account for up to 50% of the total forest photosynthesis.

The agreement between the predicted seasonal courses of the integrated photosynthesis components and the measured stand photosynthesis indicates that the annual cycle of photosynthesis is similar throughout the canopy. Chamber measurements at the top of the canopy thus seem to represent the whole canopy photosynthesis accurately. We can determine the diurnal and seasonal course of photosynthetic production of the stand by just parameterising photosynthesis model with data from a couple of shoots and integrating the instantaneous rate of shoot photosynthesis over the whole canopy and over time.

7.6.2.3 Ecosystem CO_2 Exchange

CO_2 exchange of ecosystems is determined by the uptake of CO_2 by photosynthesis and release of CO_2 by respiration. All living organisms require energy for maintenance of tissues and construction of new tissue. Respiratory release of CO_2 is, therefore, an important part of the CO_2 exchange of ecosystems. A large part of the chemical energy bound in photosynthates is at some point consumed in respiratory processes that supply energy for maintenance and growth of plants themselves (Sections 6.3.2.1 and 6.3.3.1.4) and for heterotrophic organisms decomposing plant biomass. CO_2 is subsequently released into the atmosphere.

Respiration of Foliage

Respiration of green leaves in general follows temperature and can be described fairly accurately with an exponential temperature response function (Section 6.3.2.1). The absolute level of respiration also varies during the summer, for example due to variation in the proportions of maintenance and growth respiration (Section 6.3.3.1.2). A single temperature response of respiration applied over the year yields satisfactory results when it comes to determining the stand carbon budget, but understanding the short-timescale dynamics requires more sophisticated approaches. Mechanistic modelling of respiration is, however, not a straightforward task. For example, we need detailed information on the timing of growth for predicting the proportions of growth and maintenance respiration.

Respiration of foliage in trees at SMEAR II was calculated using an empirical relationship between respiration and temperature (Section 6.3.2.1). The obtained rate of respiratory CO_2 release per unit needle surface area was then multiplied by the total needle area per square metre ground in the stand. We used a simple empirical way to take into account the seasonal variation that is not directly related to temperature. Temperature sensitivity was assumed to be constant over the year but the seasonal course in the base level of respiration was estimated daily in a seven-day moving time window of shoot CO_2 exchange data (see Section 6.3.3.1.2).

Respiration of the vascular plants growing at the forest floor was determined from measurements of aboveground parts enclosed in a chamber. Moss respiration was measured from the living (green) parts of the moss carpet that was detached from the ground. For each species, exponential temperature responses obtained from the CO_2 exchange measurements were used over the year, seasonal variation in the response was omitted.

The effect of moisture on moss respiration was also omitted. Due to the lack of a vascular system and roots, mosses are very sensitive to rainless periods. Figure 7.6.9 illustrates CO_2 exchange of a moss population and its water content measured at SMEAR II in the summer of 2005. The used chamber is described in detail in Pumpanen et al. (2001). We removed patches of *Pleurozium schreberi* and *Dicranum polysetum* from the soil one month before the measurements and placed them on pure quartz sand bed in order to exclude the CO_2 efflux from the ground. The patches on the sand substrate were placed on the forest floor in their natural temperature, humidity and light environment. After long rainless period, there were several rain showers which increased the water content of the whole population. We observed that the water content of the population affected the amplitude of the diurnal pattern: both respiration and photosynthesis increased with water content. The amplitude almost doubled after the water content had increased from circa 80% to 180% of dry moss biomass. However, the CO_2 exchange did not get notably positive (i.e. there was little net CO_2 uptake) during daytime but stayed closed to zero regardless of the water content. In night when photosynthesis is prevented, rates of CO_2 efflux from the population increased with water content. This might indicate that moisture accelerates photosynthesis but the assimilated CO_2 is released relatively fast. That is also plausible due to mosses' lack of carbohydrate storage capacity.

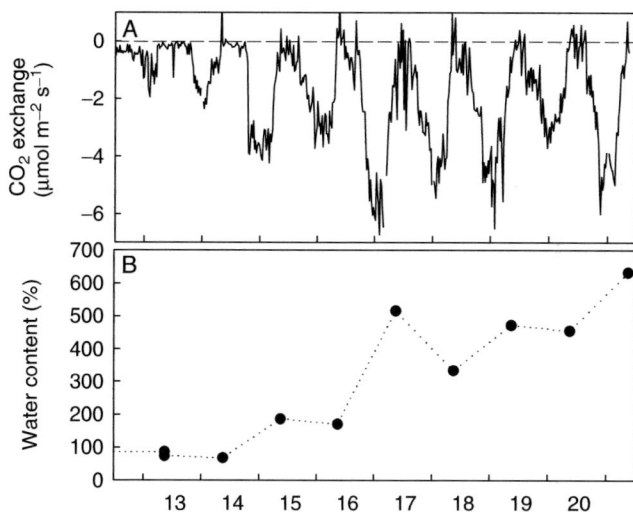

Fig. 7.6.9 CO_2 exchange per ground area (μmol m^{-2} s^{-1}, A) and water content (%, B) from a moss population in 12–21 July 2005. The detached moss patch was most of the time releasing CO_2, only in daytime there was small CO_2 uptake. Water content refers to the amount of water per unit dry moss biomass

However, the measured CO_2 exchange is a sum of uptake and release by the whole patch. Thereafter the CO_2 release from living mosses can not be easily separated from the increased CO_2 flux from dead elements: Decomposition of humus and the lowest parts of mosses increases as well after rain events.

CO_2 Efflux from Stems and Branches

Carbon efflux from woody structures consists of locally formed CO_2 from metabolic processes of living cells in the stem and bark tissues and dissolved carbon in the water of sapflow that is released within the woody tissue from soil upward (Hari et al., 1991).

The metabolic processes forming CO_2 in the stem are linked with the maintenance of the living cells in the stem xylem and bark tissue, the new growth of wood and bark as a result of cambial activity and the transport of solutes in the phloem. Despite the very large weight of woody tissue, the maintenance respiration rates of living wood remains low. This is because the living wood cells are not very active if we compare them to leaf or root tissue. This is reflected in their very low nitrogen concentration, i.e. low concentration of functional substances. Apart from the low activity, also the proportion of the living cells from the wood biomass is low, in conifers generally less than 10%. In broadleaved trees it may be larger but generally not more than 30%. Mohren (1987) estimated based on the activity and proportion of living cells that theoretically the annual stem maintenance respiration should be of the order of 1–2% of the sapwood biomass.

Stem and branch growth take up a considerable proportion of the tree photosynthetic production (up to 50% on a fertile site). It is natural that such a large share of growth is also reflected in the stem respiration rate. Although the woody tissue formed consists to a large degree of cellulose which is rather "cheap" to make, there is a considerable proportion of lignin as well. Assimilation of lignin from basic sugars releases considerable amounts of CO_2. As a general rule of thumb, Penning de Vries (1975) suggested that growth processes release about 25% of the carbon bound in biomass. Specific estimates of carbon cost of constructing different organic molecules in biomass are given in Section 9.1.5.

Since onset of growth has very fixed timing in boreal ecosystems (Section 6.3.3.1.1) and the growth process is such an important factor in stem CO_2 efflux, one might expect a stronger seasonality than just the temperature response (see Figs. 7.6.10 and 7.6.11). In the upper crown the apparent temperature response over

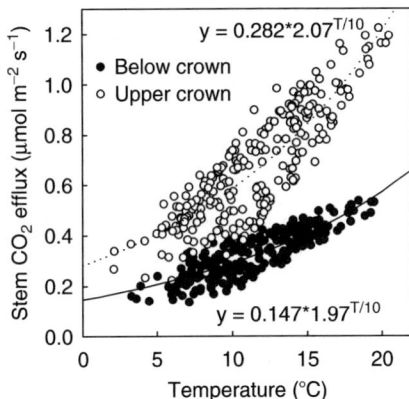

Fig. 7.6.10 The regression of stem CO_2 efflux on temperature in stem below the living crown and in the upper stem in September 2005

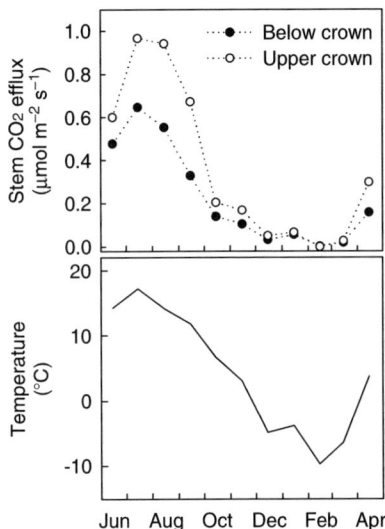

Fig. 7.6.11 Monthly means of stem CO_2 efflux in the lower part of the stem and in the middle of the living crown. In April 2006 the effluxes were higher than in November 2005 that had about the same average temperature. Also during the growing season the efflux was considerably higher in the upper part of the stem where the wood thickness growth was much higher than in the lower part

the year is actually much stronger than in the lower crown because the temperature response is confounded with the growth processes.

The respiratory CO_2 production of woody tissues can be studied by enclosing a part of the stem or branch surface inside a chamber and measuring the CO_2 efflux out of the stem. CO_2 efflux through the bark, however, is not an outcome of local respiratory activity alone but also includes diffusion of CO_2 produced in lower parts of the stem and transported by sapflow. A small amount of CO_2 in the xylem sap originates in the soil. As a result of decomposing processes and root respiration, CO_2 concentration in soil airspaces is several times higher than the atmospheric concentration (Section 7.6.1). The CO_2 concentration in the soil water equilibrates with the soil's gas concentration. When water is transported upwards, some of the dissolved gas is freed from water until an equilibrium concentration gradient is formed in the tree stem, as the permeability of living wood for gases is not very high. Naturally the concentration gradient and also the efflux would then depend on the rate of sapflow through the stem. Model analysis suggests that CO_2 originating from soil or root respiration would not be very important constituent of the stem carbon efflux, but sapflow remains an important component to consider as it is also transporting the CO_2 fluxes originating in the stem.

Respiration of woody parts (stems and branches) in the SMEAR II stand was determined from hourly stem chamber measurements. The seasonal pattern of stem CO_2 efflux was monitored with two chambers attached on the bark, one chamber in the middle of the crown and the other below the living crown. The measured fluxes were used for deriving an exponential relationship between temperature and stem CO_2 efflux (Section 6.3.2.1).

The explanatory factor in the stem CO_2 efflux model was bole temperature which was calculated as a delayed response to air temperature with a time constant τ of 4 h.

$$\frac{dT_{stem}}{dt} = \frac{T_{air} - T_{stem}}{\tau} \tag{7.6.11}$$

An exponential equation (Eq. 6.3.49) was used in modelling stem CO_2 efflux. The temperature sensitivity parameter Q_{10}, i.e., the slope of the temperature response, was first determined by fitting the respiration model to chamber measurements pooled over couple of weeks in the summer. The seasonal course in the base level of respiration was then estimated daily in a seven-day moving time window of stem CO_2 efflux data (similar analysis for shoot respiration can be found in Section 6.3.3.1.2).

The absolute level of CO_2 efflux per unit stem surface area was determined by circulating the chambers between different heights and different trees for a couple of weeks in the summer. Stem CO_2 efflux in the stand was calculated by multiplying the efflux per stem surface area by the total stem surface area ($0.5 \, m^2 \, m^{-2}$ ground) in the stand.

CO_2 Efflux from the Ground

CO_2 efflux from the ground consists of respiration of belowground parts of the plants (autotrophic respiration) and decomposition of biomass by heterotrophic organisms (heterotrophic respiration). Mechanistic understanding of the processes behind CO_2 efflux from the soil is poorer than our understanding of the factors affecting the aboveground plant processes. Several attempts have been made to empirically estimate the proportions of root respiration and microbial decomposition (e.g., Högberg et al., 2001), but linking CO_2 production in the ground mechanistically to the input of litter and root exudates, that add rapidly decaying compounds into the soil organic matter pool, is a challenging task. There are big uncertainties in determining the rates of root growth, growth respiration of the roots, root exudate release and decomposition of organic matter simply because belowground processes are harder to observe and quantify than, for example, photosynthetic production of a shoot.

Soil CO_2 efflux at SMEAR II was modelled empirically as a function of soil temperature and moisture (modelling and the processes underlying soil CO_2 effluxes are described in more detail in Section 7.6.1). The model was parameterised with measurements of CO_2 efflux from the ground with two different chamber setups. The temporal pattern was obtained from hourly measurements with three automated chambers. As the spatial variation in soil CO_2 effluxes is large, measurements at more than three locations are needed to accurately determine the absolute level of CO_2 efflux per unit ground area in the whole stand. Therefore, we made additional flux measurements with a manually operated chamber at 20–30 locations within the stand several times over the summer.

Integration of Component Fluxes

All component CO_2 fluxes at SMEAR II show systematic diurnal and seasonal variation (Figs. 7.6.12 and 7.6.13). Diurnal variation of photosynthesis within a day largely follows the changes in light. The light response is modified by stomatal responses to changes in water vapour concentration difference between leaf mesophyll and ambient air (Section 6.3.2.3).

The diurnal patterns of photosynthesis are superimposed on the temperature-driven seasonal cycle. Temperature also drives the short-term variation in respiratory fluxes, therefore, within one day, respiration follows temperature. Changes in CO_2 efflux from the soil are much slower than in photosynthesis because soil temperature changes slowly. Soil CO_2 efflux thus peaks later in the afternoon than foliage respiration because the driving factor (soil temperature) lags behind the driving factor of foliage respiration (leaf temperature). Diurnal amplitudes of stem and soil CO_2 effluxes are highest on clear spring and summer days when temperature variations are also large.

Qualitatively the diurnal courses of photosynthesis and respiration remain similar over the year but the absolute levels and the partitioning of the component CO_2

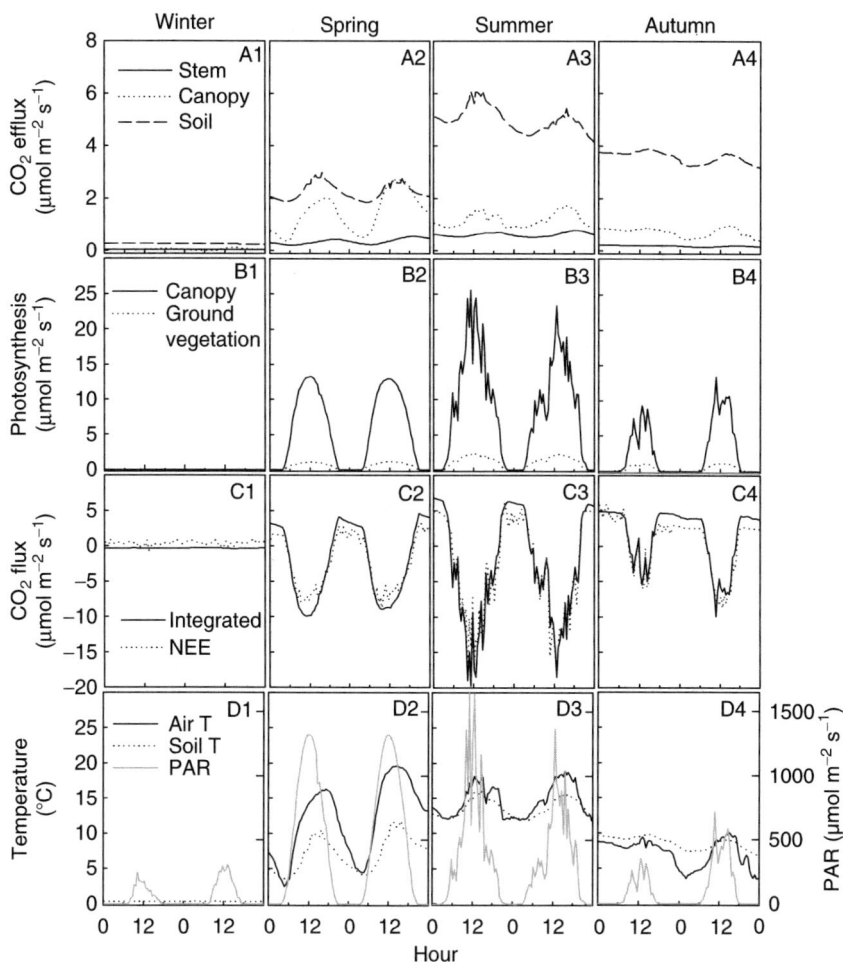

Fig. 7.6.12 Diurnal patterns of component CO_2 fluxes (A and B) and integrated component fluxes and NEE (C) on selected two-day periods in (1) winter (26–27 February), (2) spring (2–3 May), (3) summer (12–13 July) and (4) autumn (29–30 September) of 2004. Contemporary measurements of soil and air temperatures and PAR are shown in the lowest panels (D). Negative CO_2 flux values (C) denote uptake of CO_2 by the ecosystem

fluxes vary considerably (Figs. 7.6.12 and 7.6.13). The respiration component fluxes exhibit similar seasonal patterns as photosynthesis. In winter there is notable photosynthetic CO_2 uptake only during warm spells when temperature rises above 0°C. At freezing temperatures the rates of photosynthesis and aboveground respiration are very low whereas soil CO_2 efflux continues over the whole winter. Biological activity in a frozen soil is limited by the lack of liquid water, but the soil at SMEAR II never freezes completely in winter due to the insulating snowpack. Only a thin surface layer of few centimetres is frozen during cold periods. Therefore, root and microbial activity can take place all year round and CO_2 efflux from the ground

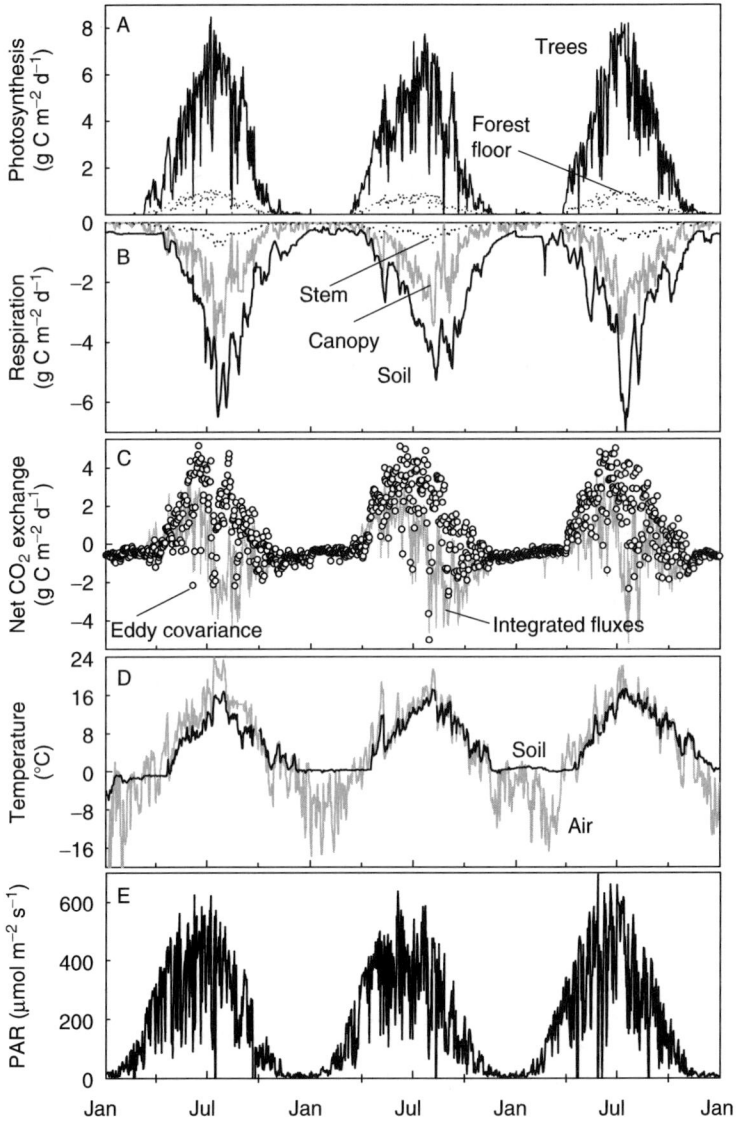

Fig. 7.6.13 Annual patterns of component CO_2 fluxes (daily means, panels A and B) over three years at SMEAR II. Positive net ecosystem CO_2 exchange (C) indicates CO_2 uptake by the stand, negative CO_2 efflux from the stand to the atmosphere. Daily means of soil and air temperatures and PAR are shown in panels D and E

never ceases totally. CO_2 efflux from the ground thus dominates the ecosystem CO_2 exchange in boreal ecosystems in winter (Fig. 7.6.13).

In spring, photosynthetic production and respiratory CO_2 effluxes increase steeply with increasing temperatures. The level of aboveground CO_2 effluxes rises more rapidly and peaks earlier in the summer than CO_2 efflux from the ground. In autumn, soil temperature declines slowly which keeps soil CO_2 efflux relatively high. The whole seasonal course of soil CO_2 efflux can be explained well with soil temperature alone. The observed temperature relationship should, however, be considered as merely apparent response because the processes and driving factors behind the observed effluxes are more complicated. Root growth and supply of root exudates into the soil are also related to photosynthetic production (Pumpanen et al., 2008), which is seasonally driven by air temperature and at a shorter timescale mainly by light.

Both in foliage and wood the respiration normalized to a standard temperature seems to be higher in the spring and early summer than in late summer and autumn (Section 6.3.3.1.2). This reflects the higher respiratory activity when the trees are recovering from winter dormancy and starting growth. There is a continuous CO_2 efflux from maintenance respiration all year round, and an additional respiration component that supplies energy for the growth processes. Attempts to separate growth and maintenance respiration from the measured fluxes have shown that the highest levels of temperature-normalised respiration coincide with the highest rates of stem diameter growth (e.g., Zha et al., 2005).

The daily ecosystem CO_2 budget is positive, i.e., the stand takes up CO_2 from the atmosphere, from approximately late April to mid-September. In the summer the stand's momentary net CO_2 balance depends on photosynthetic production, which in turn largely follows irradiance. Canopy photosynthesis is almost an order of magnitude higher than photosynthetic production by the forest floor vegetation. At night the stand is always a source of CO_2 because respiration goes on, whereas photosynthetic CO_2 fixation ceases due to darkness. From late autumn to early spring the respiratory fluxes, primarily soil CO_2 efflux, dominate the stand CO_2 exchange and the stand is a source of CO_2 to the atmosphere on a daily basis. Daily photosynthetic production in winter is very low due to low temperatures, low light and short daylight hours.

The majority, 60–70% of the stand's respiratory fluxes at SMEAR II originates below the ground surface. Annually this means an efflux of 500–600 g (C) m^{-2}. The annual respiration in Scots pine shoots was about 0.5 g (C) per gram needles, totalling about 250 g (C) per square metre ground. There is uncertainty of about 20% in the annual foliage respiration, mainly because day-time respiration is extrapolated from night-time temperature regressions. The annual respiration of woody tissues of trees was clearly smaller than foliage respiration, only about 100 g (C) m^{-2} ground. In trees, most respiration takes place in foliage and roots and only a minor part is allocated to maintenance and growth respiration of woody tissues. Because of this, respiration is rather constant fraction of photosynthesis (Landsberg and Waring, 1997). However, the relations are influenced by the tree size as the proportions between the foliage and root tissue vs. the living stem tissue change as tree size increases (Mäkelä and Valentine, 2001).

Respiration of forest floor vegetation was small but in relative terms higher than in trees. The allocation to woody structures in dwarf shrubs is lower than in trees, and nonexistent in herbs and mosses. Therefore, we estimate that respiration of forest floor vegetation is about 80% of their photosynthetic production, whereas in trees the typical ratio between respiration and photosynthesis would be roughly 50% (e.g., Waring et al., 1998).

Component Fluxes Versus Eddy Covariance

The diurnal patterns of the combined component fluxes and the measured net ecosystem exchange at SMEAR II agree well (Fig. 7.6.12). The afternoon decline in CO_2 uptake is similar in the integrated CO_2 flux and in the measured NEE. The seasonal courses are also similar (Fig. 7.6.13). There is, however, a systematic difference in the absolute levels of the fluxes. In the summer the integrated CO_2 exchange usually results in weaker carbon sink than the net ecosystem exchange measured by eddy covariance. The disagreement mainly originates in underestimation in modelled canopy photosynthesis or overestimation of daytime respiration components because at night the integrated fluxes and eddy covariance follow each other very closely (Mammarella et al., 2007). As net ecosystem exchange is a sum of two fluxes of opposite sign, small relative errors in the modelled component fluxes can result in a fairly large relative uncertainty in the sum of those fluxes.

In spring the agreement between eddy covariance and the integrated fluxes is good. On some days in early spring the integrated fluxes suggest higher CO_2 uptake than observed by eddy covariance. In those cases the integrated stand photosynthesis is probably overestimated because the simple annual cycle model cannot predict the decline of photosynthesis after freezing nights. Comparison between the integrated photosynthesis and photosynthesis derived from eddy covariance supports this (Fig. 7.6.8). This leads to overestimation in net CO_2 uptake when days are fairly warm but night frosts are frequent.

Annually the integration of the component CO_2 fluxes at SMEAR II results in a small net carbon sequestration, 50–100 g (C) m^{-2}. During years 1997–2006 the annual net ecosystem exchange measured by eddy covariance varied between 140 and 260 g (C) m^{-2} indicating that the stand clearly is a sink of carbon (e.g., Suni et al., 2003). Ideally, the sum of component fluxes, as well as net ecosystem exchange, equals net biomass accumulation, which mainly consists of the increment of the woody biomass of the trees. At SMEAR II the component fluxes do not close the carbon balance completely: the difference between the integrated photosynthetic production and respiration, 50–100 g (C) m^{-2} a^{-1}, is clearly lower than the annual net biomass increment of about 10 m^3 or about 200 g (C) m^{-2} a^{-1} (see Section 7.9). This discrepancy, however, is well within the accuracy of the integrated fluxes. The absolute level of canopy photosynthesis is especially uncertain due to, for example, the simplifications in determining light distributions over the foliage surface area and in conversion of the photosynthetic parameters from the flattened experimental shoots to free shoots in the canopy (Section 7.5.1). Integration of ground and stem CO_2 effluxes is technically a more straightforward task but soil heterogeneity that causes large spatial variation in the measured soil CO_2 effluxes must be taken into account. Therefore, we utilise additional flux measurements at more than 20 locations to reduce this uncertainty to an acceptable level. These simplifications and systematic errors of the integration process altogether create uncertainty of roughly 100–200 g (C) m^{-2} a^{-1} in the integrated ecosystem CO_2 exchange.

Carbon sequestration measured by eddy covariance is a closer match to the tree growth. The eddy covariance method has random and systematic errors (Baldocchi, 2003; Rannik et al., 2006; Richardson et al., 2005; Aubinet et al., 2005; Mammarella et al., 2007). The random component of the uncertainty originates in the stochastic nature of turbulence and more or less cancels out in annual carbon balances. Annually the random uncertainty was estimated at 50–80 g C m^{-2} a^{-1} (Baldocchi, 2003; Rannik et al., 2006).

The systematic errors in eddy covariance are more severe than the random noise. Turbulent fluxes are determined from the information on upward and downward movements of air, and corresponding concentrations in the moving air parcels. Matter or energy transported in only one direction (advection) cannot be detected. Eddy covariance thus underestimates the actual ecosystem exchange under stable atmospheric stratification when there is little turbulent vertical movement of air, especially at night. The source area of the flux detected by eddy covariance, so called footprint, varies with wind direction and turbulence. This is important if the stand is heterogeneous and the sources and the sinks of CO_2 are not evenly distributed. Topography of the site also alters flow patterns and the footprint area.

Calculating carbon budgets from eddy-covariance data always involves replacing part of the measured fluxes with model estimates that are based on the measurements considered as "good". Energy balance closure, i.e., the relationship between the measured heat fluxes and the available energy (net radiation) is often incomplete also in those measurements. This suggests that CO_2 fluxes as well might be underestimated. The exact magnitude of possible underestimation, however, cannot be concluded directly from the energy balance closure (Wilson et al., 2001). At SMEAR II, net carbon sequestration was close to tree growth, i.e., the net ecosystem exchange measured by eddy covariance is probably not severely biased.

The chamber measurements seem to represent the different components of the whole forest ecosystem CO_2 exchange surprisingly accurately. We can determine the diurnal and seasonal course of photosynthetic production of the stand by just parameterising photosynthesis model with data from a couple of shoots and integrating the instantaneous rate of shoot photosynthesis over the whole canopy and over time. Given the uncertainty arising from all the simplifications and systematic errors in the integration process, the agreement of integrated fluxes with eddy covariance (Fig. 7.6.13) and growth measurements can be considered fairly good.

CO_2 Fluxes in Different Aged Stands

The partitioning of the component CO_2 fluxes at SMEAR II is typical of managed boreal Scots pine stands with closed canopy (e.g., Wang et al., 2004). The magnitudes and the relative proportions of CO_2 exchange by different biomass compartments can, however, vary greatly during the stand development. In managed stands the course of natural succession is altered by thinnings where suppressed trees are removed, and finally the stand development is ceased by clear-felling that removes most of the aboveground woody biomass from the stand.

When a mature forest stand is clear-cut, a great amount of dead biomass is left in the stand. Unless the cutting residue is collected for fuel, the branches and foliage from the felled trees are left lying on the ground. The stumps and roots are also destined to die and be decomposed by soil fauna and microbes. The branches, stumps and dead roots mainly consist of fairly resilient compounds like cellulose and lignin. Their masses, however, can be great and therefore the effect of their decomposition on stand CO_2 exchange is considerable for several years after clear-felling.

We have previously studied carbon balances of different aged Scots pine stands with eddy covariance (Kolari et al., 2004) and found that total respiration did not vary much in the chronosequence (Fig. 7.6.14). The partitioning of decomposition and root respiration, however, changed considerably. At a previously clear-cut site, formerly a Scots pine stand, the annual photosynthetic production was about 300 g (C) m^{-2}, which means that respiration of the living vegetation must have been smaller than this. Thus the decomposition of cutting residue and dead roots contributed to more than half of the total CO_2 efflux of about 700 g (C) m^{-2} a^{-1}. Pumpanen et al. (2004) observed similar carbon losses at another clear-cut site: 350 g (C) m^{-2}, 23% of the total carbon pool in the aboveground logging residue, was released during the first year following clear-cutting.

In a developing forest stand the foliage biomass is first low and mainly consists of ground vegetation. In young stands with missing or sparse tree canopy, the field and ground layer vegetation can have considerably higher photosynthetic production than under a closed canopy. Photosynthetic production at the clear-cut site with no trees and the vegetation consisting mainly of lingonberry, grasses and mosses was more than twice the photosynthesis of the forest floor vegetation at SMEAR II. The biomasses of the field and ground layer vegetation at the clear-cut site were

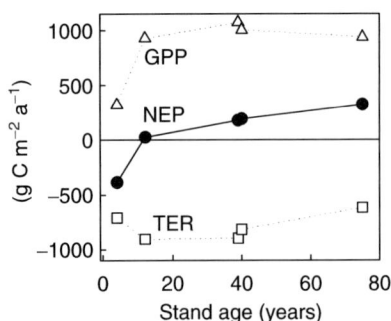

Fig. 7.6.14 Annual net ecosystem productivity (NEP), photosynthesis (GPP) and respiration (TER) as a function of stand age. NEP is the net carbon balance of the stand, positive values indicate carbon uptake by the stand, negative loss of carbon from the stand to the atmosphere. The carbon budgets are based on eddy-covariance measurements in four different aged Scots pine stands (Rannik et al., 2002; Kolari et al., 2004). The 39- and 40-year-old stands refer to flux measurements at SMEAR II in 2000 and 2001, respectively

similar to that in the SMEAR II forest, but due to the lack of shading trees their photosynthetic production was much higher. Tree foliage mass increases gradually until the canopy is closed, whereafter the leaf area stays fairly stable. Tree photosynthesis first increases with foliage biomass. The increase saturates and finally levels off when the canopy is closed. The expanding tree canopy increasingly shades the forest floor, resulting in decline of photosynthesis in the lower vegetation layers.

Due to the decaying cutting residue and the low photosynthetic production and growth, a forest stand is a net source of CO_2 to the atmosphere during the first years after clearfelling (Rannik et al., 2002; Kowalski et al., 2004). Along with increasing foliage area and photosynthetic production and decreasing pools of decaying residue, a young stand turns from a source to a sink of CO_2. In boreal Scots pine stands this happens at around 10 years of age (Karjalainen, 1996; Kolari et al., 2004). After canopy closure the component fluxes of stand CO_2 exchange are fairly stable. Photosynthetic production and soil CO_2 effluxes were similar at SMEAR II and at a nearby 75-year-old stand (Kolari et al., 2004; Fig. 7.6.14). Middle-aged forests in general are usually sinks of carbon, the annual sink strength being comparable to the net increment of woody biomass (e.g., Schelhaas et al., 2004).

In a managed stand, thinning occasionally alters the carbon balance because new woody debris (branches, stumps and roots of the felled trees) is introduced and some of the photosynthetising foliage is removed. The effect does not seem very dramatic or long-lasting though (Vesala et al., 2005). The increased light on the forest floor allows the understorey and forest floor vegetation to compensate for the decrease in canopy photosynthesis. Compared to photosynthesis and respiration of living vegetation, CO_2 efflux from decaying thinning residue, including stumps and roots, is fairly small because the decomposition is slow (Section 6.4.1). The partial thinning performed in early 2002 at SMEAR II did not significantly change the carbon balance (Vesala et al., 2005). The partitioning of the CO_2 fluxes, however, changed: tree photosynthesis declined by about 8% but the decrease was partly compensated by increased forest floor photosynthesis. Respiration of living trees also decreased simply because their total biomass decreased. It is probable that no sharp increase in CO_2 effluxes due to the newly introduced coarse woody debris can be distinguished because it takes time for the microbes to colonise all the dead wood and the rate of its decomposition is slow.

Conclusions

CO_2 exchange of a forest ecosystem is determined by carbon sinks (photosynthesis) and sources (respiration). Photosynthesis and respiration show strong seasonality related to temperature. The respiratory fluxes are also driven by the supply of photosynthates, thus they have similar seasonal patterns as photosynthesis. The annual cycle and environmental responses observed at the leaf or shoot level and integrated over space and time can accurately predict photosynthetic production of the whole forest stand.

7.6.2.4 Developing an Empirical Model Consistent with Process Knowledge for the Daily Photosynthetic Production of Coniferous Canopies

The problem with the physiologically-based canopy photosynthesis models developed to describe the metabolic processes, as well as the distribution of phytomass and light attenuation in canopies, is that they generally require very detailed input information at a high temporal resolution and thus are impractical for large scale applications.

An alternative approach has been to build models that summarise the fast, leaf-level responses at a longer-term, whole-canopy scale, and take longer-term average weather data as their input. Examples of such models are the widely-used Forest-BGC and Biome-BCG (Running and Gower, 1991; Running and Hunt, 1993; Kennedy et al., 2006) with a daily time step, and the 3-PG model (Landsberg and Waring, 1997) with a monthly time step. Moving from a fast time scale compatible with the usual shoot-level photosynthesis measurements to a coarser scale requires integration over time and space to relate the available measurements with the summary model parameters. Because of uncertainties in the methods of summarising, parameter values need usually be calibrated to make model predictions agree with observations (Law et al., 2000; Kramer et al., 2002). With the increasing availability of canopy-level estimates of GPP from eddy covariance measuring stations, this calibration process has become more feasible than ever.

On the other hand, the availability of eddy covariance canopy-level GPP estimates from different regions and biomes also opens up a third approach: developing canopy-level photosynthesis models using statistical model fitting. A few studies have been conducted using hourly (Van Dijk et al., 2005), daily (Yuan et al., 2007) and monthly (Maselli et al., 2006) time scales. However, if we want the fitted models to be applicable as generally as possible, we should base the model structure and parameters on our biological understanding of the processes. The structures of the summary-type photosynthesis models that have been shown to be consistent with canopy level measurements (Landsberg and Waring, 1997; Thornton et al., 2002) would therefore seem good candidates for empirical model fitting. One such candidate is the widely used light use efficiency (LUE) approach (McMurtrie et al., 1994; Landsberg and Waring, 1997). In the following, we summarise a study (Mäkelä et al., 2007) that applied the LUE approach for modelling daily photosynthetic production of coniferous canopies in boreal and temperate conditions.

The LUE model assumes that in theoretical optimal conditions, where no environmental factors restrict photosynthesis, canopy GPP depends linearly on absorbed photosynthetically active radiation (APAR). The proportionality constant is referred to as *potential* LUE, and it gives the amount of carbon that the canopy is able to photosynthesise with one unit of APAR within one unit of time. For real (suboptimal) conditions, the potential LUE need be modified by the effects of restricting environmental factors; as a result we get *actual* LUE, which change in time together with the environmental factors. For the daily canopy GPP, we decided to try the following model:

$$P(t) = \beta I_{APAR}(t) f_L(t) f_S(t) f_D(t) f_W(t) + \varpi(t) \qquad (7.6.12)$$

where $P(t)$ is canopy GPP (g C m^{-2}) during day t; β (g C (mol APAR)$^{-1}$) is potential LUE; $I_{APAR}(t)$ is APAR (mol m^{-2}) during day t; $f_L(t)$, $f_S(t)$, $f_D(t)$ and $f_W(t)$ are light, temperature, water vapour pressure deficit (VPD) and soil water content modifier functions assuming values in [0, 1] in day t; and $\varpi(t)$ is random error during day t. The actual LUE of the canopy in day t is the product of β, $f_L(t)$, $f_S(t)$, $f_D(t)$ and $f_W(t)$.

Although fairly linear over monthly or annual time periods, GPP has been found to be strongly nonlinear with respect to APAR at the daily scale (Medlyn et al., 2003; Turner et al., 2003). This is because the unavailability of CO_2 for dark reactions reduces photosynthesis at high photosynthetically active radiations (Section 6.3.2.3). To account for this nonlinearity, we defined the light modifier so as to yield the rectangular hyperbola when multiplied with the linear response included in the LUE model:

$$f_L(t) \equiv \frac{1}{\gamma I_{APAR}(t) + 1} \qquad (7.6.13)$$

where γ (m^2 mol^{-1}) is an empirical parameter.

In boreal and temperate conifers, the main effect of temperature has been proposed to be through the regulation system of the annual cycle that changes the concentrations of functional substances (Section 6.3.3.1.3). This effect can be described as a temperature-dependent dynamic delay process (Mäkelä et al., 2004; Van Dijk et al., 2005). Accordingly, we modelled the effect of temperature on daily GPP using a temperature transformation $S(t)$ similar to the concept of the state of the functional substances of photosynthesis S_P (Section 6.3.3.1.3): from delayed temperature, $X(t)$ ($^{\circ}$C),

$$X(t) = X(t-1) + \frac{1}{\tau}[T(t) - X(t-1)], X(1) = T(1) \qquad (7.6.14)$$

calculated from the mean daily ambient temperature, $T(t)$($^{\circ}$C), with parameter τ (d) as the time constant of the delay process, we formed a piecewise linear function

$$S(t) = \max\{X(t) - X_0, 0\} \qquad (7.6.15)$$

with X_0 ($^{\circ}$C) as a threshold value of the delayed temperature. The temperature modifier was then defined as

$$f_S(t) \equiv \min\left\{\frac{S(t)}{S_{max}}, 1\right\} \qquad (7.6.16)$$

where the empirical parameter S_{max} ($^{\circ}$C) determines the value of $S(t)$ at which the temperature modifier attains its saturating level. A saturating function was considered more appropriate than the commonly used parabolic temperature effect (McMurtrie et al., 1994; Landsberg and Waring, 1997), as the parabolic function has been observed to cause underestimating bias during the warm season (Thornton et al., 2002).

Stomata react to high water saturation deficit in the air (Sections 6.3.2.3 and 6.3.3.2.2). Following Landsberg and Waring (1997), we thus defined the VPD modifier as a decreasing exponential function:

$$f_D(t) \equiv e^{\kappa D(t)} \qquad (7.6.17)$$

where $D(t)$ (g H_2O m^{-3}) is VPD in day t and $\kappa < 0$ (m^3 g^{-1} H2O) is an empirical parameter.

Furthermore, stomata also control water loss when the water pool in the soil is small (Section 6.3.3.2.2). Following Landsberg and Waring (1997) again, we modelled the reduction caused by decreasing soil water content as

$$f_W(t) \equiv \left[1 + \left(\frac{1 - W(t)}{\alpha}\right)^v\right]^{-1} \qquad (7.6.18)$$

where $W(t)$ is the relative extractable water (REW) in day t and α and v are empirical parameters. REW was defined as

$$W(t) \equiv \min\left\{\frac{\theta(t) - \theta_{WP}}{\theta_{FC} - \theta_{WP}}, 1\right\} \qquad (7.6.19)$$

where $\theta(t)$ is volumetric soil water content (SWC) ($m^3 m^{-3}$), and θ_{WP} and θ_{FC} are SWC at permanent wilting point and at field capacity, respectively.

The model was fitted to data from the following five European coniferous forest sites with eddy covariance measurement towers in them: Sodankylä in Finland (67°22′ N, 26°38′ E; *Pinus sylvestris*), Hyytiälä in Finland (61°51′ N, 24°18′ E; *Pinus sylvestris*), Norunda in Sweden (60°05′ N, 17°29′ E; *Pinus sylvestris, Picea abies*), Tharandt in Germany (50°58′ N, 13°34′ E; *Picea abies*), and Bray in France (44°42′ N, 0°46′ W; *Pinus pinaster*). There were 2–8 years of daily data in each site, making altogether 18 site-years. The sites were contrasting not only in terms of their geographical locations, which formed a good north-south gradient, but also in terms of leaf area index, stand structure and age. After fitting, the model was tested against independent data from two coniferous American sites: Northern old black spruce (NOBS) stand in northern Manitoba, Canada (55°53′ N, 98°29′ W; *Picea mariana*), and Metolius in Oregon, USA (44°27′ N, 121°33′ W; *Pinus ponderosa*). These are not standard sites exactly: in NOBS the mosaic of vegetation within 500 m from the eddy covariance tower consists of about 75% imperfectly to very poorly drained wetlands, whereas Metolius is a dry sandy soil site at a fairly high elevation in a semi-arid region subject to summer drought. There were three years of daily data in each of the test sites.

Daily *canopy* APAR was taken to be a constant, site-specific fraction of daily total above-canopy PAR; the fraction was estimated either from below-canopy PAR measurements or from the canopy leaf area index using the Lambert-Beer law with a site-specific extinction coefficient. Daily *ecosystem* GPP was estimated from eddy flux measurements in the same way as at SMEAR II (Section 7.6.2.1). As we could not separate out the effects of soil and ground vegetation on the ecosystem GPP, we ended up in a slight inconsistency: fitting a canopy GPP model to the data of ecosystem GPP.

The model parameters were first estimated separately for each site and year (site-year-specific models) and for all years in each site (site-specific models); the idea was to study the significance of the different modifiers and to investigate the between-years and between-sites variation in the parameter estimates. A set of common parameters for all the sites and years was then sought by fitting the model with the significant modifiers over the whole data set (whole-data model). Finally, a modification of the whole-data model was fitted where the LUE parameter β was allowed to vary between sites while the other parameters were shared (variable-LUE model).

The soil water modifier improved the model fit significantly only in few site-years, whereas the other modifiers became statistically significant in practically all site-years and sites. Accordingly, we discuss below only the results from the model without the soil water modifier.

The site-year-specific models fitted fairly well in the data, however, the year-specific models in each site performed almost equally well (Figs. 7.6.15 and 7.6.17).

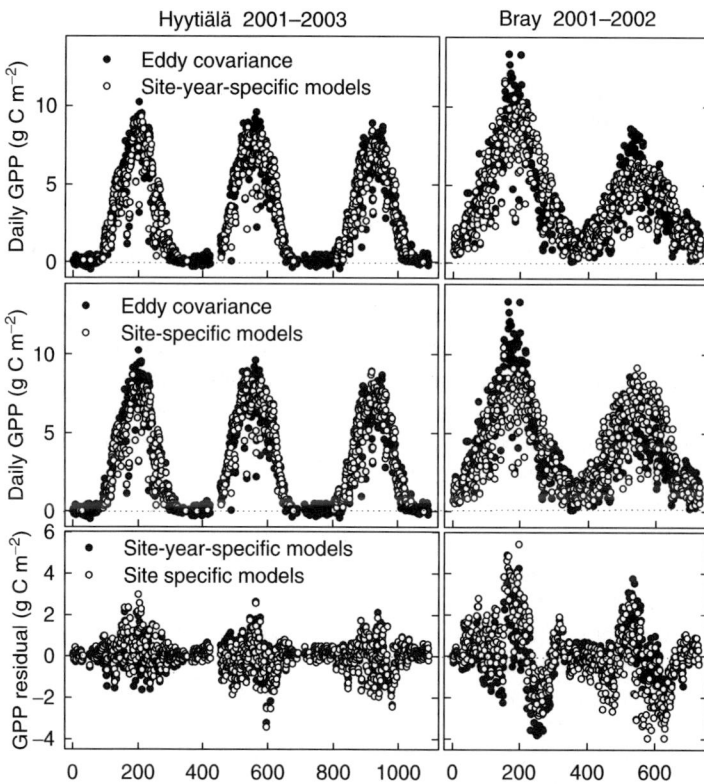

Fig. 7.6.15 The best (Hyytiälä) and the worst (Bray) fits of the site-year-specific and site-specific models containing light (f_L), temperature (f_S) and VPD (f_D) modifiers

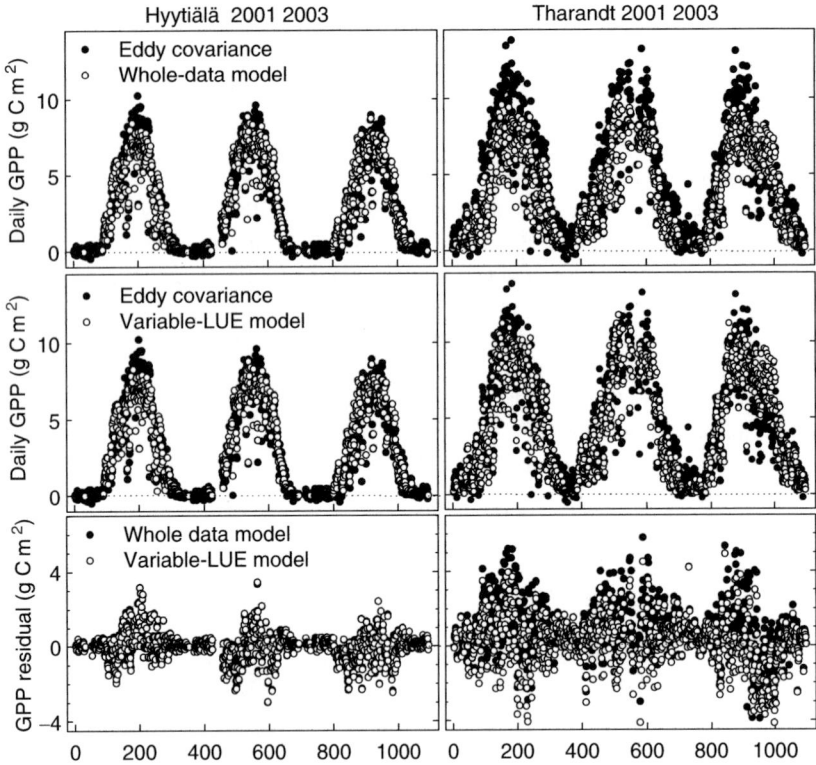

Fig. 7.6.16 The best (Hyytiälä) and the second worst (Tharandt) fits of the whole-data and variable-LUE models containing light (f_L), temperature (f_S) and VPD (f_D) modifiers

In other words, allowing between-years variation in the GPP response to APAR, temperature and VPD did not bring much benefit compared to keeping the response constant in each site. Further, the response could be assumed similar even across sites as long as the LUE parameter β was allowed to vary between sites, as the variable-LUE model fitted to the data almost as well as the site-specific models (Figs. 7.6.16 and 7.6.17). This means that the LUE parameter largely governed the level of GPP in each site and that the gain from the site-wise estimation of the other parameters was negligible.

The whole-data and variable-LUE models were assessed against the test data from the two North American sites. For the variable-LUE model, the LUE parameter β was estimated for each test site conditional on the values obtained from the whole European data for the other parameters. The prediction performance of the variable-LUE model in the test sites was comparable to its fit in the estimation sites, whereas the whole-data model did rather badly in both test sites (Figs. 7.6.15 and 7.6.18). Interestingly, the temporal pattern of the residuals in the dry Metolius

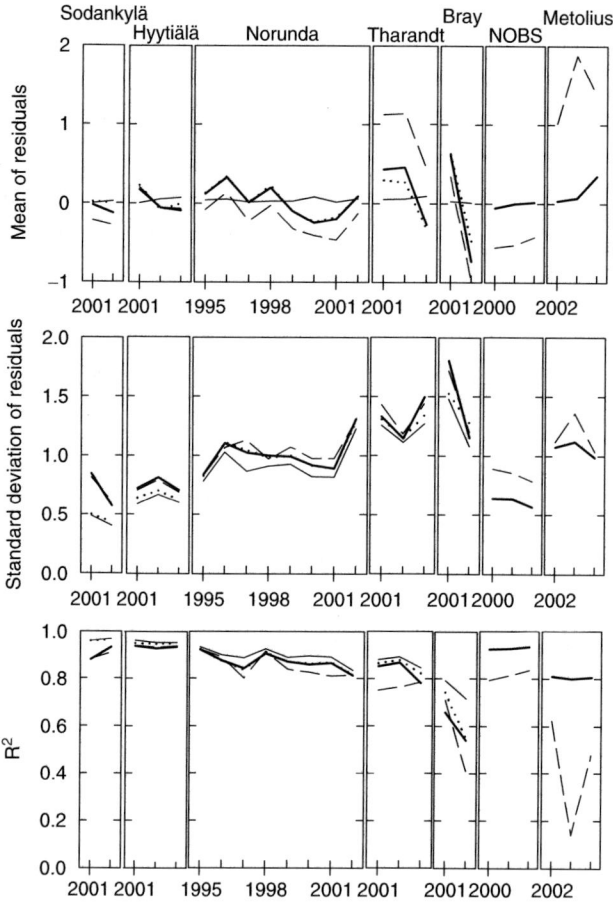

Fig. 7.6.17 GPP residual (g C m^{-2} d^{-1}) diagnostics for the models containing light (f$_L$), temperature (f$_S$) and VPD (f$_D$) modifiers in all the years in the estimation and test sites. R^2 is the unadjusted coefficient of determination. Solid line indicates Site-year-specific models, dotted line Site-specific models, thick line Variable-LUE model and dash line Whole-data model

site (Fig. 7.6.18) resembled that of Bray (Fig. 7.6.15), the only European site in this study where water was likely to be limited.

The annual GPP values summed up from the daily model estimates agreed well with the annual sums of the eddy covariance GPP values (Fig. 7.6.19). According with the daily results, the estimates from the site-specific models were not considerably better than those from the variable-LUE model, but both of these were clearly more accurate than the estimates from the whole-data model.

The main finding above was that while the same response to the environmental driving variables can be assumed across different sites and species, the level of GPP is still site-specific. If we assume that the biochemical mechanism of photosynthetic production is universal and largely independent of species (Section 5.2.1; Landsberg and Waring, 1997), this suggests that some site-specific factors affecting the level

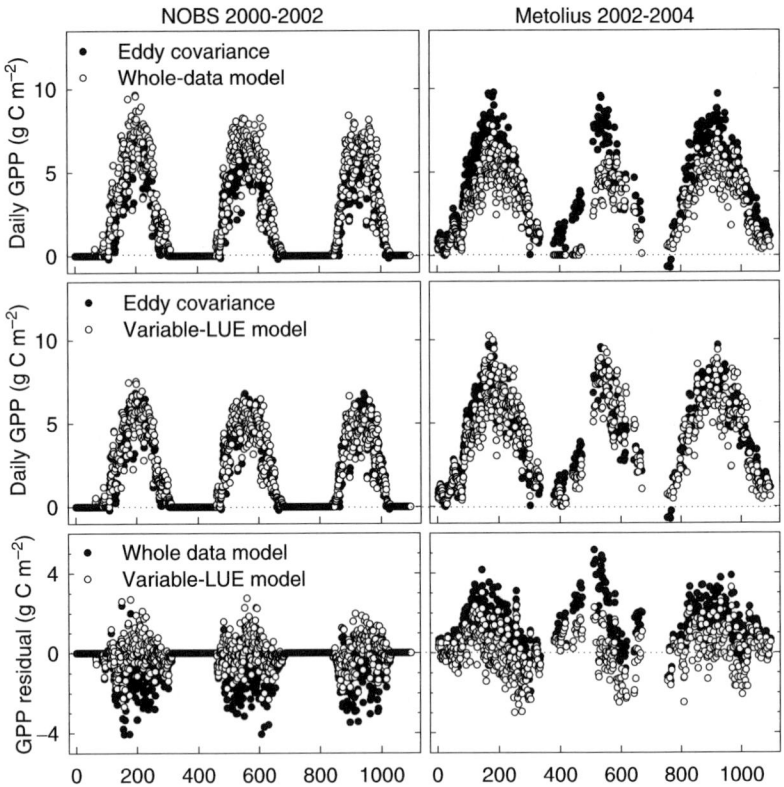

Fig. 7.6.18 Daily GPP predictions with the whole-data and the variable-LUE models containing light (f_L), temperature (f_S) and VPD (f_D) modifiers against daily eddy covariance GPP in the two test sites

of light use efficiency have not been included, or have been mis-represented, in the model. An obvious candidate for such a factor is foliar nitrogen needed to construct functional substances (Section 5.2.1) and suggested in many studies to be a key determinant of canopy photosynthesis (Ågren, 1996; Smith et al., 2002). Indeed, when plotting the variable-LUE model estimates of β against available information on foliar N per leaf area in the sites, a clear relationship was detected. Other factors causing differences in β could be differences in ground vegetation and uncertainty in estimating the fraction of absorbed PAR in the sites.

The relative importance of the significant modifying factors somewhat varied between the sites. The more northern the site the more important delayed temperature appeared to be: while the temperature modifier together with linear light explained almost 90% of the variation in daily GPP in Sodankylä and Hyytiälä and around 80% in Norunda, the effect was less pronounced in Tharandt and Bray and came up mainly through an interaction with VPD and light. In all the site-years delayed temperature was a better explanatory variable than current temperature, as

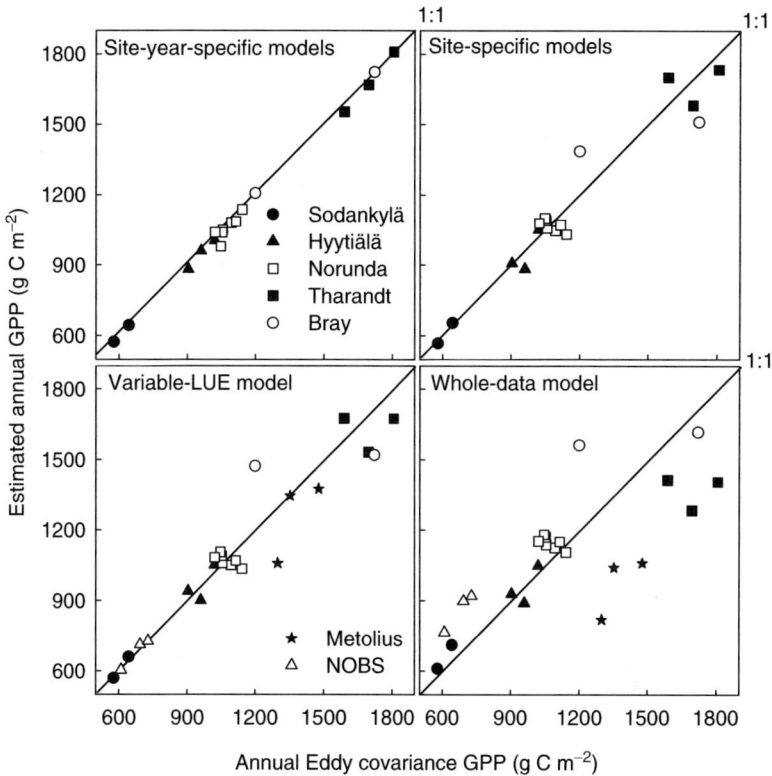

Fig. 7.6.19 Annual GPP estimated with the models containing light (f_L), temperature (f_S) and VPD (f_D) modifiers against annual eddy covariance GPP in the estimation and test sites

the time constant $\tau > 1$ in all the cases. The choice of a saturating light function instead of a linear one and the incorporation of a reducing effect of high VPD values had a smaller but still significant influence, fairly similar in all the sites. It was surprising that soil water did not become a significant explanatory factor. However, this lack of response may be not because of a lack of effect, but because of the sparse data available on soil water. Also, part of the soil water effect was probably embedded in VPD, as drought periods tend to be accompanied by high VPD.

There was variation in GPP at each site between years that was not captured by the non-year-specific models (Fig. 7.6.17). Part of this may be due to inter-annual variation of leaf area which was not included in the model. Some of it may also be due to systematic errors in the eddy covariance measurements that can only be explained by factors not entering the photosynthesis model (variation in footprint area of net ecosystem exchange with wind direction and turbulence regime, temporal and spatial variation in the proportions in which trees vs. ground vegetation and soil contribute to net ecosystem exchange, error related to the estimation of total ecosystem respiration from night-time measurements and temperature).

The statistical approach chosen here to determine the parameter values of a physiologically meaningful model naturally has its restrictions. Although the parameters have a phenomenological interpretation, the approach cannot provide information about the "true" magnitude of the parameters that are strongly correlated with each other, and the values can only be compared among models with exactly the same structure. This was clearly demonstrated in this study by the fact that the parameter values varied between the different model formulations. Parameter correlations reflect the inter-dependence between PAR and the driving variables in the modifiers. Thus, adding a new modifier affects the parameter estimates and even estimability of the other modifiers.

Although the individual parameter estimates cannot be interpreted as "true" values, the full model may still accurately represent the response of GPP to the combination of driving variables. In this sense, the result may be more reliable than in models where the response has been carefully identified for each factor at a time, as small errors in parameter estimates may lead to larger errors in the prediction when combined with other model components. The combined model will then need to be calibrated, as is the case with many physiologically based models (Law et al., 2000; Kramer et al., 2002).

In conclusion, the study showed that the day-to-day variation of GPP over a wide geographical range of temperate and boreal coniferous forests can be rather generally explained by the variation in absorbed PAR, annual cycle of photosynthesis, and VPD. The explanation seemed more robust in the more boreal, less drought-limited sites, but no definite improvement was gained in this study by including the relative soil water content as an explanatory variable. The absorbed PAR is a function of total PAR and leaf area index, possibly modified to some extent by stand structure (Duursma and Mäkelä, 2007). However, after accounting for the three modifying factors, some between-sites variation remained in the potential LUE. For practical applications of the model, the site-specific values of potential LUE can relatively easily be estimated if eddy covariance data are available.

7.6.3 Maximum Transpiration Rate and Water Tension During Drought

Remko Duursma[1,2], Pasi Kolari[3], Martti Perämäki[3], and Pertti Hari[3]

[1] University of Western Sydney, Centre for Plant and Food Science, Australia
[2] Macquarie University, Department of Biological Science, Australia
[3] University of Helsinki, Department of Forest Ecology, Finland

Transpiration often exceeds rainfall even in the humid boreal forests. Although the soil water pool filled up during autumn and winter is large, it may be depleted during long spells of high-pressure weather and drought is developed. As the soil water tension (P_S) becomes increasingly negative during a drought, water transport in the soil-plant-atmosphere system becomes increasingly difficult. In a simple one-dimensional representation of flow, water flow from the bulk soil to the leaf is proportional to the conductance of the pathway, and the tension difference between soil and leaf (Eq. 7.6.20).

$$E = k_T \cdot (P_S - P_L) \qquad (7.6.20)$$

where E is the canopy average leaf transpiration rate (mol m^{-2} (leaf) s^{-1}), k_T the leaf-specific hydraulic conductance of the whole pathway (mol m^{-2} (leaf) s^{-1} MPa^{-1}), P_S soil water tension (MPa) and P_L bulk leaf water tension. The conductance of the pathway can be split up into soil (see Section 6.3.2.8) and plant conductance (see Section 7.5.2.2), because they are coupled in series,

$$\frac{1}{k_T} = \frac{1}{k_S} + \frac{1}{k_{PL}} \qquad (7.6.21)$$

where k_S is the leaf-specific hydraulic conductance of the pathway from 'bulk soil' to the root surface (see Box 1 for estimation of this conductance), and k_{PL} hydraulic conductance of the whole plant (from fine roots to the leaf). The relative values of the two main conductances, that characterize the transport of water from the soil to the root surface and from the roots to the leaf (Section 6.3.2.8, Fig. 7.6.20), affect the response of plant water uptake during drought.

From Eq. 7.6.20 it becomes clear that as the soil water tension decreases, the leaf water tension needs to decrease to allow the same transpiration rate. Several classic studies have attempted to relate leaf water tension measurements to stomatal conductance, and have often used it as a covariate in empirical models of stomatal conductance (e.g., Jarvis, 1976). However, leaf water tension is of a different nature than evaporative demand, temperature, and light, that strongly control c_{H2O} (stomatal conductance Section 6.3.2.3), because c_{H2O} itself controls leaf water tension through transpiration, that is, there is a feedback loop. Usually poor relationships are found between P_L and c_{H2O}, except sometimes in very dry soils (Rambal et al., 2003).

In early morning, leaf water tension is more or less equilibrated with the soil water tension; this pre-dawn tension subsequently declines during a drought (Rambal et al., 2003). Towards midday, evaporative demand and light intensity

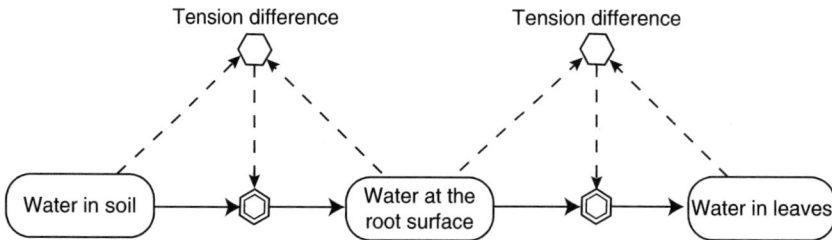

Fig. 7.6.20 Schematic of the transport of water from bulk soil to the leaves when the transport pathway is split into just two main pathways, from the soil to the root surface (soil-determined resistance) and from the roots to the leaf (plant-determined resistance). The symbols are introduced in the Fig. 1.2.1

increase, leading to increased transpiration and decreased leaf water tension. Sometime during the day, the daily maximum transpiration rate (E_{max}) is reached, corresponding to a minimum leaf water tension (P_{min}). From Eq. 7.6.21,

$$E_{max} = k_T \cdot (P_S - P_{min}) \tag{7.6.22}$$

When the drought progresses, the negative tension cannot decrease indefinitely, because water transporting elements will embolize at sufficiently low negative pressures (Section 7.5.2), depending on xylem properties (Sperry and Tyree, 1988). Many woody species are thought to limit transpiration by closing stomates before a serious loss of hydraulic conductance due to embolization occurs (Jones and Sutherland, 1991; Cochard et al., 1996; Bond and Kavanagh, 1999). This behaviour would result in a near constant minimum leaf water tension (P_{min}) when the soil water tension decreases, as has been shown experimentally (Loustau et al., 1996; Irvine et al., 1998; Delzon et al., 2004; Fisher et al., 2006). It may be possible that leaves maintain a safety margin before major embolisms occur (Cochard et al., 1996), or that embolization cannot be completely avoided, especially in a severe drought (Saliendra et al., 1995; Nardini and Salleo, 2000; Salleo et al., 2000).

This 'homeostatic' behaviour of stomata in response to dry soil results in a near constant midday leaf water tension during a drought, and explains why poor relationships between leaf water tension and stomatal conductance are usually found, making it doubtful whether to include leaf water tension as a covariate in stomatal conductance models. Even though midday bulk leaf tension does not change during a drought, it is likely that stomatal conductance is still controlled by leaf water tension, much like a thermostat keeps temperature constant in a room by sensing very small changes in room temperature (Saliendra et al., 1995). The exact mechanism by which stomata sense changes in water tension is still controversial (Buckley and Mott, 2002), as well as the question to what degree chemical signals (especially ABA) play a role in signalling drought. Both in *Betula occidentalis* (Saliendra et al., 1995) and *Pinus sylvestris* (Perks et al., 2002), it was found that chemical signals were unlikely to play a major role, as the dynamics of stomatal closure could be well explained with hydraulic signals only, and chemical signals would travel much more slowly than hydraulic signals. The pressure driven intimate linking between the water and carbon transport suggests, however, that stomata need to be responsive to both water availability from soil and to source- sink relationship in the phloem transport (see Section 7.5.2).

The optimal stomatal control model (see Section 6.3.2.3) does not incorporate the leaf water tension as a control on stomatal action, but a recent extension accounts for the gradual stomatal closure with soil drying (Duursma et al., 2008). The idea is that stomatal conductance follows the optimal stomatal control model, unless the maximum allowable transpiration rate (E_{max}) (Eq. 7.6.22) is lower. The value of E_{max} depends on soil water content, conductance of the soil-plant pathway, root to shoot ratio, and soil type (Eq. 7.6.22, and see Box 1 for the conductance of the soil to root pathway). The final estimate of the stomatal conductance is then,

$$c_{H2O} = \min(c_{OPTI}, E_{max}/VPD)$$

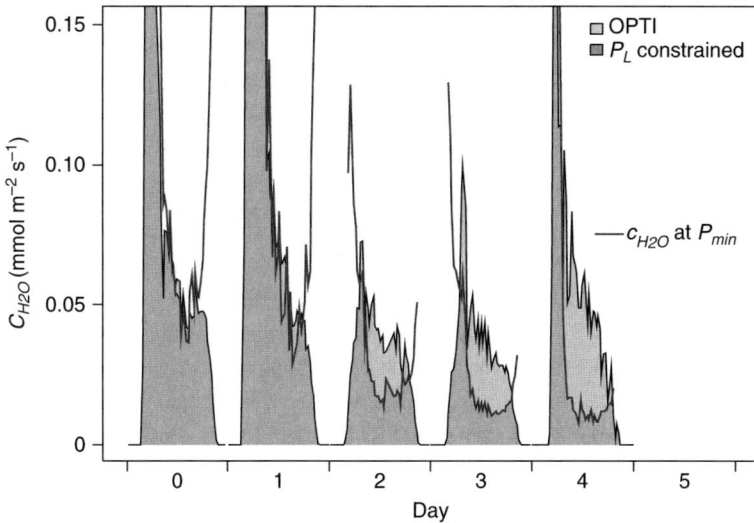

Fig. 7.6.21 Modelled stomatal conductance (c_{H2O}) during five selected (non-consecutive) days in the summer of 2006, using measured light, vapor pressure deficit, and temperature. Soil water content declined from left to right. The thick black line stomatal conductance calculated from maximum transpiration rate achievable with the current soil and plant conductance, and the minimum leaf water tension (Eqs. 7.6.21–7.6.22), the dark grey area is the achievable c_{H2O} (the minimum of c_{OPTI} and c_{Emax}). Note that even during severe water stress (day 4), stomata are still opened for a short period in the morning

where c_{OPTI} is the estimate from the optimal stomatal control model (Section 6.3.2.3), E_{max} is calculated with Eq. 7.6.22, and VPD is the vapour pressure deficit. This method is illustrated in Fig. 7.6.21 for five selected days during the 2006 drought. During the first two days, when soil water content is still relatively high, c_{H2O} estimated from E_{max} is equal or higher than stomatal conductance from the optimal control model, in other words there is no effect of soil water content on stomatal conductance. In the next three days, stomatal conductance from E_{max} is substantially lower than c_{OPTI}, and effects of water stress occur especially later in the day.

To test the prediction of E_{max} with Eq. 7.6.22 within this hydraulic framework, we used measurements of transpiration rate and sap flow during the 2006 drought (see Section 6.3.3.2.2). Using measured soil water content, we estimated the hydraulic conductance of the whole pathway (Duursma et al., 2008), and predicted E_{max} with 7.6.22. Figure 7.6.22 shows that the decline in daily maximum transpiration rate is similar to that predicted with the model.

Box 1: Transport of water to the root surface

In dry soil, the conductance to water transport from the bulk soil to the root surface can be sufficiently low that the water tension at the root surface is much lower than that in the bulk soil. For prediction of water uptake in dry soil, this conductance needs to be estimated. Because the root system and the flow paths of water are highly complex, some simplifications are needed to arrive at a practical estimate of the soil conductance. A typical simplification is to consider the root system as one long straight root with a cylinder of soil around it (Gardner, 1960; Passioura and Cowan, 1968). The amount of soil around the idealized long root depends then on the rooting density. Gardner (1960) developed a steady state approximation to the soil conductance (k_S), assuming that the soil hydraulic conductivity (K_s) does not vary within the soil cylinder. The expression for leaf-specific soil conductance ($\mathrm{mol\,m^{-2}\,s^{-1}\,MPa^{-1}}$) is

$$k_S = \frac{R_l}{L} \cdot \frac{2\pi \cdot K_S(P_S)}{\log(r_{cyl}/r_{root})} \qquad (7.6.23)$$

where R_l is root length index (m root m^{-2} soil surface), L leaf area index (m^2 leaf m^{-2} soil surface), K_s soil conductivity (mol m^{-1} s^{-1} MPa^{-1}), P_s 'bulk soil' water tension, r_{root} root radius, and r_{cyl} the radius of a cylinder of soil to which the root has access. The derivation of Eq. 7.6.23 is very involved, and not shown here (see Passioura and Cowan, 1968). The soil can be divided up into identical cylinders with a root along the middle axis, resulting in the radius of the cylinder

$$r_{cyl} = 1/\sqrt{\pi \cdot L_v} \qquad (7.6.24)$$

where L_v is the root length density (m m^{-3}) of the roots that are active in water uptake. Soil hydraulic conductivity is a function of soil water tension and soil type. Here, we use a simple empirical equation relating K_s to soil water tension from Campbell (1974),

$$K_s(P) = K_{sat}\left(\frac{P_e}{P_s}\right)^{2+3/b} \qquad (7.6.25)$$

where K_{sat} is saturated conductivity (mol m^{-1} s^{-1} MPa^{-1}), P_s soil water tension (MPa), and P_e (MPa) and b empirical coefficients that are estimated from an empirical soil water retention curve (Campbell, 1974),

$$P_s = P_e\left(\frac{\theta}{\theta_{sat}}\right)^{-b} \qquad (7.6.26)$$

where θ is volumetric water content (m^3 m^{-3}), and θ_{sat} is water content at saturation. The coefficient b takes values between ca. 2.5 (sand) and 11 (clay) (Cosby et al., 1984).

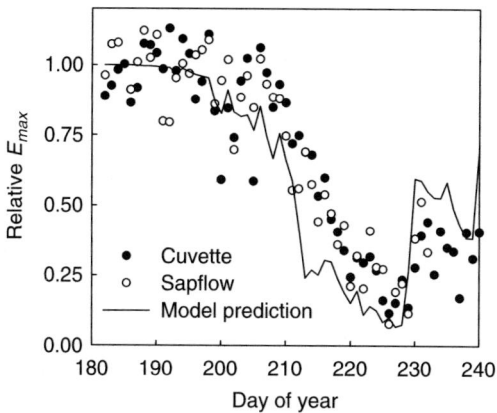

Fig. 7.6.22 Measured and modelled decline of relative daily maximum transpiration rates during the 2006 drought at the SMEAR-II station. Daily maximum transpiration rates were expressed relative to the average maximum transpiration rate in the period DOY 182–200, during which no effect of soil water content was found, and weather conditions were fairly constant

Conclusions

The decline in daily maximum transpiration during the 2006 drought in Hyytiälä could be well explained with the assumptions of a constant plant hydraulic conductance, constant daily minimum leaf water tension, but variable conductance from soil to root surface. It is concluded that the resistance from soil to the root surface is an important limiting factor in water transport during drought.

7.6.4 Fluxes and Concentrations of N_2O in Boreal Forest Soil

Mari Pihlatie[1], Jukka Pumpanen[2], Asko Simojoki[3], and Pertti Hari[2]

[1] University of Helsinki, Department of Physics, Finland
[2] University of Helsinki, Department of Forest Ecology, Finland
[3] University of Helsinki, Department of Applied Chemistry and Microbiology, Finland

N_2O is a green house gas (Section 10.1) and N_2O flux from forest soil reduces the nitrogen pool in the ecosystem. During recent years, information regarding N_2O emissions from boreal ecosystems has markedly increased. Emissions of N_2O from soils in the boreal region vary greatly from near zero to 25 kg N_2O-N ha^{-1} year^{-1} (Martikainen et al., 1993; Nykänen et al., 1995; Kasimir-Klemedtsson and Klemedtsson, 1997; Regina et al., 1999; Maljanen et al., 2003; Regina et al., 2004). Most studied boreal ecosystems are agricultural soils, and forested and natural peatlands. The highest N_2O emissions have been measured from agricultural peat soils (Kasimir-Klemedtsson et al., 1997; Regina et al., 2004) and afforested peat soils with former agricultural history (Maljanen et al., 2001; Mäkiranta et al.,

2007). The smallest N_2O emissions have been measured in natural peatlands (Martikainen et al., 1993; Regina et al., 1999).

Scattered emission measurements or laboratory studies of N_2O production in boreal upland forest soils indicate that these ecosystems are very small sources of N_2O (Schiller and Hastie, 1996; Paavolainen and Smolander, 1998; Brumme et al., 2005). However, the studies on mineral upland forest soils have been short and concentrated on investigating the effects of different forest management practices on trace gas emissions (Martikainen et al., 1993; Paavolainen et al., 2000; Maljanen et al., 2006).

Although the gaseous losses of nitrogen into the atmosphere from boreal upland forest soils are only a minor part of the nitrogen cycling within the ecosystem the gaseous losses of N_2O and NO are important to global climate change and to the photochemistry of the lower troposphere. Nitrous oxide is a strong greenhouse gas and accounts for approximately 6% of the anticipated global warming (IPCC, 2007), and nitrogen monoxide (NO) is a chemically active trace gas in the troposphere and plays an important role in atmospheric chemistry (Seinfeld and Pandis, 1998).

The production or consumption of N_2O depends on the activity of soil microbes, which in turn depends on the availability of the substrates, NH_4^+ and NO_3^- ions, and organic carbon for the microbial processes. Soil temperature in general drives the activity of soil microbes, while soil moisture controls the aeration and gas diffusion in the soil. The transport of N_2O between the soil layers is driven by the concentration gradient (Fig. 7.6.23). The physical structure of the soil and the proportion of air-filled pore space strongly influence the gas transport within the soil.

We measured soil N_2O fluxes by static chamber and soil gradient methods. The chamber measurements with six chambers were conducted weekly to fortnightly during the snow free period in 2002–2003, and once per month in the winter. The chambers were rectangular and made of stainless steel (width x length x height: $0.29 \times 0.40 \times 0.10$ m). During a measurement the collars were closed with a stainless steel chamber (width x length x height: $0.29 \times 0.40 \times 0.20$ m) and a fan mixing the headspace of the chamber. The chambers were closed for 60 min and four gas samples were drawn from the chamber at 20 min intervals. In the soil-gradient method the concentrations of N_2O were monitored at different soil depths. The gas collectors, installed in the various soil layers, were sampled for soil air every fortnight or month and analyzed for N_2O by gas chromatography. The soil gradient method relies on the knowledge of soil structure, total porosity and soil water content, which are used as parameters in the flux calculation. The fluxes between the different soil layers were calculated using the model described in Pumpanen et al. (2003). The calculations are explained in detail in Pihlatie et al. (2007).

We measured N_2O fluxes ranging from small emissions to small soil uptake with both soil chamber and soil gradient techniques. In the summer and most of

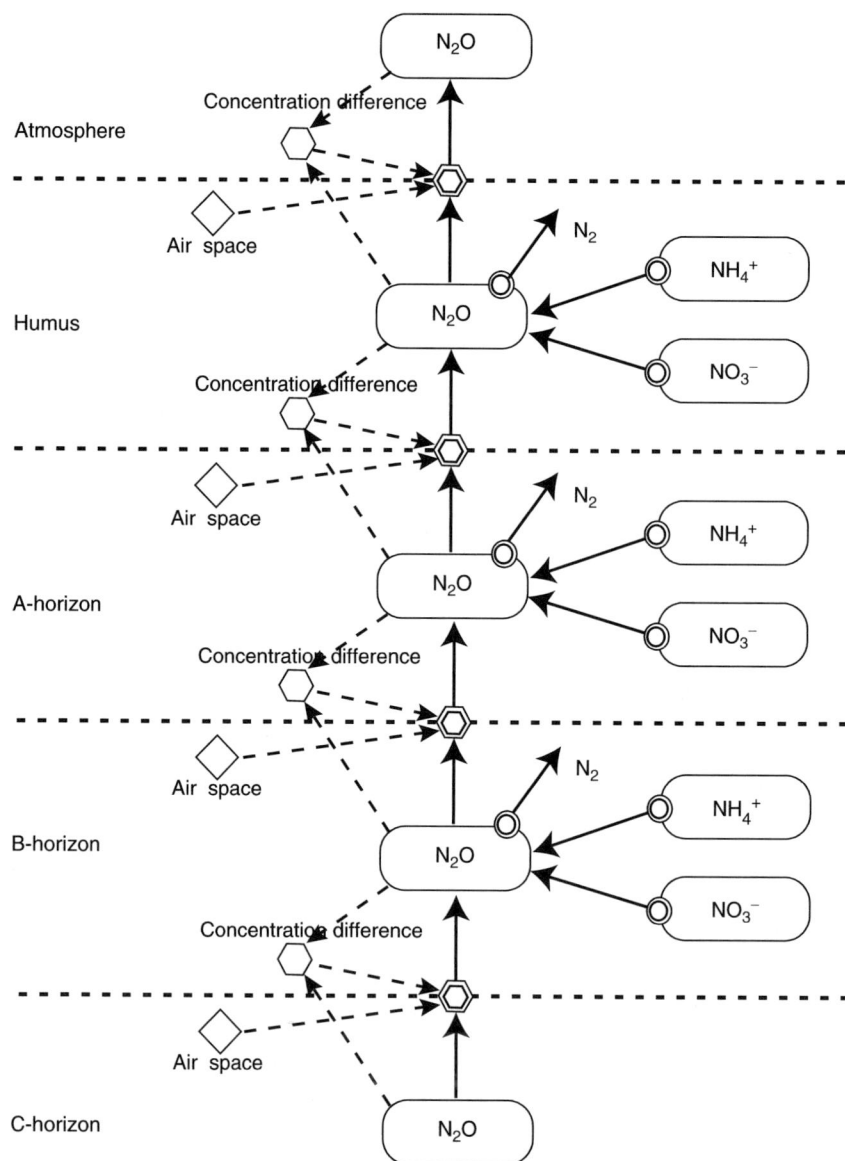

Fig. 7.6.23 Production and transport of N_2O in the soil profile of a Podzol forest soil. Boxes denote the concentration of the various nitrogen compounds (N_2O, N_2, NH_4^+, NO_3^-) in different layers of the soil, bold arrows illustrate the transport of N_2O. The symbols are introduced in the Fig. 1.2.1

the time in the spring the soil N_2O concentrations increased with depth of the soil (see Figs. 7.6.24–7.6.26). In the autumn, the N_2O concentration profile changed and the highest N_2O concentrations were measured in the O-horizon (Fig. 7.6.25). This autumn-peak in N_2O concentration in the O-horizon occurred after a litter fall,

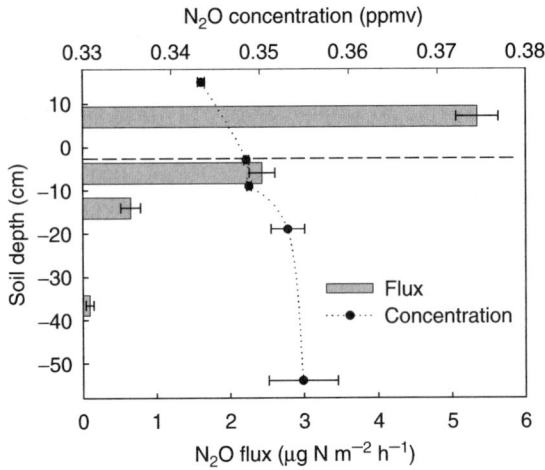

Fig. 7.6.24 Example of summer time (3 July 2003) soil N_2O concentration profile and N_2O fluxes as an average from the pits 100, 130 and 160. Error bars represent the standard errors of the mean between the pits and between the ambient air samples (n = 4), respectively. A dotted line between the concentration measurements is drawn to guide the eye of the reader (Redrawn from Pihlatie et al., 2007)

which in 2002 was exceptionally high due to the summer drought. We considered that the litter fall stimulated N_2O production in the top-soil layer, possibly due to increased organic matter mineralization in the litter and humus layer and a consequent release of mineral N into the soil (Pihlatie et al., 2007).

The fluxes between different soil horizons varied between the seasons. The O-horizon acted as a source of N_2O in the autumn but as a sink in the spring, whereas, the A-horizon acted as a sink during most of the autumn and as a source in the spring (see Figs. 7.6.24–7.6.26). The highest N_2O emissions calculated from the soil profiles occurred in July–August 2002 when the soil temperatures were high and the soil moisture was intermediate (see Fig. 7.6.25).

According to Ambus et al. (2006) the N_2O production in European forest ecosystems is mostly driven by nitrification since that is the sole source of NO_3^- for denitrification. As nitrification and denitrification are often strongly coupled, they occur simultaneously in the soil. Nitrification in the aerobic microsites provides the substrates, NO_2^- and NO_3^-, for the denitrification that takes place in the anaerobic microsites. Non-fertilized forest ecosystems receive nitrogen only from the atmosphere and hence regions with low N deposition, such as boreal forests in Finland, are largely N-limited. The mineral N in the soil is predominantly ammonium, and as a result of low nitrification activities, the soil nitrate content remains very low (Priha and Smolander, 1999; Priha et al., 1999; Ambus et al., 2006; Pihlatie et al., 2007).

The small N_2O emissions and the variability between uptake and emission from the Hyytiälä upland forest soil are in line with other measurements from upland boreal forest soils (Saari et al., 1997; Schiller and Hastie, 1996; Simpson

Fig. 7.6.25 Mean soil N_2O concentrations and N_2O fluxes in the autumn 2002. The error bars for the soil concentrations and fluxes are the standard errors of the mean between the four locations (pits). For ambient air values the error bars express the standard errors of the mean between four replicate air samples. A dotted line between the concentration measurements is drawn to guide the eye of the reader (Redrawn from Pihlatie et al., 2007)

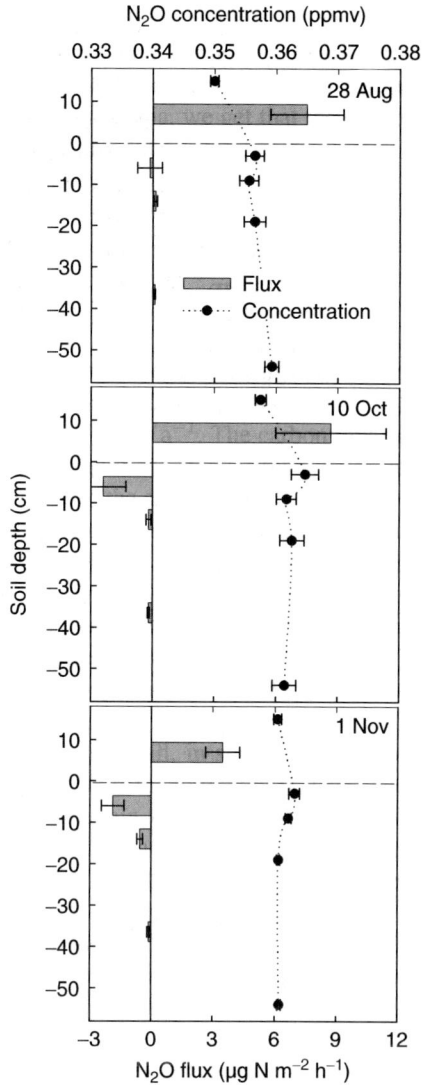

et al., 1999). Consumption of atmospheric N_2O has also been reported in some N-limited temperate and Mediterranean forest ecosystems (Butterbach-Bahl et al., 1998; Goossens et al., 2001; Rosenkranz et al., 2006). Both in the Mediterranean and the Hyytiälä boreal pine forest soil the organic top-soil layer was responsible for most of the N_2O consumption (Rosenkranz et al., 2006; Pihlatie et al., 2007). As the observations of N_2O uptake by soils have only recently gained attention, the processes responsible for N_2O consumption in soils are poorly understood (Conen and Neftel, 2007; Chapuis-Lardy et al., 2007). Anaerobic denitrification is currently the only biological process known to consume N_2O (Conrad, 1996).

Fig. 7.6.26 Mean soil N_2O concentrations and N_2O fluxes in the spring 2003. The error bars for the soil concentrations and fluxes are the standard errors of the mean between the four locations (pits). For ambient air values the error bars express the standard error of the mean between four replicate air samples. A dotted line between the concentration measurements is drawn to guide the eye of the reader (Redrawn from Pihlatie et al., 2007)

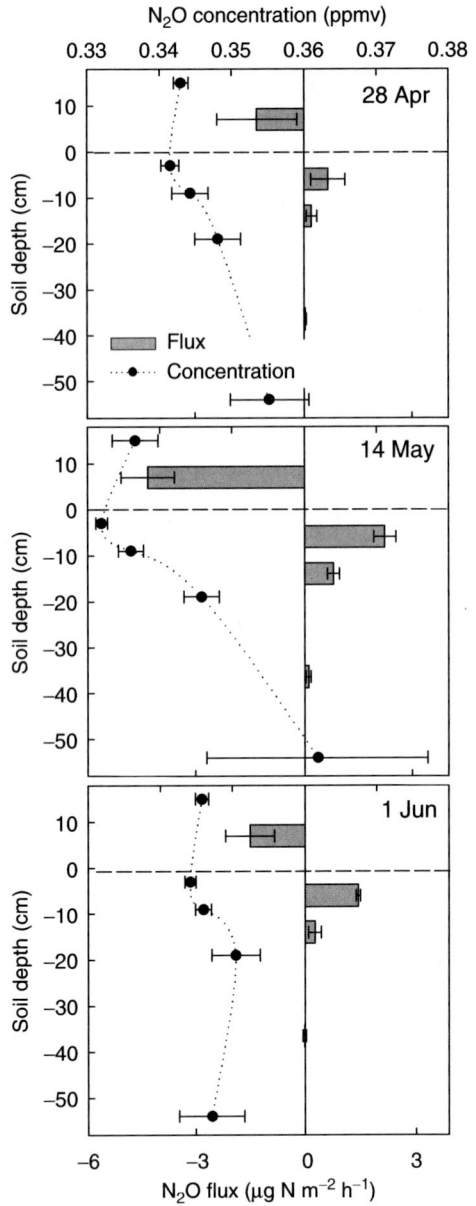

However, also consumption of N_2O by nitrifying bacteria and heterotrophic bacteria has been hypothesized to consume N_2O (Wrage et al., 2001; Rosenkranz et al., 2006; Chapuis-Lardy et al., 2007).

7.6.5 Methane Fluxes in Boreal Forest Soil

Mari Pihlatie[1], Asko Simojoki[2], Jukka Pumpanen[3], and Pertti Hari[3]

[1] University of Helsinki, Department of Physics, Finland
[2] University of Helsinki, Department of Applied Chemistry and Microbiology, Finland
[3] University of Helsinki, Department of Forest Ecology, Finland

7.6.5.1 Behaviour of CH_4 in Soil

Methane is a green house gas and its concentration in the atmosphere is increasing (Section 10.1). The ability of soils to produce or consume methane (CH_4) is widespread across different ecosystems and vegetation zones. Bacterial oxidation of atmospheric CH_4 is the main biotic sink for atmospheric methane (e.g., Conrad, 1996; Le Mer and Roger, 2001; Section 6.4.3), and is considered to account for 10% of the global CH_4 sink (Topp and Pattey, 1997). Methane is emitted primarily from wetlands, rice paddies and landfill soils, which account for approximately half of the annual global methane production. In addition, recent studies report significant CH_4 emissions from plants under aerobic conditions; however, the mechanism of the emissions and the source strength still remain unclear (Keppler et al., 2006; Kirschbaum et al., 2006; Ferretti et al., 2007).

As the CH_4 oxidation in forest soils is limited by diffusion of CH_4 into the soil, one would consider that the highest CH_4 oxidation activity should take place on the surface layer of the soil. However, several studies have shown that the most active CH_4 oxidizing soil layer, in forest soils, is located below the litter and humus layers of the soil. In studies conducted in Finnish upland forest soils the maximum CH_4 oxidation activity occurred in the uppermost 0–5 cm mineral soil (Saari et al., 1997, 1998) or at approximately 7–12 cm from the mineral soil surface (Kähkönen et al., 2002).

Gas diffusivity is the main factor controlling the rate of movement of CH_4 through the soil to the sites of microbial oxidation and thus the rate of CH_4 oxidation (Ball et al., 1997a; Smith et al., 2003). Gas diffusivity and air-filled porosity are largely determined by soil structure, bulk density and water content (Ball et al., 1997a). In forest soils the thickness and the structure of the litter layer controls the diffusion of CH_4 into the topmost mineral soil, which is the layer most active in CH_4 oxidation (Saari et al., 1997, 1998; Kähkönen et al., 2002). The litter layer in a deciduous forest is often thicker and more compact as compared to the needle litter in coniferous forest soil. Brumme and Borken (1999) found that leaves from broad deciduous forest can reduce diffusion of CH_4 into the mineral soil more than needles.

As the soil's physical properties determine the rate of diffusion of gases into and from the soil, CH_4 oxidation in general tends to be higher in well-aerated and coarse textured soils than in soils with limited aeration. Higher CH_4 oxidation has been reported from coarse texture soils or soils with variable particle sizes than from fine silt and clay soils (Dörr et al., 1993; Saari et al., 1997; Regina et al., 2007). Regina and colleagues (2007) found that CH_4 oxidation in boreal agricultural soils strongly correlate with macropores of the soil. The more there were macropores or the less there were micropores in the studied peat, loamy sand and clay soils, the higher was the mean annual CH_4 oxidation rate.

As the diffusion of CH_4 in water is several magnitudes slower than that in the air, increasing soil water content reduces the availability of atmospheric CH_4 to methane oxidizing bacteria. Numerous field studies have reported increasing CH_4 oxidation rates with decreasing water content of the soil (e.g., Mosier et al., 1991; Sitaula et al., 1995; Billings et al., 2000). In very dry soils the CH_4 oxidation process is water limited and an increase in soil water content increases the CH_4 oxidation rate (Whalen and Reegurgh, 1996). At water saturated conditions the oxidation CH_4 slows down and starts to be limited by oxygen.

Oxidation of CH_4 in boreal forest soils is rarely limited by soil water, since even prolonged dry periods do not dry the soil down to the levels that reduce CH_4 oxidation. In a laboratory study of Saari and colleagues (1998) CH_4 oxidation activity in boreal pine forest soil started to decrease only when soil gravimetric moisture content fell below 4%, and the CH_4 oxidation was inhibited only at 1% gravimetric moisture content (Saari et al., 1998). Optimum soil moisture content for CH_4 oxidation in boreal upland forest soils seems to be between 20% and 35% of water holding capacity (WHC)[6] (Whalen and Reeburgh, 1996; Gulledge and Schimel, 1998a; Saari et al., 1998). Peat soils differ from upland mineral soils as the optimum soil water content for CH_4 oxidation is considerably higher, approximately 50% WHC (Whalen and Reeburgh, 1996). The CH_4 oxidizers in peat soils and upland mineral soils are probably different since they respond differently to changes in soil moisture, and they oxidize CH_4 at different rates. It has been suggested that the CH_4 oxidizing communities differ between peat soils and soils that are commonly water saturated, and upland forest soils (Billings et al., 2000; Saari et al., 2004).

Methane oxidizing bacteria seem much less sensitive to the temperature of the soil than methane producing bacteria. Several studies have shown that the response of CH_4 oxidation to soil temperature is very small (Billings et al., 2000; Smith et al., 2003). Also, Q_{10} values for CH_4 oxidation (1.1–2.0) are much smaller than those for other soil processes such as respiration (2.1–4.4), or CH_4 production (5.3–16) (Crill, 1991; Kähkönen et al., 2002).

The response of CH_4 oxidation to temperature does not seem to differ between climatic zones, but more on the microbial community. Peat soils (forested peatlands,

---

[6] Water holding capacity (WHC) is defined as the amount of water that the soil profile can hold, whereas percent of the WHC refers to a percentage of soil water as compared to the capacity.

natural peatlands) seem to respond more to soil temperature than upland mineral soils. It is suggested that microbial communities in environments with high concentrations of CH_4 in the soil are more controlled by soil temperature than microbial communities living on atmospheric CH_4 (Whalen and Reeburgh, 1996; Billings et al., 2000).

Numerous studies have reported that nitrogen fertilization decreases CH_4 oxidation rates by soils (e.g., Mosier et al., 1991; Crill et al., 1994; Kasimir-Klemedtsson and Klemedtsson, 1997; Butterbach-Bahl et al., 1998; Whalen, 2000; Kähkönen et al., 2002; Wang and Ineson, 2003). The reduction in CH_4 oxidation is considered to result mainly from an inhibitory effect of ammonia on the first step of CH_4 oxidation (Gulledge and Schimel, 1998b), but also from toxicity of nitrite to the last step of CH_4 oxidation (Topp and Pattey, 1997; Whalen, 2000). Due to the physiological similarities between CH_4 and NH_4^+-ions, ammonia is thought to compete for the enzyme active sites of methane mono-oxygenase (MMO, Section 6.4.3). In this reaction the MMO participates in the nitrogen cycling by oxidizing ammonia to hydroxylamine, which is then oxidized to nitrite and further to nitrate by hydroxylamine oxidoreductase and nitrite oxidase, respectively (Bédard and Knowles, 1989). This reaction is, however, an immediate response and the behaviour in long-lasting elevated ammonium availability may be different.

In boreal forest ecosystems the most active CH_4 oxidation occurs in the uppermost mineral soil layer, and not in the top litter and humus (Saari et al., 1997, 1998; Kähkönen et al., 2002). In nutrient poor forest ecosystems the litter and humus layers are the most active layers for NH_3 formation and they also are susceptible for atmospheric nitrogen deposition. Hence, if we consider that the litter layer does not limit diffusion of CH_4 into the soil, one explanation for the subsurface maximum could be ammonium inhibition of CH_4 oxidation in the top litter layer (King and Schnell, 1994).

Increase in soil mineral nitrogen does not always decrease soil CH_4 oxidation. Moreover, several studies report no effects of nitrogen fertilization on CH_4 consumption rates (e.g., Gulledge et al., 1997; Steinkamp et al., 2001; Borken et al., 2002; Ambus and Robertson, 2006), or stimulation of CH_4 oxidation by nitrogen additions (Bodelier and Laanbroek, 2004).

7.6.5.2 CH_4 Fluxes

Fluxes of CH_4 were measured by soil chamber and soil gas gradient methods. The measurements were conducted during a one year period in 2002–2003. The chamber measurements with six static chambers were conducted weekly to fortnightly during the snow free period in 2002–2003, and once per month in the winter. The chambers were rectangular and made of stainless steel (width $\times$ length $\times$ height: $0.29 \times 0.40 \times 0.10$ m). During a measurement the collars were closed with a stainless steel chamber (width x length x height: $0.29 \times 0.40 \times 0.20$ m) and a fan mixing the headspace of the chamber. The chambers were closed for 60 min and four gas samples were drawn from the chamber at 20 min intervals.

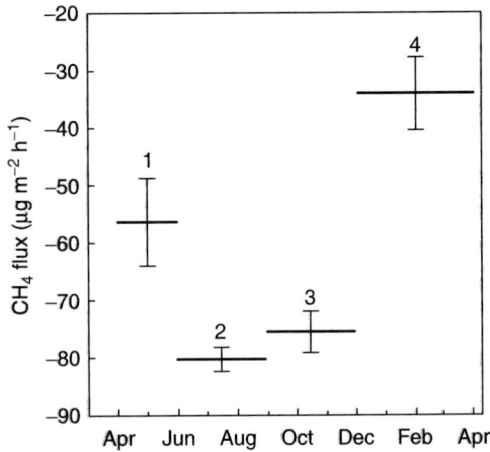

Fig. 7.6.27 Seasonal mean CH_4 fluxes at SMEAR II station in Hyytiälä measured by soil chambers during 2002–2003 (1 = Apr–May, 2 = Jun–Aug, 3 = Sep–Nov, 4 = Dec–Mar). Error bars represent standard errors of the mean

The boreal Scots pine forest soil at Hyytiälä was a net sink for CH_4 throughout the year. The CH_4 fluxes followed a clear seasonal variation with a maximum oxidation during June–July and a minimum in December–January (see Fig. 7.6.27). The mean CH_4 uptake calculated from the chamber measurements during the June 2002–July 2003 measurement period was $-72.0 \pm 4.2\,\mu g\,CH_4\,m^{-2}\,h^{-1}$ (uptake $\pm 2\times SE$), and the fluxes ranged from -145.2 to $-0.6\,\mu g\,CH_4\,m^{-2}\,h^{-1}$. The maximum CH_4 uptake from individual chamber measurements was measured in September 2002 and the minimum in March 2003. Annual cumulative CH_4 oxidation at Hyytiälä Scots pine forest soil summed up to $-0.51\,g\,CH_4\,m^{-2}$.

These flux estimates only account for the CH_4 exchange between atmosphere and soil with ground vegetation. We do not have a direct estimate of the contribution of trees to the ecosystem scale CH_4 exchange, which Keppler and colleagues (2006) suggested a significant source of CH_4. However, if we scale up the CH_4 oxidation rate at Smear II station to the whole boreal forest area and compare this to the current estimates of the CH_4 emissions from boreal forests, we can roughly compare the magnitudes of the two. We get an annual CH_4 oxidation rate of $-7.0\,Tg\,CH_4$ for boreal forests. This CH_4 uptake rate is much larger than the annual emission estimate of $2.8\,Tg\,CH_4$ from boreal forests by Kirschbaum et al. (2006). Hence, based on our measurements at Smear II station, the CH_4 oxidation in boreal forest soils seems to overcome the possible CH_4 emissions from forest canopies.

7.6.5.3 Soil CH_4 Concentration Profiles

In the soil-gradient method the concentrations of CH_4 were monitored at different soil depths. The gas collectors were installed in the O-, A-, B- and C-horizons, at approximate depths of 3, 7, 23 and 71 cm from the topsoil. The various soil layers, were sampled for soil air every fortnight or month and analyzed for CH_4 by gas chromatography. The soil gradient method relies on the knowledge of soil structure, total porosity and soil water content, which are used as parameters in the flux calculation. The fluxes between the different soil layers were calculated using the model described in Pumpanen et al. (2003). The model application for N_2O fluxes has been explained in detail by Pihlatie et al. (2007).

Methane concentration decreased with depth of the soil throughout the year (see Fig. 7.6.28). The ambient CH_4 concentrations were always higher than the soil CH_4 concentrations. The strongest change in CH_4 concentration occurred between the humus layer and the A-horizon. This was interpreted as the most active CH_4 oxidation layer in this podzol soil. Kähkönen et al. (2002) found that the most active CH_4 oxidizing soil layer in this forest soil located at approximately 10 cm below the humus layer. In our study the gas samplers were located at approximate depths of 3, 7, 23 and 71 cm from the soil surface in the O-, A-, B- and C-horizon, respectively. The soil concentrations reveal that throughout the year 2002–2003 the most active CH_4 consuming soil layer was in the top-most mineral soil, between 3 and 7 cm depth of the soil.

In several studies the most active CH_4 oxidizing layer in forest soils have been the top mineral soil underlying the organic litter and humus layer. In boreal pine forest soils the litter and humus layers are relatively thin and consisting of loose material. Such a layer allows easy diffusion for the atmospheric CH_4 into the active CH_4 oxidizing layers. Ball et al. (1997a, b) found that air diffusivity in forest soils decreases with the depth of the soil, and that the diffusivity of the soil is the key factor controlling CH_4 oxidation.

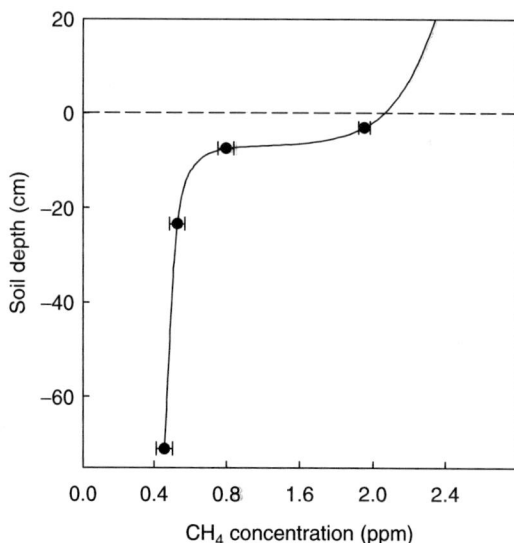

Fig. 7.6.28 Methane concentration in the soil profile of a boreal pine forest at Hyytiälä. Soil concentrations represent mean values over four locations and over one year of measurement period in 2002–2003. Error bars represent standard errors of the mean

In spring, during April–May, when snow was melting and soil water content was close to water saturation, the CH_4 concentrations increased in deep soil. During this period the conditions in the soil may have become anaerobic and enabled CH_4 production by methanogenic bacteria. This increase in CH_4 in the deep soil layers, however, was very small, indicating that simultaneous production and oxidation may have taken place.

Methane uptake rates in Hyytiälä Scots pine forest soil were generally higher than earlier reported from boreal regions (Schiller and Hastie, 1996; Burke et al., 1997; Savage et al., 1997; Saari et al., 1998) and from temperate forests (Butterbach-Bahl et al., 1998). In the boreal forest region in Scandinavia, the nitrogen deposition is at least an order of magnitude lower than that of Central Europe. As discussed in the section on nitrogen in the soil (see 7.9) this boreal forest site is nitrogen limited with a negligible amount of ammonium-N in the soil. Also, because the coarse textured sandy soil is well drained, it is rarely saturated with water. These conditions create an ideal environment for CH_4 oxidizing bacteria.

7.7 Reactive Gases in the Air

Boris Bonn[1], Michael Boy[2], and Heidi Hellén[3]

[1] Frankfurt University, Institute of Atmospheric and Environmental Sciences, Germany
[2] University of Helsinki, Department of Physics, Finland
[3] Finnish Meteorological Institute, Finland

This section focuses on reactive trace gases, their chemical processing (sources and sinks) and transport (advection, mixing and dilution), which determine their atmospheric lifetime as well as their concentration variations with time and height at boreal forest sites. In order to display the links between different atmospheric trace gases, we apply the chemical processes (reactions) described already in Section 6.2.2 and boundary layer transport (Section 5.3) to measurements, performed exemplarily at the boreal forest site SMEAR II station in Hyytiälä (Finland).

By doing so the aim is to answer the following questions: How do reactive trace gases such as ozone, OH and volatile organic compounds (VOCs) vary with height and time at a boreal forest site?

7.7.1 Ozone

Let us start with ozone, i.e. the most important trace gas for most gas-phase processes in the atmosphere. Its production and destruction are the initiation of any chemical process such as oxidation reactions, causing the formation of radicals such as OH, NO_2, NO_3 or peroxy radicals. Hence it interferes with most of the gaseous compounds, not only directly by reactions but for example by production of OH, which reacts instead. This shows the mesh of chemical reactions linking most of the chemical compounds. Only very unreactive species such as carbon dioxide are decoupled from this mesh.

Sinks

Ozone itself is destroyed by reactions with various trace gases (see Section 6.2.2): (i) hydrogen oxides OH and HO_2 (HO_x), (ii) reactive VOCs (e.g. isoprene, terpenes), (iii) nitrogen oxides such as NO and NO_2 (NO_x) and by (iv) sun light (photolysis). The contribution of individual processes to the total destruction process of ozone increases from (i) towards (iv) as shown in Table 7.7.1. About half a percent of the ozone molecules react with HO_x molecules at daylight, but significantly less during night-time, when HO_x concentrations are smallest. Ozone interacts faster with reactive VOCs such as mono- and sesquiterpenes. But any chemical reactions are insignificant with the most abundant but less reactive VOCs, i.e. methane, formaldehyde and acetone. Other reactive VOCs like isoprene are much less abundant in boreal forest environments and therefore do not affect ozone that much. However, depending on the emission strength by the forest, biogenic terpenes can significantly reduce the local concentration of ozone. For instance, if we assume individual volume mixing ratios of 100 ppt_v for both, i.e. mono- and sesquiterpenes (Bonn et al., 2006) as a mean estimate during summer time, monoterpenes are responsible for only about 0.1% (taken as α-pinene) of the ozone destruction, while sesquiterpenes cause about 14%. The latter fraction is caused by the high reactivity of sesquiterpene molecules and is therefore highly variable with distance to the emitting needles because of the change in present concentration during transport by wind. Hence, a reduction of local ozone concentration close to the emitting needles within the canopy is expected. In contrast to the reduction of ozone by emitted VOCs, vertical mixing, which is most intense during daylight (maximum of emissions), counterbalances this effect. The destruction of ozone by reactive terpenes is highest within the forest and especially in the canopy (see Section 6.3.2.6). This leads to a notable vertical pattern of ozone, which is rather constant above tenths of meters above the forest but declines somewhat below because of VOC emissions. The two remaining and major loss processes for ozone are the oxidation by nitrogen oxides and the photolysis by sunlight, which are commonly taken into account in the simplified NO_x-VOC-plots for estimations of ambient ozone concentrations from NO_x and VOC concentrations (see Fig. 7.7.1 for conditions at Hyytiälä). At maximum, sun light destroys about half of the ozone molecules and the reactions with NO consume about a third (33.3%), forming NO_2. In this context, NO_2 serves as a reservoir for ozone, producing ozone, when irradiated (day-time). By contrast in the absence of sunlight (night-time) destruction of ozone by NO_2 becomes most important ($>50\%$).

Table 7.7.1 Sink reactions for ozone and OH as well as for compounds (NO, NO_2, OH, HO_2 and SO_2) linked to their formation. The individual mean strengths are compared to the total loss of the individual compound (see Boy et al. (2006) for OH and HO_2) and is accompanied by the individual chemical lifetimes with respect to a certain loss term at Hyytiälä during day-time. Lifetimes of monoterpenes vary strongly for different structures between about 1 and 5 h. For sesquiterpenes the lifetime is nearly exclusively dependent on ozone and is between tens of seconds up to 2 min

Sink reaction	Mean contribution	Individual lifetime τ
Ozone (mean lifetime 1.55 h)		
+ hν (sunlight)	50.2%	2.5 h
+ NO		4.1 h
+ NO_2	33.3%	20.3 days
+ sesquiterpene*	14.1%	9.6 h
+ monoterpene*	0.1%	57.9 days
+ HO_2	0.4%	> 1 year
+ OH		
NO (mean lifetime ca. 20–30 s)		
+ RO_2	43.2%	44.2 s
+ HO_2	33.8%	56.4 s
+ O_3	22.9%	83.4 s
+ OH	0.2%	10,000 s
NO_2 (mean lifetime ca. 1.5 min)		
+ hν	91.4%	10.1 min
+ HO_2	4.2%	30.8 min
+ NO_3	3.9%	40 min
+ O_3	0.2%	13.9 h
+ OH	0.2%	9.3 h
HO_2 (mean lifetime ca. 2 s)		
+ NO	74.1%	2.3 s
+ O_3	14.2%	11.8 s
+ RO_2	8.9%	18.9 s
+ HO_2	2.8%	58.8 s
OH (mean lifetime ca. 1 s)		
+ CO	40.1%	3.1 s
+ NO_2	30.0%	4.1 s
+ CH_4	13.0%	9.6 s
+ NMHC	4.2%	29.4 s
+ HCHO	5.1%	24.4 s
+ O_3	3.7%	33.3 s
+ H_2	3.2%	38.5 s
+ SO_2	0.5%	250 s
SO_2 (mean lifetime ca. 10 days)		
+ OH (gas ox.)	100%	10.3 days

*Assuming sesquiterpene and monoterpene volume mixing ratio to be 100 ppt_v (2.5×10^9 molecules cm^{-3}).

Source

The source of ozone is entirely light dependent. Since the ozone photolysis and formation represents a null cycle but not a net source, the only process causing net ozone production is the photolysis of NO_2. VOCs on the other hand are required to

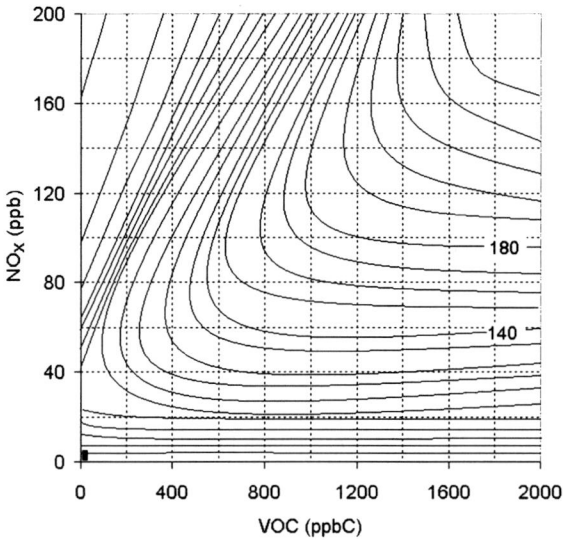

Fig. 7.7.1 Ambient maximum ozone concentration as a function of present NO_x and VOC mixing ratios during day-time. The dot marks the location of the Hyytiälä (SMEAR II, Finland) boreal forest site. It is apparent that the boreal forest site of Hyytiälä is at NOx limited conditions and a change in VOC doesn't affect the maximum ozone concentration much

provide HO_2 and initiate a 'recycling' of NO (oxidation process) forming NO_2 with continuous ozone production as long as VOC, NO_x and sunlight are available (see e.g. Jacob, 1999). Any reaction preventing the cycling between NO and NO_2 and the formation of ozone are thus a limit for ozone production (see again Fig. 7.7.1 and the limitation with respect to VOCs and NO_x). An example for a preventive reaction is the formation of nitric acid (NO_2 + OH), which deposits on surfaces of the forest. Others are nitrogen oxides such as PAN, nitrates and H_2O_2. However, less is known about their deposition on needles or leaves because of the difficulties in detection. For example surface PAN might be able to affect ozone concentration too, when decomposing to NO_2 and NO_3.

Consequently, ozone is destroyed during darkness (night-time) and produced as well as destroyed during sunlight (day time). Therefore, sunlight is the key in driving atmospheric reactions. A typical scheme displaying the dependence of ozone on NO_x and VOCs is shown in Fig. 7.7.1. It indicates the mean noon-time conditions, i.e. the cross, found in Hyytiälä as a remote background station close to the lower production limit. NO_x and VOCs are smaller than in most urban environments and a NO_x limitation is expected. This might switch to VOC limitation during winter-time too. But since the sunlight is no constant but varies with height, during day and especially during different seasons, we will tackle this in the following paragraphs.

Vertical Profile

The vertical profiles of chemical compounds depend on the differences in sources and sinks as well as on boundary layer transport (Chapter 5). With respect to NO_x the forest can act as a small source (NO emissions or photolysis of deposited nitrates, Section 6.3.2.8), while some nitrogen compounds (nitric acid, PAN type species, nitrates) will mainly deposit. The deposited compounds might re-evaporate at noon or early afternoon, when their ambient vapour pressure exceeds their saturation. Nitric acid concentration gets smallest close to the forest and within because of deposition, while NO_2 enhances slightly (reduced photolysis). In case of NO emission at the surface the same applies for NO too.

Since most of the processes are limited by solar radiation and temperature controlled, major vertical changes occur above the canopy, especially for compounds produced by reactions such as radicals and stable products like carbonyl compound and acids. Some changes, which are caused by reactive VOC emissions, can appear already in the canopy, most pronounced at the sunlit part. Therefore, any emitted VOCs display a clear vertical profile dependent on their reactivity: the higher reactive the more apparent the vertical shape. OH formation is extremely dependent on sunlight with some additional contributions from ozone reactions with reactive VOCs (dominant during night-time). This leads to a steady-state situation above and a sharp decrease within the forest, when light intensity decreases. The opposite is true for less reactive and long lived compounds. These compounds are rather homogeneously distributed with height (e.g. methane, nitrates, carbonyl compounds, SO_2), depending predominantly on the vertical stability and the mixing process.

This vertical mixing (transport) becomes important during night-time, when blocking processes at the surface from long range transported pollution above. A stable boundary layer with only minor vertical mixing will decouple the boundary layer from the air at higher altitudes and sharp concentration changes with height appear. For example pollutants cannot enter the surface air from their transport altitude in the free troposphere (above 1–2 km) and thus a potential temperature-like vertical profile can be found for those gases with no major local sources. This situation changes during day-time, when sunlight and increasing temperature weaken the vertical stability and force the air to be mixed downward from the free troposphere. This causes vertical differences in concentrations to decrease. Thus, the strongest vertical profiles will appear for gases with forest sources (e.g. terpenes) and sinks (e.g. ozone) during night-time depending on their individual atmospheric lifetime. Shorter lived ones display a stronger gradient (OH for instance) and long-lived a negligible one (e.g. CO_2).

An exemplary plot is shown for the vertical profile of sulphuric acid (Fig. 7.7.2), which indicates this. Formed from the OH reaction with SO2 and removed by aerosol as well as forest surfaces, the situation is rather constant with height above the forest at noon-time during summer. Below an additional sink (forest) and the reduction in light penetration into the vegetation layer causes the concentration to decrease. This is strongest at stable meteorological conditions, but weaker at noon. A small secondary maximum appears below the canopy close to the surface, where the sink is smaller and the residence time longer.

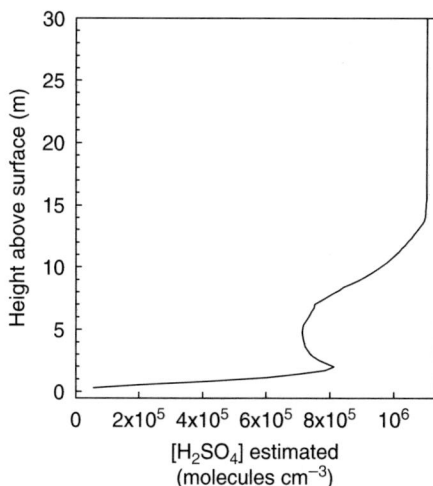

Fig. 7.7.2 Approximated noontime vertical distribution of sulphuric acid during summer

Daily Behaviour

This difference between night-time and day time conditions becomes apparent in the daily pattern of most of the trace gases (Fig. 7.7.3). For example consider the daily pattern of ozone. As mentioned before, the formation of ozone takes place only during daylight. Therefore, the concentration of ozone decreases from sunset until sunrise. The magnitude of the decrease depends on the present concentrations of NO_x, HO_2 and reactive VOCs. Once the sun rises in the morning, photolysis of NO_2 starts and the destruction is more and more compensated by the formation via NO_2 photolysis. The formation is strongest at the highest concentration of NO_2 and at highest radiation intensity (in the morning and at noon). Hence, a maximum in ozone concentration is found shortly after noon and in the early afternoon, before the loss processes become stronger than the sink. All the compounds, which are directly dependent on solar radiation, display a clear maximum at noon and decline towards the night. This applies for OH and HO_2 (HO_x). Those compounds show also a remarkable decline from above to within the canopy due to the reduced available sunlight and their short atmospheric lifetime τ (OH: tenths of a second, HO_2: several minutes).

On the other hand, VOC concentrations from the forest are destroyed most effectively during the day (OH and ozone) but less during the night. Their daily pattern is determined by their emissions aiming to compensate the loss. Anthropogenic VOCs, transported from a distant source, may display a two maximum behaviour. During night they are not transported downward from the free troposphere (travelling altitude) and thus their concentration is fairly low. At noon their destruction (mainly by OH) is most intense and their concentration small as well. Both maxima appear in between, i.e. the morning and afternoon, when vertical mixing occurs

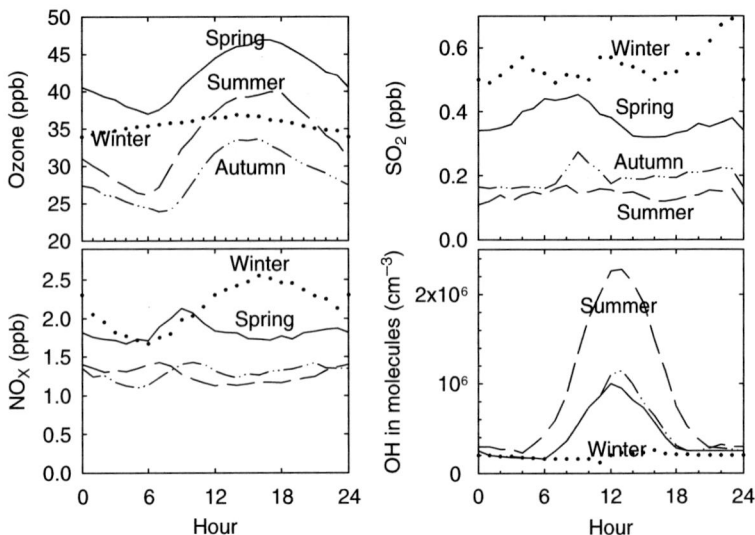

Fig. 7.7.3 The daily pattern of ozone, NO_x, SO_2 and OH during winter spring, summer and autumn. The datasets of ozone and SO_2 are based on the average of seven years measurements at SMEAR II station in Hyytiälä (Lyubovtseva et al., 2005), while OH is based on estimations by Boy et al. (2005) and NO_x is taken from measurements of 2003–2005

and destruction is slow. Nevertheless, it should be noted that the description given above applies for cloud free conditions. If a cloud blocks or reduces the radiation flux downward, the chemical processes slow down, vertical transport is weakened and a more scattered behaviour is to be observed.

A further important parameter affecting the reactions is temperature. The temperature affects so-called storage terms such as N_2O_5 and PAN as well as the emissions of biogenic VOCs. It influences anthropogenic sources (traffic and heating) too and is given by the prevalent meteorological conditions (circulation of weather systems). All of this varies clearly throughout the year during different seasons, which we will investigate next.

Seasonal Behaviour

This dependence of ozone formation on sunlight makes the daily variation of ozone rather small during winter time (SMEAR II average: 30 ± 1 ppbv $\approx 7.5 \times 10^{11}$ molecules cm^{-3}, see Fig. 7.7.3). During that time temperatures are very low and the vegetation is covered by snow resulting in very small emissions of VOCs by the forest. Mainly anthropogenic pollutants (VOCs, NO_x and SO_2), which are advected by wind, are present during that time due to the prevailing wind pattern and anthropogenic heating. However, the small amount of sunlight causes only a moderate vertical mixing and the daily pattern is usually rather constant for most of the trace gases. Because of the rather small emission of VOCs by the forest during winter, the production of ozone is VOC limited and thus small. Its sink depends mainly on reactions with NO_x. NO_x is highest during that time because of large anthropogenic sources (heating and traffic). At the same time the removal of NO_x by reactions with OH is smallest. OH needs sunlight to be formed and consequently its concentration is lowest during winter.

The situation changes during spring: Now the time of daily sunlight increases, the snow starts to melt and temperatures increase. This causes the vegetation to start photosynthesizing and releasing VOCs (see Section 6.3.2.5). By contrast, anthropogenic pollution decreases and NO_x concentrations decline due to the reduced need for heat and the changing meteorological wind pattern. Because of the increase in sunlight, the boundary layer development gets stronger (Chapter 5), vertical mixing intensifies and the daily cycle becomes more pronounced. This affects the daily pattern of VOC concentrations. Although the temperature dependent emission increases at sunrise and it is found highest at noon, the VOC concentrations are found maximum partly during night-time because of trapping in a shallow boundary layer. The drop at sunrise depends on the meteorological causing an intensified trace gas dilution the more unstable the atmosphere becomes. This dilution affects strongly the NO_x concentrations. NO_x is transported from above towards the forest, hence displaying the maximum in the morning and afternoon. At noon the concentration of destructive gases such as OH are highest (sunlight) and remove NO_x from the gas-phase as nitric acid, HONO or nitrates. The daily NO_x variation and VOC release initiate a more distinct daily pattern of ozone (mean: 36–46 ppb) than during winter, displaying a minimum during the morning just before sun-rise. Following the night with destruction only, ozone increases until the early afternoon when its concentration decreases as the sun sets and the production declines.

This situation becomes more and more pronounced until the middle of the summer. Because of the temperature VOCs are emitted at their highest, even during the night at significant levels Section 6.3.2.5). By contrast, NO_x is smallest since there is no heating during that time and vertical mixing is strongest because of the shortest nights. This causes ozone on average to range from 26 ppb (6 a.m. in the morning) to about 40 ppb (6 p.m. in the afternoon). In Fig. 7.7.3 (upper left) the change in sunrise time becomes apparent by the times ozone starting to increase after passing its minimum. During summer the production of ozone is clearly NO_x limited and its sink is caused by both reactive VOCs and NO_x. Herein VOC sink summarizes two possibilities: a) the reaction with reactive VOCs (e.g. biogenic terpenes, Fig. 7.7.4) as well as b) the reaction with HO_2, which is also formed by the first one. Both, terpenes and HO_2, are at their highest abundance during summer. By contrast, concentrations of anthropogenic VOCs and SO_2 are smallest during summer and vary mainly because of vertical mixing and destruction by OH during the middle of the day.

With respect to the 'cleaning' medium of the atmosphere (OH) the situation is extremely dependent on sunlight. Once formed the OH radical survives only milliseconds before reaction. However compared to the photo production of OH this process is much less efficient and causes a night-time concentration of several 10^4 molecules cm^{-3} at maximum compared to a summertime daily concentration between 10^6 and 10^7 molecules cm^{-3} reached at noon.

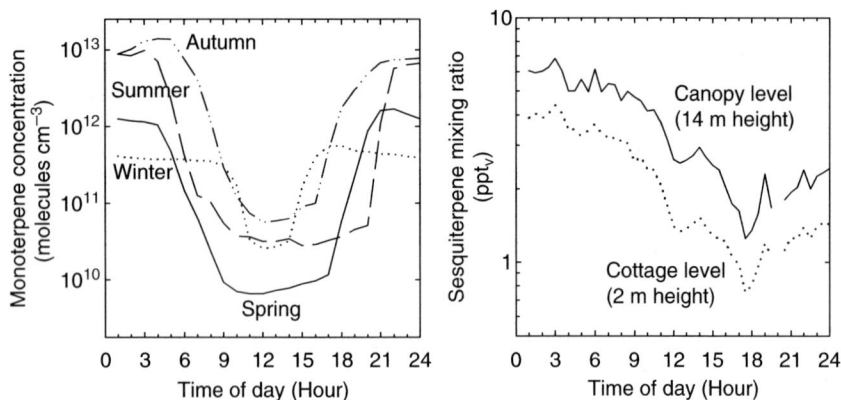

Fig. 7.7.4 VOC concentrations simulated from emissions and vertical mixing. Left: Mean monoterpene daily variations in four different seasons are displayed (Boy et al., 2005). On the right sesquiterpene volume mixing ratios (pptv) are given for an exemplary day in April 2004 at soil and top of the canopy level to show the impact on chemical reactivity and boundary layer stability on concentrations for very reactive trace gases

During autumn the situation looks rather similar to spring, but water vapour concentrations and VOC emissions are higher then. Both originate in the higher temperature, the active biosphere and the missing snow/ice cover of the surface. This causes the ambient ozone mixing ratio to be shifted to smaller values (min.: 24 ppb, max.: 34 ppb), but keeping the same daily pattern as during spring.

Daily OH maxima decrease because of the declining solar radiation. Night-time concentrations however are quite similar to those during the summer, depending on the availability of terpenes and other reactive VOCs.

The second cleansing substance nitrate radical (NO_3) behaves exactly the opposite way from OH: It is formed in reaction to NO_2 and ozone, hence mainly during the night and can be 'stored' by reacting with NO_2 as N_2O_5. Both, NO_3 and N_2O_5 are destroyed very efficiently by solar radiation and they get lost once the sun rises. During night NO_3 is the most efficient destructing agent of VOCs, thus replacing OH during the night.

Annual Behaviour

The seasonality of OH and NO_3 is complementary: OH follows nicely the solar zenith angle during the day with a maximum around 10^6 molecules per cm^3 at noon and a minimum around several 10^3 molecules cm^{-3} during night-time and displays a kind of steady state from spring time on until late autumn, because it is balanced by the reaction with the increasing amount of VOCs. NO_3 is highest during the night, destroyed by the daily sunlight. Its concentration is night-time maximum from late autumn to early spring. It is also available during summer nights. Nevertheless, because of the shortest night and its fastest destruction, the maximum NO_3 is smallest within the year.

So far only the gas-phase has been treated. But recent studies (e.g. Kulmala et al., 2004) revealed a link between the gas-phase compounds concentrations and the aerosol formation events. The compounds believed to be mainly involved in aerosol formation are (i) sulphuric acid (Fig. 7.7.5) and (ii) large reactive VOCs emitted from the biosphere (Fig. 7.7.4).

7.7.2 Sulphuric Acid

We start with sulphuric acid, known, e.g., from acid rain periods over Fennoscandia (Jacob, 1999) and atmospheric aerosol studies (Kulmala et al., 2004). It forms by the reaction of OH and SO_2 (Seinfeld and Pandis, 1998; Finnlayson-Pitts and Pitts, 2000). Therefore, its production is closely related to the seasonal and vertical profile of OH, since SO_2 is a rather long lived gas (lifetime between 3 and 10 days), which is advected from mainly anthropogenic pollutants. The sink for sulphuric acid is either a dry or a wet (highly water soluble) surface or any aerosol particles present, onto which it condenses. Assuming that its atmospheric lifetime is mainly determined by the condensation process it results in about 10–30 min. This leads to a fairly similar daily and seasonal pattern as to be observed for OH (Boy et al., 2005): The minimum is observed during the night-time and the concentration increases from sunrise until noon. Similar to OH sulphuric acid concentrations decline afterwards until sunset. However, maximum daily concentrations (noon) are expected to be rather constant at about 2×10^7 molecules cm^{-3} from late spring until early autumn (Fig. 7.7.5). The rise in OH daily maximum concentration is balanced by a decrease in SO_2 and an increased aerosol sink present during summer. Nevertheless, any clouds present will reduce OH and thus H_2SO_4 too.

Therefore the aerosol mass formation by sulphuric acid is highest at noon and its daily maximum stays fairly constant from spring until autumn.

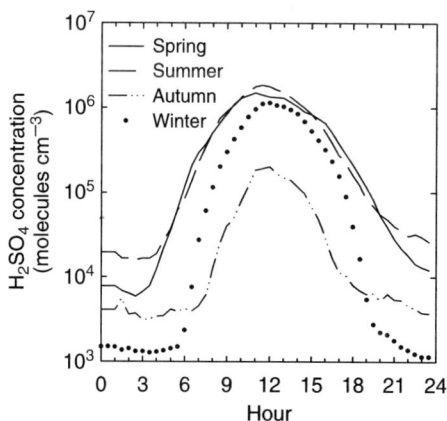

Fig. 7.7.5 Seasonally averaged daily sulphuric acid concentrations by Boy et al. (2005). The daily pattern remains rather unchanged once sufficient sunlight is available: spring, summer and autumn, while there is no notable daily cycle during winter

7.7.3 Reactive Biogenic VOCs

Reactive biogenic VOCs such as mono- and sesquiterpenes have highest emittance at the highest temperatures, thus during summer with a clear daily cycle. If there is a significant winter-time emission of very reactive terpenes remains speculative. Their concentrations depend on their atmospheric lifetimes. For reactive ones this is given by the survival time for reactions with ozone, OH and NO_3. Thus very short-lived ones, emitted by the forest, can be seen only within the forest (canopy or soil). Their concentration drops sharply above the forest height. Interesting to note in this context is the estimated daily pattern of mono- and sesquiterpenes. Their concentration is expected highest during the night because of the suppressed mixing and thus dilution, although the emission is highest during day-time (Fig. 7.7.4). Especially for the very reactive sesquiterpenes a clear vertical change can be seen, when comparing canopy level (13 m) with cottage level (2 m).

It has been known for a long time (Went, 1960) that terpenes contribute to atmospheric particle formation. However, the particular role remains still unclear. Terpenes, and especially the very reactive sesquiterpenes, are proposed by some scientists to initiate nucleation (e.g. Bonn and Moortgat, 2003). Others believe that they contribute mainly by production of condensable species with rather low volatility (Fig. 7.7.6).

The reaction causing products of lowest volatility is the one with ozone (Seinfeld and Pandis, 1998). One of the important products is anticorrelated with the amount of present water vapour (Bonn et al., 2002; Bonn and Moortgat, 2003) (highest during summer). Therefore the highest aerosol mass production is to be expected during summer but starting already in spring time next to the 'awakening' of the biosphere.

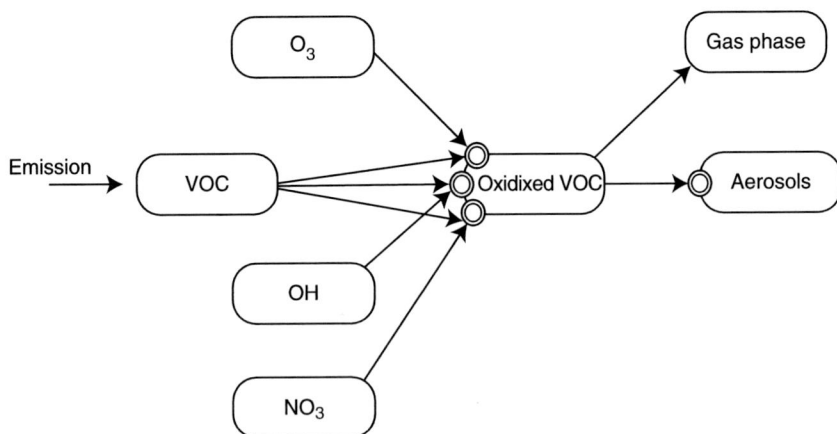

Fig. 7.7.6 The link between the emission of volatile organic compounds, their processing in the atmosphere and the suggested formation of particulate matter. At which stage they contribute, at the formation of the initial nuclei or at later growth, remains speculative. Boxes indicate amounts, double circles processes, and arrows flows

One open topic is the initiating compound for nucleation in boreal forests. Nucleation clearly depends on the present aerosol surface area, which takes up the compounds otherwise nucleating. Unfortunately, since there is no current tool to investigate the chemical composition of smallest clusters or particles, the starting compound of nucleation remains speculative and an open topic for research.

7.8 Aerosol Particles in the Air

Ilona Riipinen[1], Miikka Dal Maso[1,2], Tuukka Petäjä[1,3], Lauri Laakso[1], Tiia Grönholm[1], and Markku Kulmala[1]

[1] University of Helsinki, Department of Physics, Finland
[2] ICG-2: Troposphäre, Forschungszentrum Jülich, Germany
[3] Earth and Sun Systems Laboratory, Division of AtmosphericChemistry, National Center for Atmospheric Research, Boulder, CO, USA

7.8.1 Atmospheric Aerosol Concentrations

Aerosols absorb and reflect solar radiation as well as thermal radiation of the globe (Section 7.2). In this way, they contribute to the behaviour of the atmosphere and to the radiation available for photosynthesis. The BVOCs emitted by vegetation effect on the aerosol particle formation and growth. The BVOC emissions react to the climate change (Section 10.3.5) and their contribution to aerosol dynamics will change. Thus understanding of the formation and growth of aerosol particles is important for the analysis of feedbacks from boreal forests to climate change.

The dynamic processes affecting the atmospheric aerosol particle populations are summarized in Fig. 7.8.1 (see also Section 6.2.2) and their relative effects on the particle number, surface, and mass are presented in Table 7.8.1. The time evolution of the particle size distribution can be formulated with the general aerosol dynamic equation (GDE), which accounts for the sources and the sinks of particles of different sizes (e.g., Seinfeld and Pandis, 1998). If we flag each size section in the distribution with a subscript i, the general dynamic equation for the ith size section would read (see also Section 6.2.2):

$$\frac{dN_i}{dt} = C_i^+ - CoagS_iN_i + p_{i-1}N_{i-1} - (p_i + \gamma_i)N_i + \gamma_{i+1}N_{i+1} + J_0(t)\delta_{g^*,i} + S_i - R_i$$

$$(7.8.1)$$

where N refers to the particle number concentration, C^+ to the particle flux resulting from coagulation, and $CoagS$ is the coagulation sink for particles in the investigated size class. p and γ are the condensation and evaporation coefficients, and J_0 is the particle formation rate, and g^* is the size section to which the particles are formed. The delta function $\delta_{g*,i}$ is 1 if $i = g^*$ and 0 in all other size sections, and it takes care

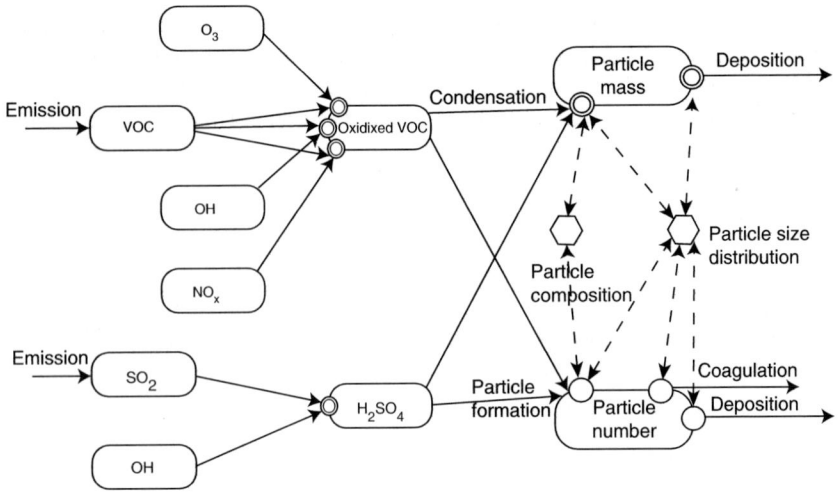

Fig. 7.8.1 A summary of the basic aerosol dynamical processes taking place in boreal forest. Particles are produced by nucleation of condensable vapours, for instance sulphuric acid and oxidation products of VOCs. The newly formed particles grow by the condensation of these/additional vapours onto particle surfaces to reach sizes where they can act as cloud condensation nuclei. The main processes removing particles from the air are dry and wet deposition, the latter referring to particle removal by rain. The total particle concentration is also affected by the coagulation of particles, which particularly decreases the concentration of freshly nucleated particles. Boxes indicate amounts, double circles processes, arrows (solid line) flows, and arrows (dashed line) effects on a given process.

Table 7.8.1 The effect of different aerosol dynamical processes on the aerosol particle total number, total surface area and total mass. The number of signs corresponds to the size of the effect. Plus signs indicate an increase; minus signs indicate a decrease

	Particle number	Particle surface	Particle mass
Particle formation	+ + + +	+	+
Condensation	No effect	+ +	+ + +
Coagulation	– – –	– –	No effect
Deposition	– –	– –	– – –

of adding new particles only to the smallest size section. S and R refer to any other aerosol number sources and sinks, including, e.g., dust emissions and deposition.

The left-hand side of Eq. 7.8.1 describes the temporal evolution of the particle number concentration in the *ith* size section. The first term on the right-hand side of the equation represents the flux of particles to the *ith* section resulting from the coagulation of smaller particles, whereas the second term describes the loss of *i*-sized particles due to their collisions and coagulation with other particles. The third, fourth and fifth terms represent the particle number gain and loss due to condensational growth and evaporation: the third term describes the growth of particles smaller than *i*, the fourth term the condensation/evaporation to/from the *i*-sized particles and the fifth term the evaporation of particles larger than size i. The last three terms in Eq. 7.8.1 correspond to particle number increase/decrease by new particle formation and other sources and sinks. Note that each size section is represented by a differential equation and an analytical solution to this set of equations is only possible in some special cases. Typically the equation is solved numerically with several simplifications.

Taking a closer look at the times of particle formation in boreal forest (see also Section 6.2.3), some clear patterns can be found. The bursts are almost exclusively observed during day-time, at least 2 h after sunrise and on average 3–4 h after sunrise but mostly before noontime. The start time of the bursts typically follows the variation of the time of sunrise (Mäkelä et al., 2000b). Incoming solar radiation is found to correlate with the growth rate of the new particles, which leads to speculation that increasing amounts of radiation increase the photochemical production rate of condensing vapour, which increases the formation and growth rates of the particles. These effects naturally increase the probability of particles surviving the early stages of growth until they are detectable.

Probable compounds participating in the atmospheric nucleation are, e.g., sulphuric acid, water, ammonia, and a variety of organic compounds which can vary in different environments. In boreal forest the effect of the organics is clearly seen, and a significant part of the condensing vapours is thought to be a mixture of oxidation products of volatile organic compounds (VOC), which in turn are produced by trees (e.g., Hakola et al., 2003). The exact role of condensable organics is not completely understood, but there are clear indications that organics are needed to explain the particle growth rates observed in the boreal forest (Kulmala et al., 2004b). Organics may also affect the very first steps of aerosol formation. However, exact information on, e.g., the composition of the smallest atmospheric particles (1–10 nm in diameter) is very difficult to obtain with current commercial instruments.

The condensing vapours affect the aerosol climatic properties in several ways. By making the small particles grow, they increase the atmospheric lifetime of those particles and cause them to them reach sizes at which they can act as cloud condensation nuclei. The vapours also naturally affect the composition and therefore the chemical and physical properties of the particles, thus affecting the limiting size at which they can activate as cloud droplets. As the particles deposit on the surfaces they carry their constituents along with them. Therefore depositing aerosol particles are also likely to contribute significantly to, e.g., the nitrogen fertilization of the forest soil.

7.8.2 Profiles

Total fine particle concentrations observed in the boreal forests typically range from 10 to 100,000 cm^{-3}, depending on the site, for instance its distance from roads, cities and other sources of anthropogenic pollution. Figure 7.8.2 shows an exemplary time series of running four-month averages of aerosol particle concentrations

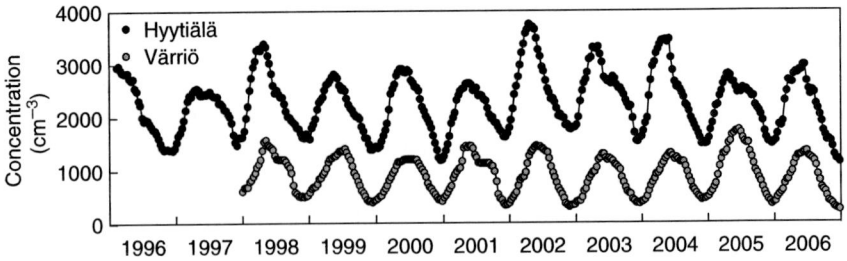

Fig. 7.8.2 The four-month running average of the total particle concentration at SMEAR II, Hyytiälä, Finland and at SMEAR I, Värriö, Finland. One can easily see the annual variation, with maxima in the summer and minima in the winter. The concentrations are clearly lower at SMEAR I in Värriö than at SMEAR II, Hyytiälä

Fig. 7.8.3 The two-week running average of the total particle number concentration at SMEAR II, Hyytiälä, Finland in the year 2000. The seasonal variation with maxima in spring and autumn and minima in summer and winter can be seen

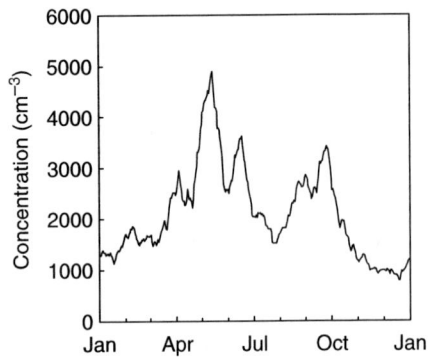

in the size range of 3–800 nm measured in 1996–2007 at the SMEAR II. The same figure presents also a similar plot for the years 1998–2007 in the cleaner environment of SMEAR I in Värriö. At remote continental sites like Hyytiälä and Värriö most of the particles are typically concentrated in the size range of ca. 50–150 nm. On average, approximately 30% of these, potentially climatically active, particles were originally formed in a nucleation event.

An exemplary plot on the seasonal variation of particle number concentrations in boreal forest is shown in Fig. 7.8.3, where the two-week average particle concentration observed at SMEAR II during the year 2000 is plotted. In spring time a clear maximum in the particle number concentration is observed, and a second, somewhat smaller, maximum is seen in the autumn. The daily variation of particle numbers on a day with clear nucleation and on the other hand on a non-event day are presented in Fig. 7.8.4. A clear increase in particle concentrations can be observed during new particle formation, compared to a day with no nucleation and growth.

The vertical profiles of the particle number concentrations vary as a function of, e.g., the height of the mixed boundary layer, and the intensity of the possible nucleation in different parts of the troposphere. Figure 7.8.6 present a vertical concentration profile measured near the SMEAR II station on 10.3.2006 and 13.3.2006 (Laakso et al., 2007). The measurements were taken from a hot air

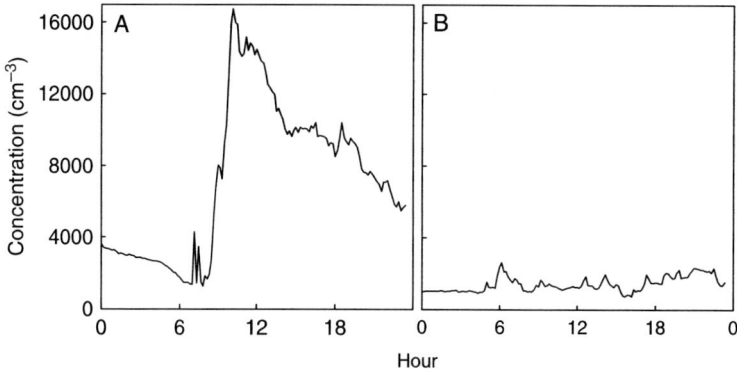

Fig. 7.8.4 The total particle number concentration during an event day (A) a non-event day (B) at SMEAR II, Hyytiälä, Finland. During the particle formation period (at ca. 10–11 o'clock) on a event day, the total particle number increases almost 10-fold. The particle number concentration fluctuates around a few thousand particles per cubic centimeter

Fig. 7.8.5 Hot air balloon taking off in front of the old dining room in Hyytiälä. The basket is full of eager researchers for boundary layer measurements

balloon (Fig. 7.8.5) due to its high lifting capacity and capability to follow air mass. The first of the days shown here corresponds to a day without new particle formation, whereas the latter is a day with a clear nucleation event. The profiles reveal that during new particle formation the particles are clearly concentrated near the

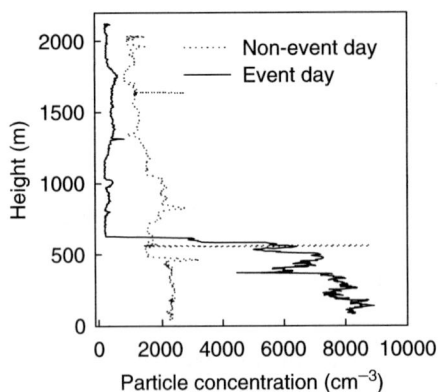

Fig. 7.8.6 The number concentration of over 10 nm particles as a function of altitude during a non-event day on 10.3.2006 and on an event day on 13.3.2006 at SMEAR II, Hyytiälä. No significant altitude dependence is observed on the non-event day. The particle number concentration has a clear maximum close to the Earth's surface on the event day

surface; whereas on the non-event day no such clear dependence on the measurement height is observed. This indicates that most observed nucleation takes place in the atmospheric boundary layer near the forest canopy, rather than at the higher altitudes (see also Section 5.3). This observation supports the idea that compounds emitted by forest vegetation have an important role in new particle formation, and particularly in particle growth.

7.8.3 Nucleation Event Annual Cycles

Classification and selection of event and non-event days reveals that tropospheric particle formation in the boreal forest is quite common, at least more common than previously assumed. In Hyytiälä, roughly every fourth day is found to contain a particle formation burst, based on the analysis of the particle size distribution data from years 1996 to 2006 (Dal Maso et al., 2005, 2007). For SMEAR I in Värriö, the corresponding frequency is one event every five days. (Dal Maso et al., 2007). The annual statistics of nucleation event days for Hyytiälä in 1996–2006 and Värriö in 1996–2004 are presented in Fig. 7.8.7. It seems that occurrence of new particle formation has an inter-annual trend, with a cycle close to 10 years. The reasons behind this trend, however, are still somewhat unclear, as the exact mechanism of new particle formation in the boreal forest is still under debate.

On the seasonal scale, two clear peaks in formation frequency are found: one in spring time (March–May) and another centred around September. During these periods, the event fraction in Hyytiälä is typically over 50% of all days, and even higher than that if undefined days are removed from the statistics. In contrast, wintertime

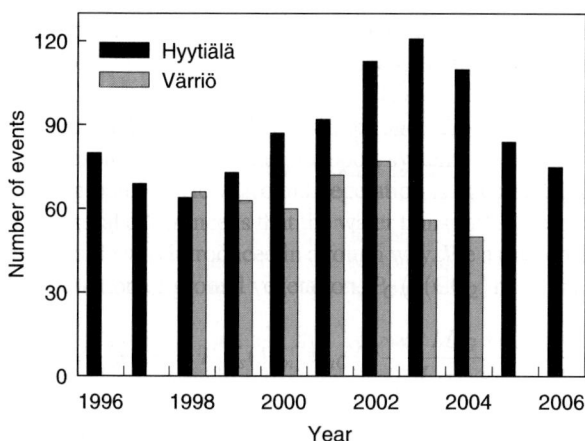

Fig. 7.8.7 The number of particle formation events each year at SMEAR I (Värriö) and SMEAR II (Hyytiälä), Finland

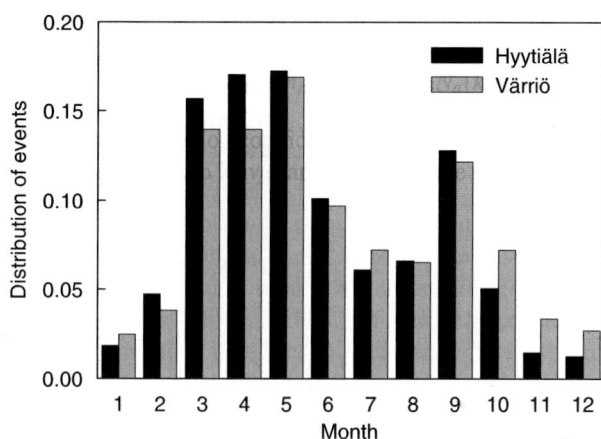

Fig. 7.8.8 The monthly distribution of particle formation events at SMEAR I (Värriö) and SMEAR II (Hyytiälä), Finland

(November to mid-February) is a very inactive time for particle formation. In summer a dip in the frequency is observed. The annual distribution of nucleation event days in the more northern Värriö station is similar, with a somewhat less clear minimum in the summer. Summaries of the annual variation of the occurrence of nucleation events for Hyytiälä and Värriö are presented in Fig. 7.8.8. The spring time peaks in particle formation are related to the spring recovery of photosynthesis (Section 6.3.2.3) the forest vegetation and related emissions of biogenic organics (Sections 6.3.2.5 and 7.7.3).

7.9 Annual Energy, Carbon, Nitrogen and Water Fluxes and Amounts at SMEAR II

Pertti Hari[1], Pasi Kolari[1], Eero Nikinmaa[1], Mari Pihlatie[2],
Jukka Pumpanen[1], Liisa Kulmala[1], Asko Simojoki[3], Timo Vesala[2],
and Markku Kulmala[2]

[1] University of Helsinki, Department of Forest Ecology, Finland
[2] University of Helsinki, Department of Physics, Finland
[3] University of Helsinki, Department of Applied Chemistry and Microbiology, Finland

Previous sections have described the main processes behind the material exchange in ecosystems. However, for their proper understanding, we also need quantitative values describing the flows of material and energy in the biogeochemical cycles. Energy, water and carbon fluxes in a forest ecosystem are large while those of nitrogen are small. The quantification of the fluxes requires versatile field stations that are able to follow continuously the material and energy flows. The planning and construction principle of our field station, SMEAR II, was to measure all relevant material and energy fluxes and amounts in the forest ecosystem and to understand the factors affecting the processes that generate these fluxes. Using the direct measurements and approaches described previously we can now summarise the fluxes. In this chapter we concentrate on the annual mean values obtained in measurements. While these values give us very good basic picture of the quantities, it need to mentioned that the interannual variation in many of the fluxes is large.

7.9.1 Energy Fluxes as Radiation and as Organic Matter

Solar radiation is the source of energy available to the vegetation (Fig. 7.9.1). Its annual input at SMEAR II is about $30{,}000\,\mathrm{MJ\ ha^{-1}\,a^{-1}}$. Photosynthesis converts solar radiation energy into chemical form as sugars: the efficiency of conversion is low, less than 3% at annual level. However, the sugars, formed in photosynthesis, are the source of energy for plant metabolism and the main source of raw material for growth of new plant structures and through their input also source of energy for all other living organisms in the ecosystem. The annual amount of energy bound in photosynthesis is $800\,\mathrm{MJ\ ha^{-1}\ a^{-1}}$. The biggest usage of sugars is respiration and root exudates and fine root turnover. The formation of new permanent structures consumes clearly less, only $200\,\mathrm{MJ\ ha^{-1}a^{-1}}$, thus only less than 1% of solar radiation is bound into growth of the trees and ground vegetation. The energy flux associated to by litter production is rather small, $80\,\mathrm{MJ\ ha^{-1}a^{-1}}$. If we assume that the soil organic matter does not accumulate, then the flux from soil organic matter to microbes is also $80\,\mathrm{MJ\ ha^{-1}a^{-1}}$.

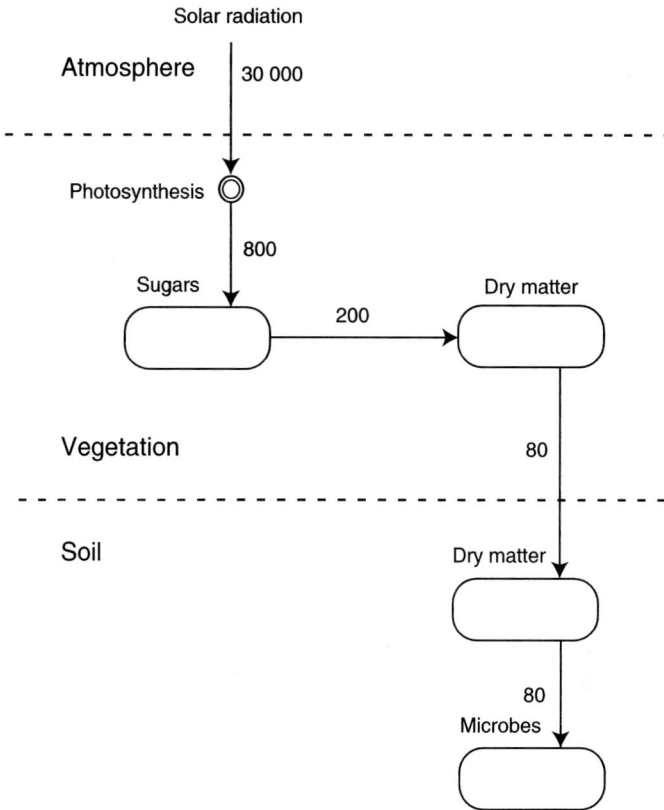

Fig. 7.9.1 The energy fluxes (MJ ha^{-1} a^{-1}) as solar radiation and as carbon compounds in the ecosystem around SMEAR II

7.9.2 Carbon Pools and Fluxes

Carbon enters the forest ecosystem in photosynthesis both in the canopy (37,000 kg CO_2 ha^{-1} a^{-1}) and in the ground vegetation (4,000 kg CO_2 ha^{-1} a^{-1}, Fig. 7.9.2. Sugars are utilised for production of ATP and formation of new structures. Vegetation respiration, formation of ATP, consumes 11,000 kg CO_2ha^{-1}a^{-1}. The remaining sugars are used for growth of new leaves, xylem elements for transporting water and fine roots and to root respiration and exudates. Decomposing microbes in the soil live on the litter of needles, wood and fine roots, that is altogether 3,200 kg dry matter ha^{-1}a^{-1}. If we assume that the soil organic matter is no longer accumulating, then the microbial respiration based on decomposition is 6,400 kg (CO_2) ha^{-1}a^{-1}.

The lifetime of fine roots is about one year, of needles three years and of wood about 100 years. The needle litter fall is approximately 3,200 kg dry weight ha^{-1}a^{-1}. The wood litter production during the life time of trees is small, and fine root litter is so difficult to measure that we do not have direct measurements of its annual production. The different litter components have specific chemical

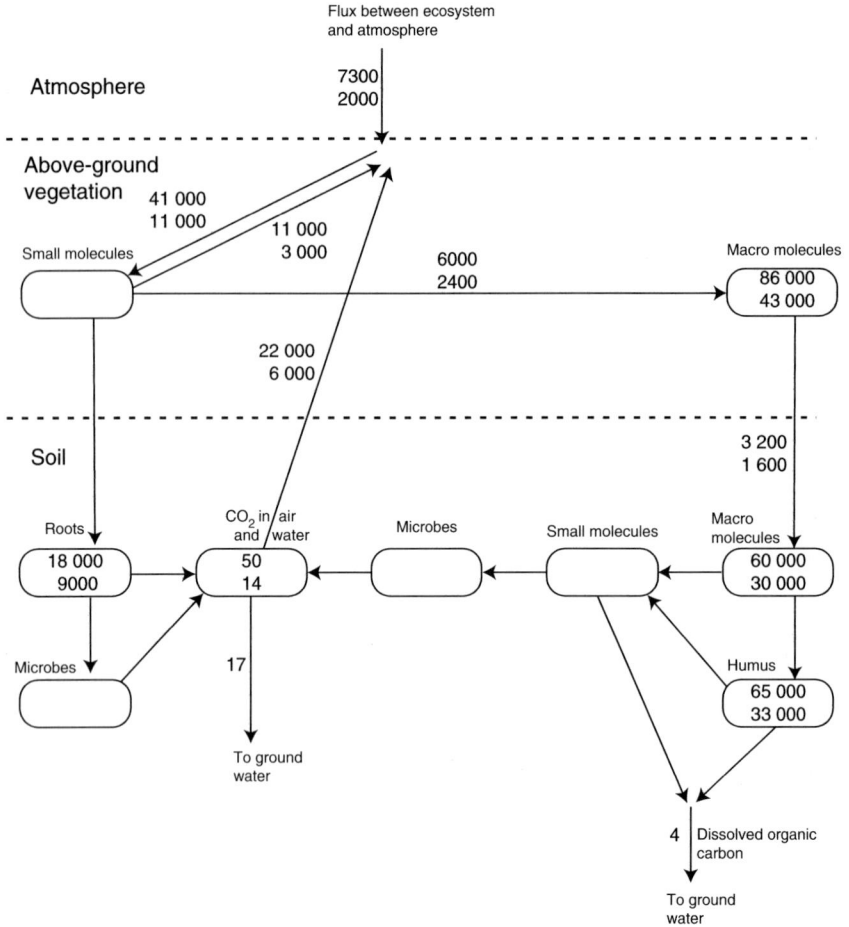

Fig. 7.9.2 The carbon amounts and fluxes in the ecosystem around SMEAR II. Upper numbers indicate the masses (kg dry weight ha^{-1}) and fluxes of the compounds (kg CO$_2$ ha^{-1} a^{-1} or kg dry weight ha^{-1} a^{-1}). Lower masses indicate the masses (kg C ha^{-1}) and fluxes (kg C ha^{-1} a^{-1}) as carbon

composition. When they are combined we get annual litter components: 80 kg proteins ha^{-1}a^{-1}, 260 kg starch ha^{-1}a^{-1}, 540 kg lipids ha^{-1}a^{-1}, lignin 1,050 kg ha^{-1}a^{-1} and 1,500 kg cellulose ha^{-1}a^{-1}.

7.9.3 Mean Residence Time of Carbon

The mean residence time in a pool/storage of any substance provides a good understanding of the dynamics of the substance in the pool. It is defined as the time required to accumulate or to deplete the pool with the mean in- or outflow. The mean residence times can be determined only for major routes in the circulation of material in the ecosystem. We can approximate the mean residence time by dividing the pool of substances with the mean in- or outflows. The structures of vegetation die and form outflow from the vegetation as litter production. The mean residence time of carbon in the vegetation is about 20 to 30 years.

Microbial decomposition determines the residence times of macromolecules in the soil. The amounts of starch, lipids, cellulose, lignin and proteins have been measured at SMEAR II and the annual inputs can be obtained from litter fall and their macromolecule concentrations. These results show that the mean residence time of organic macromolecules is high: for proteins, starches, lipids and cellulose, the means are 45, 15, 37 and 16 years, respectively. We cannot determine the mean residence time for lignin, since our method to analyse the chemical composition of soil organic matter does not separate lignin and humus (cf. Section 5.4.2).

Humus compounds dominate (65,000 kg dry weight $ha^{-1}a^{-1}$) the soil organic matter. The lifetime of humus is very long, on the scale of millennia (Liski et al., 2005). If we assume that the mean residence time is 1,000 years and that the amount of humus compounds are in steady state, we get a rough approximation for the annual formation and decomposition of humus (65 kg dry weight $ha^{-1}a^{-1}$). The CO_2 pool in soil air is small and the efflux from soil is large resulting in a small mean residence time, only a bit below a day.

7.9.4 Water Pools and Fluxes

All biological processes occur in liquid phase and water is an important transport media at long distance within trees. These facts make water necessary for living organisms. On the other hand, photosynthesis and transpiration are strongly coupled, since CO_2 dissolves in water at the mesophyll cell walls and diffusion transports CO_2 and water through stomata: CO_2 going in and H_2O coming out. Rainfall is water input in forest ecosystems and transpiration and run-off are outputs (Fig. 7.9.3). Since the rainfall is in boreal forests often around 300–800 mm in a year the annual inflow is large, at SMEAR II, 7 400,000 kg (H_2O) $ha^{-1}a^{-1}$ during the years 1998–2002. Water leaves the forest ecosystem either as evapotranspiration to the atmosphere or as lateral transport in deeper layers in the soil. Small watersheds enable measurement of the runoff flux in the soil from our stand. The evapotranspiration is 2,800,000 kg (H_2O) $ha^{-1}a^{-1}$ and runoff 3,000,000 kg (H_2O) $ha^{-1}a^{-1}$. The difference between out- and inflows at SMEAR II is 1 600,000 kg (H_2O) $ha^{-1}a^{-1}$ indicating measuring problems. The systematic errors in rainfall and evapotranspiration measurements are large, but the present imbalance is disturbing large. Most of the water in the forest ecosystem is either loosely bound on the surfaces of soil particles or in the wood. Both these amounts are large at SMEAR II, on soil particles 150,000 kg (H_2O) ha^{-1} and in the wood 800,000 kg (H_2O) ha^{-1}.

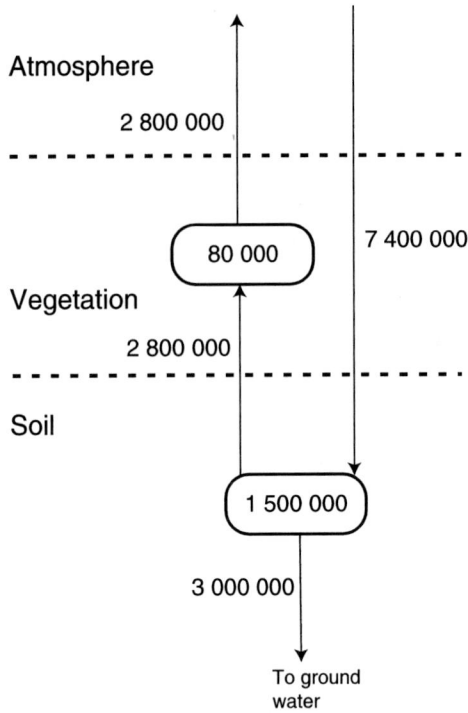

Fig. 7.9.3 Annual water fluxes (kg (H_2O) ha^{-1} a^{-1}) and amounts (kg (H_2O) ha^{-1}) in the ecosystem around SMEAR II

7.9.5 Mean Residence Time of Water

The water pools in the boreal forest ecosystem at SMEAR II are large, as they are in most boreal forests. Evapotranspiration is minimal in winter due to humid cold weather and stomatal closure. In contrast, the evapotranspiration is intensive in summer because the temperatures are often over 20°C, humidity is low and stomata open. The mean residence time is reasonable only during summer because water fluxes are negligible in winter. The mean annual residence time for water in soil is 60 days and in wood eight days. If we take into consideration that transpiration is very small in winter, we can say that the mean residence time of water is in the soil about 30 days and in the wood four days.

7.9.6 Nitrogen Pools and Fluxes

Molecular nitrogen is not available for plants, it has to be converted to ions, either ammonium or nitrate, before nitrogen can be utilised in metabolism. Nitrogen ions

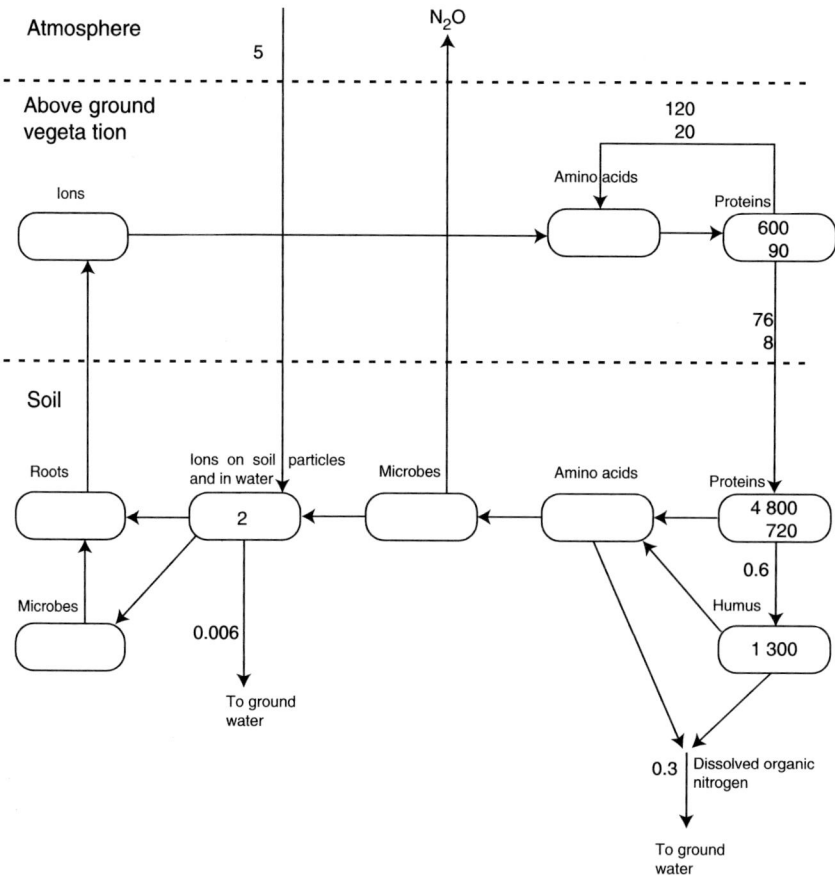

Fig. 7.9.4 The nitrogen amounts and fluxes in the ecosystem around SMEAR II. Upper numbers indicate the masses kg proteins ha^{-1} and fluxes of the compounds kg proteins ha^{-1} a^{-1} or dry weight of organic matter kg proteins ha^{-1} a^{-1}. Lower numbers indicate the masses kg N ha^{-1} and fluxes kg (N) ha^{-1} a^{-1}

enter the SMEAR II ecosystem either in deposition from the atmosphere or in microbial nitrogen fixation. Thereafter nitrogen circulates in the system either as proteins, amino acids or ions (Fig. 7.9.4). Plants take ammonium and nitrate ions from soil. The nitrate is converted to ammonium and used in amino acid synthesis together with ammonium ions taken from soil. The amino acids are utilised in protein synthesis to build the needed functional substances. When the tissue dies, then the proteins are decomposed back to amino acids, which are used for protein synthesis in some living tissue. However, about 40% of the proteins in the dying tissue are lost as litter to the soil. The proteins in the soil are decomposed to amino acids by extra cellular enzymes emitted by microbes. A small proportion is consumed in the formation of humus substances. Microbes take up the amino acids to construct proteins or utilise the energy in amino acids and release ammonium ion into soil water.

The internal cycling within the stand is about 76 kg (proteins) $ha^{-1} a^{-1}$. Fluxes with the surroundings are small when compared with internal cycling. The input is about 5 kg (N) $ha^{-1} a^{-1}$ by wet deposition. The role of nitrogen fixation is unclear; we assume that it can be neglected. The outflows are still smaller, the biggest loss is as dissolved organic nitrogen (DON) in the runoff 0.3 kg (N) $ha^{-1} a^{-1}$. There are short episodes in autumn and spring when the N_2O flow from the system is above detection limit of chamber measurements, but the flux is so small that it can be regarded as very marginal for N budget. NO_x flux between the needles and atmosphere is clearly measurable when the sun is shining. The NO_x comes evidently from a UV-radiation driven process on the needle surface and we assume that the nitrogen is not coming from the metabolism of trees and it can be omitted in the nitrogen budget of the ecosystem.

The differences between sizes of nitrogen pools in forest ecosystems are several magnitudes, reflecting very active and rather passive processes. The ion pool in the soil is small, only 2 kg (N) ha^{-1}. The amount of proteins in vegetation is large reflecting the fact that functional substances are proteins. The value is 600 kg proteins ha^{-1}. The biggest protein nitrogen pool is in the soil, 4,800 kg proteins ha^{-1}. The nitrogen in humus is in a very inactive form and it does not participate in the normal nitrogen cycle in the forest ecosystem, but its amount is rather large, 1,300 kg (N) ha^{-1}.

7.9.7 Mean Residence Time of Nitrogen

The variation in nitrogen pool sizes and in the fluxes is large in the forest ecosystem around SMEAR II, which is reflected also in mean residence times in various components of the system. The pool of nitrogen ions in the soil is small and the nitrogen uptake by vegetation is large resulting in a short mean residence time, only 60 days; if we take into consideration that nutrient uptake is very small in winter, we can say that the mean residence time for nitrogen ions is about 30 days. Most of the proteins in living tissues have rather short lifetimes, but re-use of the amino acids is efficient, resulting in rather long mean residence time in vegetation, about eight years. The decomposition of proteins by extra- cellular enzymes in the soil is a slow process. In addition, the circulation in soil generated by microbes prolongs the stay of nitrogen in the soil. The mean residence time of proteins in the soil at SMEAR II is 45 years.

The nitrogen pool in the forest ecosystem around SMEAR II is large: when expressed in the units nitrogen per hectare, we get 2,300 kg (N) ha^{-1}. There are a short and weak N_2O and N_2 formation periods when the gas diffusion is blocked by water but their emissions can be neglected. The leakage as ions is small compared with that of dissolved organic nitrogen (DON). The mean nitrogen residence time in the forest ecosystem is long, according to our measurements about 6,000 years.

Extensive forest fires are an essential part of development of boreal forests. Then all green components, small woody parts and the top layer of soil burn releasing the nitrogen that was present in proteins as NO_x. Fire frequency varies according to the location, but it can be assumed to be several times in a millennia. If we assume that proteins remaining in the unburned wood and that burned from the soil are equal and if we assume further that there are five fires in a millennia, we get that the outflow is 3 kg (proteins) ha^{-1} a^{-1} and the nitrogen residence time caused by forest fires is about 1,500 years.

7.9.8 Comparison of Fluxes, Pools and Residence Times

The differences of fluxes between water, carbon and nitrogen are large. Water flux from the atmosphere into SMEAR II ecosystem is 7,400,000 kg (H_2O) ha^{-1} a^{-1} and nitrogen flux out of the ecosystem is 0.3 kg (N) ha^{-1} a^{-1}. The carbon flux into the system (41,000 kg CO_2 ha^{-1} a^{-1}) is two magnitudes smaller than water flux and five magnitudes bigger than nitrogen flux out of the system.

Although water flux into the ecosystem is two magnitudes larger than the carbon flux, the carbon pool is clearly larger than the pool of water. The water is running fast through the system and carbon remains for long periods either in wood or in soil organic matter resulting in large pools. There are functional substances in all living cells enabling processes. The nitrogen content in vegetation and macromolecules in soil is around 1% and thus nitrogen pool in the ecosystem is about two magnitudes smaller than that of carbon.

The residence time of water in the ecosystem is small, in summer time about 30 days. In contrast, the nitrogen stays in the system for millennia if it has been synthesized into organic form. This difference is again five magnitudes reflecting the very different behaviour of water and nitrogen in the ecosystem, water is flowing through and nitrogen is accumulating. The carbon is between water and nitrogen, although it flows through the system like water. Its residence time is 20 years.

The differences in the mean residence times between different water pools in the ecosystem are small, only one magnitude between soil and wood. Carbon compartment pools have bigger variation because the CO_2 pool in the soil is very dynamic, the residence time is only a day. The humus storage in the soil is extremely stable: its residence time is millennia. The variation in residence times is about six magnitudes. Nitrogen has also big variation in the residence times, but smaller than carbon. The fastest pool is ammonium ions in the soil: they have residence time about 30 days. The residence time of organic nitrogen in the ecosystem is in the range of millennia. Thus the variation in residence times is one magnitude smaller than in the case of carbon.

The variations in the fluxes, amounts and residence times of carbon, water and nitrogen are large, often several magnitudes. The big variation hampers the analysis of the ecosystems since several time levels have to be treated simultaneously. Proper analysis of the fluxes, amounts and residence times is the key to understand the growth of forest ecosystems and their response to climate change.

Chapter 8
Connections Between Processes, Transport and Structure

Eero Nikinmaa, Pertti Hari, and Annikki Mäkelä

8.1 Emergent Properties

Tree structure forms the physical framework for processes and transport to take place. It also forms the means for the tree to reach towards resources provided by its environment. Thus structure, processes and transport, but also the mechanical stability and long term survival, are strongly interconnected, and the efficiency of whole-tree functioning in terms of energy and raw material use is dependent on how well these components are balanced with each other. Imbalance would be wasteful with respect to both energy and raw material and would result into worse performance in competitive environment. Therefore balanced connections between structure, processes and transport can be assumed to be subjected to strong selective pressures. Since long in botanical literature, number of different observations of regular structural features has been reported (MacCurdy, 2002). A claim can be made that efficient and balanced structures adapted to different environments have emerged in the course of evolution of organisms (see Section 1.2: Basic idea 6).

A balanced structure, or a balance between two parallel processes, can be regarded as an emergent property at the whole-tree level that is not directly derivable from the component processes and structures at the space and time element level. Instead, the existence of some balancing principle has to be assumed. This makes the research and prediction of whole-tree functioning more complicated than just scaling-up of phenomena from the space and time element level to the ecosystem level. In the preceding chapters we have discussed the latter as consisting of three steps; (i) analysis of structures, processes and transport phenomena, (ii) construction of mathematical models to describe the object and (iii) a test of the model against measurements. In order to include the assumption of balanced structure, an additional step has to be incorporated. We need to (iv) formulate linking between

Eero Nikinmaa, Pertti Hari, and Annikki Mäkelä
University of Helsinki, Department of Forest Ecology, Finland

P. Hari, L. Kulmala (eds.) *Boreal Forest and Climate Change,*
© Springer Science+Business Media B.V., 2009

structures, processes and transport phenomena, and finally as before, (iv) test the models and hypotheses against measurements.

It has often been suggested that evolution is the source of testable connection principles (Mäkelä et al., 2002). Any biological entity, whether plant or animal, are learning systems over time. Their current response to their surroundings, which is genetically coded in their DNA, is such that has enabled them to survive the conditions up to the present. In principle, all the information we need to predict the response of organism to their surroundings is there. However, even simple organisms, such as plants, are complex systems where large number of interactions between different plant parts takes continuously place. This makes the prediction of whole plant responses to their surrounding very difficult, as for example the phenotypic plasticity of trees shows (Messier and Nikinmaa, 2000). An alternative approach is to assume performance criteria for the whole plant that had presumably had to take place in the historical development of the organism and study their consequences and compare that with the observed behavior. The often used approach is optimality of behavior relative to critical features (Givnish, 1986). This approach has enjoyed limited interest among plant biologists and modelers, perhaps because it has been thought of as involving a fair share of intuitive and subjective elements. However, apart of being sometimes very efficient means of approximating plant behavior (Cowan and Farquhar, 1977; Hari et al., 1986; Berninger et al., 1996, Section 6.3.2.3), it will also give insight on the performance level of plants relative to the particular feature being studied.

In this chapter, we review some of the most important interactions between structure, processes and transport from the point of view of understanding whole-tree growth and efficient allocation developed in evolution. We first discuss the functional role of structure in trees, and how it relates to tree growth, and then formulate some connecting principles and present quantitative evidence related to these.

8.2 Functional Role of Structure and Growth of Trees

Whole tree architecture has a significant role in many functional traits in trees, as previous chapters have pointed out. Crown structure influences light interception and thus the photosynthesis of the foliage. The cross-sectional area of water-conducting wood, its permeability and its length all have a significant role in controlling water and assimilate transport in trees. The spatial distribution of fine roots determines how trees access nutrients and water. The interplay between these structural properties and resource acquisition results in feedbacks that influence tree shape and size. For example, the idea of the hydraulic limitation of tree growth (Yoder et al., 1994) has been well established in studies of maximum size that trees can reach (e.g., Koch et al., 2003). An important component of long term survival of trees is also the mechanical stability of the main axis of the tree, which is also linked to the material properties and dimensions of them.

There is feedback among the different hierarchical levels of tree; for example, molecular level processes determine process rates in tree organs but these can also be influenced by whole-tree structure. This is most obvious in photosynthetic production where functional substances in the chloroplasts of leaf parenchyma cells determine the CO_2-binding capacity, but the instantaneous rate of CO_2-intake also depends, through stomatal action, on the water supply capacity of the water-conducting pathway from soil to leaf (see Section 7.5.2). Thus, in living systems the whole is to some degree constrained by the parts (reductionism would claim that it is completely so) but at the same time the parts are to some degree constrained by the whole (downward causation; Campbell, 1974; Nikinmaa, 1992).

There cannot be an organism whose internal functions do not follow the laws of physics and chemistry, but regulation under the control of the genetic code is needed to determine the composition and the shape of the whole organism. Once particular composition has emerged, it very much influences the behavior of its components.

The structure of trees is constrained by their inherited growth pattern. Chapter 5 has pointed out that the growth of trees takes place at the specialised meristematic tissue, both at the ends of the growing axes, bringing about the architectural shape of the tree, and all around the formed axes, such as tree stems, where the cambium increases the tree girth with increasing tree size. This secondary growth helps trees to cope with changing transport and support requirements. These growth processes depend on a supply of the required growth resources, the carbohydrates from the primary production of photosynthesis, proteins from the nitrogen assimilation and a number of other elements either in their ion form or incorporated as functional groups into organic molecules. Furthermore, the water supply needs to be adequate to allow sufficient turgor pressure at the growing sites, which expands the newly formed structures into their final size.

Supplementary to the required resources, there is a complex regulatory system that brings out the response pattern of growth to the surrounding conditions. This regulatory system involves a number of proteins and hormones such as auxin, cytokinins and gibberellins that control the initiation and growth of the meristematic tissue relative to the season and the tissue's position within the plant. The growth potential of a tree is much higher than the phenotype at any given moment might reveal. Regulation among the growth compartments suppresses some parts while allocating to other parts more resources than would be expected on the basis of their growth environment only (for example the typical pattern of apical dominance, Nikinmaa et al., 2003).

Trees can be viewed as modular organisms with reiterative growth pattern of simple structures repeating over time, bringing about the tree crown development (e.g., Hallé et al., 1978). Many physiological studies have suggested that branches are autonomous with respect to their carbon requirement (e.g., Dickson and Isebrands, 1991), viewing individual shoot development mainly as a local process, with photosynthetic production efficiency playing a key role (Linder and Axelsson, 1982). However, as already pointed out in Section 7.5, there is an intimate link between the assimilate production at the source, assimilate use at the sink and the phloem transport connecting them. The latter is also linked with water uptake from the soil. Such a source-transport-sink approach is probably the most adequate view of the growth process at the whole-tree level. It is fairly easy to understand conceptually, and it has also been successfully applied to growth modelling at an aggregated level (e.g., Thornley, 1991). However, (i) the large number of "competing" sinks and supplying sources, (ii) the involvement of hormones that are largely produced at opposing ends of the growing system (developing shoots and leaves at one end and roots at the other end) and (iii) the combination of a "fast" bulk flow over long distances with local diffusion at the meristematic level, make the quantitative description of tree growth at the process level a very challenging task. The complexity created by the detailed description of the source-transport-sink systems seem to give rise to many more degrees of freedom in growth than are actually observed in the growth of real trees.

The study of regularities in tree structure, based on connections between processes, transport phenomena and tree structure, can help to overcome the problem of a too complicated and detailed analysis of growth and still bring us useful information about how and why trees grow as they do in different environments. During vegetative growth, trees face an allocation problem; they can use the available resources, mainly sugars and nitrogen, for building either productive structures or transport and support systems. At the life cycle level also the division between allocation to vegetative growth of reproduction becomes important. Continuous functioning of processes requires resources and products assimilated in other tree parts and also transport of the end products away from their sites of production. The capacity of the transport system easily becomes a rate limiting factor (e.g. Höltta et al., 2006). Thus an adequate balance between the assimilation and transport rates brings out performance improvement in comparison to less balanced cases. As the transport rate is intimately linked with the structural features, this should reflect on the dimensions of the structure. Thus balance is crucial for the success of trees in evolution. The regularities in tree structure may be regarded as an emergent property that could be studied in terms of an efficient or balanced allocation.

A point of view emphasizing the role of emergent properties in growth and development focuses on how well certain structural traits perform as functional wholes. This approach considers the connectivity of different tree organs and their functional balance. Watson and Casper (1984) studied tree development consisting of integrated physiological units, stressing the importance of a balanced supply and transport of both above-ground and below-ground resources for undisturbed tree functioning. According to this view, tree growth is the development of integrated leaf–wood–root units. This approach resembles the pipe model approach (Shinozaki et al., 1964a, b, see Section 5.2.5).

In the following sections, we review some approaches that take the view of different tree compartments, i.e. larger structural entities with some principal functional task performing as functional wholes, thus balancing different source-transport-sink relationships in the tree. Those balancing requirements act as constraints to whole tree shape consequently providing quantitative constraints to growth allocation.

8.3 Structural Regularities Generated by Water Transport

Transpiration is a prerequisite of photosynthesis, since diffusion transports CO_2 into and H_2O out of stomata. In addition, the mass flow of water in transpiration is much larger than that of CO_2 in photosynthesis, for example in boreal coniferous forests, the difference is nearly 100-fold. The need for water transport generates regularities in the woody components, and the tree structure can be viewed as a balanced continuum of water transport elements (Hari et al., 1986). If we assume that a constant ratio is maintained throughout tree development between foliage mass and the area of the conducting tissue below (i.e. the pipe model principle, Shinozaki et al., 1964a, b; Mäkelä, 2002; Berninger et al., 2005, see Fig. 8.3.1), and that the ratio remains fairly constant even across different local environments in the canopy, then it is straightforward to describe how the growth should be distributed to maintain such a structural balance (Hari et al., 1985; Valentine, 1985; Mäkelä, 1986).

Let M_n denote the needle mass of a tree, A_s the cross-sectional area of the stem below the living crown. The pipe model principle is

$$M_n = a_s A_s \qquad (8.3.1)$$

where a_s is a parameter.

The pipe model can be formulated also for the branches. Let A_b^i be the cross-sectional area of the branch i. Then

$$M_n = a_b \sum_i A_b^i \qquad (8.3.2)$$

where a_b is a parameter.

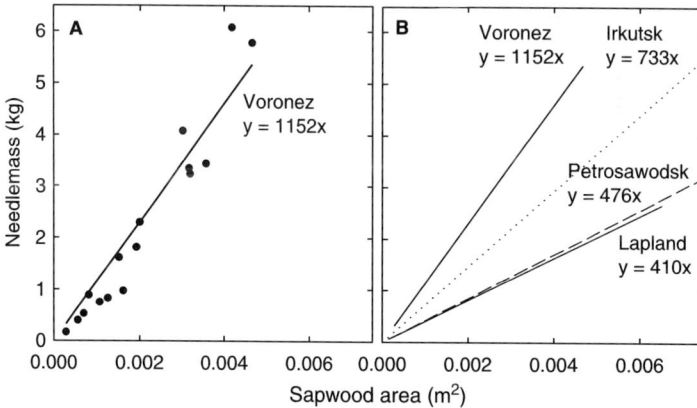

Fig. 8.3.1 (A) Measurements and fitted relationship between sapwood area at crown base and crown needle mass in Scots pine in Voronez and (B) the fitted relationships in Voronez and in Irkutsk, Petrosawodsk and Lapland

The stem sapwood biomass, M_s is obtained as function of mean average height of the needles, h_n and tree needle mass

$$M_s = \rho_s a_s M_n h_n \qquad (8.3.3)$$

where ρ_s is wood density of the stem.

Equations 8.3.1–8.3.3 demonstrate that if we know the foliage growth, we can approximate how the other biomass compartments should grow to maintain the pipe model relationships in the tree.

Above, we have discussed the pipe model in terms of water transport to the leaves. Shinozaki et al. (1964a) did not make this interpretation, but presented the model more as a description of measured data. The reasons for the pipe model structure could be either mechanical or physiological or both. However, the water transport interpretation has been extensively explored in the literature, and there is mounting evidence that balancing the water conducting pathway is far more complicated a problem than a simple balance between conducting area and transpiring surface. The water conducting "tubes" are not hollow pipes with constant diameter from top to stump, but they consist of separate conduits connected by wallcrossings both longitudinally and radially (Siau, 1984). In addition to the conduits, wood consists of fibres that provide support rather than a water pathway. Therefore the need of water transport should be considered as first approximation of the underlying complicated connections between structure and water transport in wood.

Deriving the thickness growth from the foliage growth at tree level is relatively straight-forward from these premises. However, when considering the tree length growth and the pattern of branch formation and growth, also the long-term patterns of tree success are involved. Investment in leaves at low heights allows a larger proportion of photosynthetic production to be used for foliage growth compared with investment higher up in the canopy, because wood growth is proportional to both net foliage growth and the distance between foliage and roots. However, in a competitive environment, foliage is also likely to be shaded, possibly resulting in lower net foliage growth. Depending on the competitive environment, different height growth strategies are favoured (Givnish, 1995).

Game theory has been applied to study height growth strategies (Mäkelä, 1986; Givnish, 1986), but an optimal allocation for different conditions can also be formulated (Nikinmaa, 1992).

The shoots grown in the upper or lower parts of the tree crown play different roles in crown dynamics. Because young, first-order shoots in the upper crown are unlikely to be immediately shaded, they play an important role in supporting tree structure for many years. On the other hand, shoots in the lower crown, even if at first in a strong light environment, become shaded and less able to support other shoots. Therefore, it seems that there are mechanisms facilitating the control of shoot growth in relation to shoot position for trees that normally compete for light during canopy development (Nikinmaa et al., 2003)

8.4 Balance Between Photosynthesis and Nitrogen Uptake

It has long been well known that nitrogen availability is crucial for the growth rate of crops and stands, however, the mechanisms of the impact of N are still not fully understood. One of the main theories to explain nitrogen impacts is the so-called "functional balance". The essence of the functional balance theory is that plant tissue has a characteristic nitrogen concentration determined by the need for functional substances, and this characteristic concentration is maintained through allocation of growth.

Davidson (1969) first presented a model based on the functional balance idea. The model was for a herbaceous plant consisting of a shoot and a root and undergoing exponential growth. For such a plant, the following relationship holds between the balanced nitrogen level B_N ($\mathrm{kg\,N\,kg^{-1}C}$) and shoot and root biomasses, M_{sh} and M_r

$$\sigma_N M_r = B_N \sigma_C M_{sh} \qquad (8.4.1)$$

where σ_N and σ_C are specific nitrogen and carbon uptake rates by the root and shoot (per unit time). According to this model, changes in the specific rates σ_N and σ_C lead to respective changes in the relative sizes of the root and shoot biomass, while the tissue N concentration tends to be constant. The model was supported by evidence from a series of experiments in white clover (Brouwer, 1962).

In a more complicated situation, i.e., if the plant is not undergoing exponential growth, and if it has other components in addition to shoots and roots, the functional balance requirement has to be re-phrased. In such a case, the functional balance can be derived from the requirement that the amount of N utilised in the growth of the new tissue is proportional to that of the carbon used in new growth, with tissue-specific N concentrations. Let N_G desote the amount of nitrogen used for growth of new tissue. Then

$$N_G = \sum_i B_N^i G_i \qquad (8.4.2)$$

where B_N^i is the balanced N concentration in component i and G_i is the growth of component i in dry weight units.

On the other hand, the nitrogen utilized in new growth has to equal the nitrogen taken up by the roots, U_N, plus the nitrogen recycled from the senescing components of the plant, therefore

$$N_G = U_N + \sum_i f_i B_N^i s_i \qquad (8.4.3)$$

where s_i is the loss rate of component i and f_i is the proportion of N recycled from i.

Furthermore, U_N is proportional to the fine root biomass, $M_{r,}$

$$U_N = \sigma_r M_r \qquad (8.4.4)$$

and total growth depends on all biomass components through photosynthesis and respiration Assuming that that f_i and C_{Ni} are constant, these equations allow us to solve for the amount of fine root growth required to satisfy the nitrogen demand of the plant.

In support of the theory in trees, there is evidence that the ratio of foliage to fine root biomass is strongly dependent on site quality (Santantonio, 1989; Vanninen and Mäkelä, 1999; Helmisaari et al., 2007), and that especially in conifers, the foliage N concentration varies little over a wide range of site conditions and climates (Le Maire et al., 2005; Helmisaari et al., 2007).

Modifications of the functional balance approach include models allowing for variations in the tissue N concentration. In Thornley's transport resistance model, variation is allowed in the substrate N concentration, while structural tissue is assumed fixed by stochiometry (Thornley, 1991). Other studies have assumed that N concentration has an impact on the rate of physiological processes, including growth, photosynthesis and respiration, and searched for optimum N concentrations along with the optimal allocation (Mäkelä and Sievänen, 1987; Hilbert, 1990; Ågren and Franklin, 2003). As the light conditions within tree crown vary often as much or more than the average light intercepted between trees studies have been made how nitrogen should be allocated within tree crowns. When considering only nitrogen and light received by leaves in a tree crown, an arrangement where foliar nitrogen concentration decreases linearly with available light seems to bring optimal canopy production (Sands, 1995).

Conclusions

Structural regularities, balancing processes and transport, have emerged through evolution. The efficient use of resources is the key to understanding allocation of sugars and nitrogen to new tissues.

Chapter 9
MicroForest

9.1 Growth and Development of Forest Ecosystems; the MicroForest Model

Pertti Hari[1], Mirja Salkinoja-Salonen[2], Jari Liski[3], Asko Simojoki[2], Pasi Kolari[1], Jukka Pumpanen[1], Mika Kähkönen[2], Tuomas Aakala[1], Mikko Havimo[4], Roope Kivekäs[1], and Eero Nikinmaa[1],

[1] University of Helsinki, Department of Forest Ecology, Finland
[2] University of Helsinki, Department of Applied Chemistry and Microbiology, Finland
[3] Finnish Environment Institute, Research Department, ResearchProgramme for Global Change, Finland
[4] University of Helsinki, Department of Forest Resource Management, Finland

9.1.1 Background

Clear regularities in processes, structure and transport in forest ecosystems have been highlighted in the previous chapters. The most important conclusions are (i) light, temperature and temperature history determines photosynthetic rate (Section 6.3.3.1.3), (ii) extinction of light reduces photosynthesis in the canopy (Section 7.6.2), (iii) there is a tight relationship between the needle mass of a tree and the cross-sectional area of sapwood (Chapter 8), (iv) tissue types are characterised by specific concentrations of functional substances (Section 5.2.1) and (v) decomposition of soil organic matter proceeds slowly (Sections 6.4.1 and 7.9). These results, expressed quantitatively, enable construction of a dynamic model describing the growth of a forest ecosystem.

Two main lines have emerged in the development of carbon-balance models: the model must be simple and easy to use for global applications (e.g., Landsberg and Waring, 1997) and there should be a theory-driven analysis which aims towards causal understanding of forest growth (e.g., Thornley and Johnson, 1990; Thornley,

P. Hari, L. Kulmala (eds.) *Boreal Forest and Climate Change,*
© Springer Science+Business Media B.V., 2009

1991; Bossel, 1991; Nissinen and Hari, 1998). De Wit et al. (1970) formulated the carbon-balance approach in the quantitative analysis of plant production, i.e., growth is based on carbohydrates formed in photosynthesis. The original formulation was for cereals but the same principles were later also applied in the analysis of tree growth (Hari et al., 1982; Mohren, 1987). Apart from the carbon balance, the availability of nutrients has a strong influence on stand growth and development (Ingestad and Ågren, 1991). A proper description of nitrogen in trees and microbes, releasing nitrogen from dead plant material, is needed to understand the dynamics of metabolic processes in forest ecosystems.

The introduction of the connection between tree structure and functions using emergent and empirically observed structural regularities was a clear step forward in understanding of the dynamics of resource allocation and tree growth in carbon balance models (Hari et al., 1985; Valentine, 1985; Landsberg, 1986; Mäkelä, 1986). Also the functional balance principle has been successful in predicting the relative growth between organs that take up carbon and nitrogen (Mäkelä, 1997). However, this approach has not yet been linked with models of explicit nitrogen cycling that would allow full evaluation of the relative importance of carbon and nitrogen in ecosystem dynamics.

Here we develop and test a model that links earlier tested concepts of tree and stand primary production and its allocation among tree parts (Hari et al., 1982, 1985; Nikinmaa, 1992; Perttunen et al., 1996; Mäkelä, 1997, 2002; Nissinen and Hari, 1998; Hari, 1999; Mäkelä and Mäkinen, 2003) with a novel dynamic treatment of nitrogen, especially the role of soil microbes in the decomposition of soil organic matter. We also, in order to complete the picture of nutrient cycling, introduce, ground vegetation into the analysis as a new component that has more rapid turnover rate than trees. The model is based on results obtained in Chapters 4–8.

9.1.2 Carbon and Nitrogen Fluxes in a Forest Ecosystem

The carbon cycle is driven by photosynthesis and respiration leading to a net accumulation of carbon in the form of long lasting structures through the process of growth. These eventually die and become an energy and nutrient source for soil microbes that release part of the bound carbon. Metabolic processes and senescence generate carbon flow through a forest ecosystem (Fig. 9.1.1).

Nitrogen circulates within the internal structure of plants, within the organic matter and microbes, and between the vegetation and soil. The latter cycle starts with nitrogen uptake by the roots from the pool of available nutrients in the soil and goes, via the formation of new tissue, senescence and decomposition of organic matter in the soil, back to the available pool. The within-soil cycle is generated by extra-cellular decomposition of large protein molecules where microbes take up amino acids, form enzymes and emit them and where soil proteins are broken into amino acids. The within-plant cycle results from the retranslocation of proteins as amino acids from dying plant parts (Fig. 9.1.2).

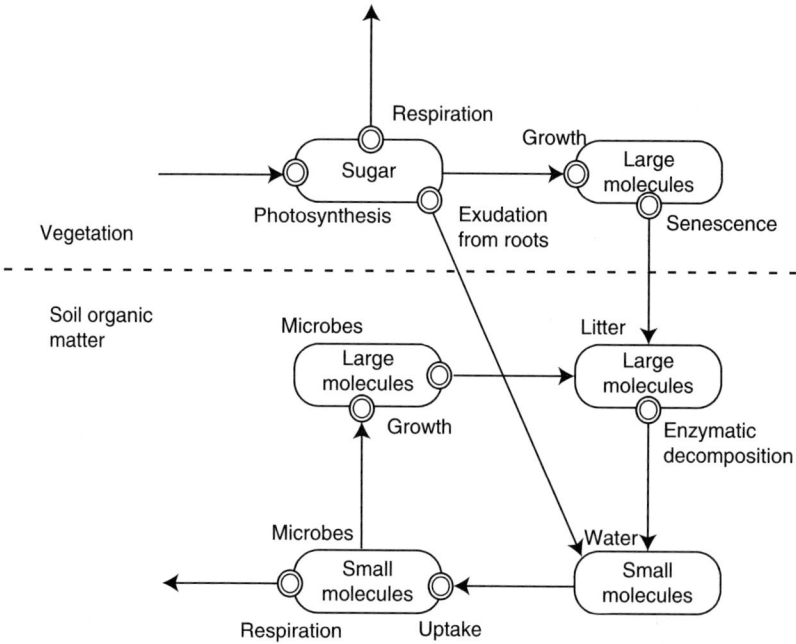

Fig. 9.1.1 Carbon compound fluxes in forest ecosystem from atmosphere via leaves and microbes back to atmosphere. The symbols are introduced in the Fig. 1.2.1

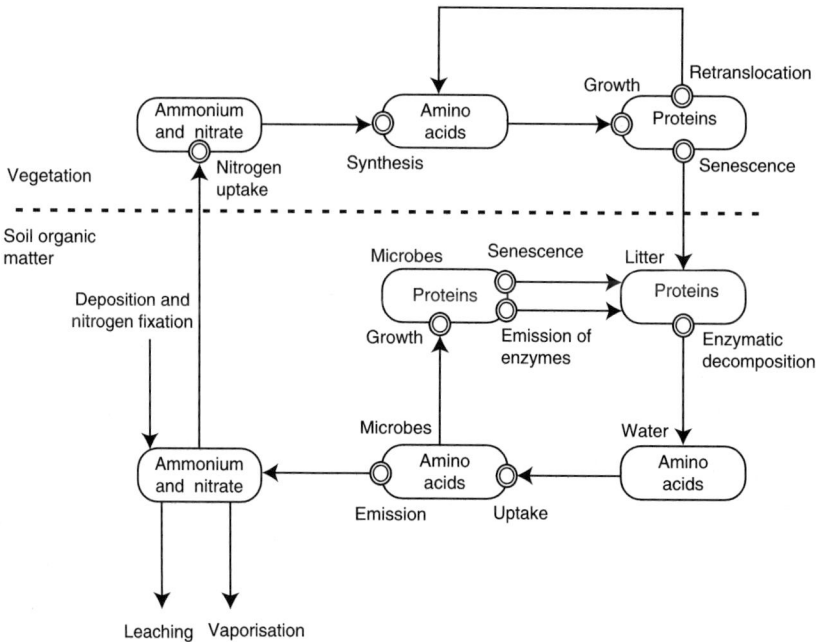

Fig. 9.1.2 Circulation of nitrogen compounds in forest ecosystem from available nitrogen in the soil into metabolism of trees and microbes back to available nitrogen. The symbols are introduced in the Fig. 1.2.1

Carbon and nitrogen fluxes generate changes in the pools of organic matter in a forest ecosystem. When the flows are combined with the pools, a dynamic model of the system can be constructed that describes the development of trees, ground vegetation and forest soil. The ecosystem model is split into three submodels; (i) trees, (ii) ground vegetation and (iii) soil.

9.1.3 Trees

The dominating component in a forest ecosystem is trees. They provide raw material for the organic material in soil and they determine the light climate of ground vegetation.

Trees and individual plants in ground vegetation are the basic units of the above ground component of a forest ecosystem. They collect material for metabolism and growth from their environment, grow and die. There are three different steps in the growth and development; (i) collection of raw material, (ii) formation of new tissues and (iii) senescence. The modelling of development of tree structure is divided similarly into the formation of raw material and the allocation of the obtained raw material to different components and finally senescence.

Photosynthesis produces sugars and nitrogen uptake and retranslocation provides amino acids for growth. These resources are utilised under the control of the regulation system. The allocation of sugars and amino acids to new leaves, to water pipes in branches, stem and transport roots to fine roots is an emergent property at tree level. The connections between structures, functions and transport phenomena are keys to understand allocation of carbon and nitrogen in a tree (Chapter 8).

Proper treatment of the regulation system to produce allocation of carbon and nitrogen to productive parts (needles and fine roots) and to supporting water transport systems in branches, stems and coarse roots is crucial for the success of a tree in a stand (Figs. 9.1.3 and 9.1.4). Allocation determines the relations between the different components and enables balanced connections among photosynthesis, nutrient and water uptake, and water transport. Thus allocation is also an important aspect in the construction of growth models based on carbon and nitrogen fluxes. The primordias for all needles to grow in the next spring are laid down in late summer in the buds. Thus the allocation is determined to great extend in the formation of the buds and Scots pine operates with annual time step.

The basic functional unit of a tree is assumed to be a whorl of branches and leaves (needles) that photosynthesises, transports water and nutrients and takes up nutrients and water (Fig. 9.1.5). Trees are more complicated than just a sum of the whorls; the regulation system of tree architecture allocates sugars from the lower part of the canopy to the top whorls of the tree. The senescence of old needles is missing in the top whorls and reuse of xylem does not contribute to the water transport of new needles. This is why construction of the water transport system for the top whorls consumes such a great quantity of sugars, that photosynthesis of the whorl can not produce and additional sugars are needed from the lower parts of the crown.

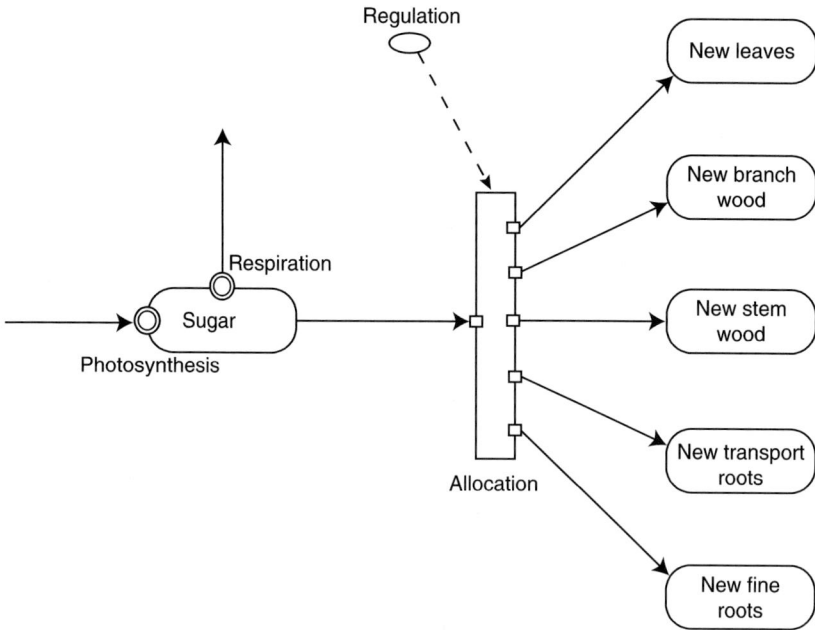

Fig. 9.1.3 Visualization of allocation of carbon compounds to new leaves, new water transport system in branches, stem, transport roots and fine roots under the control of the regulation system. A symbol for allocation is introduced here. The other symbols are introduced in the Fig. 1.2.1

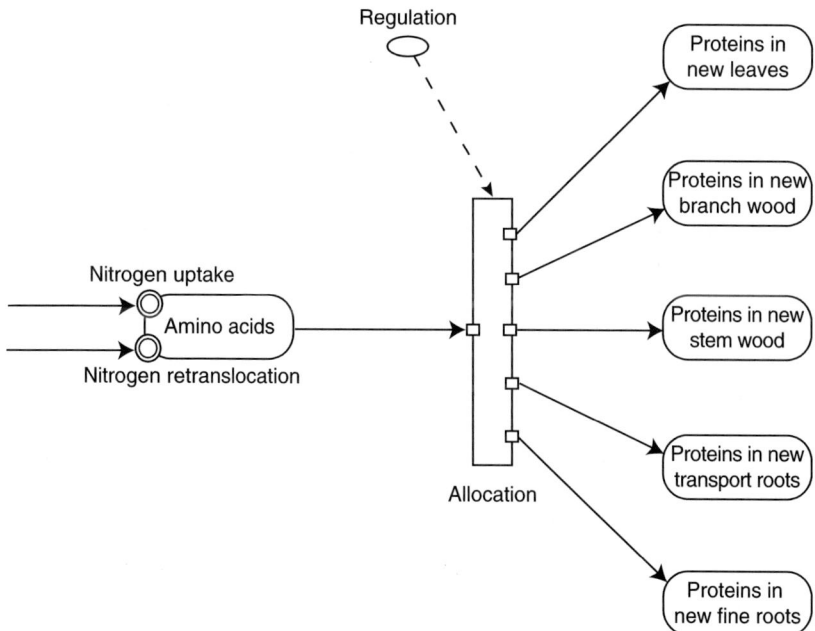

Fig. 9.1.4 Visualization of allocation of nitrogen compounds to new leaves, new water transport system in branches, stem, transport roots and fine roots under the control of the regulation system

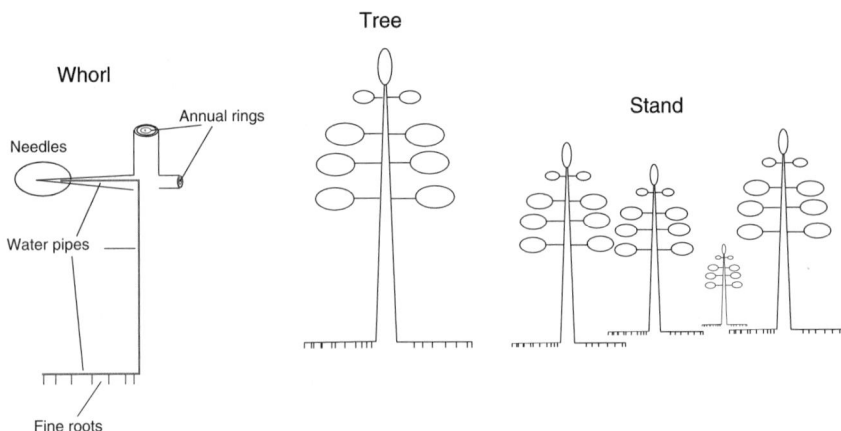

Fig. 9.1.5 Hierarchical description of stand structure in MicroForest. Most detailed level is a whorl including needles, water pipes and fine roots. Intermediate level is tree obtained as sum of whorls. Most aggregated level is stand formed by size classes

Structure of Trees

The structure of trees is an important dynamic property influencing the ecosystem fluxes. Thus a structural description needs to be sufficient to reflect both the variation of allocation of resources within a single tree and differentiation of trees within a stand (Hari et al., 1985). We describe the stand structure hierarchically: the most detailed level is the whorl, the tree being the intermediate level and the stand the most aggregated (Fig. 9.1.5). To allow for dynamic development of tree crowns, we assume that whorls are the functional units of a tree, including the water transport structures in its branches, stem, transport roots, and fine roots as suggested by Watson and Casper (1984 and Chapter 8). Pine trees form annually a whorl on the top of the tree, where new branches and a leader shoot are formed. Other branches elongate, new needles grow on them and the annual rings are formed in stem and branches. Tree growth dynamics are driven by the growth dynamics of the individual whorls and the tree is derived as their sum. A stand is a population of trees that is divided into size classes that are characterized by the number of trees in the class and the mean tree representing the average properties of each size class.

The needles, woody structures and fine roots are described with masses. This is not, however, sufficient since the chemical structure of wood differs rather considerably from that of needles and fine roots. Plants synthesize macromolecules, proteins, cellulose, lignin, lipids and starch using sugars and nutrients as raw material. Proteins are the active components in living cells: they catalyze biochemical reactions and transport substances through membranes. Cellulose and lignin form the cell wall, lipids the membranes, starch acts as reserve and proteins are functional substances (Section 5.2.1). In addition, the lifetimes in the decay process of litter are macromolecule specific. Thus all masses are treated with the five macromolecules both in trees and soil.

The nitrogen concentration in proteins is high, about 15% and in cellulose, lignin, lipids and starch it is negligible. Needles and fine roots are metabolically very active: they synthesize new molecules and move molecules and ions through membranes. In contrast, wood is rather passive, it transports water. This activity difference is reflected in the concentrations of functional substances, being high in needles and fine roots and low in wood (Chapter 5).

The stand structure is described as a discrete system where masses and dimensions can be assigned to whorls, size classes of trees and simulated years. For example, let $M_n(i, j, k)$ (g(dry matter)) describe the needle mass in the size class i, whorl j and year k. In continuation the notation follows this principle. The whorls run from the base of the tree where the whorl was formed at year 1 to the top of the tree where the whorl is formed at year k. Since the driving structural component of the model is the whorl also the masses required in other parts of the tree, i.e., $M_b(i, j, k)$, $M_s(i, j, k)$, $M_c(i, j, k)$ and $M_r(i, j, k)$ (g(dry matter)) are branch, stem, transport root and fine root growth formed to support new needles $M_n(i, j, k)$ (Fig. 9.1.5).

Apart from the masses each tree has dimensions, i.e., height, length of branches in the whorls and the diameters of stem and branches.

Processes of Trees

The annual photosynthetic production of a whorl is the starting point for analysis of the development of tree structure. The photosynthetic rate, i.e., the rate of formation of sugars, is determined by environmental factors and the annual cycle of the trees (Section 6.3.3.1.3; Hari and Mäkelä, 2003; Mäkelä et al., 2004). The annual photosynthetic production is needed, since the structure of trees operates with annual time step, and it is obtained by integrating the photosynthetic rate over the photosynthetically active period (Sections 7.1 and 7.5.1).

The reduction of photosynthetic production of a whorl by shading is important for understanding of canopy development and differentiation of size classes (Section 7.5.1). A link between the shading needles above the whorl and its photosynthetic production is needed. Let $P_w(i, j, k)$ (g[CO_2]) denote the photosynthetic production of the jth whorl in size class i during year k.

The degree of photosynthetic interaction of the whorl j in the ith size class during year k, $i_p(i, j, k)$, is defined as the change in photosynthetic production caused by shading

$$i_P(i,j,k) = \frac{P(x_{i\,j},k)}{P_o} \tag{9.1.1}$$

where x_{ij} is a representative point within the whorl and $P(x_{ij}, k)$ is annual photosynthetic production at the point x_{ij} during the year k and P_o is annual photosynthetic production in unshaded conditions.

The shading is formed by two components, the general reduction of light in the canopy and the shading within a tree. The two sources of shading have to be combined. Let $M_S(i, j, k)$ (g[dry matter]) denote the shading leaf mass of the jth whorl in the ith size class during the year k, $h(i, j)$ (cm) the height of the jth whorl in the

ith size class and $L_C(i,k)$ (cm) the crown length in the size class i during the year k. The shading mass is the sum of the leaf mass in the stand and weighted needle mass of the tree itself above the point

$$M_s(i, j, k)$$

$$= \sum_{i_a=1}^{m} \sum_{h(i_a, j_a) \geq h(i,j)}^{k} N(i_a, k) M_n(i_a, j_a, k) + a_{s1} \frac{\sum\limits_{j_a=j+1}^{k} M_n(i, j_a, k)}{L_C(i,k)} \sum_{j_a=j+1}^{k} M_n i, j_a, k$$

$$(9.1.2)$$

where m is the number of size classes and $N(i,k)$ is the number of trees in size class i during year k. The Eq. 9.1.2 looks rather complicated although its structure is rather obvious. The first term refers to the shading by surrounding trees and the second to self-shading. The summing over i_a refers to size classes and j_a to the whorls. Only the whorls above the whorl can shade it. This is taken in consideration with the lower limit of summing over height. The self-shading is assumed to be proportional to the needle mass in the tree above and the needle mass per unit of stem is used as weight.

The shading determines the degree of interaction, which is approximated as follows:

$$i_p(i, j, k) = f_s(M_s(i, j, k)) = \frac{1}{1 + (a_{s2} M_s(i, j, k))^{0.8}} \qquad (9.1.3)$$

The photosynthetic production of the jth whorl in size class i during year k is

$$P_w(i, j, k) = P_o i_p(i, j, k) M_n(i, j, k) \qquad (9.1.4)$$

When Eqs. 9.1.1–9.1.4 are combined we get

$$P_w(i, j, k) = P_o \frac{M_n(i, j, k)}{1 + (a_{s2} M_s(i, j, k))^{0.8}} \qquad (9.1.5)$$

This equation combines the structure of the stand with photosynthetic production of a whorl and it is a very aggregated formula that utilizes the understanding of photosynthetic process and extinction of light in the canopy gained in Chapters 6 and 7.

Most biochemical reactions demand energy which is provided by the conversion of energy rich ATP to ADP (Section 6.3.2.1). The energy yielding reverse reaction in branches, stems and roots are carried out by burning sugars and other carbon compounds resulting in release of CO_2, and it is called maintenance respiration. Let $R(i, j, k)$ (g(CO_2)) denote the amount of CO_2 produced in the maintenance respiration of the jth whorl in the ith size class and its water transport structures in branches, stem, transport and fine roots during year k and let $M_b(i, j, k)$ (g(dry matter)) denote the water transporting structures in branches, $M_s(i, j, k)$ in stem $M_t(i, j, k)$ in transport roots and $M_r(i, j, k)$ in fine roots, respectively (g(dry matter)). Assume that the maintenance respiration of each component is proportional to its living mass. Then

$$R(i,j,k) = a_{nr}M_n(i,j,k) + a_{br}M_b(i,j,k) + a_{sr}M_s(i,j,k) + a_{cr}M_c(i,j,k) + a_{rr}M_r(i,j,k)$$
$$(9.1.6)$$

where a_{nr}, a_{br}, a_{sr}, a_{tr} and a_{rr} are parameters. We include the root exudates in the fine root respiration (Fig. 9.1.1).

Apart from maintenance respiration, energy is needed for new tissue growth, since it involves synthesis of new macromolecules. This is proportional to the amount of new tissue formed and is called growth respiration. In material transformations considerable weight loss occurs, part of which is due to growth respiration. One kilogram of sugars produces 0.7 kg proteins, 0.8 kg cellulose, 0.44 kg lignin and 0.28 kg lipids (Penning de Vries et al., 1974). The macromolecule concentrations are tissue type specific (Section 5.2.1). Needles and fine roots are assumed to have the same concentrations. The characteristic feature of wood is low protein and lipid concentrations. The protein concentration of needles and fine roots is 0.089, starch 0.079, cellulose 0.466, lignin 0.331 and lipids 0.169. In wood the concentrations are: proteins 0.0016, starch 0.0034, cellulose 0.721, lignin 0.257 and lipids 0.017. When the tissue type specific concentrations and mass loss are combined with the synthesis of macromolecules, then the conversion coefficients a_{ngr}, a_{wgr} and a_{rgr} are obtained (Penning de Vries et al., 1974).

Conservation of Carbon

When the allometric relationships are used in dynamic studies, we need to consider also the longevity of the tissues. Let $G_n(i, j, k)$ (g(dry matter)) denote the growth of needles in the whorl during the year k in the jth whorl in the ith size class. The lifetime of sapwood is long when compared to that of needles, and this has to be introduced into the analysis. When needles die (at the age of three years in Scots pine growing in Southern Finland), the sapwood used for water transport to the dying needles is released for reuse. Let $G_{Ab}(i, j, k)$ (cm^2) denote the growth of sap wood area in branches, $G_{As}(i, j, k)$ (cm^2) in stem, and $G_{At}(i, j, k)$ (cm^2) in transport roots.

Water transport is the key to understanding the regularities in tree structure (Chapter 8). This fact is introduced into the model of tree structure with the requirement that the area of new sapwood fulfils the need of water transport capacity for the growing needles. Thus

$$G_{Ab}(i,j,k) = a_b \left(G_n(i,j,k) - G_n(i,j,k-3) \right) \qquad (9.1.7)$$
$$G_{As}(i,j,k) = a_s (G_n(i,j,k) - G_n(i,j,k-3)) \qquad (9.1.8)$$
$$G_{At}(i,j,k) = a_t \left(G_n(i,j,k) - G_n(i,j,k-3) \right) \qquad (9.1.9)$$

where parameters a_b, a_s, and a_r describe the sap wood requirement per unit leaf mass for branches, stem and coarse roots. If needle mass in the whorl is decreasing, then the extra water transport capacity is lost.

Pine trees have basal swelling at the stem base that is not explained by water transport, but which is evidently required for mechanical stability. Let $\Delta A_E(h,i,j,k)$ (cm^2) denote the additional stem area at height h grown during year k utilizing sugars from whorl j in the ith size class. A rough approximation is used to introduce the stem base expansion

$$\Delta A_E(h,i,j,k) = \begin{cases} 0, & if\, h > h(i,k)/10 \\ a_{b1}(1 - \frac{h/10}{h(i,k)})G_{As}(i,j,k), & if\, h < h(i,k)/10 \end{cases} \tag{9.1.10}$$

where a_{b1} is a parameter.

Let $M_B(i, j, k)$ (g(dry weight)) denote the mass of growth of stem base expansion supporting the jth whorl in the ith size class during the kth year. $M_B(i, j, k)$ is

$$M_B(i,j,k) = d_S \int\limits_{0}^{h(i,k)/10} \Delta A_E(h,i,j,k)dh \tag{9.1.11}$$

where d_s is wood density.

Let $M_T(i, j, k)$ (g(dry weight)) denote the mass of the growth o0f the water transport system supporting the jth whorl in the ith size class during the kth year. It is formed by the growths in the branches, stem, and transport roots. Let $l_b(i, j, k)$ (cm) denote the mean length of branches in the jth whorl during the kth year, $h(i, j)$ (cm) the height of the jth whorl in the ith size class and $b_t(i, j, k)$ (cm) the length of transport roots. Then the mass of the growth of the water transport system is

$$M_T(i,j,k) = d_b l_b(i,j,k)G_{Ab}(i,j,k) + d_s h(i,j)G_{As}(i,j,k) + d_t b_t(i,j,k)G_{At}(i,j,k) \tag{9.1.12}$$

The parameters d_b, d_s and d_t describe wood density in branches, stem and transport roots.

The top of a tree is unable to grow with its own photosynthetic production instead the regulation system of tree architecture allocates sugars from lower whorls to support the top. From all photosynthetic products a share that depends on tree size is allocated for the development of the top of the tree. Let $T_A(i, j, k)$ (g(CO_2)) denote the amount of carbohydrates allocated for the top of the tree from the jth whorl in the ith size class during the kth year. It is approximated with the following function:

$$T_A(i,j,k) = \frac{a_{t1}(P_w(i,j,k) - R(i,j,k))}{1 + h(i,k)/a_{t2}} \tag{9.1.13}$$

where a_{t1} and a_{t2} are parameters.

The preceding equations describe growth relative to other biomass compartments. To calculate actual growth, additional conditions are needed. The carbon balance equation has been the cornerstone of process-based growth models since the work by De Wit et al., in 1970 which is mass conservation at annual and tree level (see Section 1.2: Basic idea 9). The carbon balance equation states that all sugars formed during a year are used for growth, production of ATP resulting in maintenance respiration and for the top of the tree. It is

$$P_w(i,j,k) - R(i,j,k) - T_A(i,j,k) = a_{ngr}G_n(i,j,k) + a_{wgr}M_T(i,j,k) + a_{rgr}M_r(i,j,k)$$
$$(9.1.14)$$

The parameters $a_{n\,gr}$, $a_{w\,gr}$ and $a_{r\,gr}$ are chemical conversion coefficients from sugars to tissue for needles, wood and fine roots. Conservation of mass and energy was frequently used to derive equations in Chapter 7. The carbon balance equation is an application of the same principle at both tree and annual levels.

Conservation of Nitrogen

Forest stands take nitrogen from soil mainly through fine roots. Spatiotemporal complexity of the soil system, the root growth dynamic with microbial (mycorrhiza) interaction and the short lifetime of the ammonium ion in soil (Section 7.9) render the quantitative description of the system very complex. Overall, available nitrogen content and amount of fine roots seem to determine the annual uptake. Retranslocation from senescent tissues is an important source of nitrogen for bigger trees. The role of foliar nitrogen exchange with the atmosphere is an open question, but in this study the nitrogen oxide emissions and depositions are considered negligible (Hari et al., 2003).

The lifetime of cells varies, fine roots are active about one growing season, needles photosynthesize for three years and woody tracheids transport water for decades. When a cell dies, then a considerable portion of proteins are broken down and the resulting amino acids are transported and stored for later use to synthesize new proteins. This results in internal circulation of nitrogen within a tree (Section 6.3.2.4). The nitrogen content of needle litter is less than half of that in active needles (Helmisaari, 1992).

The nitrogen for protein synthesis is taken by fine roots from the soil or reused from senescent tissues. All tissues have their specific protein content, thus they have specific requirements of nitrogen for growth (e.g., Mohren, 1987). The amount of nitrogen needed for growth of a whorl is obtained by summing the products of growth and specific nitrogen content over all tissue types.

The nitrogen balance equation states that the amount of nitrogen needed for growth equals the sum of uptake and reuse of nitrogen.

$$n_n\,G_n(i,j,k) + n_w\,M_T(i,j,k) + n_r\,M_r(i,j,k) = u M_r(i,j,k)N_a(k) + c\,n_n\,G_n(i,j,k-3)$$
$$(9.1.15)$$

where $N_a(k)$ (g(N) m^{-2}) is the amount of nitrogen in the soil that is available to the tree in year k and n_n, n_w and n_r (g(N) g(dry weight)$^{-1}$) are nitrogen concentrations of needles, wood and fine roots, u is nutrient uptake and c is a reuse parameter. Again, the nitrogen balance equation is an application of the conservation principle at tree and annual levels.

Core of the Model

The fundamental assumption underlying the tree component of the model MicroForest is that the regulation system of trees is able to fulfil simultaneously the structural constraints (Eqs. 9.1.7–9.1.9), carbon (Eq. 9.1.14) and nitrogen (Eq. 9.1.15) balance equations for each whorl in the tree. Thus we assume simultaneously two functional balances: (i) between needle transpiration and water transport in the wood and (ii) demand of nitrogen and nitrogen uptake. The five equations contain five unknowns: needle growth, sap wood area growth in branches, stem and coarse roots and fine roots. Thus they can be solved from the equations. Solution of the five equations results in an allocation of carbon and nitrogen within the tree under the control of the regulation system (Figs. 9.1.3 and 9.1.4).

There is no senescence in the three uppermost whorls of a pine tree when the maximum needle age is three years. This is why the release of water transport capacity and reuse of nitrogen plays no role in the allocation of carbohydrates for the treetops. Instead, sugars from lower parts of the crown have to be used. This additional photosynthetic production is divided in equal shares between the top whorls. Thus the growths in the top whorls can be solved from structural regularities and balance equations with slight modifications.

Bookkeeping

A solution of the five equations, introducing structural regularities and carbon and nitrogen balance, results in dynamic allocation of sugars and nitrogen to needles, fine roots and the water transport system in branches stem and coarse roots. The solution also introduces the regulation of formation of tree structure into the model.

The needle mass in a whorl in the ith size class in the jth whorl during year k is increased by growth and decreased by senescence, thus

$$M_n(i,j,k+1) = M_n(i,j,k) + G_n(i,j,k) - G_n(i,j,k-3) \qquad (9.1.16)$$

The transition from whorl to tree level sums up the growths at whorl level. The area of the annual tree ring in each stem segment is determined by the water transport for the needles in the above whorls. Let A(i, j, k) (cm^2) denote the cross-sectional area of the stem segment formed in the ith size class during the jth year at age k. The cross-sectional area of the stem at the top of the tree is zero, thus $A(i, k, k) = 0$. The area, $A(i, j, k)$ is obtained recursively by adding new water pipes for the above whorls on the area during the previous year.

$$A(i,j,k) = A(i,j,k-1) + \sum_{jj=j+1}^{k} G_{As}(i,jj,l) \qquad (9.1.17)$$

The diameter of each segment of the stem can now be determined from the relationship between the area and diameter of a sphere. Let $d(i, j, k)$ (cm) denote the diameter of the segment formed during the kth year in the ith size class at age k.

$$d(i,j,k) = 2\sqrt{A(i,j,k)/\pi} \qquad (9.1.18)$$

Extension

The height growth of trees is more complicated to derive from simple functional relationships and more evolutionary arguments have been used (Mäkelä, 1985, 1988). The length and diameter are mechanistically related (Ylinen, 1952) but also a simple empirical linear relationship for open grown trees have been found (Ek, 1971).

In the model the extension is not explicitly considered in the mass balance equations. Instead, the extension of the year k is determined with diameter growth and the position of the tree in the stand, similarly to Sievänen (1992). Ek (1971) and Mäkelä and Sievänen (1992) have observed that for open grown pines the ratio between height and diameter growth is constant. The shading accelerates height growth in stands relative to diameter growth (Mäkelä and Vanninen, 1998; Ilomäki et al., 2003). The height growth of the tree in the ith size class during the year $k, \Delta h(i,k)$ (cm), is assumed to be

$$\Delta h(i,k) = a_{h1}\Delta r(i,k)(1 + a_{h2}(1 - I_{Pi})^4) \qquad (9.1.19)$$

where a_{h1} and a_{h2} are parameters, Δr (cm) is radial growth at stem base, and I_{Pi} the mean interaction is

$$I_{pi}(i,k) = \frac{\sum\limits_{j=i}^{k} i_p(i,j,k)M_n(i,j,k)}{\sum\limits_{j=i}^{k} M_n(i,j,k)} \qquad (9.1.20)$$

We assume that the needles fill space with constant volume density, ρ_n (g(dry weight) cm^{-3}) at the outer surface of the crown at each whorl. Let $L_B(i,j,k)$ denote the mean length of a branch in the ith size class in the jth whorl during year k. The above assumption determines the length growth of branches, $L_B(i,j,k)$ (cm). It is the solution of the following equation

$$\frac{G_n(i,j,k)}{2\pi\Delta L_B(i,j,k)L_B(i,j,k)\Delta h(i,k)} = \rho_n \qquad (9.1.21)$$

Volume

The treatment of trees is completed with a determination of the stem volumes on the three levels of hierarchy. Let $V_w(i, j, k)$ (cm^3) denote the volume of the stem segment formed in the ith size class during the jth year at age k. It is the cross-sectional area multiplied by the height growth during the jth year

$$V_w(i,j,k) = A(i,j,k)(h(i,k) - h(i,k-1))$$ (9.1.22)

The stem volume, $V_s(i,k)$ (m^3) in size class i during year k is obtained by summing the volumes of all stem segments,

$$V_s(i,k) = \sum_{j=1}^{k} V_b(i,j,k)/1000000$$ (9.1.23)

The transition to large units is accounted for in the above equation.

The last step in the model of a tree component is the determination of stand volume. Let $V(k)$ (m^3) denote the volume of the stand at age k. It is obtained by summing the volumes of individual trees:

$$V(k) = \sum_{i=1}^{m} N(i,k)V_s(i,k)$$ (9.1.24)

where $N(i,k)$ is the number of trees in the size class i during the year k and m is the number of size classes.

Senescence

In the model, the needles become senescent and die at the age of three years. The life-time of fine roots is still shorter, they are annual. A new whorl is formed annually at the top of the tree, thereafter it grows utilising its own photosynthesis. When the stand grows, the branches become more and more shaded and its photosynthesis is reduced resulting in a slow decline of the needle mass in the whorl and in the extension growth of the branches. We assume that the whorl finally dies when its extension growth is less than 1 cm.

The smaller trees become more and more shaded, especially after canopy closure resulting in declining trend in photosynthesis per gram of needles in the tree. Thus the carbohydrates available for growth have also a declining trend which is reflected in the growth of the tree. We assume that the probability of a tree dying is proportional to the relative decrease in its needle mass.

9.1.4 Ground Vegetation

The forest floor is usually covered by ground vegetation, which consists of dwarf shrubs, herbs and mosses. The living biomass of ground vegetation is often rather small, but when the trees are not reducing light falling on ground vegetation too strongly, then the growth of new leaves is of the same magnitude as that of trees. The chemical composition of ground vegetation is close to that of tree leaves.

There are no fundamental differences in the basic processes of ground vegetation and trees they photosynthesize and take up nitrogen and use the obtained resources in an efficient way for maintenance and growth. The only major difference is that, due to short transport distances, the dwarf shrubs and herbs do not allocate considerable amounts of resources to the water transport system.

From a modelling point of view, ground vegetation is considered analogously to trees. The most essential difference is that the water transport system can be omitted and the effect of shading was introduced in a rough way. We assumed that the annual photosynthetic production of ground vegetation, $P_G \, (\text{g}(CO_2)\,\text{m}^{-2})$, is

$$P_G(M_{nG}) = P_g \, f_s(M_{sn}) M_{nG} \left(1 - \frac{M_{nG}}{M_{nG} - g_1}\right) \qquad (9.1.25)$$

M_{nG} (g(dry weight)), is the leaf mass of ground vegetation, M_{sn} the needle mass of the tree stand and function f_s is defined in the Eq. 9.1.3 and p_g and g_1 are parameters.

The leaf and fine root growths are solved from carbon and nitrogen balance equations. We assume, however, that the annual leaves require only the amount of nitrogen that is in the leaf litter. The balance equations are:

$$(P_G - R_G) = a_{ngr} M_{nG} + a_{rgr} M_{rG} \qquad (9.1.26)$$

$$(1 - c) n_n M_{nG} + n_r M_{rG} = u M_{rG} N_a(k) \qquad (9.1.27)$$

where a_{ngo} and a_{rgo} are conversion coefficients from sugars to dry matter and n_n and n_r are nitrogen concentrations in leaves and roots, c is the nutrient recycling and u a nutrient uptake parameter.

9.1.5 Soil

The structure of soil organic matter includes its spatial distribution and chemical composition that are determined by the litter fall and microbial decomposition processes. Soil biopolymers are not available for microbes or trees. Extra cellular enzymes are needed for microbial turnover of the organic biopolymers in forest soils (Linkins et al., 1990; Criquet, 2002; Waldrop et al., 2004). Large biopolymers (proteins, lipids, starch, cellulose and lignin) cleave to small molecules such as sugars, amino acids, fatty acids and various other aliphatic and aromatic compounds (Tomme et al., 1995; Eivazi and Bayan, 1996; Pavel et al., 2004; Saito et al., 2003; Wittmann et al., 2004; Section 6.4.1). These small molecules are taken up by the microbes. In addition, small amounts of humic substances are formed. The decomposition of organic matter in soil is a slow process: it requires decades (Section 7.9). The lifetime of humus is very long and varies according to its chemical properties (Liski et al., 1998). The fluxes connected with humic substances are, however, so small that we neglect them (Section 7.9).

When microbes use amino acids to produce ATP, they emit NH_4^+ ion. Nitrogen that is liberated in the decomposition process is available for trees. Most of the available nitrogen is loosely adsorbed on the surface of soil particles as ammonium ions (Section 6.2.4). The lifetime of ammonium ion in the soil is short, only about 30 days in summer (Section 7.9)

Litter fall is the main carbon and nitrogen input into forest soil, while nutrient uptake causes the main nitrogen flow out. In addition, deposition and leaching of nitrogen and emission of nitric gases connect the forest soil with its environment. The microbes decompose larger molecules in the litter fall with extracellular enzymes (Section 6.4.1). Gaseous substances flow eventually to the atmosphere. Fig. 9.1.6 shows carbon compound fluxes between the main pools in the soil. Circulation of nitrogen within soil (Fig. 9.1.7) is additional complication in the soil nitrogen fluxes compared to carbon fluxes.

We modelled the carbon and nitrogen compound fluxes in Figs. 9.1.6 and 9.1.7. As a first approximation, simple model structures and either linear or multiplicative dependencies are used. In order to capture the qualitative difference between the rather permanent, structural carbohydrates and readily usable smaller carbon based molecules, we divide the organic matter into categories of large and small molecules.

We treat five types of biopolymers (proteins, lipids, starch, cellulose and lignin) synthesized during the growth of the tissue (Section 5.2). The letter n refers to the type of substrate (n = 1, 2, 3, 4, 5). The following symbols are used for the state variables describing the amounts of carbon rich substances in soil:

C_{Ln} = carbon compounds in soil formed by large molecules of type n $(g\,m^{-2})$
C_{Sn} = carbon compounds in the soil formed by small molecules of type n $(g\,m^{-2})$
M_{Mn} = mass of microbes emitting enzyme to break molecules of type n $(g\,m^{-2})$
E_n = enzymes breaking large carbon molecules of type n $(g\,m^{-2})$
C_{SMn} = mass of free small carbon compounds of type n in microbes $(g\,m^{-2})$

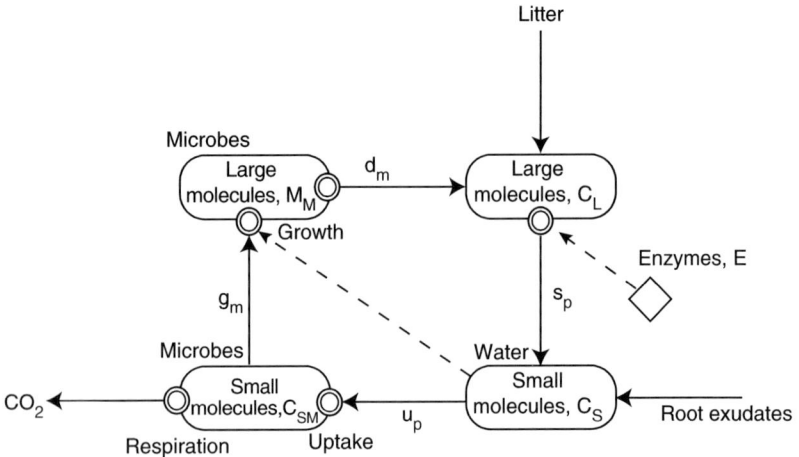

Fig. 9.1.6 Carbon compound fluxes in forest soil starting with litter fall via enzymatic splitting of large molecules to release in atmosphere in microbial metabolism. The symbols are introduced in the Figure 1.2.1

Litter

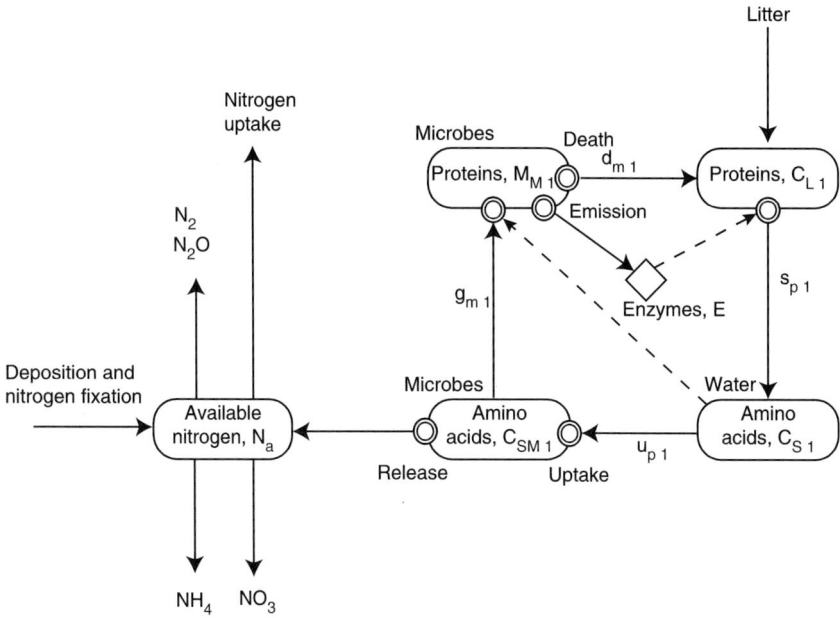

Fig. 9.1.7 Nitrogen compound fluxes in forest soil starting with litter fall via enzymatic splitting of proteins to release in plant available form. Microbial activity generates an internal loop within the forest soil

Table 9.1.1 The connection between the index and molecule types

Index	Small molecules	Large molecules	Enzyme
1	Amino acids	Proteins	Peptidase
2	Sugars	Starch	Amylase
3	Fatty acids	Lipids	Esterase
4	Sugar	Cellulose	Glucanase
5	Aliphatic acids	Lignin	Ligninase

The connection between the index and molecule types is presented in Table 9.1.1. The structure of the equations for flows can be seen from Figs. 9.1.6 and 9.1.7.

Splitting of large carbon molecules of type n, s_{pn}

$$s_{pn} = b_{1n} E_n C_{Ln} \qquad (9.1.28)$$

Uptake of small molecules of type n by microbes u_{pn}

$$u_{pn} = b_{2n} C_{Sn} \qquad (9.1.29)$$

Growth of microbes emitting enzyme-splitting macromolecules of type n, g_{mn},

$$g_{mn} = b_{4n} C_{Sn} M_{Mn} \qquad (9.1.30)$$

Death of microbes emitting enzyme-splitting macromolecules of type n, d_{mn}

$$d_{mn} = b_{5n}M_{Mn} \qquad (9.1.31)$$

The masses of microbes are treated in terms of the five chemical groups.

The protein and amino acid fluxes in the soil include additional aspects connected with the circulation of nitrogen rich compounds in the soil.

The uptake of amino acids by microbes of type n, u_{pn}

$$u_{pn} = c_1 \frac{M_{Mn}}{\sum\limits_{j=1}^{5} M_{Mj}} C_{S1} \qquad (9.1.32)$$

Production of an enzyme that splits macromolecules of type n, p_{en}

$$p_{en} = c_2 C_{SM1} \frac{M_{Mn}}{\sum\limits_{j=1}^{5} M_{Mj}} \qquad (9.1.33)$$

Release of ammonium n_r

$$n_r = c_3 C_{SM1} \qquad (9.1.34)$$

Microbial respiration r_m

$$r_m = c_4 \sum_{j=1}^{5} e_l C_{SMl} \qquad (9.1.35)$$

where chemical constants e_l convert different carbon compounds to CO_2.

Destruction of enzyme splitting macromolecules of type n, d_{en}

$$d_{en} = c_5 E_1 E_n \qquad (9.1.36)$$

The differential equations for the state variables are obvious, as demonstrated for the small carbon compounds of type n:

$$\frac{dC_{Sn}}{dt} = s_{pn} - u_{pn} \qquad (9.1.37)$$

The nitrogen in the soil that is available to plants, N_a, plays a key role in connecting trees and soil. It is obtained from the flows changing the amount of available nitrogen

$$\frac{dN_a}{dt} = -u \sum_{i=1}^{m} \sum_{j=1}^{k} M_r(i,j,k)N_a(k) - uM_{rG}(k) + r_a + d_{ep} + n_f - l_e - e_v \qquad (9.1.38)$$

where d_{ep} is nitrogen deposition, n_f nitrogen fixation, l_e is nitrogen leaching, e_v is nitrogen evaporation as NO or N_2O and the sum is the nutrient uptake term in Eq. 9.1.15 for trees.

9.1.6 Parameter Values

We have described in the previous chapters the measurements of environment, processes and tree structure made at SMEAR II. These results are the backbone of the estimation of values of the parameters in the model MicroForest.

MicroForest includes 54 parameters to be determined. We applied five approaches to determine the values: (i) chemical information, (ii) direct measurements of structure, process rates or chemical composition, (iii) requiring static behaviour of some state variables, (iv) literature values and (v) estimation from the behaviour of the model. We tried to minimize the number of parameters estimated from the behaviour of the model (Table 9.1.2).

Table 9.1.2 List of parameters

Parameter	Value	Unit	Equation, 9.1:	Source
P_o	13.6	$g(CO_2)\ g(dry\ weight)^{-1}$	1,4,5,41	SMEAR II
a_{s1}	400	$cm\ g(dry\ weight)^{-1}$	2	Mäkelä and Sievänen, 1992
a_{s2}	0.00000027	$g(dry\ weight)^{-1}$	3,5	Simulations with Micro-Forest
a_{nr}	2	$g(CO_2)\ g(dry\ weight)^{-1}$	6	SMEAR II
a_{br}	0.02	$g(CO_2)\ g(dry\ weight)^{-1}$	6	SMEAR II
a_{sr}	0.02	$g(CO_2)\ g(dry\ weight)^{-1}$	6	SMEAR II
a_{tr}	0.02	$g(CO_2)\ g(dry\ weight)^{-1}$	6	SMEAR II
a_{rr}	17	$g(CO_2)\ g(dry\ weight)^{-1}$	6	Simulations with Micro-Forest
a_b	0.03	$cm^2\ g(dry\ weight)^{-1}$	7	SMEAR II
a_s	0.033	$cm^2\ g(dry\ weight)^{-1}$	8	SMEAR II
a_t	0.015	$cm^2\ g(dry\ weight)^{-1}$	9	SMEAR II
a_{b1}	1.5		10	Simulations with Micro-Forest
d_b	0.4	$g(dry\ weight)\ cm^{-3}$	12	SMEAR II
d_s	0.4	$g(dry\ weight)\ cm^{-3}$	12	SMEAR II
d_t	0.2	$g(dry\ weight)\ cm^{-3}$	12	SMEAR II
a_{t1}	0.85		13	Simulations with Micro-Forest
a_{t2}	500	cm	13	Simulations with Micro-Forest
a_{ngr}	1.35	$g(sugar)\ g(dry\ weight)^{-1}$	14,26	Penning de Vries et al., 1974
a_{wgr}	1.3	$g(sugar)\ g(dry\ weight)^{-1}$	14	Penning de Vries et al., 1974
a_{rgr}	1.35	$g(sugar)\ g(dry\ weight)^{-1}$	14,26	Penning de Vries et al., 1974
n_n	0.0134	$g(N)\ g(dry\ weight)^{-1}$	15,27	SMEAR II
n_w	0.00024	$g(N)\ g(dry\ weight)^{-1}$	15,27	SMEAR II
n_r	0.0134	$g(N)\ g(dry\ weight)^{-1}$	15	SMEAR II
c	0.575	$g(N)\ g(dry\ weight)^{-1}$	15	SMEAR II

(Continued)

Table 9.1.2 (*Continued*)

Parameter	Value	Unit	Equation, 9.1:	Source
u	0.002	m^2 g(dry weight)$^{-1}$	15	Simulations with Micro-Forest
a_{h1}	50		19	Mäkelä and Sievänen, 1992
a_{h2}	2,400		19	Simulations with Micro-Forest
ρ_v	0.00025	g(dry weight) cm^{-3}	21	SMEAR II
p_g	30	g(sugar) g(dry weight)$^{-1}$	25	SMEAR II
g_1	80	g(dry weight) m^{-2}	25	SMEAR II
b_{11}	0.00043	m^2 a^{-1} g^{-1}	28	Stable behaviour
b_{12}	0.0035	m^2 a^{-1} g^{-1}	29	Stable behaviour
b_{13}	0.0006	m^2 a^{-1} g^{-1}	28	Stable behaviour
b_{14}	0.0012	m^2 a^{-1} g^{-1}	28	Stable behaviour
b_{15}	0.0002	m^2 a^{-1} g^{-1}	28	Stable behaviour
b_{21}	1	a^{-1}	29	Stable behaviour
b_{22}	1	a^{-1}	29	Stable behaviour
b_{23}	1	a^{-1}	29	Stable behaviour
b_{24}	1	a^{-1}	29	Stable behaviour
b_{25}	1	a^{-1}	29	Stable behaviour
b_{41}	0.004	a^{-1}	30	Stable behaviour
b_{42}	0.0003	a^{-1}	30	Stable behaviour
b_{43}	0.0003	a^{-1}	30	Stable behaviour
b_{44}	0.00006	a^{-1}	30	Stable behaviour
b_{45}	0.0003	a^{-1}	30	Stable behaviour
b_{51}	0.004	a^{-1}	31	Stable behaviour
b_{52}	0.007	a^{-1}	31	Stable behaviour
b_{53}	0.003	a^{-1}	31	Stable behaviour
b_{54}	0.004	a^{-1}	31	Stable behaviour
b_{55}	0.002	a^{-1}	31	Stable behaviour
c_1	1	a^{-1}	32	Stable behaviour
c_2	0.55	a^{-1}	33	Stable behaviour
c_3	6.7	a^{-1}	34	Stable behaviour
c_4	10	a^{-1}	35	Stable behaviour
c_5	0.00043	a^{-1}	36	Stable behaviour

The most important parameter influencing stand behaviour is the annual photosynthetic production in unshaded conditions, since photosynthesis produces most of the raw material for plant metabolism and growth. The model MicroForest includes processes in which photosynthesized carbon is converted to other forms. We used the conversion coefficients according to Penning de Vries (1975). The values of the parameters in the chemical composition of needles and wood were based on measurements of samples from the stand. The chemical composition of fine roots was assumed to be the same as that of needles. The chemical properties of microbes were based on literature.

The information for determining the values of the soil parameters is very limited since the traditional measurements do not differentiate proteins, cellulose, lignin, starch and lipids in the soil organic matter. The pools of rapidly decaying carbon compounds are, however, very large indicating slow turnover rates (Liski et al., 1998). The parameter values related to the decomposition process, i.e., those parameter values that were connected with microbes and extra cellular enzymes, were adjusted so that the pools of organic matter were close to observed values.

The values of seven parameters were fitted to give acceptable model behaviour. The criteria were: (i) the parameters in the effect of shading on canopy photosynthesis and fine root respiration (including root exudates) were fitted to yield annual photosynthetic production and soil CO_2 efflux that corresponded to soil chamber and eddy-covariance measurements, (ii) the stem base enlargement parameters and allocation to the top of the crown were selected so that the model produced the observed form of the stem base and crown, (iii) nitrogen uptake efficiency was estimated so that the model produced reasonable fine root mass, and (iv) the effect of shading on height growth was adjusted to produce measured height growth patterns.

We considered the influence of increased nitrogen deposition during the last century by increasing it from initially $1 \, kg \, ha^{-1}$ to $5 \, kg \, ha^{-1}$ at the age of 40 years. The leaching of nitrogen from the stand was assumed negligible, which is in accordance with observations at our field sites. We included nitrogen fixation into the approximation of nitrogen deposition.

The stand around SMEAR II station was sown with pine after prescribed burning 40 years ago. The time at three years after clear-cut was selected as the starting moment of the model simulations. The number of trees in each size class was derived from the standing living trees at 40 years. Apart from the thinning of the cluster of sown pine that was done during the sapling state, no tree mortality had taken place during the first 40 years of stand development.

The initial state of trees was determined retrospectively from the tree rings of the existing trees at the stem base. The initial needle mass was determined with the relationship between needle mass and sap wood area. The prescribed burning was assumed to have burned the cutting residues of the final cut of the previous tree generation, essentially maintaining the soil organic matter at the same level as just before the clear-cut. As the pools of soil organic matter change rather slightly in the late phases of stand development according to model simulations (Nissinen and Hari, 1998 Liski et al., 2005.), we assumed that the initial state of the soil was the same as that we measured at the age of 40. The initial state of microbes and extra cellular enzymes was based on measurements at SMEAR II in the year 2002.

9.1.7 Simulations

The overall development shows that carbon accumulation into the ecosystem increases over the stand development, but nitrogen cycles in the system with rather constant rates (Fig. 9.1.8). The majority of the biomass accumulates into the

Fig. 9.1.8 Simulated development of carbon compounds (upper panel) and proteins (lower panel) in trees and soil

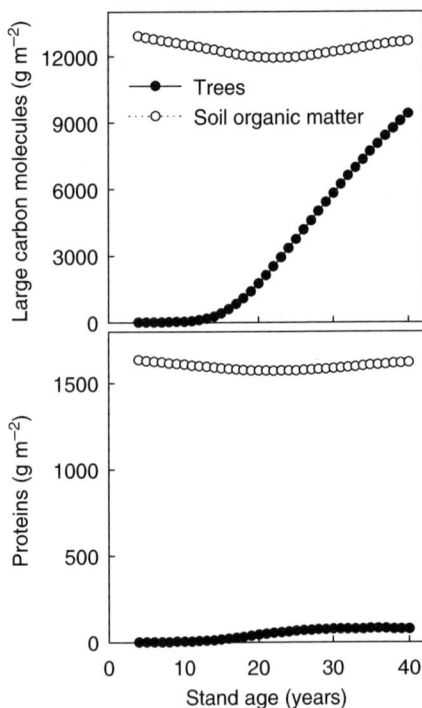

Fig. 9.1.8 Simulated development of carbon compounds (upper panel) and proteins (lower panel) in trees and soil

growing wood but there is a small increasing trend also in the soil organic matter as the branch litter starts accumulating into the stand (Fig. 9.1.8). In these simulations we assume that dead stems do not enter soil organic matter but that they are removed from the stand. The input of protein rich litter does not vary much, as opening of the stand after clear cut allows fast development of ground vegetation that takes up the nitrogen released in decomposition. Later on, the gradual development of trees directs the nitrogen to the active tree tissue in leaves and roots. As a result, the dynamics of the nitrogen cycling of the stand is very stable. The vegetation is able to take up all the nitrogen that is released in decomposition, thus the amount of plant available nitrogen in soil solution is low (Section 7.9). Some dynamics in nitrogen is introduced when the annual ground vegetation is replaced with evergreen pines.

The biomass growth of trees becomes larger than that of ground vegetation as a very large proportion of biomass is allocated to wood that has clearly a lower proportion of active tissues (Fig. 9.1.8). The released nitrogen from organic matter is sufficient in maintaining the higher biomass accumulation, as the average concentration of nitrogen in tree tissue is lower than in ground vegetation. Over the rotation, a small accumulation of nitrogen in wood can be observed (Fig. 9.1.8), but it does not reflect on plant available nitrogen as the simulations assume that the nitrogen input to the site from the atmosphere is increasing according to the trend during the last century.

The needle, fine root and wood masses in the simulations are solutions of the carbon and nitrogen balance equations and they are solved annually. Thus allocation to different components of trees becomes dynamic (Fig. 9.1.9). Especially during the first 20 years the regulation system changes allocation strongly from needles and fine roots to woody components transporting water. The increasing water transport distance causes this phenomenon when the tree is growing.

Photosynthesis drives the carbon and nitrogen cycles. The development of CO_2 flow from atmosphere to trees and ground vegetation (Fig. 9.1.10) is closely connected with the dynamics of needle mass in the stand (Fig. 9.1.11). The low needle mass of trees in the early phase of stand development reduces their photosynthetic

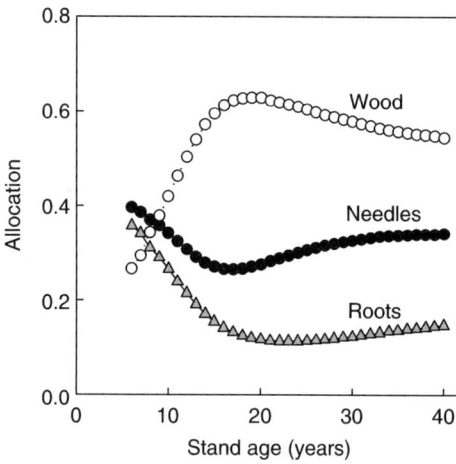

Fig. 9.1.9 Simulated development of allocation to needles, roots and to water pipes. The allocation is the solution of the five equations describing structural regularities, and carbon and nitrogen balance

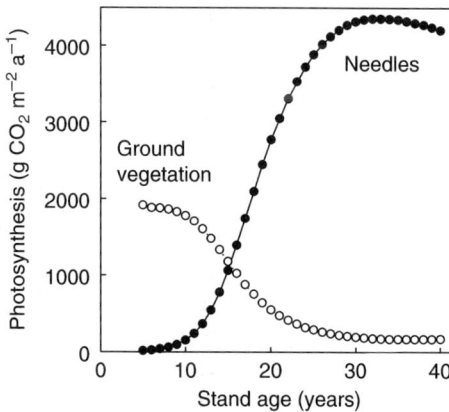

Fig. 9.1.10 Simulated annual photosynthesis (g CO_2 m^{-2} a^{-1}) as function of stand age.

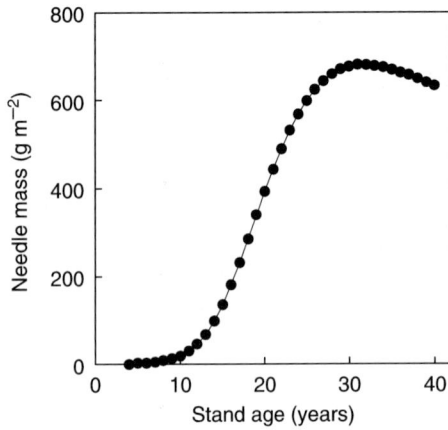

Fig. 9.1.11 Simulated needle mass (g m^{-2}) in the stand

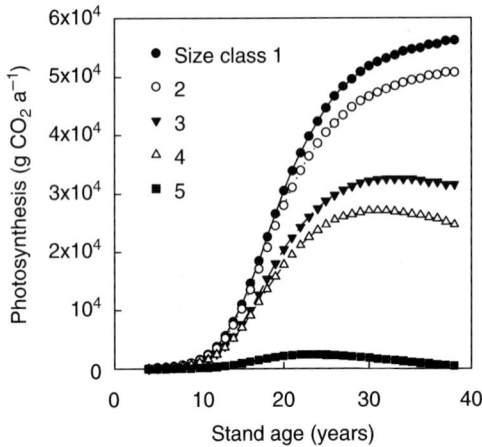

Fig. 9.1.12 Simulated development of photosynthesis per one tree (g CO_2 a^{-1}) in the five size classes

production but allows high CO_2 fixation of ground vegetation as a considerable amount of solar radiation reaches the forest floor. Later during the stand development the situation is reversed, tree foliage shades the ground vegetation and they photosynthesize very little.

There are small differences in initial values of the height, diameter and needle mass between the size classes. The seedlings grow exponentially during the first ten years. Thereafter the shading begins to reduce the photosynthesis, especially in the smaller size classes resulting in reduced growth and differentiation of size classes (Fig. 9.1.12). The differences in annual photosynthetic production generate the differentiation of the size classes into dominated and suppressed trees in simulations.

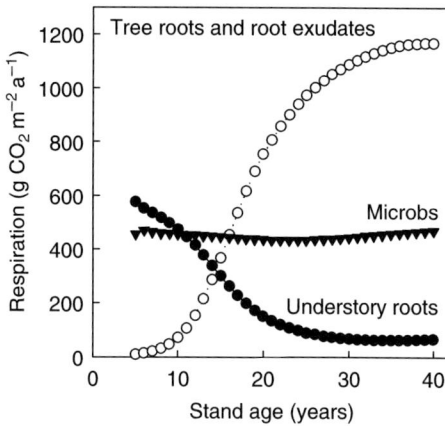

Fig. 9.1.13 Simulated annual respiration (g CO_2 m^{-2} a^{-1}) from the soil

The soil CO_2 efflux (Fig. 9.1.13) is more static during stand development than photosynthesis of the trees. The dominating component in the CO_2 flux from the soil is root respiration and root exudates of trees and ground vegetation. These two components balance each other to a great extent during the simulation period in harmony with their production. The microbial respiration depends on litter production but remains very stable over the rotation. Together these tendencies cause forests to be carbon sources when young but to become sinks as they mature.

Nitrogen fluxes from the surrounding environment into forest ecosystems are small when compared to those within the system. The amount of leaves in a forest ecosystem after canopy closure is rather stable and characteristic for the site. Thus the flow of nitrogen into the vegetation generated by root uptake and flow out by senescent leaves and fine roots are close to each other. This is why the nitrogen release into plant available form by microbes from proteins in the soil is very important for the formation of new productive tissues in leaves and roots. The fertility of the site is determined by the annual release of nitrogen in the decomposition of proteins.

The large protein pool in the soil filters out the effect of small variations in the litter input into soil and the release of ammonium ions is rather stable during stand development (Section 7.9). The ground vegetation has an important role in taking up nitrogen released in decomposition at the early phase of stand development.

The performance analysis of the model against measured tree growth shows reasonable agreement with measurements. The simulated height developments are rather close to the measured ones in each size class, the fit is demonstrated in Fig. 9.1.14. The fits between the observed and modeled height and diameter developments in the four biggest size classes are rather similar to that shown in Fig. 9.1.14. However, for the smallest size class MicroForest gives smaller heights and diameters than the observed ones. The measured and simulated development of stand volumes are rather close to each other.

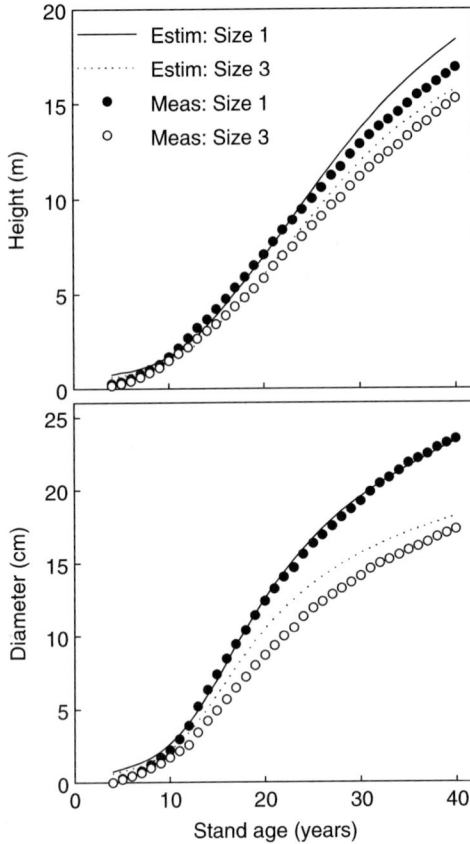

Fig. 9.1.14 Comparison of measured and simulated height (m) and diameter (cm) developments at SMEAR II in the biggest and in 3rd size class. Dots indicate measured values and lines simulated ones

9.1.8 Comments on MicroForest

Small differences between two large flows of ecosystem carbon uptake and ecosystem respiration determine whether a forest stand is a carbon source or a sink. There is quite a lot of uncertainty as regards the global carbon sink in forests depending on the uncertainty of the fate of carbon in the soil. Even small inaccuracies in determining the associated flow rates may twist our understanding of the role of forest in carbon sequestration. There is still quite a high unexplained carbon sink in the global carbon budget that has often been assumed to be in boreal forests. Micro-Forest provides a new tool to analyse the role of boreal forest in the global carbon balance.

It has been argued that the ecosystem ability to bind carbon depends on the nitrogen cycle (Ågren and Bosatta, 1996). Linkages between carbon and nitrogen cycles control the carbon uptake, storage and turnover in terrestrial ecosystems.

Considerable changes in carbon storing by an ecosystem can take place if there is a shift in the nitrogen pool from soil to woody vegetation, as the C:N ratio of the former is only about half of that of the latter (Rastetter et al., 1991). Vegetation growth, allocation of growth between different types of tissue and decomposition of nitrogen in soil will determine the balance of carbon and nitrogen in soil and in vegetation (Aber et al., 1991).

To study the linkages between carbon and nitrogen cycles in boreal Scots pine forest, we developed a model where detailed treatment of microbial decomposition of litter is linked to a well-tested tree growth modeling approach (Hari et al., 1982, 1985; Mäkelä and Hari, 1986; Nikinmaa, 1992; Perttunen et al., 1996; Mäkelä, 1997, 2002; Nissinen and Hari, 1998; Mäkelä and Mäkinen, 2003). The previous modeling that describes allocation of growth to different tissue types from a pipe model (Shinozaki et al., 1964a and b) and functional balance principle (Davidson, 1969) is complemented by explicit treatment of chemical composition of the tissues. In the present version the tissue type composition is kept constant, but as there is important variation in the growth allocation during the stand development, also the C:N ratio of trees changes over the stand rotation. The litter decomposition model fully utilizes variable chemical composition of the litter as the decomposition process is split into enzymatic action of microbes on the molecules of the organic matter.

Traditionally, decomposition models have used aggregated litter types for which they have produced aggregated decomposition rates. Some models have been only interested in carbon release (e.g., Liski et al., 1998) while in other also nitrogen release is considered (e.g., Rastetter et al., 1991; Aber et al., 1991; Running and Gower, 1991; Johnson, 1999). The crucial role of extra-cellular enzymes in the decomposing of soil organic matter was realized about ten years ago and thereafter a rather rapid development in soil microbiology has taken place (Linkins et al., 1990; Criquet, 2002; Waldrop et al., 2004). MicroForest is the first attempt to base decomposition in a dynamic model on the action of extra cellular enzymes. In principle, these developments offer very good possibilities for studying the nitrogen cycle in the natural environment.

The treatment of soil organic matter according to its chemical structure allows efficient use of quantitative methods, since all soil components can, at least in principle, be directly measured. We applied wet chemical methods of food and soil chemistry with slight modifications to determine the content of lipids, carbohydrates, lignin and proteins in plants and soil (Section 5.4.2). Some organic fractions were combined to make larger groups (sugars and starch, hemicellulose and cellulose). Similarly, all N hydrolysable in 6 M HCl was assumed to originate from amino acids and was converted to crude protein by multiplying with 6.25. Amino N was indeed the major hydrolysable N fraction (56–70%) in all samples. The analytical methods were deliberately rough and insufficient for detailed study of organic matter composition in plant and soil. The methods have to be developed to more sophisticated ones that have been intensively researched, especially with respect to nitrogenous compounds and humus (Kögel-Knabner, 2000; Chefetz et al., 2002; Poirier et al., 2003).

The analysis by Mäkelä (1997) has shown that functional balance is well suited for describing the wood production dynamics of stands of different fertility in the boreal region if the root uptake efficiency of nitrogen is changed from site type to another. It is noteworthy, that a constant uptake rate of nitrogen per unit mass of roots produced realistic values in her simulations (Mäkelä, 1997). The model MicroForest is similar to Mäkelä's model in the basic principles of structural formation of a tree but it also explicitly accounts for the molecular composition of tissues, ground vegetation as an independent factor in the ecosystem and nitrogen cycling in the soil by soil microbes. Also in our case, the simulations that described tree development most realistically, gave essentially very constant annual nitrogen release in the soil that was related to litter production with delay. In Mäkelä's work the production differences between site types would result from different uptake efficiencies. Here the same result follows from a larger rotating N-pool in the soil. If indeed the nutrient cycling is close to steady state throughout the stand development, as our model analysis and Mäkelä's results would suggest, this will greatly facilitate the process based stand production estimates.

An important aspect of the model's behaviour was the dynamic allocation of carbohydrates among tree compartments. At stand level there is differentiation of trees into different canopy layers (Vuokila, 1987) and the development of crown heights that have important implications for carbohydrate allocation (Vanninen and Mäkelä, 2000; Ilomäki et al., 2003) MicroForest calculates the crown development in a very dynamic manner that considers both local conditions and position within the crown. These have been shown to be important determinants in tree architecture development (Sievänen et al., 2000).

The behaviour of complex systems is difficult to predict and therefore credibility of the system description can only be achieved if different aspects of a model's behaviour are simultaneously projected against real world phenomena (Nikinmaa, 1992). Data used in model evaluation and testing should be coherent with the model's structure. The data are, however, often limited to growth and yield tables and to thinning experiments measured at 5–10 year intervals. Annual observations would be better in pointing out the driving factors of forest productivity (Berninger et al., 2004). Recent developments in simultaneous monitoring capabilities at organ or patch level (Hari and Mäkelä, 2003) and ecosystem level (Vesala et al., 1998) also provide valuable information for model testing. At the same time the soil analysis methods are developing and they should be fully utilized in the estimation and testing of models.

The model MicroForest and the measuring system SMEAR II have both been constructed to analyze and measure material and energy fluxes in a forest ecosystem. They are based on the same thinking and SMEAR II provides information that can be easily utilized in the determination of the parameter values and testing the model. Most of the parameter values in the tree component were determined with direct measurements, with literature and general chemical information only seven parameter values had to be estimated from the behaviour of the model. As such this number is still too high. However, most of the values are such that they can, in principle, be measured. As we reach the limits of our understanding, a close collaboration between modeling and design of measurements becomes apparent. Complex systems require modelling tools in helping to understand their dynamics but modelling that cannot be evaluated against measurements is closer to computer games than science.

The measurements at SMEAR II cover all measurable carbon and nitrogen fluxes between the forest ecosystem and its surroundings and pools in the trees and soil. The simulations with MicroForest are in agreement with these measurements. The only exception is that the simulated soil CO_2 efflux remains somewhat low. This means that our description of the carbon and nitrogen balances of the ecosystem with its surroundings are within the margins of the present measuring accuracy and precision. The estimation of parameter values improves the fit of the model with measurements and for critical evaluation additional tests are needed.

Global climate change, increasing CO_2, increasing temperature and enhanced nitrogen deposition, will have major effects upon boreal forests. Evidently, photosynthesis is accelerating, the photosynthetically active period is expanding, decomposition of soil organic matter is becoming more rapid and nitrogen deposition is increasing the nitrogen pool in soil. The causal effects of temperature and increasing atmospheric CO_2 on processes in boreal forest ecosystems can easily be introduced into the annual photosynthetic production and decomposition of soil organic matter. In addition to the other parameters discussed above, MicroForest uses nitrogen deposition as input. All these features make MicroForest a powerful tool to analyse the effect of climate change on boreal forests as will be shown in Chapter 11.

MicroForest is operational summary of the regularities in the processes, structure and transport in the forest ecosystem detected in the previous chapters. It indicates that the growth and development of forest ecosystem can be derived from these regularities with rather satisfactory accuracy.

9.2 Testing MicroForest

9.2.1 Need for Rigorous Testing

Roope Kivekäs[1], Mikko Havimo[2], and Pertti Hari[1]

[1] University of Helsinki, Department of Forest Ecology, Finland
[2] University of Helsinki, Department of Forest Resource Management, Finland

A dynamical stand model produces a highly detailed description of stand development over a rotation period. Proper testing of a detailed stand model requires data representing the same time span and the same level of detail as the output of the model. This kind of data is missing to a great extent due to high costs of compiling them and due to measuring traditions. Decent or good success in comparison to rough scale forest inventory or growth and yield table data argues for more detailed level of testing, i.e., more challenging tests in the case of stand models, forecasting the annual development of single tree dimensions for decades represents significantly more severe testing than that of standing stock or average height.

In well-formulated models, where described state variables and functions have counterparts in the real world, comparison of measurements with the model output often points out suspicious properties of the model. Conversion of the idea also works, but with limitations: reasonable output supports the theories on which the model is constructed. However, it is unable to prove them. This has an implication to the model: if a reasonable, well-measured input leads to unreasonable results, there is something wrong in the model structures. Low quality data weakens the supporting strength of the test, especially if the data is not sufficiently detailed, then several important aspects are beyond the test.

Complex dynamical models are generally sensitive to initial states high sensitivity means large effects on output, but also requires suitable input. Hence, measurements of the initial state require effort and focus on quality. As MicroForest has an extremely high level of detail, also the measurements of the initial state have to be entered in detail.

The structure of the model MicroForest was derived utilising knowledge of processes and environmental factors and of regularities in tree structure generated by evolution. The values of the parameters in the model were obtained either from measurements at SMEAR II or from model behaviour. However, the number of estimated parameters was kept as low as possible to avoid estimation bias in the results.

We can say that MicroForest is based on the results obtained at SMEAR II. This fact raises the question: Are the results valid only for the stand around SMEAR II or are they valid also for other pine stands? Models often are developed in "general" or "scalable" form, so that factors confining the model to certain locations are imbedded in the values of some parameters. This is done hoping that the model will contain some essential features of the phenomenon it describes, and given enough information about prevailing environment, would apply to the whole scale of environments, where the phenomenon is present. Moving the model from one edge in the environment of the phenomenon to quite the opposite, just by adjusting the parameter values concerned, and then putting those results against measured ones, performs a severe test for the models applicability.

MicroForest includes three features which enable its application to other and very different stands. (i) The annual photosynthetic production depends on the climate and weather of the site. This is embedded into the value of the parameter Po. (ii) The initial state for trees enables a very detailed description of the stand structure in the beginning. (iii) The amount of soil organic matter at initial state introduces the fertility of the site.

To rigorously test MicroForest, we collected detailed measurements of growth history of trees from several stands located near the SMEAR II stand and far north and far south of it. On all locations, Lapland, southern Finland and Estonia, we measured the growth histories of five size classes from five to six stands varying in fertility and density, from very fertile to very poor soils and from sparse to extremely dense stands.

9.2.2 Test with Six Stands near SMEAR II

Mikko Havimo[1], Roope Kivekäs[2], Juho Aalto[2], Pauliina Schiestl[2], and Pertti Hari[2]

[1] University of Helsinki, Department of Forest Resource Management, Finland
[2] University of Helsinki, Department of Forest Ecology, Finland

Six stands near SMEAR II station were selected to test the performance of Micro-Forest. The study stands represent typical managed forest in Finland they are even-aged, dense, and consist mainly of one tree species. The measurements from these stands forms a new data set, which has been collected after construction of Micro-Forest and which has not been used in the development of the model. Data from test stands is used only to determine the initial state for the model. Therefore MicroForest is placed under severe test, when values simulated by it are compared against the new measurements about 1 km from SMEAR II (Fig. 9.2.2).

The stands were established in 1978 by clear-cutting an old forest, and regenerating the area by burning over and ploughing. Experimental stands were chosen, having stand densities between 1,800–10,000 stems per hectare, mean breast height diameter 8–15 cm and mean height 8–11 m. All stands were on stony mineral soil. The stands were dominated by Scots pine, having only small numbers of Norway spruces (*Picea abies*) and birches (*Betula pendula* and *Betula pubescens*). The study area is located on Southern boreal vegetation zone: mean annual temperature is $+3.3°C$ and mean annual precipitation is 700 mm (Drebs et al., 2002). For testing MicroForest, the sample plots were measured in spring 2005, when the trees were 27 years old.

9.2.2.1 Test Stands

Stand 1, location 61°50'55''N24°17'22''E, mean diameter 8 cm, mean height 8 m. The sample plot consisted 3,500 stems per hectare on a Vaccinium type site according to Finnish forest type classification (Cajander, 1926). The stand was almost purely pine, but there were some small undergrowth spruces (height <1.3 m). Mosses (*Pleurozium schreberi*) dominated the ground vegetation, but there was also some *Vaccinium vitis-idea* and *Vacciniumin myrtillus* dwarf shrubs.

Stand 2, location 61°50'57'' N 24°17'31'' E, mean diameter 8 cm, mean height 8 m. The stand density was 3,800 stems per hectare, and site fertility Vaccinium type. This was pure pine stand on a very stony soil. *P. schreberi* was the dominating species of the ground vegetation, other common species being *Calluna vulgaris* and *V. vitis-idae*.

Stand 3, location 61°50'59'' N 24°17'37'' E, mean diameter 11 cm, mean height 10 m. The stand contained 1,800 stems per hectare. Along with pines, there were few birches in the dominating crown layer and vigorous spruce undergrowth. This Vaccinium type site was mainly on mineral soil, but there was also a small peatland patch inside the plot. The ground was covered mostly by *V. vitis-idae*, but *P. schreberi*, *V. myrtillus* and *Polytrichum strictum* were also common.

Stand 4, location 61°51′2″ N 24°17′26″ E, mean diameter 8 cm, mean height 7 m. This stand was exceptionally dense (10,000 stems per hectare), therefore undominant trees and ground vegetation were under dark shade. The ground was covered with *P. schreberi,* whit only few patches of *V. vitis-idae.* The number of undergrowth spruces was small. The stand type was Vaccinium.

Stand 5, location 61°51′7″ N 24°17′20″ E, mean diameter 8 cm, mean height 6 m. The stand was similar to the sample plot 4. The stand density was 8,000 stems per hectare[1], so shading of trees and ground vegetation was heavy. The ground vegetation was dominated by *P. schreberi* and the stand type was Vaccinium.

Stand 6, location 61°51′9″ N 24°17′26″ E, mean diameter 15 cm, mean height 11 m. This was the most fertile sample plot, where the stand type was Myrtillus. The stand density was only 2,100 stems per hectare, which was the second lowest among the sample plots, therefore the mean diameter and height were high. The ground vegetation was mainly different kind of herbs, *V. myrtillus* being the dominating species. *V. vitis-idae* and *Deschampsia flexuosa.*

For obtaining the sample trees, a sample plot consisting about 200 Scots pines was established on each stand. Trees damaged by elk (*Alces alces*) or snow were excluded from population and sampling. Measured trees were divided into five size classes by breast height diameter in such a way that size classes consisted of 5%, 15%, 30%, 30% and 20% of the trees. Six trees were systematically sampled from each size class for measurements. Whorl heights were measured from sample trees, and a sample disk was cut on stump height. Disks were photographed, and widths of annual rings were measured with an image analysis program.

In MicroForest, the initial state describing the stand is very brief, including only stand structure and soil fertility. All parameters in the model are obtained from the stand around SMEAR II station. The initial state includes three size class specific components: (i) diameter of seedlings, (ii) height of seedlings, (iii) number of stems per hectare. In principle, the initial state should also include measured soil fertility, but the fertility for initial state should have been measured when the stands were established in 1978. When the stands were established, the fertilities were determined with the traditional forest type classification system (Cajander, 1926). Micro-Forest uses amount of proteins in soil as a measure of fertility, and this measure is not, at the moment, comparable to the old classification. The method to determine soil fertility is described below.

The sizes of seedlings at the initial state can be determined from the diameters of annual rings and heights of whorls. However, it is difficult to separate annual rings of first three or four years, so there is some noise in diameter measurements of 3–4 years old seedlings. Since initial state is the only stand information that MicroForest uses, the model is sensitive to measurement errors in seedling sizes. When seedling age is five years, noise in annual ring measurements is considerably smaller than in 3–4 year old seedlings. This is why five-year old seedlings are used in the initial state (Fig. 9.2.1).

Tree photosynthesis depends largely on needle mass. Therefore, the needle mass of initial state seedlings has a major effect on simulation results. The needle mass of five-year old seedlings was predicted from stem diameter with a regression model.

Fig. 9.2.1 Five years old seedlings similar to those used to determine the initial state

This was done for each size class and the obtained masses were used to determine the initial states.

The amount of proteins in soil at the initial state is the measure of fertility in the model. As was discussed before, amount of proteins at the initial state was not measured. However, we can estimate it from the model behaviour. Estimation is done by choosing a property, which depends on the amount of proteins in soil. We chose diameter on stump height of the second size class as the suitable characteristic. The estimation procedure was based on adjusting the initial amount of proteins in such a way that the model produced the measured diameter of the second size class at the end of the study period. The diameters of the second size class were rejected in the evaluation of the model, since we wanted to avoid estimation bias in the tests.

We assumed that nitrogen deposition mainly coming from traffic and industry was $1\,kg\,(N)\,ha^{-1}\,a^{-1}$ in the beginning of the study period and increased linearly to $5\,kg\,ha^{-1}\,a^{-1}$. These values are in agreement with measurements and have been used when MicroForest was fitted to SMEAR II data.

Of the six measured stands, we present here results of the two stands, where the modelled and measured values were in best or in weakest agreement. The ability of the model to predict diameter on stump height was very good in all size classes in the stand 6 (Fig. 9.2.3) which represents the best agreement. The situation with the tree height was not as good as with diameter. The model overestimated the height of largest size class, but the correspondence was pretty good in third and fifth size classes (Fig. 9.2.3).

Fig. 9.2.2 Wintery aspect in one of the test stands in Southern Finland

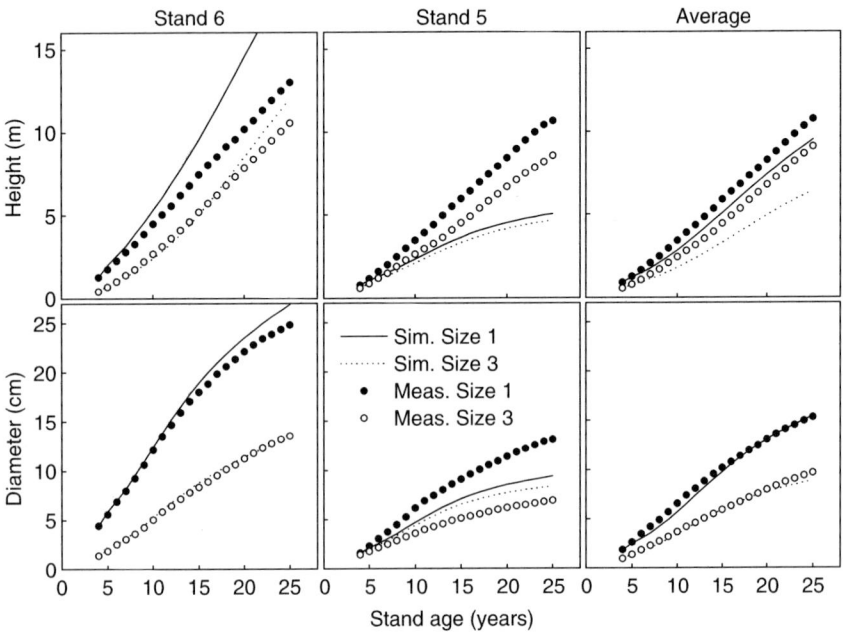

Fig. 9.2.3 Simulated and measured height and diameter in Stand number 6 (Myrtillus type site, 2,000 stems ha^{-1}), in Stand number 5 (Vaccinium type, 4,400 stems ha^{-1}) and in average in the test performed in Hyytiälä

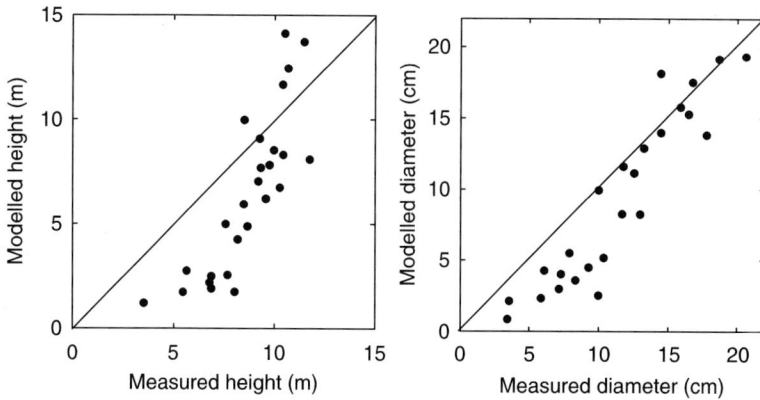

Fig. 9.2.4 Comparison of measured and simulated heights and diameters in all size classes and stands at the age of 27 years in Hyytiälä

In the stand no 5 (Fig. 9.2.2) the model and measurements were in weakest agreement. The model underestimated the diameter on stump height in the largest and smallest size classes, but overestimated it in the third size class (Fig. 9.2.3). The modelled height was also unsatisfactory, since model overestimated the height in first and third size classes, and underestimated it in the smallest size class (Fig. 9.2.3).

The overall prediction power of the model can be found in Fig. 9.2.4, where modelled and measured diameters and heights of all size classes and stands at the end of the study period are compared. The modelled diameters are in good agreement with measured ones, but modelled heights lack behind measured ones, especially in the smaller size classes.

We calculated the averages over stands for each size class to study the model performance in describing the development of size differences. The agreement in the average diameter in the biggest size classes is very good (Fig. 9.2.3), but in the smallest one, the modelled diameter lacks behind the measured one. Similar way, the modelled height is in good agreement with the measured ones in the largest size class (Fig. 9.2.3), but in the third and smallest size classes the modelled height are clearly smaller than the measured ones.

The reason for problems in modelled heights is likely in the model structure the method of calculating height growth does not work properly. In principle this is not surprising, since in MicroForest height growth is not connected to tree photosynthesis and to structural regularities as strongly as diameter growth. At the moment our understanding on height growth is clearly limited. This is a major finding, because the idea behind the model testing is to detect weakest points in the model, and enable the development of understanding.

9.2.3 Test in Estonia at the Southern Border of the Boreal Zone

Kajar Köster[1], Ahto Kangur[1], Pertti Hari[2], and Kalev Jõgiste[1]

[1] Estonian University of Life Sciences, Institute of Forestry and Ryral Engineering, Estonia
[2] University of Helsinki, Department of Forest Ecology, Finland

The results obtained with MicroForest seem to have explaining power also in other stands around SMEAR II. This raises again the question: Can MicroForest be applied to Scots pine stands growing in other locations than Hyytiälä in boreal forests? We selected five stands in Estonia at the southern border of boreal forests about 300–400 km south from Hyytiälä. The altitude in the selected stands varies between 32 to 56 m above the sea level. Selected stands represent Scots pine dominated hemiboreal coniferous forests on sandy soil with different site fertilities. The climatic conditions are rather similar in all locations. Mean annual temperature is 4–6°C. Annual precipitation is 500–750 mm, about 40–80 mm of which falls as snow. The length of the growing period (daily air temperature above 5°C) is between 170 and 180 days.

The measurements in the stands were done as in Hyytiälä.

Plot 1 in Lahemaa is representing Scots pine forest on poor sandy soil conditions on *Cladonia* site type from Lahemaa National Park (North of Estonia (59°32′ N, 25°56′ E, Fig. 9.2.5). Although the stand was established by planting in rows with pine seedlings on furrows (Table 9.2.1), the area now is strictly protected and no subject for forest management. The forest is Scots pine monoculture with almost no understorey and with poor shrub and herb layer characterized by *Vaccinium vitis-idea, Calluna vulgaris, Melampyrum pratense, Carex ericetorum* and *Diphasiastrum complanatum*. Forest floor has continuous moss and lichens cover characterized by *Cladina stellaris, Cladina rangiferina, Cladina mitis, Cladonia furcata, Pleurozium schreberi, Dicranum polysetum, Dicranum spurium* and *Polytrichum juniperinum*.

The plots 2, 3 and 4 are in North-western part of Estonia (59°21′ N, 23°81′ E, Fig. 9.2.6) in Vihterpalu and are growing on poor sandy soils on *Calluna* (plot 2 and plot 3) and *Vaccinium* (plot 4) site types. The sampling site is located on the Vihterpalu fire hazard area, which was burned 60 years ago. Post fire the area was afforested by planting Scots pine in rows (plot 3, Fig. 9.2.6). The site conditions in the area vary due to varying microrelief: there are small hills on the poor sandy soil. Groundwater table is high and in lower places thick raw humus layer has formed indicating changes in growth conditions towards *Vaccinum uliginosum* site type (plot 4). In plot 4 the variation in tree diameter and height was higher due to uneven tree spacing on the plot. The stands in sample area are dominated by Scots pine, understorey and shrub layer is very poor or even missing. In ground vegetation *Calluna vulgaris, Vaccinium vitis-idea, Empetrum nigrum, Vaccinium myrtillus, Ledum palustre, Vaccinium uliginosum* and *Eriophorum vaginatum* can be found. The moss and lichens cover is continuous and characterized by *Pleurozium schreberi, Dicranum* spp., *Sphagnum* spp., and *Cladina* spp.

Fig. 9.2.5 One of the test stands was in Lahemaa National Park

Plot 5 is located in Järvselja Training and Experimental Forest of Estonian University of Life Sciences (58°25′ N, 27°46′ E). The sample area represents Scots pine dominated forests on sandy soil on *Oxalis-Rhodococcum* site type with comparatively high fertility. The stand in sample area represents a well managed second Scots pine generation in commercial forest on a former agricultural land. Although the management effect, the stand is shoving a relatively high stand density: 1,570

Table 9.2.1 Stand characteristics on sample plots

Location	Plot no	Site type	H	$D_{1.3}$	Age	Trees ha^{-1}
Lahemaa	1	*Cladonia*	9.7	10.4	34	2,900
Vihterpalu	2	*Calluna*	8.9	9.8	61	2,100
Vihterpalu	3	*Calluna*	8.9	10.6	61	2,200
Vihterpalu	4	*Vaccinium uliginosum*	8.5	9.3	50	1,800
Järvselja	5	*Oxalis-Rhodococcum*	18	19.8	32	1,570

Fig. 9.2.6 Scots pines Vihterpalu test site in Estonia

trees per hectare. The stand is Scots pine dominated having Norway spruce and deciduous tree species like *Betula pendula, Sorbus aucuparia* in understorey. Shrub and bush layer is not very dense characterized by *Rubus idaeus, Vaccinium vitis-idea, Vaccinium myrtillus.*

Diameter growth was measured using CORIM Maxi device (PREISSER Messtechnik GmbH, Germany). Height growth was measured as heights of whorls starting from the top of the tree.

Nitrogen deposition is in Estonia clearly higher than in Hyytiälä. We assumed that deposition was $1\,kg\,ha^{-1}\,a^{-1}$ in the beginning of the study period and increased linearly to $9\,kg\,ha^{-1}\,a^{-1}$ in Estonia. These values are in agreement with measurements.

The initial state was determined as in Hyytiälä.

The model prediction succeeded best in Järvselja the development of diameter and height were close to each other in the size classes 1, 2, 3 and 4 (Fig. 9.2.7). It was surprising that the model based on measurements in the stand around SMEAR II predicted the growth of the stand growing in Järvseljä without estimation of the initial protein pool in the soil within measuring accuracy and precision. However the smallest size class grew clearly less in the model than in reality.

Vihterpalu growth condition (plots 2, 3 and 4) differ much compared to Järvselja and Lahemaa, the soil is much poorer. Again in the smaller size classes the model under predicts the growths (Fig. 9.2.7). The model is unable to predict the growth pattern: the model overestimates the growth in young stands (10–20 years) and underestimates at the end of test period.

The model MicroForest predicted rather unsatisfactorily the mean diameter and height growth patterns in the size classes: the modelled growths were too large in the middle of the study period and too little at the end (Fig. 9.2.7). The model failed

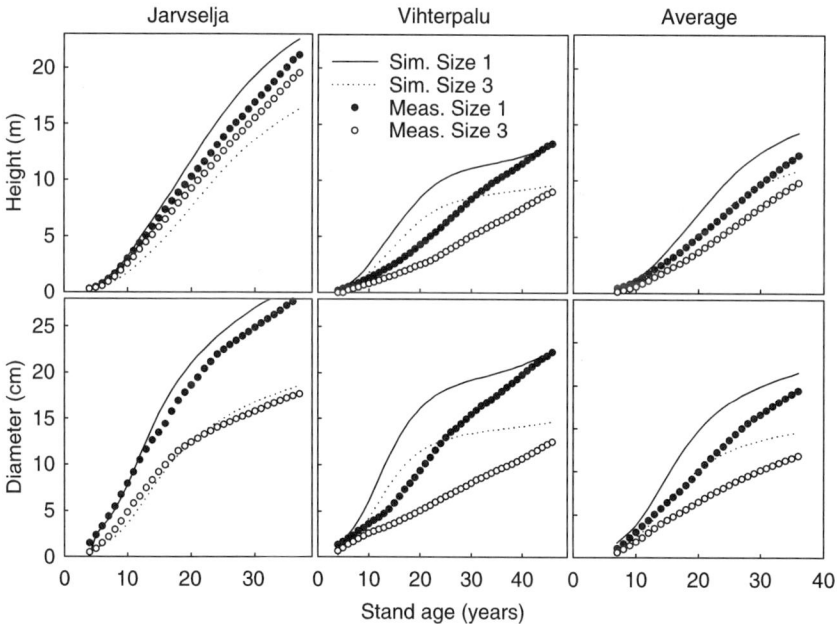

Fig. 9.2.7 Simulated and measured height and diameter in Jarvselja, Vihterpalu and in average in the test performed in Estonia

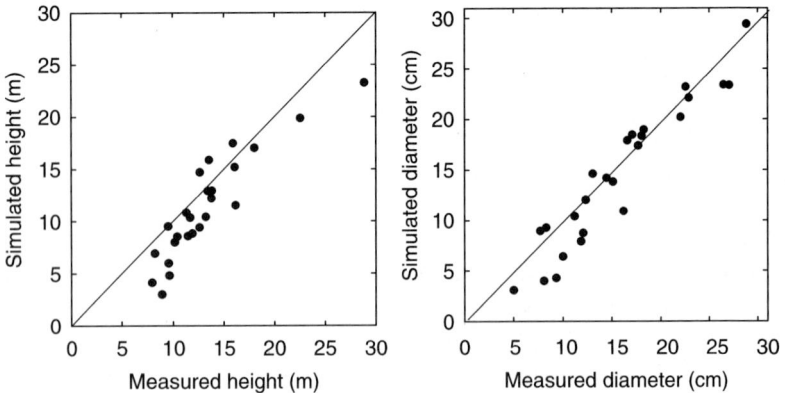

Fig. 9.2.8 Measured and simulated heights and diameters in all size classes and stands at the end of simulations of the test forests in Estonia

on these poor sites to predict the dynamics of nitrogen properly, the availability was too large at the age 10–20 years and thereafter too small. The model was recovering from the nitrogen shortage at the end of the study period. The nitrogen dynamics in soil explains the failure in the prediction of growths.

MicroForest was able to predict the diameters and the heights at the end of the study period rather satisfactorily, although the height growth of the small trees had tendency to grow too little (Fig. 9.2.8).

Acknowledgments We thank Mr. Loic Gruson for the analysis of the annual rings.

9.2.4 Test at Northern Timber Line

Veli Pohjonen[1], Petteri Mönkkönen[1], and Pertti Hari[2]

[1] Värriö Subarctic Research Station, Finland
[2] University of Helsinki, Department of Forest Ecology, Finland

The sample forests at northern timber line were selected inside relatively even, young Scot pine stands in five different locations: Kotovaara (Fig. 9.2.9), Pulkkatunturi, Papulampi, Oulanka and Aitaselka. Geographically the study area belongs to Eastern Finnish Lapland, in the municipality of Salla. The sample plots are situated between the latitudes 66°30 and 67°46 North, between the longitudes 28°14 and 29°50 East. The altitude in the sample forests varies between 210 and 400 m above the sea level. In these latitudes the alpine timber line of Scots pine and Norway spruce is at 420–430 m above sea level. The area is 700 km to north-east from SMEAR II.

All these forests represent northern boreal coniferous forests. They are situated north of Arctic Circle, near to the northern timber line, but well inside large forest zone where also regulated commercial forestry is practiced. The mean annual

Fig. 9.2.9 Kotovaara test forest. The big trees were excluded from the population and also from sampling. Photo: Teuvo Hietajärvi

temperature in the study area is about −1°C and annual rainfall is about 550 mm. Growing season lasts on average 105–120 calendar days and the snow covers the ground for 200–225 days per year. Due to very northern location the sun never sets at midsummer and photosynthesis occurs also at nights (Section 6.3.2.3).

In Northern Finland Scots pine is the most common tree species: 76% of forests in Lapland are dominated by Scots pine. Because of large clear felling in mid 1900s and subsequent re-establishment to Scots pine, young forests of less than 50–70 years are typical all over Northern Finland. All sample plots are dominated by Scots pine mixed with more or less pubescent birch (*Betula pubescens*) or its bush-like subspecies *Betula pubescens* ssp. *czerepanowii*.

9.2.4.1 Sample Plots

Kotovaara hill. The sample plot is situated at 67°46 N and 29°35′ E, 400 m above sea level, inside Värriö Strict Nature Reserve. The plot is on the top of a hill, close to the alpine timberline near SMEAR I measuring station. The stocking is 770 stems per hectare and the growing volume 57 m³ per hectare. Kotovaara hill has never been subject to fellings. There are two dominating age classes. However, most trees were born about in late 1930s or in mid 1950s. In addition some scattered older Scots pines with light mixture of pubescent birches as understorey are found. The soil on the site is moraine with many stones. The typical vegetation: *Empetrum nigrum*, *Vaccinium myrtillus*, *Vaccinium vitis-ideae* and *Pleurozium schreberi*, tells also about rather poor soil conditions. Some pines are damaged by snow.

Pulkkatunturi fell. The sample plot is situated at $67°46'$ N and $29°43'$ E, 231 m above sea level on the western slope of an open fell. This plot is also inside Varrio Strict Nature Reserve, and the forest has never been subject to fellings. The stocking of the pine forest is 1,240 stems per hectare and growing volume 31 m^3 per hectare. The dominating species is Scots pine intermixed with some pubescens birches. The trees on the sample plot forest were naturally regenerated after forest fires. The dominating age classes were in late 1950s or in late 1960s. The soil in the site is poor and stony and bare rock surface is in sight in small spots in the site. *Vaccinium uligonosum*, *Vaccinium myrtillus*, *Empetrum nigrum* and *Ledum palustre* characterize the ground vegetation.

Papulampi forest. The sample plot is situated at $67°44'$ N and $29°50'$ E, 240 m above sea level, in commercial forest zone. The site is located in northern part of the large forest fire area of Tuntsa, which burned in 1960. After the fire the areas were planted with Scots pine. The stocking of the rather well-growing pine forest is 1,470 stems per hectare and growing volume 106 m^3 per hectare. Some intermixed pubescent birch is found. The soil is sandy without stones. The ground water is quite near to the surface soil. Typical plants on the site are *Vaccinium myrtillus*, *V. vitisideae*, *V. uliginosum*, *Empetrum nigrum*, *Ledum palustre* and *Pleurozium schreberi*.

Oulanka forest. The sample plot is situated at $66°30'$ N and $28°40'$ E, 300 m above sea level, in an area of regular commercial forestry. The area has been clear felled in the 1960s, prepared via controlled burning, and planted with Scots pine seedlings in the late 1960s. The stocking of relatively dense pine forest is 1,640 stems per hectare and the growing volume 171 m^3 per hectare. Some pubescent birches are found as undergrowth. The soil is sandy without stones. The dominating ground vegetation is made by *Vaccinium myrtillus* and *V. uliginosum*.

Aitaselka forest. The sample plot is situated at $66°39'$ N and $28°14'$ E, 213 m above sea level. The young Scots pine forest is uneven by age due to natural regeneration after previous clear felling. The dominating age classes were born in 1960s. The stocking is 1,500 stems per hectare and growing stock 68 m^3 per hectare. The sample plot is situated on the western slope of a small hill. The soil is poor moraine with many stones. *Vaccinium uligonosum*, *V. myrtillus*, *V. vitis-ideae* and lichens from *Cladina* sp. form typical vegetation on this site.

The five size classes were determined and measured as at SMEAR II. We assumed that deposition in Lappland was 1 kg ha^{-1} a^{-1} in the beginning of the study period and increased linearly to 3 kg ha^{-1} a^{-1}. These values are in agreement with measurements.

The prediction by MicroForest was most successfully at Oulanka (Fig. 9.2.10): the measured and predicted diameters and heights in the size classes were nearly as close as in Hyytiälä and also the diameter and height patterns are rather similar in each size class. Again the smallest trees grow too little in the simulations.

The prediction of the growths of the biggest size classes was quite successful in Kotovaara (Fig. 9.2.10), although this stand is growing on the timber line. The smallest trees are clearly younger than the dominating trees and MicroForest is unable to predict the growths of the smallest size classes, the trees grow rather well and the model predicts very small diameter and height increments.

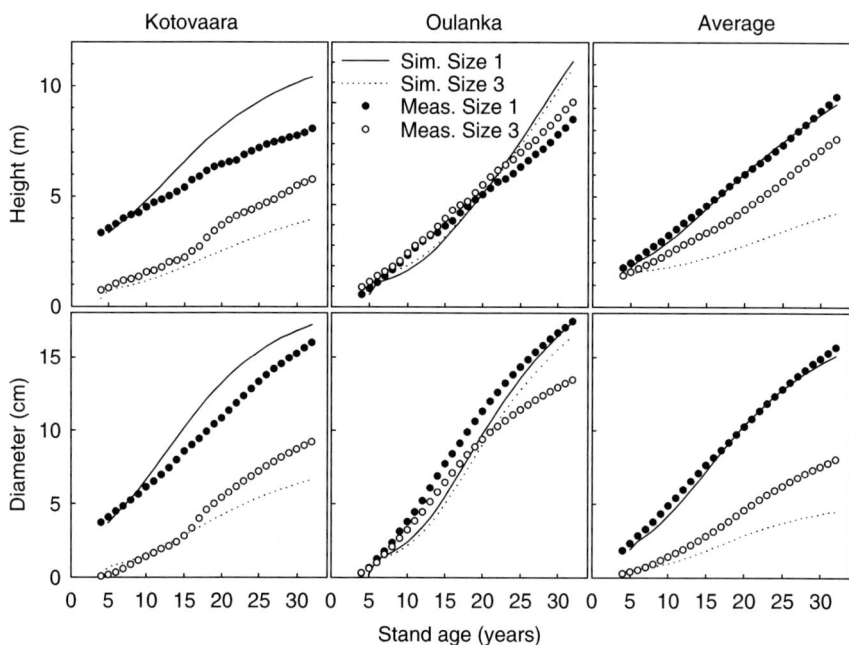

Fig. 9.2.10 Simulated (lines) and measured (dots) heights and diameters of two size classes in Kotovaara, Oulanka and in average in the test performed in northern timberline

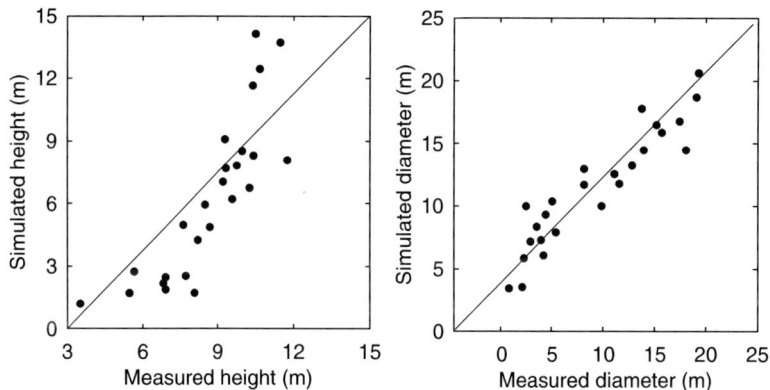

Fig. 9.2.11 Simulated and simulated heights and diameters in all size classes in the current stage of the forests in northern timberline

The prediction of the mean diameter and height growths in each size class was successful in the biggest size classes, and prediction failed rather clearly for the smallest size classes (Fig. 9.2.10). This result is rather similar as in Hyytiälä.

The prediction of the size-class-specific diameters and heights at the end of the study period indicated rather good agreement, especially in the biggest size classes (Fig. 9.2.11).

9.2.5 Evaluation of the Performance of MicroForest

Pertti Hari[1], Mikko Havimo[2], Roope Kivekäs[1], and Eero Nikinmaa[1]

[1] University of Helsinki, Department of Forest Ecology, Finland
[2] University of Helsinki, Department of Forest Resource Management, Finland

The model MicroForest is based on theory dealing with metabolic basis of growth and development of forest ecosystems and regularities in tree structure. The theoretical ideas underlying the model in a nutshell are:

1. Processes produce and consume material for living organisms.
2. The environmental factors and the amounts of active functional substances determine process rates.
3. The amounts of active substances are under the control of regulation system.
4. Active substances are proteins, cell walls cellulose and lignin, membranes lipids and storages starch.
5. Microbes decompose with extracellular enzymes macro molecules.
6. Strong structural regularities have developed in evolution connecting processes transport and structure.
7. Processes generate carbon and nitrogen fluxes within forest ecosystems.
8. Different processes within a tree are combined with conservation of carbon and nitrogen resulting in carbon and nitrogen balance equations.

In addition, the operational model includes numerous technical details smearing rather clear theoretical basis of MicroForest. The present version is result of long lasting development (Hari et al., 1982, 1985 Mäkelä and Hari, 1986 Nikinmaa, 1992 Perttunen et al., 1996 Mäkelä, 1997, 2002 Nissinen and Hari, 1998 Mäkelä and Mäkinen, 2003).

The measuring station SMEAR II was planned and implemented to measure processes and mass, energy and momentum fluxes in a forest ecosystem. Thus the measurements speak the same language as MicroForest. This fact makes the obtained data very useful in determining the parameter values in the model. Mikro-Forest include rather many parameters to be determined. Most of the values can be obtained from direct measurements, as for example, the annual photosynthetic production in unshaded conditions.

MicroForest is in agreement within the measuring accuracy and precision with all measurements at SMEAR II. For example, the carbon budget is closed, the difference between in and out fluxes is accumulated into wood in the stand and the carbon storage of the soil is stable. However, there is evident estimation bias in the agreement between the model and measurements, since values of many parameters have been obtained from the same measurements. Thus additional measurements from different stands are needed to get proper picture of the performance of the model.

We selected six stands 1 km north from SMEAR II to test the prediction power of MicroForest. We measured annual diameter and height down to age five years in five size classes. The measured size-class specific diameters and heights determined the seedlings in the initial state. The needle mass at five years age was calculated from empirical regression. Number of trees per hectare described the density.

MicroForest requires the amount of proteins in soil in the initial state. This could not be determined from direct measurements instead it was estimate from the diameter of the second biggest size class at the end of the study period. Thus the remaining measurements do not affect the model behaviour and we have no estimation bias in diameter growths of the size classes 1, 3, 4, and 5 as well as in all heights. This strict testing really differs from the normal modelling praxis and makes our test very demanding.

The measurements indicated extremely clearly that the biggest trees at the year of measurements were also biggest at the age of five years and correspondingly for the small trees. The size classes were formed already at the age of five years. Micro-Forest was able to produce the differentiation of the size classes into dominating and suppressed trees.

The prediction of diameter and height growths was rather successful when the stands were similar to that around SMEAR II. Thus model predicts properly near SMEAR II, if the stand is to some extend similar with that one used in the construction of the model and MicroForest fails in predicting height growths at high densities on poor sites. At high densities, the predicted height growths were clearly too small.

Scots pine stands look rather similar to each other in the south-north direction in the boreal zone. This gives rise to the question: What is the prediction power of MicroForest in other locations than Hyytiälä? We selected five stands in Estonia at the southern border of boreal forests and in Lapland to test the prediction power in the boreal zone.

The results from Estonia were rather surprising since in the stand in Järvselja, the prediction was as close as at SMEAR II stand. In Vihterpalu there is some discrepancy in the diameter patterns of prediction and measurements. MicroForest seems to be able predict stand development also at the southern border of the boreal zone. There are, however, problems on poor sites since the nitrogen dynamics generates to large growths at the age 10–20 years.

SMEAR I is located at northern timber line about 700 km to north-east from SMEAR II. The measuring station is at elevation of 400 m and there are no larger Scots pines above 420 m a.s.l. We measured five stands around and south from SMEAR I within distance of 100 km.

MicroForest was able to predict all stands having reasonable homogenous stand structure in the early phases of their development. If the stand included smaller trees clearly younger than the dominating trees, as around SMEAR I, then MicroFores failed in the prediction of the small trees. Again, the prediction was especially successful when the stand was similar to the one used in the construction of the model, i.e. soil is fertile enough and stand density around 2,000 trees per hectare.

The model MicroForest predicted development of diameter and heights well in all homogenous stands which are of medium fertility. This type of forests dominates the Finnish and Estonian forests. On the other hand, the prediction showed evident shortcomings in the modelled growths, especially in height, in smallest size class, in low fertility sites and at extremely high densities. These findings are important for the development of MicroForest.

The structure of the model MicroForest was derived using theoretical ideas which are not connected with special features of Scots pine they are general for all coniferous trees. This raises the question: Can MicroForest be applied for all coniferous trees? The parameter values were mainly determined from species-specific measurements and they are evidently characteristic for each species or ecotype. Thus we can hypothesize with good reasons that MicroForest can be applied to other species in other countries, if the species-specific parameter values are determined utilising measurements of the species in question.

Testing of the hypothesis of general applicability of MicroForest is rather easy and straightforward: we need 5–10 test stands that should be measured as was done in the tests near SMEAR II. In addition, we need some measurements dealing with photosynthesis and tree structure. We are ready to help in the arrangement of the measurements and in computation.

We do not know any similar test of growth models, thus we can not compare the results with others. The stand volume, the most important characteristic in applications, depends strongly on the diameters of the biggest trees and, not so strongly on heights. Thus the failures in height increments in smallest size classes are not so crucial for the applications in the research of the impacts of climate change on boreal forests and in practical forestry.

Conclusion

The model MicroForest has rather strong prediction power when tested in a wide range in south-north direction. MicroForest predicts diameters, especially of the biggest trees, rather well. The prediction of height development fails, especially in dense stands on poor soil. However, the prediction of stand volume, the most important characteristics, is rather successful. The height growths and nitrogen dynamics on poor soil should be improved.

Chapter 10
Interactions Between Boreal Forests and Climate Change

10.1 Climate Change

Jouni Räisänen[1] and Heikki Tuomenvirta[2]

[1] University of Helsinki, Department of Physics, Finland
[2] Finnish Meteorological Institute, Finland

10.1.1 Mechanisms of Climate Change

The increasing concentrations of CO_2 and other greenhouse gases (Chapter 1) change the behaviour of radiation energy in the atmosphere. Several processes (Section 6.2) respond to this redistribution of radiation. The conservation equations of mass, energy and momentum described in Section 7.2 can be solved numerically to study the resulting effects on climate.

Changes in climate occur both as a result of natural variability and as a response to anthropogenic forcing. Some part of the natural variability is forced, that is, caused by external factors such as solar variability and volcanic eruptions. Another part is unforced, associated with the nonlinear internal dynamics of the climate system such as interactions between the atmosphere and the oceans. Because of these natural mechanisms, climate has always varied, on time scales ranging from years to millions of years (Jansen et al., 2007), and it would continue to vary in the future regardless of what mankind is doing. Nevertheless, when we focus on this and the following centuries, the effects of natural variability will likely be secondary when compared with anthropogenic changes in the global climate (Meehl et al., 2007).

Anthropogenic climate change is often referred to simply as *global warming* (e.g., Houghton, 1997; Harvey, 2000). This refers to the expected tendency of increased greenhouse gas concentrations to warm up the surface and the lower atmosphere in all or nearly all parts of the world. Nevertheless, other anthropogenic forcing mechanisms such as changes in land use and, in particular, increased aerosol concentrations complicate this picture (Forster et al., 2007). Furthermore, although temperature is the climate variable in which the anthropogenic changes are expected to be strongest when compared with natural variability, many other aspects of climate will also change.

The Earth has a *natural greenhouse effect* which helps to keep the surface of our planet much warmer than it would otherwise be. The surface of the Earth emits thermal radiation almost like a black body, with the intensity of radiation increasing with the fourth power of the surface temperature (Jin and Liang, 2006; Section 6.2.1). On the other hand, considering the energy budget of the Earth and its atmosphere as a whole, there must be a close balance between the solar radiation absorbed by our planet and the thermal radiation escaping to space. From this condition it follows that, if the atmosphere were transparent to thermal radiation, the average surface temperature of the Earth would be only about $-19°C(254\,K)$. The observed mean surface temperature is about $33°$C higher. This difference is due to the fact that most of the thermal radiation emitted by the surface is absorbed by the atmosphere, and the radiation that actually escapes to space therefore has its origin in atmospheric layers much cooler than the surface. Of the so-called *greenhouse gases* that are responsible for this effect, the most important is water vapour (with a contribution of about 60% in clear-sky conditions), followed by carbon dioxide CO_2 (26%) and ozone $O_3(8\%)$. Methane (CH_4) and nitrous oxide (N_2O) together account for the remaining 6% (Kiehl and Trenberth, 1997). Clouds also amplify the greenhouse effect, with a contribution comparable to that of CO_2, but this is more than compensated by their ability to reflect solar radiation to space.

Because of human activities, the concentrations of many greenhouse gases have increased and are projected to increase further in the future. The effect that a given change in the atmospheric composition has on the planetary energy budget is commonly quantified by *radiative forcing*. This measures, in approximate terms, the change in the energy balance of the Earth-atmosphere system that a given change in external conditions would induce with no compensating changes in climate (for the exact definition, see Houghton et al., 2001, p. 795). For the increase in CO_2 concentration from pre-industrial time to the present (280–380 ppmv), the radiative forcing is approximately $1.66\,W\,m^{-2}$ (Forster et al., 2007). Smaller positive forcing contributions have come from increases in $CH_4(0.48\,W\,m^{-2})$, $N_2O(0.16\,W\,m^{-2})$, various halocarbons $(0.34\,W\,m^{-2})$ and tropospheric $O_3(0.35\,W\,m^{-2})$. With the exception of O_3, all these forcings can be estimated with a good accuracy $(\pm10\%)$, because the radiative properties of gas molecules are well known and the lifetime of these greenhouse gases (excluding O_3) is long enough to make them almost evenly distributed in the lower atmosphere. All in all, the present-day radiative forcing from anthropogenic increases in greenhouse gas concentrations approaches $3\,W\,m^{-2}$. This number does not include changes in water vapour because direct anthropogenic sources of water vapour are negligible when compared with the natural hydrological cycle. However, observations suggest that water vapour is increasing in response to observed atmospheric warming and is thereby providing a positive feedback to ongoing climate change (Trenberth et al., 2007).

Greenhouse gas forcing has, however, been partly compensated by a net negative forcing from increased aerosol loading. The direct effect of anthropogenic aerosols, associated with the scattering and absorption of solar radiation by aerosol particles, varies in sign between different aerosol types, but the net forcing is very likely negative. Forster et al. (2007) give it a best estimate of $-0.5\,\mathrm{W\,m^{-2}}$ and a wide estimated uncertainty range of -0.1 to $-0.9\,\mathrm{W\,m^{-2}}$. The anthropogenic increase in aerosols also results in an increase in available cloud condensation nuclei, which has the potential to increase cloud albedo. The negative forcing due to this indirect aerosol effect is estimated as $-0.7\,\mathrm{W\,m^{-2}}$, but with a very wide uncertainty range (-0.3 to $-1.8\,\mathrm{W\,m^{-2}}$). Furthermore, aerosols may also affect the lifetime of clouds and hence the total cloud cover, but this effect is even more poorly quantified.

As the anthropogenic increase in aerosol loading has resulted in a negative radiative forcing, it has most likely slowed down the warming that would have resulted from increases in greenhouse gas concentrations alone. However, the magnitude of the aerosol forcing and hence the degree of this compensation is poorly known (e.g., Andreae et al., 2005). Also in contrast to the more or less evenly distributed greenhouse gas forcing, the aerosol forcing is concentrated in the vicinity of areas with the largest aerosol-generating emissions (e.g., Forster et al., 2007). Thus, greenhouse-gas induced and aerosol-induced climate changes are expected to have partly different geographical distributions. Note, however, that these differences are moderated by atmospheric circulation, which tends to spread the effects of even localised forcings to the global scale (e.g., Boer and Yu, 2003).

Positive radiative forcing tends to increase and negative forcing decrease the global mean temperature, but the magnitude of the forcing does not directly tell the magnitude of the temperature response. The latter depends on several feedbacks acting in the climate system and it therefore needs to be estimated with climate models. However, model simulations suggest that the ratio of the response to the magnitude of forcing is approximately the same for different forcing agents (e.g., Joshi et al., 2003; Hansen et al., 2005).

In the absence of amplifying feedbacks, the global climate would be quite resistant to radiative forcing. For example, a hypothetical doubling of the atmospheric CO_2 concentration, which gives a radiative forcing of about $3.7\,\mathrm{W\,m^{-2}}$, would result in a global mean warming of only about $1.2°\,C$, once the climate has had sufficient time[1] to reach a new equilibrium (e.g., Hansen et al., 1984). In fact, this number, known as *climate sensitivity,* is most likely higher but its exact value is very hard to determine. According to the latest estimates, which combine evidence from model simulations and observations (Meehl et al., 2007), it is probably within the range 2–$4.5°\,C$.

The water vapour holding capacity of air increases quasi-exponentially with increasing temperature, by approximately 7% for each $1°\,C$ of warming. Assuming that there will be no large changes in relative humidity, which is supported by both observations and model simulations (Held and Soden, 2006), the actual concentration of water vapour will also increase. Because water vapour is the most important

---

[1] Due to the large heat capacity of the oceans, this equilibrium time is of the order of centuries.

greenhouse gas in the atmosphere, the increase in its concentration will result in a strong positive feedback. Another most likely positive feedback is associated with changes in snow and ice cover. With a warming of climate, ice and snow cover will be most likely reduced, which increases the solar radiation absorbed at the surface and therefore acts to amplify the warming. This feedback is particularly important in high latitudes, but its role in amplifying the global mean warming is expected to be smaller than the role of increasing water vapour (Webb et al., 2006).

Clouds are important for the global climate, as they both reflect solar radiation to space (a cooling effect) and absorb thermal radiation emitted by the surface (a warming effect). Changes in cloudiness may thus also result in an important feedback effect. However, the complexity of cloud dynamics and microphysical processes makes this feedback very difficult to determine, although the most recent model estimates suggest that it will also probably be positive (Colman, 2003; Webb et al., 2006). As a result, changes in cloudiness appear to be the most important uncertainty in the response of climate to increasing greenhouse gas concentrations (e.g., Webb et al., 2006).

The interaction between biosphere and climate may also result in feedbacks. Focusing, in particular, on the role of boreal forests, there are several possible feedbacks:

1. The model simulations discussed in Section 10.2.4 suggest that the growth of boreal forests will increase with increasing temperature. This will cause the trees to sequester more CO_2 from the atmosphere, which represents a negative feedback. On the other hand, decomposition of organic matter in the soil is also expected to accelerate with warming. The net effect of boreal forests on the CO_2 concentration will depend on which of these competing feedbacks dominates. As globally averaged, however, model results indicate a positive feedback between climate and the carbon cycle. If acting alone, warming of climate would reduce the total amount of carbon stored in the global soil-biosphere system (Friedlingstein et al., 2006).

2. Model simulations also point to an increase in the leaf area index of the trees in a warmer climate, which has the potential to increase transpiration. This may have multiple effects on climate. Increased transpiration is expected to increase water vapour concentration in the atmosphere, thus amplifying the greenhouse effect. On the other hand, the resulting increase in humidity is also expected to promote formation of low clouds, which will reduce the solar radiation reaching the surface. Finally, an increase in transpiration would leave a smaller fraction of the available radiation energy for heating the surface that, like increases in cloudiness, would tend to reduce the surface temperature.

3. Increased activity of the forest canopy is also likely to lead to an increase in the formation of biogenic aerosols, with implications on the scattering of solar radiation and cloud formation.

4. Increases in the leaf area index and the density of the forest canopy may reduce the surface albedo, particularly during the snow season when the albedo contrast between mostly snow-free tree crowns and open snow-covered land is large. This positive feedback is potentially important, particularly during late winter

and early spring when there is still snow on the ground but a lot of solar radiation is already available. Simulations with coupled climate-vegetation models suggest that this feedback may have played an important role in explaining the relative warmth of the northern hemisphere high-latitude continents during the mid-Holocene 6,000 years ago (e.g., Wohlfahrt et al., 2004).

10.1.2 Observed Changes in Climate

The longest series of direct meteorological observations in the boreal region cover over 200 years. For example, in Stockholm (Moberg et al., 2002) and St. Petersburg (Jones and Lister, 2002) air temperature and pressure have been measured since the mid-18th century. However, only from the latter part of the 19th century have there been enough stations to calculate the global and boreal zone area-averaged temperatures. For the other parameters describing climatic conditions, such as precipitation amount, data coverage and its accuracy is limited until the 20th century. Increasingly comprehensive observations are only available, e.g., for snow cover and water vapour, during the recent decades. In this section, we will describe succinctly what kind of changes has been observed in the boreal climate.

The global mean temperature has increased $+0.8 \pm 0.2°$ C since the late 19th century (Trenberth et al., 2007). The warming has been larger over land areas than over the oceans. In the Northern Hemisphere the warming has been largest during winter and spring. In the boreal zone, the annual mean temperatures have warmed more than in the global mean (Fig. 10.1.1). The 20th century warming has occurred in three phases: warming until the 1940s, followed by a slight cooling and a rapid warming from about the 1970s.

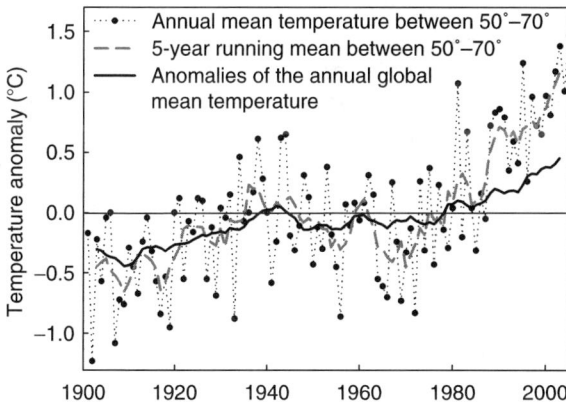

Fig. 10.1.1 Anomalies of annual mean temperature (1901–2005) in the land areas between latitudes 50° N and 70° N (baseline period 1961–1990). Annual values are in a dotted line with spheres and the 5-year running mean is a grey dash line. The 5-year running mean of the anomalies of the annual global mean temperature is shown as a black line. The data is from NCDC (Smith and Reynolds, 2005)

In the boreal zone (land areas between the latitudes 50° and 70° N), mean temperature during the period from May to August (hereafter MJJA) has increased in a similar three-phase behaviour as the annual mean temperature. Warming since the mid-1970s has been 0.40° C/decade (period 1976–2005) that is about triple the rate during the whole 20th century. Although this linear trend is a rough estimate of the 20th century warming, it allows an easy way to characterize spatial differences. From Fig. 10.1.2A, B, it can be seen that warming has not been spatially uniform. In fact, there are some limited regions in the boreal zone that show no warming or may

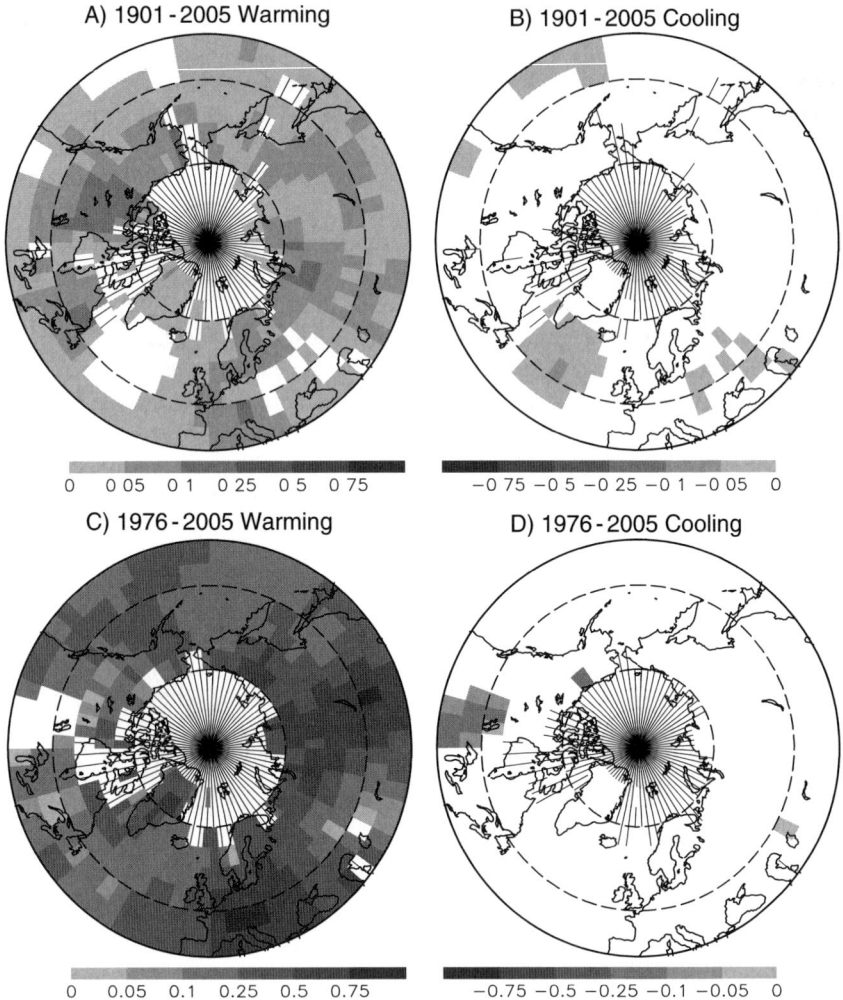

Fig. 10.1.2 Linear trend of MJJA temperatures (° C/decade) for (A)–(B) 1901–2005 and (C)–(D) 1976–2005. Areas with warming are shown in (A) and (C), areas with cooling in (B) and (D). Areas with missing data are indicated with north-south-oriented lines (The data is from NCDC; Smith and Reynolds, 2005)

even have cooled. However, the vast majority of the boreal area shows an increasing temperature trend over the period 1901–2005. The largest positive trends are about $+1.5°$ C per century.

Since the 1970s, MJJA temperature trends are characterized by widespread warming (Fig. 10.1.2C, D). The largest trends of up to $+0.7°$ C per decade can be found in central Siberia. However, there are regions that have cooled, e.g., in central North America. Changes in atmospheric circulation modify the regional temperature trends and natural climatic variability affects the statistical significance of the trends. A majority of the positive trends in Fig. 10.1.2A are statistically significant at 5% level, but due to the large natural variability and the short time period just the largest trends are statistically significant over the land areas. One clear implication of temperature increase is the lengthening of the growing season which can be seen as advanced bud burst of Rowan (Section 6.3.3.1.1).

Effects of urbanisation on the land-based temperature record are negligible as far as hemispheric- and continental-scale averages are concerned, because the very real but local effects are avoided or accounted for in the data sets. Trenberth et al. (2007) estimate the error due to urbanisation to be less than $0.01°$ C/decade (best estimate $0.006°$ C/decade) over land areas. Similarly, there may exist some random like error sources, like changes in observation sites, that tend to cancel out when averaging is done.

Area averaged temperature variations and changes are known with good accuracy during the 20th century. However, precipitation variations are known with spatially varying accuracy. There are two basic difficulties that cause inaccuracies to precipitation estimates. Firstly, measurements of precipitation suffer from undercatch due to wind and other error sources. Local conditions have a large effect on the measured amount, so any changes, e.g. in wind exposure or improvements in the instrument, may break the homogeneity of measurement series. Secondly, global datasets have too few stations to catch the large spatial variability of precipitation, thus, in some regions of the world the sample is too small to cover the spatially complex phenomena. For these reasons precipitation trends are known less accurately than temperature trends.

There are several global land precipitation data sets but they do not show consistent long-term trends over global land areas (Trenberth et al., 2007). In the NCDC dataset used in this book, boreal zone annual, as well as MJJA, precipitation shows some increase during the 20th century (Peterson and Vose, 1997).

The small overall increase in MJJA precipitation amounts since the 1970s consists of opposite trends within the boreal zone (Fig. 10.1.3). Precipitation has increased in Alaska, Northern Europe and parts of Northern Asia while decreases have been observed in western and eastern parts of North America as well as in Eastern Eurasia. From spatially detailed analysis it is known that the pattern of changes can be more complex than shown in Fig. 10.1.3. For example the increase in Northern Europe breaks into regions of increase and decrease in a more detailed analysis (BACC Author Team, 2008). Due to large natural variability, regional trends may lack coherence and trend analysis is sensitive to selection of the time period.

486 J. Räisänen, H. Tuomenvirta

A) 1976-2005 Increase   B) 1976-2005 Decrease

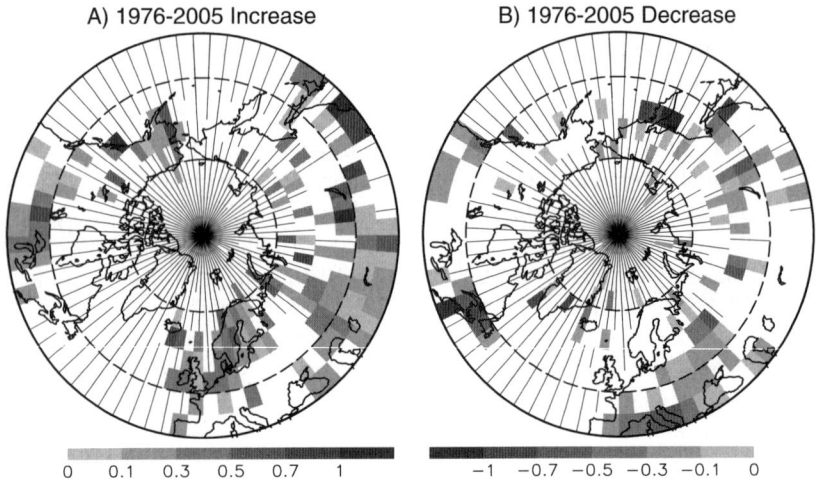

| 0 | 0.1 | 0.3 | 0.5 | 0.7 | 1 | | −1 | −0.7 | −0.5 | −0.3 | −0.1 | 0 |

Fig. 10.1.3 Linear trend of MJJA precipitation amount for 1976–2005 (mm day^{-1}/decade). Areas with increasing precipitation are shown in (A) and areas with decreasing precipitation in (B). Areas with missing data are indicated with north-south-oriented lines (The data is from NCDC)

The moisture holding capacity of the atmosphere increases at a rate of about 7% per degree Celsius. Temperature increase alone may therefore alter the hydrological cycle, especially the characteristics of precipitation (amount, frequency, intensity, duration, type). A warmer climate is expected to increase the risk for both drought and floods. Furthermore, changes in precipitation characteristics are complicated by aerosols that affect cloud formation processes. The changes in water that is available for the trees, i.e. soil moisture, are determined by changes in precipitation, evapotranspiration and runoff. There are studies utilising observations and model calculations (e.g., Dai et al., 2004; Huntington, 2006; Groisman et al., 2007) but a comprehensive view of changes in soil moisture conditions is lacking in the boreal zone.

Intensification of westerlies since the 1960s has increased air flow from oceans to continents and its effect on precipitation and temperature anomalies can be seen during the winter half-year (Trenberth et al., 2007). Although the most notable climatic changes in the boreal zone have occurred during the winter and spring seasons, some of these changes also affect conditions at the beginning of the growing season that has shifted earlier. Snow covered area has declined in the Northern Hemisphere and the reduction has been largest during the months of March and April (Lemke et al., 2007). Earlier snowmelt occurs near the edges of the snow covered area, thus lengthening the period of evaporation from soil and increasing the risk for drought. In April the Northern Hemisphere snow cover extent is strongly correlated with the 40–60° N April temperature reflecting the feedback between snow and temperature.

As noted in the previous subsection, the concentration of greenhouse gases and aerosols has increased markedly during the industrial period. The globally averaged net radiative forcing since 1750, due to both human-induced warming and cooling effects, is estimated to be +1.6W m^{-2}, with uncertainty range from +0.6

to $+2.4\,\mathrm{W\,m^{-2}}$ (Forster et al., 2007). Hegerl et al. (2007) conclude that most of the observed increase in global mean temperature during the last 50 years is very likely due to the increase of anthropogenic greenhouse gas concentrations. However, in the boreal zone natural climatic variability is large and modes of variability such as the Artic Oscillation (or its "European" manifestation, the North Atlantic Oscillation) provide large regional differences in the observed climatic changes.

10.1.3 Global Climate Models

The primary tool used for understanding past climate changes and making projections of future climate change are three-dimensional Global Climate Models (GCMs), also known as General Circulation Models. These models attempt to explicitly simulate the main physical phenomena that affect atmospheric weather and ocean circulation, land surface conditions and the state of sea ice and snow cover. The models are built on well-known hydrodynamic and thermodynamic principles that describe the conservation of mass, water, energy and momentum within the climate system, rather than, for example empirical correlations between temperature and greenhouse gas concentrations. The basic equations used in atmospheric models, that is, two equations for the horizontal wind components, the thermodynamic equation, the hydrostatic balance equation, the continuity equation, and conservation equations for water vapour and other trace substances, were given in Section 7.2. A similar set of equations is also used in ocean models, with the main difference that the equation for water vapour is replaced by an equation for salinity. This solid physical basis gives a strong reason to think that these models are useful tools for exploring the behaviour of the climate system and its response to changes in external forcing such as increases in greenhouse gas concentrations.

On the other hand, limitations in computing power make it impossible to represent the full complexity of the real world in any model. This, together with the uncertainty associated with future greenhouse gas and aerosol emissions and natural climate variability, makes it impossible to give exact predictions of future climate.

The atmospheric components in current GCMs typically have a grid spacing of 200–300 km in the horizontal direction. In the vertical direction there are typically about 30 levels between the surface and the model top at 30–50 km height, the spacing of levels increasing from a few hundred meters in the boundary layer to several kilometres in the stratosphere. Resolution of the oceanic model components is similar or slightly better. Phenomena acting on scales smaller than the grid spacing cannot be resolved explicitly. The impact of these phenomena needs to be parameterized, that is, estimated indirectly from the grid scale conditions simulated by the model.

The most important parameterized phenomena in the atmospheric components of climate models include the transfer of solar and thermal radiation, small-scale mixing in the atmospheric boundary layer, vertical convection and, in particular, the formation of clouds and precipitation. The other model components, representing the ocean, sea ice and land surface have their own parameterized phenomena, and parameterizations are also needed when describing exchange of heat, water and momentum between the atmosphere and other components. Some of the feedbacks that affect the magnitude and in some cases directions of climate changes are strongly sensitive to the parameterization of sub-grid scale phenomena, changes in cloudiness being the most notorious example.

In addition to the parameterization problem, uncertainty in the results of climate models also arises from the neglect of some potentially important phenomena. For example, climate-induced changes in vegetation have this far been excluded in most model simulations, including those used in Section 10.1.4 below.

The atmospheric components of GCMs are, except for a coarser horizontal and vertical resolution, very similar to the models used in operational weather forecasting. The raw model output thus consists of daily or sub-daily time series of weather, for each grid point at the surface and higher in the atmosphere. Obviously, however, the daily evolution of weather is not predictable for several decades in the future. Useful information in climate model output therefore resides in the statistical properties (long-term means, measures of variability, etc.) rather than in the daily details of simulation. For a sufficiently strong external forcing such as a large increase in greenhouse gas concentrations, the changes in these statistical properties become large enough to be discernible from the internally generated chaotic variability in the simulations.

GCMs have been developed and are used at a large number of research institutions. An intercomparison between the climate changes simulated by different models provides one way of estimating the uncertainty in future climate changes in the real world. However, this measure should be used with some care. While a disagreement between model simulations demonstrates that at least some of the models give false projections, an agreement between them does not necessarily prove that the models are right.

10.1.4 Projected Climate Changes for the Rest of the 21st Century

Climate modellers often prefer to talk about projections rather than predictions or forecasts of future climate change. This is because the actual evolution of climate will be partly dependent on future human activities, in particular the magnitude of greenhouse gas and aerosol emissions. To characterize the uncertainty associated with emissions, a number of different emission scenarios have been developed. The most widely used set of these are the IPCC (Intergovernmental Panel on Climate Change) SRES (Special Report on Emissions Scenarios) scenarios (Nakienovi and Swart, 2000).

The SRES scenarios are based on alternative but plausible and internally consistent sets of assumptions about the demographic, socioeconomic and technological changes that together determine the evolution of emissions in the future. For a more detailed discussion of these underlying assumptions, the reader is referred to

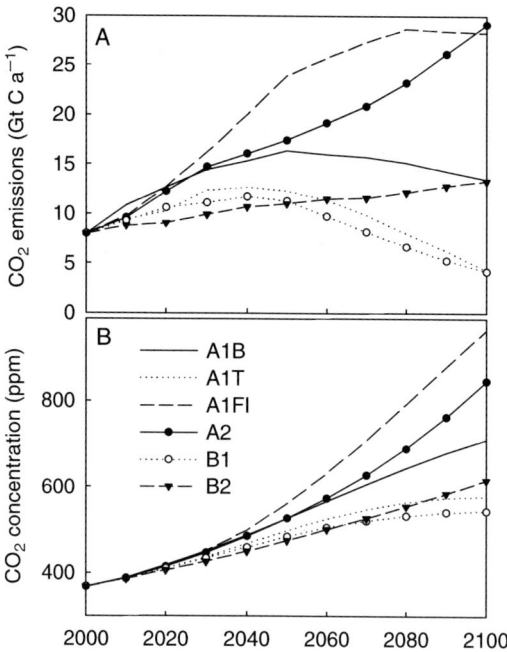

Fig. 10.1.4 (A) CO_2 emissions and (B) the resulting best-estimate CO_2 concentrations for the six SRES scenarios

Nakienovi and Swart (2000). The scenarios remain reasonably similar in the early 21st century; for example, all of them include an initial increase in CO_2 emissions in the next few decades (Fig. 10.1.4). Towards the end of the century, the scenarios diverge, with some of them indicating a continuing rapid increase of greenhouse gas emissions driven by the increasing energy demand of the world, and others pointing towards a gradual decline in emissions allowed by reduced population growth, alternative energy sources and a shift towards a less energy-intensive economy. Nevertheless, the concentrations of the most long-lived greenhouse gases, particularly CO_2 and N_2O, keep rising even in those scenarios in which emissions are reduced. The best-estimate CO_2 concentration in the year 2100, as derived by an off-line carbon cycle model, varies from ca. 540–960 ppmv between the lowest (B1) and highest (A1FI) SRES scenarios (Houghton et al., 2001), to be compared with the present-day concentration of 380 ppmv.

Anthropogenic increases in atmospheric aerosol loading have suppressed greenhouse-induced warming during the 20th century but will not necessarily continue to do so in the future. The main contributor to anthropogenic aerosol-induced cooling, SO_2 emissions, are still projected to increase in a global mean sense during the first decades of the 21st century in many SRES scenarios, which would suppress warming during this period. By the end of the century, however, the world-wide introduction of cleaner technologies is projected to reduce global SO_2 emissions well below present-day levels. Due to the very short lifetime of tropospheric aerosols, this would result in an immediate decrease in sulphate aerosol concentrations. Thus, greenhouse-gas induced warming is projected to become increasingly dominant over aerosol-induced cooling.

In its Fourth Assessment Report (Meehl et al., 2007), the IPCC estimates that the global mean warming that would occur during this century under the lowest SRES scenario (B1) would be within the range $1.1-2.9°$ C, with a best estimate of $1.8°$ C. The corresponding range for the highest scenario (A1FI) is $2.4-6.4°$ C, with a best estimate of $4.0°$ C. In both cases, the quoted uncertainty ranges take into account both the variation in warming under the same scenario of greenhouse gas concentrations between different GCMs and the uncertainty in modelling the feedback from climate change to the atmospheric CO_2 concentration. The within-scenario uncertainty associated with the modelling of feedbacks in the climate system is thus comparable with the differences between the lowest and highest emission scenarios. However, this should not obscure the fact that the magnitude of the warming will depend strongly on the magnitude of greenhouse gas emissions. Thus, as mankind has little control of feedbacks in the natural climate system, our primary possibility to alleviate future climate changes is to reduce the emissions of CO_2 and other greenhouse gases.

Below, the typical features and uncertainties of GCM-simulated climate change are illustrated for the SRES A1B scenario, which is, in terms of greenhouse gas emissions, in the midrange of the SRES scenarios (Fig. 10.1.5). For example, CO_2 emissions are projected to approximately double by 2050, declining slightly thereafter, and CO_2 concentration is projected to rise approximately to 710 ppmv by the year 2100. The analysis is based on the results of 21 recent GCMs contributing to the World Climate Research Programme 3rd Coupled Model Intercomparison Project (WCRP CMIP3; http://www-pcmdi.llnl.gov/ipcc/about_ipcc.php; see also

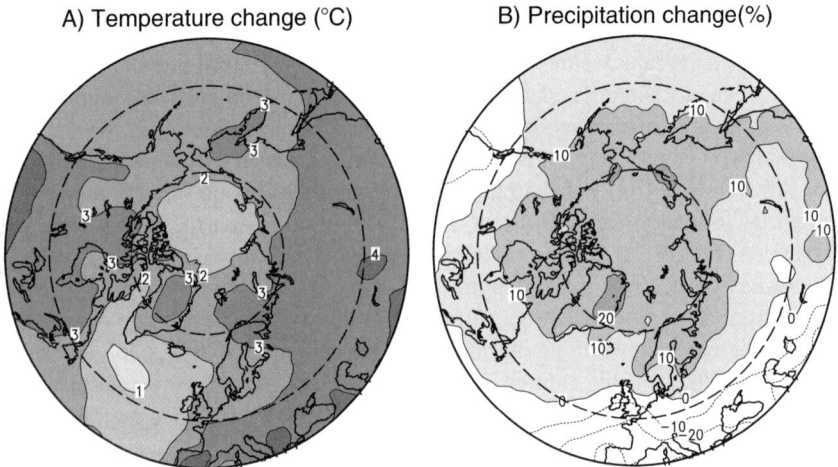

A) Temperature change (°C)   B) Precipitation change(%)

Fig. 10.1.5 (A) Simulated changes in MJJA mean temperature (° C) and (B) precipitation (%) from the recent past (1971–2000) to the late 21st century (2070–2099), as averaged over 21 climate models. All simulations are based on the SRES A1B emissions scenario

Räisänen, 2007) and climate changes are calculated as differences between the mean values for the years 2070–2099 and 1971–2000. The focus is on the MJJA season that is most important for growth of boreal forests.

Maps of temperature and precipitation change, as averaged over the 21 models, are shown in Fig. 10.1.5. The average simulated warming in the boreal forest zone is slightly over 3° C, being comparable to but slightly higher than the 21-model annual mean global warming (2.6° C) in the same simulations. The warming is geographically relatively uniform but has, in this season, a tendency to increase from north to south. Precipitation is simulated to increase slightly in almost all of the boreal forest zone, but with larger increases in the north than in the south. The opposite patterns of temperature and precipitation change may be physically related via the surface energy balance. The risk that warming would lead to a drying of the soil is larger in the southern areas where the increase in precipitation is smaller and, if the soil becomes sufficiently dry, then the ability of evapotranspiration to cool the surface will be reduced. Nevertheless, as illustrated by Figs. 10.1.6 and 10.1.7 below, the

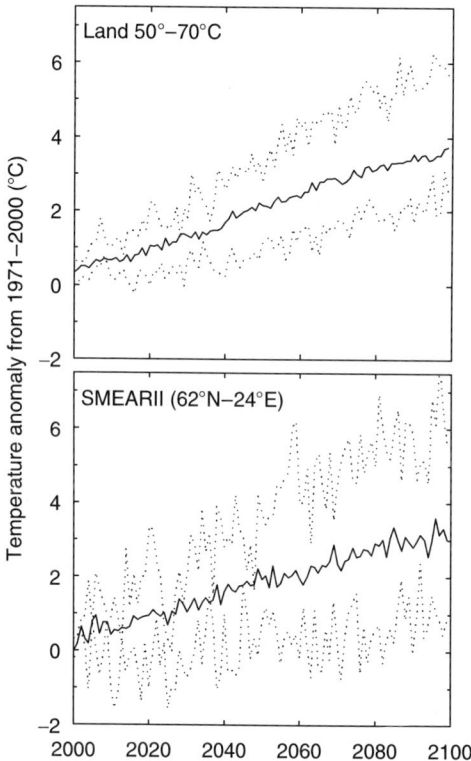

Fig. 10.1.6 Simulated time series of MJJA mean temperature, as averaged over all land areas at 50°–70° N (upper panel), and in an individual location (lower panel). Temperatures are given as differences from the mean value in the years 1971–2000. In each case, the central line represents the mean of 21 climate models and the other two lines the individual models with the smallest and the largest warming

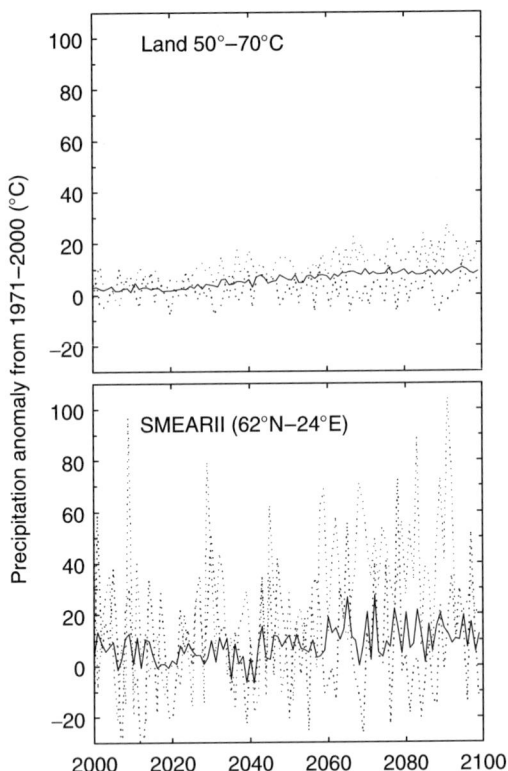

Fig. 10.1.7 As Fig. 10.1.6, but for changes in precipitation

geographical variation of these simulated climate changes is relatively small when compared with the variation between individual models.

Climate is expected to warm gradually throughout the 21st century, although the warming is punctuated by substantial interannual variability (Fig. 10.1.6). The details of this variability are not predictable and they are therefore largely smoothed out when averaging the temperatures over a large number of individual models. However, the differences in the rate of warming between individual models are also substantial, and more so for individual grid boxes than for averages taken over the whole boreal zone (Fig. 10.1.6). Still, all models simulate some warming in practically the whole boreal forest zone.

For precipitation, the signal-to-noise ratio between the simulated anthropogenic climate change and natural variability is much lower than for temperature (Fig. 10.1.7), particularly when we consider the evolution of climate on the scale of individual grid boxes. Anthropogenic changes in precipitation are thus likely to become more slowly discernible from natural variability than changes in temperature. The variation between the individual models is also larger for precipitation than for temperature changes. Nevertheless, in most parts of the boreal forest zone, a large majority of the models agree on a long-term increase in precipitation.

Acknowledgments We acknowledge the modelling groups for making their model output available for analysis, the Program for Climate Model Diagnosis and Intercomparison (PCMDI) for collecting and archiving this data, and the WCRP's Working Group on Coupled Modelling (WGCM) for organizing the model data analysis activity. The WCRP CMIP3 multi-model dataset is supported by the Office of Science, U.S. Department of Energy.

10.2 Climate Change and Boreal Forests

10.2.1 The Approach

Pertti Hari

University of Helsinki, Department of Forest Ecology, Finland

Evidence is accumulating, and being better understood, that we are experiencing climate change caused by emissions from the use of fossil fuels and other industrial activities. In addition, the large-scale use of forested land for agriculture has caused significant release of carbon from trees and forest soils into the atmosphere (Section 10.1). The concentrations of so-called greenhouse gases in the atmosphere are increasing and they absorb more than previously the globe's radiation of heat into space. Thus the energy balance of the globe has been disturbed and more energy is accumulating on the surface of the globe, resulting in a temperature increase (Section 10.1).

The overall pattern of climate change is clear and the physical basis is non-problematic. However, there are several aspects of these changes that cause uncertainties in the conclusions. The measures required for adaptation to climate change are expensive and the political decisions to bring them about have been continuously postponed, although some first steps have been accepted. The main argument against clear reduction of emissions has been that the scientific evidence is not sufficient and the uncertainties in our understanding of the nature of climate change are too large to warrant action. Thus a better understanding of the nature of climate change and its effects on forests is urgently needed.

The statistical approach has made good contributions to science during recent decades. Often it is implicitly assumed that we should primarily use statistical methods in the analysis of the expected effects of climate change on forests. When the terms "representative sample" or "unbiased estimates" are used, they refer to statistical generalisations from a sample to a population. We face two problems when applying a statistical approach to the effects of climate change on boreal forests. The practical problem is that it is impossible to get real representative samples from boreal forests because of their huge area and weak transportation systems, caused to a great extent by non-existence of roads. A more important problem is principal to our quandary; we can use statistical generalisation only from sample to population.

However, we want to say something about boreal forests in the future, but we cannot sample future forests. Thus statistical generalisation cannot be applied for prediction purposes.

The emissions from burning of fossil fuels and from expanding agriculture change the material, energy and momentum fluxes in the atmosphere and between the atmosphere and forest ecosystems. Dynamic modelling can be used to analyse material and energy flows, as shown in the Eq. 2.2.3. Thus dynamic modelling based on material, energy and momentum fluxes seems to be the adequate methodological approach to analyse the interactions between boreal forests and climate change.

We use theory-driven analysis, which covers the relevant causal mechanisms of climate change on boreal forests. Then we proceed step-wise starting from the analysis of changes in environmental factors, especially CO_2 concentration and temperature. The causal relationships generated by the structures (Chapter 5), processes (Chapter 6) and transport phenomena (Chapter 4) determine the responses of the boreal forests to the changing environmental factors. The transition to ecosystem and boreal forest level is done with aggregating models. The obtained results are neither representative nor unbiased. As stated above, these requirements for results, dealing with future forests, are unrealistic and show a weak knowledge of the nature of statistical methods.

10.2.2 Process Responses to Climate Change

Pertti Hari, Jaana Bäck, and Eero Nikinmaa

University of Helsinki, Department of Forest Ecology, Finland

Atmospheric CO_2 concentration and temperature are changing environmental factors that directly affect biosphere processes. Carbon dioxide is an important raw material for sugar formation (Section 6.3.2.3) and temperature influences the rate of biochemical reactions carried by functional substances, especially enzymatic decomposition in the soil (Section 6.4). Thus the climate change has caused and will cause responses in metabolic processes in boreal forests.

The CO_2 flux to leaf limits photosynthesis, especially at high photon flux densities. The rate of light reactions, and resulting formation of ATP and NADPH for carbon reactions, is rather independent of temperature in normal summer conditions. The availability of light energy and acclimation of light reactions (Section 6.3.3.2.1) to the consumption of the carbon reactions determine the flow of energy to the carbon reactions. In carbon reactions the end products of light reactions are used to bind CO_2 into carbohydrate chains (Section 6.3.2.3). At high photon flux densities the capacity of light and carbon reactions may exceed the influx of CO_2 that depend on its atmospheric concentration and diffusivity of transport pathway into the chloroplasts.

In Section 6.3.2.3 atmospheric CO_2 concentration was introduced into the model that study how photosynthetic rate depend on environmental factors since it determines together with leaf and leaf boundary layer transport processes the availability of CO_2 to the carbon reactions. Thus the model (Eq. 6.3.15) can be applied to introduce the effect of increasing CO_2 on photosynthesis. The ambient carbon dioxide concentration is included specifically also in the model of degree of stomatal opening. Thus the models for specific photosynthetic rate provide a natural tool to approximate the effect of atmospheric CO_2 concentration on sugar formation. The annual production is obtained by integration over active period in summer.

Acclimations of the photosynthetic machinery in leaves and the leaf structure cause uncertainties in the approximation of photosynthetic response to increasing CO_2. Experiments have shown that the stomatal fine structure of leaves, as well as the enzymes and pigments involved in photosynthesis acclimate to changing CO_2 levels (Section 6.3.3.2.3). The experimental results are contradictory, which is rather natural since the acclimations to high CO_2 concentrations have not been tested during the last 600,000 years, when the atmospheric concentration has been below 300 ppm. The atmospheric concentration has varied in the time scale of 10,000 years, but it has been rather stable in the time scale of millennia. Thus the concentration has been stable during lifetime of individuals. Proper testing of acclimations in evolution requires variation in the driving environmental factor during the lifetime of individuals of the species in consideration. Evidently, the acclimations to elevated CO_2 concentrations can be even harmful for photosynthetic production. Another problem with the estimates is that plants may acclimate at the individual level leading to variable allocation of both carbon and nitrogen between plant parts that are difficult to account for in experiments. The importance of acclimations cause large uncertainties in the values of annual photosynthetic production at high CO_2 concentrations.

The annual cycle of metabolism is a characteristic feature of vegetation in the boreal zone. The metabolism is slowly activated in the spring; for example, the transition of photosynthesis from the inactive winter state to full activity in summer takes at least one month at our field station. The concentrations of functional substances are low or they are inactivated in winter. The regulation system of the annual cycle activates the functional substances (Section 6.3.3.1.3), which results in faster metabolism. The regulation of the annual cycle is based largely on temperature (Section 6.3.3.1.3). In northern boreal conditions earlier springs, in particular, increase the amounts of materials produced in processes as then also plenty of light energy is available in contrast to autumn.

Microbes emit extra cellular enzymes that decompose the macromolecules in the soil. Thereafter, microbial metabolism utilises the resulting small molecules (Section 6.4.2). Thus decomposition of organic matter in the soil is an enzymatic process. Since all enzymatic reactions depend strongly on temperature, also decomposition in the soil is temperature dependent. The increase of microbial decomposition along temperature is also evident in the measured microbial respiration in the soil (Section 7.6.1). The release of ammonium ions in microbial decomposition will increase nitrogen availability in soil and accelerate tree growth. In addition, the emissions of N_2O will increase and also methane oxidation may decrease.

10.2.3 Ecosystem Responses to Climate Change

Pertti Hari and Eero Nikinmaa

University of Helsinki, Department of Forest Ecology, Finland

There are several changes in the specific physiological processes caused by the climate change. These responses have to be converted to stand level to get the understanding about the expected changes in boreal forests. For example, maximum photosynthetic rate can be assumed to accelerate due to the increased availability of CO_2. Is the growth increased in the same way as photosynthesis or are there internal feedbacks in a tree or in forests that modify the response? The ecosystem model MicroForest (Chapter 9) combines several processes and regularities in tree structure to produce ecosystem development. Simulations with MicroForest give evident insight into the responses of boreal forest ecosystems.

The increasing atmospheric CO_2 concentration increases maximum photosynthetic rate and increases the annual amount of sugars available for growth and metabolism. The annual amount of photosynthesis in unshaded conditions is a key parameter in the model. This fact opens a possibility to introduce the effect of increasing CO_2 concentration on photosynthesis into model simulations.

We analysed the effect of increasing CO_2 concentration on stand development with SMEAR II stand. The basic run in the Chapter 9 assumes implicitly stable CO_2 concentration. For comparison, we introduced the effect of increasing CO_2 by assuming that the parameter values are estimated at 380 ppm i.e., this concentration is the reference CO_2. The simulated stand photosynthesis had, as expected, a clear increasing trend (Fig. 10.2.1A). In contrast, the simulated stand volume reacted very weakly to the accelerated photosynthesis (Fig. 10.2.1C): the stand volume was even smaller at the end of the simulation than without increasing CO_2 concentration. This result is caused by the change in allocation to favour roots. As only CO_2 concentration was modified, the release of nitrogen from proteins in soil by microbes did not change and annual nitrogen uptake remained stable. The trees were unable to produce additional functional substances and this is why the response of the stand to elevated CO_2 is very small. The model assumes no acclimation in foliar nitrogen content. If trees were to acclimate their leaf nitrogen content to the availability of nitrogen, the growth of trees would have reacted too along increasing atmospheric CO_2 concentration.

Expected temperature increase has two effects; it prolongs the metabolically active season and accelerates decomposition of soil organic matter. Extension of the active season increases annual photosynthetic production and its effects are rather similar to those obtained for increasing CO_2 concentration, i.e., the effect is rather small if it is not associated with increased nitrogen uptake from soil.

The decomposition of soil organic matter is slow in boreal forests since the pools of proteins – lignin, cellulose and lipids – are over ten fold when compared with the annual litter input. The model MicroForest has a special sub model for decomposition of soil organic matter. The value of the parameter b_{1n} in Eq. 9.1.28 determines the decomposition rate in the simulations. The effect of increasing temperatures on

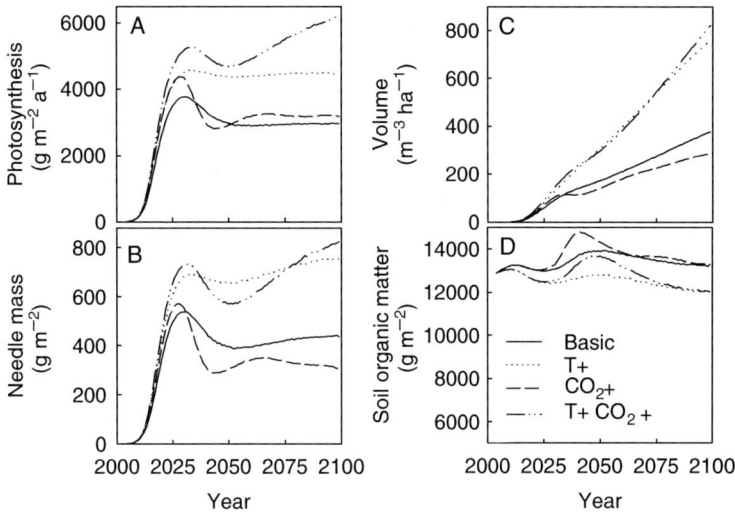

Fig. 10.2.1 Simulated (A) photosynthesis (expressed in units of CO_2), (B) needle mass, (C) volume and (D) mass of soil organic matter (dry weight) by MicroForest for the years 2000–2100 with different scenarios: no change in temperature or atmospheric CO_2 concentration, increase in temperature ($0.4°C/$decade), increase in CO_2 concentration ($1.5\,ppm\,a^{-1}$) and increase both in temperature and CO_2 concentration

stand growth was introduced by assuming that the dependence of the value of the parameter b_{1n} for temperature increase ΔT is

$$b_{1n}(\Delta T) = q_{10}{}^{\Delta T} b_{1n}^{0} \qquad (10.2.1)$$

where q_{10} is parameter and b_{1n}^{0} is the parameter value used in the Chapter 9. The value of the parameter q_{10} was assumed to be 3 in accordance with the Section 7.6.1.

Temperature records show an increasing trend in southern Finland, although the interannual variation nearly hides the signal. We assumed that temperature will increase $0.4°C/$decade which is the observed warming in the boreal zone (Section 10.1.2) When this temperature increase is combined with the dependence of the value of the parameter on temperature, we get that decomposition of soil organic matter will be 20% faster in the year 2020 than in the year 1980, and 70% faster in the year 2100.

The simulations assuming accelerated decomposition of soil organic matter due to temperature increase indicated a strong response. The needle mass increased 70% at the end of the simulation in the year 2100 (Fig. 10.2.1B) and the annual photosynthetic production of the stand reflected the increase in needle mass (Fig. 10.2.1A). Annual wood increment responded also strongly (Fig. 10.2.1C). The increased availability of nitrogen in the soil due to accelerated decomposition of proteins in soil generates these large responses in simulations (Section 9.1.7).

The increasing decomposition of soil organic matter generated large responses in the simulations. It is surprising that changes in the pools of soil organic matter are small, having only a very small declining trend (Fig. 10.2.1D). The accelerated decomposition of proteins in soil enables a clear increase in nitrogen uptake and synthesis of functional substances, especially for leaves and fine roots. The accelerated decomposition is compensated to a great extent by the increased litter input on the soil. Thus the weak response of soil organic matter can be explained by the compensating effect of increased litter fall.

When the accelerated decomposition due to temperature increase and increasing CO_2 concentration were combined in the simulation, still larger responses was obtained. The needle mass increased 90%, photosynthesis 110%, and stand volume 120%. The large organic molecules in soil decreased with 10%. This small decrease can be explained with increased litter input that compensates the faster decomposition. The strong response can be explained with high availability of nitrogen for the synthesis of functional substances and with accelerated photosynthesis due to increased availability of CO_2.

The annual increase in release of ammonium from proteins in the soil is used to synthesize proteins in leaves, water transport system and fine roots. The lifetime of needles and fine roots is short thus nitrogen returns quickly into soil as proteins. In contrast most of the woody structures live the whole lifetime of the trees and proteins accumulate into woody structures. Thus proteins are moved from soil into the pool of woody structures in trees. Since the annual release of nitrogen determines the fertility of the site (Section 9.1.7) and the protein concentration in wood is over one magnitude smaller than in soil, the increase in wood mass is about one magnitude larger than the loss in soil organic matter.

Anthropogenic nitrogen deposition is large in the south-western corner of European boreal forests; it can be over ten fold when compared with natural deposition. The leaching and evaporation of nitrogen are very small at SMEAR II (Section 7.9) thus the nitrogen circulates efficiently within the ecosystem. The accumulation is, however, very slow; it can last for centuries. Magnani et al. (2007) reported clear relationship between nitrogen deposition and carbon fixation by ecosystems. Eddy-covariance measurements in stands forming chronosequences were used to obtain the net primary production. The material consists of 20 stands growing in Europe, Siberia and North America. Most of the stands are coniferous but four of them are hardwood. Scots pine is the dominating tree species on six sites (including SMEAR II stand). The analysis by Magnani et al. (2007) resulted a clear relationship between nitrogen deposition and net primary production as shown in Fig. 10.2.2.

MicroForest is based on carbon and nitrogen fluxes in a forest ecosystem and nitrogen deposition is included as input flux in the ecosystem (Fig. 9.1.2). We simulate the effect of nitrogen deposition on net primary production with MicroForest using as nitrogen deposition inputs wet nitrogen depositions used in the study by Magnani et al. (2007). We applied linearly increasing nitrogen depositions in the simulations of SMEAR II stand in such a way that at the end of the simulations at the age of 40 years the deposition was as used in the study by Magnani et al. (2007) and we determined the net primary production of the stand as the increase in wood and soil organic material mass. The simulated relationship between nitrogen deposition and net primary production is rather close to that reported by Magnani et al. (2007). However, the shape of the relationship is different, since in the original study

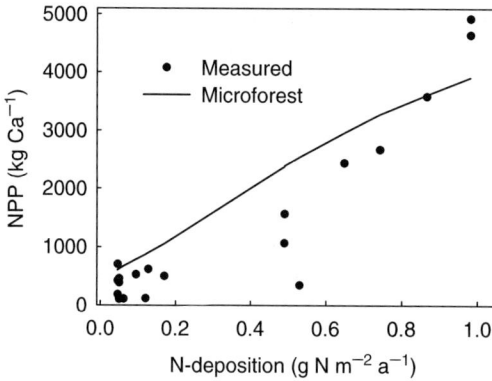

Fig. 10.2.2 Relationship between nitrogen deposition and net primary production according to measurements by Magnani et al. 2007 and according to simulations with MikroForest

the regression is nonlinear and simulations result in linear relationship. The nitrogen impact reported by Magnani et al. (2007) has been argued to be too large. The simulations would, however, suggest that slowly increasing deposition and accumulation into the ecosystem might bring about observed pattern.

10.2.4 Response of Boreal Forests to Climate Change

Pertti Hari and Eero Nikinmaa

University of Helsinki, Department of Forest Ecology, Finland

Boreal forests are versatile ecosystems; there are, however, similarities among them. The most important similarity is that the processes are the same; all forests photosynthesise, take up nutrients, grow and finally die. Thus behind the versatile species, the same processes and their dependence on environmental factors are generating regularities in the flows of material and energy. We urgently need understanding of the changes in the biogeochemical cycles of boreal forests, not only because the boreal forest soils contain large pools of carbon that may be released into atmosphere with climatic warming.

Boreal forests have most tendency of all forests to grow as even aged stands; they are born after some catastrophe, such as clear cut, fire or storm and they develop until a new catastrophe occurs, often at a rather high age of the stand. Also more complicated stand structures exist, but for the sake of simplicity they are not considered in this exercise. The natural fire cycle for the boreal biome is estimated to be approximately 100–150 years (Gauthier et al., 1996). The rotation period in forestry is also about 100 years.

Boreal forests include several carbon storages that are relevant in the connection of climate change. Let $M_B(t)$ denote the amount of a carbon component in boreal forests as a function of time t and $F_B(t)$ the flow of a compound between the boreal-forests and the atmosphere. The component can be needle mass, amount of wood, amount of proteins, lignin, cellulose, lipids and starch in soil and the flow of CO_2, VOCs or N_2O. Changes in these masses or flows reflect the response of boreal forest to climate change. Let $M_S(t, \tau_S)$ denote the corresponding masses in a stand at stand age τ_S as a function of time t and $F_S(t, \tau_S)$ correspondingly. If the boreal forests are formed by even aged stands, then we can approximate the masses and fluxes by summing the stand specific values over the whole area. Assume that

$$M_B(t) \approx \sum_{\tau_S=1}^{\infty} A(t, \tau_S) M_S(t, \tau_S) \qquad (10.2.2)$$

$$F_B(t) \approx \sum_{\tau_S=1}^{\infty} A(t, \tau_S) F_S(t, \tau_S) \qquad (10.2.3)$$

where $A(t, \tau_S)$ is the area of stands having age τ_S in the boreal forests as a function of time t.

Approximation for any of the masses $M_S(t, \tau_S)$ at moment t and age τ_S can be obtained from simulations with MicroForest. The development of the areas of stands at a given age trough the centuries is difficult to obtain. This problem can be avoided by assuming that all stands have the same maximum age, often called as rotation time, and that the areas of stand age classes are the same. Let τ_R denote the rotation time. Then we get

$$M_B(t) \approx \sum_{\tau_S=1}^{\tau_R} \frac{A_B}{\tau_R} M_S(t, \tau_S) \qquad (10.2.4)$$

$$F_B(t) \approx \sum_{\tau_S=1}^{\tau_R} \frac{A_B}{\tau_R} F_S(t, \tau_S) \qquad (10.2.5)$$

where A_B is the total area of boreal forests.

The boreal forests cover a huge area between $50°$ and $70°$ N: their total area is about $1.66 * 10^9$ ha. There are several species forming stands and the variation in fertility and the length of growing season is large. On the other hand, we have available very detailed information from one stand (SMEAR II) and dynamic model (MicroForest), based on carbon and nitrogen flows and structural regularities, tuned to the SMEAR II stand and tested with stands at the southern and northern border of boreal forests. We need urgently to understand the effects of climate change on boreal forests. Thus the question rises: Can we apply the model MicroForest to all boreal forests?

We are facing two unpleasant alternatives, either we introduce strong assumptions and apply MicroForest or we state that it is premature to say anything about the behaviour of boreal forests under climate change. We think that valuable insight can be gained when applying MicroForest on all boreal forests even though the analysis is rough. The analysis should be tested with new material and the detected shortcomings should be improved. In this way we can improve our understanding in the long run. The other alternative is a blind alley and will not result in progress in the field.

We assume that the material and energy fluxes in boreal forests will respond to climate change as described in MicroForest and we apply MicroForest to all boreal forests. In addition, we normalise the results in such a way that the biomass is $4.2*10^7$ g ha^{-1} which value has been measured for boreal forests in North America (Botkin and Simpson, 1990).

We analyse the responses of boreal forests to increasing CO_2, to nitrogen deposition and to accelerated decomposition due to temperature increase.

The increasing atmospheric CO_2 concentration has a strong effect on the simulated photosynthesis of boreal forests. The soil CO_2 efflux also responds strongly to the increasing CO_2 concentration. In contrast the volume of boreal forests reacts very little to the increased availability of carbon dioxide. Thus the main result in simulations is that the increased photosynthesis is compensated with increased root respiration and exudates. This result could be expected from the simulation in the Section 10.2.3. As mentioned earlier, the increase in productivity is most probably an overestimate as the acclimation processes of the maximum photosynthetic rate are still difficult to consider and we did not consider any reducing factors that e.g. periods of increasing water shortages could cause.

Next we studied the impact of increasing temperature. We simulated the expected temperature increase on soil decomposition in such a way that the temperature increases 0.4° C/decade after the year 1965 as in the Fig. 10.1.1. The acceleration of decomposition of large carbon molecules in soil by temperature increase (Section 6.4.1) is introduced according to Eq. 10.2.1. The temperature increase caused large responses in all characteristics, the needle mass increased, annual photosynthesis increased also but to a lesser extent than needle mass; soil CO_2 efflux showed rather weak response and the largest response was in the stem volume. The temperature rise increased decomposition of proteins in the soil enabling increased nitrogen uptake and synthesis of functional substances.

Finally, we analysed the combined effect of increasing CO_2 concentration and temperature increase. This alternative resulted in large responses since the restriction caused by the synthesis of functional substances did not anymore reduce the effect of accelerated photosynthesis on growth (Fig. 10.2.3). Photosynthesis, stem growth and volume tripled and needle mass doubled in the simulation by the year 2100 and the response was still increasing. In contrast the macromolecules in soil had slow declining trend.

The boreal forests were clear carbon sinks in the simulation assuming temperature increase 0.4° C/decade starting 1965, which is the observed mean summer time temperature increase. The increase in the mass of wood in boreal forests during the year 2005 was 0.73 Pg (dry weight) a^{-1} and decrease of organic matter in soil was 0.21 Pg (dry weight) a^{-1} (1 Pg = 10^{15} g). Thus the mass of carbon compounds increased in boreal forests 0.52 Pg (dry weight) a^{-1}. The observed annual

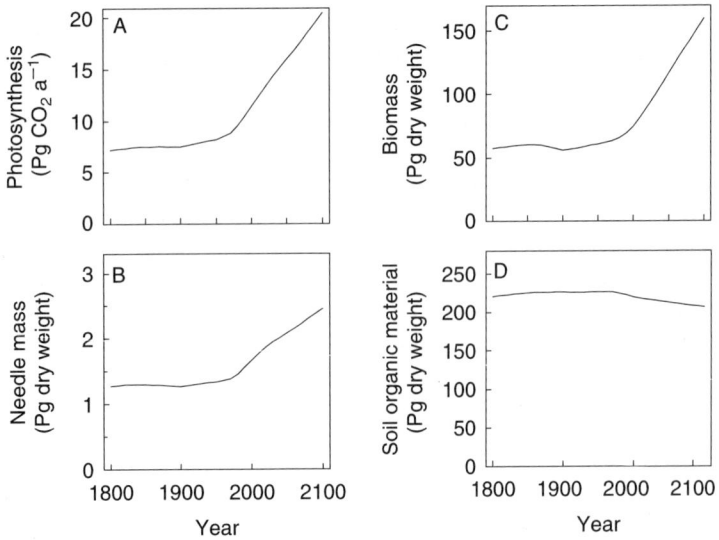

Fig. 10.2.3 Simulated (A) photosynthesis, (B) needle mass, (C) biomass and (D) organic material in soil from the year 1800 till 2100 when temperature and atmospheric CO_2 concentration increases

temperature increase is, however, larger than that in summer time, $0.5°\,C/$decade as can be seen from the Fig. 10.1.1. The simulations are rather sensitive to the temperature increase rate and the annual value results in $0.90\,Pg$ (dry weight) a^{-1} increase in the mass of vegetation and $0.33\,Pg$ (dry weight) a^{-1} loss of organic matter in soil, if the temperature increase is $0.5°\,C/$decade. The boreal forests are stronger carbon sinks in this case.

The nitrogen deposition is high in those areas of boreal forests, which are close to large industrialised centres. However, most boreal forests are remote and their nitrogen deposition is still close to natural one (Dentener, 2006). We assumed that 30% of boreal forests are clearly affected by anthropogenic nitrogen deposition and we assumed that the annual increase in nitrogen deposition is $0.015\,g\,(N)m^{-2}\,a^{-1}$ starting from the year 1950 and the deposition will stabilise in the year 2050 at the value $1.5\,g\,(N)\,m^{-2}\,a^{-1}$ (cf. Dentener, 2006). These assumptions are crude but they enable approximation of the magnitude of the effect of nitrogen deposition on boreal forests.

The response of boreal forests to increasing nitrogen deposition is rather similar to that caused by accelerated decomposition of soil organic matter caused by temperature increase. The increased availability of nitrogen changes allocation between fine roots and needles. The increased nitrogen availability also enables construction of functional substances for additional needles. These two phenomena result in increasing needle mass and wood growth to transport water for the needles. The macro molecules in soil increase because of increasing litter fall. This increase, and decrease, when temperature is increasing, is the only major difference between the behavior of the carbon pools in boreal forests in the two simulations.

We assumed that the nitrogen deposition covers only 30% of the boreal forests that decreases the effect of nitrogen deposition. The simulated annual increase in biomass in the year 2005 is about $0.1\,Pg\,(C)\,a^{-1}$ and $0.01\,Pg(C)\,a^{-1}$ in soil organic matter. These values are clearly smaller than in the case of accelerated decomposition of soil organic matter.

The simulations did not consider the influence of climate change on the water cycle that is due to increased evapotranspiration in warmer climate and alterations in the rainfall pattern. As the boreal forests are currently predominantly moist, water is not limiting productivity in current situation. With climate change the situation may change, at least regionally and therefore the estimates presented here can be considered as upper range values. Another important and linked phenomena is how the forest fires are changing. With warmer weather their frequency probably increases that will accelerate the carbon return into atmosphere.

10.3 Feedback from Boreal Forests to Climate Change

10.3.1 Carbon Sequestration

Pertti Hari and Eero Nikinmaa

University of Helsinki, Department of Forest Ecology, Finland

The role of boreal forest, either as sink or source of atmospheric carbon, has received considerable attention during the last decade. Several large projects have aimed to quantify the carbon fluxes between forests and atmosphere (e.g., LTEEF, LTEEFII, Euroflux, CarboMont, CarboAge and CarboEurope). These studies have produced valuable results but still the role of boreal forests in the atmospheric carbon balance is unclear.

Increasing atmospheric carbon dioxide accelerates photosynthesis and temperature increase prolongs the photosynthetically active period and accelerates decomposition of soil organic matter, especially of proteins. According to the simulations with MicroForest, the carbon stock in trees (wood and needles) in boreal forests responds clearly after 1990 (Fig. 10.2.3) and the increase is rather linear. The increased availability of nitrogen for synthesis of proteins dominates the response. The annual simulated increase in biomass in boreal forests due to the accelerated decomposition in the year 2000 is about $0.8\,Pg$ (dry weight) a^{-1} (Fig. 10.2.3). The carbon concentration in plant material is about 50%, thus the boreal forest vegetation bounds about $0.4\,Pg\,(C)\,a^{-1}$. The carbon accumulation to boreal forests due to nitrogen deposition is about $0.1\,Pg\,(C)\,a^{-1}$. Thus the carbon sequestration by vegetation in boreal forests is according our simulations about $0.5\,Pg\,(C)\,a^{-1}$.

The amounts of macromolecules in soil (proteins, cellulose, lignin, lipids and starch) are rather stable in the soil, only a rather slight decline can been seen (Fig. 10.2.3), $0.15\,Pg$ (C) a^{-1}. This somewhat strange result is caused by two compensating effects, (i) accelerated decomposition and (ii) increased litter input. Humus is the largest component in soil organic matter. It decomposes very slowly, in the time scale of millennia. Its decomposition is, of course also accelerated, but its formation is also increased and these changes compensate each other to great extent. The fluxes connected with humus are very small when compared with annual photosynthetic production or annual decomposition of organic matter in soil and their neglect does not introduce any major inaccuracies.

10.3.2 N_2O Emissions from Boreal Forests

Mari Pihlatie[1], Jukka Pumpanen[2], and Pertti Hari[2]

[1] University of Helsinki, Department of. Physics, Finland
[2] University of Helsinki, Department of Forest Ecology, Finland

Climate change may lead to changes in nitrogen (N) cycling and the linkages between carbon and nitrogen cycling in boreal forest ecosystems. There is very little information available on the effects of increasing temperature on nitrogen (N) dynamics in boreal forests. If we assume that the effect of accelerated decomposition of proteins in soil organic matter, and a release of ammonium (NH_4^+)-ions, is similar to that caused by anthropogenic or artificial N deposition onto forest ecosystems, we can roughly estimate the possible changes in N cycling in the boreal zone.

The expected temperature increase will enhance the decomposition of proteins in the soil and result in an increase in ammonium flux from microbes into soil water. This will activate nitrifying bacteria in the soil to utilise the newly formed ammonium in their metabolism and turn it into nitrate (Section 6.4.2). The nitrate is then taken up either by vegetation or by denitrifying bacteria, which at low oxygen concentration may convert nitrate to gaseous forms of nitrogen such as N_2O (Section 6.4.2). Diffusion transports N_2O from the soil into the atmosphere, generating a flux from the forest soil to the atmosphere.

Experiments with forest N fertilization have shown that forest ecosystems with different fertility status respond differently to the additional N input. In general, N fertilization tends to increase soil N_2O and NO emissions (e.g., Bowden et al., 1991; Brumme and Beese, 1992; Matson et al., 1992; Sitaula et al., 1995; Butterbach-Bahl et al., 1998; Rennenberg et al., 1998; Venterea et al., 2003, 2004; Ambus and Robertson, 2006). Strong responses have been found either in short-term laboratory experiments with high N additions or field experiments with N fertilization rates similar to the rates applied to agricultural soils (up to $20\,g\,(N)\,m^{-2}\,a^{-1}$). The emissions of N_2O and NO seem to largely depend on whether the ecosystem N limited or not at the time of N additions. Nitrogen cycling in a forest ecosystem, which already exhibits large N oxide emissions, responds strongly to the added N, whereas an ecosystem that has been limited by N uses up the added fertilizer N rapidly and the N oxide emissions are elevated only for a short period of time, if at all. For instance, Maljanen and colleagues (2006) measured N_2O emissions from a fertile

N fertilized and unfertilized spruce forest soil in Finland. Despite elevated mineral N concentrations in the soil, they found no increase in N_2O emissions in the field, whereas in the laboratory the N additions increased nitrification and N_2O production in the soil. The effects of N additions may also differ between forest types. Ambus and Robertson (2006) discovered that soil N fertilization increased N_2O emissions and soil mineral N content in a deciduous forest ecosystem, whereas there was no response in a coniferous forest soil.

Effects of smaller N applications or the long-term effects of exposure to elevated N deposition on nitrogen cycling are much less studied. Venterea and colleagues (2003) reported elevated NO emissions but non-affected N_2O emissions from forest soils with a chronic (12–13 years) N fertilization up to $15\,g\,(N)\,m^{-2}\,a^{-1}$. They also found that net nitrification rates and nitrate concentrations had increased in forest soil with N fertilization rates of 5 and $15\,g\,(N)\,m^{-2}\,a^{-1}$.

As the boreal forests are strongly N limited ecosystems, a small increase in N deposition may have a beneficial effect on forest growth. Recent work by Magnani and colleagues (2007) suggests that carbon sequestration by temperate and boreal forest is strongly dependent on the atmospheric N deposition. The forest ecosystems studied showed no signs of N saturation with the total N deposition rates up to $15\,g\,(N)\,m^{-2}\,a^{-1}$. This indicates that majority of the boreal and temperate forest ecosystems, outside the most industrialized regions, are able to retain most of the additional N and hence benefit from the increased N deposition (Gaige et al., 2007; Magnani et al., 2007). This may, however, not be true for the forest ecosystems in most industrial regions of Central Europe and North-America, where the atmospheric N deposition approaches to the rates normally applied onto agricultural fields ($>5\,g\,(N)\,m^{-2}\,a^{-1}$). Aber and colleagues (1998) suggested that forest ecosystems in the most industrialized regions have become N saturated and are prone to high losses of N such as nitrate leaching or N_2O and NO emissions.

In order to estimate the N_2O emissions from boreal forests in changing climate, we make a general assumption that the expected temperature increase will accelerate decomposition of proteins in the soil resulting in increase in production of NH_4^+-ions. Then we assume that the effect of accelerated decomposition of proteins is similar to that caused by anthropogenic or artificial nitrogen deposition onto forest ecosystems, which allows us to roughly estimate the changes in N_2O emissions from boreal forests. As mentioned above, the data on N_2O emission rates from coniferous forest ecosystems with different N deposition or N fertilization rates is limited. In the Fig. 10.3.1 the emissions of N_2O from forest ecosystems are plotted against N deposition. Figure 10.3.1A shows the direct N_2O emissions against atmospheric N deposition, whereas the Fig. 10.3.1B shows the relative increase in N_2O emission in relation to added N as a fertilizer.

The enhanced decomposition of proteins will produce annually about $1\,g\,(N)\,m^{-2}\,a^{-1}$ of newly available nitrogen in the year 2100 according to simulations by MicroForest. If we sum this to the current N deposition rate of $0.5\,g\,(N)\,m^{-2}\,a^{-1}$ at Smear II station, the available N in boreal forest soil would increase up to $1.5\,g\,(N)\,m^{-2}\,a^{-1}$. From the regression of observed N_2O emissions in

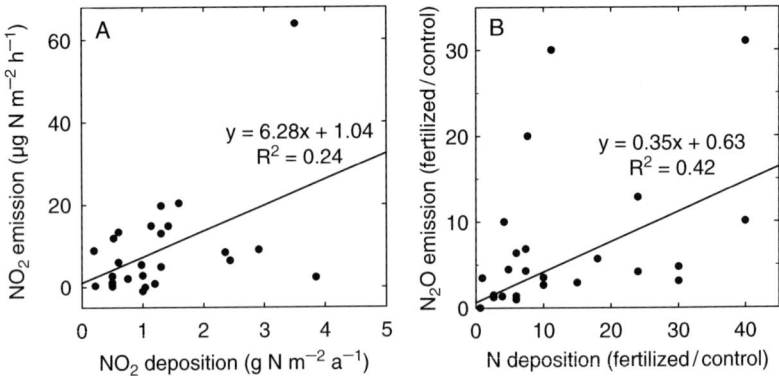

Fig. 10.3.1 N_2O emissions from coniferous and deciduous forest soils in Europe and United States as function of nitrogen (N) deposition and nitrogen fertilization (Data from Bowden et al., 1991; Brumme and Beese, 1992; Matson et al., 1992; Sitaula et al., 1995; Klemedtsson et al., 1997; Papen et al., 2001; Ambus and Robertson, 2006; Maljanen et al., 2006; Pilegaard et al., 2006; and measurements from Smear II station). (A) N_2O emissions against atmospheric N deposition, (B) N_2O emissions from N fertilization experiments, y-axis: N_2O emission from fertilized/control plots, and x-axis: N deposition in fertilized/control plots. Control plots receive N from the atmosphere only, whereas N fertilized plots received additions of N from fertilizers

relation to nitrogen deposition in forests (Fig. 10.3.1) we get that the N_2O emissions from boreal forest soils could be $10\mu g\,m^{-2}a^{-1}$ in the year 2100.

When this emission rate is scaled up to the whole boreal forests with the area of 1.7×10^9 ha, the magnitude of N_2O emissions from boreal forests in the year 2100 could be $1.5\,Tg\,(N)\,a^{-1}$ ($1.5 \times 10^{12}\,g\,(N)a^{-1}$). This is approximately three times as much as the current N_2O emissions of $0.5\,Tg\,(N)\,a^{-1}$ from boreal forests based on emission measurements at Smear II station in Hyytiälä. Compared to the global N_2O emissions of approximately $17.7\,Tg\,(N)a^{-1}$ (IPCC, 2007), the contribution of boreal forests may markedly increase in within the next 100 years.

10.3.3 CH_4 *Fluxes in Changing Climate*

Mari Pihlatie

University of Helsinki, Department of. Physics, Finland

As the temperature increase will accelerate the rate of organic matter decomposition and consequent release of ammonium into boreal forest soil, this may influence the CH_4 oxidation capacity of the soil. Several studies have reported that forest fertilization decreases the CH_4 oxidation capacity (e.g. Ojima et al., 1993; Bodelier and Laanbroek, 2004; Chan et al., 2005). However, as the climate change induced changes in the soil available ammonium are slow, the impact of additional nitrogen on CH_4 oxidation probably remains small.

Ojima and colleagues (1993) estimated that the increased nitrogen deposition onto temperate forests and grasslands have decreased the CH_4 oxidation by 30–70%. So far, however, there is no estimate on how much nitrogen deposition has already decreased CH_4 oxidation in boreal forest soils (Brumme et al., 2005). The effect of additions of different N-salts on the CH_4 oxidation of forest soils has been studied in laboratory incubations and often with additions of high levels of N. These laboratory experiments have shown that high levels of N strongly inhibit CH_4 oxidation, whereas low levels of N additions have no or only a small influence on CH_4 oxidation. Calculations of the effect of current atmospheric N deposition onto boreal upland forest soil in Hyytiälä indicate that the current low N deposition does not suppress CH_4 oxidation (Kähkönen et al., 2002). According to Whalen (2000) it is unlikely that the inhibitory levels of N via atmospheric N deposition can be achieved, and hence an increase in the atmospheric load of N will probably not diminish CH_4 oxidation capability of the boreal soils.

It is worth noticing that most experiments studying the effects of N additions in CH_4 oxidation are short-term laboratory experiments with high-levels of N additions. In addition, these studies rarely consider the influence of increased N deposition on N turnover, which according to Mosier and colleagues (1991) may be more important in determining the sensitivity of atmospheric CH_4 oxidation. Chan and colleagues (2005) investigated the consequence of long-term N additions on soil CH_4 dynamics in a temperate deciduous forest ecosystem. They found that CH_4 consumption was reduced by 35% after 8 years of N fertilization at an annual rate of $100 \, kg \, (N) \, ha^{-1} (10 \, g \, (N) \, m^{-2})$. They suggested that N cycling strongly controls the CH_4 consumption and hence changes in the N cycle due to N deposition or fertilization may alter rates of CH_4 consumption in forest soils.

The uncertainties about the effects of individual factors and their interactions on CH_4 oxidation, and coupling between N and C cycles underline the need for further studies. Large uncertainties still remain concerning the in-situ effect of additions of N on CH_4 oxidation in different ecosystems. According to Bodelier and Laanbroek (2004) the research community seems to have accepted that N fertilization decreases CH_4 oxidation, whereas a large number of publications report opposite effects of N additions. One additional factor adding uncertainties to future projections regarding CH_4 consumption is the changing climate with respect to temperature and precipitation, and further changes in the coupling between C and N cycles. The projected increase in temperature and precipitation due to global climate change may change also CH_4 oxidation in boreal forest soils.

Overall, the slow increase of mineral N in the soil probably does not markedly affect CH_4 oxidation rate in boreal forest soil. The decrease in CH_4 oxidation rate of boreal forest soils will evidently be of minor importance to the global atmospheric concentrations of CH_4.

10.3.4 Climatic Effects of Increased Leaf Area: Reduced Surface Albedo and Increased Transpiration

Jouni Räisänen and Sampo Smolander

University of Helsinki, Department of Physics, Finland

Green needles absorb (Section 6.2.1) strongly solar radiation and convert the radiation energy to heat; in contrast, snow reflects most of the radiation. The temperature increase will accelerate protein decomposition in the soil (Section 6.4.1) that is reflected in increasing needle mass in the stand (Section 10.2.3). Thus the increasing needle mass may generate positive feedback to the climate change.

As postulated in Section 10.2.4, climate warning should lead to an increase of leaf area index (LAI) in boreal forests. This could result in several climatic feedbacks (Section 10.1.1). In this subsection we focus on two of them: the impact of the LAI change on surface albedo, and its impact on transpiration. After a brief physical discussion of these feedbacks, their potential climatic importance is evaluated in simulations conducted with a state-of-the art atmospheric general circulation model.

To calculate the total albedo of a land surface, both (i) the spectral composition of the incoming below-atmosphere radiation, and (ii) the spectral albedo of the surface, need to be known. For the below-atmosphere solar radiation spectra, the 25-band model of Freidenreich and Ramaswamy (1999) was employed to compute gaseous absorption, together with the DISORT radiative transfer code (Stamnes et al., 1988, 2000) to account for atmospheric multiple scattering. Cloud-free subarctic winter atmospheric conditions (McClatchey et al., 1971) were assumed with a visible optical depth of 0.2 for continental aerosols and a solar zenith angle of 60°. The resulting spectra is shown in Fig. 10.3.2.

To calculate the total albedo of a land surface, the land surface spectral albedo in different wavelengths needs to be weighted according to the incoming below-atmosphere solar energy. To calculate these weights, the 25-band model

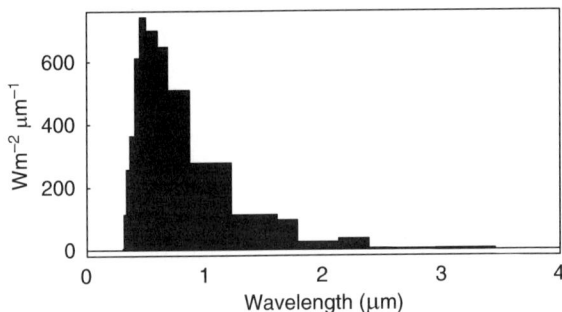

Fig. 10.3.2 Modelled 25-band spectral composition of the radiation penetrating atmosphere, as used in calculating the surface albedo

of Freidenreich and Ramaswamy (1999) was employed to compute gaseous absorption, together with the DISORT radiative transfer code (Stamnes et al., 1988, 2000) to account for atmospheric multiple scattering. Cloud-free subarctic winter atmospheric conditions (McClatchey et al., 1971) were assumed with a visible optical depth of 0.2 for continental aerosols and a solar zenith angle of 60°. The resulting weighting is shown in Fig. 10.3.2.

The spectral albedo of snow was calculated using routines available from a NASA GSFC FTP site (ftp://climate1.gsfc.nasa.gov/wiscombe/). The snowbed was modelled as an optically thick layer of spherical ice crystals (150 μm diameter was assumed). The ice refractive index routine (REFICE) is based on Warren (1984) and the particle Mie scattering routine (MIEV) on Wiscombe (1980). The snowbed albedo, taking into account multiple scattering, was solved with the DISORT plane-parallel radiative transfer code (Stamnes et al., 1988, 2000). A test of this approach was presented by Green et al. (2002). The resulting spectra are shown in Fig. 10.3.3. When calculating the snow-covered land surface albedo with these spectra, the result is 0.78, close to the value 0.8 used in the ECHAM5 model (Roeckner et al., 2003) as the maximum albedo of unvegetated snow-covered areas.

Coniferous needle spectral albedo (needle spectral reflectance + transmittance) was calculated with the LIBERTY model[2] by Dawson et al. (1998), using the default values for parameters as provided with the model. The LIBERTY model gives results only for 0.4−2.5 μm range, so the values at these endpoints were extended to cover the wider range. As most of the weight for albedo calculations is contained in shorter wavelengths than 2 μm (see Fig. 10.3.2), the resulting error is minimal. The needle spectral albedo is shown in Fig. 10.3.3. (Calculating a weighted needle spectral albedo using the weights of Fig. 10.3.2 results to 0.42.)

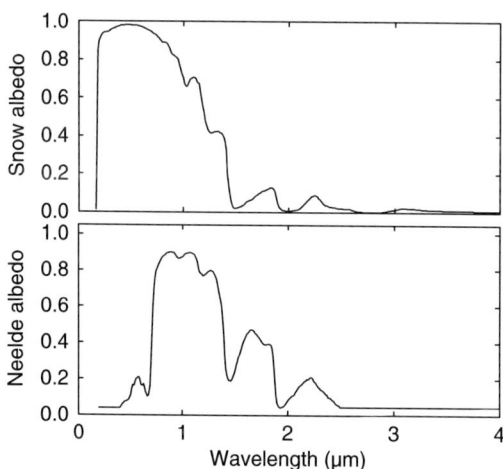

Fig. 10.3.3 Spectral albedos for snow and conifer needles

---

[2] http://www.geos.ed.ac.uk/homes/s0455489/LIBERTY.html

The spectral albedo of a coniferous forest on a snow-covered ground was calculated based on the spectral albedos of snow and coniferous needles. We used a two-stream canopy radiative transfer model by Ross (1981), modified for coniferous canopies as proposed by Smolander and Stenberg (2005). In the model, the needles are clumped together as coniferous shoots, and the resulting effects for radiative transfer (self-shadowing, multiple scattering) are accounted for (shoot self-shadowing parameter STAR = 0.125, corresponding to a shoot level recollision probability of 0.5, was used). The canopy consisted of randomly located shoots, no branch or crown structure is included in the model. The lack of crown is a somewhat unrealistic assumption in the model, but in this case we opted for simplicity.

The average spectral albedo of the system consisting of snow-covered ground and canopy was calculated in the 25 wavebands, and average weighted albedo was calculated. These calculations were repeated for canopies of different leaf area indices (LAIs). The resulting canopy albedo as a function of canopy LAI is shown in Fig. 10.3.4. As needles are non-flat objects, the coniferous canopy leaf (needle) area index was defined on a half of the total surface area basis (Chen and Black, 1992).

All other factors being the same, transpiration from vegetation should be directly proportional to the transpiring surface area. Thus, for example, a doubling of LAI should lead to a doubling of transpiration. In reality, this is a gross overestimate. This is because transpiration is limited by the availability of water and, in particular, by the availability of energy. Furthermore, an initial increase in transpiration tends to increase boundary-layer relative humidity. This reduces the vapour pressure difference between plant stomata and the near-surface air, thus acting as a negative feedback to transpiration. On the other hand, any changes in transpiration may affect other aspects of the hydrological cycle, such as cloudiness and precipitation.

To estimate the impact of the LAI-albedo and LAI-transpiration feedbacks in more quantitative terms, sensitivity experiments were conducted with the ECHAM5 atmospheric general circulation model (GCM) (Roeckner et al., 2003, 2006) coupled to a simple model of the upper ocean. ECHAM5 is a state-of-the-art atmospheric GCM developed by the Max Planck Institute for Meteorology in Hamburg. For the simulations described here, ECHAM5 was run with a relatively low horizontal resolution (spectral truncation to total wave number 31, corresponding to a grid spacing of approximately 3.75° in latitude and longitude)

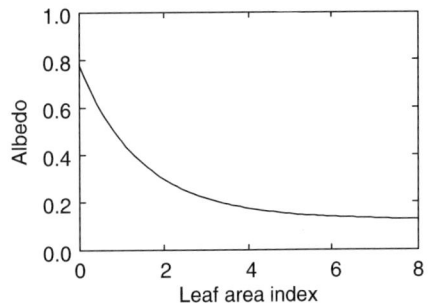

Fig. 10.3.4 Albedo of coniferous canopy on a snow-covered ground, as a function of canopy leaf area index

and with 19 levels in the vertical, with the model top at the 10 hPa pressure surface (approximately 30 km). The atmospheric model was connected to a simple model of the upper ocean, consisting of a heat balance equation for a 50-meter thick water layer and a prognostic equation for sea ice thickness. To compensate for the lack of ocean currents in the thermodynamic ocean model, flux adjustments (Sausen et al., 1988) were used to keep sea surface temperatures near observed present-day values in the control simulation. An evaluation of the simulated present-day climate with observational data revealed a level of skill well comparable with other global climate models (Räisänen, 2007). In a standard experiment with a doubling of the atmospheric CO_2 concentration, the model simulated a global mean warming of about 3° C, in good agreement with other models and the Intergovernmental Panel on Climate Change best estimate of climate sensitivity (Meehl et al., 2007).

In the ECHAM5 standard set-up, the evolution of LAI follows a prescribed seasonal cycle based on the data set compiled by Hagemann (2002). The growing and dormant season values of LAI are derived from ecosystem type classifications from the U.S. Geological Survey (2001), and the seasonal cycle between these extremes is based at high latitudes on observed monthly mean temperatures. In all months with an observed mean temperature of less than 5° C, LAI retains its dormant season (winter) value. The January and July LAI values in the ECHAM5 control run are shown in Fig. 10.3.5.

To study the sensitivity of the simulated climate to changes in LAI, two 30-year model simulations with increased LAI were made and compared with a control simulation (CTRL) with standard LAI values. In one (LAI_PLUS), a seasonally invariant increase was added to the standard LAI values at latitudes 50−70° N. The magnitude of this increase was such that the dormant season LAI was doubled; in the growing season the absolute increase was equally large but the relative increase was smaller. In the other experiment (LAI_DOUBLE), LAI was doubled in each month. Thus, LAI_PLUS is equivalent to doubling the LAI of

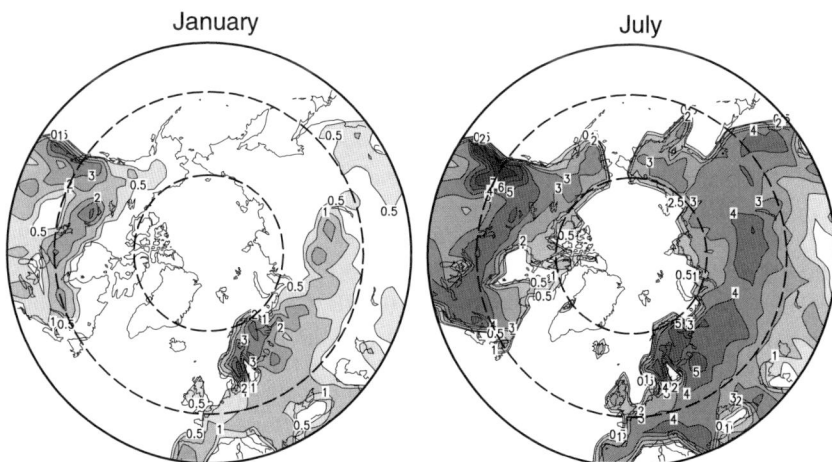

Fig. 10.3.5 Leaf Area Index (LAI) in the ECHAM5 control run in January and July

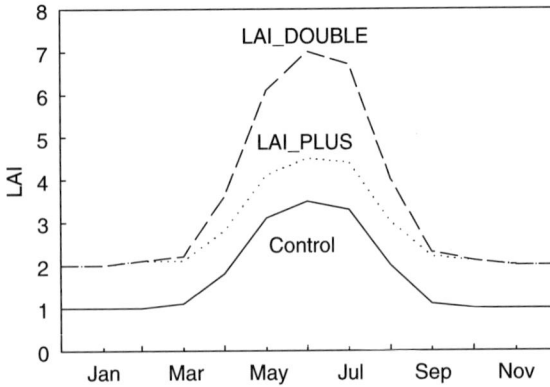

Fig. 10.3.6 Seasonal cycles of LAI, as averaged over the Eurasian and North American continents at latitudes $50°-70°$ N, in the control simulation (CTRL) and the two sensitivity experiments LAI_PLUS and LAI_DOUBLE

evergreen vegetation (in particularly, coniferous trees) whereas in LAI_PLUS the LAI of all vegetation types is doubled. All three simulations assumed the same near-present concentrations of greenhouse gases. The seasonal cycles of LAI in the three simulations, averaged over the Eurasian and North American continents at latitudes $50°-70°$ N, are shown in Fig. 10.3.6. Note that LAI_PLUS and LAI_DOUBLE are essentially identical from October to April, so that they only differ from late spring to early autumn.

All three simulations were run for 30 years. To avoid eventual spin-up problems in the beginning of the simulations, the first five years were discarded, and the mean values shown here were calculated over the last 25 years of the simulations.

The increase in LAI affects the simulated climate in ECHAM5 by two primary mechanisms. First, it acts to reduce the surface albedo when there is snow on the ground. Second, it affects the calculation of surface fluxes, including the partitioning of available energy between sensible and latent heat fluxes. Other potential consequences of increased vegetation activity, such as changes in carbon cycle and in the production of biogenic aerosols, are not presented in these simulations.

The average seasonal evolution of the simulated surface albedo in the boreal forest zone is shown in Fig. 10.3.7A. Because LAI only affects the albedo in ECHAM5 when there is snow on the ground, there is no difference between the three simulations in summer. In winter and early spring, the average surface albedo is about 3% lower in LAI_PLUS and LAI_DOUBLE than in CTRL. The geographical distribution of the LAI_PLUS – CTRL difference during late winter and early spring (February–March–April) is shown in Fig. 10.3.7B. Because surface albedo is also affected by changes in snow and ice conditions, the changes are not exclusively limited to those areas where LAI was increased.

The energy balance equation for the ground surface layer, including soil and vegetation, can be written with sufficient accuracy as

$$E_S = I_A + I_T + g_{Heat}^S + g_{Heat}^L - S_M \qquad (10.3.1)$$

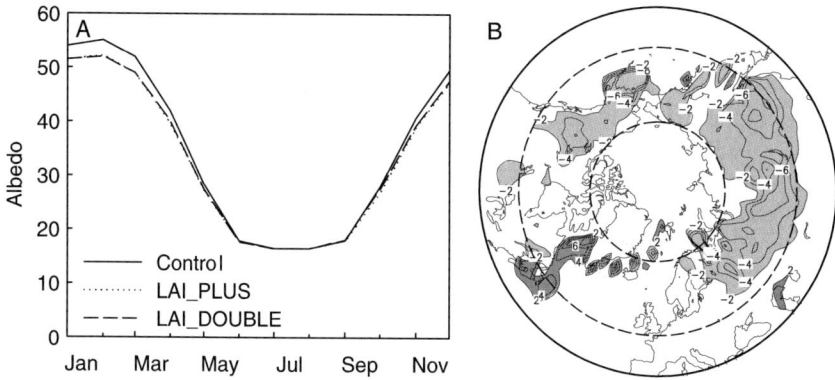

Fig. 10.3.7 (A) Seasonal cycles of surface albedo, as averaged over the Eurasian and North American continents at latitudes $50° - 70°$ N, in the three simulations. (B) The difference LAI_PLUS minus CTRL in February–March–April mean surface albedo. Contour interval 2%, zero contour omitted. Light (dark) shading indicates areas where the albedo has decreased (increased) by at least 2%

Here E_S represents the energy stored in unit time as heat in soil or via photosynthesis in vegetation. I_A is the solar radiation absorbed at the surface, i.e., the difference between the incoming and reflected solar radiation. I_T, which is generally negative, is the net thermal radiation, i.e., the difference between the thermal radiation received by the surface from the atmosphere and the thermal radiation emitted upward by the surface. The sensible heat flux g_{Heat}^S represents direct, mainly turbulent, heat transport from the surface to the atmosphere, whereas the latent heat flux g_{Heat}^L gives the heat consumed to evapotranspiration. S_M is the energy consumed by the melting of snow.

As averaged over the annual cycle, E_S is very small in land areas. S_M is also small excluding the periods of most rapid snowmelt. In ECHAM5, it has in the boreal forest zone an annual area mean of only $2\,\mathrm{W\,m^{-2}}$. Furthermore, the differences in both E_S and S_M between the three simulations are negligible. Thus, to a good approximation, the changes LAI_PLUS – CTRL and LAI_DOUBLE – CTRL fulfil the equation

$$\Delta I_A + \Delta I_T \approx \Delta g_{Heat}^S + \Delta g_{Heat}^L \qquad (10.3.2)$$

Thus, changes in the sum of net solar and net thermal radiation are balanced by changes in the sum of the sensible heat flux and the latent heat flux.

Figure 10.3.9 shows the four terms of (10.3:2) for the differences LAI_PLUS – CTRL and LAI_DOUBLE – CTRL, as averaged over the whole boreal forest zone at $50° - 70°$ N. The decrease in surface albedo shown in Fig. 10.3.7 acts to increase the amount of solar radiation absorbed at the surface in winter and spring (Fig. 10.3.9A). The increase is largest (on the average about $2\,\mathrm{W\,m^{-2}}$) in March and April when there is much more solar radiation available than earlier in winter.

In summer, absorbed solar radiation decreases, particularly in LAI_DOUBLE, in which the increase in LAI is largest. At the same time, there is an increase in I_T

(Fig. 10.3.9), although this does not fully compensate the decrease in I_A. Both of these changes appear to be caused by increased cloudiness (Fig. 10.3.9B below), which is most likely a consequence of the LAI-transpiration feedback.

Boreal forest zone mean changes in the sensible and latent heat fluxes are shown in Fig. 10.3.9C, D. For most of the year, the latent heat flux is larger in LAI_PLUS and LAI_DOUBLE than in CTRL. The sensible heat flux increases relative to CTRL in March and April, but decreases in the summer, particularly in LAI_DOUBLE. Thus, as expected, the increase in LAI requires the model to consume a larger fraction of the available radiation energy to the latent heat flux at the expense of the sensible heat flux.

The increase in latent heat flux means an increase in evapotranspiration. However, the increase is relatively modest: in both LAI_PLUS and LAI_DOUBLE, the annual area mean evapotranspiration is only 2% (6 mm per year) higher than in CTRL. The relative change is thus over an order of magnitude smaller than the change of LAI applied in the experiments. This might partly result from compensation between increasing transpiration and decreasing surface evaporation; unfortunately this cannot be verified because these two components are not separated in the ECHAM5 output. More importantly, evapotranspiration is limited by the availability of energy (Eq. 10.3.2). Because the change in the net radiation balance $(I_A + I_T)$ is quite small, a substantial increase in g_{Heat}^L would require a substantial decrease in g_{Heat}^S. Although the sensible heat flux decreases in summer, particularly in LAI_DOUBLE, this decrease is small compared with the area mean of g_{Heat}^L in CTRL (about $25\,\mathrm{W\,m^{-2}}$ in the annual mean and $60\,\mathrm{W\,m^{-2}}$ in summer). Thus, energy is only available for a rather small relative increase in g_{Heat}^L.

Another constraint to evapotranspiration is the availability of water. This factor likely limits the simulated increase in evapotranspiration in summer, particularly in the southern parts of the boreal forest zone.

The increase in evapotranspiration and the decrease in sensible heat flux (that reduces the heating of the lower atmosphere) together act to increase the near-surface relative humidity in summer, particularly in LAI_DOUBLE (Fig. 10.3.8). On the other hand, moistening of the surface layer is one of the factors that limit the increase in evapotranspiration. The increase in relative humidity also leads to an increase in cloudiness in summer (Fig. 10.3.8B). Therefore less solar radiation reaches the surface (Fig. 10.3.9A), although this is partly compensated by an increase in the atmospheric counterradiation in the thermal part of the electromagnetic spectrum (Fig. 10.3.9B). In summary, the changes in the hydrological cycle and in the surface energy balance are strongly interconnected.

The increase in LAI leads to a small overall increase (about 5 mm per year) of precipitation in the boreal forest zone (Fig. 10.3.10B). In LAI_DOUBLE, in particular, the largest increase occurs in summer.

The impacts of increased LAI on simulated surface air temperature are twofold (Figs. 10.3.8D and 10.3.10). From January to April, the temperature response is dominated by the decrease in surface albedo. Thus, temperatures are slightly higher in LAI_PLUS and LAI_DOUBLE than in the control simulation. The largest area mean warming of about $1°$C occurs in March in LAI_DOUBLE and in February in

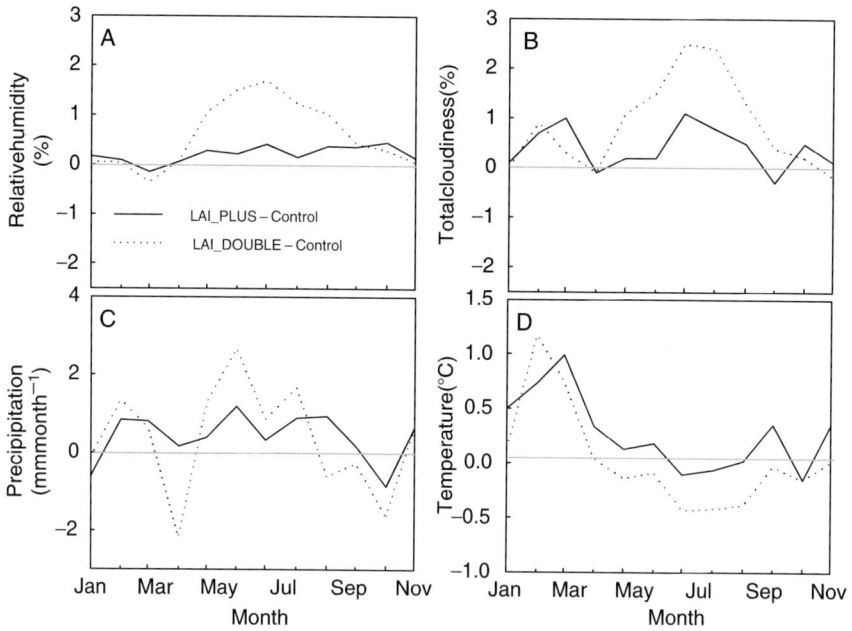

Fig. 10.3.8 Differences LAI_PLUS – CTRL and LAI_DOUBLE – CTRL in (A) near-surface relative humidity (in %), (B) total cloudiness (in % of sky), (C) precipitation (in mm per month) and (D) surface air temperature (in ° C), as averaged over the Eurasian and North American continents at latitudes $50° - 70°$ N

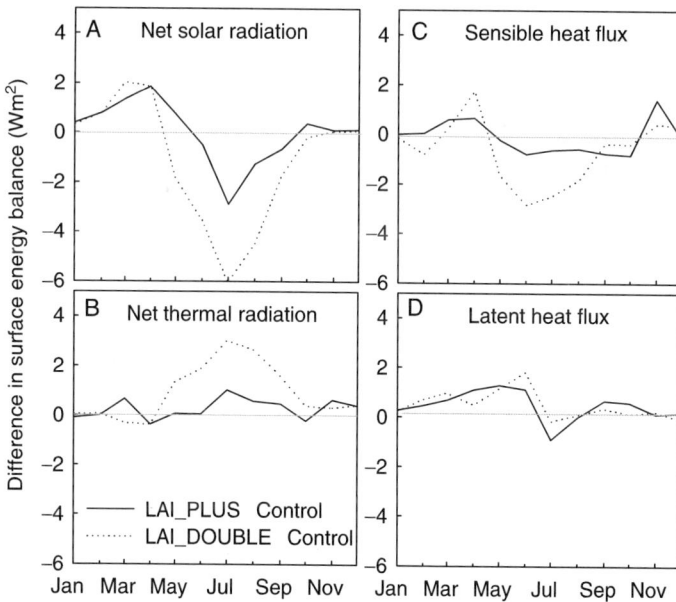

Fig. 10.3.9 Differences in surface energy balance $(W m^{-2})$ between LAI_PLUS and LAI_DOUBLE and the control simulation, as averaged over the Eurasian and North American continents at latitudes $50° - 70°$ N. (A) Net solar radiation, (B) net thermal radiation, (C) sensible heat flux, (D) latent heat flux

A) Feb–Mar–Apr, LAI_PLUS–CTRL   B) Jun–Jul–Aug, LAI_DOUBLE–CTRL

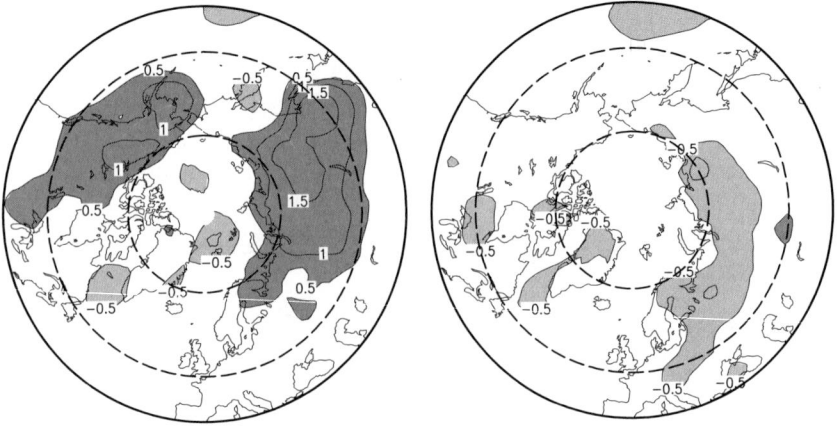

Fig. 10.3.10 Differences in surface air temperature (A) between LAI_PLUS and CTRL in February–March–April, and (B) between LAI_DOUBLE and CTRL in June–July–August. Contour interval is 0.5° C, zero contour is omitted. Dark (light) shading indicates areas where temperature has decreased (increased) by at least 0.5° C

LAI_PLUS. The warming is greatest in Siberia (Fig. 10.3.10A). By contrast, the summer mean temperature decreases slightly in most of the boreal forest zone, particularly in LAI_DOUBLE, at least in part due to the increase in cloudiness (Fig. 10.3.10B). However, the sign of the summertime temperature changes might be sensitive to the GCM-specific parameterizations of albedo and other surface properties. In an earlier model study by Bonan et al. (1992), a total removal of boreal forests also led to a cooling in summer, resulting from a much later snowmelt in spring.

Acknowledgments We thank Dr. Petri Räisänen for the incoming solar radiation computations and Dr. Tristan Quaife for the snow albedo computations.

10.3.5 BVOC Emissions from Boreal Forests

Jaana Bäck and Pertti Hari

University of Helsinki, Department of Forest Ecology, Finland

Increasing atmospheric CO_2 concentration accelerates photosynthesis; the relationship is nearly linear at present atmospheric concentration (Section 6.3.2.3). The acclimation of photosynthesis to elevated CO_2 concentration can not be predicted since it has not been tested in evolution, at least during the last million years (Section 6.3.3.2.3). In addition, the increase in needle mass (Section 10.2.3), caused by accelerated decomposition of proteins due to temperature increase (Section 6.4.1), will still enhance the photosynthesis of boreal forests.

Despite of insufficient understanding of the BVOC emission responses to climate change, we know that the synthesis of many compounds is closely linked with photosynthesis at present concentration ranges (Section 6.3.3.2.4). Enhanced photosynthetic production, in connection with increasing leaf biomass will strongly influence the BVOC emissions from boreal forest stands in the future climate. Positive effects on emissions are also expected to follow from changes in water and nutrient availability, and increased temperature, in particular for those compounds that are having permanent storage pools in the foliage and trunks, such as monoterpenes. Taking these factors into account, we can roughly assume that the response of BVOC emissions to climatic change is resembling that of photosynthesis.

The photosynthetic production was doubled in simulations with MicroForest (Fig. 10.2.3). This is why we assume that the BVOC emissions will also double in the year 2100 when compared with present emissions. However, we realize that this is an assumption where potential significant interactions, e.g. increased respiratory demands for substrates involved in BVOC biosynthesis (Rosenstiel et al., 2003), may partially cancel the effects of increased photosynthetic production on emissions. In future, when more field measurements will be available for testing and developing of models, the present emission approximation should be improved.

10.3.6 Climatic Effects of Increasing Aerosols

Markku Kulmala[1], Ilona Riipinen[1], Miikka Dal Maso[1], and Tuukka Petäjä[1,2]

[1] University of Helsinki, Department of. Physics, Finland
[2] Earth and Sun Systems Laboratory, Division of Atmospheric Chemistry, National Center for Atmospheric Research, Boulder, CO, USA

Condensable vapour concentration effects on aerosol particle formation and growth (Section 6.2.3). Anthropogenic emissions of CO_2 change the atmospheric CO_2 concentration; accelerate photosynthesis and consequently BVOC emissions from boreal forests (Section 10.3.5). The emitted volatile carbon compounds react in the air and lose their volatility and condense to form and grow aerosol particles (Section 7.8). Thus changes in boreal forests are reflected into the aerosols in the air.

Atmospheric aerosols affect the climate in two distinct ways. First, they affect the Earth's radiation budget directly by scattering the solar radiation (direct effect). Second, they act as condensation nuclei for cloud droplets (CCN) having therefore a significant impact on the radiative properties and lifetimes of clouds (indirect aerosol effect). According to the latest report by the Intergovernmental Panel on Climate Change (IPCC, 2007), the total climatic effect of atmospheric aerosols is estimated to be cooling: the radiative forcing resulting from the direct effect is estimated to be between -0.1 and $-0.9\,Wm^{-2}$, whereas the corresponding estimate for the indirect effect is $-1.8\ldots-0.3\,Wm^{-2}$. However, the IPCC also reports that the radiative forcing of the aerosols is currently subject to the largest uncertainties of all individual components in their radiative forcing calculations. In this work we provide rough estimates on how an increased aerosol number concentration resulting from increased biogenic activity would affect the aerosol radiative forcing.

According to recent global model calculations by Spracklen et al. (2006), approximately 30% of the global particle concentration originates from biogenic secondary aerosol formation events described in Section 6.2.3. However, the newly formed particles need to grow fast enough to survive to sizes (ca. 50–100 nm in diameter) at which they can act as cloud condensation nuclei and have a significant climatic effect – otherwise these small particles will be lost by coagulation and deposition processes (see e.g., Kerminen and Kulmala, 2002; Section 6.2.3). Competition between the growth and loss processes is therefore among the main factors determining the magnitude of the particle source provided by the boreal forests.

Condensation of the oxidation products of volatile organic compounds (VOCs) emitted by boreal forests seems to explain a significant fraction of the observed 3–25 nm particle growth rates at the SMEAR I and II sites (Kulmala et al., 1998, 2004a; Hirsikko et al., 2005). Besides their evident role in particle growth above 3 nm, it is possible that the organics emitted by the forests affect also the very first steps of the particle formation and growth processes. The atmospheric particle formation takes place close to 2 nm in diameter and the growth rates observed for the 2–3 nm particles (ca. $1.5\,mn\,h^{-1}$, see e.g., Hirsikko et al., 2005) are typically approximately half of the values reported for particles larger than 3 nm (ca. $3\,nm\,h^{-1}$, e.g. Dal Maso et al., 2005).

Because of the aforementioned links between the VOCs and the particle formation, an increase in VOC emissions is likely to result in an increase of the particle source provided by the boreal forests (see also Kulmala et al., 2004c). The change in the produced climatically active aerosol particle numbers will affect the climatic role of boreal forests: an increase in the cloud condensation nuclei number would lead to more long-lived clouds and therefore add to a negative feedback between climate warming and particle production (IPCC, 2007).

To assess this feedback between climate change and boreal forests, we have estimated an increase of the production of 150 nm particles resulting from doubling the emissions of VOCs (see Section 10.3.5). We assumed that the particle formation rate at 2 nm (J_2) stays constant over time, being approximately $1\,cm^{-3}s^{-1}$. We then calculated the corresponding formation rates at 3 and 150 nm (J_3 and J_{150}) now and in the future when VOC emissions are doubled compared to the current situation. The calculations were made taking into account an increase of growth rates caused by an increasing amount of VOCs, and their effect on the competition of growth and coagulation losses (see Kerminen and Kulmala, 2002 for details).

In the estimations the growth rate of 2–3 nm particles was assumed to be $1.5\,nm^{-1}$, and the growth rate of particles larger than 3 nm was set to $3\,nm^{-1}$. We first assumed that the organics start to contribute to growth from the very beginning, i.e., from 2 nm, and second that they start to condense on the particles only when they have reached sizes above 3 nm, the latter being the more likely case. The ratios of the 150 nm particle production rates in the future (with doubled VOC emissions) and at the present time are $130-300\%$ in the case when organics contribute in the growth from 2 nm and $103-110\%$ in the case when the organics contribute to the growth from 3 nm. The ranges for the ratios have been estimated by assuming first that the increase in VOC concentrations does not affect the coagulation sink (see Section 6.2.3) of the freshly nucleated particles (maximum ratio $J_{150,future}/J_{150,now}$) and second that it maximally enhances the sink provided by the background particle concentration (minimum ratio $J_{150,future}/J_{150,now}$). The obtained ratio in the 150 nm particle production rates ranges from $103-300\%$, which would enhance the climate cooling effect of the forest-produced aerosol at least with the same factor compared to the current radiative forcing (estimated to be between -0.03 and $-1.1\,\mathrm{W\,m^{-2}}$, see Kurtén et al., 2003). However, the response between the aerosol source strength and the radiative transfer is highly nonlinear, so these estimations are still subject to significant uncertainties and should be treated as order of magnitude estimations only.

10.4 Evaluation of the Connections Between Boreal Forests and Climate Change

Pertti Hari[1], Jouni Räisänen[2], Eero Nikinmaa[1], Timo Vesala[2], and Markku Kulmala[2]

[1] University of Helsinki, Department of Forest Ecology, Finland
[2] University of Helsinki, Department of Physics, Finland

10.4.1 Climate Change

Extensive use of forested areas for agriculture starting in the 18th century released CO_2 from forests and forest soil into the atmosphere, resulting in an increasing trend in CO_2 concentration. The increasing use of fossil fuels in the 20th century generated an accelerating concentration increase in the atmosphere. This increasing trend is expected to continue, since mankind can not manage without fossil energy. The first global attempts to reduce CO_2 emissions were made in the Kyoto agreement. This is a good start, but the agreement can, in the best case, only cut the accelerating trend, and atmospheric concentration will continue to increase at a stable rate. However, the CO_2 emissions during the last years have exceeded those in the Fig. 10.1.4 (Raupach et al., 2007) and the increase in the atmospheric CO_2 concentration is accelerating (Canadell et al., 2007).

Direct measurements of atmospheric CO_2 concentration started in 1958 in Mauna Loa, Hawaii. The measurements indicate a clear annual cycle and an accelerating increase. The bubbles in ice in Antarctica enable construction of atmospheric CO_2 concentration time series over 600,000 years. The development of atmospheric CO_2 concentration is well documented and the causes of the present increase are well understood.

Atmospheric CO_2 is a very important environmental factor for vegetation since it affects photosynthesis (Section 6.3.2.3). The increasing atmospheric concentration will have great effects on the plant kingdom. The biggest uncertainty is involved in the acclimation differences between species which makes the development of ecosystems and their species composition problematic to predict.

Use of fossil fuels in industry and intensive agriculture emit many substances into the atmosphere. If the substances have long life times in the atmosphere, then they accumulate and the concentrations increase. Methane, CO_2 and N_2O are, for example such gases and they clearly accumulate in the atmosphere. Emitted, reactive gases, such as SO_2 and NO_x, do not accumulate; instead they start an air chemical reaction chain and are removed from the atmosphere, often in connection with rain (Section 7.8). The life times of reactive emissions in the atmosphere are so short that they are deposited rather close to the source, usually within some thousands of kilometres. Thus most boreal forests are so remote that the reactive gases are of minor importance.

Emissions of SO_2 and NO_x from industry and from nitrogen fertilizer applied in agriculture increase aerosol particle formation and growth (Section 7.8). The increased aerosol particles concentrations and changes in optical properties of tropospheric clouds lead to higher reflection of solar radiation back into space and hinder radiation from reaching the surface of the globe, thus cooling the climate, but the effects in quantitative terms are rather weakly known at present.

The above mentioned changes in the atmosphere are well documented; rather non-problematic measurements have provided their development for decades, as much as 600,000 years for CO_2 and some other trace gases. Thus the change in concentrations is a clear fact and it has to be taken as the starting point when analysing the behaviour of the atmosphere.

A huge amount of solar radiation energy falls on the globe, and this radiation is either absorbed or scattered into space. The absorbed energy runs a versatile system of processes and energy transport. Classical physics has dealt with energy conversion and transport for 200 years, at least (Chapter 4 and Sections 6.2.1 and 7.2). The description of processes and transport phenomena is very well tested and the theory has passed the tests successfully. We can say that classical physics describes the behaviour in the atmosphere very well.

The fundamental equations for the behaviour of the atmosphere were derived in the Section 7.2. They are based on processes and transport phenomena which are combined with conservation of mass, energy and momentum. Thus the model structure is derived from the knowledge of classical physics. The resulting equations describe the behaviour of the atmosphere when solar energy is the driving force of the system.

The equations describing the behaviour of the atmosphere can only be solved numerically by dividing the atmosphere into boxes and determining the fluxes of mass, energy and momentum between the boxes. There are a large number of details which have to be specified for numerical solutions. The GCM models (General Circulation Model or Global Climate Model) compute numerical solutions of the atmospheric equations. They are generally thought to be our most useful tool in the analysis of present climate change.

Vegetation interacts in a complicated manner with the atmosphere, by transpiration, photosynthesis, absorption of solar radiation, heat exchange etc. These processes have been introduced as rough approximations in global climate models. Also the effect of interactions between aerosol particles and solar radiation is complicated, because particle size and composition is important for absorption and reflection of solar radiation and for cloud formation. The necessary idealising assumptions (simplifications) introduce inaccuracies in the calculations. In addition the box size in GCMs has to be rather large to limit computation time, which again introduces additional inaccuracies.

There are several factors capable of affecting the global energy balance and hence climate, the uncertainties of which have been carefully analysed. The magnitude of the contribution to the global energy balance of anthropogenic forcings is illustrated in Fig. 10.4.1. Several factors increase the global greenhouse effect; CO_2 has the

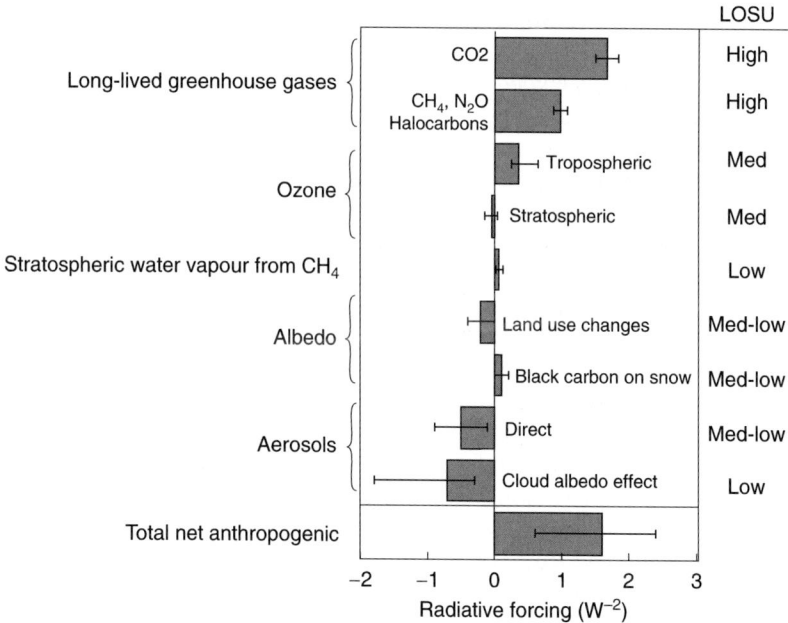

Fig. 10.4.1 Anthropogenic factors causing radiative forcing of climate, their magnitude, estimates of uncertainty and level of their scientific understanding (LOSU) (Redrawn from IPCC Fourth Assessment; IPCC, 2007)

biggest anthropogenic effect and CH_4 the second. In addition, many other gases contribute to the warming of our climate. Albedo changes due to land use changes and aerosols have cooling effects on the climate. The biggest cooling is caused by aerosols and their effect is also the most weakly known.

Global Climate Models have been used already for 30 years. They are under development and improvement, but the basic behaviour of the models has remained rather stable. They produce rather well the present climate everywhere on the globe, which is a rather large achievement. In addition, they have been able to simulate quite well the warming observed during the 20th century (Hegerl et al., 2007) and predict the warmer climate during the early 21st century (Rahmstorf et al., 2007) which indicate prediction power. There are good reasons to believe that models give broadly reliable forecasts of future climate change. Still, there are large differences in the details of the changes between different models, and it is hard to distinguish between the more and less likely forecasts (Räisänen, 2007).

The view that at least most of the recent warming has been caused by increasing greenhouse gas concentrations (Hegerl et al., 2007) has been continuously challenged by alternative explanations. Many of the "climate skeptics" who advocate these views base their argumentation on the fact that climate has varied before. As there have been warmer periods in Earth's geological past, is it then not reasonable to assume that the recent warming is just part of a natural cycle that would have occurred regardless of what mankind was doing? Such natural variations certainly take place, for causes ranging from variations in Earth's orbit and solar and volcanic activity to internal dynamics of the atmosphere-ocean system (e.g., Jansen et al., 2007).

Nevertheless, there are good reasons to think that the global warming observed during the past half-century or so has not been natural. First, climate model simulations indicate that the observed change has been too large and rapid to be explained by internal variability alone – particularly at the global scale on which regional warmings and coolings caused by variations in atmospheric circulation and ocean currents largely cancel each other. Note that, although there have been abrupt and dramatic changes in climate during glacial periods, particularly in the North Atlantic region (Jansen et al., 2007), it is still unlikely that internal dynamics of the climate system would induce variations of this magnitude in the present, fundamentally different conditions. Second, while increased solar activity and lack of explosive volcanism probably accelerated global warming in the early 20th century, there is no evidence that these factors would have contributed to the more recent warming. In fact, model simulations suggest that the net effect of these factors would have been to cool the global climate during the past half-century, which was characterized by several large volcanic eruptions and a rather stable level of solar activity (Hegerl et al., 2007). Third, and most importantly, the observed warming agrees well with GCM experiments that simulate the effects of increased greenhouse gas concentrations based on our best physical understanding of the climate system.

The present understanding of the factors that affect climate is, of course, still imperfect. It is thus possible that some pieces of the puzzle are still missing, for example regarding highly controversial indirect effects of solar activity (e.g.,

Svensmark, 2007). If such factors can be substantiated by physical evidence, then there is no reason against including them in climate models, although it seems very unlikely that they would substantially alter our projections of 21st century climate change. However, claiming that increased greenhouse gas concentrations will not warm the climate because climate has varied before due to other causes represents a very illogical argumentation and it omits the energy balance in the atmosphere. Thus it implicitly assumes that some rather unspecific variation is more relevant than physical knowledge.

Mankind has to decide either to continue present emissions into the atmosphere or to reduce them. The reduction is very expensive because it means a strong decrease in the use of fossil fuels. If scientific knowledge is used in the decision making, then global climate models should be applied because they give now the most relevant information about climate change.

The GCMs have been applied to make scenarios of the climate development during the 21st century that clearly indicate warming (Section 10.1). There are some differences in the idealising assumptions between different versions of the model resulting in some scatter of the results. The main argument is clear. As a result of increased greenhouse gas concentrations, a smaller fraction of the thermal radiation emitted by the surface and the warm lower troposphere escapes directly to space. Thus, to maintain a balance between absorbed solar radiation and outgoing thermal radiation, the surface and the lower atmosphere must warm so that the resulting increase in the emission of thermal radiation compensates the reduction due to increasing greenhouse gases.

In summary, we are facing global climate change which is very dangerous for mankind. Urgent measures are needed to stop negative development and emissions have to be strongly and rapidly reduced.

10.4.2 Responses of Boreal Forests

Photosynthesis uses as raw material solar energy, atmospheric CO_2 and water. The present low atmospheric carbon dioxide concentrations clearly limit photosynthesis, especially at high light intensities, and increase in atmospheric CO_2 results in acceleration of photosynthesis (Section 6.3.2.3). The relationship between specific photosynthetic rate and CO_2 concentration is linear in the models in Section 6.3.2.3. Transpiration depends primarily on the water vapour concentration difference between stomatal cavity and ambient air, not on CO_2 difference, thus the ratio between transpiration and photosynthesis is changing. This change is evidently not small and the ratio can be doubled in the year 2100 when compared with the present one. Thus the structural regularities based on need of water transport are no longer favourable for trees (Chapter 8). In addition, although drought events may become more frequent, their severity may be mitigated since plants reduce transpiration levels at higher CO_2 concentration.

Plants acclimate to elevated CO_2 concentrations; experiments have shown changes in the concentrations of functional substances in plants grown in increased CO_2 concentrations (Section 6.3.3.2.3). The test in evolution makes the acclimations predictable, since only favourable acclimations can pass successfully the competition in evolution. The atmospheric CO_2 concentration has been rather stable and below 300 ppm at least for 600,000 years. Thus the acclimations to present CO_2 concentration have not been tested in evolution for a very long time and we can not predict acclimations to increasing atmospheric CO_2 concentration and they can be even harmful for some species. Those plant species having favourable acclimations will benefit greatly and species composition will change.

The magnitude of the contribution to the global energy balance of anthropogenic forcings is illustrated in Fig. 10.4.1. Several factors increase the global greenhouse effect; CO_2 has the biggest anthropogenic effect and CH_4 the second. In addition, many other gases contribute to the warming of our climate. Albedo changes due to land use changes and aerosols have cooling effects on the climate. The biggest cooling is caused by aerosols and their effect is also the most weakly known.

The decomposition of macromolecules entering the organic matter pool in the soil as litter is very slow because the organic matter pool in the soil is large. Most of the energy rich macromolecules are important only for microbes in the soil as a source of raw material and energy. The nitrogen in functional substances in the litter is also important for microbes but it is crucial for the growth of vegetation, since microbes break proteins with extracellular enzymes and thereafter utilise the amino acids in their metabolism, releasing some of the nitrogen as NH_4 into soil water. In addition, the decomposition produces potassium, phosphorus, calcium and other nutrients in soil water and releases CO_2. The nitrogen in soil organic matter is either in proteins or in humus. The lifetime of proteins is decades and of humus millennia. This is why these two macromolecule types have very different roles in the soil nitrogen dynamics and they should be treated separately. The C:N ratio is commonly used to describe the soil organic matter. Then the proteins and humus compounds are pooled together and rather uninformative concept is obtained.

The crucial step in decomposition of functional substances in the soil is splitting of proteins into amino acids by extracellular enzymes (Section 6.4.1). Because temperature dependence is a characteristic feature of enzymatic reactions, splitting of proteins into amino acids also depends strongly on temperature. The mean residence time of proteins in the soil is large, 50 years (Section 7.9). When the decomposition of proteins is accelerated by temperature, then increased flow of NH_4 becomes from proteins to the soil ammonium. The large pool in the soil enables big amounts of additional nitrogen and also large effects in growth.

MicroForest simulated clear growth increases along with temperature increase. When, in addition, the CO_2 increase was introduced, the response was large. This result comes from the fundamental logic of the model, since in this case both growth limiting factors; availability of nitrogen and sugars were removed. Strong responses have been observed in FACE experiments to combined nitrogen fertilization and CO_2 enrichment (Oren et al., 2001; McCarthy et al., 2006).

Concentrations of several atmospheric trace gases are changing. These gases do not have clear links with processes in vegetation, thus their direct effects on the plant kingdom are evidently of minor importance. Also aerosols have no direct effect on vegetation processes; they can, however, change the properties of light and in this way have an effect on photosynthesis. This effect is small when compared with that of CO_2. Thus we can conclude that the main direct effects of climate change on boreal ecosystems is (i) via acceleration of photosynthesis by increasing CO_2 and temperature and (ii) accelerating of decomposition of proteins in soil caused by increasing temperature.

10.4.3 Feedbacks from Boreal Forests

Most of the boreal forests are remote and access is very difficult because roads are often missing. Only areas of intensive forestry can easily be visited and measured. On the other hand, we urgently need information dealing with the response of boreal forest to climate change. We have to choose between two unpleasant alternatives, we either state that we have no information or understanding and refuse to say anything about the response of boreal forests to climate change or we utilise our limited information and understanding to get some rough picture of the responses combined with the available results on remote sensing observations and CO_2 profiles.

We need information about the response of boreal forests to climate change now for political decision making. On the other hand, we have very detailed information and knowledge from one stands and the tests of the knowledge in north-south direction indicates that the results can be applied at the southern and northern border of Scots pine (Section 9.2). This raises the question: Can this knowledge be applied to the whole boreal zone?

The basic processes, structures, transport phenomena and regularities in forest structure are the same in all trees and evidently also in the whole plant kingdom. Thus the overall responses of trees to climate change are rather similar, but there are some differences in the quantitative responses. The model structures applied in this book are derived from regularities in processes, structures and transport phenomena, often using conservation of mass, energy or momentum, fundamental features for all ecosystems in the boreal forests. Thus model structures are the same for all ecosystems in boreal forests. There are, however, differences, but they are in the values of the parameters involved (see Section 7.6.2). Thus we have good reasons to assume that the results obtained with Scots pine can give indication on the patterns of change in the whole boreal zone.

The limited knowledge of the structure of boreal forests has to be compensated with strong assumptions. This, of course, introduces additional inaccuracies but we have no other choice. The applied assumption, all stands in boreal forests are even aged, have the same rotation period and the age distribution is even, are highly idealising, i.e., contrary to reality, but evidently this idealisations do not introduce considerable inaccuracies in the development pattern of boreal forests during the period 1800–2100.

Satellite images combined with forest inventories and measurements of CO_2 profiles in connection with atmospheric models provide estimates of carbon uptake by boreal forests. There are, however, problems in interpreting these results to consider boreal forests since their coverage is different. In addition they do not treat the same characteristic of boreal forests; satellite image based estimates deal with carbon stock in vegetation and CO_2 profiles with carbon uptake. The flow of carbon in fellings from forests is the most important difference.

Increasing atmospheric carbon dioxide accelerates photosynthesis and temperature increase prolongs the photosynthetically active period and accelerates decomposition of soil organic matter, especially of proteins. According to the simulations with MicroForest, the carbon stock in trees (wood and needles) in boreal forests responds clearly after 1990 (Fig. 10.2.3) and the increase is rather linear. The increased availability of nitrogen for synthesis of proteins dominates the response. The annual simulated increase in biomass in boreal forests due to the accelerated decomposition in the year 2000 is about $0.8\,Pg\,a^{-1}$ (Section 10.2.4). The carbon concentration in plant material is about 50%, thus the boreal forest vegetation bounds about $0.4\,Pg\,(C)\,a^{-1}$. The carbon accumulation to boreal forests due to nitrogen deposition is about $0.1\,Pg\,(C)\,a^{-1}$. Thus the carbon sequestration by vegetation in boreal forests is according our simulations about $0.5\,Pg\,(C)\,a^{-1}$.

Analysis of satellite images together with forest inventories by Myneni et al. (2001) results that the carbon sink in biomass in Northern forests is $0.68\,Pg\,(C)\,a^{-1}$. The analysis by Myneni et al. (2001) does not cover only the boreal forests, large areas of temperate forests are included and also some Northern boreal forests are missing. Thus direct comparison of the two estimates of carbon sequestration is problematic, however, we can state that the results are in reasonable agreement with each other.

The estimates obtained with CO_2 profiles deal with the carbon uptake or sink. They also include carbon transported from boreal forests in wood products. Thus we have to estimate the amount of wood utilised in forestry. The official statistics by FAO (UN-ECE/FAO Forest Resources Assessment) provide annual fellings of coniferous trees as 979 million cubic metre per acre. When this is converted to units $Pg\,(C)\,a^{-1}$ we obtain 0.2. Thus the carbon flow away from boreal forests caused by forestry is a considerable component in the carbon budget.

The approximation of carbon uptake of northern terrestrial uptake based on CO_2 profiles by Stephens et al. (2007) covered a still larger area, the whole northern terrestrial region. Their estimate for the northern terrestrial carbon sequestration is $1.5\,Pg\,(C)\,a^{-1}$. There is rather large variation between different inverse models, based on CO_2 profiles, estimates for northern terrestrial carbon sink from 0.5 to $4\,Pg\,(C)\,a^{-1}$ (Stephens et al., 2007). However, the result by Stephens et al. (2007) is the most recent and it is based on latest measurements. The magnitude of sink in boreal forest, based on CO_2 profiles, is evidently $1\,Pg\,(C)\,a^{-1}$.

We can now estimate the carbon sink in boreal forests with three approaches:

1. Dynamic modelling based on theoretical understanding of photosynthesis, growth and soil processes together with field measurements enables simulation of carbon stocks in boreal forests during 1800–2100. Then the estimate for increase in biomass is $0.5 \, Pg \, (C) \, a^{-1}$, decrease in soil organic matter $0.15 \, Pg \, (C) \, a^{-1}$ and the carbon flow from forests caused by forestry is $0.2 \, Pg \, (C) \, a^{-1}$. When these estimates are combined we get that the carbon sink in boreal forests is $0.55 \, Pg \, (C) \, a^{-1}$.

2. Satellite images combined with forestry inventory results resulted $0.68 \, Pg \, (C) \, a^{-1}$ annual accumulation in biomass in the boreal forests. The decrease in soil organic matter we have to take from the previous theoretical analysis $0.15 \, Pg \, (C) \, a^{-1}$ and the outflow in forest products is $0.2 \, Pg \, (C) \, a^{-1}$. Combining these results we get $0.738 \, Pg \, (C) \, a^{-1}$ as the estimate for the carbon sink in the boreal forests.

3. Measurements of CO_2 profiles combined with atmospheric models resulted in the magnitude estimate $1 \, Pg \, (C) \, a^{-1}$.

The three rather different approaches gave rather similar results. Although the methods are rough the similarity of the obtained results increase confidence on them. Anyway, the obtained result that the carbon sinks in boreal forests is a bit below $1 \, Pg \, (C) \, a^{-1}$ is the most reliable one at the moment.

Our estimate of the carbon sink in boreal forests is so high that the boreal forests have evidently an important role in the global carbon budget. The sink in coniferous forests seems to be about 10% of the global carbon emissions. According simulations (Section 10.2.4) the carbon sequestration will continue and even increase to some extend during the 21st century. This raises the question: Can be thrust on this result.

The prediction of carbon sequestration for 100 years is problematic since then we have to assume that there are no fundamental changes in the behaviour of the boreal forests. The vegetation and soil microbes and fauna have adapted to the rather stable environmental conditions after ice age. The CO_2 concentration has increased from 280 to 380 ppm and the increase is accelerating, temperature and nitrogen availability in soil have increased thus the environmental factors have changed and will still change outside the range of environmental factors during adaptation of the boreal forests to the prevailing environment.

The protection of vegetation against fungal and insect damage, the over wintering of pathogens and insects and the annual cycle of vegetation are examples of the potential critical features in the ecosystems and they will in the long run cause large scale damage which will make our prediction too optimistic, i.e. the carbon sequestration is too large. In this way the adaptation to the new environment begins.

The increased decomposition of proteins due to climate change will enhance the availability of ammonium to vegetation and microbes. It is rather possible that also nitrification of ammonium (Section 6.4.2) will start to convert ammonium to nitrate and denitrification will convert further nitrate to gaseous nitrogen. As a side product, N_2O is released. We approximated the possible N_2O release from boreal forests. The obtained total emission ($1.5 \, Tg \, (C) \, a^{-1}$) were so small that they have only minor importance for atmospheric concentration since the present emissions are $17.7 \, Tg \, (C) \, a^{-1}$.

Increased decomposition will also enhance the nitrogen uptake of vegetation and trees will have more nitrogen for synthesis of functional substances in leaves, thus the leaf mass will increase. Conifer leaves effectively absorb solar radiation while, in contrast, snow reflects nearly all radiation (Section 10.3.4). Thus in the spring, increasing needle mass will enhance the capture of solar energy by boreal forests. We analysed this effect with global climatic models and they indicated a rather clear increase in humidity and cloudiness in summer and a temperature increase up to $1°C$ in the spring. Changes in humidity and cloudiness are caused by larger transpiration in summer due to a larger transpiring area.

The effect of climate change on volatile carbohydrate (BVOC) emissions is poorly understood at the moment. The reasonable assumption is that BVOC emissions behave like photosynthesis in the long run, since synthesis of BVOCs in the leaves is connected with photosynthesis. The contribution of BVOC to aerosol formation and growth is better understood, although additional knowledge is needed. The effect of increasing aerosols in the atmosphere is cooling due to enhanced reflection of solar radiation back to space. However, the obtained result is quite uncertain, which is seen in the big difference of maximum and minimum estimates of change in radiation forcing.

Boreal forests produce a number of evident feedbacks on climate change. An increasing leaf mass amplifies climate change, especially in the spring. Carbon sequestration in the increased wood mass reduces increase in the atmospheric CO_2 concentration and enhanced aerosol formation and growth amplify reflection of solar energy. When these effects are combined, then evidently the negative feedbacks are stronger than the positive ones and boreal forests slightly hinder climate change.

Chapter 11
Concluding Remarks

Pertti Hari, Eero Nikinmaa, and Markku Kulmala

Our book covers a wide range of phenomena in boreal forests and in the atmosphere. We combine biological, physical, meteorological, microbiological and chemical knowledge to get a holistic picture of the interactions between boreal forests and climate change. Close interactions between ecologists and physicists have been characteristic of the work done in our research team, called APFE (Aerosol Physics and Forest Ecology). Ecological and physical phenomena have similarities and specific features that make this cooperation fruitful. In addition, modern instrumentation is based on physics and without understanding the basis of measurements, it is impossible to properly utilise new capabilities to gain information from forests.

Processes are the engine of interaction between boreal forests and the atmosphere. The processes generate concentration, temperature and pressure differences which are necessary for transport. Physical and chemical processes are spontaneous, i.e., they occur whenever the necessary substances and energy are available. Biological processes, in contrast, need also active substances, enzymes, pigments or membrane pumps which enable or greatly accelerate chemical reactions. Processes in connection with weak transport also accumulate material, such as stable organic compounds in vegetation, aerosols in the air, water in clouds and ammonium ions on soil particles.

Most physical processes treated in our book have been studied for centuries and they are well understood; the underlying theories have been tested with extensive experiments and measurements. Only aerosol processes are not yet fully understood, evidently because the field is rather new. However, the rapid increase in knowledge of the behaviour of aerosols is evident.

Our knowledge of biological processes is more uncertain. Large spatial variations in environmental factors and the combination of data from several objects have often blurred the information in the measurements. Our results indicate that biological

P. Hari and E. Nikinmaa
University of Helsinki, Department of Forest Ecology, Finland

M. Kulmala
University of Helsinki, Department of Physics, Finland

P. Hari, L. Kulmala (eds.) *Boreal Forest and Climate Change,*
© Springer Science+Business Media B.V., 2009

processes are very regular and clear results can be obtained with good quality of measurement and by utilising the different time and spatial scales of processes and regulation of functional substances involved.

Transport is a physical phenomenon and it should be treated within the physical framework. Concentration, temperature and pressure differences generate the material and energy fluxes. Transport has the tendency to level out differences and it ends without processes generating new substances or energy to be transported. Continuous flow requires stable processes to provide the material for transport. The biological transport phenomena are based on physical mechanisms.

Conservation of mass, energy and momentum is the key to combining processes and transport. The same conservation principle was repeatedly utilised when models were constructed to describe phenomena in trees, ecosystems and the atmosphere. In this way, understanding at a more aggregated level can be derived from basic process and transport phenomena. The inputs and outputs of the resulting combined models can often be measured in the field, which enables testing of the basis of the approach.

Ecosystem development is clearly a biological phenomenon, within the framework of physics and chemistry. Vegetation produces large numbers of macromolecules for cell walls, membranes and functional substances. Very stable macromolecules, such as cellulose and lignin, are important for carbon sequestration in ecosystems, and nitrogen rich functional substances enable processes in living cells and in soil solution. The formation of complicated structure in cells, organelles and trees, developed in evolution, is a characteristic feature for biological objects; no correspondence can be found in the fields of physics and chemistry.

The chemical composition of the atmosphere is changing due to anthropogenic influence, use of fossil fuels, destruction of forests and other industrial emissions. The processes in the air react to these changes. When the changing processes are combined with transport in the air utilising the conservation of energy, mass and momentum we get Global Climatic Models (GCM), called also global circulation models. Thus the physical basis of GCM is very clear and solid. However, the idealising assumptions needed for the numerical solution of GCMs introduce inaccuracies in the calculations. More attention should be paid to alternative idealising assumptions and on the inaccuracies they generate.

Boreal forests react to changes in atmospheric composition, since photosynthesis is accelerated by an increase in the raw material, CO_2. Vegetation acclimates to increased CO_2 by changing fine structure and functional substances. These acclimations have not been tested in evolution and they can not be predicted. In addition the changes in temperature have an effect on several processes in vegetation and soil. The accelerated decomposition of proteins in soil is evidently the most important response since it provides additional nitrogen for vegetation to construct functional substances. This link seems to result in a large increase in stem volume, thus the amount of carbon in boreal forests will increase.

The changes in boreal forests generate feedbacks to climate change, both accelerating and damping ones. The most important negative feedbacks are: Accumulation of stem wood slows down the increase in atmospheric CO_2 concentration and an increased number of aerosols in the air reflects more solar radiation back to space. The positive feedbacks are: Increasing needle mass which decreases reflected radiation especially in spring, and an increased flow of N_2O from forest soils which increases the capture of thermal radiation from the globe. These feedbacks are in general poorly understood and their studies require a combination of ecological and physical knowledge.

The rapid development of instrumentation has opened several novel measuring possibilities. For example, it is possible to construct a system that measures simultaneously CO_2 exchange utilising both stable carbon isotopes, transpiration based on two stable oxygen isotopes, ozone deposition, NO_x exchange and the emissions of volatile carbon compounds (VOC). This kind of versatile data opens several interesting possibilities for research.

Our data on boreal forests and especially on boreal forest–climate change interactions are weak. However, the present instrumentation enables high quality comprehensive measurements to fill the gap in information about current changes in our environment and in living nature. We need a comprehensive and hierarchical system of stations covering the entire area of boreal forests. Advanced measuring stations, like SMEAR II, are complex to construct and tedious to run, which is why only a limited number of flagship stations can be constructed. Flux stations that concentrate on the fluxes between boreal forests and the atmosphere are easier to build and manage, which makes it possible to establish a rather large number of such stations. Basic stations that measure the state of the forests and environment are technically simpler and their number can be so large that they can cover sufficiently the large spatial variation in forests and environment. The analysis of structures, processes and transport phenomena, utilising data from a network of such measuring stations, would greatly improve our understanding of the interactions between boreal forests and climate change.

The main objective in scientific research is to discover permanent and general regularities in nature. There are evident regularities in non-living nature which physical and chemical sciences, ever since Newton, have successfully detected. The behaviour of aerosols in the atmosphere is not yet fully understood, but rapid progress is going in the field. The success has been more limited in the ecological research with only rather weak regularities been found.

The proposed metatheory described in our opening chapter has been applied throughout the entire book. Our existing physical knowledge can be expressed within it and it can be applied to aerosol research. In addition, eco-physiological knowledge can be formulated and developed within it. Our metatheory is consistent with the known principles of physics, which enables the utilisation of modern measuring techniques in producing data to test subtheories and to obtain descriptions of the environment as demonstrated at SMEAR II and I. The approach is based on material and energy fluxes and the conservation of mass, energy and momentum which make dynamic modelling a powerful tool in developing and testing the ideas. Statistical methods are needed to treat measuring errors and great spatial and temporal variation in the phenomena.

Several clear regularities, such as the dependence of photosynthesis on environmental factors and the regulation of annual cycle, water and sugar transport within trees as driven by water tension and pressure and the development of forest ecosystems based on photosynthesis, nitrogen uptake and structural regularities have been detected and successfully tested in field conditions. Several aspects, however, still remain open for further research, especially the regularities in tree structure that are generated in evolution to combine processes, transport and structure.

Our metatheory enables transitions from detailed to more aggregated levels; for example, (i) the measured ecosystem CO_2 flux can be explained with processes at space element level and transport within the ecosystem, (ii) ecosystem growth and development with carbon and nitrogen processes in trees and soil, and regularities in tree structure and (iii) carbon sequestration of boreal forests can be explained with ecosystem dynamics and release of nitrogen by accelerated decomposition of proteins in soil. The emergent properties may generate additional complications in transitions to more aggregated levels and they should always be analysed in the connection of climbing up in the hierarchy.

The present climate change is basically a physical phenomenon, since increasing atmospheric concentrations change the energy balance of the globe, resulting in responses in temperature, rain fall, etc. Boreal forests react to the changes in such factors as atmospheric CO_2 concentration, rain-fall and temperature. These responses generate feedbacks from boreal forests to the climate. Thus physical and biological aspects have to be combined, which requires coherent treatment by practitioners of different disciplines. In this book, we have combined tree ecophysiological, soil scientific, microbiological, physical, meteorological and aerosol physical knowledge utilising modern instrumentation and modelling to analyse interactions between boreal forests and climate change. The metatheory has clearly contributed to the studies of interactions between disciplines. The negative feedbacks from boreal forests to climate change are evidently stronger than positive ones, thus boreal forests will probably slightly slow down the climate change.

Our message in this book in a nutshell is: (i) The fluxes of material, energy and momentum and the conservation principles of mass, energy and momentum form a sound basis for biological, physical, meteorological and chemical research of boreal forests and climate change. (ii) There are strong and clear regularities in boreal forests and in the atmosphere. (iii) Modern measuring techniques are capable of providing accurate and precise measurements about regularities. (iv) Dynamic modelling, based on the conservation of mass, energy and momentum is an effective tool for studying the regularities. (v) Anthropogenic emissions into the atmosphere are changing the climate in a harmful manner; strong and rapid reduction of emissions is needed.

References

Aalto, P. 2004. *Atmospheric ultrafine particle measurements.* (Diss.) Department of Physical Sciences. University of Helsinki.

Aalto, T. and Juurola, E. 2001. Parameterization of a biochemical CO_2 exchange model for birch (*Betula pendula* Roth.). *Boreal Environment Research* 6: 53–64.

Aalto, T., Hari, P., and Vesala, T. 2002a. Comparison of an optimal regulation model and biochemical model in explaining CO_2 exchange in field conditions. *Silva Fennica* 36: 615–623.

Aalto, T., Hatakka, J., Paatero, I., Tuovinen, J.-P., Aurela, M., Laurila, T., Holmén, K., Trivett, N., and Viisanen, Y. 2002b. Tropospheric carbon dioxide concentrations at a northern boreal site in Finland: basic variations and source areas. *Tellus B*54: 110–126.

Aaslyng, J.M., Rosenqvist, E., and Høgh-Schmidt, K. 1999. A sensor for microclimatic measurements of photosynthetically active radiation in a plant canopy. *Agricultural and Forest Meteorology* 96: 189–197.

Aber, J., Mellilo, J., Nadelhofer, J., Pastor, J., and Boone, D. 1991. Factors controlling nitrogen cycling and nitrogen saturation in northern temperate forest ecosystem. *Ecological Applications* 1(3): 303–315.

Aber, J., McDowell, K., Nadelhoffer, K., Magill, A., Berntson, G., Kamakea, M., McNulty, S., Currie, W., Rustad, L., and Fernandez, I. 1998. Nitrogen Saturation in Temperate Forest Ecosystems. Hypothesis revisited. *BioScience* 48(11): 921–934.

Adamson, A.W. and Gast, A.P. 1997. *Physical Chemistry of Surfaces*, 6th ed. Wiley. New York, USA. 784 p.

Ågren, G.I. 1996. Nitrogen productivity or photosynthesis minus respiration to calculate plant growth. *Oikos* 76: 529–535.

Ågren, G.I. and Bosatta, E. 1996. Quality: a bridge between theory and experiment in soil organic matter studies. *Oikos* 76: 522–528.

Ågren, G.I. and Franklin, O. 2003. Root: shoot ratios, optimization and nitrogen productivity. *Annals of Botany* 92: 795–800.

Ahti, T., Hämet-Ahti, L., and Jalas, J. 1968. Vegetation zones and their sections in northwestern Europe. *Annales Botanici Fennici* 5: 169–211.

Aitken, J.A. 1897. On some nuclei of cloudy condensation. *Transactions of the Royal Society* XXXIX: 15–25.

Aitken, S.N. and Hannerz, M. 2001. Genecology and gene resource management strategies for conifer cold hardiness. In: Bigras, F.J. and Columbo, S.J. (Ed.). *Conifer Cold Hardiness.* Kluwer. Dordrecht, the Netherlands. p. 25–53.

Allen, J.F. and Forsberg, J. 2001. Molecular recognition in thylakoid structure and function. *Trends Plant Science* 6: 317–326.

Allen, L.H., Stewart, D.W., and Lemon, E.R. 1974. Photosynthesis in plant canopies: effect of light response curves and radiation source geometry. *Photosynthetica* 8: 184–207.

Altimir, N., Vesala, T., Keronen, P., Kulmala, M., and Hari, P. 2002. Methodology for direct field measurements of ozone flux to foliage with shoot chambers. *Atmospheric Environment* 36: 19–29.

Altimir, N., Tuovinen, J.-P., Vesala, T., Kulmala, M., and Hari, P. 2004. Measurements of ozone removal to Scots pine shoots: calibration of a stomatal uptake model including the non-stomatal component. *Atmospheric Environment* 38: 2387–2398.

Altimir, N., Kolari, P., Tuovinen, J.-P., Vesala, T., Bäck, J., Suni, T., Kulmala, M., and Hari, P. 2006. Foliage surface ozone deposition: a role for surface moisture? *Biogeosciences* 3: 209–228.

Ambus, P. and Robertson, G.P. 2006. The effect of increased N deposition on nitrous oxide, methane and carbon dioxide fluxes from unmanaged forest and grassland communities in Michigan. *Biogeochemistry* 79: 315–337.

Ambus, P., Zechmeister-Boltenstern, S., and Butterbach-Bahl, K. 2006. Sources of nitrous oxide emitted from European forest soils. *Biogeosciences* 3: 135–145.

Amthor, J.S. 1991. Respiration in a future, higher-CO_2 world. *Plant Cell and Environment* 14: 13–20.

Amthor, J.S. 1994. Plant respiratory responses to the environment and their effects on the carbon balance. In: Wilkinson, R.E. (Ed.). *Plant–Environment Interactions*. Marcel Dekker. New York, USA. p. 501–554.

Andreae, M.O., Jones, C.D., and Cox, P.M. 2005. Strong present-day aerosol cooling implies a hot future. *Nature* 435: 1187–1190.

Angers, D.A. and Caron, J. 1998. Plant-induced changes in soil structure: processes and feedbacks. *Biogeochemistry* 42: 55–72.

Arya, S.P. 2001. *Introduction to Micrometeorology*, 2nd ed. Academic Press. London, UK. 420 p.

Atkin, O.K. and Tjoelker, M.G. 2003. Thermal acclimation and the dynamic response of plant respiration to temperature. *Trends in Plant Science* 8: 343–351.

Aubinet, M., Grelle, A., Ibrom, A., Rannik, Ü., Moncrieff, J., Foken, T., Kowalski, A.S., Martin, P.H., Berbigier, P., Bernhofer, C., Clement, R., Elbers, J., Granier, A., Grunwald, T., Morgenstern, K., Pilegaard, K., Rebmann, C., Snijders, W., Valentini, R., and Vesala, T. 2000. Estimates of the annual net carbon and water exchange of forests: the EUROFLUX methodology. *Advances in Ecological Research* 30: 113–176.

Aubinet, M., Berbigier, P., Bernhofer, C., Cescatti, A., Feigenwinter, C., Granier, A., Grünwald, T., Havrankova, K., Heinesch, B., Longdoz, B., Marcolla, B., Montagnani, L., and Sedlak, P. 2005. Comparing CO_2 storage and advection conditions at night at different CARBOEUROFLUX sites. *Boundary-Layer Meteorology* 43: 345–364.

Babst, B.A., Ferrieri, R.A., Gray, D.W., Lerdau, M., Schlyer, D.J., Schueller, M., Thorpe, M.R., and Orians, C.M. 2005. Jasmonic acid induces rapid changes in carbon transport and partitioning in Populus. *New Phytologist* 167: 63–72.

BACC Author Team. 2008. *Assessment of Climate Change for the Baltic Sea Basin, Regional Climate Studies*. Springer. Heidelberg, 474 pp.

Bachmann, J. and van der Ploeg, R.R. 2002. A review on recent developments in soil water retention theory: interfacial tension and temperature effects. *Journal of Plant Nutrition and Soil Science* 165: 468–478.

Bäck, J., Neuvonen, S., and Huttunen, S. 1994. Pine needle growth and fine structure after prolonged acid rain treatment in the subarctic. *Plant, Cell & Environment* 17: 1009–1021.

Bäck, J., Hari, P., Hakola, H., Juurola, E., and Kulmala, M. 2005. Dynamics of monoterpene emissions in Pinus sylvestris during early spring. *Boreal Environment Research* 10: 409–424.

Baggs, E.M., Cadisch, G., Stevenson, M., Pihlatie, M., Regar, A., and Cook, H. 2003. Nitrous oxide emissions resulting from interactions between cultivation technique, residue quality and fertiliser application. *Plant and Soil* 254: 361–370.

Baldocchi, D.D., Falge, E., Gu, L.H., Olson, R., Hollinger, D., Running, S., Anthoni, P., Bernhofer, C., Davis, K., Evans, R., Fuentes, J., Goldstein, A., Katul, G., Law, B., Lee, X.H., Malhi, Y., Meyers, T., Munger, W., Oechel, W., U, K.T.P., Pilegaard, K., Schmid, H.P., Valentini, R., Verma, S., Vesala, T., Wilson, K., Wofsy, S., 2001. Fluxnet: a new tool to study the temporal and spatial variability of ecosystem-scale carbon dioxide, water vapor, and energy flux densities. *Bull. Am. Meteorol. Soc.* 82: 2415–2434.

Baldocchi, D.D. 2003. Assessing the eddy covariance technique for evaluating carbon dioxide exchange rates of ecosystems: past, present and future. *Global Change Biology* 9: 479–492.

Baldock, J.A. and Nelson, P.N. 2000. Soil organic matter. In: Sumner, M.E. (Ed.). *Handbook of Soil Science*. CRC Press. Boca Raton, Florida, USA. p. B25–B84.

Ball, B.C., Dobbie, K.E., Parker, J.P., and Smith, K.A. 1997a. The influence of gas transport and porosity on methane oxidation in soils. *Journal of Geophysical Research* 102(D19): 23301–23308.

Ball, B.C., Smith, K.A., Klemedtsson, L., Brumme, R., Sitaula, B.K., Hansen, S., Priemé, A., MacDonald, J., and Horgan, G.W. 1997b. The influence of soil gas transport properties on methane oxidation in a selection of northern European soils. *Journal of Geophysical Research* 102(D19): 23309–23317.

Baraldi, R., Rapparini, F., Oechel, W.C., Hastings, S.J., Bryant, P., Cheng, Y., and Miglietta, F. 2004. Monoterpene emission responses to elevated CO2 in a Mediterranean-type ecosystem. *New Phytologist* 161: 17–21.

Barber, J., Malkin, S., and Telfer, A. 1989. The origin of chlorophyll fluorescence in vivo and its quenching by the photosystem II reaction centre. *Philosophical Transactions of the Royal Society of London B* 323: 227–239.

Barnola, J.-M., Raynaud, D., Lorius, C., and Barkov, N.I. 2003. Historical CO_2 record from the Vostok ice core. In Trends: A Compendium of Data on Global Change. Carbon Dioxide Information Analysis Center, Oak Ridge National.

Laboratory, U.S. Department of Energy. Oak Ridge, Tennessee, USA.

Baron, P.A. and Willeke, K. (Ed.). 2001. *Aerosol Measurement: Principles, Techniques, and Applications*, 2nd ed. Wiley. New York, USA.

Bédard, C. and Knowles, R. 1989. Physiology, biochemistry, and specific inhibitors of CH_4, NH_4^+, and CO oxidation by methanotrophs and nitrifiers. *Microbiological Reviews* 53: 68–84.

Bendall, D.S. 2006. Photosynthesis: light reactions. *Encyclopaedia of Life Sciences*. Wiley. New York, USA.

Bender, M. and Conrad, R. 1992. Kinetics of CH_4 oxidation in oxic soils exposed to ambient air or high CH_4 mixing ratios. *FEMS Microbiology Ecology* 101: 687–696.

Bengough, A.G., Bransby, M.F., Hans, J., McKenna, S.J., Roberts, T.J., and Valentine, T.A. 2006. Root responses to soil physical conditions; growth dynamics from field to cell. *Journal of Experimental Botany* 57: 437–447.

Berg, B. 2000. Litter decomposition and organic matter turnover in northern forest soils. *Forest Ecology and Management* 133: 13–22.

Berg, B. and Dise, N. 2004. Calculating the long-term stable nitrogen sink in northern European forests. *Acta Oecologica* 26: 15–21.

Berg, B., Booltink, H., Breymeyer, A., Ewertson, A., Gallardo, A., Holm, B., Johansson, M.-B., Koivuoja, S., Meentemeyer, V., Nyman, P., Olofsson, J., Petterson, A.-S., Reurslag, A., Staaf, H., Staaf, I., and Uba, L. 1991. *Data on Needle Litter Decomposition and Soil Climate as Well as Site Characteristics for Some Coniferous Forest Sites. Part 2. Decomposition Data*, 2nd ed. Swedish University of Agricultural Sciences, Department of Ecology and Environmental Research. Report 42.

Bergh. J., McMurtrie, R.E. and Linder, S.1998. Climatic factors controlling the productivity of Norway spruce: a model based analysis.*Forest Ecology and Management* 110: 127–139.

Bernacchi, C.J., Pimentel, C., and Long, S.P. 2003. In vivo temperature response functions of parameters required to model RuBP-limited photosynthesis. *Plant, Cell and Environment* 26: 1419–1430.

Berninger, F. and Nikinmaa, E. 1994. Geographical variation in the foliage mass - wood cross-sectional area ratios in young Scots Pine stands. *Canadian Journal of Forest Research* 24: 2263–2268.

Berninger, F., Mäkelä, A., and Hari, P. 1996. Optimal control of gas exchange during drought: empirical evidence. *Annals of Botany* 77: 469–476.

Berninger, F., Nikinmaa, E., Sievänen, R., and Nygren, P.A. 2000. Modelling of reserve carbohydrate dynamics, regrowth and nodulation in a N2-fixing tree managed by periodic prunings. *Plant, Cell & Environment* 23: 1025–1040.

Berninger, F., Nikinmaa, E., Hari, P., Lindholm, M., and Meriläinen, J. 2004. Simulation of tree ring growth using process based approaches. *Tree Physiology* 24: 193–204.

Berninger, F., Coll, L., Vanninen, P., Mäkelä, A., Palmroth, S., and Nikinmaa, E. 2005. Effects of tree size and position on pipe model ratios in Scots pine. *Canadian Journal of Forest Research* 35: 1294–1305.

Beyer, L. 1996. The chemical composition of soil organic matter in classical humic compound fractions and in bulk samples – a review. *Zeitschrif für Pflanzenernährung und Bodenkunde* 159: 527–539.

Beyer, L., Vogt, B., and Kobbemann, C. 1996. A simple wet chemical extraction procedure to characterize soil organic matter (SOM): 2. Reproducibility and verification. *Communications in Soil Science and Plant Analysis* 27: 2229–2241.

Bhiry, N., Payette, S., and Robert, E.C. 2007. Peatland development at the arctic tree line (Quebec, Canada) influenced by flooding and permafrost. *Quaternary Research* 67: 426–437.

Billings, S.A., Richter, D.D., and Yarie, J. 2000. Sensitivity of soil methane fluxes to reduced precipitation in boreal forest soils. *Soil Biology & Biochemistry* 32: 1431–1441.

Bird, R.B., Stewart, W.E., and Lightfoot, E.N. 2002. *Transport Phenomena*, 2nd ed. Wiley. New York, USA. 895 p.

Bisbee, K.E., Gower, S.T., Norman, J.M., and Nordheim, E.V. 2001. Environmental controls on ground cover species composition and productivity in a boreal black spruce forest. *Oecologia* 129: 261–270.

Björkman, O. 1981a. Responses to different quantum flux densities. In: Lange, O.L., Nobel, P.S., Osmond, C.B., and Ziegler, H. (Eds.). *Encyclopedia of Plant Physiology*, New Series. Vol. 12A. Springer. Heidelberg, Berlin. p. 57–107.

Björkman, O. 1981b. The response of photosynthesis to temperature. In: Grace, J., Ford E.D., and Jarvis, P.G. (Eds.). *Plants and Their Atmospheric Environment. The 21st Symposium of the British Ecological Society, Edinburgh 1979.* Blackwell Scientific. Oxford. p. 273–301.

Bligh, E.G. and Dyer, D.J. 1959. A rapid method of total lipid extraction and purification. *Canadian Journal of Biochemistry and Physiology* 37: 911–917.

Blume, H.-P., Brümmer, G.W., Schwertmann, U., Horn, R., Kögel-Knabner, I., Stahr, K., Auerswald, K., Beyer, L., Hartmann, A., Litz, N., Scheinost, A., Stanjek, H., Welp, G., and Wilke, B.-M. 2002. Organische Substanz und Bodenbiologie. In: Scheffer/Schachtschabel. *Lehrbuch der Bodenkunde.* 15th Auflage. Spektrum Akademischer Verlag. Heidelberg, Berlin. p. 51–102.

Bodelier, P.L.E. and Laanbroek, H.J. 2004. Nitrogen as a regulatory factor of methane oxidation in soils and sediments. *FEMS Microbiology Ecology* 47: 265–277.

Boer, G.J. and Yu, B. 2003. Climate sensitivity and response. *Climate Dynamics* 20: 415–429.

Bohlmann, J., Meyer-Gauen, G., and Croteau, R. 1998. Plant terpenoid synthases: molecular biology and phylogenetic analysis. *Proceedings of the National Academy of Sciences USA* 95: 4126–4133.

Bonan, G.B., Pollard, D., and Thompson, S.L. 1992. Effects of boreal forest vegetation on global climate. *Nature* 359: 716–718.

Bond, B.J. and Kavanagh, K.L. 1999. Stomatal behavior of four woody species in relation to leaf-specific hydraulic conductance and threshold water potential. *Tree Physiology* 19: 503–510.

Bonn, B. and Moortgat, G.K. 2003. Sesquiterpene ozonolysis: origin of new particle formation. *Geophysical Research Letters* 30: 1585.

Bonn, B., Schuster, G., and Moortgat, G.K. 2002. Influence of water vapor on the process of new particle formation during monoterpene ozonolysis. *Journal of Physical Chemistry A* 106: 2869–2881.

Bonn, B., Hirsikko, A., Hakola, H., Kurtén, T., Laakso, L., Boy, M., Dal Maso, M., Mäkelä, J.M., and Kulmala, M. 2006. Ambient sesquiterpene concentration and its link to air ion measurements. *Atmospheric Chemistry and Physics Discussions* 6: 13165–13224.

Boone, R.D., Nadelhoffer, K.J., Canary, J.D., and Kaye, J.P. 1998. Roots exert a strong influence on the temperature sensitivity of soil respiration. *Nature* 396: 570–572.

Borken, W., Beese, F., Brumme, R., and Lamersdorf, N. 2002. Long-term reduction in nitrogen and proton inputs did not affect atmospheric methane uptake and nitrous oxide emission from a German spruce forest soil. *Soil Biology & Biochemistry* 34: 1815–1819.

Bossel, H. 1991. Modelling forest dynamics: moving from description to explanation. *Forest Ecology and Management* 42: 129–142.

Bothe, H., Jost, G., Schloter, M., Ward, B.B., and Witzel, K.-P. 2000. Molecular analysis of ammonia oxidation and denitrification in natural environments. *FEMS Microbiology Reviews* 24: 673–690.

Botkin, D.B. and Simpson, L. 1990. The first statistically valid estimate of biomass for a large region. *Biogeochemistry* 9: 161–174.

Bowden, R.D., Melillo, J.M., Steudler, P.A., and Aber, J.D. 1991. Effects of nitrogen additions on annual nitrous-oxide fluxes from temperate forest soils in the Northeastern United-States. *Journal of Geophysical Research* 96: 9321–9328.

Bowes, G. 1991. Growth at elevated CO_2. Photosynthetic responses mediated through Rubisco. *Plant Cell and Environment* 14: 795–806.

Boy, M., Kulmala, M., Ruuskanen, T.M., Pihlatie, M., Reissell, A., Aalto, P.P., Dal Maso, M., Hellen, H., Hakola, H., Jansson, R., Hanke, M., and Arnold, F. 2005. Sulphuric acid closure and contribution to nucleation mode particle growth. *Atmospheric Chemistry and Physics* 5: 863–878.

Brassard, B.W. and Chen, H.Y.H. 2006. Stand structural dynamics of North American boreal forest.*Critical Reviews in Plant Sciences* 25: 115–137.

Bremner, J.M. 1965. Organic forms of nitrogen. In: Black, C.A. (Ed.). *Methods of Soil Analysis. Part 2. Chemical and Microbiological Properties.* Agronomy 9. American Society of Agronomy. Madison, Wisconsin, USA. p. 1238–1255.

Bremner, J.M. 1997. Sources of nitrous oxide in soils. *Nutrient Cycling in Agroecosystems* 49: 7–16.

Brennen, C.E. 1995. *Cavitation and Bubble Dynamics*. Oxford University Press. New York, USA. 291 p.

Bricard, J., Delattre, P., Madelaine, G., and Pourprix, M. 1976. Detection of ultra-fine particles by means of a continuous flux condensation nucleus counter. In: Liu, B.Y.H. (Ed.). *Fine Particles: Aerosol Generation, Measurement, Sampling, and Analysis.* Academic Press. New York, USA. p. 565–580.

Britto, D.T., Glass, A.D.M., Kronzucker, H.J., and Siddiqi, M.Y. 2001. Cytosolic concentrations and transmembrane fluxes of NH_4^+/NH_3. An evaluation of recent proposals. *Plant Physiology* 125: 523–526.

Britto, D.T., Glass, A.D.M., and Kronzucker, H.J. 2002. NH_4^+ toxicity in higher plants: a critical review. *Journal of Plant Physiology* 159: 567–584.

Brouwer, R. 1962. Distribution of dry matter in the plant. *Netherlands Journal of Agricultural Science* 10: 361–376.

Brownlee, C. 2001. The long and the short of stomatal density signals. *Trends in Plant Science* 6: 441–442.

Brumme, R. and Beese, F. 1992. Effect of liming and nitrogen fertilization on emissions of CO_2 and N_2O from a temperate forest. *Journal of Geophysical Research* 97(D12): 12851–12858.

Brumme, R. and Borken, W. 1999. Site variation in methane oxidation as affected by atmospheric deposition and type of temperate forest ecosystem. *Global Biogeochemical Cycles* 13: 493–502.

Brumme, R., Verchot, L.V., Martikainen, P.J., and Potter, C.S. 2005. Contribution of trace gases nitrous oxide (N_2O) and methane (CH_4) to the atmospheric warming balance of forest biomes. In: Griffiths, H. and Jarvis, P. (Eds.). *The Carbon Balances of Forest Biomes.* Garland Science, BIOS Scientific Publishers, Cromwell Press. Trowbridge, UK.

Brunekreef, B. and Holgate, S. 2002. Air pollution and health. *Lancet* 360: 1233–1242.

538   References

Brusa, R.W. and Fröhlich, C. 1986. Absolute radiometers (PMO6) and their experimental characterization. *Applied Optics* 25: 4173–4180.

Buchanan, B.B., Gruissem, W., and Jones, R.L. (Eds.). 2000. *Biochemistry and Molecular Biology
of Plants*. American Society of Plant Physiologists. Rockville, Maryland, USA. 1367 p.

Buchanan-Wollaston, V. 1997. The molecular biology of leaf senescence. *J Exp Bot* 48: 181–199.

Buchmann, N. 2000. Biotic and abiotic factors controlling soil respiration rates in *Picea abies*
stands. *Soil Biology and Biochemistry* 32: 1625–1635.

Buckley, T.N. and Mott, K.A. 2002. Dynamics of stomatal water relations during the humidity
response: implications of the two hypothetical mechanisms. *Plant Cell and Environment* 25:
407–419.

Bunce, J.A. 2000. Acclimation to temperature of the response of photosynthesis to increased carbon dioxide concentration in Taraxacum officinale. *Photosynthesis Research* 64: 89–94.

Bunge, M. 1996. *Finding Philosophy in Social Science*. Yale University Press. New Haven, USA,
and London, UK.

Buol, S.W., Hole, F.D., McCracken, R.J., and Southard, R.J. 1997. *Soil Genesis and Classification*,
4th ed. Iowa State University Press. Ames, Iowa, USA. 527 p.

Burdon, J. 2001. Are the traditional concepts of the structures of humic substances realistic? *Soil
Science* 166: 752–769.

Burke, R.A., Zepp, R.G., Tarr, M.A., Miller, W.L., and Stocks, B.J. 1997. Effect of fire on soil-
atmosphere exchange of methane and carbon dioxide in Canadian boreal forest sites. *Journal
of Geophysical Research* 102(D24): 29289–29300.

Burton, D.L., and Beauchamp, E.G. 1994. Profile nitrous oxide and carbon dioxide concentrations
in a soil subject to freezing. *Soil Science Society of America Journal* 58: 115–122.

Busch, F., Hüner, N.P.A., and Ensminger, I. 2007. Increased air temperature during simulated
autumn conditions does not increase photosynthetic carbon gain but affects the dissipation
of excess energy in seedlings of the evergreen conifer Jack Pine. *Plant Physiology* 143:
1242–1251.

Butterbach-Bahl, K., Gasche, R., Huber, C.H., Kreutzer, K., and Papen, H. 1998. Impact of N-input
by wet deposition on N-trace gas fluxes and CH$_4$-oxidation in spruce forest ecosystems of the
temperate zone in Europe. *Atmospheric Environment* 32(3): 559–564.

Buurman, P. and Jongmans, A.G. 2005. Podzolization and soil organic matter dynamics. *Geoderma*
125: 71–83.

Cabada, J., Khlystov, A., Wittig, A., Pilinis, C., and Pandis, S. 2004. Light scattering by fine particles during the Pittsburgh Air Quality Study: measurements and modeling. *Journal of Geophysical Research* 109: D16S03.

Cajander, A.K. 1926. The theory of forest types. *Acta Forestalia Fennica* 29: 1–108.

Campbell, D.T. 1974. Downward causation in hierarchically organised biological systems. In:
Ayala, F.J. and Dobzhansky, T. (Eds.). *Studies in the Philosophy of Biology: Reduction and
Related Problems*. Macmillan. Basingstoke, London. p. 179–186.

Campbell, R.K. and Sugano, A.I. 1975. Phenology and bud burst in Douglas-fir related to provenance, photoperiod, chilling, and flushing temperature. *Botanical Gazette* 136: 290–298.

Campbell, R.K. and Sugano, A.I. 1979. Genecology of bud-burst phenology in Douglas-fir:
response to flushing temperature and chilling. *Botanical Gazette*. 140: 223–31.

Canadell, J.G., Le Quéré, C., Raupach, M.R., Field, C.B., Buitenhuis, E.T., Ciais, P., Conway,
T.J., Gillett, N.P., Houghtonh, R.A., and Marland, G. 2007. Contributions to accelerating
atmospheric CO2 growth from economic activity, carbon intensity, and efficiency of natural
sinks. *Proceedings of the National Academy of Sciences* 104: 18866–18870.

Canham, C.D. 1985. Suppression and release during canopy recruitment in *Acer saccharum*. *Bulletin of the Torrey Botanical Club* 112: 134–145.

Cannell, M.R.G. and Thornley, J.H.M. 2000. Modelling the components of plant respiration: some
guiding principles. *Annals of Botany* 85: 45–54.

Caraglio, Y. and Barthélémy, D. 1997. Revue critique des termes relatifs à la croissance et à la
ramification des tiges des végétaux vasculaires. In: Bouchon, J., de Reffye, Ph., and Barthélémy,

D. (Eds.). *Modélisation et simulation de l'architecture des végétaux*, INRA Éditions. Versailles, France. p. 11–87.

Castillo, F.J. and Greppin, H. 1988. Extracellular ascorbic acid and enzyme activities related to ascorbic acid metabolism in Sedum album L. leaves after ozone exposure. *Experimental and Environmental Botany* 28: 231–238.

Cescatti, A. 1997. Modelling the radiative transfer in discontinuous canopies of asymmetric crowns I. Model structure and algorithms. *Ecological Modelling* 101: 263–274.

Chalot, M., Javelle, A., Blaudez, D., Lambilliote, R., Cooke, R., Sentenac, H., Wipf, D., and Botton, B. 2002. An update on nutrient transport processes in ectomycorrhizas. *Plant and Soil* 244: 165–175.

Chalot, M., Blaudez, D., and Brun, A. 2006. Ammonia: a candidate for nitrogen transfer at the mycorrhizal interface. *Trends in Plant Science* 11: 263–266.

Chan, A.S.K., Steudler, P.A., Bowden, R.D., Gulledge, J., and Cavanaugh, C.M. 2005. Consequences of nitrogen fertilization on soil methane consumption in a productive temperate deciduous forest. *Biology Fertility of Soils* 41: 182–189.

Lardy, L., Wrage, N., Metay, A., Chotte, J.-L., and Bernoux, M. 2007. Soils, a sink for N_2O? A review. *Global Change Biology*. 13: 1–17.

Chefetz, B., Salloum, M.J., Deshmukh, A.P. and Hatcher, P.G. 2002. Structural components of humic acids as determined by chemical modifications and carbon-13 NMR, pyrolysis-, and thermochemolysis-gas chromatography/mass spectrometry. *Soil Science Society of America Journal* 66: 1159–1171.

Chen, H.H. and Li, P.H. 1978. Interactions of low temperature, water stress, and short days in the induction of stem frost hardiness in red osier dogwood. *Plant Physiology* 62: 833–835.

Chen, J.M. and Black, T.A. 1992. Defining leaf area index for non-flat leaves. *Plant, Cell and Environment* 15: 421–429.

Chen, J.M., Rich, P.M., Gower, S.T., Norman, J.M., and Plummer, S. 1997. Leaf area index of boreal forests: theory, techniques, and measurements. *Journal of Geophysical Research-Atmospheres* 102: 29429–29443.

Chen, Z. and Gallie, D.R. 2004. The ascorbic acid redox state controls guard cell signalling and stomatal movements. *Plant Cell* 16: 1143–1162.

Chitnis, P.R. 2001. Photosystem I: function and physiology. *Annual Review of Plant Physiology and Plant Molecular Biology* 52: 593–626.

Clapham, D., Ekberg, I., Little, C.H.A., and Savolainen, O. 2001. Molecular biology of conifer frost tolerance and potential applications to tree breeding. In: Bigras, F.J. and Columbo, S.J. (Eds.). *Conifer Cold Hardiness*. Kluwer. Dordrecht, The Netherlands. p. 187–219.

Cochard, H., Ewers, F., and Tyree, M.T. 1994. *Journal of Experimental Botany* 45: 1085–1089.

Cochard, H., Bréda, N., and Granier, A. 1996. Whole tree hydraulic conductance and water loss regulation in Quercus during drought: evidence for stomatal control of embolism? *Annales des Sciences Forestieres* 53: 197–206.

Coe, H., Gallagher, M.W., Choularton, T.W., and Dore, C. 1995. Canopy scale measurements of stomatal and cuticular O_3 uptake by Sitka spruce. *Atmospheric Environment* 29: 1413–1423.

Colman, R. 2003. A comparison of climate feedbacks in general circulation models. *Climate Dynamics* 20: 865–873.

Conen, F. and Neftel, A. 2007. Do increasingly depleted $\delta^{15}N$ values of atmospheric N_2O indicate a decline in soil N_2O reduction? *Biogeochemistry* 82: 321–326.

Conrad, R. 1996. Soil microorganisms as controllers of atmospheric trace gases (H_2, CO, CH_4, OCS, N_2O, and NO). *Microbiological Reviews* 60: 609–640.

Conrad, R. 1999. Contribution of hydrogen to methane production and control of hydrogen concentrations in methanogenic soils and sediments. *FEMS Microbiology Ecology* 28: 193–202.

Constable, J.V.H., Litvak, M.E., Greenberg, J.P., and Monson, R.K. 1999. Monoterpene emission from coniferous trees in response to elevated CO_2 and climate warming. *Global Change Biology* 5: 255–267.

Corey, A.T. and Logsdon, S.D. 2005. Limitations of the chemical potential. *Soil Science of America Journal* 69: 976–982.

Cosby, B.J., Hornberger, G.M., Clapp, R.B., and Ginn, T.R. 1984. A statistical exploration of the relationships of soil-moisture characteristics to the physical properties of soils. *Water Resources Research* 20: 682–690.

Cowan, I.R. 1977. Stomatal behaviour and environment. *Advances in Botanical Research* 4: 117–228.

Cowan, I.R. and Farquhar, G.D. 1977. Stomatal function in relation to leaf metabolism and environment. In: Jennings, D.H. (Ed.). *Integration of Activity in the Higher Plant*. Cambridge University Press. Cambridge. p. 471–505.

Coyle, M. 2005. *The Gaseous Exchange of Ozone at Terrestrial Surfaces: Non-stomatal Deposition to Grassland*. (Diss.) School of Geosciences, Faculty of Science and Engineering, The University of Edinburgh.

CRC. 2003. *CRC Handbook of chemistry and physics*, 84th edition. CRC Press, FL, USA. 2496 p.

Crill, P.M. 1991. Seasonal patterns of methane uptake and carbon dioxide release by a temperate woodland soil. *Global Biogeochemical Cycles* 5: 319–334.

Crill, P.M., Martikainen, P.J., Nykänen, H., and Silvola, J. 1994. Temperature and N fertilization effects on methane oxidation in a drained peatland soil. *Soil Biology and Biochemistry* 26(10): 1331–1339.

Criquet, S. 2002. Measurement and characterization of cellulase activity in sclerophyllous forest litter. *Journal of Microbiological Methods* 50: 165–173.

Dahl, E. and Mork, E. 1959. Om sambandet mellom temperatur, ånding og vekst hos gran. *Picea abies* (L.) Karst. Medd. *Det Norske Skogforsøksvesen* 53: 83–93.

Dai, A., Trenberth, K.E., and Qian, T. 2004. A global data set of Palmer Drought Severity Index for 1870–2002: relationship with soil moisture and effects of surface warming. *Journal of Hydrometeorology* 5: 1117–1130.

Dal Maso, M., Kulmala, M., Riipinen, I., Wagner, R., Hussein, T., Aalto, P.P., and Lehtinen, K.E.J. 2005. Formation and growth of fresh atmospheric aerosols: eight years of size distribution data from SMEAR II, Hyytiälä, Finland. *Boreal Environment Research* 10: 323–336.

Dal Maso, M., Sogacheva, L., Aalto, P.P., Riipinen, I., Komppula, M., Tunved, P., Korhonen, L., Suur-Uski, V., Hirsikko, A., Kurtén, T., Kerminen, V.-M., Lihavainen, H., Viisanen, Y., Hansson, H.-C., and Kulmala, M. 2007. Aerosol size distribution measurements at four Nordic field stations: identification, analysis and trajectory analysis of new particle formation bursts. *Tellus* B59: 350–361.

Dalton, F.N. and Van Genuchten, M.Th. 1986. The time-domain reflectometry method for measuring soil water content and salinity. *Geoderma* 38: 237–250.

Darrigol, O. 2005. *Worlds of Flow. A History of Hydrodynamics from the Bernoullis to Prandtl*. Oxford University Press. Oxford, UK. 356 p.

Dau, H. 1994. Molecular mechanisms and quantitative models of variable photosystem II fluorescence. *Photochemistry and Photobiology* 60: 1–23.

Davidson, E.A. 1991. Fluxes of nitrous oxide and nitric oxide from terrestrial ecosystems. In: Rogers, J.E. and Whitman, W.B. (Eds.). *Microbial Production and Consumption of Greenhouse Gases: Methane, Nitrogen Oxides, and Halomethanes*. American Society for Microbiology. Washington, USA. p. 219–235.

Davidson, E.A. 1993. Soil water content and the ratio of nitrous oxide to nitric oxide emitted from soil. In: Oremland, R.S. (Ed.). *Biogeochemistry of Global Change Radiatively Active Trace Gases*. Chapman & Hall, New York, USA. p. 369–386.

Davidson, E.A., Belk, E., and Boone, R.D. 1998. Soil water content and temperature as independent or confounded factors controlling soil respiration in a temperate mixed hardwood forest. *Global Change Biology* 4: 217–227.

Davidson, E.A., Janssens, I.A., and Luo, Y. 2006. On the variability of respiration in terrestrial ecosystems: moving beyond Q_{10}. *Global Change Biology* 12: 154–164.

Davidson, R.L. 1969. Effect of root/leaf temperature on root/shoot ratios in some pasture grasses and clover. *Annals of Botany* 33: 561–569.

Davies, P.J. 1995. The plant hormones: their nature, occurrence, and functions. In: Davies, P.J. (Ed.).*Plant Hormones: Physiology, Biochemistry and Molecular Biology.* Kluwer. Boston, Massachussetts, USA. p. 1–12.

De Boer, W. and Kowalchuk, G.A. 2001. Nitrification in acid soils: micro-organisms and mechanisms. Review. *Soil Biology & Biochemistry* 33: 853–866.

De Boer, A.H. and Volkov, V. 2003. Logistics of water and salt transport through the plant: structure and functioning of the xylem. *Plant, Cell & Environment* 26: 87–101.

Coninck, F. 1980. Major mechanisms in formation of spodic horizons. *Geoderma* 24: 101–128.

DeLuca, T.H., Zackrisson, O., Nilsson, M.-C., and Sellstedt, A. 2002. Quantifying nitrogen-fixation in feather moss carpets of boreal forests. *Nature* 419: 917–920.

Delwiche, C.F. and Sharkey, T.D. 1993. Rapid appearance of ^{13}C in biogenic isoprene when $^{13}CO_2$ is fed to intact leaves. *Plant Cell and Environment* 16: 587–591.

Delzon, S., Sartore, M., Burlett, R., Dewar, R., and Loustau, D. 2004. Hydraulic responses to height growth in maritime pine trees. *Plant Cell and Environment* 27: 1077–1087.

Demmig-Adams, B. and Adams, W.W. 1996. The role of xanthophyll cycle carotenoids in the protection of photosynthesis. *Trends in Plant Science* 1: 21–26.

Demmig-Adams, B., Adams, III, W.W. Barker, B.A., Logan, B.A., Bowling, D.R., and Verhoeven, A.S. 1996. Using Chlorophyll fluorescence to assess the fraction of absorbed light allocated to thermal dissipation of excess excitation. *Physiologia Plantarum* 98: 253–264.

den Camp, H.J.M.O., Kartal, B., Guven, D., van Niftrik, L.A.M.P., Haaijer, S.C.M., van der Star, W.R.L., van de Pas-Schoonen, K.T., Cabezas, A., Ying, Z., Schmid, M.C., Kuypers, M.M.M., van de Vossenberg, J., Harhangi, H.R., Picioreanu, C., van Loosdrecht, M.C.M., Kuenen, J.G., Strous, M., and Jetten, M.S.M. 2006. Global impact and application of the anaerobic ammonium-oxidizing (anammox) bacteria. *Biochemical Society Transactions* 34: 17–178.

Dentener, F. 2006. Global Maps of Atmospheric Nitrogen Deposition, 1860, 1993, and 2050. Data Set from Oak Ridge National Laboratory Distributed Active Archive Center, Oak Ridge, Tennessee, USA. Available at http://www.daac.ornl.gov/.

Derenne, S. and Largeau, C. 2001. A review of some important families of refractory macromolecules: composition, origin and fate in soils and sediments. *Soil Science* 166: 833–847.

Dewar, R.C., Medlyn, B.E. and McMurtrie, R.E. 1998. A mechanistic analysis of light and carbon use efficiencies. *Plant, Cell and Environment* 21: 573–588.

Dibb, J.E., Arsenault, M., Peterson, M.C., and Honrath, R.E. 2002. Fast nitrogen oxide photochemistry in Summit, Greenland snow. *Atmospheric Environment* 36: 2501–2511.

Dickson, R.E. and Isebrands, J.G. 1991. Leaves as regulators of stress responses. In: Mooney, H.A., Winner, W.E., and Pell, E.J. (Eds.). *Response of Plants to Multiple Stresses.* Academic Press. San Diego, California. p. 4–34.

Donaldson, L.A. 2001. Lignification and lignin topochemistry - an ultrastructural view. *Phytochemistry* 57: 859–873.

Doran, J.W., Mielke, L.N., and Stamatiadis, S. 1988. Microbial activity and N cycling as regulated by soil water-filled pore space. Proceedings of the 11th ISTRO Conference, Edinburgh, UK: 49–54. Conference Organizing Committee, Edinburgh, UK.

Dörr, H., Katruff, L., and Levin, I. 1993. Soil texture parameterization of the methane uptake in aerated soils. *Chemosphere* 26: 697–713.

Drake, B.G., Gonzàlez-Meler, M.A., and Long, S.P. 1997. More efficient plants: a consequence of rising atmospheric CO_2? *Annual Review of Plant Physiology and Plant Molecular Biology* 48: 609–639.

Drebs, A., Nordlund, A., Karlsson, P., Helminen, J., and Rissanen, P. 2002. Climatological Statistics of Finland 1971–2000, Climatological statistics of Finland 2001, Finnish Meteorological Institute, Helsinki.

Dubois, M., Gilles, K.A., Hamilton, J.K., Rebers, P.A., and Smith, F. 1956. Colorimetric method for determination of sugars and related substances. *Analytical Chemistry* 28: 350–356.

Dudareva, N., Pichersky, E., and Gerschenzon, J. 2004. Biochemistry of plant volatiles. *Plant Physiology* 135: 1893–1902.

Duursma, R. and Mäkelä, A. 2007. Summary models for light interception and light-use efficiency of non-homogeneous canopies. *Tree Physiology* 27: 859–870.

Duursma, R., Kolari, P., Perämäki, M., Nikinmaa, E., Hari, P., Delzon, S., Loustau, D., Ilvesniemi, H., Pumpanen, J., and Mäkelä, A. 2008. Predicting the decline in daily maximum transpiration rate of two pinestands during drought based on constant minimum leaf water potentialand plant hydraulic conductance. *Tree Physiology* 28: 265–276.

Eamus, D. and Jarvis, P. 1989. The direct effects of increase in the global atmospheric CO_2 concentration on natural and commercial temperate trees and forests.*Advances in Ecological Research* 19: 1–55.

Einsle, O. and Kroneck, P.M.H. 2004. Structural basis of denitrification. *Biological Chemistry* 385: 875–883.

Einstein, A. 1905. Über einen die Erzeugung und Verwandlung des Lichtes betreffenden heuristischen Gesichtspunkt. *Annalen der Physik* 17: 132–148.

Eisele, F.L. and Tanner, D.J. 1993. Measurement of the gas phase concentration of H_2SO_4 and methane sulfonic acid and estimates of H_2SO_4 production and loss in the atmosphere. *Journal of Geophysical Research* 98: 9001–9010.

Eivazi, F. and Bayan, M.R. 1996. Effects of long-term prescribed burning on the activity of select soil enzymes in an oak-hickory forest. *Canadian Journal of Forest Research* 2: 1799–1804.

Ek, A.R. 1971. *Size-Age Relationships for Open Grown Red Pine*. The University of Wisconsin, Forestry Research Notes 156. 4p.

Ekblad, A., Wallander, H., and Näseholm, T. 1998. Chitin and ergosterol combined to measure total and living fungal biomass in ectomycorrhizas. *New Phytologist* 138: 143–149.

Eskling, M., Emanuelsson, A., and Åkerlund, H.-E. 2001. Enzymes and mechanisms for violaxanthin-zeaxanthin conversion. In: Aro, E.-M., and Andersson, B. (Eds.). *Regulation of Photosynthesis*. Kluwer. Dordrecht, The Netherlands. p. 433–452.

Evert, R.F. 2006. *Esau's Plant Anatomy: Meristems, Cells, and Tissues of the Plant Body: Their Structure, Function, and Development*, 3rd ed. Wiley. Holboken, New Jersey, USA.

Falge, E., Baldocchi, D., Olson, R.J., Anthoni, P., Aubinet, M., Bernhofer, C., Burba, G., Ceulemans, R., Clement, R., Dolman, H., Granier, A., Gross, P., Grünwald, T., Hollinger, D., Jensen, N.-O., Katul, G., Keronen, P., Kowalski, A., Ta Lai, C., Law, B.E., Meyers, T., Moncrieff, J., Moors, E., Munger, J.W., Pilegaard, K., Rannik, Ü., Rebmann, C., Suyker, A., Tenhunen, J., Tu, K., Verma, S., Vesala, T., Wilson, K., and Wofsy, S. 2001. Gap filling strategies for defensible annual sums of net ecosystem exchange. *Agricultural and Forest Meteorology* 107: 43–69.

Fall, R. 2003. Abundant oxygenates in the atmosphere: a biochemical perspective.*Chemical Reviews* 103: 4941–4951.

Fang, C. and Moncrieff, J.P. 1999. A model for soil CO_2 production and transport 1: model development. *Agricultural and Forest Meteorology* 95: 225–236.

Farmer, V.C. and Lumsdon, D.G. 2001. Interactions of fulvic acid with aluminium and a proto-imogolite sol: the contribution of E-horizon eluates to podzolization. *European Journal of Soil Science* 52: 177–188.

Farquhar, G.D., Caemmerer von, S., and Berry, J.A. 1980. A biochemical model of photosynthetic CO2 assimilation in leaves of C3 species. *Planta* 149: 78–90.

Feigenwinter, C., Bernhofer, C., and Vogt, R. 2004. The influence of advection on short term CO_2 budget in and above a forest canopy. *Boundary-Layer Meteorology* 113: 201–224.

Fernandez, I.J. and Kosian, P.A. 1987. Soil air carbon dioxide concentrations in a New England spruce-fir forest. *Soil Science Society of America Journal* 51: 261–263.

Ferretti, D.F., Miller, J.B., White, J.W.C., Lassey, K.R., Lowe, D.C., and Etheridge, D.M. 2007. Stable isotopes provide revised global limits of aerobic methane emissions from plants. *Atmospheric Chemistry and Physics* 7: 237–241.

Ferrier, J.M., Tyree, M.T., and Christy, A.L. 1975. The theoretical time-dependent behavior of a Münch pressure-flow system: the effect of sinusoidal time variation in sucrose loading and water potential. *Canadian Journal of Botany* 53: 1120–1127.

Ferry, J.G. 2002. *Methanogenesis Biochemistry. Encyclopedia of Life Sciences*. Wiley. Chichester, UK.

Finér, L., Messier, C., and De Grandpré, L. 1997. Fine-root dynamics in mixed boreal conifer - broad-leafed forest stands at different successional stages after fire. *Canadian Journal of Forest Research* 27: 304–314.

Finnigan, J.J., Clement, R., Malhi, Y., Leuning, R., and Cleugh, H.A. 2003. A re-evaluation of long-term flux measurement techniques. Part I: averaging and coordinate rotation. *Boundary-Layer Meteorology* 107: 1–48.

Finnlayson-Pitts, B. and Pitts, J.N. 2000. *Chemistry of the Upper and Lower Atmosphere*, 2nd ed. Academic Press. New York, USA.

Fischbach, R.J., Staudt, M., Zimmer, I., Rambal, S., and Schnitzler, J.-P. 2002. Seasonal pattern of monoterpene synthase activities in leaves of the evergreen tree Quercus ilex. *Physiologia Plantarum* 114: 354–360.

Fisher, J.B., Angeles, G.A., Ewers, F.W., and Lopez-Portillo, J. 1997. Survey of root pressure in tropical vines and woody species. *International Journal of Plant Science* 158: 44–50.

Fisher, R.A., Williams, M., Do Vale, R.L., Da Costa, A.L., and Meir, P. 2006. Evidence from Amazonian forests is consistent with isohydric control of leaf water potential. *Plant Cell and Environment*. 29: 151–165.

Flagan, R.C. 1998. History of electrical aerosol measurements. *Aerosol Science and Technology* 28: 301–380.

Flexas, J. and Medrano, H. 2002. Drought-inhibition of photosynthesis in C3 plants: stomatal and nonstomatal limitations revisited. *Annals of Botany* 89: 183–189.

Flügge, U.-I. 1999. Phosphate translocators in plastids. *Annual Review of Plant Physiology and Plant Molecular Biology* 50: 27–45.

Ford, D. 2000. *Scientific Method for Ecological Research*. Cambridge University Press. UK. 564 p.

Forster, P., Ramaswamy, V., Artaxo, P., Berntsen, T., Betts, R.A., Fahey, D.W., Haywood, J., Lean, J., Lowe, D.C., Myhre, G., Nganga, J., Prinn, R., Raga, G., Schulz, M., and van Dorland, R. 2007. Changes in atmospheric constituents and in radiative forcing. In: Solomon, S., Qin, D., Manning, M., Chen, Z., Marquis, M., Averyt, K.B., Tignor M., and Miller H.L. (Eds.). *Climate Change 2007: The Physical Science Basis*. Cambridge University Press. Cambridge, UK and New York, USA. p. 129–234.

Fowler, D., Flechard, C., Cape, J.N., Storeton-West, R.L., and Coyle, M. 2001. Measurements of ozone deposition to vegetation quantifying the flux, the stomatal and non-stomatal components. *Water, Air and Soil Pollution* 130: 63–74.

Francis, C.A., Beman, J.M., and Kuypers, M.M.M. 2007. New processes and players in the nitrogen cycle: the microbial ecology of anaerobic and archaeal ammonia oxidation. MINI-REVIEW. *The ISME Journal* 1: 19–27.

Freidenreich, S.M. and Ramaswamy, V. 1999. A new multiple-band solar radiative paramaterization for general circulation models. *Journal of Geophysical Research*. 104: 31389–31409.

Freijer, J.I. and Leffelaar, P.A. 1996. Adapted Fick's law applied to soil respiration. *Water Resources Research* 32: 791–800.

Frey, T.E.A. 1978. The Finnish school and forest site-types. In: Whittaker, R.H. (Ed.). *Classification of Plant Communities*. Dr. W. Junk. Hague, The Netherlands. p. 81–110.

Friedli, H., Lötscher, H., Oeschger, H., Siegenthaler, U., and Stauffer, B. 1986. Ice core record of 13C/12C ratio of atmospheric CO_2 in the past two centuries. *Nature* 324: 237–238.

Friedrich, M.W. 2005. Methyl-coenzyme M reductase genes - unique functional markers for methanogenic and anaerobic methane-oxidizing Archaea. *Methods in Enzymology* 397: 428–442.

Fry, S.C. 2001. Plant cell walls. In: *Encyclopedia of Life Sciences*. Wiley. Chichester, UK. www.els.net.

Fuchigami, L.H., Weiser, C.J., Kobayashi, K., Timmis, R., and Gusta, L.V. 1982. A degree growth stage (°GS) model and cold acclimation in temperate woody plants. In: Li, P.H. and Sakai, A. (Eds.). *Plant Cold Hardiness and Freezing Stress*, Vol. 2. Academic Press. New York, USA. p. 93–116.

Gaige, E., Dail, D.B., Hollinger, D.Y., Davidson, E.A., Fernandez, I.J., Sievering, H., White, A., and Halteman, W. 2007. Changes in canopy processes following whole-forest canopy nitrogen fertilization of a mature spruce-hemlock forest. *Ecosystems* 10: 1133–1147.

Gardner, W.R. 1960. Dynamic aspects of water availability to plants.*Soil Science* 89: 63–73.

Garrett, P.W., and Zahner, R. 1973. Fascicle density and needle growth responses of red pine to water supply over two seasons. *Ecology* 54: 1328–1334.

Gates, D.M., Keegan, H.J., Schleter, J.C., and Weidner, V.R. 1965. Spectral properties of plants. *Applied Optics* 4: 11–20.

Gauthier, S., Bergman, A., and Bergeron, Y. 1996. Forest dynamics modelling under natural fire cycles: a tool to define natural mosaic diversity for forest management. *Environmental Monitoring and Assessment* 39(1–3): 417–434.

Genty, B., Briantais, J.-M., and Baker, N.R. 1989. The relationship between the quantum yield of photosynthetic electron transport and quenching of chlorophyll fluorescence. *Biochimica et Biophysica Acta* 990: 87–92.

Geßler, A., Rienks, M., and Rennenberg, H. 2002. Stomatal uptake and cuticular adsorption contribute to dry deposition of NH_3 and NO_2 to needles of adult spruce (*Picea abies*) trees. *New Phytologist* 156: 179–194.

Ghashghaie, J. and Cornic, G. 1994. Effect of temperature on partitioning of photosynthetic electron flow between CO_2 assimilation and O_2 reduction and the CO_2/O_2 specifity of Rubisco. *Journal of Plant Physiology* 143: 643–650.

Gielen, B., Jach, M.E., and Ceulemans, R. 2000. Effects of season, needle age and elevated atmospheric CO_2 on chlorophyll fluorescence parameters and needle nitrogen concentration in Scots pine (*Pinus sylvestris*). *Photosynthetica* 38: 13–21.

Gilmore, A. and Ball, M.C. 2000. Protection and storage of chlorophyll in overwintering evergreens. *Proceedings of the National Academy of Sciences of the USA* 97: 11098–11101.

Gilmore, A.M. 1997. Mechanistic aspects of xanthophyll cycle-dependent photoprotection in higher plant chloroplasts and leaves. *Physiologia Plantarum* 99: 197–209.

Givnish, T.J. 1985. On the use of optimality arguments. In: Givnish, T.J. (Ed.). *On the Economy of Plant Form and Function*. Cambridge University Press. Cambridge, UK. p. 3–9.

Givnish, T.J. 1986. Biomechanical constraints on crown geometry in forest herbs. In: Givnish, T.J. (Ed.). *On the Economy of Plant Form and Function*. Cambridge University Press. Cambridge, UK. p. 525–583.

Givnish, T.J. 1995. Plant stems: biomechanical adaptation for energy capture and influence on species distributions. In: Gartner, B.L. (Ed.). *Plant Stems: Physiology and Functional Morphology*. Academic Press. San Diego, California, USA. p. 3–49.

Glass, A.D.M., Britto, D.T., Kaiser, B.N., Kinghorn, J.R., Kronzucker, H.J., Kumar, A., Okamoto, M., Rawat, S., Siddiqi, M.Y., Unkles, S.E., and Vidmar, J.J. 2002. The regulation of nitrate and ammonium transport systems in plants. *Journal of Experimental Botany* 53: 855–864.

Glinski, J. and W. Stepniewski. 1985. *Soil Aeration and Its Role for Plants*. CRC Press. Boca Raton, Florida, USA.

Goldstein, A.H., McKay, M., Kurpius, M.R., Schade, G.W., Lee, A., Holzinger, R., and Rasmussen, R. 2004. Forest thinning experiment confirms ozone deposition to forest canopy is dominated by reaction with biogenic VOCs. *Geophysical Research Letters* 31: L22106.

Goossens, A., de Visscher, A., Boeckz, P., and Van Cleemput, O. 2001. Two-year field study on the emission of N_vO from coarse and middle-textured Belgian soils with different land use. *Nutrient Cycling in Agroecosystems* 60: 23–34.

Goulden, M.L. and Crill, P.M. 1997. Automated measurements of CO_2 exchange at the moss surface of a black spruce forest. *Tree Physiology* 17: 537–542.

Govindarajulu, M., Pfeffer, P.E., Jin, H., Abubaker, J., Douds, D.D., Allen, J.W., Bückling, H., Lammers, P.J., and Shachar-Hill, Y. 2005. Nitrogen transfer in the arbuscular mycorrhizal symbiosis. *Nature* 435: 819–823.

Gower, S.T. and Norman, J.M. 1991. Rapid estimation of leaf area index in conifer and broad-leaf plantations. *Ecology* 72: 1896–1900.

Graham, J.H. and Miller, R.M. 2005. Mycorrhizas: gene to function. *Plant and Soil* 274: 79–100.

Greacen, E.L. and Oh, J.S. 1972. Physics of root growth. *Nature, New Biology* 235: 24–25.

Greaves, J.R. and Carter, E.G. 1920. Influence of moisture on the bacterial activities of the soil. *Soil Science* 10: 361–387.

Green, R.O., Dozier, J., Robert, D., and Painter, T. 2002. Spectral snow-reflectance models for grain-size and liquid-water fraction in melting snow for the solar-reflected spectrum. *Annals of Glaciology* 34: 71–73.

Greer, D.H. 1983. Temperature regulation of the development of frost hardiness in *Pinus radiate* D. Don. *Australian Journal of Plant Physiology* 10: 539–547.

Gregory, P.J. 2006. Roots, rhizosphere and soil: the route to a better understanding of soil science? *European Journal of Soil Science* 57: 2–12.

Groffman, P. 1991. Ecology of nitrification and denitrification in soil evaluated at scales relevant to atmospheric chemistry. In: Rogers, J.E. and Whitman, W.B. (Eds.). *Microbial Production and Consumption of Greenhouse Gases: Methane, Nitrogen Oxides, and Halomethanes*. American Society for Microbiology. Washington, USA. p. 201–217.

Grogan, P. and Chapin, III, F.S. 1999. Arctic soil respiration: effects of climate and vegetation depend on season. *Ecosystems* 2: 451–459.

Groisman, P.Ya., Sherstyukov, B.G., Razuvaev, V.N., Knight, R.W., Enloe, J.G., Stroumentova, N.S., Whitfield, P.H., Førland, E., Hannsen–Bauer, I., Tuomenvirta, H., Alexandersson, H., Mescherskaya, A.V., and Karl, T.R. 2007. Potential forest fire danger over Northern Eurasia: changes during the 20th century. *Global and Planetary Change* 56: 371–386.

Guenther, A.B., Monson, R.K., and Fall, R. 1991. Isoprene and monoterpene emission rate variability: observations with eucalyptus and emission rate algorithm development. *Journal of Geophysical Research* 96: 10.799–10.808.

Gulledge, J. and Schimel, J.P. 1998a. Moisture control over atmospheric CH_4 consumption and CO_2 production in diverse Alaskan soils. *Soil Biology and Biochemistry* 30(8/9): 1127–1132.

Gulledge, J. and Schimel, J.P. 1998b. Low-concentration kinetics of atmospheric CH_4 oxidation in soil and mechanisms of NH_4^+ inhibition. *Applied and Environmental Microbiology* 64(11): 4291–4298.

Gulledge, J., Doyle, A.P., and Schimel, J.P. 1997. Different NH_4^+-inhibition patterns of soil CH_4 consumption: a result of distinct CH_4 –oxidizer populations across sites? *Soil Biology and Biochemistry* 29(1): 13–21.

Gunderson, C.A. and Wullschleger, S.D. 1994. Photosynthetic acclimation in trees to rising atmospheric CO_2: a broader perspective. *Photosynthetic Research* 39: 369–388.

Gustafsson, J.P., Berggren, D., Simonsson, M., Zysset, M., and Mulder, J. 2001. Aluminium solubility mechanism in moderately acid Bs horizons of podzolized soils. *European Journal of Soil Science* 52: 655–665.

Hacke, U.G., Sperry, J.S., Pockman, W.T., Davis, S.D., and McCulloh, K.A. 2001. Trends in wood density and structure are linked to prevention of xylem implosion by negative pressure. *Oecologia* 126: 457–461.

Hagemann, S. 2002. *An Improved Land Surface Parameter Data Set for Global and Regional Climate Models*. Max-Planck-Institute for Meteorology, Report 336. Hamburg, Germany. 21 p.

Hakola, H., Laurila, T., Hiltunen, V., Hellen, H., and Keronen, P. 2003. Seasonal variation of VOC concentrations above a boreal coniferous forest. *Atmospheric Environment* 37: 1623–1634.

Hakola, H., Tarvainen, V., Bäck, J., Ranta, H., Bonn, B., Rinne, J., and Kulmala, M., 2006. Seasonal variation of mono- and sesquiterpene emission rates of Scots pine. *Biogeosciences* 3: 93–101.

Haldrup, A., Jensen, P.E., Lunde, C., and Scheller, H.V. 2001. Balance of power: a view of the mechanism of photosynthetic state transitions. *Trends in Plant Science* 6: 301–305.

Hallé, F., Oldeman, R.A.A., and Tomlinson, P.B. 1978. *Tropical Trees and Forests: An Architectural Analysis*. Springer. Berlin, Germany. 441 p.

Hämeri, K., Väkevä, M., Hansson, H.-C., and Laaksonen, A. 2000. Hygroscopic growth of ultrafine ammonium sulphate aerosol measured using an ultrafine tandem differential mobility analyzer. *Journal of Geophysical Research* 105: 22231–22242.

Handisides, G.M., Plass-Dülmer, C., Gilge, S., Bingemer, H., and Berresheim, H. 2003. Hohen-peissenberg Photochemical Experiment (HOPE 2000): measurements and photostationary state calculations of OH and peroxy radicals. *Atmospheric Chemistry and Physics* 3: 1565–1588.

Hänninen, H. 1995. Effects of climatic change on trees from cool and temperate regions: an eco-physiological approach to modelling of bud burst phenology. *Canadian Journal of Botany* 73: 183–199.

Hänninen, H. 2006. Climate warming and the risk of frost damage to boreal forest trees: identifi-cation of critical ecophysiological traits. *Tree Physiology* 26: 889–898.

Hänninen, H. and Kramer, K. 2007. A framework for modelling the annual cycle of trees in boreal and temperate regions. *Silva Fennica* 41: 167–205.

Hänninen, H., Kolari, P., and Hari, P. 2005. Seasonal development of Scots pine under climatic warming: effects on photosynthetic production. *Canadian Journal of Forest Research* 35: 2092–2099.

Hansen, J., Lacis, A., Rind, D., Russell, G., Stone, P., Fung, I., Ruedy R., and Lerner, J. 1984. Cli-mate sensitivity: analysis of feedback mechanisms. *Meteorological Monograph* 29: 130–163.

Hansen, J., Sato, M., Ruedy, R., Nazarenko, L., Lacis, A., Schmidt, G., Russell, G., Aleinov, I., Bauer, M., Bauer, S., Bell, N., Cairns, B., Canuto, V., Chandler, M., Cheng, Y., Genio, A.D., Faluvegi, G., Fleming, E., Friend, A., Hall, T., Jackman, C., Kelley, M., Kiang, N., Koch, D., Lean, J., Lerner, J., Lo, K., Menon, S., Miller, R., Minnis, P., Novakov, T., Oinas, V., Perlwitz, J., Perlwitz, J., Rind, D., Romanou, A., Shindell, D., Stone, P., Sun, S., Tausnev, N., Thresher, D., Wielicki, B., Wong, T., Yao, M., and Zhang, S. 2005. Efficacy of climate forcings. *Journal of Geophysical Research* 110: D18104.

Hari, P. 1972. Physiological stage of development in biological models of growth and maturation. *Annales Botanici Fennici* 9: 107–115.

Hari, P. 1999. Towards a quantitative theory in the research of plant production. In: Purohit, S., Agarwal, S., Vyas, S., and Gehlot, H. (Eds.). *Agro's Annual Review of Plant Physiology IV*. Agrobios. India. p. 1–45.

Hari, P. and Häkkinen, R. 1991. The utilization of old phenological time series of budburst to compare models describing annual cycles of plants. *Tree Physiology* 8: 281–287.

Hari, P. and Kulmala, M. 2005. Station for measuring ecosystems–atmosphere relations (SMEAR II). *Boreal Environment Researach* 10: 315–322.

Hari, P. and Mäkelä, A. 2003. Annual pattern of photosynthesis of Scots pine in the boreal zone. *Tree Physiology* 23: 145–155.

Hari, P., Leikola, M., and Räsänen, P. 1970. A dynamic model of the daily high increment of plants. *Annales Botanici Fennici* 7: 375–378.

Hari, P., Kellomäki, S., Mäkelä, A., Ilonen, P., Kanninen, M., Korpilaahti, E., and Nygren, M. 1982. Dynamics of early development of tree stand (in Finnish with English summary). *Acta Forestalia Fennica* 177: 39 p.

Hari, P., Kaipiainen, L., Korpilahti, E., Mäkelä, A., Nilsson, T., Oker–Blom, P., Ross, J., and Salminen, R. 1985. *Structure, Radiation and Photosynthetic Production in Coniferous Stands*. University of Helsinki, Department of Silviculture, research notes. 54: 1–233.

Hari, P., Heikinheimo, P., Mäkelä, A., Kaipiainen, L., Korpilahti, E., and Salmela, J. 1986. Trees as a water transport system. *Silva Fennica* 20: 205–210.

Hari, P., Nygren, P., and Korpilahti, E. 1991. Internal circulation of carbon within a tree. *Canadian Journal of Forest Research* 21: 514–515.

Hari, P., Keronen, P., Bäck, J., Altimir, N., Linkosalo, T., Pohja, T., Kulmala, M., and Vesala, T. 1999. An improvement of the method for calibrating measurements of photosynthetic CO_2 flux. *Plant, Cell and Environment* 22: 1297–1301.

Hari, P., Raivonen, M., Vesala, T., Munger, J.W., Pilegaard, K., and Kulmala, M. 2003. Ultraviolet light and leaf emission of NOx. *Nature* 422: 134.

Harju, A.M., Venäläinen, M., Anttonen, S., Viitanen, H., Kainulainen, P., Saranpää, P., and Vapaavuori, E. 2003. Chemical factors affecting the brown-rot decay resistance of Scots pine heartwood. *Trees* 17: 263–268.

Harvey, L.D.D. 2000. *Global Warming: The Hard Science*. Pearson Education. Harlow, UK. 336 p.

He, J.-Z., Shen, J.-P., Zhang, L.-M., Zhu, Y-G., Zheng, Y.-M., Xu, M.-G., and Di, H. 2007. Quantitative analyses of the abundance and composition of ammonia-oxidizing bacteria and ammonia-oxidizing archaea of a Chinese upland red soil under long-term fertilization practices. *Environmental Microbiology* 9: 2364–2374.

Heard, D.E. (Ed.). 2006. *Analytical Techniques for Atmospheric Measurement*. Blackwell. Oxford. 510 p.

Hegerl, G.C., Zwiers, F.W., Braconnot, P., Gillett, N.P., Luo, Y., Marengo, J., Nicholls N., Penner, J.E., and Stott, P.A. 2007. Understanding and attributing climate change. In: Solomon, S., Qin, D., Manning, M., Chen, Z., Marquis, M., Averyt, K.B., Tignor, M., and Miller, H.L. (Eds.). *Climate Change 2007: The Physical Science Basis*. Cambridge University Press. Cambridge, UK and New York, NY, USA. 337–383.

Held, I.M. and B.J. Soden. 2006. Robust responses of the hydrological cycle to global warming. *Journal of Climate* 19: 5686–5699.

Helmisaari, H. 1992. Nutrient retranslocation within the foliage of *Pinus sylvestris*. *Tree Physiology* 10: 45–58.

Helmisaari, H.-S., Derome, J., Nöjd, P., and Kukkola, M. 2007. Fine root biomass in relation to site and stand characteristics in Norway spruce and Scots pine. *Tree Physiology* 27: 1493–1504.

Hendrickson, L., Furbank, R.T., and Chow, W.S. 2004. A simple alternative approach to assessing the fate of absorbed light energy using chlorophyll fluorescence. *Photosynthesis Research* 82: 73–81.

Hering, S., Stolzenburg, M., Quant, F., Oberreit, D., and Keady, P. 2005. A laminar-flow, water-based condensation particle counter WCPC. *Aerosol Science and Technology* 39: 659–672.

Hikosaka, K. and Hirose, T. 1998. Leaf and canopy photosynthesis of C_3 plants at elevated CO_2 in relation to optimal partitioning of nitrogen among photosynthetic components: theoretical prediction. *Ecological Modelling* 106: 247–259.

Hilbert, D.W. 1990. Optimization of plant root: shoot ratios and internal nitrogen concentration. *Annales of Botany* 66: 91–99.

Hillel, D. 1998. *Environmental Soil Physics*. Academic Press. Boston, USA. 771 p.

Hinds, W.C. 1982. *Aerosol Technology: Properties, Behavior, and Measurement of Airborne Particles*, 1st ed. Wiley. New York. 424 pp.

Hinds, W.C. 1999. *Aerosol Technology: Properties, Behavior, and Measurements of Airborne Particles*, 2nd ed. Wiley-Interscience. New York, USA. 504 p.

Hinsinger, P., Gobran, G.R., Gregory, P.J., and Wenzel, W.W. 2005. Rhizosphere geometry and heterogeneity arising from root mediated physical and chemical processes. *New Phytologist* 168: 293–303.

Hirsikko, A., Laakso, L., Hõrrak, U., Aalto, P.P., Kerminen, V.-M., and Kulmala, M. 2005. Annual and size dependent variation of growth rates and ion concentrations in boreal forest. *Boreal Environment Research* 10: 357–370.

Hishi, T. 2007. Heterogeneity of individual root within the fine root architecture: causal links between physiological and ecosystem functions. *Journal of Forest Research* 12: 126–133.

Hoefnagel, M.H.N., Atkin, O.K., and Wiskich, J.T. 1998. Interdependence between chloroplasts and mitochondria in the light and the dark. *Biochimica et Biophysica Acta/Bioenergetics* 1366: 235–255.

Högberg, P., Nordgren, A., Buchmann, N., Taylor, A.F.S., Ekblad, A., Högberg, M.N., Nyberg, G., Ottosson-Löfvenius, M., and Read, D.J. 2001. Large-scale forest girdling shows that current photosynthesis drives soil respiration. *Nature* 411: 789–792.

Holbrook, N.M. and Zwieniecki, M.A. 1999. Embolism repair and xylem tension: do we need a miracle? *Plant Physiology* 120: 7–10.

Holton, J.R. 2004. *An Introduction to Dynamic Meteorology*. Elsevier Academic Press. New York, USA. 535 p.

Hölttä, T., Vesala, T., Nikinmaa, E., Perämäki, M., Siivola, E., and Mencuccini, M. 2005. Field measurements of ultrasonic acoustic emissions and stem diameter variations. New insight into the relationship between xylem tensions and embolism. *Tree Physiology* 25: 237–243.

Hölttä, T., Vesala, T., Sevanto, S., Perämäki, M., and E. Nikinmaa. 2006. Modeling xylem and phloem water flows in trees according to cohesion theory and Münch hypothesis. *Trees — Structure and Function* 20: 67–78.

Horn, H.S. 1971. *The Adaptive Geometry of Trees*. Princeton University Press. Princeton, New Jersey, USA. 144 p.

Horton, P., Ruban, A.V., and Walters, R.G. 1996. Regulation of light harvesting in green plants. *Annual Review of Plant Physiology and Plant Molecular Biology* 47: 655–684.

Houghton, J.T. 1997. *Global Warming: The Complete Briefing*, 2nd ed. Cambridge University Press, Cambridge, UK. 251 p.

Houghton, J.T., Ding, Y., Griggs, D.J., Noguer, M., van der Linden, P.J., and Xiaosu, D. (Eds.). 2001. *Climate Change 2001. The Scientific Basis. Contribution of Working group I to the Third Assessment Report of the Intergovernmental Panel on Climate Change (IPCC)*. Cambridge University Press. Cambridge, UK. 944 p.

Howard, D.M. and Howard, P.J.A. 1993. Relationships between CO_2 evolution, moisture content and temperature for a range of soil types. *Soil Biology Biochemistry* 25: 1537–1546.

Huntington, T.G. 2006. Evidence for intensification of the global water cycle: review and synthesis. *Journal of Hydrology* 319: 83–95.

Hytteborn, H., Maslov, A.A., Nazimova, D.I., and Rysin, L.P. 2005. Boreal forests of Eurasia. In: Andersson, F. (Ed.). *Coniferous Forests. Ecosystems of the World, Vol. 6*. Elsevier. Amsterdam, The Netherlands. p. 23–99.

Ilomäki, S., Mäkelä, A., and Nikinmaa, E. 2003. Crown rise due to competition drives biomass allocation in silver birch (*Betula pendula* L.). *Canadian Journal of Forest Research* 33: 2395–2404.

Ilvesniemi, H. and Liu, C. 2001. Biomass distribution in a young Scots pine stand. *Boreal Environment Research* 6: 3–8.

Ingestad, T. and Ågren, G. 1991. The influence of plant nutrition on biomass allocation. *Ecological Applications* 1: 168–174.

IPCC. 2007. *Contribution of Working Group I to the Fourth Assessment Report of the Intergovernmental Panel on Climate Change*. In: Solomon, S., Qin, D., Manning, M., Chen, Z., Marquis, M., Averyt, K., Tignor, M.M.B., and Miller, H.L. (Eds.). Cambridge University Press, Cambridge, UK and New York, USA, 800 pp.

Irvine, J. and Grace, J. 1997. Continuous measurements of water tensions in the xylem of trees based on the elastic properties of wood. *Planta* 202: 455–461.

Irvine, J., Perks, M.P., Magnani, F., and Grace, J. 1998. The response of *Pinus sylvestris* to drought: stomatal control of transpiration and hydraulic conductance. *Tree Physiology* 18: 393–402.

Jach, M.E. and Ceulemans, R. 2000. Effects of season, needle age and elevated atmospheric CO_2 on photosynthesis in Scots pine (*Pinus sylvestris*). *Tree Physiology* 20: 145–157.

Jackson, G.E. and Grace, J. 1996. Field measurements of xylem cavitation: are acoustic emissions useful? *Journal of Experimental Botany* 47: 1643–1650.

Jacob, D.J. 1999. *Introduction to Atmospheric Chemistry*. Princeton University Press. Princeton, New Jersey, USA.

Jansen, B., Nierop, K.G.J., and Verstraten, J.M. 2005. Mechanisms controlling the mobility of dissolved organic matter, aluminium and iron in podzol B horizons. *European Journal of Soil Science* 56: 537–550.

Jansen, E., Overpeck, J., Briffa, K.R., Duplessy, J.-C., Joos, F., Masson-Delmotte, V., Olago, D., Otto-Bliesner, B., Peltier, W.R., Rahmstorf, S., Ramesh, R., Raynaud, D., Rind, D., Solomina, O., Villalba, R., and Zhang, D. 2007. Palaeoclimate. In: Solomon, S., Qin, D., Manning, M., Chen, Z., Marquis, M., Averyt, K.B., Tignor, M., and Miller, H.L. (Eds.). *Climate Change 2007: The Physical Science Basis. Contribution of Working Group I to the Fourth Assessment Report of the Intergovernmental Panel on Climate Change*. Cambridge University Press. Cambridge, UK and New York, USA. p. 433–497.

Jarvis, P.G. 1976. The interpretation of the variations in leaf water potential and stomatal conductance found in canopies in the field. *Philosophical Transactions Of the Royal Society London Series* B273(927): 593–610.

Jayne, J.T., Leard, D.C., Zhang, X.F., Davidovits, P., Smith, K.A., Kolb, C.E., and Worsnop, D.R. 2000. Development of an aerosol mass spectrometer for size and composition analysis of sub-micron particles. *Aerosol Science and Technology* 33: 49–70.

Jenkins, G.M. and Watts, D.G. 1968. *Spectral Analysis and Its Applications.* Holden-Day. San Francisco, USA. 525 p.

Jiang, Q.-Q. and Bakken L.R. 1999. Nitrous oxide production and methane oxidation by different ammonia-oxidizing bacteria. *Applied and Environmental Microbiology* 65(6): 2679–2684.

Jin, M. and Liang, S. 2006. An improved land surface emissivity parameter for land surface models using global remote sensing observations. *Journal of Climate* 19: 2867–2881.

Johnson, D. 1999. Simulated nitrogen cycling response to elevated CO_2 in *Pinus taeda* and mixed deciduous forests. *Tree Physiology* 19: 321–327.

Jones, D.L., Healey, J.R., Willett, V.B., Farrar, J.F., and Hodge, A. 2005. Dissolved organic nitrogen uptake by plants – an important N uptake pathway? *Soil Biology & Biochemistry* 37: 413–423.

Jones, H.G. and Sutherland, R.A. 1991. Stomatal control of xylem embolism. *Plant Cell and Environment* 14: 607–612.

Jongmans, A.G., van Breemen, N., Lundström, U., van Hees, P.A.W., Finlay, R.D., Srinivasan, M., Unestam, T., Giesler, R., Melkerud, P.-A., and Olsson, M. 1997. Rock-eating fungi. *Nature* 389: 682–683.

Joshi, M., Shine, K., Ponater, M., Stuber, N., Sausen, R., and Li, L. 2003. A comparison of climate response to different radiative forcings in three general circulation models: towards an improved metric of climate change. *Climate Dynamics* 20: 843–854.

Jouzel, J., Lorius, C., Petit, J.R., Genthon, C., Barkov, N.I., Kotlyakov, V.M., and Petrov, V.M. 1987. Vostok ice core: a continuous isotope temperature record over the last climatic cycle (160,000years). *Nature* 329: 403–408.

Jouzel, J., Barkov, N.I., Barnola, J.M., Bender, M., Chappellaz, J., Genthon, C., Kotlyakov, V.M., Lipenkov, V., Lorius, C., Petit, J.R., Raynaud, D., Raisbeck, G., Ritz, C., Sowers, T., Stievenard, M., Yiou, F., and Yiou, P. 1993. Extending the Vostok ice-core record of palaeoclimate to the penultimate glacial period. *Nature* 364: 407–412.

Jouzel, J., Waelbroeck, C., Malaize, B., Bender, M., Petit, J.R., Stievenard, M., Barkov, N.I., Barnola, J.M., King, T., Kotlyakov, V.M., Lipenkov, V., Lorius, C., Raynaud, D., Ritz, C., and Sowers, T. 1996. Climatic interpretation of the recently extended Vostok ice records. *Climate Dynamics* 12: 513–521.

Jurgens, G., Lindstrom, K., and Saano, A. 1997. Novel group within the kingdom Crenarchaeota from boreal forest soil. *Applied and Environmental Microbiology* 63: 803–805.

Juurola, E. 2003. Biochemical acclimation patterns of *Betula pendula* and *Pinus sylvestris* seedling to elevated carbon dioxide concentration. *Tree Physiology* 23: 85–95.

Juurola, E. 2005. *Photosynthesis, CO_2 and temperature – an approach to analyse the constraints to acclimation of trees to increasing CO_2 concentration.* (Diss.) Dissertationes Forestales 4. 47 p.

Kähkönen, M.A., Wittmann, C., Kurola, J., Ilvesniemi, H., and Salkinoja-Salonen, M.S. 2001. Microbial activity of boreal forest soil in a cold climate. *Boreal Environment Research* 6: 19–28.

Kähkönen, M., Wittmann, C., Ilvesniemi, H., Westman, C.J., and Salkinoja-Salonen, M. 2002. Mineralization of detritus and oxidation of methane in acid boreal coniferous forest soils: seasonal and vertical distribution and effects of clear-cut. *Soil Biology & Biochemistry* 34: 1191–1200.

Kaimal, J.C. and Finnigan, J.J. 1994. *Atmospheric Boundary Layer Flows. Their Structure and Measurement.* Oxford University Press. New York, USA. 289 p.

Kaiser, E.-A., Kohrs, K., Kücke, M., Schnug, E., Heinemeyer, O., and Munch, J.C. 1998. Nitrous oxide release from arable soil: importance of N-fertilization, crops and season. *Soil Biology and Biochemistry* 30: 1553–1563.

Kaiser, W.M., Weiner, H., and Huber, S.C. 1999. Nitrate reductase in higher plants: a case study for transduction of environmental stimuli into control of catalytic activity. *Physiologia Plantarum* 105: 385–390.

Kaldenhoff, R., Ribas Carbo, M., Flexas Sans, J., Lovisolo, C., Heckwolf, M., and Uehlein, N. 2008. Aquaporins and plant water balance. *Plant, Cell & Environment* 31: 658–666.

Kanninen, M., Hari, P., and Kellomäki, S. 1982. A dynamic model for above ground growth of dry matter production in a forest community. *Journal of Applied Ecology* 19: 465–476.

Karjalainen, T. 1996. Dynamics and potentials of carbon sequestration in managed stands and wood products in Finland under changing climatic conditions. *Forest Ecology and Management* 80: 113–132.

Karkalas, J. 1985. An improved method for the determination of native and modified starch. *Journal of the Science in Food and Agriculture* 36: 1019–1027.

Karl, T., Fall, R., Rosenstiel, T.N., Prazeller, P., Larsen, B., Seufer, G., and Lindinger, W. 2002. On-line analysis of the $^{13}CO_2$ labeling of leaf isoprene suggests multiple subcellular origins of isoprene precursors. *Planta* 215: 894–905.

Kasimir-Klemedtsson, Å. and Klemedtsson, L. 1997. Methane uptake in Swedish forest soil in relation to liming and extra N-deposition. *Biology and Fertility of Soils* 25: 296–301.

Kaurichev, I.S., Panov, N.P., Rozov, N.N., Stratonovich, M.V., and Fokin, A.D. 1989. *Soil Science*, 4th ed.. Agropromizdat, Moscow. (In Russian).

Keeling, C.D. and Whorf, T.P. 2005. Atmospheric CO_2 records from sites in the SIO air sampling network. In: *Trends: A Compendium of Data on Global Change*. Carbon Dioxide Information Analysis Center, Oak Ridge National Laboratory, U.S. Department of Energy, Oak Ridge, Tennessee, USA.

Kellomäki, S. and Oker-Blom, P. 1983. Canopy structure and light climate in a young Scots Pine stand. *Silva Fennica* 17: 1–21.

Kennedy, R.E., Turner, D.P., Cohen, W.B., and Guzy, M. 2006. A method to efficiently apply a biogeochemical model to a landscape. *Landscape Ecology* 21: 213–224.

Keppler, F., Hamilton, J.T.G., Braß, M., and Röckmann, R. 2006. Methane emissions from terrestrial plants under aerobic conditions. *Nature* 439: 187–191.

Kerminen, V.-M. and Kulmala, M. 2002. Analytical formulae connecting the "real" and the "apparent" nucleation rate and the nuclei number concentration for atmospheric nucleation events. *Journal of Aerosol Science* 33: 609–622.

Keronen, P., Reissell, A., Rannik, Ü., Pohja, T., Siivola, E., Hiltunen, V., Hari, P., Kulmala, M., and Vesala, T. 2003. Ozone flux measurements over a Scots pine forest using eddy covariance method: performance evaluation and comparison with flux-profile method. *Boreal Environment Research* 8: 425–443.

Keskitalo, J., Bergquist, G., Gardeström, P., Jansson, S. 2005. A cellular timetable of autumn senescence. *Plant Physiol.* 139: 1635–1648.

Kesselmeier, J. and Staudt, M. 1999. Biogenic volatile organic compounds (VOC): an overview on emission, physiology and ecology. *Journal of Atmospheric Chemistry* 33: 23–88.

Kiehl, J.T. and Trenberth, K.E. 1997. Earth's annual global mean energy budget. *Bulletin of the American Meteorological Society* 78: 197–208.

Killham, K. 1990. Nitrification in coniferous forest soils. *Plant and Soil* 128: 31–44.

Killham, K. 1994. *Soil Ecology*. Cambridge University Press. Cambridge, UK. 242 p.

Kimball, B.A., Idso, S.B., Johnson, S., and Rillig, M.C. 2007. Seventeen years of carbon dioxide enrichment of sour orange trees: final results. *Global Change Biology* 13: 2171–2183,.

King, G.M. and Schnell, S. 1994. Effect of increasing atmospheric methane concentration on ammonium inhibition of soil methane consumption. *Nature* 370: 282–284.

Kirschbaum, M.U.F. 1995. The temperature dependence of soil organic matter decomposition, and the effect of global warming on soil organic C storage. *Soil Biology and Biochemistry* 27(6): 753–760.

Kirschbaum, M.U.F., Bruhn, D., Etheridge, D.M., Evans, J.R., Farquhar, G.D., Gifford, R.M., Paul, K.I., and Winters, A.J. 2006. A comment on the quantitative significance of aerobic methane release by plants. *Functional Plant Biology* 33: 521–530.

Kitajima, M. and Butler, W.L. 1975. Quenching of chlorophyll fluorescence and primary photochemistry in chloroplasts by dibromothymoquinone. *Biochimica et Biophysica Acta* 376: 105–115.

Klemedtsson, L., Kasimir Klemedtsson, Å., Moldan, F., and Weslien, P. 1997. Nitrous oxide emission from Swedish forest soils in relation to liming and simulated increased N-deposition. *Biology and Fertility of Soils*. 25: 290–295.

Klemedtsson, L., Jiang, Q., Klemedtsson, A.K., and Bakken, L. 1999. Autotrophic ammonium-oxidising bacteria in Swedish mor humus. *Soil Biology & Biochemistry* 31(6): 839–847.

Knoblauch, M. and van Bel, A.J.E. 1998. Sieve tubes in action. *Plant Cell* 10: 35–50.

Knowles, R. 1982. Denitrification. *Microbiological Reviews* 46: 43–70.

Koch, G.W., Sillet, S.C., Jennings, G.M., and Davis, S.D. 2004. The limits to tree height. *Nature* 428: 851–854.

Kögel-Knabler, I. 1995. Composition of soil organic matter. In: Alef, K. and Nannipieri, P. (Eds.). *Methods in Applied Soil Microbiology and Biochemistry*. Academic Press. London, UK. p. 66–78.

Kögel-Knabner, I. 2000. Analytical approaches for characterizing soil organic matter. *Organic Geochemistry* 31: 609–625.

Kögel-Knabner, I. 2002. The macromolecular organic composition of plant and microbial residues as inputs to soil organic matter. *Soil Biology and Biochemistry* 34: 139–162.

Kolari, P., Keronen, P., and Hari, P. 2004a. The accuracy of transpiration measirements with a dynamic chamber system. *Report Series in Aerosol Science* 68: 112–114.

Kolari, P., Pumpanen, J., Rannik, U., Ilvesniemi, H., Hari, P., and Berninger, F. 2004b. Carbon balance of different aged Scots pine forests in Southern Finland. *Global Change Biology* 10: 1106–1119.

Kolari, P., Pumpanen, J., Kulmala, L., Ilvesniemi, H., Nikinmaa, E., Grönholm T., and Hari, P. 2006. Forest floor vegetation plays an important role in photosynthetic production of boreal forests. *Forest Ecology and Management* 221: 241–248.

Kolari, P., Lappalainen, H.K., Hänninen, H., and Hari, P. 2007. Relationship between temperature and the seasonal course of photosynthesis in Scots pine at northern timberline and in southern boreal zone. *Tellus* 59B: 542–552.

Kontunen-Soppela, S. and Laine, K. 2001. Seasonal fluctuation of dehydrins is related to osmotic status in Scots pine needles.*Trees* 15: 425–430.

Körner, C. 2006. Plant CO_2 responses: an issue of definition, time and resource supply.*New Phytologist* 172: 393–411.

Koski, V. and Sievänen, R. 1985. Timing of growth cessation in relation to the variations in the growing season. In: Tigerstedt, P.M.A., Puttonen, P., and Koski, V. (Eds.). *Crop Physiology of Forest Trees*. Helsinki University Press. Helsinki, Finland. p. 167–193.

Kousaka, Y., Niida, T., Okuyama, K., and Tanaka, H. 1982. Development of a mixing type condensation nucleus counter. *Journal of Aerosol Science* 13: 231–240.

Kowalchuk, G.A. and Stephen, J.R. 2001. Ammonia-oxidizing bacteria: a model for molecular microbial ecology. *Annual Reviews Microbiology* 55: 485–529.

Kowalski, A.S., Loustau, D., Berbigier, P., Manca, G., Tedeschi, V., Borghetti, M., Valentini, R., Kolari, P., Berninger, F., Rannik, Ü., Hari, P., Rayment, M., Mencuccini, M., Moncrieff, J., and Grace, J. 2004. Paired comparisons of carbon exchange between undisturbed and regenerating stands in four managed forests in Europe. *Global Change Biology* 10: 1707–1723.

Kramer, K., Leinonen, I., Bartelink, H.H., Berbigier, P., Borghetti, M., Bernhofer, C.H., Cienciala, E., Dolman, A.J., Froer, O., Gracia, C.A., Granier, A., Grünwald, T., Hari, P., Jans, W., Kellomäki, S., Loustau, D., Magnani, F., Markkanen, T., Matteucci, G., Mohren, G.M.J., Moors, E., Nissinen, A., Peltola, H., Sabaté, S., Sanchez, A., Sontag, M., Valentini, R., and Vesala, T. 2002. Evaluation of six process-based forest growth models using eddy-covariance measurements of CO_2 and H_2O fluxes at six forest sites in Europe. *Global Change Biology* 8: 213–230.

Krause, G.H. and Weis, E. 1991. Chlorophyll fluorescence and photosynthesis: the basics. *Annual Review of Plant Physiology and Plant Molecular Biology* 42: 313–349.

Kreuzwieser, J., Cojocariu, C., Jüssen, V., and Rennenberg, H. 2002. Elevated atmospheric CO_2 causes seasonal changes in carbonyl emissions from *Quercus ilex*. *New Phytologist* 154: 327–333.

Kronzucker, H.J., Siddiqi, M.Y., and Glass, A.D.M. 1996. Kinetics of NH_4^+ influx in spruce. *Plant Physiology* 110: 773–779.

Kruse, J. and Adams, M.A. 2008. Sensitivity of respiratory metabolism and efficiency to foliar nitrogen during growth and maintenance. *Global Change Biology* 14: 1–19.

Kull, O. and Niinemets, Ü. 1998. Distribution of leaf photosynthetic properties in tree canopies: comparison of species with different shade tolerance. *Functional Ecology* 12: 472–479.

Kulmala, L., Launiainen, S., Pumpanen, J., Lankreijer, H., Lindroth, A., Hari, P., and Vesala, T. 2008. H_2O and CO_2 fluxes at the floor of a boreal pine forest. *Tellus* 60B: 167–178.

Kulmala, M., Toivonen, A., Mäkelä, J.M., and Laaksonen, A. 1998. Analysis of the growth of nucleation mode particles observed in Boreal forest. *Tellus* 50B: 449–463.

Kulmala, M., Hienola, J., Pirjola, L., Vesala, T., Shimmo, M., Altimir, N., and Hari, P. 1999. A model for NO_x-O_3-terpene chemistry in chamber measurements of plant gas exchange. *Atmospheric Environment* 33: 2145–2156.

Kulmala, M., Hämeri, K., Aalto, P.P., Mäkelä, J.M., Pirjola, L., Nilsson, E.D., Buzorius, G., Rannik, Ü., Dal Maso, M., Seidl, W., Hoffman, T., Janson, R., Hansson, H.-C., Viisanen, Y., Laaksonen, A., and O'Dowd, C.D. 2001a. Overview of the international project on biogenic aerosol formation in the boreal forest (BIOFOR). *Tellus* 53B: 423–343.

Kulmala, M., Dal Maso, M., Mäkelä, J.M., Pirjola, L., Väkevä, M., Aalto, P., Miikkulainen, P., Hämeri, K., and O'Dowd, C.D. 2001b. On the formation, growth and composition of nucleation mode particles. *Tellus* 53B: 479–490.

Kulmala, M., Vehkamäki, H., Petäjä, T., Dal Maso, M., Lauri, A., Kerminen, V.-M., Birmili, W., and McMurry, P.H. 2004a. Formation and growth rates of ultrafine atmospheric particles: a review of observations. *Journal of Aerosol Science*. 35: 143–176.

Kulmala, M., Laakso, L., Lehtinen, K.E.J., Riipinen, I., Dal Maso, M., Anttila, T., Kerminen, V.-M., Hõrrak, U., Vana, M., and Tammet, H. 2004b. Initial steps of aerosol growth. *Atmospheric Chemistry and Physics* 4: 2553–1560.

Kulmala, M., Suni, T., Lehtinen, K.E.J., Dal Maso, M., Boy, M., Reissell, A., Rannik, Ü., Aalto, P.P., Keronen, P., Hakola, H., Bäck, J., Hoffmann, T., Vesala, T., and Hari, P. 2004c. A new feedback mechanism linking forests, aerosols, and climate. *Atmospheric Chemistry and Physics* 4: 557–562.

Kumar, P. and Foufoula-Georgiou, E., 1997. Wavelet analysis for geophysical applications. *Reviews of Geophysics* 35: 385–412.

Küppers, M. 1989. Ecological significance of above ground patterns in woody plants: a question of cost-benefit relationships. *Tree* 4: 375–379.

Kürten, A., Curtius, J., Nillius, B., and Borrmann, S. 2005. Characterization of an automated water-based expansion condensation nucleus counter for ultrafine particles. *Aerosol Science and Technology* 39: 1174–1183.

Kurtén, T., Kulmala, M., Dal Maso, M., Suni, T., Reissell, A., Vehkamäki, H., Hari, P., Laaksonen, A., Viisanen, Y., and Vesala, T. 2003. Estimation of different forest-related contributions to the radiative balance using observations in southern Finland, *Boreal Environment Research* 8: 275–285.

Kuuluvainen, T. 2002. Natural variability of forests as a reference for restoring and managing biological diversity in Boreal Fennoscandia. *Silva Fennica* 36: 97–125.

Kuuluvainen, T. and Pukkala, T. 1987. Effect of crown shape and tree distribution on the spatial distribution of shade. *Agricultural and Forest Meteorology* 40: 215–231.

Kuuluvainen, T. and Pukkala, T. 1989. Simulation of within-tree and between-tree shading of direct radiation in a forest canopy: effect of crown shape and sun elevation. *Ecological Modelling* 49: 89–100.

Laakso, L., Grönholm, T., Kulmala, L., Haapanala, S., Hirsikko, A., Lovejoy, E.R., Kazil, J., Kurtén, T., Boy, M., Nilsson, E.D., Sogachev, A., Riipinen, I., Stratmann, F. and Kulmala, M. 2007: Hot-air balloon as a platform for boundary layer profile measurements during particle formation. *Boreal Env. Res.* 12: 279–294.

Lalonde, S., Wipf, D., and Frommer, W.B. 2004. Transport mechanisms for organic forms of carbon and nitrogen between source and sink. *Annual Review of Plant Biology* 55: 341–372.

Lam, P., Jensen, M.M., Lavik, G., McGinnis, D.F., Müller, B., Schubert, C.J., Amann, R., Thamdrup, B., and Kuypers, M.M.M. 2007. Linking crenarchaeal and bacterial nitrification to anammox in the Black Sea. *Proceedings of National Academy of Sciences* 104: 7104–7109.

Landsberg, J.J. 1986. *Physiological Ecology of Forest Production.* Academic Press. London, UK. 198 p.

Landsberg, J.J. and R.H. Waring. 1997. A generalised model of forest productivity using simplified concepts of radiation-use efficiency, carbon balance and partitioning. *Forest Ecology and Management* 95: 209–228.

Langebartels, C., Kerner, K., Leonardi, S., Schrauder, M., Trost, M., Heller, W., and Sandermann, H. 1991. Biochemical plant responses to ozone. I. Differential induction of polyamine and ethylene biosynthesis in tobacco. *Plant Physiology* 95: 882–889.

Law, B.E., Waring, R.H., Anthoni, P.M., and Aber, J.D. 2000. Measurements of gross and net ecosystem productivity and water vapour exchange of a Pinus ponderosa ecosystem, and an evaluation of two generalized models. *Global Change Biology* 6: 155–168.

Lawlor, D.W. 1993. *Photosynthesis. Molecular, Physiological and Environmental Processes.* Longman Scientific and Technical. Essex, UK.

Lawlor, D.W. 2001. *Photosynthesis.* BIOS Scientific Publishers. Oxford, UK. p. 386.

Lawlor, D.W., Delgado, E., Habash, D.Z., Driscoll, S.P., Mitchell, V.J., Mitchell, R.A.C., and Parry, M.A.J. 1995. Photosynthetic acclimation of winter wheat to elevated CO_2 and temperature. In: Mathis, P. (Ed.). *Photosynthesis: From Light to Biosphere,* Vol. V. Kluwer. The Netherlands. p. 989–992.

Le Maire, G., Davi, H., Soudani, K., François, C., Le Dantec, V., and Dufrêne, E. 2005. Modelling annual production and carbon dioxide fluxes of a large managed temperate forest using forest inventories, satellite data and field measurements. *Tree Physiology* 25: 859–872.

Le Mer, J. and Roger, P. 2001. Production, oxidation, emission and consumption of methane by soils. A review.*European Journal of Soil Biology* 37: 25–50.

Lee, K.E. and Foster, R.C. 1991. Soil fauna and soil structure. *Australian Journal of Soil Research* 29: 745–775.

Leininger, S., Urich, T., Schloter, M., Schwark, L., Qi J., Nicol, G.W., Prosser, J.I., Schuster, S.C., and Schleper, C. 2006. Archaea predominate among ammonia-oxidizing prokaryotes in soils. *Nature* 442: 806–809.

Leinonen, I. 1996. A simulation model for the annual frost hardiness and freeze damage of Scots pine. *Annals of Botany* 78: 687–693.

Lemke, P., Ren, J., Alley, R., Allison, I., Carrasco, J., Flato, G., Fuiji, Y., Kaser, G., Mote, P., Thomas, R., and Zhang, T. 2007. *Observations: Changes in Snow, Ice and Frozen Ground.* In: Solomon, S., Qin, D., Manning, M., Chen, Z., Marquis, M., Averyt, K.B., Tignor, M., and Miller, H.L. (Eds.). *Climate Change 2007: The Physical Science Basis. Contribution of Working Group I to the Fourth Assessment Report of the Intergovernmental Panel on Climate Change.* Cambridge University Press. Cambridge, UK and New York, USA. p. 337–383.

Leverenz, J.W. and Öquist, G. 1987. Quantum yields of photosynthesis at temperatures between $-2°C$ and $35°C$ in cold-tolerant C3 plant (*Pinus sylvestris*) during the course of one year. *Plant, Cell and Environment* 10: 287–295.

Lewis, J.D., Tissue, D.T., and Strain, B.D. 1996. Seasonal response of photosynthesis to elevated CO_2 in loblolly pine (*Pinus taeda* L.) over two growing seasons. *Global Change Biology* 2: 103–114.

Lezica, R.F. and Quesada-Allue, L. 1990. Chitin. *Methods in Plant Biochemistry* 2: 443–481.

Lichtenthaler, H.K. 1999. The 1-deoxy-D-xylulose-5-phosphate pathway of isoprenoid biosynthesis in plants. *Annual Review of Plant Physiology and Plant Molecular Biology* 150: 47–65.

Lin, J., Jach, M.E., and Ceulemans, R. 2001. Stomatal density and needle anatomy of Scots pine (*Pinus sylvestris*) are affected by elevated CO_2.*New Phytologist* 150: 665–674.

Linder, S. and Axelsson, B. 1982. Changes in carbon uptake and allocation patterns as a result of irrigation and fertilization in young *Pinus sylvestris* stands. In: Waring, R.H. (Ed.). *Carbon Uptake and Allocation in Subalpine Ecosystems as a Key to Management.* Forestry Research Laboratory. Corvallis. Oregon, USA. p. 38–44.

Lindinger, W., Hansel, A., and Jordan, A. 1998. On-line monitoring of volatile organic compounds at ppt levels by means of Proton Transfer Reaction Mass Spectrometry (PTR-MS). Medical applications, food control and environmental research. *International Journal of Mass Spectrometry Ion Processes* 173: 191–241.

Linkins, A.E., Sinsabaugh, R.L., NcClugherty, C.A., and Melillo, J.M. 1990. Comparison of cellulase activity in decomposing leaves in a hardwood forest and woodland stream. *Soil Biology & Biochemistry* 22: 423–425.

Linkosalo, T. 2000. Mutual regularity of spring phenology of some boreal tree species: predicting with other species and phenological models. *Canadian Journal of Forest Research* 30: 667–673.

Linn, D.M. and Doran, J.W. 1984. Effect of water-filled pore space on carbon dioxide and nitrous oxide production in tilled and nontilled soils. *Soil Science Society of America Journal* 48: 1267–1272.

Liski, J., Ilvesniemi, H., Mäkelä, A., and Starr, M. 1998. Model analysis of the effects of soil age, fires and harvesting on the carbon storage of boreal forest soils. *European Journal of Forest Science* 49(3): 407–416.

Liski, J., Palosuo, T., Peltoniemi M., and Sievänen, R. 2005. Carbon and decomposition model Yasso for forest soils. *Ecological Modelling* 189: 168–182.

Lloyd, J. and Taylor, J.A. 1994. On the temperature dependence of soil respiration. *Functional Ecology* 8: 315–323.

Long, S.P. 1991. Modification of the response of photosynthetic productivity to rising temperature by atmospheric CO_2 concentration: has its importance been underestimated? *Plant, Cell and Environment* 14: 729–739.

Longeutaud, F., Mothe, F., Leban, J.-M., and Mäkelä, A. 2006. Picea abies sapwood width: variations within and between trees. *Scandinavian Journal of Forest Research* 21: 41–53.

Loqué, D. and Wirén, N. von. 2004. Regulatory levels for the transport of ammonium in plant roots. *Journal of Experimental Botany* 55: 1293–1305.

Loreto, F. and Sharkey, T.D. 1990. A gas exchange study of photosynthesis and isoprene emission in red oak (Quercus rubra L.). *Planta* 182: 523–531.

Loreto, F. and Velikova, V. 2001. Isoprene produced by leaves protects the photosynthetic apparatus against ozone damage, quenches ozone products, and reduces lipid peroxidation of cellular membranes. *Plant Physiology* 127: 1781–1787.

Loreto, F., Ciccioli, P., Cecinato, A., Brancaleoni, E., Frattoni, M., and Tricoli, D. 1996. Influence of environmental factors and air composition on the emission of a-pinene from Quercus ilex leaves. *Plant Physiology* 110: 267–275.

Loreto, F., Fischbach, R.J., Schnitzler, J.-P., Ciccioli, P., Brancaleoni, E., Calfapietra, C., and Seufert, G. 2001. Monoterpene emission and monoterpene synthase activities in the Mediterranean evergreen oak *Quercus ilex* L. grown at elevated CO2 concentrations. *Global Change Biology* 7: 709–717.

Loustau, D., Berbigier, P., Roumagnac, P., Arruda-Pacheco, C., David, J.S., Ferreira, M.I., Pereira, J.S., and Tavares, R. 1996. Transpiration of a 64-year-old maritime pine stand in Portugal 1. Seasonal course of water flux through maritime pine. *Oecologia* 107: 33–42.

Lundmark, T., Bergh, J., Strand, M., and Koppel, A. 1998. Seasonal variation of maximum photochemical efficiency in boreal Norway spruce stands. *Trees* 13: 63–67.

Lundström, U.S., van Breemen, N., and Bain, D. 2000a. The pozolization process. A review. *Geoderma* 94: 91–107.

Lundström, U.S., van Breemen, N., Bain, D., van Hees, P.A.W., Giesler, R., Gustafsson, J.P., Ilvesniemi, H., Karltun, E., Melkerud, P.-A., Olsson, M., Riise, G., Wahlberg, O., Bergelin, A., Bishop, K., Finlay, R., Jongmans, A.G., Magnusson, T., Mannerkoski, H., Nordgren, A., Nyberg, L., Starr, M., and Strand, L.T. 2000b. Advances in understanding the podzolization process resulting from a multidisciplinary study of three coniferous forest soils in the Nordic Countries. *Geoderma* 94: 335–353.

Luo, Y., Sims, D.A., and Griffin, K.L. 1998. Nonlinearity of photosynthetic responses to growth in rising atmospheric CO_2: an experimental and modelling study. *Global Change Biology* 4: 173–183.

Luomala, E.-M., Laitinen, K., Sutinen, S., Kellomäki, S., and Vapaavuori, E. 2005. Stomatal density, anatomy and nutrient concentrations of Scots pine needles are affected by elevated CO_2 and temperature. *Plant, Cell & Environment* 28: 733–749.

Luu, D.-T. and Maurel, C. 2005. Aquaporins in a challenging environment: molecular gears for adjusting plant water status. *Plant, Cell and Environment* 28: 85–96.

Lyubovtseva, Y.S., Sogacheva, L., Dal Maso, M., Bonn, M., Keronen, P., and Kulmala, M. 2005. Seasonal variations of trace gases, meteorological parameters, and formation of aerosols in boreal forests. *Boreal Environment Research* 6: 493–510.

MacCurdy, E. (Ed.). 2002. *The Notebooks of Leonardo Da Vinci. Definitive edition in one volume.* Konecky & Konecky. Old Saybrook, Connecticut, USA. 1180 p.

Magnani, F., Mencuccini, M., and Grace, J. 2000. Age-related decline in stand productivity: the role of structural acclimation under hydraulic constraints. *Plant, Cell and Environment* 23: 251–263.

Magnani, F., Mencuccini, M., Borghetti, M., Berbigier, P., Berninger, F., Delzon, S., Grelle, A., Hari, P., Jarvis, P.G., Kolari, P., Kowalski, A.S., Lankreijer, H., Law, B.E., Lindroth, A., Loustau, D., Manca, G., Moncrieff, J.B., Rayment, M., Tedeschi, V., Valentini, R., and Grace, J. 2007. The human footprint in the carbon cycle of temperate and boreal forests. *Nature* 447: 848–850.

Mäkelä, A. 1985. Differential games in evolutionary theory: height growth strategies of trees. *Theoretical Population Biology.* 27: 239–267.

Mäkelä, A. 1986. Implications of the pipe model theory on dry matter partitioning and height growth of trees. *Journal of Theoretical Biology* 123: 103–120.

Mäkelä, A. 1988. *Models of Pine Stand Development: An Eco-physiological Systems Analysis.* University of Helsinki, Department of Silviculture. Research Notes 62: 1–267.

Mäkelä, A. 1997. A carbon balance model of growth and self-pruning in trees based on structural relationships. *Forest Science* 43: 7–24.

Mäkelä, A. 2002. Derivation of stem taper from the pipe theory in a carbon balance framework. *Tree Physiology* 22: 891–905.

Mäkelä, A. and Hari, P. 1986. Stand growth model based on carbon uptake and allocation in individual trees. *Ecological Modelling* 33: 205–229.

Mäkelä, A. and Mäkinen, H. 2003. Generating 3D sawlogs with a process-based growth model. *Forest Ecology and Management* 184: 337–354.

Mäkelä, A. and Sievänen, R. 1987. Comparison of two shoot-root partitioning models with respect to substrate utilization and functional balance. *Annals of Botany* 59: 129–140.

Mäkelä, A. and Sievänen, R. 1992. Height growth strategies in open-grown trees. *Journal of Theoretical Biology* 159: 443–467.

Mäkelä, A. and Valentine, H.T. 2001. The ratio of NPP to GPP: evidence of change over the course of stand development. *Tree Physiology* 21: 1015–1030.

Mäkelä, A. and Vanninen, P. 1998. Impacts of size and competition on tree form and distribution of above ground biomass in Scots pine. *Canadian Journal of Forest Research* 28: 216–227.

Mäkelä, A., Givnish, T.J., Berninger, F., Buckley, T.N., Farquhar, G.D., and Hari, P. 2002. Challenges and opportunities of the optimality approach in plant ecology. *Silva Fennica* 36: 605–614.

Mäkelä, A., Hari, P., Berninger, F., Hänninen, H., and Nikinmaa, E. 2004. Acclimation of photosynthetic capacity in Scots pine to the annual cycle of temperature. *Tree Physiology* 24: 369–376.

Mäkelä, A., Kolari, P., Karimäki, J., Nikinmaa, E., Perämäki, M., and Hari, P. 2006. Modelling five years of weather-driven variation of GPP in a boreal forest. *Agricultural and Forest Meteorology* 139: 382–398.

Mäkelä, A., Pulkkinen, M., Kolari, P., Lagergren, F., Berbigier, P., Lindroth, A., Loustau, D., Nikinmaa, E., Vesala, T., and Hari, P. 2007. Developing an empirical model of stand GPP with the LUE approach: analysis of eddy covariance data at five contrasting conifer sites in Europe. *Global Change Biology* 14: 92–108.

Mäkelä, J.M., Aalto, P., Jokinen, V., Pohja, T., Nissinen, A., Palmroth, S., Markkanen, T., Seit-sonen, K., Lihavainen, H., and Kulmala, M. 1997. Observations of ultrafine aerosol formation and growth in boreal forest. *Geophysical Research Letters* 24: 1219–1222.

Mäkelä, J.M., Koponen, I.K., Aalto, P.P., and Kulmala, M. 2000a. One-year data of submicron size modes of tropospheric background aerosol in Southern Finland. *Journal of Aerosol Science* 31: 595–611.

Mäkelä, J.M., Dal Maso, M., Pirjola, L., Keronen, P., Laakso, L., Kulmala, M., and Laaksonen, A. 2000b. Characteristics of the atmospheric particle formation events observed at a boreal forest site in southern Finland. *Boreal Environment Research* 5: 299–313.

Mäkiranta, P., Hytönen, J., Aro, L., Maljanen, M., Pihlatie, M., Potila, H., Shurpali, N., Laine, J., Lohila, A., Martikainen P.J., and Minkkinen, K. 2007. Soil greenhouse gas emissions from afforested organic soil croplands and cutaway peatlands. *Boreal Environment Research* 12: 159–175.

Maljanen, M., Hytönen, J., and Martikainen, P.J. 2001. Fluxes of N_2O, CH_4 and CO_2 on afforested boreal agricultural soils. *Plant and Soil* 231: 113–121.

Maljanen, M., Liikanen, A., Silvola, J., and Martikainen, P.J. 2003. Nitrous oxide emissions from boreal organic soil under different land-use. *Soil Biology & Biochemistry* 35: 689–700.

Maljanen, M., Jokinen, H., Saari, A., Strömmer, R., and Martikainen, P. 2006. Methane and nitrous oxide fluxes, and carbon dioxide production in boreal forest soil fertilized with wood ash and nitrogen. *Soil Use and Management* 22: 151–157.

Malkin, S. and Fork, D.C. 1981. Photosynthetic units of sun and shade plants. *Plant Physiology* 67: 580–583.

Mammarella, I., Kolari, P., Rinne, J., Keronen, P., Pumpanen J., and Vesala, T. 2007. Determining the contribution of vertical advection to the net ecosystem exchange at Hyytiälä forest (Finland). *Tellus* 59B: 900–909.

Mandl, F. 1971. *Statistical Physics*, John Wiley, London. 379 p.

Mandl, F. 1988. *Statistical Physics*, 2nd ed. Wiley. Chichester, UK. 395 p.

Manninen, S., Huttunen, S., and Perämäki, P. 1997. Needle S fractions and S to N ratios as indices of SO_2. *Water, Air, and Soil Pollution* 95: 277–228.

Mäntylä, E., Lång V., and Palva, E.T. 1995. Role of abscisic acid in drought-induced freezing tolerance, cold acclimation, and accumulation of LTI78 and RAB18 proteins in Arabidopsis thaliana. *Plant Physiology* 107: 141–148.

Martikainen, P.J. 1984. Nitrification in two coniferous forest soils after different fertilization treatments. *Soil Biology and Biochemistry* 16: 577–582.

Martikainen, P.J. 1985. Nitrous oxide emission associated with autotrophic ammonium oxidation in acid coniferous forest soil. *Applied Environmental Microbiology* 50: 1519–1525.

Martikainen, P.J., Lehtonen, M., Lång, K., De Boer, W., and Ferm, A. 1993. Nitrification and nitrous oxide production potentials in aerobic soil samples from the soil profile of a Finnish coniferous site receiving high ammonium deposition. *FEMS Microbiology Ecology* 13: 113–122.

Martin, F., Kohler, A., and Duplessis, S. 2007. Living in harmony in the wood underground: ectomycorrhizal genomics. *Current Opinion in Plant Biology* 10: 204–210.

Maselli, F., Barbati, A., Chiesi, M., Chirici, G., and Corona, P. 2006. Use of remotely sensed and ancillary data for estimating forest gross primary productivity in Italy *Remote Sensing of Environment* 100: 563–575.

Matson, P., Gower, S.T., Volkmann, C., Billow, C., and Grier, C.C. 1992. Soil nitrogen cycling and nitrous oxide flux in a Rocky Mountain Douglas-fir forest: effects of fertilization, irrigation and carbon addition. *Biogeochemistry* 18: 101–117.

Mauldin, III, R.L., Frost, G.J., Chen, G., Tanner, D.J., Prevot, A.S.H., Davis, D.D., and Eisele, F.L. 1998. OH measurements during the First Aerosol Characterization Experiment (ACE 1): observations and model comparisons. *Journal Geophysical Research* 103: 16713–16729.

Maurel, C. 1997. Aquaporins and water permeability of plant membranes. *Annual Review of Plant Physiology and Plant Molecular Biology* 48: 399–429.

Mazhitova, G. 2006. Soils of the boreal forest. In: *Encyclopedia of Soil Science*. Taylor & Francis. New York. p. 183–187.

McCarthy, H., Oren, R., Finzi, A., and Johnsen, K. 2006. Canopy Leaf area constrains [CO_2]-induced enhancement of productivity and partitioning among aboveground carbon pools. *PNAS* 103: 19356–19361.

McCarty, G.W. 1999. Modes of action of nitrification inhibitors. *Biology and Fertility of Soils* 29: 1–9.

McClatchey, R.A., Fenn, R.W., Selby, J.E.A., Volz, F.E., and Garing, J.S. 1971. *Optical properties of the atmosphere*. Report AFCRL-71-0279 (available from Air Force Geophysics Laboratory, Hanscom Air Force Base, MA 01731, USA).

McCree, K.J. 1970. An equation for the respiration of white clover plants grown under controlled conditions. In: Setlik, I. (Ed.). *Prediction and Measurement of Photosynthetic Productivity*. Pudoc. Wageningen, The Netherlands. p. 221–229.

McCully, M.E., Huang, C.X., and Ling, L.E.C. 1998. Daily embolism and refilling of xylem vessels in the roots of field-grown maize. *New Phytologist* 138: 327–342.

McMurry, P.H. 2000. The history of condensation nucleus counters. *Aerosol Science and Technology* 33: 297–322.

McMurry, P.H. and Stolzenburg, M.R. 1989. On the sensitivity of particle size to relative humidity for Los Angeles Aerosols. *Atmospheric Environment* 23: 497–507.

McMurtrie, R.E., Gholz, H.L., Linder, S., and Gower, S.T. 1994. Climatic factors controlling the productivity of pine stands: a model-based analysis. *Ecological Bulletin (Copenhagen)* 43: 173–188.

McTiernan, K.B., Coûteaux, M.-M., Berg, B., Berg, M.P., de Anta, R.C., Gallardo, A., Kratz, W., Piussi, P., Remacle, J., and De Santo, A.V. 2003. Changes in chemical composition of *Pinus sylvestris* needle litter during decomposition along a European coniferous forest climatic transect. *Soil Biology & Biochemistry* 35: 801–812.

Mecke, M. and Ilvesniemi, H. 1999. Near-saturated hydraulic conductivity and water retention in coarse podzol profiles. *Scand. J. For. Res.* 14: 391–401.

Medlyn, B., Barrett, D., Landsberg, J., Sands, P., and Clement, R. 2003. Conversion of canopy intercepted radiation to photosynthate: review of modelling approaches for regional scales. *Functional Plant Biology* 30: 153–169.

Meehl, G.A., Stocker, T.F., Collins, W., Friedlingstein, P., Gaye, A., Gregory, J., Kitoh, A., Knutti, R., Murphy, J., Noda, A., Raper, S., Watterson, I., Weaver A., and Zhao, Z.-C. 2007. Global climate projections. In: Solomon, S., Qin, D., Manning, M., Chen, Z., Marquis, M., Averyt, K.B., Tignor, M., and Miller, H.L. (Eds.). *Climate Change 2007: The Physical Science Basis*. Cambridge University Press. Cambridge, UK and New York, USA. 747–785.

Melcher, P.J., Goldstein, G., Meinzer, F.C., Yount, D.E., Jones, T.J, Holbrook, N.M., and Huang, C.X. 2001. Water relations of coastal and estuarine *Rhizophora mangle*: xylem pressure potential dynamics of embolism formation and repair. *Oecologia* 126: 182–192.

Mencuccini, M., Grace, J., and Fiovaranti, M. 1997. Biomechanical and hydraulic determinants of tree structure in Scots pine: anatomical characteristics. *Tree Physiology* 17: 105–113.

Mengel, K. and Kirkby, E.A. 2001. *Principles of Plant Nutrition*, 5th ed. Kluwer. Dordrecht, The Netherlands. 849 p.

Mertes, S., Schröder, F., and Wiedensohler, A. 1995. The particle detection efficiency curve of the TSI-3010 CPC as a function of temperature difference between saturator and condenser. *Aerosol Science and Technology* 23: 257–261.

Messier, C. and Nikinmaa, E. 2000. Effects of light availability and sapling size on the growth, biomass allocation and crown morphology of understory sugar maple, yellow birch and American beech. *Ecoscience* 7: 345–356.

Miller, A.J. and Cramer, M.D. 2004. Root nitrogen acquisition and assimilation. *Plant and Soil* 274: 1–36.

Moberg, A., Bergström, H., Krigsman, J.R., and Svanered, O. 2002. Daily air temperature and pressure series for Stockholm (1756–1998). *Climatic Change* 53: 171–212.

Mohren, G.M.J. 1987. *Simulation of Forest Growth, Applied to Douglas Fir Stand in the Nether-lands.* (Diss.) Agricultural University of Wageningen, The Netherlands. 184 p.

Mokma, D. 2006. Spodosols. In:*Encyclopedia of Soil Science.* Taylor & Francis. New York. p. 1682–1684.

Mokma, D. and Evans, C.V. 2000. Spodosols. In: Sumner, M.E. (Ed.). *Handbook of Soil Science.* CRC Press. Boca Raton, FL. p. E307–E321.

Mokma, D.L., Yli-Halla, M., and Lindqvist, K. 2004. Podzol formation in sandy soils of Finland. *Geoderma* 120: 259–272.

Moncrieff, J.B. and Fang, C. 1999. A model for soil CO_2 production and transport 2: application to a florida Pinus elliotte plantation. *Agricultural and Forest Meteorology* 95: 237–256.

Monson, R.K. and Fall, R. 1989. Isoprene emission from aspen leaves. *Plant Physiology* 90: 267–274.

Monteith, J.L. and Unsworth, M. 1990. *Principles of Environmental Physics.* Edward Arnold. London, UK, 304 p.

Monteith, J.L. 1977. Climate and the efficiency of crop production in Britain. *Philosophical Trans-actions of the Royal Society of London B* 281: 277–294.

Morgan, P.B., Bollero, G.A., Nelson, R.L., Dohleman, F.G., and Long, S.P. 2005. Smaller than predicted increase in aboveground net primary production and yield of field-grown soybean under fully open-air [CO_2] elevation. *Global Change Biology* 11: 1856–1865.

Morison, J.I.L. and Lawson, T. 2007. Does lateral gas diffusion in leaves matter? *Plant, Cell and Environment* 30: 1072–1085.

Morrell, J.J. and Gartner, B.L. 1998. Wood as a material. In: Bruce, A. and Palfreyman, J.W. (Eds.). *Forest Products Biotechnology.* Taylor & Francis. London, UK, p. 1–14.

Mosier, A., Schimel, D., Valentine, D., Bronson, K., and Parton, W. 1991. Methane and nitrous oxide fluxes in native, fertilized and cultivated grasslands. *Nature* 350: 330–332.

Müller, P., Li, X.-P., and Niyogi, K.K. 2001. Non-photochemical quenching. A response to excess light energy. *Plant Physiology* 125: 1558–1566.

Müller, T., Avolio, M., Olivi, M., Benjdia, M., Rikirsch, E., Kasaras, A., Fitz, M., Chalot, M., and Wipf, D. 2007. Nitrogen transport in the ectomycorrhiza association: the *Hebeloma cylindrosporum-Pinus pinaster* model. *Phytochemistry* 68: 41–51.

Myneni, R.B., Dong, J., Tucker, C.J., Kaufmann, R.K., Kauppi, P.E., Liski, J., Zhou, L., Alexeyev, V., and Hughes, M.K. 2001. A large carbon sink in the woody biomass of Northern forests. *Proceedings of the National Academy of Sciences USA* 98(26): 14784–14789.

Nakienovi, N. and Swart, R. (Eds.). 2000. *Emissions Scenarios. A Special Report of Working Group III of the Intergovernmental Panel on Climate Change.* Cambridge University Press. Cambridge, UK, and New York, USA. 599 p.

Nardini, A. and Salleo, S. 2000. Limitation of stomatal conductance by hydraulic traits: sensing or preventing xylem cavitation. *Trees* 15: 14–24.

Näsholm, T. 1994. Removal of nitrogen during needle senescence in Scots pine (*Pinus sylvestris* L.). *Oecologia* 99: 290–296.

Nehls, U., Mikolajewski, S., Magel, E., and Hampp, R. 2001. Carbohydrate metabolism in ectomy-corrhizas: gene expression, monosaccharide transport and metabolic control. *New Phytologist* 150: 533–541.

Nelson, D.W. and Sommers, L.E. 1996. Total carbon, organic carbon, and organic matter. In: Bigham, J.M. (Ed.). *Methods of Soil Analysis.* Part 3. Chemical Methods. SSSA Book Series: 5. Madison, Wisconsin, USA. p. 961–1010.

Nelson, P.N. and Baldock, J.A. 2005. Estimating the molecular composition of a diverse range of natural organic materials from solid-state 13C NMR and elemental analysis. *Biogeochemistry* 72: 1–34.

Nicol, G.W. and Schleper, C. 2006. Ammonia-oxidising Crenarchaeota: important players in the nitrogen cycle? *Trends in Microbiology* 14: 207–212.

Niinemets, Ü. and Reichstein, M. 2002. A model analysis of the effects of nonspecific monoter-penoid storage in leaf tissues on emission kinetics and composition in Mediterranean sclero-phyllous Quercus species. *Global Biogeochemical Cycles* 16: 1110.

Nikinmaa, E. 1992. Analyses of the growth of Scots Pine; matching structure with function. *Acta Forestalia Fennica* 235. 68 p.

Nikinmaa, E., Goulet, J., Messier, C., Sievänen, R., Perttunen, J., and Lehtonen, M. 2003. Shoot growth and crown development; the effect of crown position in 3D simulations. *Tree Physiology* 23: 129–136.

Nissinen, A. and Hari, P. 1998. Effects of nitrogen deposition on tree growth and soil nutrients in boreal Scots pine stands. *Environmental Pollution* 102: 61–68.

Nixon, P.J. and Mullineaux, C.W. 2001. Regulation of photosynthetic electron transport. In: Aro, E.-M. and Andersson, B. (Eds.). *Regulation of Photosynthesis*. Kluwer. Dordrecht, The Netherlands. p 533–555.

Nobel, P.S. 1991. *Physicochemical and Environmental Plant Physiology*. Academic Press. San Diego, Florida, USA. 635 p.

Nobel, P.S. 2005. *Physicochemical and Environmental Plant Physiology*, 3rd ed. Academic Press/Elsevier. Burlington, Massachusetts, USA. 540 p.

Noctor, G. and Foyer, C.H. 1998. A re-evaluation of the ATP:NADPH budget during C3 photosynthesis: a contribution from nitrate assimilation and its associated respiratory activity. *Journal of Experimental Botany* 49: 1895–1908.

Nykänen, H., Alm, J., Lång, K., Silvola, J., and Martikainen, P.J. 1995. Emissions of CH_4, N_2O and CO_2 from a virgin fen and a fen drained for grassland in Finland. *Journal of Biogeography* 22: 351–357.

Oades, J.M. 1993. The role of biology in the formation, stabilization and degradation of soil structure. *Geoderma* 56: 377–400.

Oades, J.M., Kirkham, M.A., and Wagner, G.H. 1970. The use of gas-liquid chromatography for the determination of sugars extracted from soils by sulfuric acid. *Soil Science Society of America Proceedings* 34: 230–235.

Ogren, W.L. 1984. Photorespiration: pathways, regulation and modification. *Annual Review of Plant Physiology* 35: 415–422.

Ojima, D.S., Valentine, D.W., Mosier, A.R., Parton, W.J., and Schimel, D.S. 1993. Effect of land use change on methane oxidation in temperate forest and grassland soils. *Chemosphere* 26: 675–685.

Oker-Blom, P., Kellomäki, S., Valtonen, E., and Väisänen, H. 1988. Structural development of *Pinus sylvestris* stands with varying initial density: a simulation model, *Scand. J. Forest Res.* 3: 185–200.

Oker-Blom, P. and Smolander, H. 1988. The ratio of shoot silhouette area to total needle area in Scots Pine. *Forest Science* 34: 894–906.

Oker-Blom, P., Pukkala, T., and Kuuluvainen, T. 1989. Relationships between radiation interception and photosynthesis in forest canopies — effect of stand structure and latitude. *Ecological Modelling* 49: 73–87.

Onsager, L. 1931a. Reciprocal relations in irreversible processes, I. *Physical Review* 37: 405–426.

Onsager, L. 1931b. Reciprocal relations in irreversible processes II. *Physical Review* 38: 2265–2279.

Öquist, G., Mårtensson, O., Martin, B., and Malmberg, G. (1978). Seasonal effects on chlorophyll-protein complexes isolated from Pinus sylvestris. *Physiologia Plantarum* 44: 187–192.

Öquist, G., Gardeström, P., and Huner, N.P.A. 2001. Metabolic changes during cold acclimation and subsequent freezing and thawing. In: Bigras, F.J. and Colombo, S.J. (Eds.). *Conifer Cold Hardiness*. Kluwer. Dordrecht, The Netherlands. p. 137–163.

Öquist, G. and Huner, N.P.A. 2003. Photosynthesis of overwintering evergreen plants. *Review of Plant Biology* 54: 329–355.

Oren, R., Ellsworth, D., Johnsen, K., Phillips, N., Ewers, b., Maier, C., Shäfer, K., McCarthy, G., McNulty, S., and Katul, G. 2001. Soil fertility limits carbon sequestration by forest ecosystems in a CO_2-enriched atmosphere. *Nature* 411: 469–472.

Ott, R.L. and Mendenhall, W. 1995. *Understanding Statistics*, 6th ed. Duxbury Press. Belmont, California, USA. 716 p.

Ottander, C. and Öquist, G. 1991. Recovery of photosynthesis in winter stressed Scots pine. *Plant Cell and Environment* 14: 345–349.

Ottander, C., Campbell, D., and Öquist, G. 1995. Seasonal changes in photosystem II organisation and pigment composition in Pinus sylvestris. *Planta* 197: 176–183.

Paavolainen, L. and Smolander, A. 1998. Nitrification and denitrification in soil from a clear-cut Norway spruce (Picea abies) stand. *Soil Biology and Biochemistry* 30: 775–781.

Paavolainen, L., Fox, M., and Smolander, A. 2000. Nitrification and denitrification in forest soil subjected to sprinkling infiltration. *Soil Biology and Biochemistry* 32: 669–678.

Palmroth, S. and Hari, P. 2001. Evaluation of the importance of acclimation of needle structure, photosynthesis, and respiration to available photosynthetically active radiation in a Scots pine canopy. *Canadian Journal of Forest Research* 31: 1235–1243.

Palva, L., Garam, E., Manoochehri, F., Hari, P., Rajala, K., Ruotoistenmäki, H., and Seppälä, I. 1998a. A novel multipoint measuring system of photosynthetically active radiation. *Agricultural and Forest Meteorology* 89: 141–147.

Palva, L., Garam, E., Siivola, E., Sepponen, R., and Hari,P. 1998b. Quantifying spatial variability of photosynthetically active radiation within canopies using a multipoint measuring system. *Agricultural and Forest Meteorology* 92: 163–171.

Palva, L., Markkanen, T., Siivola, E., Garam, E., Linnavuo, M., Nevas, S., Manoochehri, F., Palmroth, S., Rajala, K., Ruotoistenmäki, H., Vuorivirta, T., Seppälä, I., Vesala, T., Hari, P., and Sepponen, R. 2001. Tree scale multipoint measuring system of photosynthetically active radiation. *Agricultural and Forest Meteorology* 106: 71–80.

Papen, H., Daum, M., Steinkamp, R., and Butterbach-Bahl, K. 2001. N$_2$O and CH$_4$-fluxes from soils of a N-limited and N-fertilized spruce forest ecosystem of the temperate zone. *Journal of Applied Botany – Angewandte botanik* 75(3–4): 159–163.

Parson, W.W. and Nagarajan, V. 2003. Optical spectroscopy in photosynthetic antennas. In: Green, B.R. and Parson, W.W. (Eds.). *Light-Harvesting Antennas in Photosynthesis.* Kluwer. Dordrecht, The Netherlands. p. 83–127.

Passioura, J.B. and Cowan, I.R. 1968. On solving the non-linear diffusion equation for the radial flow of water to roots. *Agricultural Meteorology* 5: 129–134.

Patrick, J.W. 1997. Phloem unloading: sieve element unloading and post-sieve element transport. *Annual Review of Plant Physiology and Plant Molecular Biology* 48: 191–222.

Paul, E.A. and Clark, F.E. 1989. *Soil Microbiology and Biochemistry.* Academic Press. San Diego, California, USA. 273 p.

Pavel, R., Doyle, J., and Steinberger, Y. 2004. Seasonal patterns of cellulase concentration in desert soil. *Soil Biology & Biochemistry* 36: 549–554.

Paw, U.K.T., Baldocchi, D.D., Meyers, T.P., and Wilson, K.B. 2000. Correction of eddy-covariance measurements incorporating both advective effects and density fluxes. *Boundary-Layer Meteorology* 97: 487–511.

Pearcy, R.W., Ehleringer, J.R., Mooney, H.A., and Rundel, P.W. (Eds.). 1989. *Plant Physiological Ecology: Field Methods and Instrumentation.* Chapman & Hall. New York, USA.

Pelkonen, P. 1980. The uptake of carbon dioxide in Scots pine during spring. *Flora* 169: 386–397.

Pelkonen, P. and Hari, P. 1980. The dependence of the springtime recovery of CO$_2$ uptake in Scots pine on temperature and internal factors. *Flora* 169: 398–404.

Penning de Vries, F.W.T. 1975. The cost of maintenance processes in plant cells. *Annals of Botany* 39: 77–92.

Penning de Vries, F.W.T., Brunsting, A.H.M., and Van Laar, H.H. 1974. Products, requirements and efficiency of biosynthesis: a quantitative approach. *Journal of Theoretical Biology* 45: 339–377.

Pensa, M. and Sellin, A. 2002. Needle longevity of Scots pine in relation to foliar nitrogen content, specific leaf area, and shoot growth in different forest types. *CanadianJournal of Forest Research* 32: 1225–1231.

Peñuelas, J. and Munne-Bosch, S. 2005. Isoprenoids: an evolutionary pool for photoprotection. *Trends in Plant Science* 10: 166–169.

Perämäki, M., Nikinmaa, E., Sevanto, S., Ilvesniemi, H., Siivola, E., Hari, P., and Vesala, T. 2001. Tree stem diameter variation and transpiration in Scots pine; an analysis using dynamic sap flow model. *Tree Physiology* 21: 889–897.

Perämäki, M., Vesala, T., and Nikinmaa, E. 2005. Modeling the dynamics of pressure propagation and diameter variation in tree. *Tree Physiology* 25: 1091–1099.

Perks, M.P., Irvine, J., and Grace, J. 2002. Canopy stomatal conductance and xylem sap abscisic acis (ABA) in mature Scots pine during a gradually imposed drought. *Tree Physiology* 22: 877–883.

Persson, J., Gardeström, P., and Näsholm, T. 2006. Uptake, metabolism and distribution of organic and inorganic nitrogen sources by *Pinus sylvestris*. *Journal of Experimental Botany* 57: 2651–2659.

Perttunen, J., Sievänen, R., Nikinmaa, E., Salminen, H., Saarenmaa, H., and Väkevä, J. 1996. LIGNUM: a tree model based on simple structural units.*Annals of Botany* 77: 87–98.

Perttunen, J., Nikinmaa, E., Lechowicz, M.J., Sievänen, R., and Messier, C. 2005. Application of the functional-structural tree model LIGNUM to sugar maple saplings (Acer saccharum Marsh) growing in forest gaps. *Annals of Botany* 88: 471–481.

Petäjä, T., Mordas, G., Manninen, H., Aalto, P., Hämeri, K., and Kulmala, M. 2006. Detection efficiency of water-based TSI Condensation Particle Counter 3785. *Aerosol Science and Technology* 40: 1090–1097.

Peterson, C.A., Enstone, D.E., and Taylor, J.H. 1999. Pine root structure and its potential significance for root functions. *Plant and Soil* 217: 205–213.

Peterson, T.C. and Vose, R.S. 1997. An overview of the Global Historical Climatology Network temperature database. *Bulletin of the American Meteorological Society* 78: 2837–2848.

Petit, J.R., Jouzel, J., Raynaud, D., Barkov, N.I., Barnola, J.-M., Basile, I., Bender, M., Chappellaz, J., Davis, M., Delayque, G., Delmotte, M., Kotlyakov, V.M., Legrand, M., Lipenkov, V.Y., Lorius, C., Pepin, L., Ritz, C., Saltzman, E., and Stievenard, M. 1999. Climate and atmospheric history of the past 420,000 years from the Vostok ice core, Antarctica. *Nature* 399: 429–436.

Pettersson, R., McDonald, A.J.S., and Stadenberg. I. 1993. Response of small birch plants (*Betula pendula* Roth.) to elevated CO_2 and nitrogen supply. *Plant Cell and Environment* 16: 1115–1121.

Phillips, N.G., Ryan, M.G., Bond, B.J., McDovell, N.G., Hinckley, T.M., and ermák, J. 2003. Reliance on stored water increases with tree size in three species in the Pacific Northwest. *Tree Physiology* 23: 237–245.

Phillips, R.J. and Dungan, S.R. 1993. Asymptotic analysis of flow in sieve tubes with semipermeable walls. *Journal of Theoretical Biology* 162: 465–485.

Piccolo, A. 2001. The supramolecular structure of humic substances. *Soil Science* 166: 810–832.

Pichersky, E. and Gershenzon, J. 2002. The formation and function of plant volatiles: perfumes for pollinator attraction and defense. *Current Opinion in Plant Biology* 5: 237–243.

Pickard, W.F. 1981. The ascent of sap in plants. *Progress in Biophysics and Molecular Biology* 37: 181–229.

Pietarinen, I., Kaninen, M., Hari, P., and Kellomäki, S. 1982. A simulation model for daily growth of shoots, needles and stem diameter in Scots pine trees. *Forest Science* 28: 573–581.

Pietikäinen, J., Vaijärvi, E., Ilvesniemi, H., Fritze, H., and Westman, C.J. 1999. Carbon storage of microbes and roots and the flux of CO_2 across a moisture gradient. *Canadian Journal of Forest Research* 29: 1197–1203.

Pihlatie, M., Pumpanen, J., Rinne, J., Ilvesniemi, H., Simojoki, A., Hari, P., and Vesala, T. 2007. Gas concentration driven fluxes of nitrous oxide and carbon dioxide in boreal forest soil. *Tellus* 59B: 458–469.

Pilegaard, K., Skiba, U., Ambus, P., Beier, C., Brüggemann, N., Butterbach-Bahl, K., Dick, J., Dorsey, J., Duyzer, J., Gallagher, M., Gasche, R., Horvath, L., Kitzler, B., Leip, A., Pihlatie, M.K., Rosenkranz, P., Seufert, G., Vesala, T., Westrate, H., and Zechmeister-Boltenstern, S. 2006. Factors controlling regional differences in forest soil emission of nitrogen oxides (NO and N_2O). *Biogeosciences* 3: 651–661.

References

Pinelli, P. and Loreto, F., 2003. (CO_2)–C12 emission from different metabolic pathways measured in illuminated and darkened C3 and C4 leaves at low, atmospheric and elevated CO_2 concentration. *Journal of Experimental Botany* 54: 1761–1769.

Poirier, N., Derenne, S., Balesdent, J., Mariotti, A., Massiot D., and Largeau, C. 2003. Isolation and analysis of the non-hydrolysable fraction of a forest soil and an arable soil (Lacadée, southwest France). *European Journal of Soil Science* 54: 243–255.

Porcar-Castell, A., Bäck, J., Juurola, E., and Hari, P. 2006. Dynamics of the energy flow through PSII under changing light conditions: a model approach. *Functional Plant Biology* 33: 229–239.

Possell, M., Heath, J., Hewitt, C.N., Ayres, E., and Kerstiens, G. 2004. Interactive effects of elevated CO_2 and soil fertility on isoprene emissions from Quercus robur.*Global Change Biology* 10: 1835–1843.

Possell, M., Hewitt, C.N., and Beerling, D.J. 2005. The effects of glacial atmospheric CO_2 concentrations and climate on isoprene emissions by vascular plants. *Global Change Biology* 11: 60–69.

Poth, M. and Focht, D.D. 1985. ^{15}N kinetic analysis of N_2O production by Nitrosomonas europaea: an examination of nitrifier denitrification. *Applied Environmental Microbiology* 49: 1134–1141.

Pothier, D., Margolis, H.A., and Waring, R.H. 1989. Patterns of change of saturated sapwood permeability and sapwood conductance with stand development. *Canadian Journal of Forest Research* 19: 432–439.

Preining, O., Wagner, P.E., Pohl, F.G., and Szymanski, W. 1981. *Heterogeneous Nucleation and Droplet Growth*. University of Vienna, Institute of Experimental Physics, Vienna, Austria.

Prescott, C.E., Zabek, L.M., Staley, C.L., and Kabzems, R. 2000. Decomposition of broadleaf and needle litter in forests of British Columbia: influences of litter type, forest type, and litter mixtures. *Canadian Journal of Forest Research* 30: 1742–1750.

Priha, O. and Smolander, A. 1999. Nitrogen transformations in soil under *Pinus sylvestris*, *Picea abies* and *Betula pendula* at two forest sites. *Soil Biology and Biochemistry* 31: 965–977.

Priha, O., Grayston, S.J., Pennanen, T., and Smolander, A. 1999. Microbial activity related to C and N cycling and microbial community structure in the rhizospheres of *Pinus sylvestris*, *Picea abies* and *Betula pendula* seedlings in an organic and mineral soil. *FEMS Microbiology Ecology* 30: 187–199.

Pumpanen, J., Ilvesniemi, H., Keronen, P., Nissinen, A., Pohja, T., Vesala, T., and Hari, P. 2001. An open chamber system for measuring soil surface CO_2 efflux: analysis of error sources related to the chamber system. *Journal of Geophysical Research* 106 (D8): 7985–7992.

Pumpanen, J., Ilvesniemi, H., and Hari, P. 2003. A process-based model for predicting soil carbon dioxide efflux and concentration. *Soil Science Society of America Journal* 67: 402–413.

Pumpanen, J., Westman, C.-J., and Ilvesniemi, H. 2004. Soil CO_2 efflux from a podsolic forest soil before and after clear-cutting and site preparation. *Boreal Environment Research* 9: 199–212.

Pumpanen, J., Ilvesniemi, H., Kulmala, L., Siivola, E., Laakso, H., Kolari, P., Helenelund, C., Laakso, M., Uusimaa, M., and Hari, P. 2008. Respiration in boreal forest soil as determined from carbon dioxide concentration profile. *Soil Science Society of America Journal*, in print.

Raghothama, K.G. and Karthikeyan, A.S. 2005. Phosphate acquisition. *Plant and Soil* 274: 37–49.

Rahmstorf, S., Cazenave, A., Church, J.A., Hansen, J.E., Keeling, R.F., Parker, F.E., and Somerville, R.C.J. 2007. Recent climate observations compared to projections. *Science* 316: 709.

Räisänen, J. 2007. How reliable are climate models? *Tellus* 59A: 2–29.

Raivonen, M., Keronen, P., Vesala, T., Kulmala, M., and Hari, P. 2003. Measuring shoot-level NO_x flux in field conditions: the role of blank chambers. *Boreal Environment Research* 8: 445–455.

Raivonen, M., Bonn, B., Sanz, M.J., Vesala, T., Kulmala, M., and Hari, P. 2006. UV-induced NO_y emissions from Scots pine: could they originate from photolysis of deposited HNO_3? *Atmospheric Environment* 40: 6201–6213.

Rambal, S., Ourcival, J., Joffre, R., Mouillot, F., Nouvellon, Y., Reichstein, M., and Rocheteau, A. 2003. Drought controls over conductance and assimilation of a Mediterranean evergreen ecosystem: scaling from leaf to canopy. *Global Change Biology* 9: 1813–1824.

Rannik, Ü., Altimir, N., Raittili, J., Suni, T., Gaman, A., Hussein, T., Hölttä, T., Lassila, H., Latokartano, M., Lauri, A., Natsheh, A., Petäjä, T., Sorjamaa, R., Ylä-Mella, H., Keronen, P., Beringer, F., Velala, T., Hari, P., and Kulmala, M. 2002. Fluxes of carbon dioxide and water vapour over Scots pine forest and clearing. *Agricultural and Forest Meteorology* 111: 187–202.

Rannik, Ü., Keronen, P., Hari, P., and Vesala, T. 2004. Estimation of forest-atmosphere CO_2 exchange by direct and profile techniques. *Agricultural and Forest Meteorology.* 126: 141–155.

Rannik, Ü., Kolari, P., Vesala, T., and Hari, P. 2006. Uncertainties in measurement and modelling of net ecosystem exchange of a forest. *Agricultural and Forest Meteorology* 138: 244–257.

Rapparini, F., Baraldi, R., Miglietta, F., and Loreto, F. 2004. Isoprenoid emission in trees of Quercus pubescens and Quercus ilex with lifetime exposure to naturally high CO_2 environment. *Plant, Cell & Environment* 27: 381–391.

Rastetter, E., Ryan, M., Shaver, G., Mellilo, J., Nadelhoffer, K., Hobbie, J., and Aber, J. 1991. A general biochemical model describing the responses of the C and N cycles in terrestrial ecosystems to changes in CO_2, climate and N deposition. *Tree Physiology* 9: 101–126.

Raupach, M.R., Marland, G., Ciais, P., Le Quéré, C., Canadell, J.G., Klepper, G., and Field, C.B. 2007. Global and regional drivers of accelerating CO_2 emissions. *PNAS* 104: 10288–10293.

Rautio, P., Huttunen, S., and Lamppu, J. 1998. Element concentrations in Scots pine needles on radial transects across a subarctic area. *Water, Air, and Soil Pollution* 102: 389–405.

Regina, K., Silvola, J., and Martikainen, P.J. 1999. Short-term effects of changing water table on N_2O fluxes from peat monoliths from natural and drained boreal peatlands. *Global Change Biology* 5(2): 183–189.

Regina, K., Syväsalo, E., Hannukkala, A., and Esala, M. 2004. Fluxes of N_2O from farmed peat soils in Finland. *European Journal of Soil Science* 55: 591–599.

Regina, K., Pihlatie, M., Esala, M., and Alakukku, L. 2007. Methane fluxes on boreal arable soils. *Agriculture, Ecosystems and Environment* 119: 346–352.

Reich, P.B., Walters, M.B., Tjoelker, M.G., Vanderklein, D., and Buschena, C. 1998. Photosynthesis and respiration rates depend on leaf and root morphology and nitrogen concentration in nine boreal tree species differing in relative growth rate. *Functional Ecology* 12: 395–405.

Rennenberg, H., Kreutzer, K., Papen, H., and Weber, P. 1998. Consequences of high loads of nitrogen for spruce (*Picea abies*) and beech (*Fagus sylvatica*) forests. *New Phytologist* 139: 71–86.

Repo, T. and Pelkonen, P. 1986. Temperature step response of hardening in *Pinus sylvestris* seedlings. *Scandinavian Journal of Forest Research* 1: 271–284.

Repo, T., Leinonen, I., Wang, K.-Y., and Hänninen, H. 2006. Relation between photosynthetic capacity and cold hardiness in Scots pine. *Physiologia Plantarum* 126: 224–231.

Rey, A. and Jarvis, P.G. 1998. Long-term photosynthetic acclimation to increased atmospheric CO_2 concentration in young birch trees (Betula pendula). *Tree Physiology* 18: 441–450.

Richardson, A.D., Hollinger, D.Y., Burba, G.G., Davis, K.J., Flanagan, L.B., Katul, G.G., Munger, J.W., Ricciuto, D.M., Stoy, P.C., Suyker, A.E., Verma, S.B., and Wofsy, S.C. 2006. A multi-site analysis of random error in tower-based measurements of carbon and energy fluxes. *Agricultural and Forest Meteorology* 136: 1–18.

Riikonen, J., Holopainen, T., Oksanen, E., and Vapaavuori, E. 2005. Leaf photosynthetic characteristics of silver birch during three years of exposure to elevated concentrations of CO_2 and O_3 in the field. *Tree Physiology* 25: 549–560.

Ritchie, R.J. 2006. Estimation of cytoplasmic nitrate and its electrochemical potential in barley roots using $^{13}NO_3^-$ and compartmental analysis. *New Phytologist* 171: 643–655.

Roberntz, P. and Stockfors, J. 1998. Effects of elevated CO_2 concentration and nutrition on net photosynthesis, stomatal conductance and needle respiration of field-grown Norway spruce trees. *Tree Physiology* 18: 233–241.

Roeckner, E., Bäuml, G., Bonaventura, L., Brokopf, R., Esch, M., Giorgetta, M., Hagemann, S., Kirchner, I., Kornbleuh, L., Manzini, E., Rhodin, A., Schlese, U., Schulzweida U., and Tomkins, A. 2003. *The Atmospheric General Circulation Model ECHAM5, Part I: Model Description.* Max-Planck-Institute for Meteorology Report 349. Hamburg, Germany, 127 p.

Roeckner, E., Brokopf, R., Esch, M., Giorgetta, M., Hagemann, S., Kornblueh, L., Manzini, E., Schlese, U., and Schulzweida, U. 2006. Sensitivity of simulated climate to horizontal and vertical resolution in the ECHAM5 atmosphere model. *Journal of Climate* 19: 3771–3791.

Rondón, A., Johansson, C., and Granat, L. 1993. Dry deposition of nitrogen dioxide and ozone to coniferous forests. *Journal of Geophysical Research* 98: 5159–5172.

Rondón, A. and Granat, L. 1994. Studies on the dry deposition of NO_2 to coniferous species at low NO_2 concentrations. *Tellus* 46B: 339–352.

Room, P.M., Maillette, L., and Hanan, J.S. 1994. Module and metamer dynamics and virtual plants. *Advances in Ecological Research* 25: 105–157.

Rosenkranz, P., Brüggemann, N., Papen, H., Xu, Z., Seufert, G., and Butterbach-Bahl, K. 2006. N_2O, NO and CH_4 exchange, and microbial N turnover over a Mediterranean pine forest soil. *Biogeosciences* 3: 121–133.

Rosenstiel, T.N., Potosnak, M.J., Griffin, K.L., Fall, R., and Monson, R. 2003. Increased CO_2 uncouples growth from isoprene emission in an agriforest ecosystem. *Nature* 421: 256–259.

Ross, J. 1981. *The Radiation Regime and Architecture of Plant Stands*. Kluwer. The Hague, The Netherlands. 391 p.

Ross, J., Sulev, M., and Saarelaid, P. 1998. Statistical treatment of the PAR variability and its application to willow coppice. *Agricultural and Forest Meteorology* 91: 1–21.

Running, S.W. and Hunt, E.R. 1993. Generalization of a forest ecosystem process model for other biomes, BIOME-BGC, and an application for global-scale models. In: Ehleringer, J.R. and Field, C.B. (Eds.). *Scaling Physiological Processes: Leaf to Globe*. Academic Press. San Diego, CA, USA. p. 141–158.

Running, S. and Gover, S. 1991. FOREST-BGC, A general model of forest ecosystem processes for regional applications. II Dynamic carbon allocation and nitrogen budgets. *Tree Physiology* 9: 147–160.

Ruuskanen, T.M., Kolari, P., Bäck, J., Kulmala, M., Rinne, J., Hakola, H., Taipale, R., Raivonen, M., Altimir, N., and Hari, P. 2005. On-line field measurements of monoterpene emissions from Scots pine by proton transfer reaction - mass spectrometry. *Boreal Environment Research* 10: 553–567.

Ryan M.G. 1991. Effects of climate change on plant respiration. *Ecological Applications* 1: 157–167.

Ryan, M.G., Lavigne, M.B., and Gower, S.T. 1997. Annual carbon cost of autotrophic respiration in boreal forest ecosystems in relation to species and climate. *Journal of Geophysical Research* 102: 28871–28883.

Saari, A. and Martikainen, P.J. 2001. Differential inhibition of methane oxidation and nitrification in forest soils by dimethyl sulfoxide (DMSO). *Soil Biology & Biochemistry* 33: 1567–1570.

Saari, A., Martikainen, P.J., Ferm, A., Ruuskanen, J., De Boer, W., Troelstra, S.R., and Laanbroek, H.J. 1997. Methane oxidation in soil profiles of Dutch and Finnish coniferous forests with different soil texture and atmospheric nitrogen deposition. *Soil Biology & Biochemistry* 29: 1625–1632.

Saari, A., Heiskanen J., and Martikainen, P.J. 1998. Effect of the organic horizon on methane oxidation and uptake in soil of a boreal Scots pine forest. *FEMS Microbiology Ecology* 26: 245–255.

Saari, A., Rinnan, R., and Martikainen, P.J. 2004. Methane oxidation in boreal forest soils: kinetics and sensitivity to pH and ammonium. *Soil Biology & Biochemistry* 36: 1037–1046.

Saarikoski, S., Mäkelä, T., Hillamo, R., Aalto, P.P., Kerminen, V.-M., and Kulmala, M. 2005. Physico-chemical characterization and mass closure of size-segregated atmospheric aerosols in Hyytiälä, Finland. *Boreal Environment Research* 10: 385–400.

Saito, T., Hong, P., Kato, K., Okazaki, M., Inagaki, H., Maeda, S., and Yokogava, Y. 2003. Purification and characterization of an extracellular laccase of fungus (family Chaetomiaceae) isolated from soil. *Enzyme and Microbial Technology* 33: 520–526.

Sakai, A. and Larcher, W. 1987. *Frost Survival of Plants. Responses and Adaptation to Freezing Stress*. Springer. Berlin, Germany. 321 p.

Sakr, S., Alves, G., Morillon, R., Maurel, K., Decourteix, M., Guilliot, A., Fleurat-Lessard, P., Julien, J.-L., and Chrispeels, M.J. 2003. Plasma membrane aquaporins are involved in winter embolism recovery in walnut tree. *Plant Physiology* 133: 630–641.

Saliendra, N.Z., Sperry, J.S., and Comstock, J.P. 1995. Influence of leaf water status on stomatal response to humidity, hydraulic conductance, and soil drought in Betula occidentalis. *Planta* 196: 357–366.

Salleo, S., LoGullo, M., De Paoli, D., and Zippo, M. 1996. Xylem recovery from cavitation-induced embolism in young plants of *Laurus nobilis*: a possible mechanism. *New Phytologist* 132: 47–56.

Salleo, S., Nardini, A., Pitt F., and Lo Gullo, M.A. 2000. Xylem cavitation and hydraulic control of stomatal conductance in Laurel (*Laurus nobilis* L.). *Plant, Cell & Environment* 23: 71–79.

Salleo, S., Lo Gullo, M., Trifilo, P., and Nardini, A. 2004. New evidence for a role of vessel-associated cells and phloem in the rapid xylem refilling of cavitated stems of *Laurus nobilis*. *Plant, Cell & Environment* 27: 1065–1076.

Salleo, S., Trifilò, P., and Lo Gullo, M. 2006. Phloem as a possible major determinant of rapid cavitation reversal in stems of *Laurus nobilis* (laurel). *Functional Plant Biology* 33: 1063–1074.

Salminen, R. (Ed.). 2005. *Geochemical Atlas of Europe. Part 1: Background Information, Methodology and Maps.* Geological Survey of Finland. Espoo, Finland. 526 p.

Salo-Väänänen, P. 1996. *Determination of Protein Content in Foods by the Amount of Total Nitrogen or Amino Acids.* (Diss.) EKT series 1050. University of Helsinki. Department of Applied Chemistry and Microbiology. 195 p.

Sandermann, H., Wellburn, A.R., and Heath, R.L. 1997. *Forest decline and ozone. A comparison of controlled chamber experiments and field experiments.* Ecological Studies 127. Springer Verlag. Berlin. 400 p.

Sands, P.J. 1995. Modeling canopy production. 1. Optimal distribution of photosynthetic resources. *Australian Journal of Plant Physiology* 22: 593–601.

Santantonio, D. 1989. Dry-matter partitioning and fine-root production in forests: new approaches to a difficult problem. In: Pereira, J.S. and Landsberg, J.J. (Eds.). *Biomass Production by Fast-Growing Trees.* Kluwer. Norwell, Massachusetts, USA. p. 57–72.

Sarvas, R. 1967. The annual period of development of forest trees. *Proceedings of the Finnish Academy of Science and Letters* 1965: 211–231.

Sarvas, R. 1972. Investigations on the annual cycle of development of forest trees. Active period. *Communicationes Instituti Forestalis Fenniae* 76: 1–110.

Sarvas, R. 1974. Investigations on the annual cycle of development of forest trees. II. Autumn dormancy and winter dormancy. *Communicationes Instituti Forestalis Fenniae* 84: 1–101.

Sausen, R., Barthel, K., and Hasselman, K. 1988. Coupled ocean–atmosphere models with flux-correction. *Climate Dynamics* 2: 145–163.

Savage, K., Moore, T.R., and Crill, P.M. 1997. Methane and carbon dioxide exchanges between the atmosphere and northern boreal forest soils. *Journal of Geophysical Research* 102(D24): 29279–29288.

Scarascia-Mugnozza, G., De Angelis, P., Matteucci, G., and Valentini, R. 1996. Long-term exposure to elevated [CO_2] in a natural *Quercus ilex* L. community: net photosynthesis and photochemical efficiency of PSII at different levels of water stress. *Plant Cell and Environment* 19: 43–654.

Scheel, T., Dörfler, C., and Kalbitz, K. 2007. Precipitation of dissolved organic matter by aluminum stabilizes carbon in acidic forest soils. *Soil Science Society of America Journal* 71: 64–74.

Schelhaas, M.J., Nabuurs, G.J., Jans, W., Moors, E., Sabaté, S., and Daamen, W.P. 2004. Closing the carbon budget of a Scots pine forest in the Netherlands. *Climatic Change* 67: 309–328.

Schiller, D.L. and Hastie, D.R. 1996. Nitrous oxide and methane fluxes from perturbed and unperturbed boreal forest sites in northern Ontario. *Journal of Geophysical Research* 101: 22767–22774.

Schimel, J.P. and Bennett, J. 2004. Nitrogen mineralization: challenges of a changing paradigm. *Ecology* 85: 591–602.

Schindlbacher, A., Zechmeister-Bolternstern, S., and Butterbach-Bahl, K. 2004. Effects of soil moisture and temperature on NO, NO_2, and N_2O emissions from European forest ecosystems. *Journal of Geophysical Research* 109: D17302.

Schlentner, R.E. and van Cleve, K. 1985. Relationships between CO_2 evolution from soil, substrate temperature, and substrate moisture in four mature forest types in interior Alaska. *Canadian Journal of Forest Research* 15: 97–106.

Schmid, M.C., Risgaard-Petersen, N., van de Vossenberg, J., Kuypers, M.M.M., Lavik, G., Petersen, J., Hulth, S., Thamdrup, B., Canfield, D., Dalsgaard, T., Rysgaard, S., Sejr, M.K., Strous, M., Op den Camp, H.J.M., and Jetten, M.S.M. 2007. Anaerobic ammonium-oxidizing bacteria in marine environments: widespread occurrence but low diversity. *Environmental Microbiology* 9: 2364–2374.

Schmidt, I., Sliekers, O., Schmid, M., Cirpus, I., Strous, M., Bock, E., Kuenen, J.G., and Jetten, M.S.M. 2002. Aerobic and anaerobic ammonia oxidizing bacteria competitors or natural partners? – minireview. *FEMS Microbiology Ecology* 39: 175–181.

Schnitzer, M. 2000. A lifetime perspective on the chemistry of soil organic matter. *Advances in Agronomy* 68: 1–58.

Scholefield, P.A, Doick, K.J., Herbert, B.M.J., Hewitt, C.N.S., Sshnitzler, J.-P., Pinelli, P., and Loreto, F. 2004. Impact of rising CO_2 on emissions of volatile organic compounds: isoprene emission from Phragmites australis growing at elevated CO_2 in a natural carbon dioxide spring. *Plant, Cell and Environment* 27: 393–401.

Schüßler, A., Martin, H., Cohen, D., Fitz, M., and Wipf, D. 2006. Characterization of a carbohydrate transporter from symbiotic glomeromycotan fungi. *Nature* 444: 933–936.

Scurlock, J.M.O., Asner, G.P., and Gower, S.T. 2001. *Worldwide Historical Estimates and Bibliography of Leaf Area Index, 1932–2000.* ORNL Technical Memorandum TM-2001/268, Oak Ridge National Laboratory, Oak Ridge, Tennessee, USA.

Segers, R. 1998. Methane production and methane consumption: a review of processes underlying wetland methane fluxes. *Biogeochemistry* 41: 25–51.

Seager, R., Battisti, D.S., Yin, J., Gordon, N., Naik, N., Clement, A.C., and Cane, M.A. 2002. Is the Gulf Stream responsible for Europe's mild winters? *Quarterly Journal of the Royal Meteorological Society* 128: 2563–2586.

Seinfeld, J.H. and Pandis, S.N. 1998. *Atmospheric Chemistry and Physics: From Air Pollution to Climate Change.* J. Wiley, New York, USA. 1326 p.

Sem, G.J. 2002. Design and performance characteristics of three continuous-flow condensation particle counters: a summary. *Atmospheric Research* 62: 267–294.

Shao, M., Czapiewski, K.V., Heiden, A.C., Kobel, K., Komenda, M., Koppmann, R., and Wildt, J. 2001. Volatile organic compound emissions from Scots pine: mechanisms and description by algorithms. *Journal of Geophysical Research* 106: 20483–20491.

Sharkey, T.D. 2005. Effects of moderate heat stress on photosynthesis: importance of thylakoid reactions, rubisco deactivation, reactive oxygen species, and thermotolerance provided by isoprene. *Plant, Cell and Environment* 28: 269–277.

Sharkey, T.D., Loreto, F., and Delwiche, C.F. 1991. High carbon dioxide and sun/shade effect on isoprene emissions from oak and aspen tree leaves. *Plant, Cell and Environment* 14: 333–338.

Shinozaki, K., Yoda, K., Hozumi, K., and Kira, T. 1964a. A quantitative analysis of plant form - The pipe model theory. I. Basic analyses. *Japanese Journal of Ecology* 14: 97–105.

Shinozaki, K., Yoda, K., Hozumi, K., and Kira, T. 1964b. A quantitative analysis of plant form - the Pipe Model theory. II. Further evidence of the theory and its application in forest ecology. *Japanese Journal of Ecology* 14: 133–139.

Siau, J.F. 1984. *Transport Processes in Wood.* Springer. New York, USA. 211 p.

Sievänen, R. 1992. *Construction and Identification of Models for Tree and Stand Growth.* (Diss.) Helsinki University of Technology, Automation Technology Laboratory, Research Reports. 9: 1–52.

Sievänen, R., Lindner, M., Mäkelä, A., and Lash, P. 2000. Volume growth and survival graphs: a method for evaluating process-based models. *Tree Physiology* 20: 357–365.

Simek, M. 2000. Nitrification in soil – terminology and methodology. *Rostlinna Vyroba* 46: 385–395.

Simojoki, A. 2001. *Oxygen supply to plant roots in cultivated mineral soils.* (Diss.) Department of Applied Chemistry and Microbiology, University of Helsinki. Pro Terra No. 7. Helsinki. 1–59.

Simpson, I.J., Edwards, G.C., and Thurtell, G.W. 1999. Variations in methane and nitrous oxide mixing ratios at the southern boundary of a Canadian boreal forest. *Atmospheric Environment* 33: 1141–1150.

Šimůnek, J. and Suarez, D.L. 1993. Modeling of carbon dioxide transport and production in soil 1. Model development. *Water Resources Research* 29: 487–497.

Singh, U. and Uehara, G. 1999. Electrochemistry of the double layer: principles and applications to soils. In: Sparks, D.L. (Ed.). *Soil Physical Chemistry*, 2nd ed. CRC Press. Boca Raton, Florida, USA.

Sinsabaugh, R.L., Carreiro, M.M., and Alvarez, S. 2002. Enzyme and microbiological dynamics of litter composition. In: Burns, R.G. and Dick, R.P. (Eds.). *Enzymes in the Environment*. Marcel Dekker. New York, Basel. p. 249–265.

Sitaula, B.K., Bakken, L., and Abrahamsen, G. 1995. N-fertilization and soil acidification effects on N_2O and CO_2 emission from temperate pine forest soil. *Soil Biology & Biochemistry* 27: 1401–1408.

Sjöström, E. 1993. *Wood Chemistry. Fundamentals and Applications*, 2nd ed. Academic Press. San Diego, California, USA, 249 p.

Skopp, J., Jawson, M.D., and Doran, J.W. 1990. Steady-state aerobic microbial activity as a function of soil water content. *Soil Science Society of America Journal* 54: 1619–1625.

Skre, O. 1975. Exchange in Norwegian Tundra plants studied by infrared gas analyzer technique. In: Wielgolaski, F.E. (Ed.). *Fennoscandian Tundra Ecosystems. Part 1: Plants and Microorganisms*. Springer. Heidelberg, Berlin, and New York, USA. p. 168–183.

Smith, J.N., Moore, K.F., Eisele, F.L., Voisin, D., Ghimire, A.K., Sakurai, H., and McMurry, P.H. 2005. Chemical composition of atmospheric nanoparticles during nucleation events in Atlanta. *Journal of Geophysical Research* 110: D22S03.

Smith, K.A., Ball, T., Conen, F., Dobbie, K.E., Massheder, J., and Rey, A. 2003. Exchange of greenhouse gases between soil and atmosphere: interactions of soil physical factors and biological processes. *European Journal of Soil Science* 54(4): 779–791.

Smith, K.C., Magnuson, C.E., Goeschl, J.D., and DeMichele, D.W. 1980. A time-dependent mathematical expression of the Münch hypothesis of phloem transport. *Journal of Theoretical Biology* 86: 493–505.

Smith, M.L., Ollinger, S.V., Martin, M.E., Aber, J.D., Hallett, R.A., and Goodale, C.L. 2002. Direct estimation of aboveground forest productivity through hyperspectral remote sensing of canopy nitrogen. *Ecological Applications* 12: 1286–1302.

Smith, T.M. and Reynolds, R.W. 2005. A global merged land and sea surface temperature reconstruction based on historical observations (1880–1997). *Journal of Climate* 18: 2021–2036.

Smits, M.M., Hoffland, E., Jongmans, A.G., and van Breemen, N. 2005. Contribution of mineral tunneling to total feldspar weathering. *Geoderma* 125: 59–69.

Smolander, S. and Stenberg, P. 2005. Simple parameterizations of the radiation budget of uniform broadleaved and coniferous canopies. *Remote Sensing of Environment* 94: 355–363.

Sparks, J.P., Monson, R.K., Sparks, K.L., and Lerdau, M. 2001. Leaf uptake of nitrogen dioxide (NO_2) in a tropical wet forest: implications to tropospheric chemistry. *Oecologia* 127: 214–221.

Sperry, J.S. 2003. Evolution of water transport and xylem structure. *International Journal of Plant Science* 164: 115–127.

Sperry, J.S. and Tyree, M.T. 1988. Mechanism of water stress-induced xylem embolism. *Plant Physiology* 88: 581–587.

Spracklen, D.V., Carslaw, K.S., Kulmala, M., Kerminen, V.-M., Mann, G.W., and Sihto, S.-L. 2006. The contribution of boundary layer nucleation events to total particle concentrations on regional and global scales. *Atmospheric Chemistry and Physics* 6: 5631–5648.

Stafsform, J.P. 1995. Developmental potential of shoot buds. In: Gartner, B.L. (Ed.). *Plant Stems: Physiology and Functional Morphology*. Academic Press. San Diego, California, USA. p. 257–279.

Stamnes, K., Tsay, S.C., Wiscombe, W., and Jayaweera, K. 1988. A numerically stable algorithm for discrete-ordinate-method radiative transfer in multiple scattering and emitting layered media. *Applied Optics* 27: 2502–2509.

Stamnes, K., Tsay, S.C., Wiscombe W., and Laszlo, I. 2000. A general-purpose numerically stable computer code for discrete-ordinate-method radiative transfer in scattering and emitting layered media, DISORT Report v1.1.

Staudt, M., Joffre, R., Rambal, S., and Kesselmeier, J. 2001. Effect of elevated CO_2 on monoterpene emission of young *Quercus ilex* trees and its relation to structural and ecophysiological parameters. *Tree Physiology* 21: 437–445.

Steinkamp, R., Butterbach-Bahl, K., and Papen, H. 2001. Methane oxidation by soils of an N limited and N fertilized spruce forest in the Black Forest, Germany. *Soil Biology & Biochemistry* 33: 145–153.

Stenberg, P. 1995. Penumbra in within-shoot and between-shoot shading in conifers and its significance for photosynthesis. *Ecological Modelling* 77: 215–231.

Stenberg, P. 1996. Simulations of the effects of shoot structure and orientation on vertical gradients in intercepted light by conifer canopies. *Tree Physiology* 16: 99–108.

Stenberg, P., Palmroth, S., Bond, B.J., Sprugel, D.G., and Smolander, H. 2001. Shoot structure and photosynthetic efficiency along the light gradient in a Scots pine canopy. *Tree Physiology* 21: 805–814.

Stephens, B.B., Gurney, K., Tans, P., Sweeney, C., Peters, W., Bruhwiler, L., Ciais, P., Ramonet, M., Bousquet, P., Nakazawa, T., Aoki, S., Machida, T., Inoue, G., Vinnichenko, N., Lloyd, J., Jordan, A., Heimann, M., Shibistova, O., Langenfelds, R.L., Steele, L.P., Francey, R.J., and Denning, A.S. 2007. Weak northern and strong tropical land carbon uptake from vertical profiles of atmospheric CO_2. *Science* 316: 1732–1735.

Stevenson, F.J. 1996. Nitrogen – organic forms. In: Sparks, D.L. (Ed.). *Methods of Soil Analysis. Part 3. Chemical Methods*. SSSA Book Series no. 5. Madison, Wisconsin, USA. p. 1185–1200.

Stitt, M. 1990. Fructose-2,6-bisphosphate as a regulatory molecule in plants. *Annual Review of Plant Physiology and Plant Molecular Biology* 41: 153–185.

Stitt, M. 1991. Rising CO_2 levels and their potential significance for carbon flow in photosynthetic cells. *Plant Cell and Environment* 14: 741–762.

Stocks, B.J., Fosberg, M.A., Lynham, T.J., Mearns, L., Wotton, B.M., Yang, Q., Jin, J.-Z., Lawrence, K., Harley, G.R., Mason, J.A., and McKenney, D.W. 1998. Climate change and fire potential in Russian and Canadian boreal forests. *Climate Change* 38: 1–13.

Stolzenburg, M. and McMurry, P. 1991. An ultrafine aerosol condensation nucleus counter. *Aerosol Science and Technology* 14: 48–65.

Strangeways, I. 2003. *Measuring the Natural Environment*. Cambridge University Press. New York, USA. 544 p.

Strous, M. and Jetten, M.S.M. 2004. Anaerobic oxidation of methane and ammonium. *Annual Review of Microbiology* 58: 99–117.

Suarez, D.L. and Šimnek, J. 1993. Modeling of carbon dioxide transport and production in soil 2. Parameter selection, sensitivity analysis and comparison of model predictions to field data. *Water Resources Research* 29: 499–513.

Suni, T., Rinne, J., Reissell, A., Altimir, N., Keronen, P., Rannik, Ü., Dal Maso, M., Kulmala, M., and Vesala, T. 2003. Long-term measurements of surface fluxes above a Scots pine forest in Hyytiälä, southern Finland, 1996–2001. *Boreal Environment Research* 8: 237–301.

Sutinen, M-J., Arora, R., Wisniewski, M., Ashworth, E., Strimbeck, R., and Palta, J. 2001. Mechanisms of frost survival and freeze-damage in nature. In: Bigras, F.J. and Colombo, S.J. (Eds.). *Conifer Cold Hardiness*. Kluwer. Dordrecht, The Netherlands.

Sutton, R. and Sposito, G. 2005. Molecular structure in soil humic substances: the new view. *Environmental Science & Technology* 39: 9009–9015.

Svensmark, H. 2007. Cosmoclimatology: a new theory emerges. *Astronomy and Geophysics* 48: 1.18–1.24.

Swift, R.S. 1996. Organic matter characterization. In: Sparks, D.L. (Ed.). *Methods of Soil Analysis. Part 3. Chemical Methods* – SSSA Book Series no. 5. Madison, Wisconsin, USA, SSSA, ASA. p. 1011–1069.

Taiz, L. and Zeiger, E. 2002. *Plant Physiology*, 3rd ed. Sinauer Associates. Sunderland, Massachusetts, USA. 690 p.

Taiz, L. and Zeiger, E. 2006. *Plant Physiology*, 4th ed. Sinauer Associates. Sunderland, Massachusetts, USA. 764 p.

Tanskanen, N. 2006. *Aluminium Chemistry in Ploughed Podzolic Forest Soils.* (Diss.) University of Helsinki, Department of Forest Ecology. Available at www.metla.fi/dissertationes

Tarvainen, V., Hakola, H., Hellen, H., Bäck, J., Hari, P., and Kulmala, M. 2005. Temperature and light dependence of the VOC emissions of Scots pine. *Atmospheric Chemistry and Physics* 5: 6691–6718.

Tate, III, R.L. 2002. Microbiology and enzymology of carbon and nitrogen cycling. In: Burns, R.G. and Dick, R.P. (Eds.). *Enzymes in the Environment.* Marcel Dekker. New York, Basel. p. 227–248.

Taylor, J.H. and Peterson, C.A. 2005. Ectomycorrhizal impacts on nutrient uptake pathways in woody roots. *New Forests* 30: 203–214.

Thoene, B., Rennenberg, H., and Weber, P. 1996. Absorption of atmospheric NO_2 by spruce (*Picea abies*) trees. II. Parameterization of NO_2 fluxes by controlled dynamic chamber experiments. *New Phytologist* 134: 257–266.

Thomas, G.E. and Stamnes, K. 1999. Radiative transfer in the atmosphere and ocean.Cambridge University Press. Cambridge, UK. 517 p.

Thomashow, M.F. 1998. Role of cold-responsive genes in plant freezing tolerance. *Plant Physiology* 118: 1–7.

Thomashow, M.F. 2001. So What's new in the field of plant cold acclimation? lots! *Plant Physiology* 125: 89–93.

Thompson, M.V. and Holbrook, N.M. 2004. Scaling phloem transport: information transmission. *Plant, Cell & Environment* 27: 509–519.

Thompson, N.M. and Holbrook, N.M. 2003. Scaling phloem transport water potential equilibrium and osmoregulatory flow. *Plant Cell and Environment* 26: 1561–1577.

Thornley, J.H.M. 1972. A balanced quantitative model for root:shoot ratios in vegetative plants. *Annals of Botany* 36: 431–441.

Thornley, J.H.M. 1991. A Transport-resistance model of Forest Growth and Partitioning. *Annals of Botany* 68: 211–226.

Thornley, J.H.M. 2002. Instantaneous canopy photosynthesis: analytical expressions for sun and shade leaves based on exponential light decay down the canopy and an acclimated non-rectangular hyperbola for leaf photosynthesis. *Annals of Botany* 89: 451–458.

Thornley, J.H.M. and Johnson, I.R. 1990. *Plant and Crop Modelling. A Mathematical Approach to Plant and Crop Physiology.* Clarendon Press, Oxford, UK. 669 p.

Thornton, P.E., Law, B.E., Gholz, H.L., Clark, K.L., Falge, E., Ellsworth, D.S., Goldstein, A.H., Monson, R.K., Hollinger, D., Falk, M., Chen, J., and Sparks, J.P. 2002. Modeling and measuring the effects of disturbance history and climate on carbon and water budgets in evergreen needleleaf forests. *Agricultural and Forest Meteorology* 113: 185–222.

Tietema, A., De Boer, W., Riemer, L., and Verstraten, J.M. 1992. Nitrate production in nitrogen-saturated acid forest soils: vertical distribution and characteristics. *Soil Biology and Biochemistry* 16: 577–582.

Tietema, A., Riemer, L., Verstraten, J.M., van der Maas, M.P., van Wijk, A.J., and van Voorhuyzen, I. 1993. Nitrogen cycling in acid forest soils subject to increased atmospheric nitrogen input. *Forest Ecology and Management* 57: 29–44.

Tisdall, J.M. and Oades, J.M. 1982. Organic matter and water-stable aggregates in soils. *Journal of Soil Science* 33: 141–163.

Tjoelker, M.G., Oleksyn, J., and Reich, P.B. 1998. Seedlings of five boreal tree species differ in acclimation of net photosynthesis to elevated CO_2 and temperature. *Tree Physiology* 18: 715–726.

Tognetti, R., Johnson, J.D., Michelozzi, M., and Raschi, A. 1998. Response of foliar metabolism in mature trees of *Quercus pubescens* and *Quercus ilex* to long-term elevated CO_2. *Environmental and Experimental Botany* 39: 233–245.

Tomme, P., Warren, A.J., and Gilkes, N.R. 1995. Cellulose hydrolysis by bacteria and fungi. *Advances in Microbial Physiology* 37: 2–81.

Topp, G.C., Davis, J.L., and Annan, A.P. 1980. Elctromagnetic determination of soil water content: measurements in coaxial transmission lines. *Water Resources Research* 16: 574–582.

Topp, E. and Pattey, E. 1997. Soils as sources and sinks for atmospheric methane. *Canadian Journal of Soil Sciences* 77: 167–178.

Tournaire-Roux, C., Sutka, M., Javot, H., Gout, E., Gerbeau, P., Luu, D-T., Bligny, R., and Maurel, C. 2003. Cytosolic pH regulates root water transport during anoxic stress through gating of aquaporins. *Nature* 425: 393–397.

Trenberth, K.E., Jones, P.D., Ambenje, P.G., Bojariu, R., Easterling, D.R., Klein Tank, A.M.G., Parker, D.E., Renwick, J.A., Rahimzadeh, F., Rusticucci, M.M., Soden, B.J., and Zhai, P.-M. 2007. Observations: surface and atmospheric climate change. In: Solomon, S., Qin, D., Manning, M., Chen, Z., Marquis, M., Averyt, K.B., Tignor, M., and Miller, H.L. (Eds.). *Climate Change 2007: The Physical Science Basis* Cambridge University Press. Cambridge, UK and New York, USA, 235–336.

Treusch, A.H., Leininger S., Kleutzin A., Schuster, S.C., Klenk, H.-P., and Schleper, C. 2005. Novel genes for nitrite reductase and Amo-related proteins indicates a role of uncultivated mesophilic crenarchaeota in nitrogen cycling. *Environmental Microbiology* 7: 1985–1955.

Tritton, D.J. 1988. *Physical Fluid Dynamics*. Clarendon Press. Oxford, UK. 519 p.

Troeh, F.R., Jabro, J.D., and Kirkham, D. 1982. Gaseous diffusion equations for porous materials. *Geoderma* 27: 239–253.

Tunved, P., Hansson, H.-C., Kerminen, V.-M., Ström, J., Dal Maso, M., Lihavainen, H., Viisanen, Y., Aalto, P.P., Komppula, M., and Kulmala, M. 2006. High natural aerosol loading over boreal forests. *Science* 312: 261–263.

Tuomivaara, T., Hari, P., Rita, H., and Häkkinen, R. 1994. *The Guide-Dog Approach: A Methodology for Ecology*. University of Helsinki, Department of Forest Ecology publications 11.

Turner, D.P., Urbanski, S., Bremer, D., Wofsy, S.C., Meyers, T., Gower, S.T., and Gregory, M. 2003. A cross-biome comparison of daily light use efficiency for gross primary production. *Global Change Biology* 9: 383–395.

Turunen, M. and Huttunen, S. 1996. Scots pine needle surfaces on radial transects across the north boreal area of Finnish Lapland and the Kola Peninsula of Russia. *Environmental Pollution* 93: 175–194.

Tuskan, G.A., DiFazio, S., Jansson, S., Bohlmann, J., Grigoriev, I., Hellsten, U., Putnam, N., Ralph, S., Rombauts, S., Salamov, A., Schein, J., Sterck, L., Aerts, A., Bhalerao, R.R., Bhalerao, R.P., Blaudez, D., Boerjan, W., Brun, A., Brunner, A., Busov, V., Campbell, M., Carlson, J., Chalot, M., Chapman, J., Chen, G.-L., Cooper, D., Coutinho, P.M., Couturier, J., Covert, S., Cronk, Q., Cunningham, R., Davis, J., Degroeve, S., Déjardin, A., dePamphilis, C., Detter, J., Dirks, B., Dubchak, I., Duplessis, S., Ehlting, J., Ellis, B., Gendler, K., Goodstein, D., Gribskov, M., Grimwood, J., Groover, A., Gunter, L., Hamberger, B., Heinze, B., Helariutta, Y., Henrissat, B., Holligan, D., Holt, R., Huang, W., Islam-Faridi, N., Jones, S., Jones-Rhoades, M., Jorgensen, R., Joshi, C., Kangasjärvi, J., Karlsson, J., Kelleher, C., Kirkpatrick, R., Kirst, M., Kohler, A., Kalluri, U., Larimer, F., Leebens-Mack, J., Leplé, J.-C., Locascio, P., Lou, Y., Lucas, S., Martin, F., Montanini, B., Napoli, C., Nelson, D.R., Nelson, C., Nieminen, K., Nilsson, O., Pereda, V., Peter, G., Philippe, R., Pilate, G., Poliakov, A., Razumovskaya, J., Richardson, P., Rinaldi, C., Ritland, K., Rouzé, P., Ryaboy, D., Schmutz, J., Schrader, J., Segerman, B., Shin, H., Siddiqui, A., Sterky, F., Terry, A., Tsai, C.-J., Uberbacher, E., Unneberg, P., Vahala, J., Wall, K., Wessler, S., Yang, G., Yin, T., Douglas, C., Marra, M., Sandberg, G., Van de Peer, Y. and Rokhsar, D.

2006. The genome of black cottonwood, *Populus trichocarpa* (Torr. & Gray). *Science* 313: 1596–1604.

Tyree, M.T. and Dixon, M.A. 1983. Cavitation events in *Thuja occidentalis* L. Ultrasonic acoustic emissions from the sapwood can be measured. *Plant Physiology* 72: 1094–1099.

Tyree, M.T. and Sperry, J.S. 1988. Do woody plants operate near the point of catastrophic xylem dysfunction caused by dynamic water stress? Answers from a model. *Plant Physiology* 88: 574–580.

Tyree, M.T. and Zimmermann, M.H. 2002. *Xylem Structure and the Ascent of Sap*, 2nd ed. Springer, New York, USA.

Tyree, M.T., Christy, A.L., and Ferrier, J.M.A. 1974. Simpler iterative steady state solution of Münch pressure-flow systems applied to long and short translocation paths. *Plant Physiology* 54: 589–600.

Tyree, M.T., Fiscus, E.L., and Wullschleger, S.D., and Dixon, M.A. 1986. Detection of xylem cavitation in corn under field conditions. *Plant Physiology* 82: 597–599.

Tyree, M.T., Salleo, S., Nardini, A., Lo Gullo, M.A., and Mosca, R. 1999. Refilling of embolized vessels in young stems of laurel: do we need a new paradigm? *Plant Physiology* 120: 11–21.

Underwood, A.J. 1997. *Experiments in Ecology: Their Logical Design and Interpretation Using Analysis of Variance*. Cambridge University Press, Cambridge, UK.

USDA-NRCS, Soil Survey Division, World Soil Resources. 2005. Global Soil Regions Map. Washington DC. Available at http://soils.usda.gov/use/worldsoils/mapindex/order.html.

Valentine, D.L. 2007. Adaptations to energy stress dictate the ecology and evolution of the Archaea. *Nature* 5: 316–323.

Valentine, H.T. 1985. Tree-growth models: derivations employing the pipe model theory. *Journal of Theoretical Biology* 117: 579–584.

Van Breemen, N. and Buurman, P. 2002. *Soil Formation*, 2nd ed. Kluwer. Dordrecht, The Netherlands. 404 p.

Van Breemen, N., Mulder, J., and van Grinsven, J.J.M. 1987. Impacts of acid atmospheric deposition on woodland soils in the Netherlands. II. Nitrogen transformations. *Soil Science Society of America Journal* 51: 1634–1640.

Van Breemen, N., Finlay, R., Lundström, U., Jongmans, A.G., Giesler, R., and Olsson, M. 2000. Mycorrhizal weathering: a true case of mineral plant nutrition? *Biogeochemistry* 49: 53–67.

Van Dijk, A.I.J.M., Dolman, A.J., and Schulze, E.D. 2005. Radiation, temperature, and leaf area explain ecosystem carbon fluxes in boreal and temperate European forests. *Global Biogeochemical Cycles* 19: GB2029.

Van Hees, P.A.W., Jones, D.L., and Godbold, D.L. 2002. Biodegradation of low molecular weight organic acids in coniferous forest pozolic soils. *Soil Biology & Biochemistry* 34: 1261–1272.

Van Rees, K. 2006. Soils Associated with major forest ecosystems. In: *Encyclopedia of Soil Science*. Taylor and Francis, Boca Raton, FL. p. 722–724.

Vanninen, P. and Mäkelä, A. 1999. Fine root biomass of Scots pine stands differing in age and soil fertility in southern Finland. *Tree Physiology* 19: 823–830.

Vanninen, P. and Mäkelä, A. 2000. Needle and stem wood production in Scots pine (*Pinus sylvestris*) trees of different age, size and competitive status. *Tree Physiology* 20: 527–533.

Vanninen, P. and Mäkelä, A. 2005. Carbon budget for Scots pine trees: effects of size, competition and site fertility on growth allocation and production. *Tree Physiology* 25: 17–30.

Vanninen, P., Ylitalo, H., Sievänen, R., and Mäkelä, A. 1996. Effects of age and site quality on the distribution of biomass in Scots pine (*Pinus sylvestris* L.). *Trees* 10: 231–238.

Venterea, R.T., Groffman, P.M., Verchot, L.V., Magill, A.H., and Aber, J.D. 2004. Gross nitrogen process rates in temperate forest soils exhibiting symptoms of nitrogen saturation. *Forest Ecology and Management* 196: 129–142.

Véry, A.-A. and Sentenac, H. 2003. Molecular mechanisms and regulation of K + transport in higher plants. *Annual Reviews of Plant Biology* 54: 575–603.

Vesala, T., Haataja, J., Aalto, P., Altimir, N., Buzorius, G., Garam, E., Hämeri, K., Ilvesniemi, H., Jokinen, V., Keronen, P., Lahti, T., Markkanen, T., Mäkelä, J.M., Nikinmaa, E., Palmroth, S., Palva, L., Pohja, T., Pumpanen, J., Rannik, Ü., Siivola, E., Ylitalo, H., Hari, P., and Kulmala, M.

1998. Long-term field measurements of atmosphere-surface interactions in boreal forest combining forest ecology, micrometeorology, aerosol physics and atmospheric chemistry. *Trends in Heat, Mass and Momentum Transfer* 4: 17–35.

Vesala, T., Markkanen, T., Palva, L., Siivola, E., Palmroth, S., and Hari, P. 2000. Effects of variations of PAR on CO_2 estimation for Scots pine. *Agricultural and Forest Meteorology* 100: 337–347.

Vesala, T., Hölttä, T., Perämäki, M., and Nikinmaa, E. 2003. Refilling of hydraulically isolated embolised vessels: model calculations. *Annals of Botany* 91: 419–428.

Vesala, T., Suni, T., Rannik, U., Keronen, P., Markkanen, T., Sevanto, S., Grönholm, T., Smolander, S., Kulmala, M., Ilvesniemi, H., Ojansuu, R., Uotila, A., Levula, J., Mäkelä, A., Pumpanen, J., Kolari, P., Kulmala, L., Altimir, N., Berninger, F., Nikinmaa, E., and Hari, P. 2005. Effect of thinning on surface fluxes in a boreal forest. *Global Biogeochemical Cycles* 19: GB2001.

Vierstra, R.D. 1993. Protein degradation in plants. *Annual Review of Plant Physiology and Plant Molecular Biology* 44: 385–410.

Vitousek, P.M. and Hobbie, S. 2000. Heterotrophic nitrogen fixation in decomposing litter: patterns and regulation. *Ecology* 81(9): 2366–2376.

Vogg, G., Heim, R., Hansen, J., Schäfer, C., and Beck, E. 1998. Frost hardening and photosynthetic performance in Scots pine (*Pinus sylvestris* L.) needles. I. Seasonal changes in the photosynthetic apparatus and its function. *Planta* 204: 193–200.

Voisin, D., Smith, J.N, Sakurai, H., McMurry, P.H., and Eisele, F.L. 2003. Thermal desorption chemical ionization mass spectrometer for ultrafine particle chemical composition. *Aerosol Science and Technology* 37: 471–475.

Von Klot, S., Peters, A., Aalto, P., Bellander, T., Berglind, N., D'Ippoliti, D., Elosua, R., Hörmann, A., Kulmala, M., Lanki, T., Löwel, H., Pekkanen, J., Picciotto, S.,Sunyer, J., Forastriere, F., and the HEAPSS study group. 2005. Ambient air pollution is associated with increased risk of hospital cardiac readmission of myocardial infarction survivors in five European cities. *Circulation* 112: 3073–3079.

Vuokila, Y. 1987. *Metsänkasvatuksen perusteet ja menetelmät*. WSOY. Porvoo, Finland. p. 258.

Vuokko, R., Kellomäki, S., and Hari, P. 1977. The inherent growth rhythm and its effect on the daily height increment of plants. *Oikos* 29: 137–142.

Wahlström, E., Reinikainen, T., and Hallanaro, E.-L. (Eds.). 1992. *Ympäristön tila Suomessa*. Forssan Kirjapaino Oy, Forssa, Finland. p. 114–117.

Waldrop, M.P., Zak, D.R., and Sinsabaugh, R.L. 2004. Microbial community response to nitrogen deposition in northern forest ecosystems. *Soil Biology & Biochemistry* 36: 1443–1451.

Walker, J., Sharpe, P.J.H., Penridge, L.K., and Wu, H. 1989. Ecological field theory: the concept and field tests. *Vegetatio* 83: 81–95.

Wang, J.Y. 1960. A critique of the heat unit approach to plant response studies. *Ecology* 41: 785–790.

Wang, K.-Y., Kellomäki, S., Zha, T.S., and Peltola, H. 2004 Component carbon fluxes and their contribution to ecosystem carbon exchange in a pine forest: an assessment based on eddy covariance measurements and an integrated model. *Tree Physiology* 24: 19–34.

Wang, S., Zordan, C., and Johnston, M. 2006. Chemical characterization of individual, airborne sub-10-nm particles and molecules.*Analytical Chemistry* 78: 1750–1754.

Wang, S.C. and Flagan, R.C. 1990. Scanning electrical mobility spectrometer. *Aerosol Science and Technology* 13: 230–240.

Wang, Z.P. and Ineson, P. 2003. Methane oxidation in a temperate coniferous forest soil: effect of inorganic N. *Soil Biology & Biochemistry* 35: 427–433.

Waring, R.H. and Running, S.W. 1978. Sapwood water storage: Its contribution to transpiration and effect on water conductance through the stems of old-growth Douglas fir. *Plant, Cell and Environment* 1: 131–140.

Waring, R.H., Whitehead, D., and Jarvis, P.G. 1979. The contribution of stored water to transpiration in Scots pine. *Plant, Cell and Environment* 2: 309–317.

Waring, R.H., Landsberg, J.J., and Williams, M. 1998. Net primary production of forests: a constant fraction of gross primary production. *Tree Physiology* 18: 129–134.

Warren, C.R. and Adams, M.A. 2004. Evergreen trees do not maximize instantaneous photosynthesis. *Trends in Plant Science* 9: 270–274.

Warren, C.R., Dreyer, E., and Adams, M.A. 2003. Photosynthesis-Rubisco relationships in foliage of *Pinus sylvestris* in response to nitrogen supply and the proposed role of Rubisco and amino acids as nitrogen stores. *Trees* 17: 359–366.

Warren, S.G. 1984. Optical constants of ice from the ultraviolet to the microwave, *Applied Optics* 23(8): 218–229.

Warren, R.A.J. 1996. Microbial hydrolysis of polysaccharides. Annual Reviews in Microbiology 50: 183–212.

Watson, M. and Casper, B.B. 1984. Morphogenetic constraints on patterns of carbon distribution in plants. *Annual Review of Ecological Systems* 15: 233–258.

Webb, E.K., Pearman, G.I., and Leuning, R. 1980. Correction of flux measurements for density effects due to heat and water vapor transfer. *Quarterly Journal of the Royal Meteorological Society* 106: 85–100.

Webb, M.J., Senior, C.A., Sexton, D.M.H., Ingram, W.J., Williams, K.D., Ringer, M.A., McAvaney, B.J., Colman, R., Soden, B.J., Gudgel, R., Knutson, T., Emori, S., Ogura, T., Tsushima, Y., Andronova, N., Li, B., Musat, I., Bony, S., and Taylor, K.E. 2006. On the contribution of local feedback mechanisms to the range of climate sensitivity in two GCM ensembles. *Climate Dynamics* 27: 17–38.

Weber, A.P.M., Schwacke, R., and Flügge, U.-I. 2005. Solute Transporters of the Plastid Envelope Membrane. *Annual Review of Plant Biology* 56: 133–164.

Weber, R.J., Orsini, D., Duan, Y., Lee, Y.-N., Klotz, P.J., and Brechtel, F. 2001. A particle-into-liquid collector for rapid measurement of aerosol bulk chemical composition. *Aerosol Science and Technology* 35: 718–727.

Wehner, B., Philippin, S., and Wiedensohler, A. 2002. Design and calibration of a thermodenuder with an improved heating unit to measure the size-dependent volatile fraction of aerosol particles. *Journal of Aerosol Science* 33: 1087–1093.

Weiser, C.J. 1970. Cold resistance and injury in woody plants. *Science* 169: 1269–1278.

Went, F.W. 1960. Blue hazes in the atmosphere. *Nature* 187: 641–645.

West, G.B., Brown, J.H., and Enquist, B.J. 1997. A general model for the origin of allometric scaling laws in biology. *Science* 276: 122–126.

WGBH. 2003. *Cell Membrane: Just Passing Through.* Available at www.teachersdomain.org/resources/tdc02/sci/life/cell/membrane.web [accessed 27 Nov 2007].

Whalen, S.C. 2000. Influence of N and non-N salts on atmospheric methane oxidation by upland boreal forest and tundra soils. *Biology and Fertility of Soils* 31: 279–287.

Whalen, S.C. and Reeburgh, W.S. 1996. Moisture and temperature sensitivity of CH_4 oxidation in boreal soils. *Soil Biology & Biochemistry* 28(10/11): 1271–1281.

Widén, B. and Majdi, H. 2001. Soil CO_2 efflux and root respiration at three sites in a mixed pine and spruce forest: seasonal and diurnal variation. *Canadian Journal of Forest Research* 31: 786–796.

Widén, B. 2002. Seasonal variation in forest-floor CO_2 exchange in a Swedish coniferous forest. *Agricultural and Forest Meteorology* 111: 283–297.

Wielicki, B.A., Harrison, E.F., Cess, R.D., King, M.D., and Randall, D.A. 1995. Mission to planet earth: role of clouds and radiation in climate. *Bulletin of the American Meteorological Society* 76: 2125–2153.

Wiese, C.B. and Pell, E.J. 2003. Oxidative modification of the cell wall in tomato plants exposed to ozone. *Plant Physiology and Biochemistry* 41: 375–382.

Wildt, J., Kley, D., Rockel, A., Rockel, P., and Segschneider, H.J. 1997. Emission of NO from several higher plant species. *Journal of Geophysical Research* 102: 5919–5927.

Wilen, R.W., Fu, P., Robertson, A.J., Abrams, S.R., Low, N.H., and Gusta, L.V. (1996). An abscisic acid analog inhibits abscisic acid-induced freezing tolerance and protein accumulation, but not abscisic acid-induced sucrose uptake in a bromegrass (*Bromus inermis Leyss*) cell culture. *Planta* 200: 138–143.

Wilkins, D., Van Oosten, J.-J., and Besford, R.T. 1994. Effects of elevated CO_2 on growth and chloroplast proteins in *Prunus avium*. *Tree Physiology* 14: 769–779.

Wilmer, C. and Fricker, M. 1996. *Stomata*, 2nd ed. Chapman & Hall. London, UK.

Wilson, K.B., Goldstein, A.H., Falge, E., Aubinet, M., Baldocchi, D., Berbigier, P., Bernhofer, Ch., Ceulemans, R., Dolman, H., Field, C., Grelle, A., Law, B., Meyers, T., Moncrieff, J., Monson, R., Oechel, W., Tenhunen, J., Valentini, R., and Verma, S. 2002. Energy balance closure at FLUXNET sites. *Agricultural and Forest Meteorology* 113: 223–243.

Winklmayr, W., Reischl, G., Lindner, A., and Berner, A. 1991. A new electromobility spectrometer for the measurement of aerosol size distributions in the size range from 1 to 1000 nm. *Journal of Aerosol Science* 22: 289–296.

Wiscombe, W. 1980. Improved Mie scattering algorithms. *Applied Optics* 19: 1505–1509.

Wit, C.T. de, Brouwer, R., and Penning de Vries, F.W.T. 1970. The simulation of photosynthetic systems. In: *Prediction and Measurement of Photosynthetic Production*. Pudoc. Wageningen, The Netherlands. p. 47–70.

Wittmann, C., Kähkönen, M.A., Ilvesniemi, H., Kurola, J., and Salkinoja-Salonen, M.S. 2004. Areal activities and stratification of hydrolytic enzymes involved in the biochemical cycles of carbon, nitrogen, sulphur and phosphorus in podsolized boreal forest soils. *Soil Biology & Biochemistry* 36 425–433.

Woebken, D., Fuchs, B.M., Kuypers, M.M.M., and Amann, R. 2007. Potential interactions of particle-associated anammox bacteria with bacterial and archaeal partners in the Namibian upwelling system. *Applied and Environmental Microbiology* 73: 4648–4657.

Woese, C.R., Magrum, L.J., and Fox, G.E. 1978. Archaebacteria. *Journal of Molecular Evolution*, 11(3): 245–252.

Wood, T. and Bhat, K.M. 1988. Methods for measuring cellulase activities. *Methods in Enzymology* 160: 87–112.

Wood, W.W. and Petraitis, M.J. 1984. Origin and distribution of carbon dioxide in the unsaturated zone of the southern high plains of Texas. *Water Resources Research* 20: 1193–1208.

Woodrow, I.E. 1994. Optimal acclimation of the photosynthetic system under enhanced CO_2. *Photosynthesis Research* 39: 401–412.

Woodward, F.I. 1987. Stomatal numbers are sensitive to increases in CO_2 from pre-industrial levels. *Nature* 327: 617–618.

Woodward, F.I. and Kelly, C.K. 1995. The influence of CO_2 concentration on stomatal density. *New Phytologist* 131: 311–327.

Wrage, N., Velthof, G.L., van Beusichem, M.L., and Oenema, O. 2001. Role of nitrifier denitrification in the production of nitrous oxide. *Soil Biology & Biochemistry* 33: 1723–1732.

Wullscheleger, S.D., Meinzer, F.C., and Vertessy, R.A. 1998. A review of whole-plant water use studies in trees. *Tree Physiology* 18: 499–512.

Yamasaki, H. and Sakihama, Y. 2000. Simultaneous production of nitric oxide and peroxynitrite by plant nitrate reductase: in vitro evidence for the NR-dependent formation of active nitrogen species. *FEBS Letters* 468: 89–92.

Ylinen, A. 1952. Über die mechanische Schaftformtheorie der Bäume. *Silva Fennica*. 76: 1–52.

Yoder, B.J., Ryan, M.G., Waring, R.H., Schoettle, A.W., and Kaufmann, M.R. 1994. Evidence of reduced photosynthetic rates in old trees. *Forest Science* 40: 513–527.

Yonebashi, H. and Hattori, T. 1980. Improvements in the method for fractional determination of soil organic nitrogen. *Soil Science and Plant Nutrition* 32: 189–200.

Yuan, W., Liu, S., Zhou, Gua., Zhou, Guo., Tieszen, L.L, Baldocchi, D., Bernhofer, C., Gholz, H., Goldstein, AH., Goulden, M.L., Hollinger, D.Y., Hu, Y., Law, B.E., Stoy, P.C., Vesala, T., and Wofsy, S.C. 2007. Deriving a light use efficiency model from eddy covariance flux data for predicting daily gross primary production across biomes. *Agricultural and Forest Meteorology* 143: 189–207.

Zeide, B. 1987. Analysis of the 3/2 power law of self-thinning. *Forest Science* 33: 517–537.

Zha, T.-S., Kellomäki, S., Wang, K.-Y., and Ryyppö, A. 2005. Respiratory responses of Scots pine stems to 5 years of exposure to elevated CO_2 concentration and temperature. *Tree Physiology* 25: 49–56.

Zhou, X., He, Y., Huang, G., Thornberry, T.D., Carroll, M.A., and Bertman, S.B. 2002. Photochemical production of HONO on glass sample manifold wall surface. *Geophysical Research Letters* 29(14): 1681.

Zimmermann, M.H. 1983. *Xylem Structure and the Ascent of Sap*. Springer. Berlin. 143 p.

Zwieniecki, M.A. and Holbrook, N.M. 1998. Short term changes in xylem water conductivity in white ash, red maple and sitka spruce. *Plant, Cell and Environment* 21: 1173–1180.

Zwieniecki, M.A. and Hollbrook, N.M. 2000. Bordered pit structure and vessel wall surface properties. Implications for embolism repair. *Plant Physiology* 123: 1015–1020.

Zwieniecki, M.A., Melcher, P.J., and Holbrook, N.M. 2001. Hydraulic properties of individual xylem vessels of *Fraxinus americana*. *Journal of Experimental Botany* 52: 257–264.

Index

DATE DUE

GAYLORD			PRINTED IN U.S.A.